2015

MASONRY

Codes and Specifications

COMPILATION

ICC
INTERNATIONAL
CODE COUNCIL

THE MASONRY SOCIETY
Advancing the knowledge of masonry

MIA
MASONRY INSTITUTE
OF AMERICA

2015 MASONRY CODES AND SPECIFICATIONS COMPILATION

First Printing, July, 2015

ISBN: 978-0-940116-22-1

COPYRIGHT © 2015 by **INTERNATIONAL CODE COUNCIL, INC.,**
THE MASONRY SOCIETY and
MASONRY INSTITUTE OF AMERICA

Advancing the knowledge of masonry

ALL RIGHTS RESERVED. This *2015 Masonry Codes and Specifications Compilation* is a copyrighted book owned by the International Code Council, Inc., The Masonry Society and the Masonry Institute of America. Without advance written permission from the copyright owners, no part of this book may be reproduced, distributed or transmitted in any form or by any means, including, without limitation, electronic, optical or mechanical means (by way of example, and not limitation, photocopying or recording by or in an information storage retrieval system). For information on permission to copy material exceeding fair use, please contact: ICC Publications, 4051 West Flossmoor Road, Country Club Hills, IL 60478. Phone 1-888-ICC-SAFE (422-7233).

Portions of this publication are reproduced, with permission, from the *2015 International Building Code* (IBC) copyright © 2014, from the *Building Code Requirements and Specification for Masonry Structures* (TMS 402-13/ACI 530-13/ASCE 5-13 and TMS 602-13/ACI 530.1-13/ASCE 6-13) copyright © 2013; from *Direct Design Handbook for Masonry Structures (TMS 403-13);* from the *Code Requirements for Determining Fire Resistance of Concrete and Masonry Construction Assemblies* (ACI 216.1-14/TMS 0216-14) copyright © 2014; and from the *Standard Method for Determining Sound Transmission Ratings for Masonry Walls* (TMS 0302-12) copyright © 2012. The *2015 International Building Code* (IBC) is a copyrighted work of the International Code Council. The *Building Code Requirements and Specification for Masonry Structures* (TMS 402-13/ACI 530-13/ASCE 5-13 and TMS 602-13/ACI 530.1-13/ASCE 6-13) are copyrighted works of The Masonry Society, American Concrete Institute and the Structural Engineering Institute of the American Society of Civil Engineers. The *Code Requirements for Determining Fire Resistance of Concrete and Masonry Construction Assemblies* (ACI 216.1-14/TMS 0216-14) is a copyrighted work of the American Concrete Institute and The Masonry Society. The *Standard Method for Determining Sound Transmission Ratings for Masonry Walls* (TMS 0302-12), *Direct Design Handbook for Masonry Structures* (TMS 403-13) and *Masonry Inspection Checklist* are copyrighted works of The Masonry Society.

This book was prepared in keeping with current information and practice for the present state of the art of masonry design and construction.

The publishers and all organizations and individuals who have contributed to this book cannot assume or accept any responsibility or liability, including liability for negligence, for errors or oversights in this data and information and in the use of such information.

Trademarks: "International Code Council", the "International Code Council" logo and the "International Building Code" are trademarks of the International Code Council, Inc.

PRINTED IN THE U.S.A.

Table of Contents

INTRODUCTION v

INTERNATIONAL BUILDING CODE, 2015 EDITION 1
   Preface 2
   2015 IBC Chapter 1, *Scope and Administration* 5
   2015 IBC Chapter 7, *Fire and Smoke Protection Features* 7
   2015 IBC Chapter 12, *Interior Environment* 21
   2015 IBC Chapter 14, *Exterior Walls* 23
   2015 IBC Chapter 16, *Structural Design* 29
   2015 IBC Chapter 17, *Special Inspections and Tests* 33
   2015 IBC Chapter 18, *Soils and Foundations* 37
   2015 IBC Chapter 21, *Masonry* 47
   2015 IBC Chapter 23, *Wood* 63

2013 BUILDING CODE REQUIREMENTS FOR MASONRY STRUCTURES 67
   Chapter 1, *General Requirements* 81
   Chapter 2, *Notation and Definitions* 87
   Chapter 3, *Quality and Construction* 105
   Chapter 4, *General Analysis and Design Considerations* 115
   Chapter 5, *Structural Elements* 127
   Chapter 6, *Reinforcement, Metal Accessories, and Anchor Bolts* 147
   Chapter 7, *Seismic Design Requirements* 159
   Chapter 8, *Allowable Stress Design of Masonry* 175
   Chapter 9, *Strength Design of Masonry* 203
   Chapter 10, *Prestressed Masonry* 233
   Chapter 11, *Strength Design of Autoclaved Aerated Concrete (AAC) Masonry* 243
   Chapter 12, *Veneer* 263
   Chapter 13, *Glass Unit Masonry* 277
   Chapter 14, *Masonry Partition Walls* 283
   Appendix A, *Empirical Design of Masonry* 289
   Appendix B, *Design of Masonry Infill* 307
   Appendix C, *Limit Design Method* 315
   Equation Conversions 317
   Conversion of Inch-Pound Units to SI Units 329

2013 SPECIFICATION FOR MASONRY STRUCTURES 331
   Part 1, *General* 337
   Part 2, *Products* 367
   Part 3, *Execution* 389

DIRECT DESIGN HANDBOOK FOR MASONRY STRUCTURES 417

CODE REQUIREMENTS FOR DETERMINING FIRE RESISTANCE OF CONCRETE AND MASONRY CONSTRUCTION ASSEMBLIES 545

STANDARD METHOD FOR DETERMINING SOUND TRANSMISSION RATINGS FOR MASONRY WALLS 559

TMS MASONRY INSPECTION CHECKLIST 581

BIA TECHNICAL NOTES ON BRICK CONSTRUCTION: SUBJECT INDEX 603

TEK MANUAL FOR CONCRETE MASONRY DESIGN AND CONSTRUCTION - TABLE OF CONTENT 607

INDEX 611

INTRODUCTION

The Masonry Society (TMS), the International Code Council (ICC®), and The Masonry Institute of America (MIA) are proud to make available this 2015 Masonry Codes and Specification Compilation which provides the most current and most used codes and specifications that are needed by masonry designers, contractors, and inspectors on a daily basis. The intent of this Compilation is to put all these needed resources in one place for easy access and use.

TMS, ICC and MIA believe that the best way to extend and improve the use of masonry is through education and the dissemination of information. Accordingly, they have joined together to develop this new resource, which will make the jobs of everyone involved in masonry design and construction easier and more consistent.

This manual, published by The Masonry Society, the International Code Council and Masonry Institute of America, provides various code requirements for masonry from the International Building Code (IBC), the Building Code Requirements for Masonry Structures (TMS 402-13/ACI 530-13/ASCE 5-13), Specification for Masonry Structures (TMS 602-13/ACI 530.1-13/ASCE 6-13), referenced as TMS 402 and TMS 602 throughout this book, and many other standards and specifications. TMS 402 and TMS 602 are developed and maintained by the Masonry Standards Joint Committee (MSJC). The book is divided into several major sections as listed in the Table of Contents. The first major section contains the 2015 IBC Masonry Provisions from Chapters 21, 14, and 17 which reference provisions from the 2013 TMS 402 and TMS 602. This section also contains masonry-related information from IBC Chapters 1, 7, 12, 16, 18 and 23 intended to provide users with the combined information needed from these resources which contain the bulk of masonry design and construction requirements needed under the 2015 IBC. When developing this section, the publishers used the following as a guide for showing the MSJC provisions with the IBC text:

- Since the IBC governs, the first section contains IBC text which extensively references the TMS 402 and TMS 602 for additional provisions.

- Section 2 contains text of TMS 402, Building Code Requirements for Masonry Structures. The TMS 402 and TMS 602 provisions are in a two-column format with the Code and Specification contained in the left column and the Commentary, which provides explanation of the provisions, in the right column.

    - In some places (e.g. IBC Section 2107.2.1 and TMS 402 Section 8.1.6.7.1.1), the IBC and MSJC contain different requirements. In such cases, the IBC requirements prevail where the IBC code has been adopted.

    - The user may also notice that the TMS 402 has been significantly reformatted. The previous Chapter 1 has been divided into seven logical design sections and more of the common design provisions have been moved into the first seven chapters.

- Section 3 contains text of TMS 602, Specification for Masonry Structures. This section is of particular interest to contractors, inspectors and anyone else related in the actual construction of masonry.

The remainder of the book includes additional references that are commonly needed by designers, contractors, inspectors, and others in the masonry industry. Some, but certainly not all, of these Codes and Specifications reference masonry requirements from Code Requirements for Determining Fire Resistance of Concrete and Masonry Construction Assemblies (ACI 216.1-14/TMS 0216-14), Standard Method for Determining Sound Transmission Ratings for Masonry Walls (TMS 0302-12), and the Direct Design Handbook for Masonry Structures (TMS 403-13). The publishers hope that by combining these resources together, application of these provisions will be easier for everyone using masonry.

This "2015 Masonry Codes and Specifications Compilation" was prepared by the Masonry Institute of America, Torrance, California in cooperation with International Code Council and The Masonry Society. Every effort has been taken to ensure that all data and information furnished is as accurate as possible. The editors and publishers cannot assume or accept any responsibility or liability, including liability for negligence, for errors or oversights in this data and information, and the use of such information.

This book contains information obtained from authentic and highly regarded sources. Reprinted material is quoted with permission, and sources are indicated. A wide variety of references are listed.

Contributing staff included John Chrysler, PE, MIA Executive Director; Thomas Escobar, Assoc. AIA, MIA Design Director; Luis Dominguez and Debby Chrysler, MIA Production Specialists. Design guidance was provided by Phillip Samblanet, PE, TMS Executive Director and Mark Johnson, ICC Executive Vice-President, Director of Business Development.

INTRODUCTION

EXCERPTS FROM

"Portions of this work are reproduced from the 2015 *International Building Code* (IBC), copyright © 2015 with the permission of the publisher, the International Code Council. The 2015 *International Building Code* (IBC), is a copyrighted work of the International Code Council."

International Code Council
500 NEW JERSEY AVENUE, NW, 6TH FLOOR
WASHINGTON, DC 20001-2070
(888) 422-7233
www.iccsafe.org

Deletion Arrow and Revision Bar
➡ Large arrow refers to text deleted from 2012 IBC.
▏ Thick bar refers to modified text from 2012 IBC.

PREFACE

Introduction

Internationally, code officials recognize the need for a modern, up-to-date building code addressing the design and installation of building systems through requirements emphasizing performance. The *International Building Code®*, in this 2015 edition, is designed to meet these needs through model code regulations that safeguard the public health and safety in all communities, large and small.

This comprehensive building code establishes minimum regulations for building systems using prescriptive and performance-related provisions. It is founded on broad-based principles that make possible the use of new materials and new building designs. This 2015 edition is fully compatible with all the *International Codes®* (I-Codes®) published by the International Code Council (ICC®), including the *International Energy Conservation Code®*, *International Existing Building Code®*, *International Fire Code®*, *International Fuel Gas Code®*, *International Green Construction Code®*, *International Mechanical Code®*, *ICC Performance Code®* for Buildings and Facilities, *International Plumbing Code®*, *International Private Sewage Disposal Code®*, *International Property Maintenance Code®*, *International Residential Code®* For One- and Two-Family Dwellings, *International Swimming Pool and Spa Code™*, *International Wildland-Urban Interface Code®* and *International Zoning Code®*.

The *International Building Code* provisions provide many benefits, among which is the model code development process that offers an international forum for building professionals to discuss performance and prescriptive code requirements. This forum provides an excellent arena to debate proposed revisions. This model code also encourages international consistency in the application of provisions.

Development

The first edition of the *International Building Code* (2000) was the culmination of an effort initiated in 1997 by the ICC. This included five drafting subcommittees appointed by ICC and consisting of representatives of the three statutory members of the International Code Council at that time, including: Building Officials and Code Administrators International, Inc. (BOCA), International Conference of Building Officials (ICBO) and Southern Building Code Congress International (SBCCI). The intent was to draft a comprehensive set of regulations for building systems consistent with and inclusive of the scope of the existing model codes. Technical content of the latest model codes promulgated by BOCA, ICBO and SBCCI was utilized as the basis for the development, followed by public hearings in 1997, 1998 and 1999 to consider proposed changes. This 2015 edition presents the code as originally issued, with changes reflected in the 2003, 2006, 2009 and 2012 editions and further changes approved by the ICC Code Development Process through 2014. A new edition such as this is promulgated every 3 years.

This code is founded on principles intended to establish provisions consistent with the scope of a building code that adequately protects public health, safety and welfare; provisions that do not unnecessarily increase construction costs; provisions that do not restrict the use of new materials, products or methods of construction; and provisions that do not give preferential treatment to particular types or classes of materials, products or methods of construction.

Adoption

The International Code Council maintains a copyright in all of its codes and standards. Maintaining copyright allows the ICC to fund its mission through sales of books, in both print and electronic formats. The *International Building Code* is designed for adoption and use by jurisdictions that recognize and acknowledge the ICC's copyright in the code, and further acknowledge the substantial shared value of the public/private partnership for code development between jurisdictions and the ICC.

The ICC also recognizes the need for jurisdictions to make laws available to the public. All ICC does and ICC standards, along with the laws of many jurisdictions, are available for free in a nondownloadable form on the ICC's website. Jurisdictions should contact the ICC at adoptions@iccsafe.org to learn how to adopt and distribute laws based on the *International Building Code* in a manner that provides necessary access, while maintaining the ICC's copyright.

Deletion Arrow and Revision Bar

 Large arrow refers to text deleted from 2012 IBC.

| Thick bar refers to modified text from 2012 IBC.

Maintenance

The *International Building Code* is kept up to date through the review of proposed changes submitted by code enforcing officials, industry representatives, design professionals and other interested parties. Proposed changes are carefully considered through an open code development process in which all interested and affected parties may participate.

The contents of this work are subject to change both through the code development cycles and the governmental body that enacts the code into law. For more information regarding the code development process, contact the Codes and Standards Development Department of the International Code Council.

While the development procedure of the *International Building Code* assures the highest degree of care, ICC, its members and those participating in the development of this code do not accept any liability resulting from compliance or noncompliance with the provisions because ICC does not have the power or authority to police or enforce compliance with the contents of this code. Only the governmental body that enacts the code into law has such authority.

Code Development Committee Responsibilities (Letter Designations in Front of Section Numbers)

In each code development cycle, code proposals to this code are considered at the Code Development Hearings by 11 different code development committees. Four of these committees have primary responsibility for designated chapters and appendices as follows:

IBC - Fire Safety
    Code Development Committee [BF]:   Chapters 7, 8, 9, 14, 26

IBC – General
    Code Development Committee [BG]:   Chapters 2, 3, 4, 5, 6, 12, 27, 28, 29, 30, 31, 32, 33, Appendices A, B, C, D, K

IBC – Means of Egress
    Code Development Committee [BE]:   Chapters 10, 11, Appendix E

IBC – Structural
    Code Development Committee [BS]:   Chapters 15, 16, 17, 18, 19, 20, 21, 22, 23, 24, 25, Appendices F, G, H, I, J, L, M

Code change proposals to sections of the code that are preceded by a bracketed letter designation, such as [A], will be considered by a committee other than the building code committee listed for the chapter or appendix above. For example, proposed code changes to Section [F] 307.1.1 will be considered by the international Fire Code Development Committee during the Committee Action Hearing in the 2016 (Group B) code development cycle.

Another example is Section [BF] 1505.2. While code change proposals to Chapter 15 are primarily the responsibility of the IBC – Structural Code Development Committee, which considers code change proposals during the 2016 (Group B) code development cycle. Section 1505.2 is the responsibility of the IBC – Fire Safety Code Development Committee, which considers code change proposals during the 2015 (Group A) code development cycle.

The bracketed letter designations for committees responsible for portions of this code are as follows:

    [A]   =   Administrative Code Development Committee;
    [BE]   =   IBC – Means of Egress Code Development Committee;
    [BF]   =   IBC – Fire Safety Code Development Committee;
    [BG]   =   IBC – General Code Development Committee;
    [BS]   =   IBC – Structural Code Development Committee;
    [E]   =   International Energy Conservation Code Development Committee (Commercial Energy Committee or Residential Energy Committee, as applicable);
    [EB]   =   International Existing Building Code Development Committee;

Deletion Arrow and Revision Bar

➡ Large arrow refers to text deleted from 2012 IBC.

❘ Thick bar refers to modified text from 2012 IBC.

[F] = International Fire Code Development Committee;
[FG] = International Fuel Gas Code Development Committee;
[M] = International Mechanical Code Development Committee; and
[P] = International Plumbing Code Development Committee.

Marginal Markings

Solid vertical lines in the margins within the body of the code indicate a technical change from the requirements of the 2012 edition. Deletion indicators in the form of an arrow (➡) are provided in the margin where an entire section, paragraph, exception or table has been deleted or an item in a list of items or a table has been deleted.

A single asterisk [*] placed in the margin indicates that text or a table has been relocated within the code. A double asterisk [**] placed in the margin indicates that the text or table immediately following it has been relocated there from elsewhere in the code. The following table indicates such relocations in the 2015 edition of the *International Building Code*.

Deletion Arrow and Revision Bar

➡ Large arrow refers to text deleted from 2012 IBC.

| Thick bar refers to modified text from 2012 IBC.

2015 IBC
CHAPTER 1*
SCOPE AND ADMINISTRATION

PART 2 — ADMINISTRATION AND ENFORCEMENT

SECTION 104
DUTIES AND POWERS OF BUILDING OFFICIAL

[A] 104.1 General. The building official is hereby authorized and directed to enforce the provisions of this code. The building official shall have the authority to render interpretations of this code and to adopt policies and procedures in order to clarify the application of its provisions. Such interpretations, policies and procedures shall be in compliance with the intent and purpose of this code. Such policies and procedures shall not have the effect of waiving requirements specifically provided for in this code.

[A] 104.10 Modifications. Where there are practical difficulties involved in carrying out the provisions of this code, the building official shall have the authority to grant modifications for individual cases, upon application of the owner or owner's authorized agent, provided that the building official shall first find that special individual reason makes the strict letter of this code impractical, and the modification is in compliance with the intent and purpose of this code and that such modification does not lessen health, accessibility, life and fire safety or structural requirements. The details of action granting modifications shall be recorded and entered in the files of the department of building safety.

[A] 104.10.1 Flood hazard areas. The building official shall not grant modifications to any provision required in flood hazard areas as established by Section 1612.3 unless a determination has been made that:

1. A showing of good and sufficient cause that the unique characteristics of the size, configuration or topography of the site render the elevation standards of Section 1612 inappropriate.

2. A determination that failure to grant the variance would result in exceptional hardship by rendering the lot undevelopable.

3. A determination that the granting of a variance will not result in increased flood heights, additional threats to public safety, extraordinary public expense, cause fraud on or victimization of the public, or conflict with existing laws or ordinances.

4. A determination that the variance is the minimum necessary to afford relief, considering the flood hazard.

5. Submission to the applicant of written notice specifying the difference between the design flood elevation and the elevation to which the building is to be built, stating that the cost of flood insurance will be commensurate with the increased risk resulting from the reduced floor elevation, and stating that construction below the design flood elevation increases risks to life and property.

[A] 104.11 Alternative materials, design and methods of construction and equipment. The provisions of this code are not intended to prevent the installation of any material or to prohibit any design or method of construction not specifically prescribed by this code, provided that any such alternative has been approved. An alternative material, design or method of construction shall be approved where the building official finds that the proposed design is satisfactory and complies with the intent of the provisions of this code, and that the material, method or work offered is, for the purpose intended, not less than the equivalent of that prescribed in this code in quality, strength, effectiveness, fire resistance, durability and safety. Where the alternative material, design or method of construction is not approved, the building official shall respond in writing, stating the reasons why the alternative was not approved.

[A] 104.11.1 Research reports. Supporting data, where necessary to assist in the approval of materials or assemblies not specifically provided for in this code, shall consist of valid research reports from approved sources.

[A] 104.11.2 Tests. Whenever there is insufficient evidence of compliance with the provisions of this code, or evidence that a material or method does not conform to the requirements of this code, or in order to substantiate claims for alternative materials or methods, the building official shall have the authority to require tests as evidence of compliance to be made at no expense to the jurisdiction. Test methods shall be as specified in this code or by other recognized test standards. In the absence of recognized and accepted test methods, the building official shall approve the testing procedures. Tests shall be performed by an approved agency. Reports of such tests shall be retained by the building official for the period required for retention of public records.

---

* Only portions of this section are shown which are particularly applicable to masonry construction. For additional information see the IBC

Deletion Arrow and Revision Bar

➡ Large arrow refers to text deleted from 2012 IBC.

| Thick bar refers to modified text from 2012 IBC.

Deletion Arrow and Revision Bar

➡ Large arrow refers to text deleted from 2012 IBC.

| Thick bar refers to modified text from 2012 IBC.

2015 IBC
CHAPTER 7*
FIRE AND SMOKE PROTECTION FEATURES

SECTION 701
GENERAL

701.1 Scope. The provisions of this chapter shall govern the materials, systems and assemblies used for structural fire resistance and fire-resistance-rated construction separation of adjacent spaces to safeguard against the spread of fire and smoke within a building and the spread of fire to or from buildings.

SECTION 704
FIRE-RESISTANCE RATING OF STRUCTURAL MEMBERS

704.1 Requirements. The fire-resistance ratings of structural members and assemblies shall comply with this section and the requirements for the type of construction as specified in Table 601. The fire-resistance ratings shall be not less than the ratings required for the fire-resistance-rated assemblies supported by the structural members.

> **Exception:** Fire barriers, fire partitions, smoke barriers and horizontal assemblies as provided in Sections 707.5, 708.4, 709.4 and 711.2, respectively.

704.7 Reinforcing. Thickness of protection for concrete or masonry reinforcement shall be measured to the outside of the reinforcement except that stirrups and spiral reinforcement ties are permitted to project not more than 0.5-inch (12.7 mm) into the protection.

SECTION 706
FIRE WALLS

706.1 General. Each portion of a building separated by one or more fire walls that comply with the provisions of this section shall be considered a separate building. The extent and location of such fire walls shall provide a complete separation. Where a fire wall separates occupancies that are required to be separated by a fire barrier wall, the most restrictive requirements of each separation shall apply.

706.7 Combustible framing in fire walls. Adjacent combustible members entering into a concrete or masonry fire wall from opposite sides shall not have less than a 4-inch (102 mm) distance between embedded ends. Where combustible members frame into hollow walls or walls of hollow units, hollow spaces shall be solidly filled for the full thickness of the wall and for a distance not less than 4 inches (102 mm) above, below and between the structural members, with noncombustible materials approved for fireblocking.

SECTION 712
VERTICAL OPENINGS

712.1 General. Each vertical opening shall comply in accordance with one of the protection methods in Sections 712.1.1 through 712.1.16.

> **712.1.8 Masonry chimney.** Approved vertical openings for masonry chimneys shall be permitted where the annular space is fireblocked at each floor level in accordance with Section 718.2.5.

SECTION 713
SHAFT ENCLOSURES

713.1 General. The provisions of this section shall apply to shafts required to protect openings and penetrations through floor/ceiling and roof/ceiling assemblies. Interior exit stairways and ramps shall be enclosed in accordance with Section 1023.

713.2 Construction. Shaft enclosures shall be constructed as fire barriers in accordance with Section 707 or horizontal assemblies in accordance with Section 711, or both.

SECTION 714
PENETRATIONS

714.1 Scope. The provisions of this section shall govern the materials and methods of construction used to protect through penetrations and membrane penetrations of horizontal assemblies and fire-resistance-rated wall assemblies.

714.3 Fire-resistance-rated walls. Penetrations into or through fire walls, fire barriers, smoke barrier walls and fire partitions shall comply with Sections 714.3.1 through 714.3.3. Penetrations in smoke barrier walls shall also comply with Section 714.4.4.

---

* Only portions of this section are shown which are particularly applicable to masonry construction. For additional information see the IBC.

Deletion Arrow and Revision Bar

→ Large arrow refers to text deleted from 2012 IBC.

| Thick bar refers to modified text from 2012 IBC.

7

714.3.1 Through penetrations. Through penetrations of fire-resistance-rated walls shall comply with Section 714.3.1.1 or 714.3.1.2.

Exception: Where the penetrating items are steel, ferrous or copper pipes, tubes or conduits, the annular space between the penetrating item and the fire-resistance-rated wall is permitted to be protected by either of the following measures:

1. In concrete or masonry walls where the penetrating item is a maximum 6-inch (152 mm) nominal diameter and the area of the opening through the wall does not exceed 144 square inches (0.0929 m^2), concrete, grout or mortar is permitted where installed the full thickness of the wall or the thickness required to maintain the fire-resistance rating.

2. The material used to fill the annular space shall prevent the passage of flame and hot gases sufficient to ignite cotton waste when subjected to ASTM E 119 or UL 263 time-temperature fire conditions under a minimum positive pressure differential of 0.01 inch (2.49 Pa) of water at the location of the penetration for the time period equivalent to the fire-resistance rating of the construction penetrated.

714.4 Horizontal assemblies. Penetrations of a fire-resistance-rated floor, floor/ceiling assembly or the ceiling membrane of a roof/ceiling assembly not required to be enclosed in a shaft by Section 712.1 shall be protected in accordance with Sections 714.4.1 through 714.4.4.

714.4.1 Through penetrations. Through penetrations of horizontal assemblies shall comply with Section 714.4.1.1 or 714.4.1.2.

Exceptions:

1. Penetrations by steel, ferrous or copper conduits, pipes, tubes or vents or concrete or masonry items through a single fire-resistance-rated floor assembly where the annular space is protected with materials that prevent the passage of flame and hot gases sufficient to ignite cotton waste when subjected to ASTM E 119 or UL 263 time-temperature fire conditions under a minimum positive pressure differential of 0.01 inch (2.49 Pa) of water at the location of the penetration for the time period equivalent to the fire-resistance rating of the construction penetrated. Penetrating items with a maximum 6-inch (152 mm) nominal diameter shall not be limited to the penetration of a single fire-resistance-rated floor assembly, provided the aggregate area of the openings through the assembly does not exceed 144 square inches (92 900 mm^2) in any 100 square feet (9.3 m^2) of floor area.

714.4.2 Membrane penetrations. Penetrations of membranes that are part of a horizontal assembly shall comply with Section 714.4.1.1 or 714.4.1.2. Where floor/ceiling assemblies are required to have a fire-resistance rating, recessed fixtures shall be installed such that the required fire resistance will not be reduced.

Exceptions:

1. Membrane penetrations by steel, ferrous or copper conduits, pipes, tubes or vents, or concrete or masonry items where the annular space is protected either in accordance with Section 714.4.1 or to prevent the free passage of flame and the products of combustion. The aggregate area of the openings through the membrane shall not exceed 100 square inches (64 500 mm^2) in any 100 square feet (9.3 m^2) of ceiling area in assemblies tested without penetrations.

SECTION 718
CONCEALED SPACES

718.1 General. Fireblocking and draftstopping shall be installed in combustible concealed locations in accordance with this section. Fireblocking shall comply with Section 718.2. Draftstopping in floor/ceiling spaces and attic spaces shall comply with Sections 718.3 and 718.4, respectively. The permitted use of combustible materials in concealed spaces of buildings of Type I or II construction shall be limited to the applications indicated in Section 718.5.

718.2.7 Concealed sleeper spaces. Where wood sleepers are used for laying wood flooring on masonry or concrete fire-resistance-rated floors, the space between the floor slab and the underside of the wood flooring shall be filled with an approved material to resist the free passage of flame and products of combustion or fireblocked in such a manner that there will be no open spaces under the flooring that will exceed 100 square feet (9.3 m2) in area and such space shall be filled solidly under permanent partitions so that there is no communication under the flooring between adjoining rooms.

Deletion Arrow and Revision Bar

➡ Large arrow refers to text deleted from 2012 IBC.

❙ Thick bar refers to modified text from 2012 IBC.

SECTION 719
FIRE-RESISTANCE REQUIREMENTS FOR PLASTER

719.3 Noncombustible furring. In buildings of Type I and II construction, plaster shall be applied directly on concrete or masonry or on *approved* noncombustible plastering base and furring.

SECTION 721
PRESCRIPTIVE FIRE RESISTANCE

721.1 General. The provisions of this section contain prescriptive details of fire-resistance-rated building elements, components or assemblies. The materials of construction listed in Tables 721.1(1), 721.1(2), and 721.1(3) shall be assumed to have the fire-resistance ratings prescribed therein. Where materials that change the capacity for heat dissipation are incorporated into a fire-resistance-rated assembly, fire test results or other substantiating data shall be made available to the building official to show that the required fire-resistance-rating time period is not reduced.

721.1.2 Unit masonry protection. Where required, metal ties shall be embedded in bed joints of unit masonry for protection of steel columns. Such ties shall be as set forth in Table 721.1(1) or be equivalent thereto.

SECTION 722
CALCULATED FIRE RESISTANCE

722.1 General. The provisions of this section contain procedures by which the fire resistance of specific materials or combinations of materials is established by calculations. These procedures apply only to the information contained in this section and shall not be otherwise used. The calculated fire resistance of concrete, concrete masonry and clay masonry assemblies shall be permitted in accordance with ACI 216.1/TMS 0216. The calculated fire resistance of steel assemblies shall be permitted in accordance with Chapter 5 of ASCE 29. The calculated fire resistance of exposed wood members and wood decking shall be permitted in accordance with Chapter 16 of ANSI/AF&PA *National Design Specification for Wood Construction* (*NDS*).

722.2.4.4 Columns built into walls. The minimum dimensions of Table 722.2.4 do not apply to a reinforced concrete column that is built into a concrete or masonry wall provided all of the following are met:

1. The *fire-resistance rating* for the wall is equal to or greater than the required rating of the column;

2. The main longitudinal reinforcing in the column has cover not less than that required by Section 722.2.4.2; and

3. Openings in the wall are protected in accordance with Table 716.5.

Where openings in the wall are not protected as required by Section 716.5, the minimum dimension of columns required to have a *fire-resistance rating* of 3 hours or less shall be 8 inches (203 mm), and 10 inches (254 mm) for columns required to have a *fire-resistance rating* of 4 hours, regardless of the type of aggregate used in the concrete.

722.3 Concrete masonry. The provisions of this section contain procedures by which the fire-resistance ratings of concrete masonry are established by calculations.

722.3.1 Equivalent thickness. The equivalent thickness of concrete masonry construction shall be determined in accordance with the provisions of this section.

722.3.1.1 Concrete masonry unit plus finishes. The equivalent thickness of concrete masonry assemblies, T_{ea}, shall be computed as the sum of the equivalent thickness of the concrete masonry unit, T_e, as determined by Section 722.3.1.2, 722.3.1.3, or 722.3.1.4, plus the equivalent thickness of finishes, T_{ef}, determined in accordance with Section 722.3.2:

$$T_{ea} = T_e + T_{ef} \quad \text{(Equation 7-6)}$$

722.3.1.2 Ungrouted or partially grouted construction. T_e shall be the value obtained for the concrete masonry unit determined in accordance with ASTM C 140.

722.3.1.3 Solid grouted construction. The equivalent thickness, T_e, of solid grouted concrete masonry units is the actual thickness of the unit.

722.3.1.4 Airspaces and cells filled with loose-fill material. The equivalent thickness of completely filled hollow concrete masonry is the actual thickness of the unit where loose-fill materials are: sand, pea gravel, crushed stone, or slag that meet ASTM C 33 requirements; pumice, scoria, expanded shale, expanded clay, expanded slate, expanded slag, expanded fly ash, or cinders that comply with ASTM C 331; or perlite or vermiculite meeting the requirements of ASTM C 549 and ASTM C 516, respectively.

722.3.2 Concrete masonry walls. The fire-resistance rating of walls and partitions constructed of concrete masonry units shall be determined from Table 722.3.2. The rating shall be based on the equivalent thickness of the masonry and type of aggregate used.

Deletion Arrow and Revision Bar

➡ Large arrow refers to text deleted from 2012 IBC.

| Thick bar refers to modified text from 2012 IBC.

TABLE 721.1(2)
RATED FIRE-RESISTANCE PERIODS FOR VARIOUS WALLS AND PARTITIONS[a, o, p]

MATERIAL	ITEM NUMBER	CONSTRUCTION	MINIMUM FINISHED THICKNESS FACE-TO-FACE[b] (inches) 4 hours	3 hours	2 hours	1 hour
1. Brick of clay or shale	1-1.1	Solid brick of clay or shale[c].	6	4.9	3.8	2.7
	1-1.2	Hollow brick, not filled.	5.0	4.3	3.4	2.3
	1-1.3	Hollow brick unit wall, grout or filled with perlite vermiculite or expanded shale aggregate.	6.6	5.5	4.4	3.0
	1-2.1	4" nominal thick units not less than 75 percent solid backed with a hat-shaped metal furring channel $^3/_4$" thick formed from 0.021" sheet metal attached to the brick wall on 24" centers with approved fasteners, and $^1/_2$" Type X gypsum wallboard attached to the metal furring strips with 1"-long Type S screws spaced 8" on center	—	—	5[d]	—
2. Combination of clay brick and load-bearing hollow clay tile	2-1.1	4" solid brick and 4" tile (not less than 40 percent solid).	—	8	—	—
	2-1.2	4" solid brick and 8" tile (not less than 40 percent solid).	12	—	—	—
3. Concrete masonry units	3-1.1[f, g]	Expanded slag or pumice.	4.7	4.0	3.2	2.1
	3-1.2[f, g]	Expanded clay, shale or slate.	5.1	4.4	3.6	2.6
	3-1.3[f]	Limestone, cinders or air-cooled slag.	5.9	5.0	4.0	2.7
	3-1.4[f, g]	Calcareous or siliceous gravel.	6.2	5.3	4.2	2.8
4. Solid concrete[h, i]	4-1.1	Siliceous aggregate concrete.	7.0	6.2	5.0	3.5
		Carbonate aggregate concrete.	6.6	5.7	4.6	3.2
		Sand-lightweight concrete.	5.4	4.6	3.8	2.7
		Lightweight concrete.	5.1	4.4	3.6	2.5
5. Glazed or unglazed facing tile, nonload-bearing	5-1.1	One 2" unit cored 15 percent maximum and one 4" unit cored 25 percent maximum with $^3/_4$" mortar-filled collar joint. Unit positions reversed in alternate courses.	—	$6^3/_8$	—	—
	5-1.2	One 2" unit cored 15 percent maximum and one 4" unit cored 40 percent maximum with $^3/_4$" mortar-filled collar joint. Unit positions side with $^3/_4$" gypsum plaster. Two wythes tied together every fourth course with No. 22 gage corrugated metal ties.	—	$6^3/_4$	—	—
	5-1.3	One unit with three cells in wall thickness, cored 29 percent maximum.	—	—	6	—
	5-1.4	One 2" unit cored 22 percent maximum and one 4" unit cored 41 percent maximum with $^1/_4$" mortar-filed collar joint. Two wythes tied together every third course with 0.030" (No. 22 galvanized sheet steel gage) corrugated metal ties.	—	—	6	—
	5-1.5	One 4" unit cored 25 percent maximum with $^3/_4$" gypsum plaster on one side.	—	—	$4^3/_4$	—
	5-1.6	One 4" unit with two cells in wall thickness, cored 22 percent maximum.	—	—	—	4
	5-1.7	One 4" unit cored 30 percent maximum with $^3/_4$" vermiculite gypsum plaster on one side.	—	—	$4^1/_2$	—
	5-1.8	One 4" unit cored 39 percent maximum with $^3/_4$" gypsum plaster on one side.	—	—	—	$4^1/_2$

For SI: 1 inch = 25.4 mm, 1 square inch = 645.2 mm², 1 cubic foot = 0.0283 m³.

a. Staples with equivalent holding power and penetration shall be permitted to be used as alternate fasteners to nails for attachment to wood framing.
b. Thickness shown for brick and clay tile is nominal thicknesses unless plastered, in which case thicknesses are net. Thickness shown for concrete masonry and clay masonry is equivalent thickness defined in Section 722.3.1 for concrete masonry and Section 722.4.1.1 for clay masonry. Where all cells are solid grouted or filled with silicone-treated perlite loose-fill insulation; vermiculite loose-fill insulation; or expanded clay, shale or slate lightweight aggregate, the equivalent thickness shall be the thickness of the block or brick using specified dimensions as defined in Chapter 21. Equivalent thickness shall include the thickness of applied plaster and lath or gypsum wallboard, where specified.
c. For units in which the net cross-sectional area of cored brick in any plane parallel to the surface containing the cores is not less than 75 percent of the gross cross-sectional area measured in the same plane.
d. Shall be used for nonbearing purposes only.
e. For all of the construction with gypsum wallboard described in this table, gypsum base for veneer plaster of the same size, thickness and core type shall be permitted to be substituted for gypsum wallboard, provided attachment is identical to that specified for the wallboard, and the joints on the face layer are reinforced and the entire surface is covered with not less than $^1/_{16}$-inch gypsum veneer plaster.
f. The fire-resistance time period for concrete masonry units meeting the equivalent thicknesses required for a 2-hour fire-resistance rating in Item 3, and having a thickness of not less than $7^5/_8$ inches is 4 hours when cores that are not grouted are filled with silicone-treated perlite loose-fill insulation; vermiculite loose-fill insulation; or expanded clay, shale or slate lightweight aggregate, sand or slag having a maximum particle size of $^3/_8$ inch.
g. The fire-resistance rating of concrete masonry units composed of a combination of aggregate types or where plaster is applied directly to the concrete masonry shall be determined in accordance with ACI 216.1/TMS 0216. Lightweight aggregates shall have a maximum combined density of 65 pounds per cubic foot.
h. See Note b. The equivalent thickness shall be permitted to include the thickness of cement plaster or 1.5 times the thickness of gypsum plaster applied in accordance with the requirements of Chapter 25.
i. Concrete walls shall be reinforced with horizontal and vertical temperature reinforcement as required by Chapter 19.

Deletion Arrow and Revision Bar

➤ Large arrow refers to text deleted from 2012 IBC.

▎ Thick bar refers to modified text from 2012 IBC.

Table 721.1(2) Notes - continued

j. Studs are welded truss wire studs with 0.18 inch (No. 7 B.W. gage) flange wire and 0.18 inch (No. 7 B.W. gage) truss wires.
k. Nailable metal studs consist of two channel studs spot welded back to back with a crimped web forming a nailing groove.
l. Wood structural panels shall be permitted to be installed between the fire protection and the wood studs on either the interior or exterior side of the wood frame assemblies in this table, provided the length of the fasteners used to attach the fire protection is increased by an amount not less than the thickness of the wood structural panel.
m. For studs with a slenderness ratio, l_e/d, greater than 33, the design stress shall be reduced to 78 percent of allowable F'_c. For studs with a slenderness ratio, l_e/d, not exceeding 33, the design stress shall be reduced 78 percent of the adjusted stress F'_c calculated for studs having a slenderness ratio l_e/d of 33.
n. For properties of cooler or wallboard nails, see ASTM C 514, ASTM C 547 or ASTM F 1667.
o. Generic fire-resistance ratings (those not designated as PROPRIETARY* in the listing) in the GA 600 shall be accepted as if herein listed.
p. NCMA TEK 5-8A, shall be permitted for the design of fire walls.
q. The design stress of studs shall be equal to a maximum of 100 percent of the allowable F'_c calculated in accordance with Section 2306.

TABLE 722.2.1.4(1)
MULTIPLYING FACTOR FOR FINISHES ON NONFIRE-EXPOSED SIDE OF WALL

TYPE OF FINISH APPLIED TO CONCRETE OR CONCRETE MASONRY WALL	TYPE OF AGGREGATE USED IN CONCRETE OR CONCRETE MASONRY			
	Concrete: siliceous or carbonate Concrete Masonry: Siliceous or carbonate; Solid clay brick	Concrete: sand-lightweight Concrete Masonry: clay tile; hollow clay brick; concrete masonry units of expanded shale and < 20% sand	Concrete: lightweight Concrete Masonry: concrete masonry units of expanded shale, expanded clay, expanded slag, or pumice < 20% sand	Concrete Masonry: concrete masonry units of expanded slag, expanded clay, or pumice
Portland cement-sand plaster	1.00	0.75[a]	0.75[a]	0.50[a]
Gypsum-sand plaster	1.25	1.00	1.00	1.00
Gypsum-vermiculite or perlite plaster	1.75	1.50	1.25	1.25
Gypsum wallboard	3.00	2.25	2.25	2.25

For SI: 1 inch = 25.4 mm.
a. For Portland cement-sand plaster 5/8 inch or less in thickness and applied directly to the concrete or concrete masonry on the nonfire-exposed side of the wall, the multiplying factor shall be 1.00.

TABLE 722.2.1.4(2)
TIME ASSIGNED TO FINISH MATERIALS ON FIRE-EXPOSED SIDE OF WALL

FINISH DESCRIPTION	TIME (minutes)
Gypsum wallboard	
³/₈ inch	10
¹/₂ inch	15
⁵/₈ inch	20
2 layers of ³/₈ inch	25
1 layer of ³/₈ inch, 1 layer of ¹/₂ inch	35
2 layers of ¹/₂ inch	40
Type X gypsum wallboard	
¹/₂ inch	25
⁵/₈ inch	40
Portland cement-sand plaster applied directly to concrete masonry	See Note a
Portland cement-sand plaster on metal lath	
³/₄ inch	20
⁷/₈ inch	25
1 inch	30
Gypsum sand plaster on ³/₈-inch gypsum lath	
¹/₂ inch	35
⁵/₈ inch	40
³/₄ inch	50
Gypsum sand plaster on metal lath	
³/₄ inch	50
⁷/₈ inch	60
1 inch	80

For SI: 1 inch = 25.4 mm.
a. The actual thickness of Portland cement-sand plaster, provided it is ⁵/₈ inch or less in thickness, shall be permitted to be included in determining the equivalent thickness of the masonry for use in Table 722.3.2.

Deletion Arrow and Revision Bar

➡ Large arrow refers to text deleted from 2012 IBC.

❙ Thick bar refers to modified text from 2012 IBC.

TABLE 722.3.2
MINIMUM EQUIVALENT THICKNESS (inches) OF BEARING OR NONBEARING CONCRETE MASONRY WALLS[a,b,c,d]

TYPE OF AGGREGATE	FIRE-RESISTANCE RATING (hours)														
	$1/2$	$3/4$	1	$1^1/_4$	$1^1/_2$	$1^3/_4$	2	$2^1/_4$	$2^1/_2$	$2^3/_4$	3	$3^1/_4$	$3^1/_2$	$3^3/_4$	4
Pumice or expanded slag	1.5	1.9	2.1	2.5	2.7	3.0	3.2	3.4	3.6	3.8	4.0	4.2	4.4	4.5	4.7
Expanded shale, clay or slate	1.8	2.2	2.6	2.9	3.3	3.4	3.6	3.8	4.0	4.2	4.4	4.6	4.8	4.9	5.1
Limestone, cinders or unexpanded slag	1.9	2.3	2.7	3.1	3.4	3.7	4.0	4.3	4.5	4.8	5.0	5.2	5.5	5.7	5.9
Calcareous or siliceous gravel	2.0	2.4	2.8	3.2	3.6	3.9	4.2	4.5	4.8	5.0	5.3	5.5	5.8	6.0	6.2

For SI: 1 inch = 25.4 mm.
a. Values between those shown in the table can be determined by direct interpolation.
b. Where combustible members are framed into the wall, the thickness of solid material between the end of each member and the opposite face of the wall, or between members set in from opposite sides, shall be not less than 93 percent of the thickness shown in the table.
c. Requirements of ASTM C 55, ASTM C 73, ASTM C 90 or ASTM C 744 shall apply.
d. Minimum required equivalent thickness corresponding to the hourly fire-resistance rating for units with a combination of aggregate shall be determined by linear interpolation based on the percent by volume of each aggregate used in manufacture.

722.3.2.1 Finish on nonfire-exposed side. Where plaster or gypsum wallboard is applied to the side of the wall not exposed to fire, the contribution of the finish to the total fire-resistance rating shall be determined as follows: The thickness of gypsum wallboard or plaster shall be corrected by multiplying the actual thickness of the finish by applicable factor determined from Table 722.2.1.4(1). This corrected thickness of finish shall be added to the equivalent thickness of masonry and the fire-resistance rating of the masonry and finish determined from Table 722.3.2.

722.3.2.2 Finish on fire-exposed side. Where plaster or gypsum wallboard is applied to the fire-exposed side of the wall, the contribution of the finish to the total fire-resistance rating shall be determined as follows: The time assigned to the finish as established by Table 722.2.1.4(2) shall be added to the fire-resistance rating determined in Section 722.3.2 for the masonry alone, or in Section 722.3.2.1 for the masonry and finish on the nonfire-exposed side.

722.3.2.3 Nonsymmetrical assemblies. For a wall having no finish on one side or having different types or thicknesses of finish on each side, the calculation procedures of this section shall be performed twice, assuming either side of the wall to be the fire-exposed side. The fire-resistance rating of the wall shall not exceed the lower of the two values calculated.

> **Exception:** For exterior walls with a fire separation distance greater than 5 feet (1524 mm), the fire shall be assumed to occur on the interior side only.

722.3.2.4 Minimum concrete masonry fire-resistance rating. Where the finish applied to a concrete masonry wall contributes to its fire-resistance rating, the masonry alone shall provide not less than one-half the total required fire-resistance rating.

722.3.2.5 Attachment of finishes. Installation of finishes shall be as follows:

1. Gypsum wallboard and gypsum lath applied to concrete masonry or concrete walls shall be secured to wood or steel furring members spaced not more than 16 inches (406 mm) on center (o.c.).

2. Gypsum wallboard shall be installed with the long dimension parallel to the furring members and shall have all joints finished.

3. Other aspects of the installation of finishes shall comply with the applicable provisions of Chapters 7 and 25.

722.3.3 Multiwythe masonry walls. The fire-resistance rating of wall assemblies constructed of multiple wythes of masonry materials shall be permitted to be based on the fire-resistance rating period of each wythe and the continuous airspace between each wythe in accordance with the following formula:

$$R_A = (R_1^{0.59} + R_2^{0.59} + ... + R_n^{0.59} + A_1 + A_2 + ... + A_n)^{1.7}$$

(Equation 7-7)

where:

R_A = Fire-resistance rating of the assembly (hours).

$R_1, R_2, ..., R_n$ = Fire-resistance rating of wythes for 1, 2, n (hours), respectively.

$A_1, A_2, ..., A_n$ = 0.30, factor for each continuous airspace for 1, 2, ...n, respectively, having a depth of $1/2$ inch (12.7 mm) or more between wythes.

Deletion Arrow and Revision Bar
Large arrow refers to text deleted from 2012 IBC.

Thick bar refers to modified text from 2012 IBC.

722.3.4 Concrete masonry lintels. Fire-resistance ratings for concrete masonry lintels shall be determined based upon the nominal thickness of the lintel and the minimum thickness of concrete masonry or concrete, or any combination thereof, covering the main reinforcing bars, as determined in according with Table 722.3.4, or by approved alternate methods.

TABLE 722.3.4
MINIMUM COVER OF LONGITUDINAL REINFORCEMENT IN FIRE-RESISTANCE-RATED REINFORCED CONCRETE MASONRY LINTELS (inches)

NOMINAL WIDTH OF LINTEL (inches)	FIRE-RESISTANCE RATING (hours)			
	1	2	3	4
6	1 1/2	2	—	—
8	1 1/2	1 1/2	1 3/4	3
10 or greater	1 1/2	1 1/2	1 1/2	1 3/4

For SI: 1 inch = 25.4 mm.

722.3.5 Concrete masonry columns. The fire-resistance rating of concrete masonry columns shall be determined based upon the least plan dimension of the column in accordance with Table 722.3.5 or by approved alternate methods.

TABLE 722.3.5
MINIMUM DIMENSION OF CONCRETE MASONRY COLUMNS (inches)

FIRE-RESISTANCE RATING (hours)			
1	2	3	4
8 inches	10 inches	12 inches	14 inches

For SI: 1 inch = 25.4 mm.

722.4 Clay brick and tile masonry. The provisions of this section contain procedures by which the fire-resistance ratings of clay brick and tile masonry are established by calculations.

722.4.1 Masonry walls. The fire-resistance rating of masonry walls shall be based upon the equivalent thickness as calculated in accordance with this section. The calculation shall take into account finishes applied to the wall and airspaces between wythes in multiwythe construction.

722.4.1.1 Equivalent thickness. The fire-resistance ratings of walls or partitions constructed of solid or hollow clay masonry units shall be determined from Table 722.4.1(1) or 722.4.1(2). The equivalent thickness of the clay masonry unit shall be determined by Equation 7-8 where using Table 722.4.1(1). The fire-resistance rating determined from Table 722.4.1(1) shall be permitted to be used in the calculated fire-resistance rating procedure in Section 722.4.2.

$$T_e = V_n/LH \qquad \text{(Equation 7-8)}$$

where:

T_e = The equivalent thickness of the clay masonry unit (inches).

V_n = The net volume of the clay masonry unit (inch^3).

L = The specified length of the clay masonry unit (inches).

H = The specified height of the clay masonry unit (inches).

TABLE 722.4.1(1)
FIRE-RESISTANCE PERIODS OF CLAY MASONRY WALLS

MATERIAL TYPE	MINIMUM REQUIRED EQUIVALENT THICKNESS FOR FIRE RESISTANCE [a, b, c] (inches)			
	1 hour	2 hours	3 hours	4 hours
Solid brick of clay or shale[d]	2.7	3.8	4.9	6.0
Hollow brick or tile of clay or shale, unfilled	2.3	3.4	4.3	5.0
Hollow brick or tile of clay or shale, grouted or filled with materials specified in Section 722.4.1.1.3	3.0	4.4	5.5	6.6

For SI: 1 inch = 25.4 mm.
a. Equivalent thickness as determined from Section 722.4.1.1.
b. Calculated fire resistance between the hourly increments listed shall be determined by linear interpolation.
c. Where combustible members are framed in the wall, the thickness of solid material between the end of each member and the opposite face of the wall, or between members set in from opposite sides, shall be not less than 93 percent of the thickness shown.
d. For units in which the net cross-sectional area of cored brick in any plane parallel to the surface containing the cores is not less than 75 percent of the gross cross-sectional area measured in the same plane.

Deletion Arrow and Revision Bar

➜ Large arrow refers to text deleted from 2012 IBC.

❘ Thick bar refers to modified text from 2012 IBC.

TABLE 722.4.1(2)
FIRE-RESISTANCE RATINGS FOR BEARING STEEL FRAME BRICK VENEER WALLS OR PARTITIONS

WALL OR PARTITION ASSEMBLY	PLASTER SIDE EXPOSED (hours)	BRICK FACED SIDE EXPOSED (hours)
Outside facing of steel studs: $1/2$" wood fiberboard sheathing next to studs, $3/4$" airspace formed with $3/4$" x $1^5/8$" wood strips placed over the fiberboard and secured to the studs; metal or wire lath nailed to such strips, $3^3/4$" brick veneer held in place by filling $3/4$" airspace between the brick and lath with mortar. Inside facing of studs: $3/4$" unsanded gypsum plaster on metal or wire lath attached to $5/16$" wood strips secured to edges of the studs.	1.5	4
Outside facing of steel studs: 1" insulation board sheathing attached to studs, 1" airspace, and $3^3/4$" brick veneer attached to steel frame with metal ties every 5th course. Inside facing of studs: $7/8$" sanded gypsum plaster (1:2 mix) applied on metal or wire lath attached directly to the studs.	1.5	4
Same as above except use $7/8$" vermiculite-gypsum plaster or 1" sanded gypsum plaster (1:2 mix) applied to metal or wire.	2	4
Outside facing of steel studs: $1/2$" gypsum sheathing board, attached to studs, and $3^3/4$" brick veneer attached to steel frame with metal ties every 5th course. Inside facing of studs: $1/2$" sanded gypsum plaster (1:2 mix) applied to $1/2$" perforated gypsum lath securely attached to studs and having strips of metal lath 3 inches wide applied to all horizontal joints of gypsum lath.	2	4

For SI: 1 inch = 25.4 mm.

722.4.1.1.1 Hollow clay units. The equivalent thickness, T_e, shall be the value obtained for hollow clay units as determined in accordance with Equation 7-8. The net volume, V_n, of the units shall be determined using the gross volume and percentage of void area determined in accordance with ASTM C 67.

722.4.1.1.2 Solid grouted clay units. The equivalent thickness of solid grouted clay masonry units shall be taken as the actual thickness of the units.

722.4.1.1.3 Units with filled cores. The equivalent thickness of the hollow clay masonry units is the actual thickness of the unit where completely filled with loose-fill materials of: sand, pea gravel, crushed stone, or slag that meet ASTM C 33 requirements; pumice, scoria, expanded shale, expanded clay, expanded slate, expanded slag, expanded fly ash, or cinders in compliance with ASTM C 331; or perlite or vermiculite meeting the requirements of ASTM C 549 and ASTM C 516, respectively.

722.4.1.2 Plaster finishes. Where plaster is applied to the wall, the total fire-resistance rating shall be determined by the formula:

$$R = (R_n^{0.59} + pl)^{1.7}$$ (Equation 7-9)

where:

R = The fire-resistance rating of the assembly (hours).

R_n = The fire-resistance rating of the individual wall (hours).

pl = Coefficient for thickness of plaster.

Values for $R_n^{0.59}$ for use in Equation 7-9 are given in Table 722.4.1(3). Coefficients for thickness of plaster shall be selected from Table 722.4.1(4) based on the actual thickness of plaster applied to the wall or partition and whether one or two sides of the wall are plastered.

722.4.1.3 Multiwythe walls with airspace. Where a continuous airspace separates multiple wythes of the wall or partition, the total fire-resistance rating shall be determined by the formula:

$$R = (R_1^{0.59} + R_2^{0.59} + ... + R_n^{0.59} + as)^{1.7}$$

(Equation 7-10)

where:

R = The fire-resistance rating of the assembly (hours).

R_1, R_2 and R_n = The fire-resistance rating of the individual wythes (hours).

as = Coefficient for continuous airspace.

Values for $R_n^{0.59}$ for use in Equation 7-10 are given in Table 722.4.1(3). The coefficient for each continuous airspace of $1/2$ inch to $3^1/2$ inches (12.7 to 89 mm) separating two individual wythes shall be 0.3.

TABLE 722.4.1(3)
VALUES OF $R_n^{0.59}$

$R_n^{0.59}$	R (hours)
1	1.0
2	1.50
3	1.91
4	2.27

Deletion Arrow and Revision Bar

➡ Large arrow refers to text deleted from 2012 IBC.

❘ Thick bar refers to modified text from 2012 IBC.

TABLE 722.4.1(4)
COEFFICIENTS FOR PLASTER, pl[a]

THICKNESS OF PLASTER (inch)	ONE SIDE	TWO SIDES
1/2	0.3	0.6
5/8	0.37	0.75
3/4	0.45	0.90

For SI: 1 inch = 25.4 mm.
a. Values listed in the table are for 1:3 sanded gypsum plaster.

722.4.1.4 Nonsymmetrical assemblies. For a wall having no finish on one side or having different types or thicknesses of finish on each side, the calculation procedures of this section shall be performed twice, assuming either side to be the fire-exposed side of the wall. The fire resistance of the wall shall not exceed the lower of the two values determined.

Exception: For exterior walls with a fire separation distance greater than 5 feet (1524 mm), the fire shall be assumed to occur on the interior side only.

TABLE 722.4.1(5)
REINFORCED MASONRY LINTELS

NOMINAL LINTEL WIDTH (inches)	MINIMUM LONGITUDINAL REINFORCEMENT COVER FOR FIRE RESISTANCE (inches)			
	1 hour	2 hours	3 hours	4 hours
6	1 1/2	2	NP	NP
8	1 1/2	1 1/2	1 3/4	3
10 or more	1 1/2	1 1/2	1 1/2	1 3/4

For SI: 1 inch = 25.4 mm.
NP = Not permitted

TABLE 722.4.1(6)
REINFORCED CLAY MASONRY COLUMNS

COLUMN SIZE	FIRE-RESISTANCE RATING (hours)			
	1	2	3	4
Minimum column dimension (inches)	8	10	12	14

For SI: 1 inch = 25.4 mm.

722.4.2 Multiwythe walls. The fire-resistance rating for walls or partitions consisting of two or more dissimilar wythes shall be permitted to be determined by the formula:

$$R = (R_1^{0.59} + R_2^{0.59} + ... + R_n^{0.59})^{1.7} \quad \textbf{(Equation 7-11)}$$

where:

R = The fire-resistance rating of the assembly (hours).

R_1, R_2 and R_n = The fire-resistance rating of the individual wythes (hours).

Values for $R_n^{0.59}$ for use in Equation 7-11 are given in Table 722.4.1(3).

722.4.2.1 Multiwythe walls of different material. For walls that consist of two or more wythes of different materials (concrete or concrete masonry units) in combination with clay masonry units, the fire-resistance rating of the different materials shall be permitted to be determined from Table 722.2.1.1 for concrete; Table 722.3.2 for concrete masonry units or Table 722.4.1(1) or 722.4.1(2) for clay and tile masonry units.

722.4.3 Reinforced clay masonry lintels. Fire-resistance ratings for clay masonry lintels shall be determined based on the nominal width of the lintel and the minimum covering for the longitudinal reinforcement in accordance with Table 722.4.1(5).

722.4.4 Reinforced clay masonry columns. The fire-resistance ratings shall be determined based on the last plan dimension of the column in accordance with Table 722.4.1(6). The minimum cover for longitudinal reinforcement shall be 2 inches (51 mm).

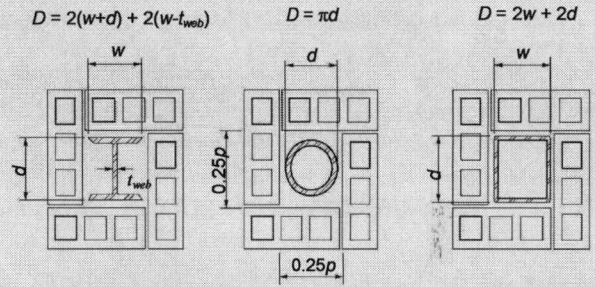

For SI: 1 inch = 25.4 mm.
d = Depth of a wide flange column, outside diameter of pipe column, or outside dimension of structural tubing column (inches).
t_{web} = Thickness of web of wide flange column (inches).
w = Width of flange of wide flange column (inches).

FIGURE 722.5.1(7)
CONCRETE OR CLAY MASONRY PROTECTED STRUCTURAL STEEL COLUMNS

722.5.1.4.5 Masonry protection. The fire resistance of structural steel columns protected with concrete masonry units or clay masonry units as illustrated in Figure 722.5.1(7), shall be permitted to be determined from the following expression:

$$R = 0.17\,(W/D)^{0.7} + [0.285\,(T_e^{1.6}/K^{0.2})]\,[1.0 + 42.7\,\{(A_s/d_m T_e)/(0.25p + T_e)\}^{0.8}]$$

(Equation 7-16)

where:

$R =$ Fire-resistance rating of column assembly (hours).

$W =$ Average weight of structural steel column (pounds per foot).

$D =$ Heated perimeter of structural steel column (inches) [see Figure 722.5.1(7)].

$T_e =$ Equivalent thickness of concrete or clay masonry unit (inches) (see Table 722.3.2 Note a or Section 722.4.1).

$K =$ Thermal conductivity of concrete or clay masonry unit (Btu/hr ft °F) [see Table 722.5.1(3)].

$A_s =$ Cross-sectional area of structural steel column (square inches).

$d_m =$ Density of the concrete or clay masonry unit (pounds per cubic foot).

$p =$ Inner perimeter of concrete or clay masonry protection (inches) [see Figure 722.5.1(7)].

722.5.1.4.6 Equivalent concrete masonry thickness. For structural steel columns protected with concrete masonry, Table 722.5.1(5) gives the equivalent thickness of concrete masonry required for various fire-resistance ratings for typical column shapes. For structural steel columns protected with clay masonry, Table 722.5.1(6) gives the equivalent thickness of concrete masonry required for various fire-resistance ratings for typical column shapes.

TABLE 722.5.1(3)
THERMAL CONDUCTIVITY OF CONCRETE OR CLAY MASONRY UNITS

DENSITY (d_m) OF UNITS (lb/ft³)	THERMAL CONDUCTIVITY (K) OF UNITS (Btu/hr · ft · °F)
Concrete Masonry Units	
80	0.207
85	0.228
90	0.252
95	0.278
100	0.308
105	0.340
110	0.376
115	0.416
120	0.459
125	0.508
130	0.561
135	0.620
140	0.685
145	0.758
150	0.837
Clay Masonry Units	
120	1.25
130	2.25

For SI: 1 pound per cubic foot = 16.0185 kg/m³, Btu/hr · ft · °F = 1.731 W/(m · K).

TABLE 722.5.1(5)
FIRE RESISTANCE OF CONCRETE MASONRY PROTECTED STEEL COLUMNS

COLUMN SIZE	CONCRETE MASONRY DENSITY POUNDS PER CUBIC FOOT	MINIMUM REQUIRED EQUIVALENT THICKNESS FOR FIRE-RESISTANCE RATING OF CONCRETE MASONRY PROTECTION ASSEMBLY, T_e (inches)				COLUMN SIZE	CONCRETE MASONRY DENSITY POUNDS PER CUBIC FOOT	MINIMUM REQUIRED EQUIVALENT THICKNESS FOR FIRE-RESISTANCE RATING OF CONCRETE MASONRY PROTECTION ASSEMBLY, T_e (inches)			
		1-hour	2-hours	3-hours	4-hours			1-hour	2-hours	3-hours	4-hours
W 14 x 82	80	0.74	1.61	2.36	3.04	W 10 x 68	80	0.72	1.58	2.33	3.01
	100	0.89	1.85	2.67	3.40		100	0.87	1.83	2.65	3.38
	110	0.96	1.97	2.81	3.57		110	0.94	1.95	2.79	3.55
	120	1.03	2.08	2.95	3.73		120	1.01	2.06	2.94	3.72
W 14 x 68	80	0.83	1.70	2.45	3.13	W 10 x 54	80	0.88	1.76	2.53	3.21
	100	0.99	1.95	2.76	3.49		100	1.04	2.01	2.83	3.57
	110	1.06	2.06	2.91	3.66		110	1.11	2.12	2.98	3.73
	120	1.14	2.18	3.05	3.82		120	1.19	2.24	3.12	3.90
W 14 x 53	80	0.91	1.81	2.58	3.27	W 10 x 45	80	0.92	1.83	2.60	3.30
	100	1.07	2.05	2.88	3.62		100	1.08	2.07	2.90	3.64
	110	1.15	2.17	3.02	3.78		110	1.16	2.18	3.04	3.80
	120	1.22	2.28	3.16	3.94		120	1.23	2.29	3.18	3.96
W 14 x 43	80	1.01	1.93	2.71	3.41	W 10 x 33	80	1.06	2.00	2.79	3.49
	100	1.17	2.17	3.00	3.74		100	1.22	2.23	3.07	3.81
	110	1.25	2.28	3.14	3.90		110	1.30	2.34	3.20	3.96
	120	1.32	2.38	3.27	4.05		120	1.37	2.44	3.33	4.12
W 12 x 72	80	0.81	1.66	2.41	3.09	W 8 x 40	80	0.94	1.85	2.63	3.33
	100	0.91	1.88	2.70	3.43		100	1.10	2.10	2.93	3.67
	110	0.99	1.99	2.84	3.60		110	1.18	2.21	3.07	3.83
	120	1.06	2.10	2.98	3.76		120	1.25	2.32	3.20	3.99
W 12 x 58	80	0.88	1.76	2.52	3.21	W 8 x 31	80	1.06	2.00	2.78	3.49
	100	1.04	2.01	2.83	3.56		100	1.22	2.23	3.07	3.81
	110	1.11	2.12	2.97	3.73		110	1.29	2.33	3.20	3.97
	120	1.19	2.23	3.11	3.89		120	1.36	2.44	3.33	4.12
W 12 x 50	80	0.91	1.81	2.58	3.27	W 8 x 24	80	1.14	2.09	2.89	3.59
	100	1.07	2.05	2.88	3.62		100	1.29	2.31	3.16	3.90
	110	1.15	2.17	3.02	3.78		110	1.36	2.42	3.28	4.05
	120	1.22	2.28	3.16	3.94		120	1.43	2.52	3.41	4.20
W 12 x 40	80	1.01	1.94	2.72	3.41	W 8 x 18	80	1.22	2.20	3.01	3.72
	100	1.17	2.17	3.01	3.75		100	1.36	2.40	3.25	4.01
	110	1.25	2.28	3.14	3.90		110	1.42	2.50	3.37	4.14
	120	1.32	2.39	3.27	4.06		120	1.48	2.59	3.49	4.28

(Continued)

Deletion Arrow and Revision Bar

 Large arrow refers to text deleted from 2012 IBC.

| Thick bar refers to modified text from 2012 IBC.

TABLE 722.5.1(5) — continued
FIRE RESISTANCE OF CONCRETE MASONRY PROTECTED STEEL COLUMNS

COLUMN SIZE	CONCRETE MASONRY DENSITY, POUNDS PER CUBIC FOOT	1-hour	2-hours	3-hours	4-hours	COLUMN SIZE	CONCRETE MASONRY DENSITY, POUNDS PER CUBIC FOOT	1-hour	2-hours	3-hours	4-hours
4 x 4 x 1/2 wall thickness	80	0.93	1.90	2.71	3.43	4 double extra strong 0.674 wall thickness	80	0.80	1.75	2.56	3.28
	100	1.08	2.13	2.99	3.76		100	0.95	1.99	2.85	3.62
	110	1.16	2.24	3.13	3.91		110	1.02	2.10	2.99	3.78
	120	1.22	2.34	3.26	4.06		120	1.09	2.20	3.12	3.93
4 x 4 x 3/8 wall thickness	80	1.05	2.03	2.84	3.57	4 extra strong 0.337 wall thickness	80	1.12	2.11	2.93	3.65
	100	1.20	2.25	3.11	3.88		100	1.26	2.32	3.19	3.95
	110	1.27	2.35	3.24	4.02		110	1.33	2.42	3.31	4.09
	120	1.34	2.45	3.37	4.17		120	1.40	2.52	3.43	4.23
4 x 4 x 1/4 wall thickness	80	1.21	2.20	3.01	3.73	4 standard 0.237 wall thickness	80	1.26	2.25	3.07	3.79
	100	1.35	2.40	3.26	4.02		100	1.40	2.45	3.31	4.07
	110	1.41	2.50	3.38	4.16		110	1.46	2.55	3.43	4.21
	120	1.48	2.59	3.50	4.30		120	1.53	2.64	3.54	4.34
6 x 6 x 1/2 wall thickness	80	0.82	1.75	2.54	3.25	5 double extra strong 0.750 wall thickness	80	0.70	1.61	2.40	3.12
	100	0.98	1.99	2.84	3.59		100	0.85	1.86	2.71	3.47
	110	1.05	2.10	2.98	3.75		110	0.91	1.97	2.85	3.63
	120	1.12	2.21	3.11	3.91		120	0.98	2.02	2.99	3.79
6 x 6 x 3/8 wall thickness	80	0.96	1.91	2.71	3.42	5 extra strong 0.375 wall thickness	80	1.04	2.01	2.83	3.54
	100	1.12	2.14	3.00	3.75		100	1.19	2.23	3.09	3.85
	110	1.19	2.25	3.13	3.90		110	1.26	2.34	3.22	4.00
	120	1.26	2.35	3.26	4.05		120	1.32	2.44	3.34	4.14
6 x 6 x 1/4 wall thickness	80	1.14	2.11	2.92	3.63	5 standard 0.258 wall thickness	80	1.20	2.19	3.00	3.72
	100	1.29	2.32	3.18	3.93		100	1.34	2.39	3.25	4.00
	110	1.36	2.43	3.30	4.08		110	1.41	2.49	3.37	4.14
	120	1.42	2.52	3.43	4.22		120	1.47	2.58	3.49	4.28
8 x 8 x 1/2 wall thickness	80	0.77	1.66	2.44	3.13	6 double extra strong 0.864 wall thickness	80	0.59	1.46	2.23	2.92
	100	0.92	1.91	2.75	3.49		100	0.73	1.71	2.54	3.29
	110	1.00	2.02	2.89	3.66		110	0.80	1.82	2.69	3.47
	120	1.07	2.14	3.03	3.82		120	0.86	1.93	2.83	3.63
8 x 8 x 3/8 wall thickness	80	0.91	1.84	2.63	3.33	6 extra strong 0.432 wall thickness	80	0.94	1.90	2.70	3.42
	100	1.07	2.08	2.92	3.67		100	1.10	2.13	2.98	3.74
	110	1.14	2.19	3.06	3.83		110	1.17	2.23	3.11	3.89
	120	1.21	2.29	3.19	3.98		120	1.24	2.34	3.24	4.04
8 x 8 x 1/4 wall thickness	80	1.10	2.06	2.86	3.57	6 standard 0.280 wall thickness	80	1.14	2.12	2.93	3.64
	100	1.25	2.28	3.13	3.87		100	1.29	2.33	3.19	3.94
	110	1.32	2.38	3.25	4.02		110	1.36	2.43	3.31	4.08
	120	1.39	2.48	3.38	4.17		120	1.42	2.53	3.43	4.22

For SI: 1 inch = 25.4 mm, 1 pound per cubic feet = 16.02 kg/m^3.
Note: Tabulated values assume 1-inch air gap between masonry and steel section.

Deletion Arrow and Revision Bar

➡ Large arrow refers to text deleted from 2012 IBC.

| Thick bar refers to modified text from 2012 IBC.

TABLE 722.5.1(6)
FIRE RESISTANCE OF CLAY MASONRY PROTECTED STEEL COLUMNS

COLUMN SIZE	CLAY MASONRY DENSITY, POUNDS PER CUBIC FOOT	1-hour	2-hours	3-hours	4-hours	COLUMN SIZE	CLAY MASONRY DENSITY, POUNDS PER CUBIC FOOT	1-hour	2-hours	3-hours	4-hours
W 14 x 82	120	1.23	2.42	3.41	4.29	W 10 x 68	120	1.27	2.46	3.26	4.35
	130	1.40	2.70	3.78	4.74		130	1.44	2.75	3.83	4.80
W 14 x 68	120	1.34	2.54	3.54	4.43	W 10 x 54	120	1.40	2.61	3.62	4.51
	130	1.51	2.82	3.91	4.87		130	1.58	2.89	3.98	4.95
W 14 x 53	120	1.43	2.65	3.65	4.54	W 10 x 45	120	1.44	2.66	3.67	4.57
	130	1.61	2.93	4.02	4.98		130	1.62	2.95	4.04	5.01
W 14 x 43	120	1.54	2.76	3.77	4.66	W 10 x 33	120	1.59	2.82	3.84	4.73
	130	1.72	3.04	4.13	5.09		130	1.77	3.10	4.20	5.13
W 12 x 72	120	1.32	2.52	3.51	4.40	W 8 x 40	120	1.47	2.70	3.71	4.61
	130	1.50	2.80	3.88	4.84		130	1.65	2.98	4.08	5.04
W 12 x 58	120	1.40	2.61	3.61	4.50	W 8 x 31	120	1.59	2.82	3.84	4.73
	130	1.57	2.89	3.98	4.94		130	1.77	3.10	4.20	5.17
W 12 x 50	120	1.43	2.65	3.66	4.55	W 8 x 24	120	1.66	2.90	3.92	4.82
	130	1.61	2.93	4.02	4.99		130	1.84	3.18	4.28	5.25
W 12 x 40	120	1.54	2.77	3.78	4.67	W 8 x 18	120	1.75	3.00	4.01	4.91
	130	1.72	3.05	4.14	5.10		130	1.93	3.27	4.37	5.34

STEEL TUBING | **STEEL PIPE**

NOMINAL TUBE SIZE (inches)	CLAY MASONRY DENSITY, POUNDS PER CUBIC FOOT	1-hour	2-hours	3-hours	4-hours	NOMINAL PIPE SIZE (inches)	CLAY MASONRY DENSITY, POUNDS PER CUBIC FOOT	1-hour	2-hours	3-hours	4-hours
4 x 4 x 1/2 wall thickness	120	1.44	2.72	3.76	4.68	4 double extra strong 0.674 wall thickness	120	1.26	2.55	3.60	4.52
	130	1.62	3.00	4.12	5.11		130	1.42	2.82	3.96	4.95
4 x 4 x 3/8 wall thickness	120	1.56	2.84	3.88	4.78	4 extra strong 0.337 wall thickness	120	1.60	2.89	3.92	4.83
	130	1.74	3.12	4.23	5.21		130	1.77	3.16	4.28	5.25
4 x 4 x 1/4 wall thickness	120	1.72	2.99	4.02	4.92	4 standard 0.237 wall thickness	120	1.74	3.02	4.05	4.95
	130	1.89	3.26	4.37	5.34		130	1.92	3.29	4.40	5.37
6 x 6 x 1/2 wall thickness	120	1.33	2.58	3.62	4.52	5 double extra strong 0.750 wall thickness	120	1.17	2.44	3.48	4.40
	130	1.50	2.86	3.98	4.96		130	1.33	2.72	3.84	4.83
6 x 6 x 3/8 wall thickness	120	1.48	2.74	3.76	4.67	5 extra strong 0.375 wall thickness	120	1.55	2.82	3.85	4.76
	130	1.65	3.01	4.13	5.10		130	1.72	3.09	4.21	5.18
6 x 6 x 1/4 wall thickness	120	1.66	2.91	3.94	4.84	5 standard 0.258 wall thickness	120	1.71	2.97	4.00	4.90
	130	1.83	3.19	4.30	5.27		130	1.88	3.24	4.35	5.32
8 x 8 x 1/2 wall thickness	120	1.27	2.50	3.52	4.42	6 double extra strong 0.864 wall thickness	120	1.04	2.28	3.32	4.23
	130	1.44	2.78	3.89	4.86		130	1.19	2.60	3.68	4.67
8 x 8 x 3/8 wall thickness	120	1.43	2.67	3.69	4.59	6 extra strong 0.432 wall thickness	120	1.45	2.71	3.75	4.65
	130	1.60	2.95	4.05	5.02		130	1.62	2.99	4.10	5.08
8 x 8 x 1/4 wall thickness	120	1.62	2.87	3.89	4.78	6 standard 0.280 wall thickness	120	1.65	2.91	3.94	4.84
	130	1.79	3.14	4.24	5.21		130	1.82	3.19	4.30	5.27

For SI: 1 inch = 25.4 mm, 1 pound per cubic foot = 16.02 kg/m³.

Deletion Arrow and Revision Bar

➤ Large arrow refers to text deleted from 2012 IBC.

| Thick bar refers to modified text from 2012 IBC.

IBC Chapter 7

Deletion Arrow and Revision Bar

➡ Large arrow refers to text deleted from 2012 IBC.

▎ Thick bar refers to modified text from 2012 IBC.

2015 IBC
CHAPTER 12*
INTERIOR ENVIRONMENT

SECTION 1201
GENERAL

1201.1 Scope. The provisions of this chapter shall govern ventilation, temperature control, lighting, yards and courts, sound transmission, room dimensions, surrounding materials and rodent-proofing associated with the interior spaces of buildings.

SECTION 1207
SOUND TRANSMISSION

1207.1 Scope. This section shall apply to common interior walls, partitions and floor/ceiling assemblies between adjacent dwelling units and sleeping units or between dwelling units and sleeping units and adjacent public areas such as halls, corridors, stairs or service areas.

1207.2 Air-borne sound. Walls, partitions and floor/ceiling assemblies separating dwelling units and sleeping units from each other or from public or service areas shall have a sound transmission class of not less than 50, or not less than 45 if field tested, for air-borne noise when tested in accordance with ASTM E 90. Penetrations or openings in construction assemblies for piping; electrical devices; recessed cabinets; bathtubs; soffits; or heating, ventilating or exhaust ducts shall be sealed, lined, insulated or otherwise treated to maintain the required ratings. This requirement shall not apply to entrance doors; however, such doors shall be tight fitting to the frame and sill.

1207.2.1 Masonry. The sound transmission class of concrete masonry and clay masonry assemblies shall be calculated in accordance with TMS 0302 or determined through testing in accordance with ASTM E 90.

---

* Only portions of this section are shown which are particularly applicable to masonry construction. For additional information see the IBC.

Deletion Arrow and Revision Bar

➡ Large arrow refers to text deleted from 2012 IBC.

❘ Thick bar refers to modified text from 2012 IBC.

IBC Chapter 12

Deletion Arrow and Revision Bar
Large arrow refers to text deleted from 2012 IBC.
Thick bar refers to modified text from 2012 IBC.

2015 IBC
CHAPTER 14*
EXTERIOR WALLS

SECTION 1401
GENERAL

1401.1 Scope. The provisions of this chapter shall establish the minimum requirements for exterior walls; exterior wall coverings; exterior wall openings; exterior windows and doors; architectural trim; balconies and similar projections; and bay and oriel windows.

SECTION 1402
DEFINITIONS

1402.1 Definitions. The following terms are defined in Chapter 2:

SECTION 202
DEFINITIONS

ADHERED MASONRY VENEER. Veneer secured and supported through the adhesion of an approved bonding material applied to an approved backing.

ANCHORED MASONRY VENEER. Veneer secured with approved mechanical fasteners to an approved backing.

BACKING. The wall or surface to which the veneer is secured.

EXTERIOR WALL. A wall, bearing or nonbearing, that is used as an enclosing wall for a building, other than a fire wall, and that has a slope of 60 degrees (1.05 rad) or greater with the horizontal plane.

EXTERIOR WALL COVERING. A material or assembly of materials applied on the exterior side of exterior walls for the purpose of providing a weather-resisting barrier, insulation or for aesthetics, including but not limited to, veneers, siding, exterior insulation and finish systems, architectural trim and embellishments such as cornices, soffits, facias, gutters and leaders.

EXTERIOR WALL ENVELOPE. A system or assembly of exterior wall components, including exterior wall finish materials, that provides protection of the building structural members, including framing and sheathing materials, and conditioned interior space, from the detrimental effects of the exterior environment.

VENEER. A facing attached to a wall for the purpose of providing ornamentation, protection or insulation, but not counted as adding strength to the wall.

WATER-RESISTIVE BARRIER. A material behind an exterior wall covering that is intended to resist liquid water that has penetrated behind the exterior covering from further intruding into the exterior wall assembly.

SECTION 1403
PERFORMANCE REQUIREMENTS

1403.1 General. The provisions of this section shall apply to exterior walls, wall coverings and components thereof.

1403.2 Weather protection. Exterior walls shall provide the building with a weather-resistant exterior wall envelope. The exterior wall envelope shall include flashing, as described in Section 1405.4. The exterior wall envelope shall be designed and constructed in such a manner as to prevent the accumulation of water within the wall assembly by providing a water-resistive barrier behind the exterior veneer, as described in Section 1404.2, and a means for draining water that enters the assembly to the exterior. Protection against condensation in the exterior wall assembly shall be provided in accordance with Section 1405.3.

Exceptions:

1. A weather-resistant exterior wall envelope shall not be required over concrete or masonry walls designed in accordance with Chapters 19 and 21, respectively.

2. Compliance with the requirements for a means of drainage, and the requirements of Sections 1404.2 and 1405.4, shall not be required for an exterior wall envelope that has been demonstrated through testing to resist wind-driven rain, including joints, penetrations and intersections with dissimilar materials, in accordance with ASTM E 331 under the following conditions:

   2.1. Exterior wall envelope test assemblies shall include at least one opening, one control joint, one wall/eave interface and one wall sill. Tested openings and penetrations shall be representative of the intended end-use configuration.

   2.2. Exterior wall envelope test assemblies shall be at least 4 feet by 8 feet (1219 mm by 2438 mm) in size.

---

* Only portions of this section are shown which are particularly applicable to masonry construction. For additional information see the IBC.

Deletion Arrow and Revision Bar

➜ Large arrow refers to text deleted from 2012 IBC.

| Thick bar refers to modified text from 2012 IBC.

2.3. Exterior wall envelope assemblies shall be tested at a minimum differential pressure of 6.24 pounds per square foot (psf) (0.297 kN/m^2).

2.4. Exterior wall envelope assemblies shall be subjected to a minimum test exposure duration of 2 hours.

The exterior wall envelope design shall be considered to resist wind-driven rain where the results of testing indicate that water did not penetrate control joints in the exterior wall envelope, joints at the perimeter of openings or intersections of terminations with dissimilar materials.

3. Exterior insulation and finish systems (EIFS) complying with Section 1408.4.1.

1403.3 Structural. Exterior walls, and the associated openings, shall be designed and constructed to resist safely the superimposed loads required by Chapter 16.

1403.4 Fire resistance. Exterior walls shall be fire-resistance rated as required by other sections of this code with opening protection as required by Chapter 7.

1403.5 Vertical and lateral flame propagation. Exterior walls on buildings of Type I, II, III or IV construction that are greater than 40 feet (12 192 mm) in height above grade plane and contain a combustible water-resistive barrier shall be tested in accordance with and comply with the acceptance criteria of NFPA 285. For the purposes of this section, fenestration products and flashing of fenestration products shall not be considered part of the water-resistive barrier.

Exceptions:

1. Walls in which the water-resistive barrier is the only combustible component and the exterior wall has a wall covering of brick, concrete, stone, terra cotta, stucco or steel with minimum thicknesses in accordance with Table 1405.2.

2. Walls in which the water-resistive barrier is the only combustible component and the water-resistive barrier has a peak heat release rate of less than 150 kW/m^2, a total heat release of less than 20 MJ/m^2 and an effective heat of combustion of less than 18 MJ/kg as determined in accordance with ASTM E 1354 and has a flame spread index of 25 or less and a smoke-developed index of 450 or less as determined in accordance with ASTM E 84 or UL 723. The ASTM E 1354 test shall be conducted on specimens at the thickness intended for use, in the horizontal orientation and at an incident radiant heat flux of 50 kW/m^2.

1403.6 Flood resistance. For buildings in flood hazard areas as established in Section 1612.3, exterior walls extending below the elevation required by Section 1612 shall be constructed with flood-damage-resistant materials.

1403.7 Flood resistance for coastal high-hazard areas and coastal A zones. For buildings in coastal high-hazard areas and coastal A zones as established in Section 1612.3, electrical, mechanical and plumbing system components shall not be mounted on or penetrate through exterior walls that are designed to break away under flood loads.

SECTION 1404
MATERIALS

1404.1 General. Materials used for the construction of exterior walls shall comply with the provisions of this section. Materials not prescribed herein shall be permitted, provided that any such alternative has been approved.

1404.4 Masonry. Exterior walls of masonry construction shall be designed and constructed in accordance with this section and Chapter 21. Masonry units, mortar and metal accessories used in anchored and adhered veneer shall meet the physical requirements of Chapter 21. The backing of anchored and adhered veneer shall be of concrete, masonry, steel framing or wood framing. Continuous insulation meeting the applicable requirements of this code shall be permitted between the backing and the masonry veneer.

1404.7 Glass-unit masonry. Exterior walls of glass-unit masonry shall be designed and constructed in accordance with Chapter 21.

SECTION 1405
INSTALLATION OF WALL COVERINGS

1405.1 General. Exterior wall coverings shall be designed and constructed in accordance with the applicable provisions of this section.

1405.2 Weather protection. Exterior walls shall provide weather protection for the building. The materials of the minimum nominal thickness specified in Table 1405.2 shall be acceptable as approved weather coverings.

Deletion Arrow and Revision Bar

➔ Large arrow refers to text deleted from 2012 IBC.

| Thick bar refers to modified text from 2012 IBC.

TABLE 1405.2 (Partial)
MINIMUM THICKNESS OF WEATHER COVERINGS

COVERING TYPE	MINIMUM THICKNESS (inches)
Adhered masonry veneer	0.25
Anchored masonry veneer	2.625
Marble slabs	1
Porcelain tile	0.025
Stone (cast artificial, anchored)	1.5
Stone (natural)	2
Stucco or exterior cement plaster Three-coat work over: Unit masonry	0.625[b]
Two-coat work over: Unit masonry	0.5[b]
Terra cotta (anchored)	1
Terra cotta (adhered)	0.25

For SI: 1 inch = 25.4 mm, 1 ounce = 28.35 g, 1 square foot = 0.093 m².
b. Exclusive of texture.

1405.4 Flashing. Flashing shall be installed in such a manner so as to prevent moisture from entering the wall or to redirect that moisture to the exterior. Flashing shall be installed at the perimeters of exterior door and window assemblies, penetrations and terminations of exterior wall assemblies, exterior wall intersections with roofs, chimneys, porches, decks, balconies and similar projections and at built-in gutters and similar locations where moisture could enter the wall. Flashing with projecting flanges shall be installed on both sides and the ends of copings, under sills and continuously above projecting trim.

1405.4.2 Masonry. Flashing and weep holes in anchored veneer shall be located in the first course of masonry above finished ground level above the foundation wall or slab, and other points of support, including structural floors, shelf angles and lintels where anchored veneers are designed in accordance with Section 1405.6.

1405.6 Anchored masonry veneer. Anchored masonry veneer shall comply with the provisions of Sections 1405.6, 1405.7, 1405.8 and 1405.9 and Sections 6.1 and 6.2 of TMS 402/ACI 530/ASCE 5.

1405.6.1 Tolerances. Anchored masonry veneers in accordance with Chapter 14 are not required to meet the tolerances in Article 3.3 F1 of TMS 602/ACI 530.1/ASCE 6.

1405.6.2 Seismic requirements. Anchored masonry veneer located in Seismic Design Category C, D, E or F shall conform to the requirements of Section 12.2.2.10 of TMS 402/ACI 530/ASCE 5.

1405.7 Stone veneer. Anchored stone veneer units not exceeding 10 inches (254 mm) in thickness shall be anchored directly to masonry, concrete or to stud construction by one of the following methods:

1. With concrete or masonry backing, anchor ties shall be not less than 0.1055-inch (2.68 mm) corrosion-resistant wire, or approved equal, formed beyond the base of the backing. The legs of the loops shall be not less than 6 inches (152 mm) in length bent at right angles and laid in the mortar joint, and spaced so that the eyes or loops are 12 inches (305 mm) maximum on center in both directions. There shall be provided not less than a 0.1055-inch (2.68 mm) corrosion-resistant wire tie, or approved equal, threaded through the exposed loops for every 2 square feet (0.2 m²) of stone veneer. This tie shall be a loop having legs not less than 15 inches (381 mm) in length bent so that the tie will lie in the stone veneer mortar joint. The last 2 inches (51 mm) of each wire leg shall have a right-angle bend. One-inch (25 mm) minimum thickness of cement grout shall be placed between the backing and the stone veneer.

2. With wood stud backing, a 2-inch by 2-inch (51 by 51 mm) 0.0625-inch (1.59 mm) zinc-coated or nonmetallic coated wire mesh with two layers of water-resistive barrier in accordance with Section 1404.2 shall be applied directly to wood studs spaced not more than 16 inches (406 mm) on center. On studs, the mesh shall be attached with 2-inch-long (51 mm) corrosion-resistant steel wire furring nails at 4 inches (102 mm) on center providing a minimum 1.125-inch (29 mm) penetration into each stud and with 8d annular threaded nails at 8 inches (203 mm) on center into top and bottom plates or with equivalent wire ties. There shall be not less than a 0.1055-inch (2.68 mm) zinc-coated or nonmetallic coated wire, or approved equal, attached to the stud with not smaller than an 8d (0.120 in. diameter) annular threaded nail for every 2 square feet (0.2 m²) of stone veneer. This tie shall be a loop having legs not less than 15 inches (381 mm) in length, so bent that the tie will lie in the stone veneer mortar joint. The last 2 inches (51 mm) of each wire leg shall have a right-angle bend. One-inch (25 mm) minimum thickness of cement grout shall be placed between the backing and the stone veneer.

3. With cold-formed steel stud backing, a 2-inch by 2-inch (51 by 51 mm) 0.0625-inch (1.59 mm) zinc-coated or nonmetallic coated wire mesh with two layers of water-resistive barrier in accordance with Section 1404.2 shall be applied directly to steel studs spaced not more than 16 inches (406 mm) on center. The mesh shall be attached with corrosion-resistant #8 self-drilling, tapping screws at 4 inches (102 mm) on center, and at 8 inches (203 mm) on

---

Deletion Arrow and Revision Bar

 Large arrow refers to text deleted from 2012 IBC.

| Thick bar refers to modified text from 2012 IBC.

center into top and bottom tracks or with equivalent wire ties. Screws shall extend through the steel connection not fewer than three exposed threads. There shall be not less than a 0.1055-inch (2.68 mm) corrosion-resistant wire, or approved equal, attached to the stud with not smaller than a #8 self-drilling, tapping screw extending through the steel framing not fewer than three exposed threads for every 2 square feet (0.2 m^2) of stone veneer. This tie shall be a loop having legs not less than 15 inches (381 mm) in length, so bent that the tie will lie in the stone veneer mortar joint. The last 2 inches (51 mm) of each wire leg shall have a right-angle bend. One-inch (25 mm) minimum thickness of cement grout shall be placed between the backing and the stone veneer. The cold-formed steel framing members shall have a minimum bare steel thickness of 0.0428 inches (1.087 mm).

1405.8 Slab-type veneer. Anchored slab-type veneer units not exceeding 2 inches (51 mm) in thickness shall be anchored directly to masonry, concrete or light-frame construction. For veneer units of marble, travertine, granite or other stone units of slab form, ties of corrosion-resistant dowels in drilled holes shall be located in the middle third of the edge of the units, spaced not more than 24 inches (610 mm) apart around the periphery of each unit with not less than four ties per veneer unit. Units shall not exceed 20 square feet (1.9 m^2) in area. If the dowels are not tight fitting, the holes shall be drilled not more than 0.063 inch (1.6 mm) larger in diameter than the dowel, with the hole countersunk to a diameter and depth equal to twice the diameter of the dowel in order to provide a tight-fitting key of cement mortar at the dowel locations where the mortar in the joint has set. Veneer ties shall be corrosion-resistant metal capable of resisting, in tension or compression, a force equal to two times the weight of the attached veneer. If made of sheet metal, veneer ties shall be not smaller in area than 0.0336 by 1 inch (0.853 by 25 mm) or, if made of wire, not smaller in diameter than 0.1483-inch (3.76 mm) wire.

1405.9 Terra cotta. Anchored terra cotta or ceramic units not less than $1^5/_8$ inches (41 mm) thick shall be anchored directly to masonry, concrete or stud construction. Tied terra cotta or ceramic veneer units shall be not less than $1^5/_8$ inches (41 mm) thick with projecting dovetail webs on the back surface spaced approximately 8 inches (203 mm) on center. The facing shall be tied to the backing wall with corrosion-resistant metal anchors of not less than No. 8 gage wire installed at the top of each piece in horizontal bed joints not less than 12 inches (305 mm) nor more than 18 inches (457 mm) on center; these anchors shall be secured to $^1/_4$-inch (6.4 mm) corrosion-resistant pencil rods that pass through the vertical aligned loop anchors in the backing wall. The veneer ties shall have sufficient strength to support the full weight of the veneer in tension. The facing shall be set with not less than a 2-inch (51 mm) space from the backing wall and the space shall be filled solidly with Portland cement grout and pea gravel. Immediately prior to setting, the backing wall and the facing shall be drenched with clean water and shall be distinctly damp when the grout is poured.

1405.10 Adhered masonry veneer. Adhered masonry veneer shall comply with the applicable requirements in this section and Sections 12.1 and 12.3 of TMS 402/ACI 530/ASCE 5.

1405.10.1 Exterior adhered masonry veneer. Exterior adhered masonry veneer shall be installed in accordance with Section 1405.10 and the manufacturer's instructions.

1405.10.1.1 Water-resistive barriers. Water-resistive barriers shall be installed as required in Section 2510.6.

1405.10.1.2 Flashing. Flashing shall comply with the applicable requirements of Section 1405.4 and the following.

1405.10.1.2.1 Flashing at foundation. A corrosion-resistant screed or flashing of a minimum 0.019-inch (0.48 mm) or 26 gage galvanized or plastic with a minimum vertical attachment flange of $3^1/_2$ inches (89 mm) shall be installed to extend not less than 1 inch (25 mm) below the foundation plate line on exterior stud walls in accordance with Section 1405.4. The water-resistive barrier shall lap over the exterior of the attachment flange of the screed or flashing.

1405.10.1.3 Clearances. On exterior stud walls, adhered masonry veneer shall be installed not less than 4 inches (102 mm) above the earth, or not less than 2 inches (51 mm) above paved areas, or not less than $^1/_2$ inch (12.7 mm) above exterior walking surface that are supported by the same foundation that supports the exterior wall.

1405.10.1.4 Adhered masonry veneer installed with lath and mortar. Exterior adhered masonry veneer installed with lath and mortar shall comply with the following.

1405.10.1.4.1 Lathing. Lathing shall comply with the requirements of Section 2510.

1405.10.1.4.2 Scratch coat. A nominal $^1/_2$-inch-thick (12.7 mm) layer of mortar complying with the material requirements of Sections 2103 and 2512.2 shall be applied, encapsulating the lathing. The surface of this mortar shall be scored horizontally, resulting in a scratch coat.

1405.10.1.4.3 Adhering veneer. The masonry veneer units shall be adhered to the mortar scratch coat with a nominal $^1/_2$-inch-thick (12.7 mm) setting bed of mortar complying with Sections 2103 and 2512.2 applied to create a full setting bed for the back of the masonry veneer units. The masonry veneer units shall be worked into the setting bed resulting in a nominal $^3/_8$-inch (9.5 mm) setting bed after the masonry veneer units are applied.

1405.10.1.5 Adhered masonry veneer applied directly to masonry and concrete. Adhered masonry veneer applied directly to masonry or concrete shall comply with the applicable requirements of Section 1405.10 and with the requirements of Section 1405.10.1.4 or 2510.7.

1405.10.1.6 Cold weather construction. Cold weather construction of adhered masonry veneer shall comply with the requirements of Sections 2104 and 2512.4.

1405.10.1.7 Hot weather construction. Hot weather construction of adhered masonry veneer shall comply with the requirements of Section 2104.

1405.10.2 Exterior adhered masonry veneers — porcelain tile. Adhered units shall not exceed $^5/_8$ inch (15.8 mm) thickness and 24 inches (610 mm) in any face dimension nor more than 3 square feet (0.28 m^2) in total face area and shall not weigh more than 9 pounds psf (0.43 kN/m^2). Porcelain tile shall be adhered to an approved backing system.

1405.10.3 Interior adhered masonry veneers. Interior adhered masonry veneers shall have a maximum weight of 20 psf (0.958 kg/m^2) and shall be installed in accordance with Section 1405.10. Where the interior adhered masonry veneer is supported by wood construction, the supporting members shall be designed to limit deflection to $^1/_{600}$ of the span of the supporting members.

1405.11 Metal veneers. Veneers of metal shall be fabricated from approved corrosion-resistant materials or shall be protected front and back with porcelain enamel, or otherwise be treated to render the metal resistant to corrosion. Such veneers shall be not less than 0.0149-inch (0.378 mm) nominal thickness sheet steel mounted on wood or metal furring strips or approved sheathing on light-frame construction.

1405.11.1 Attachment. Exterior metal veneer shall be securely attached to the supporting masonry or framing members with corrosion-resistant fastenings, metal ties or by other approved devices or methods. The spacing of the fastenings or ties shall not exceed 24 inches (610 mm) either vertically or horizontally, but where units exceed 4 square feet (0.4 m^2) in area there shall be not less than four attachments per unit. The metal attachments shall have a cross-sectional area not less than provided by W 1.7 wire. Such attachments and their supports shall be designed and constructed to resist the wind loads as specified in Section 1609 for components and cladding.

1405.11.2 Weather protection. Metal supports for exterior metal veneer shall be protected by painting, galvanizing or by other equivalent coating or treatment. Wood studs, furring strips or other wood supports for exterior metal veneer shall be approved pressure-treated wood or protected as required in Section 1403.2. Joints and edges exposed to the weather shall be caulked with approved durable waterproofing material or by other approved means to prevent penetration of moisture.

1405.11.3 Backup. Masonry backup shall not be required for metal veneer unless required by the fire-resistance requirements of this code.

IBC Chapter 14

Deletion Arrow and Revision Bar

➤ Large arrow refers to text deleted from 2012 IBC.

| Thick bar refers to modified text from 2012 IBC.

2015 IBC
CHAPTER 16*
STRUCTURAL DESIGN

SECTION 1601
GENERAL

1601.1 Scope. The provisions of this chapter shall govern the structural design of buildings, structures and portions thereof regulated by this code.

SECTION 1602
DEFINITIONS AND NOTATIONS

1602.1 Definitions. The following terms are defined in Chapter 2:

SECTION 202
DEFINITIONS

STRENGTH (For Chapter 16).

Strength design. A method of proportioning structural members such that the computed forces produced in the members by factored loads do not exceed the member design strength [also called "*load and resistance factor design*" (LRFD)]. The term "strength design" is used in the design of concrete and masonry structural elements.

SECTION 1604
GENERAL DESIGN REQUIREMENTS

1604.1 General. Building, structures and parts thereof shall be designed and constructed in accordance with strength design, load and resistance factor design, allowable stress design, empirical design or conventional construction methods, as permitted by the applicable material chapters.

1604.3.4 Masonry. The deflection of masonry structural members shall not exceed that permitted by TMS 402/ACI 530/ASCE 5.

1604.8 Anchorage. Buildings and other structures, and portions thereof, shall be provided with anchorage in accordance with Sections 1604.8.1 through 1604.8.3, as applicable.

1604.8.1 General. Anchorage of the roof to walls and columns, and of walls and columns to foundations, shall be provided to resist the uplift and sliding forces that result from the application of the prescribed loads.

1604.8.2 Structural walls. Walls that provide vertical load-bearing resistance or lateral shear resistance for a portion of the structure shall be anchored to the roof and to all floors and members that provide lateral support for the wall or that are supported by the wall. The connections shall be capable of resisting the horizontal forces specified in Section 1.4.5 of ASCE 7 for walls of structures assigned to Seismic Design Category A and to Section 12.11 of ASCE 7 for walls of structures assigned to all other seismic design categories. Required anchors in masonry walls of hollow units or cavity walls shall be embedded in a reinforced grouted structural element of the wall. See Sections 1609 for wind design requirements and 1613 for earthquake design requirements.

SECTION 1605
LOAD COMBINATIONS

1605.3 Load combinations using allowable stress design.

1605.3.1 Basic load combinations. Where allowable stress design (working stress design), as permitted by this code, is used, structures and portions thereof shall resist the most critical effects resulting from the following combinations of loads:

$D + F$ (Equation 16-8)

$D + H + F + L$ (Equation 16-9)

$D + H + F + (L_r \text{ or } S \text{ or } R)$ (Equation 16-10)

$D + H + F + 0.75(L) + 0.75(L_r \text{ or } S \text{ or } R)$ **(Equation 16-11)**

$D + H + F + (0.6W \text{ or } 0.7E)$ (Equation 16-12)

$D + H + F + 0.75(0.6W) + 0.75L + 0.75(L_r \text{ or } S \text{ or } R)$
                                                     (Equation 16-13)

$D + H + F + 0.75(0.7E) + 0.75L + 0.75S$
                                                       (Equation 16-14)

$0.6D + 0.6W + H$ **(Equation 16-15)**

$0.6(D + F) + 0.7E + H$ **(Equation 16-16)**

Exceptions:

1. Crane hook loads need not be combined with roof live load or with more than three-fourths of the snow load or one-half of the wind load.

---

* Only portions of this section are shown which are particularly applicable to masonry construction. For additional information see the IBC.

 Deletion Arrow and Revision Bar
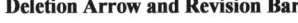 Large arrow refers to text deleted from 2012 IBC.

| Thick bars refers to modified text from 2012 IBC.

2. Flat roof snow loads of 30 psf (1.44 kN/m^2) or less and roof live loads of 30 psf (1.44 kN/m^2) or less need not be combined with seismic loads. Where flat roof snow loads exceed 30 psf (1.44 kN/m^2), 20 percent shall be combined with seismic loads.

3. Where the effect of H resists the primary variable load effect, a load factor of 0.6 shall be included with H where H is permanent and H shall be set to zero for all other conditions.

4. In Equation 16-15, the wind load, W, is permitted to be reduced in accordance with Exception 2 of Section 2.4.1 of ASCE 7.

5. In Equation 16-16, 0.6 D is permitted to be increased to 0.9 D for the design of special reinforced masonry shear walls complying with Chapter 21.

SECTION 1609
WIND LOADS

1609.1 Applications. Buildings, structures and parts thereof shall be designed to withstand the minimum wind loads prescribed herein. Decreases in wind load shall not be made for the effect of shielding by other structures.

TABLE 1609.1.2
WIND-BORNE DEBRIS PROTECTION FASTENING SCHEDULE FOR WOOD STRUCTURAL PANELS[a, b, c, d]

FASTENER TYPE	FASTENER SPACING (inches)		
	Panel Span ≤ 4 feet	4 feet< Panel Span ≤ 6 feet	6 feet< Panel Span ≤ 8 feet
No. 8 wood-screw-based anchor with 2-inch embedment length	16	10	8
No. 10 wood-screw-based anchor with 2-inch embedment length	16	12	9
$1/4$-inch diameter lag-screw-based anchor with 2-inch embedment length	16	16	16

For SI: 1 inch = 25.4 mm, 1 foot = 304.8 mm, 1 pound = 4.448 N, 1 mile per hour = 0.447 m/s.

a. This table is based on 140 mph wind speeds and a 45-foot mean roof height.
b. Fasteners shall be installed at opposing ends of the wood structural panel. Fasteners shall be located a minimum of 1 inch from the edge of the panel.
c. Anchors shall penetrate through the exterior wall covering with an embedment length of 2 inches minimum into the building frame. Fasteners shall be located a minimum of $2^1/_2$ inches from the edge of concrete block or concrete.
d. Where panels are attached to masonry or masonry/stucco, they shall be attached using vibration-resistant anchors having a minimum ultimate withdrawal capacity of 1,500 pounds.

SECTION 1613
EARTHQUAKE LOADS

1613.1 Scope. Every structure, and portion thereof, including nonstructural components that are permanently attached to structures and their supports and attachments, shall be designed and constructed to resist the effects of earthquake motions in accordance with ASCE 7, excluding Chapter 14 and Appendix 11A. The seismic design category for a structure is permitted to be determined in accordance with Section 1613 or ASCE 7.

SECTION 1615
STRUCTURAL INTEGRITY

1615.1 General. High-rise buildings that are assigned to Risk Category III or IV shall comply with the requirements of this section. Frame structures shall comply with the requirements of Section 1615.3. Bearing wall structures shall comply with the requirements of Section 1615.4.

1615.4 Bearing wall structures. Bearing wall structures shall have vertical ties in all load-bearing walls and longitudinal ties, transverse ties and perimeter ties at each floor level in accordance with this section and as shown in Figure 1615.4.

1615.4.2 Other bearing wall structures. Ties in bearing wall structures other than those covered in Section 1615.4.1 shall conform to this section.

1615.4.2.1 Longitudinal ties. Longitudinal ties shall consist of continuous reinforcement in slabs; continuous or spliced decks or sheathing; continuous or spliced members framing to, within or across walls; or connections of continuous framing members to walls. Longitudinal ties shall extend across interior load-bearing walls and shall connect to exterior load-bearing walls and shall be spaced at not greater than 10 feet (3038 mm) on center. Ties shall have a minimum nominal tensile strength, T_T, given by Equation 16-41. For ASD the minimum nominal tensile strength shall be permitted to be taken as 1.5 times the allowable tensile stress times the area of the tie.

$$T_T = wLS \leq \alpha_T S \quad \text{(Equation 16-41)}$$

where:

L = The span of the horizontal element in the direction of the tie, between bearing walls, feet (m).

w = The weight per unit area of the floor or roof in the span being tied to or across the wall, psf (N/m^2).

S = The spacing between ties, feet (m).

α_T = A coefficient with a value of 1,500 pounds per foot (2.25 kN/m) for masonry bearing wall structures and a value of 375 pounds per foot

(0.6 kN/m) for structures with bearing walls of cold-formed steel light-frame construction.

1615.4.2.2 Transverse ties. Transverse ties shall consist of continuous reinforcement in slabs; continuous or spliced decks or sheathing; continuous or spliced members framing to, within or across walls; or connections of continuous framing members to walls. Transverse ties shall be placed no farther apart than the spacing of load-bearing walls. Transverse ties shall have minimum nominal tensile strength T_T, given by Equation 16-41. For ASD the minimum nominal tensile strength shall be permitted to be taken as 1.5 times the allowable tensile stress times the area of the tie.

1615.4.2.3 Perimeter ties. Perimeter ties shall consist of continuous reinforcement in slabs; continuous or spliced decks or sheathing; continuous or spliced members framing to, within or across walls; or connections of continuous framing members to walls. Ties around the perimeter of each floor and roof shall be located within 4 feet (1219 mm) of the edge and shall provide a nominal strength in tension not less than T_p, given by Equation 16-42. For ASD the minimum nominal tensile strength shall be permitted to be taken as 1.5 times the allowable tensile stress times the area of the tie.

$$T_p = 200w \leq \beta_T \quad \text{(Equation 16-42)}$$

For SI: $T_p = 90.7w \leq \beta_T$

where:

w = As defined in Section 1615.4.2.1.

β_T = A coefficient with a value of 16,000 pounds (7200 kN) for structures with masonry bearing walls and a value of 4,000 pounds (1300 kN) for structures with bearing walls of cold-formed steel light-frame construction.

1615.4.2.4 Vertical ties. Vertical ties shall consist of continuous or spliced reinforcing, continuous or spliced members, wall sheathing or other engineered systems. Vertical tension ties shall be provided in bearing walls and shall be continuous over the height of the building. The minimum nominal tensile strength for vertical ties within a bearing wall shall be equal to the weight of the wall within that story plus the weight of the diaphragm tributary to the wall in the story below. No fewer than two ties shall be provided for each wall. The strength of each tie need not exceed 3,000 pounds per foot (450 kN/m) of wall tributary to the tie for walls of masonry construction or 750 pounds per foot (140 kN/m) of wall tributary to the tie for walls of cold-formed steel light-frame construction.

Deletion Arrow and Revision Bar

Large arrow refers to text deleted from 2012 IBC.

Thick bars refers to modified text from 2012 IBC.

Deletion Arrow and Revision Bar

➤ Large arrow refers to text deleted from 2012 IBC.

| Thick bars refers to modified text from 2012 IBC.

2015 IBC
CHAPTER 17*
SPECIAL INSPECTIONS AND TESTS

SECTION 1701
GENERAL

1701.1 Scope. The provisions of this chapter shall govern the quality, workmanship and requirements for materials covered. Materials of construction and tests shall conform to the applicable standards listed in this code.

1701.2 New materials. New building materials, equipment, appliances, systems or methods of construction not provided for in this code, and any material of questioned suitability proposed for use in the construction of a building or structure, shall be subjected to the tests prescribed in this chapter and in the approved rules to determine character, quality and limitations of use.

SECTION 1702
DEFINITIONS

1702.1 General. The following terms are defined in Chapter 2:

SECTION 202
DEFINITIONS

SPECIAL INSPECTION. Inspection of construction requiring the expertise of an approved special inspector in order to ensure compliance with this code and the approved construction documents.

Continuous special inspection. Special inspection by the special inspector who is present when and where the work to be inspected is being performed.

Periodic special inspection. Special inspection by the special inspector who is intermittently present where the work to be inspected has been or is being performed.

SPECIAL INSPECTOR. A qualified person employed or retained by an approved agency and approved by the building official as having the competence necessary to inspect a particular type of construction requiring special inspection.

STRUCTURAL OBSERVATION. The visual observation of the structural system by a registered design professional for general conformance to the approved construction documents.

SECTION 1704
SPECIAL INSPECTIONS AND TESTS, CONTRACTOR RESPONSIBILITY AND STRUCTURAL OBSERVATION

1704.1 General. Special inspections and tests, statement of special inspections, responsibilities of contractors, submittals to the building official and structural observations shall meet the applicable requirements of this section.

1704.2 Special Inspections and tests. Where application is made to the building official for construction as specified in Section 105, the owner or the owner's authorized agent, other than the contractor, shall employ one or more approved agencies to provide special inspections and tests during construction on the types of work specified in Section 1705 and identify the approved agencies to the building official. These special inspections and tests are in addition to the inspections by the building official that are identified in Section 110.

Exceptions:

1. Special inspections and tests are not required for construction of a minor nature or as warranted by conditions in the jurisdiction as approved by the building official.

2. Unless otherwise required by the building official, special inspections and tests are not required for Group U occupancies that are accessory to a residential occupancy including, but not limited to, those listed in Section 312.1.

3. Special inspections and tests are not required for portions of structures designed and constructed in accordance with the cold-formed steel light-frame construction provisions of Section 2211.7 or the conventional light-frame construction provisions of Section 2308.

4. The contractor is permitted to employ the approved agencies where the contractor is also the owner.

1704.2.1 Special inspector qualifications. Prior to the start of the construction, the approved agencies shall provide written documentation to the building official demonstrating the competence and relevant experience or training of the special inspectors who will perform

---

* Only portions of this section are shown which are particularly applicable to masonry construction. For additional information see the IBC.

Deletion Arrow and Revision Bar
➡ Large arrow refers to text deleted from 2012 IBC.

| Thick bar refers to modified text from 2012 IBC.

the special inspections and tests during construction. Experience or training shall be considered relevant where the documented experience or training is related in complexity to the same type of special inspection or testing activities for projects of similar complexity and material qualities. These qualifications are in addition to qualifications specified in other sections of this code.

The registered design professional in responsible charge and engineers of record involved in the design of the project are permitted to act as the approved agency and their personnel are permitted to act as special inspectors for the work designed by them, provided they qualify as special inspectors.

1704.2.2 Access for special inspection. The construction or work for which special inspection or testing is required shall remain accessible and exposed for special inspection or testing purposes until completion of the required special inspections or tests.

1704.2.3 Statement of special inspections. The applicant shall submit a statement of special inspections in accordance with Section 107.1 as a condition for permit issuance. This statement shall be in accordance with Section 1704.3.

> **Exception:** A statement of special inspections is not required for portions of structures designed and constructed in accordance with the cold-formed steel light-frame construction provisions of Section 2211.7 or the conventional light-frame construction provisions of Section 2308.

1704.2.4 Report requirement. Approved agencies shall keep records of special inspections and tests. The approved agency shall submit reports of special inspections and tests to the building official and to the registered design professional in responsible charge. Reports shall indicate that work inspected or tested was or was not completed in conformance to approved construction documents. Discrepancies shall be brought to the immediate attention of the contractor for correction. If they are not corrected, the discrepancies shall be brought to the attention of the building official and to the registered design professional in responsible charge prior to the completion of that phase of the work. A final report documenting required special inspections and tests, and correction of any discrepancies noted in the inspections or tests, shall be submitted at a point in time agreed upon prior to the start of work by the owner or the owner's authorized agent to the building official.

1704.3 Statement of special inspections. Where special inspection or tests are required by Section 1705, the registered design professional in responsible charge shall prepare a statement of special inspections in accordance with Section 1704.3.1 for submittal by the applicant in accordance with Section 1704.2.3.

> **Exception:** The statement of special inspections is permitted to be prepared by a qualified person approved by the building official for construction not designed by a registered design professional.

1704.3.1 Content of statement of special inspections. The statement of special inspections shall identify the following:

1. The materials, systems, components and work required to have special inspections or tests by the building official or by the registered design professional responsible for each portion of the work.
2. The type and extent of each special inspection.
3. The type and extent of each test.
4. Additional requirements for special inspections or tests for seismic or wind resistance as specified in Sections 1705.11, 1705.12 and 1705.13.
5. For each type of special inspection, identification as to whether it will be continuous special inspection, periodic special inspection or performed in accordance with the notation used in the referenced standard where the inspections are defined.

1704.3.2 Seismic requirements in the statement of special inspections. Where Section 1705.12 or 1705.13 specifies special inspections or tests for seismic resistance, the statement of special inspections shall identify the designated seismic systems and seismic force-resisting systems that are subject to the special inspections or tests.

1704.3.3 Wind requirements in the statement of special inspections. Where Section 1705.11 specifies special inspection for wind resistance, the statement of special inspections shall identify the main windforce-resisting systems and wind-resisting components that are subject to special inspections.

1704.4 Contractor responsibility. Each contractor responsible for the construction of a main wind- or seismic force-resisting system, designated seismic system or a wind- or seismic force-resisting component listed in the statement of special inspections shall submit a written statement of responsibility to the building official and the owner or the owner's authorized agent prior to the commencement of work on the system or component. The contractor's statement of responsibility shall contain acknowledgement of awareness of the special requirements contained in the statement of special inspections.

1704.5 Submittals to the building official. In addition to the submittal of reports of special inspections and tests in accordance with Section 1704.2.4, reports and certificates shall be submitted by the owner or the owner's authorized agent to the building official for each of the following:

1. Certificates of compliance for the fabrication of structural, load-bearing or lateral load-resisting members or assemblies on the premises of a registered and approved fabricator in accordance with Section 1704.2.5.1.

2. Certificates of compliance for the seismic qualification of nonstructural components, supports and attachments in accordance with Section 1705.13.2.

3. Certificates of compliance for designated seismic systems in accordance with Section 1705.13.3.

4. Reports of preconstruction tests for shotcrete in accordance with Section 1908.5.

5. Certificates of compliance for open web steel joists and joist girders in accordance with Section 2207.5.

6. Reports of material properties verifying compliance with the requirements of AWS D1.4 for weldability as specified in Section 26.5.4 of ACI 318 for reinforcing bars in concrete complying with a standard other than ASTM A 706 that are to be welded; and

7. Reports of mill tests in accordance with Section 20.2.2.5 of ACI 318 for reinforcing bars complying with ASTM A 615 and used to resist earthquake-induced flexural or axial forces in the special moment frames, special structural walls or coupling beams connecting special structural walls of seismic force-resisting systems in structures assigned to Seismic Design Category B, C, D, E or F.

1704.6 Structural observations. Where required by the provisions of Section 1704.6.1 or 1704.6.2, the owner or the owner's authorized agent shall employ a registered design professional to perform structural observations. Structural observation does not include or waive the responsibility for the inspections in Section 110 or the special inspections in Section 1705 or other section of this code.

Prior to the commencement of observations, the structural observer shall submit to the building official a written statement identifying the frequency and extent of structural observations.

At the conclusion of the work included in the permit, the structural observer shall submit to the building official a written statement that the site visits have been made and identify any reported deficiencies that, to the best of the structural observer's knowledge, have not been resolved.

1704.6.1 Structural observations for seismic resistance. Structural observations shall be provided for those structures assigned to Seismic Design Category D, E or F where one or more of the following conditions exist:

1. The structure is classified as Risk Category III or IV.

2. The height of the structure is greater than 75 feet (22 860 mm) above the base as defined in ASCE 7.

3. The structure is assigned to Seismic Design Category E, is classified as Risk Category I or II, and is greater than two stories above grade plane.

4. When so designated by the registered design professional responsible for the structural design.

5. When such observation is specifically required by the building official.

1704.6.2 Structural observations for wind requirements. Structural observations shall be provided for those structures sited where V_{asd} as determined in accordance with Section 1609.3.1 exceeds 110 mph (49 m/sec), where one or more of the following conditions exist:

1. The structure is classified as Risk Category III or IV.

2. The building height is greater than 75 feet (22 860 mm).

3. When so designated by the registered design professional responsible for the structural design.

4. When such observation is specifically required by the building official.

SECTION 1705
REQUIRED SPECIAL INSPECTIONS AND TESTS

1705.1 General. Special inspections and tests of elements and nonstructural components of buildings and structures shall meet the applicable requirements of this section.

1705.1.1 Special cases. Special inspections and tests shall be required for proposed work that is, in the opinion of the building official, unusual in its nature, such as, but not limited to, the following examples:

1. Construction materials and systems that are alternatives to materials and systems prescribed by this code.

2. Unusual design applications of materials described in this code.

3. Materials and systems required to be installed in accordance with additional manufacturer's instructions that prescribe requirements not contained in this code or in standards referenced by this code.

Deletion Arrow and Revision Bar

➡ Large arrow refers to text deleted from 2012 IBC.

| Thick bar refers to modified text from 2012 IBC.

1705.4 Masonry construction. Special inspections and tests of masonry construction shall be performed in accordance with the quality assurance program requirements of TMS 402/ACI 530/ASCE 5 and TMS 602/ACI 530.1/ASCE 6.

Exception: Special inspections and tests shall not be required for:

1. Empirically designed masonry, glass unit masonry or masonry veneer designed in accordance with Section 2109, 2110 or Chapter 14, respectively, where they are part of a structure classified as Risk Category I, II or III.

2. Masonry foundation walls constructed in accordance with Table 1807.1.6.3(1), 1807.1.6.3(2), 1807.1.6.3(3) or 1807.1.6.3(4).

3. Masonry fireplaces, masonry heaters or masonry chimneys installed or constructed in accordance with Section 2111, 2112 or 2113, respectively.

1705.4.1 Empirically designed masonry, glass unit masonry and masonry veneer in Risk Category IV. Special inspections and tests for empirically designed masonry, glass unit masonry or masonry veneer designed in accordance with Section 2109, 2110 or Chapter 14, respectively, where they are part of a structure classified as Risk Category IV shall be performed in accordance with TMS 402/ACI 530/ASCE 5, Level B Quality Assurance.

1705.4.2 Vertical masonry foundation elements. Special inspections and tests of vertical masonry foundation elements shall be performed in accordance with Section 1705.4.

1705.12 Special inspections for seismic resistance. Special inspections for seismic resistance shall be required as specified in Sections 1705.12.1 through 1705.12.9, unless exempted by the exceptions of Section 1704.2.

Exception: The special inspections specified in Sections 1705.12.1 through 1705.12.9 are not required for structures designed and constructed in accordance with one of the following:

1. The structure consists of light-frame construction; the design spectral response acceleration at short periods, S_{DS}, as determined in Section 1613.3.4, does not exceed 0.5; and the building height of the structure does not exceed 35 feet (10 668 mm).

2. The seismic force-resisting system of the structure consists of reinforced masonry or reinforced concrete; the design spectral response acceleration at short periods, S_{DS}, as determined in Section 1613.3.4, does not exceed 0.5; and the building height of the structure does not exceed 25 feet (7620 mm).

3. The structure is a detached one- or two-family dwelling not exceeding two stories above grade plane and does not have any of the following horizontal or vertical irregularities in accordance with Section 12.3 of ASCE 7:

    3.1. Torsional or extreme torsional irregularity.

    3.2. Nonparallel systems irregularity.

    3.3. Stiffness-soft story or stiffness-extreme soft story irregularity.

    3.4. Discontinuity in lateral strength-weak story irregularity.

1705.16 Exterior insulation and finish systems (EIFS). Special inspections shall be required for all EIFS applications.

Exceptions:

1. Special inspections shall not be required for EIFS applications installed over a water-resistive barrier with a means of draining moisture to the exterior.

2. Special inspections shall not be required for EIFS applications installed over masonry or concrete walls.

1705.16.1 Water-resistive barrier coating. A water-resistive barrier coating complying with ASTM E 2570 requires special inspection of the water-resistive barrier coating when installed over a sheathing substrate.

1705.17 Fire-resistant penetrations and joints. In high-rise buildings or in buildings assigned to Risk Category III or IV, special inspections for through-penetrations, membrane penetration firestops, fire-resistant joint systems and perimeter fire barrier systems that are tested and listed in accordance with Sections 714.3.1.2, 714.4.2, 715.3 and 715.4 shall be in accordance with Section 1705.17.1 or 1705.17.2.

1705.17.1 Penetration firestops. Inspections of penetration firestop systems that are tested and listed in accordance with Sections 714.3.1.2 and 714.4.2 shall be conducted by an approved agency in accordance with ASTM E 2174.

1705.17.2 Fire-resistant joint systems. Inspection of fire-resistant joint systems that are tested and listed in accordance with Sections 715.3 and 715.4 shall be conducted by an approved inspection agency in accordance with ASTM E 2393.

Deletion Arrow and Revision Bar

→ Large arrow refers to text deleted from 2012 IBC.

| Thick bar refers to modified text from 2012 IBC.

2015 IBC
CHAPTER 18*
SOILS AND FOUNDATIONS

SECTION 1801
GENERAL

1801.1 Scope. The provisions of this chapter shall apply to building and foundation systems.

SECTION 1805
DAMPPROOFING AND WATERPROOFING

1805.1 General. Walls or portions thereof that retain earth and enclose interior spaces and floors below grade shall be waterproofed and dampproofed in accordance with this section, with the exception of those spaces containing groups other than residential and institutional where such omission is not detrimental to the building or occupancy.

Ventilation for crawl spaces shall comply with Section 1203.4.

1805.2 Dampproofing. Where hydrostatic pressure will not occur as determined by Section 1803.5.4, floors and walls for other than wood foundation systems shall be dampproofed in accordance with this section. Wood foundation systems shall be constructed in accordance with AWC PWF.

1805.2.2 Walls. Dampproofing materials for walls shall be installed on the exterior surface of the wall, and shall extend from the top of the footing to above ground level.

Dampproofing shall consist of a bituminous material, 3 pounds per square yard (16 N/m^2) of acrylic modified cement, $^1/_8$ inch (3.2 mm) coat of surface-bonding mortar complying with ASTM C 887, any of the materials permitted for waterproofing by Section 1805.3.2 or other approved methods or materials.

1805.2.2.1 Surface preparation of walls. Prior to application of dampproofing materials on concrete walls, holes and recesses resulting from the removal of form ties shall be sealed with a bituminous material or other approved methods or materials. Unit masonry walls shall be parged on the exterior surface below ground level with not less than $^3/_8$ inch (9.5 mm) of Portland cement mortar. The parging shall be coved at the footing.

Exception: Parging of unit masonry walls is not required where a material is approved for direct application to the masonry.

1805.3 Waterproofing. Where the ground-water investigation required by Section 1803.5.4 indicates that a hydrostatic pressure condition exists, and the design does not include a ground-water control system as described in Section 1805.1.3, walls and floors shall be waterproofed in accordance with this section.

1805.3.2 Walls. Walls required to be waterproofed shall be of concrete or masonry and shall be designed and constructed to withstand the hydrostatic pressures and other lateral loads to which the walls will be subjected.

Waterproofing shall be applied from the bottom of the wall to not less than 12 inches (305 mm) above the maximum elevation of the ground-water table. The remainder of the wall shall be dampproofed in accordance with Section 1805.2.2. Waterproofing shall consist of two-ply hot-mopped felts, not less than 6-mil (0.006 inch; 0.152 mm) polyvinyl chloride, 40-mil (0.040 inch; 1.02 mm) polymer-modified asphalt, 6-mil (0.006 inch; 0.152 mm) polyethylene or other approved methods or materials capable of bridging nonstructural cracks. Joints in the membrane shall be lapped and sealed in accordance with the manufacturer's installation instructions.

1805.3.2.1 Surface preparation of walls. Prior to the application of waterproofing materials on concrete or masonry walls, the walls shall be prepared in accordance with Section 1805.2.2.1.

SECTION 1807
FOUNDATION WALLS, RETAINING WALLS AND EMBEDDED POSTS AND POLES

1807.1 Foundation walls. Foundation walls shall be designed and constructed in accordance with Sections 1807.1.1 through 1807.1.6. Foundation walls shall be supported by foundations designed in accordance with Section 1808.

---

* Only portions of this section are shown which are particularly applicable to masonry construction. For additional information see the IBC.

Deletion Arrow and Revision Bar

➔ Large arrow refers to text deleted from 2012 IBC.

| Thick bar refers to modified text from 2012 IBC.

1807.1.3 Rubble stone foundation walls. Foundation walls of rough or random rubble stone shall not be less than 16 inches (406 mm) thick. Rubble stone shall not be used for foundation walls of structures assigned to Seismic Design Category C, D, E, or F.

1807.1.5 Concrete and masonry foundation walls. Concrete and masonry foundation walls shall be designed in accordance with Chapter 19 or 21, as applicable.

> **Exception:** Concrete and masonry foundation walls shall be permitted to be designed and constructed in accordance with Section 1807.1.6.

1807.1.6 Prescriptive design of concrete and masonry foundation walls. Concrete and masonry foundation walls that are laterally supported at the top and bottom shall be permitted to be designed and constructed in accordance with this section.

1807.1.6.1 Foundation wall thickness. The thickness of prescriptively designed foundation walls shall not be less than the thickness of the wall supported, except that foundation walls of at least 8-inch (203 mm) nominal width shall be permitted to support brick-veneered frame walls and 10-inch-wide (254 mm) cavity walls provided the requirements of Section 1807.1.6.2 or 1807.1.6.3 are met.

1807.1.6.3 Masonry foundation walls. Masonry foundation walls shall comply with the following:

1. The thickness shall comply with the requirements of Table 1807.1.6.3(1) for plain masonry walls or Table 1807.1.6.3(2), 1807.1.6.3(3) or 1807.1.6.3(4) for masonry walls with reinforcement.

2. Vertical reinforcement shall have a minimum yield strength of 60,000 psi (414 MPa).

3. The specified location of the reinforcement shall equal or exceed the effective depth distance, d, noted in Tables 1807.1.6.3(2), 1807.1.6.3(3) and 1807.1.6.3(4) and shall be measured from the face of the exterior (soil) side of the wall to the center of the vertical reinforcement. The reinforcement shall be placed within the tolerances specified in TMS 602/ACI 530.1/ASCE 6, Article 3.4 B.11 of the specified location.

4. Grout shall comply with Section 2103.3.

5. Concrete masonry units shall comply with ASTM C 90.

6. Clay masonry units shall comply with ASTM C 652 for hollow brick, except compliance with ASTM C 62 or ASTM C 216 shall be permitted where solid masonry units are installed in accordance with Table 1807.1.6.3(1) for plain masonry.

7. Masonry units shall be laid in running bond and installed with Type M or S mortar in accordance with Section 2103.2.1.

8. The unfactored axial load per linear foot of wall shall not exceed $1.2\, t\, f'_m$ where t is the specified wall thickness in inches and f'_m is the specified compressive strength of masonry in pounds per square inch.

9. At least 4 inches (102 mm) of solid masonry shall be provided at girder supports at the top of hollow masonry unit foundation walls.

10. Corbeling of masonry shall be in accordance with Section 2104.1. Where an 8-inch (203 mm) wall is corbeled, the top corbel shall not extend higher than the bottom of the floor framing and shall be a full course of headers at least 6 inches (152 mm) in length or the top course bed joint shall be tied to the vertical wall projection. The tie shall be W2.8 (4.8 mm) and spaced at a maximum horizontal distance of 36 inches (914 mm). The hollow space behind the corbelled masonry shall be filled with mortar or grout.

1807.1.6.3.1 Alternative foundation wall reinforcement. In lieu of the reinforcement provisions for masonry foundation walls in Table 1807.1.6.3(2), 1807.1.6.3(3) or 1807.1.6.3(4), alternative reinforcing bar sizes and spacings having an equivalent cross-sectional area of reinforcement per linear foot (mm) of wall shall be permitted to be used, provided the spacing of reinforcement does not exceed 72 inches (1829 mm) and reinforcing bar sizes do not exceed No. 11.

TABLE 1807.1.6.3(1)
PLAIN MASONRY FOUNDATION WALLS[a,b,c]

MAXIMUM WALL HEIGHT (feet)	MAXIMUM UNBALANCED BACKFILL HEIGHT[e] (feet)	MINIMUM NOMINAL WALL THICKNESS (inches) Design lateral soil load[a] (psf per foot of depth) 30[f]	45[f]	60
7	4 (or less)	8	8	8
	5	8	10	10
	6	10	12	10 (solid[c])
	7	12	10 (solid[c])	10 (solid[c])
8	4 (or less)	8	8	8
	5	8	10	12
	6	10	12	12 (solid[c])
	7	12	12 (solid[c])	Note d
	8	10 (solid[c])	12 (solid[c])	Note d
9	4 (or less)	8	8	8
	5	8	10	12
	6	12	12	12 (solid[c])
	7	12 (solid[c])	12 (solid[c])	Note d
	8	12 (solid[c])	Note d	Note d
	9[f]	Note d	Note d	Note d

For SI: 1 inch = 25.4 mm, 1 foot = 304.8 mm, 1 pound per square foot per foot = 0.157 kPa/m.
a. For design lateral soil loads, see Section 1610.
b. Provisions for this table are based on design and construction requirements specified in Section 1807.1.6.3.
c. Solid grouted hollow units or solid masonry units.
d. A design in compliance with Chapter 21 or reinforcement in accordance with Table 1807.1.6.3(2) is required.
e. For height of unbalanced backfill, see Section 1807.1.2.
f. Where unbalanced backfill height exceeds 8 feet and design lateral soil loads from Table 1610.1 are used, the requirements for 30 and 45 psf per foot of depth are not applicable (see Section 1610).

1807.1.6.3.2 Seismic requirements. Based on the seismic design category assigned to the structure in accordance with Section 1613, masonry foundation walls designed using Tables 1807.1.6.3(1) through 1807.1.6.3(4) shall be subject to the following limitations:

1. Seismic Design Categories A and B. No additional seismic requirements.
2. Seismic Design Category C. A design using Tables 1807.1.6.3(1) through 1807.1.6.3(4) is subject to the seismic requirements of Section 7.4.3 of TMS 402/ACI 530/ASCE 5.
3. Seismic Design Category D. A design using Tables 1807.1.6.3(2) through 1807.1.6.3(4) is subject to the seismic requirements of Section 7.4.4 of TMS 402/ACI 530/ASCE 5.
4. Seismic Design Categories E and F. A design using Tables 1807.1.6.3(2) through 1807.1.6.3(4) is subject to the seismic requirements of Section 7.4.5 of TMS 402/ACI 530/ASCE 5.

1807.2 Retaining walls. Retaining walls shall be designed in accordance with Sections 1807.2.1 through 1807.2.3.

1807.2.1 General. Retaining walls shall be designed to ensure stability against overturning, sliding, excessive foundation pressure and water uplift. Where a keyway is extended below the wall base with the intent to engage passive pressure and enhance sliding stability, lateral soil pressures on both sides of the keyway shall be considered in the sliding analysis.

1807.2.2 Design lateral soil loads. Retaining walls shall be designed for the lateral soil loads set forth in Section 1610.

1807.2.3 Safety factor. Retaining walls shall be designed to resist the lateral action of soil to produce sliding and overturning with a minimum safety factor of 1.5 in each case. The load combinations of Section 1605 shall not apply to this requirement. Instead, design shall be based on 0.7 times nominal earthquake loads, 1.0 times other nominal loads, and investigation with one or more of the variable loads set to zero. The safety factor against lateral sliding shall be taken as the available soil resistance at the base of the retaining wall foundation divided by the net lateral force applied to the retaining wall.

Exception: Where earthquake loads are included, the minimum safety factor for retaining wall sliding and overturning shall be 1.1.

Deletion Arrow and Revision Bar

➡ Large arrow refers to text deleted from 2012 IBC.

❙ Thick bar refers to modified text from 2012 IBC.

TABLE 1807.1.6.3(2)
8-INCH MASONRY FOUNDATION WALLS WITH REINFORCEMENT WHERE d ≥ 5 INCHES[a,b,c]

MAXIMUM WALL HEIGHT (feet-inches)	MAXIMUM UNBALANCED BACKFILL HEIGHT[d] (feet-inches)	MINIMUM VERTICAL REINFORCEMENT-BAR SIZE AND SPACING (inches) Design lateral soil load[a] (psf per foot of depth) 30[e]	45[e]	60
7-4	4-0 (or less)	#4 at 48	#4 at 48	#4 at 48
	5-0	#4 at 48	#4 at 48	#4 at 48
	6-0	#4 at 48	#5 at 48	#5 at 48
	7-4	#5 at 48	#6 at 48	#7 at 48
8-0	4-0 (or less)	#4 at 48	#4 at 48	#4 at 48
	5-0	#4 at 48	#4 at 48	#4 at 48
	6-0	#4 at 48	#5 at 48	#5 at 48
	7-0	#5 at 48	#6 at 48	#7 at 48
	8-0	#5 at 48	#6 at 48	#7 at 48
8-8	4-0 (or less)	#4 at 48	#4 at 48	#4 at 48
	5-0	#4 at 48	#4 at 48	#5 at 48
	6-0	#4 at 48	#5 at 48	#6 at 48
	7-0	#5 at 48	#6 at 48	#7 at 48
	8-8[e]	#6 at 48	#7 at 48	#8 at 48
9-4	4-0 (or less)	#4 at 48	#4 at 48	#4 at 48
	5-0	#4 at 48	#4 at 48	#5 at 48
	6-0	#4 at 48	#5 at 48	#6 at 48
	7-0	#5 at 48	#6 at 48	#7 at 48
	8-0	#6 at 48	#7 at 48	#8 at 48
	9-4[e]	#7 at 48	#8 at 48	#9 at 48
10-0	4-0 (or less)	#4 at 48	#4 at 48	#4 at 48
	5-0	#4 at 48	#4 at 48	#5 at 48
	6-0	#4 at 48	#5 at 48	#6 at 48
	7-0	#5 at 48	#6 at 48	#7 at 48
	8-0	#6 at 48	#7 at 48	#8 at 48
	9-0[e]	#7 at 48	#8 at 48	#9 at 48
	10-0[e]	#7 at 48	#9 at 48	#9 at 48

For SI: 1 inch = 25.4 mm, 1 foot = 304.8 mm, 1 pound per square foot per foot = 0.157 kPa/m.
a. For design lateral soil loads, see Section 1610.
b. Provisions for this table are based on design and construction requirements specified in Section 1807.1.6.3.
c. For alternative reinforcement, see Section 1807.1.6.3.1.
d. For height of unbalanced backfill, see Section 1807.1.2.
e. Where unbalanced backfill height exceeds 8 feet and design lateral soil loads from Table 1610.1 are used, the requirements for 30 and 45 psf per foot of depth are not applicable. See Section 1610.

1807.3 Embedded posts and poles. Designs to resist both axial and lateral loads employing posts or poles as columns embedded in earth or in concrete footings in earth shall be in accordance with Sections 1807.3.1 through 1807.3.3.

1807.3.1 Limitations. The design procedures outlined in this section are subject to the following limitations:

1. The frictional resistance for structural walls and slabs on silts and clays shall be limited to one-half of the normal force imposed on the soil by the weight of the footing or slab.

2. Posts embedded in earth shall not be used to provide lateral support for structural or nonstructural materials such as plaster, masonry or concrete unless bracing is provided that develops the limited deflection required.

Deletion Arrow and Revision Bar

➤ Large arrow refers to text deleted from 2012 IBC.

| Thick bar refers to modified text from 2012 IBC.

TABLE 1807.1.6.3(3)
10-INCH MASONRY FOUNDATION WALLS WITH REINFORCEMENT WHERE d ≥ 6.75 INCHES[a,b,c]

MAXIMUM WALL HEIGHT (feet-inches)	MAXIMUM UNBALANCED BACKFILL HEIGHT[d] (feet-inches)	MINIMUM VERTICAL REINFORCEMENT-BAR SIZE AND SPACING (inches) Design lateral soil load[a] (psf per foot of depth) 30[e]	45[e]	60
7-4	4-0 (or less)	#4 at 56	#4 at 56	#4 at 56
	5-0	#4 at 56	#4 at 56	#4 at 56
	6-0	#4 at 56	#4 at 56	#5 at 56
	7-4	#4 at 56	#5 at 56	#6 at 56
8-0	4-0 (or less)	#4 at 56	#4 at 56	#4 at 56
	5-0	#4 at 56	#4 at 56	#4 at 56
	6-0	#4 at 56	#4 at 56	#5 at 56
	7-0	#4 at 56	#5 at 56	#6 at 56
	8-0	#5 at 56	#6 at 56	#7 at 56
8-8	4-0 (or less)	#4 at 56	#4 at 56	#4 at 56
	5-0	#4 at 56	#4 at 56	#4 at 56
	6-0	#4 at 56	#4 at 56	#5 at 56
	7-0	#4 at 56	#5 at 56	#6 at 56
	8-8[e]	#5 at 56	#7 at 56	#8 at 56
9-4	4-0 (or less)	#4 at 56	#4 at 56	#4 at 56
	5-0	#4 at 56	#4 at 56	#4 at 56
	6-0	#4 at 56	#5 at 56	#5 at 56
	7-0	#4 at 56	#5 at 56	#6 at 56
	8-0	#5 at 56	#6 at 56	#7 at 56
	9-4[e]	#6 at 56	#7 at 56	#7 at 56
10-0	4-0 (or less)	#4 at 56	#4 at 56	#4 at 56
	5-0	#4 at 56	#4 at 56	#4 at 56
	6-0	#4 at 56	#5 at 56	#5 at 56
	7-0	#5 at 56	#6 at 56	#7 at 56
	8-0	#5 at 56	#7 at 56	#8 at 56
	9-0[e]	#6 at 56	#7 at 56	#9 at 56
	10-0[e]	#7 at 56	#8 at 56	#9 at 56

For SI: 1 inch = 25.4 mm, 1 foot = 304.8 mm, 1 pound per square foot per foot = 0.157 kPa/m.

a. For design lateral soil loads, see Section 1610.
b. Provisions for this table are based on design and construction requirements specified in Section 1807.1.6.3.
c. For alternative reinforcement, see Section 1807.1.6.3.1.
d. For height of unbalanced backfill, see Section 1807.1.2.
e. Where unbalanced backfill height exceeds 8 feet and design lateral soil loads from Table 1610.1 are used, the requirements for 30 and 45 psf per foot of depth are not applicable. See Section 1610.

Deletion Arrow and Revision Bar

➡ Large arrow refers to text deleted from 2012 IBC.

| Thick bar refers to modified text from 2012 IBC.

TABLE 1807.1.6.3(4)
12-INCH MASONRY FOUNDATION WALLS WITH REINFORCEMENT WHERE d ≥ 8.75 INCHES[a,b,c]

MAXIMUM WALL HEIGHT (feet-inches)	MAXIMUM UNBALANCED BACKFILL HEIGHT[d] (feet-inches)	MINIMUM VERTICAL REINFORCEMENT-BAR SIZE AND SPACING (inches) Design lateral soil load[a] (psf per foot of depth) 30[e]	45[e]	60
7-4	4 (or less)	#4 at 72	#4 at 72	#4 at 72
	5-0	#4 at 72	#4 at 72	#4 at 72
	6-0	#4 at 72	#4 at 72	#5 at 72
	7-4	#4 at 72	#5 at 72	#6 at 72
8-0	4 (or less)	#4 at 72	#4 at 72	#4 at 72
	5-0	#4 at 72	#4 at 72	#4 at 72
	6-0	#4 at 72	#4 at 72	#5 at 72
	7-0	#4 at 72	#5 at 72	#6 at 72
	8-0	#5 at 72	#6 at 72	#8 at 72
8-8	4 (or less)	#4 at 72	#4 at 72	#4 at 72
	5-0	#4 at 72	#4 at 72	#4 at 72
	6-0	#4 at 72	#4 at 72	#5 at 72
	7-0	#4 at 72	#5 at 72	#6 at 72
	8-8[e]	#5 at 72	#7 at 72	#8 at 72
9-4	4 (or less)	#4 at 72	#4 at 72	#4 at 72
	5-0	#4 at 72	#4 at 72	#4 at 72
	6-0	#4 at 72	#5 at 72	#5 at 72
	7-0	#4 at 72	#5 at 72	#6 at 72
	8-0	#5 at 72	#6 at 72	#7 at 72
	9-4[e]	#6 at 72	#7 at 72	#8 at 72
10-0	4 (or less)	#4 at 72	#4 at 72	#4 at 72
	5-0	#4 at 72	#4 at 72	#4 at 72
	6-0	#4 at 72	#5 at 72	#5 at 72
	7-0	#4 at 72	#6 at 72	#6 at 72
	8-0	#5 at 72	#6 at 72	#7 at 72
	9-0[e]	#6 at 72	#7 at 72	#8 at 72
	10-0[e]	#7 at 72	#8 at 72	#9 at 72

For SI: 1 inch = 25.4 mm, 1 foot = 304.8 mm, 1 pound per square foot per foot = 0.157 kPa/m.

a. For design lateral soil loads, see Section 1610.
b. Provisions for this table are based on design and construction requirements specified in Section 1807.1.6.3.
c. For alternative reinforcement, see Section 1807.1.6.3.1.
d. For height of unbalanced backfill, see Section 1807.1.2.
e. Where unbalanced backfill height exceeds 8 feet and design lateral soil loads from Table 1610.1 are used, the requirements for 30 and 45 psf per foot of depth are not applicable. See Section 1610.

SECTION 1808 FOUNDATIONS

1808.1 General. Foundations shall be designed and constructed in accordance with Sections 1808.2 through 1808.9. Shallow foundations shall also satisfy the requirements of Section 1809. Deep foundations shall also satisfy the requirements of Section 1810.

1808.9 Vertical masonry foundation elements. Vertical masonry foundation elements that are not foundation piers as defined in Section 202 shall be designed as piers, walls or columns, as applicable, in accordance with TMS 402/ACI 530/ASCE 5.

Deletion Arrow and Revision Bar

➤ Large arrow refers to text deleted from 2012 IBC.

| Thick bar refers to modified text from 2012 IBC.

SECTION 1809
SHALLOW FOUNDATIONS

1809.1 General. Shallow foundations shall be designed and constructed in accordance with Sections 1809.2 through 1809.13.

1809.7 Prescriptive footings for light-frame construction. Where a specific design is not provided, concrete or masonry unit footings supporting walls of light-frame construction shall be permitted to be designed in accordance with Table 1809.7.

TABLE 1809.7
PRESCRIPTIVE FOOTINGS SUPPORTING WALLS OF LIGHT-FRAME CONSTRUCTION[a,b,c,d,e]

NUMBER OF FLOORS SUPPORTED BY THE FOOTING[f]	WIDTH OF FOOTING (inches)	THICKNESS OF FOOTING (inches)
1	12	6
2	15	6
3	18	8[g]

For SI: 1 inch = 25.4 mm, 1 foot = 304.8 mm.

a. Depth of footings shall be in accordance with Section 1809.4.
b. The ground under the floor shall be permitted to be excavated to the elevation of the top of the footing.
c. Interior stud-bearing walls shall be permitted to be supported by isolated footings. The footing width and length shall be twice the width shown in this table, and footings shall be spaced not more than 6 feet on center.
d. See section 1905 for additional requirements for concrete footings of structures assigned to Seismic Design Category C, D, E or F.
e. For thickness of foundation walls, see Section 1807.1.6.
f. Footings shall be permitted to support a roof in addition to the stipulated number of floors. Footings supporting roof only shall be as required for supporting one floor.
g. Plain concrete footings for Group R-3 occupancies shall be permitted to be 6 inches thick.

1809.9 Masonry-unit footings. The design, materials and construction of masonry-unit footings shall comply with Section 1809.9.1 and 1809.9.2, and the provisions of Chapter 21.

Exception: Where a specific design is not provided, masonry-unit footings supporting walls of light-frame construction shall be permitted to be designed in accordance with Table 1809.7.

1809.9.1 Dimensions. Masonry-unit footings shall be laid in Type M or S mortar complying with Section 2103.2.1 and the depth shall not be less than twice the projection beyond the wall, pier or column. The width shall not be less than 8 inches (203 mm) wider than the wall supported thereon.

1809.9.2 Offsets. The maximum offset of each course in brick foundation walls stepped up from the footings shall be $1\frac{1}{2}$ inches (38 mm) where laid in single courses, and 3 inches (76 mm) where laid in double courses.

1809.10 Pier and curtain wall foundations. Except in Seismic Design Categories D, E and F, pier and curtain wall foundations shall be permitted to be used to support light-frame construction not more than two stories above grade plane, provided the following requirements are met:

1. All load-bearing walls shall be placed on continuous concrete footings bonded integrally with the exterior wall footings.

2. The minimum actual thickness of a load-bearing masonry wall shall not be less than 4 inches (102 mm) nominal or $3\frac{5}{8}$ inches (92 mm) actual thickness, and shall be bonded integrally with piers spaced 6 feet (1829 mm) on center (o.c.).

3. Piers shall be constructed in accordance with Chapter 21 and the following:

    3.1 The unsupported height of the masonry piers shall not exceed 10 times their least dimension.

    3.2 Where structural clay tile or hollow concrete masonry units are used for piers supporting beams and girders, the cellular spaces shall be filled solidly with concrete or Type M or S mortar.

    Exception: Unfilled hollow piers shall be permitted where the unsupported height of the pier is not more than four times its least dimension.

    3.3 Hollow piers shall be capped with 4 inches (102 mm) of solid masonry or concrete or the cavities of the top course shall be filled with concrete or grout.

4. The maximum height of a 4-inch (102 mm) load-bearing masonry foundation wall supporting wood frame walls and floors shall not be more than 4 feet (1219 mm) in height.

5. The unbalanced fill for 4-inch (102 mm) foundation walls shall not exceed 24 inches (610 mm) for solid masonry, nor 12 inches (305 mm) for hollow masonry.

Deletion Arrow and Revision Bar

→ Large arrow refers to text deleted from 2012 IBC.

| Thick bar refers to modified text from 2012 IBC.

Deletion Arrow and Revision Bar

→ Large arrow refers to text deleted from 2012 IBC.

| Thick bar refers to modified text from 2012 IBC.

Summary of Chapter 21 Changes from 2012 IBC to 2015 IBC

Chapter 21 of the International Building Code® continues to evolve by removing duplicity where identical provisions are contained in both the IBC and the reference standard. In 2015 the Masonry chapter made significant progress in this evolution. For the user that may not be able to find provisions in the 2015 IBC, the table below shows where the changes occurred and generally where the provisions are now located.

2012 IBC Sections	2015 IBC Sections
2101.2 Deleted listing of most design methods (design of masonry veneer retained) in favor of simply referencing TMS 402 **2101.3** Deleted listing of construction documents since they are listed in TMS 402 Section 1.2.1 **2103.1 through 2103.7** Combined all masonry units and simply referenced TMS 602 **2104** Deleted listing of most sections in favor of referencing TMS 602 **2105** Deleted listing of most sections in favor of referencing TMS 602	**2103.2.4** Added section for mortar for adhered masonry veneer

Deletion Arrow and Revision Bar

➡ Large arrow refers to text deleted from 2012 IBC.

❘ Thick bar refers to modified text from 2012 IBC.

Deletion Arrow and Revision Bar

➡ Large arrow refers to text deleted from 2012 IBC.

| Thick bar refers to modified text from 2012 IBC.

2015 IBC
CHAPTER 21
MASONRY

SECTION 2101
GENERAL

2101.1 Scope. This chapter shall govern the materials, design, construction and quality of masonry.

2101.2 Design methods. Masonry shall comply with the provisions of TMS 402/ACI 530/ASCE 5 or TMS 403 as well as applicable requirements of this chapter.

2101.2.1 Masonry veneer. Masonry veneer shall comply with the provisions of Chapter 14.

2101.3 Special inspection. The special inspection of masonry shall be as defined in Chapter 17, or an itemized testing and inspection program shall be provided that meets or exceeds the requirements of Chapter 17.

SECTION 2102
DEFINITIONS AND NOTATIONS

2102.1 General. The following terms are defined in Chapter 2:

SECTION 202
DEFINITIONS

AAC MASONRY. Masonry made of autoclaved aerated concrete (AAC) units, manufactured without internal reinforcement and bonded together using thin- or thick-bed mortar.

ADOBE CONSTRUCTION. Construction in which the exterior load-bearing and nonload-bearing walls and partitions are of unfired clay masonry units, and floors, roofs and interior framing are wholly or partly of wood or other approved materials.

Adobe, stabilized. Unfired clay masonry units to which admixtures, such as emulsified asphalt, are added during the manufacturing process to limit the units' water absorption so as to increase their durability.

Adobe, unstabilized. Unfired clay masonry units that do not meet the definition of "Adobe, stabilized."

AREA (for masonry).

Gross cross-sectional. The area delineated by the out-to-out specified dimensions of masonry in the plane under consideration.

Net cross-sectional. The area of masonry units, grout and mortar crossed by the plane under consideration based on out-to-out specified dimensions.

AUTOCLAVED AERATED CONCRETE (AAC). Low density cementitious product of calcium silicate hydrates, whose material specifications are defined in ASTM C 1386.

BED JOINT. The horizontal layer of mortar on which a masonry unit is laid.

BRICK.

Calcium silicate (sand lime brick). A pressed and subsequently autoclaved unit that consists of sand and lime, with or without the inclusion of other materials.

Clay or shale. A solid or hollow masonry unit of clay or shale, usually formed into a rectangular prism, then burned or fired in a kiln; brick is a ceramic product.

Concrete. A concrete masonry unit made from Portland cement, water, and suitable aggregates, with or without the inclusion of other materials.

CAST STONE. A building stone manufactured from Portland cement concrete precast and used as a trim, veneer or facing on or in buildings or structures.

CELL (masonry). A void space having a gross cross-sectional area greater than $1^1/_2$ square inches (967 mm^2).

CHIMNEY. A primarily vertical structure containing one or more flues, for the purpose of carrying gaseous products of combustion and air from a fuel-burning appliance to the outdoor atmosphere.

Factory-build chimney. A listed and labeled chimney composed of factory-made components, assembled in the field in accordance with manufacturer's instructions and the conditions of the listing.

Masonry chimney. A field-constructed chimney composed of solid masonry units, bricks, stones, or concrete.

Metal chimney. A field-constructed chimney of metal.

CHIMNEY TYPES.

High-heat appliance type. An approved chimney for removing the products of combustion from fuel-burning, high-heat appliances producing combustion gases in excess of 2000°F (1093°C) measured at the appliance flue outlet (see Section 2113.11.3).

Deletion Arrow and Revision Bar

➤ Large arrow refers to text deleted from 2012 IBC.

| Thick bar refers to modified text from 2012 IBC.

Low-heat appliance type. An approved chimney for removing the products of combustion from fuel-burning, low-heat appliances producing combustion gases not in excess of 1000°F (538°C) under normal operating conditions, but capable of producing combustion gases of 1400°F (760°C) during intermittent forces firing for periods up to 1 hour. Temperatures shall be measured at the appliance flue outlet.

Masonry type. A field-constructed chimney of solid masonry units or stones.

Medium-heat appliance type. An approved chimney for removing the products of combustion from fuel-burning, medium-heat appliances producing combustion gases not exceeding 2000°F (1093°C) measured at the appliance flue outlet (see Section 2113.11.2).

COLLAR JOINT. Vertical longitudinal space between wythes of masonry or between masonry wythe and backup construction that is permitted to be filled with mortar or grout.

DIMENSIONS (for Chapter 21).

Nominal. The specified dimension plus an allowance for the joints with which the units are to be laid. Nominal dimensions are usually stated in whole numbers. Thickness is given first, followed by height and then length.

Specified. Dimensions specified for the manufacture or construction of a unit, joint or element.

FIREPLACE. A hearth and fire chamber or similar prepared place in which a fire may be made and which is built in conjunction with a chimney.

FIREPLACE THROAT. The opening between the top of the firebox and the smoke chamber.

FOUNDATION PIER (for Chapter 21). An isolated vertical foundation member whose horizontal dimension measured at right angles to its thickness does not exceed three times its thickness and whose height is equal to or less than four times its thickness.

HEAD JOINT. Vertical mortar joint placed between masonry units within the wythe at the time the masonry units are laid.

MASONRY. A built-up construction or combination of building units or materials of clay, shale, concrete, glass, gypsum, stone or other approved units bonded together with or without mortar or grout or other accepted methods of joining.

Glass unit masonry. Masonry composed of glass units bonded by mortar.

Plain masonry. Masonry in which the tensile resistance of the masonry is taken into consideration and the effects of stresses in reinforcement are neglected.

Reinforced masonry. Masonry construction in which reinforcement acting in conjunction with the masonry is used to resist forces.

Solid masonry. Masonry consisting of solid masonry units laid contiguously with the joints between the units filled with mortar.

Unreinforced (plain) masonry. Masonry in which the tensile resistance of masonry is taken into consideration and the resistance of the reinforcing steel, if present, is neglected.

MASONRY UNIT. Brick, tile, stone, glass block or concrete block conforming to the requirements specified in Section 2103.

Hollow. A masonry unit whose net cross-sectional area in any plane parallel to the load-bearing surface is less than 75 percent of its gross cross-sectional area measured in the same plane.

Solid. A masonry unit whose net cross-sectional area in every plane parallel to the load-bearing surface is 75 percent or more of its gross cross-sectional area measured in the same plane.

MORTAR. A mixture consisting of cementitious materials, fine aggregates, water, with or without admixtures, that is used to construct unit masonry assemblies.

MORTAR, SURFACE-BONDING. A mixture to bond concrete masonry units that contains hydraulic cement, glass fiber reinforcement with or without inorganic fillers or organic modifiers and water.

PRESTRESSED MASONRY. Masonry in which internal stresses have been introduced to counteract potential tensile stresses in masonry resulting from applied loads.

RUNNING BOND. The placement of masonry units such that head joints in successive courses are horizontally offset at least one-quarter the unit length.

SPECIFIED COMPRESSIVE STRENGTH OF MASONRY f'_m. Minimum compressive strength, expressed as force per unit of net cross-sectional area, required of the masonry used in construction by the approved construction documents, and upon which the project design is based. Whenever the quantity f'_m is under the radical sign, the square root of numerical value only is intended and the result has units of pounds per square inch (psi) (MPa).

STONE MASONRY. Masonry composed of field, quarried or cast stone units bonded by mortar.

STRENGTH (For Chapter 21).

Design strength. Nominal strength multiplied by a strength reduction factor.

Nominal strength. Strength of a member or cross section calculated in accordance with these provisions before application of any strength-reduction factors.

Deletion Arrow and Revision Bar

➤ Large arrow refers to text deleted from 2012 IBC.

| Thick bar refers to modified text from 2012 IBC.

Required strength. Strength of a member or cross section required to resist factored loads.

TIE, WALL. Metal connector that connects wythes of masonry walls together.

TILE, STRUCTURAL CLAY. A hollow masonry unit composed of burned clay, shale, fire clay or mixture thereof, and having parallel cells.

WALL (for Chapter 21). A vertical element with a horizontal length-to-thickness ratio greater than three, used to enclose space.

  Cavity wall. A wall built of masonry units or of concrete, or a combination of these materials, arranged to provide an airspace within the wall, and in which the inner and outer parts of the wall are tied together with metal ties.

  Dry-stacked, surface-bonded walls. A wall built of concrete masonry units where the units are stacked dry, without mortar on the bed or head joints, and where both sides of the wall are coated with a surface-bonding mortar.

  Parapet wall. The part of any wall entirely above the roof line.

WYTHE. Each continuous, vertical section of a wall, one masonry unit in thickness.

NOTATIONS.

d_b = Diameter of reinforcement, inches (mm).

F_s = Allowable tensile or compressive stress in reinforcement, psi (MPa).

f_r = Modulus of rupture, psi (MPa).

f'_{AAC} = Specified compressive strength of AAC masonry, the minimum compressive strength for a class of AAC masonry as specified in ASTM C 1386, psi (MPa). (ASTM C1386 Replaced by ASTM C1693).

f'_m = Specified compressive strength of masonry at age of 28 days, psi (MPa).

f'_{mi} = Specified compressive strength of masonry at the time of prestress transfer, psi (MPa).

K = The lesser of the masonry cover, clear spacing between adjacent reinforcement, or five times d_b, inches (mm).

L_s = Distance between supports, inches (mm).

l_d = Required development length or lap length of reinforcement, inches (mm).

P = The applied load at failure, pounds (N).

S_t = Thickness of the test specimen measured parallel to the direction of load, inches (mm).

S_w = Width of the test specimen measured parallel to the loading cylinder, inches (mm).

SECTION 2103
MASONRY CONSTRUCTION MATERIALS

2103.1 Masonry units. Concrete masonry units, clay or shale masonry units, stone masonry units, glass unit masonry and AAC masonry units shall comply with Article 2.3 of TMS 602/ACI 503.1/ASCE 6. Architectural cast stone shall conform to ASTM C 1364.

  Exception: Structural clay tile for nonstructural use in fireproofing of structural members and in wall furring shall not be required to meet the compressive strength specifications. The fire-resistance rating shall be determined in accordance with ASTM E 119 or UL 263 and shall comply with the requirements of Table 602.

2103.1.1 Second-hand units. Second-hand masonry units shall not be reused unless they conform to the requirements of new units. The units shall be of whole, sound materials and free from cracks and other defects that will interfere with proper laying or use. Old mortar shall be cleaned from the unit before reuse.

2103.2 Mortar. Mortar for masonry construction shall comply with Section 2103.2.1, 2103.2.2, 2103.2.3 or 2103.2.4.

2103.2.1 Masonry mortar. Mortar for use in masonry construction shall conform to Articles 2.1 and 2.6 A of TMS 602/ACI 530.1/ASCE 6.

2103.2.2 Surface-bonding mortar. Surface-bonding mortar shall comply with ASTM C887. Surface bonding of concrete masonry units shall comply with ASTM C 946.

2103.2.3 Mortars for ceramic wall and floor tile. Portland cement mortars for installing ceramic wall and floor tile shall comply with ANSI A108.1A and ANSI A108.1B and be of the compositions indicated in Table 2103.2.3.

TABLE 2103.2.3
CERAMIC TILE MORTAR COMPOSITIONS

LOCATION	MORTAR	COMPOSITION
Walls	Scratchcoat	1 cement; $1/5$ hydrated lime; 4 dry or 5 damp sand
Walls	Setting bed and leveling coat	1 cement; $1/2$ hydrated lime; 5 damp sand to 1 cement 1 hydrated lime, 7 damp sand
Floors	Setting bed	1 cement; $1/10$ hydrated lime; 5 dry or 6 damp sand; or 1 cement; 5 dry or 6 damp sand
Ceilings	Scratchcoat and sand bed	1 cement; $1/2$ hydrated lime; $2 1/2$ dry sand or 3 damp sand

---

Deletion Arrow and Revision Bar

→ Large arrow refers to text deleted from 2012 IBC.

| Thick bar refers to modified text from 2012 IBC.

2103.2.3.1 Dry-set Portland cement mortars. Premixed prepared Portland cement mortars, which require only the addition of water and are used in the installation of ceramic tile, shall comply with ANSI A118.1. The shear bond strength for tile set in such mortar shall be as required in accordance with ANSI A118.1. Tile set in dry-set Portland cement mortar shall be installed in accordance with ANSI A108.5.

2103.2.3.2 Latex-modified Portland cement mortar. Latex-modified Portland cement thin-set mortars in which latex is added to dry-set mortar as a replacement for all or part of the gauging water that are used for the installation of ceramic tile shall comply with ANSI A118.4. Tile set in latex-modified Portland cement shall be installed in accordance with ANSI A108.5.

2103.2.3.3 Epoxy mortar. Ceramic tile set and grouted with chemical-resistant epoxy shall comply with ANSI A118.3. Tile set and grouted with epoxy shall be installed in accordance with ANSI A108.6.

2103.2.3.4 Furan mortar and grout. Chemical-resistant furan mortar and grout that are used to install ceramic tile shall comply with ANSI A118.5. Tile set and grouted with furan shall be installed in accordance with ANSI A108.8.

2103.2.3.5 Modified epoxy-emulsion mortar and grout. Modified epoxy-emulsion mortar and grout that are used to install ceramic tile shall comply with ANSI A118.8. Tile set and grouted with modified epoxy-emulsion mortar and grout shall be installed in accordance with ANSI A108.9.

2103.2.3.6 Organic adhesives. Water-resistant organic adhesives used for the installation of ceramic tile shall comply with ANSI A136.1. The shear bond strength after water immersion shall be not less than 40 psi (275 kPa) for Type I adhesive and not less than 20 psi (138 kPa) for Type II adhesive when tested in accordance with ANSI A136.1. Tile set in organic adhesives shall be installed in accordance with ANSI A108.4.

2103.2.3.7 Portland cement grouts. Portland cement grouts used for the installation of ceramic tile shall comply with ANSI A118.6. Portland cement grouts for tile work shall be installed in accordance with ANSI A108.10.

2103.2.4 Mortar for adhered masonry veneer. Mortar for use with adhered masonry veneer shall conform to ASTM C 270 for Type N or S, or shall comply with ANSI A118.4 for latex-modified Portland cement mortar.

2103.3 Grout. Grout shall comply with Article 2.2 of TMS 602/ACI 530.1/ASCE 6.

2103.4 Metal reinforcement and accessories. Metal reinforcement and accessories shall conform to Article 2.4 of TMS 602/ACI 530.1/ASCE 6. Where unidentified reinforcement is approved for use, not less than three tension and three bending tests shall be made on representative specimens of the reinforcement from each shipment and grade of reinforcing steel proposed for use in the work.

SECTION 2104
CONSTRUCTION

2104.1 Masonry construction. Masonry construction shall comply with the requirements of Sections 2104.1.1 through 2104.1.2 and with TMS 602/ACI 530.1/ASCE 6.

2104.1.1 Support on wood. Masonry shall not be supported on wood girders or other forms of wood construction except as permitted in Section 2304.12.

2104.1.2 Molded cornices. Unless structural support and anchorage are provided to resist the overturning moment, the center of gravity of projecting masonry or molded cornices shall lie within the middle one-third of the supporting wall. Terra cotta and metal cornices shall be provided with a structural frame of approved noncombustible material anchored in an approved manner.

SECTION 2105
QUALITY ASSURANCE

2105.1 General. A quality assurance program shall be used to ensure that the constructed masonry is in compliance with the approved construction documents.

The quality assurance program shall comply with the inspection and testing requirements of Chapter 17 and TMS 602/ACI 530.1/ASCE 6.

SECTION 2106
SEISMIC DESIGN

2106.1 Seismic design requirements for masonry. Masonry structures and components shall comply with the requirements in Chapter 7 of TMS 402/ACI 530/ASCE 5 depending on the structure's seismic design category.

SECTION 2107
ALLOWABLE STRESS DESIGN

2107.1 General. The design of masonry structures using allowable stress design shall comply with Section 2106 and the requirements of Chapters 1 through 8 of TMS 402/ACI 530/ASCE 5 except as modified by Sections 2107.2 through 2107.4.

Deletion Arrow and Revision Bar

➤ Large arrow refers to text deleted from 2012 IBC.

▌ Thick bar refers to modified text from 2012 IBC.

2107.2 TMS 402/ACI 530/ASCE 5, Section 8.1.6.7.1.1, lap splices. As an alternative to Section 8.1.6.7.1.1, it shall be permitted to design lap splices in accordance with Section 2107.2.1.

2107.2.1 Lap splices. The minimum length of lap splices for reinforcing bars in tension or compression, l_d, shall be:

$$l_d = 0.002 d_b f_s \qquad \text{(Equation 21-1)}$$

For SI: $l_d = 0.29 d_b f_s$

but not less than 12 inches (305 mm). In no case shall the length of the lapped splice be less than 40 bar diameters.

where:

d_b = Diameter of reinforcement, inches (mm).

f_s = Computed stress in reinforcement due to design loads, psi (MPa).

In regions of moment where the design tensile stresses in the reinforcement are greater than 80 percent of the allowable steel tension stress, F_s, the lap length of splices shall be increased not less than 50 percent of the minimum required length. Other equivalent means of stress transfer to accomplish the same 50 percent increase shall be permitted. Where epoxy coated bars are used, lap length shall be increased by 50 percent.

2107.3 TMS 402/ACI 530/ASCE 5, Section 8.1.6.7, splices of reinforcement. Modify Section 8.1.6.7 as follows:

8.1.6.7 – Splices of reinforcement. Lap splices, welded splices or mechanical splices are permitted in accordance with the provisions of this section. All welding shall conform to AWS D1.4. Welded splices shall be of ASTM A706 steel reinforcement. Reinforcement larger than No. 9 (M #29) shall be spliced using mechanical connections in accordance with Section 8.1.6.7.3.

2107.4 TMS 402/ACI 530/ASCE 5, Section 8.3.6, maximum bar size. Add the following to Chapter 8:

8.3.6 – Maximum bar size. The bar diameter shall not exceed one-eighth of the nominal wall thickness and shall not exceed one-quarter of the least dimension of the cell, course or collar joint in which it is placed.

SECTION 2108
STRENGTH DESIGN OF MASONRY

2108.1 General. The design of masonry structures using strength design shall comply with Section 2106 and the requirements of Chapters 1 through 7 and Chapter 9 of TMS 402/ACI 530/ASCE 5, except as modified by Sections 2108.2 through 2108.3.

Exception: AAC masonry shall comply with the requirements of Chapters 1 through 7 and 11 of TMS 402/ACI 530/ASCE 5.

2108.2 TMS 402/ACI 530/ASCE 5, Section 9.3.3.3 development. Modify the second paragraph of Section 9.3.3.3 as follows:

The required development length of reinforcement shall be determined by Equation (9-16), but shall not be less than 12 inches (305 mm) and need not be greater than 72 d_b.

2108.3 TMS 402/ACI 530/ASCE 5, Section 9.3.3.4, splices. Modify items (c) and (d) of Section 9.3.3.4 as follows:

9.3.3.4 (c) – A welded splice shall have the bars butted and welded to develop at least 125 percent of the yield strength, f_y, of the bar in tension or compression, as required. Welded splices shall be of ASTM A 706 steel reinforcement. Welded splices shall not be permitted in plastic hinge zones of intermediate or special reinforced walls.

9.3.3.4 (d) – Mechanical splices shall be classified as Type 1 or 2 in accordance with Section 18.2.7.1 of ACI 318. Type 1 mechanical splices shall not be used within a plastic hinge zone or within a beam-column joint of intermediate or special reinforced masonry shear walls. Type 2 mechanical splices are permitted in any location within a member.

SECTION 2109
EMPIRICAL DESIGN OF MASONRY

2109.1 General. Empirically designed masonry shall conform to the requirements of Appendix A of TMS 402/ACI 530/ASCE 5, except where otherwise noted in this section.

2109.1.1 Limitations. The use of empirical design of masonry shall be limited as noted in Section A.1.2 of TMS 402/ACI 530/ASCE 5. The use of dry-stacked, surface-bonded masonry shall be prohibited in Risk Category IV structures. In buildings that exceed one or more of the limitations of Section A.1.2 of TMS 402/ACI 530/ASCE 5, masonry shall be designed in accordance with the engineered design provisions of Section 2101.2 or the foundation wall provisions of Section 1807.1.5.

Section A.1.2.3 of TMS 402/ACI 530/ASCE 5 shall be modified as follows:

A.1.2.3 *Wind* – Empirical requirements shall not apply to the design or construction of masonry for buildings, parts of buildings, or other structures to be located in areas where V_{asd} as determined in accordance with Section 1609.3.1 of the *International Building Code* exceeds 110 mph.

Deletion Arrow and Revision Bar

→ Large arrow refers to text deleted from 2012 IBC.

| Thick bar refers to modified text from 2012 IBC.

2109.2 Surface-bonded walls. Dry-stacked, surface-bonded concrete masonry walls shall comply with the requirements of Appendix A of TMS 402/ACI 530/ASCE 5, except where otherwise noted in this section.

2109.2.1 Strength. Dry-stacked, surface-bonded concrete masonry walls shall be of adequate strength and proportions to support all superimposed loads without exceeding the allowable stresses listed in Table 2109.2.1. Allowable stresses not specified in Table 2109.2.1 shall comply with the requirements of TMS 402/ACI 530/ASCE 5.

TABLE 2109.2.1
ALLOWABLE STRESS GROSS CROSS-SECTIONAL AREA FOR DRY-STACKED, SURFACE-BONDED CONCRETE MASONRY WALLS

DESCRIPTION	MAXIMUM ALLOWABLE STRESS (psi)
Compression standard block	45

Flexural tension
  Horizontal span
  Vertical span
	
	30
	18

Shear	10

For SI: 1 pound per square inch = 0.006895 MPa.

2109.2.2 Construction. Construction of dry-stacked, surface-bonded masonry walls, including stacking and leveling of units, mixing and application of mortar and curing and protection shall comply with ASTM C 946.

2109.3 Adobe construction. Adobe construction shall comply with this section and shall be subject to the requirements of this code for Type V construction, Appendix A of TMS 402/ACI 530/ASCE 5, and this section.

2109.3.1 Unstabilized adobe. Unstabilized adobe shall comply with Sections 2109.3.1.1 through 2109.3.1.4.

2109.3.1.1 Compressive strength. Adobe units shall have an average compressive strength of 300 psi (2068 kPa) when tested in accordance with ASTM C 67. Five samples shall be tested and no individual unit is permitted to have a compressive strength of less than 250 psi (1724 kPa).

2109.3.1.2 Modulus of rupture. Adobe units shall have an average modulus of rupture of 50 psi (345 kPa) when tested in accordance with the following procedure. Five samples shall be tested and no individual unit shall have a modulus of rupture of less than 35 psi (241 kPa).

2109.3.1.2.1 Support conditions. A cured unit shall be simply supported by 2-inch-diameter (51 mm) cylindrical supports located 2 inches (51 mm) in from each end and extending the full width of the unit.

2109.3.1.2.2 Loading conditions. A 2-inch-diameter (51 mm) cylinder shall be placed at midspan parallel to the supports.

2109.3.1.2.3 Testing procedure. A vertical load shall be applied to the cylinder at the rate of 500 pounds per minute (37 N/s) until failure occurs.

2109.3.1.2.4 Modulus of rupture determination. The modulus of rupture shall be determined by the equation:

$$f_r = 3PL_s/2S_w(S_t^2) \qquad \text{(Equation 21-2)}$$

where, for the purposes of this section only:

S_w = Width of the test specimen measured parallel to the loading cylinder, inches (mm).

f_r = Modulus of rupture, psi (MPa).

L_s = Distance between supports, inches (mm).

S_t = Thickness of the test specimen measured parallel to the direction of load, inches (mm).

P = The applied load at failure, pounds (N).

2109.3.1.3 Moisture content requirements. Adobe units shall have a moisture content not exceeding 4 percent by weight.

2109.3.1.4 Shrinkage cracks. Adobe units shall not contain more than three shrinkage cracks and any single shrinkage crack shall not exceed 3 inches (76 mm) in length or $^1/_8$ inch (3.2 mm) in width.

2109.3.2 Stabilized adobe. Stabilized adobe shall comply with Section 2109.3.1 for unstabilized adobe in addition to Sections 2109.3.2.1 and 2109.3.2.2.

2109.3.2.1 Soil requirements. Soil used for stabilized adobe units shall be chemically compatible with the stabilizing material.

2109.3.2.2 Absorption requirements. A 4-inch (102 mm) cube, cut from a stabilized adobe unit dried to a constant weight in a ventilated oven at 212°F to 239°F (100°C to 115°C), shall not absorb more than $2^1/_2$ percent moisture by weight when placed upon a constantly water-saturated, porous surface for seven days. A minimum of five specimens shall be tested and each specimen shall be cut from a separate unit.

2109.3.3 Allowable stress. The allowable compressive stress based on gross cross-sectional area of adobe shall not exceed 30 psi (207 kPa).

2109.3.3.1 Bolts. Bolt values shall not exceed those set forth in Table 2109.3.3.1.

TABLE 2109.3.3.1
ALLOWABLE SHEAR ON BOLTS IN ADOBE MASONRY

DIAMETER OF BOLTS (inches)	MINIMUM EMBEDMENT (inches)	SHEAR (pounds)
1/2	—	—
5/8	12	200
3/4	15	300
7/8	18	400
1	21	500
1 1/8	24	600

For SI: 1 inch = 25.4 mm, 1 pound = 4.448 N.

2109.3.4 Detailed requirements. Adobe construction shall comply with Sections 2109.3.4.1 through 2109.3.4.9.

2109.3.4.1 Number of stories. Adobe construction shall be limited to buildings not exceeding one story, except that two-story construction is allowed when designed by a registered design professional.

2109.3.4.2 Mortar. Mortar for adobe construction shall comply with Sections 2109.3.4.2.1 and 2109.3.4.2.2.

2109.3.4.2.1 General. Mortar for stabilized adobe units shall comply with this chapter or adobe soil. Adobe soil used as mortar shall comply with material requirements for stabilized adobe. Mortar for unstabilized adobe shall be Portland cement mortar.

2109.3.4.2.2 Mortar joints. Adobe units shall be laid with full head and bed joints and in full running bond.

2109.3.4.3 Parapet walls. Parapet walls constructed of adobe units shall be waterproofed.

2109.3.4.4 Wall thickness. The minimum thickness of exterior walls in one-story buildings shall be 10 inches (254 mm). The walls shall be laterally supported at intervals not exceeding 24 feet (7315 mm). The minimum thickness of interior load-bearing walls shall be 8 inches (203 mm). In no case shall the unsupported height of any wall constructed of adobe units exceed 10 times the thickness of such wall.

2109.3.4.5 Foundations. Foundations for adobe construction shall be in accordance with Sections 2109.3.4.5.1 and 2109.3.4.5.2.

2109.3.4.5.1 Foundation support. Walls and partitions constructed of adobe units shall be supported by foundations or footings that extend not less than 6 inches (152 mm) above adjacent ground surfaces and are constructed of solid masonry (excluding adobe) or concrete. Footings and foundations shall comply with Chapter 18.

2109.3.4.5.2 Lower course requirements. Stabilized adobe units shall be used in adobe walls for the first 4 inches (102 mm) above the finished first-floor elevation.

2109.3.4.6 Isolated piers or columns. Adobe units shall not be used for isolated piers or columns in a load-bearing capacity. Walls less than 24 inches (610 mm) in length shall be considered isolated piers or columns.

2109.3.4.7 Tie beams. Exterior walls and interior load-bearing walls constructed of adobe units shall have a continuous tie beam at the level of the floor or roof bearing and meeting the following requirements.

2109.3.4.7.1 Concrete tie beams. Concrete tie beams shall be a minimum depth of 6 inches (152 mm) and a minimum width of 10 inches (254 mm). Concrete tie beams shall be continuously reinforced with a minimum of two No. 4 reinforcing bars. The specified compressive strength of concrete shall be at least 2,500 psi (17.2 MPa).

2109.3.4.7.2 Wood tie beams. Wood tie beams shall be solid or built up of lumber having a minimum nominal thickness of 1 inch (25 mm), and shall have a minimum depth of 6 inches (152 mm) and a minimum width of 10 inches (254 mm). Joints in wood tie beams shall be spliced a minimum of 6 inches (152 mm). No splices shall be allowed within 12 inches (305 mm) of an opening. Wood used in tie beams shall be approved naturally decay-resistant or preservative-treated wood.

2109.3.4.8 Exterior finish. Exterior walls constructed of unstabilized adobe units shall have their exterior surface covered with a minimum of two coats of Portland cement plaster having a minimum thickness of 3/4 inch (19.1 mm) and conforming to ASTM C 926. Lathing shall comply with ASTM C 1063. Fasteners shall be spaced at 16 inches (406 mm) on center maximum. Exposed wood surfaces shall be treated with an approved wood preservative or other protective coating prior to lath application.

2109.3.4.9 Lintels. Lintels shall be considered structural members and shall be designed in accordance with the applicable provisions of Chapter 16.

SECTION 2110
GLASS UNIT MASONRY

2110.1 General. Glass unit masonry construction shall comply with Chapter 13 of TMS 402/ACI 530/ASCE 5 and this section.

2110.1.1 Limitations. Solid or hollow approved glass block shall not be used in fire walls, party walls, fire barriers, fire partitions or smoke barriers, or for load-bearing construction. Such blocks shall be erected with mortar and reinforcement in metal channel-type frames,

---

Deletion Arrow and Revision Bar

➡ Large arrow refers to text deleted from 2012 IBC.

▎ Thick bar refers to modified text from 2012 IBC.

structural frames, masonry or concrete recesses, embedded panel anchors as provided for both exterior and interior walls or other approved joint materials. Wood strip framing shall not be used in walls required to have a fire-resistance rating by other provisions of this code.

Exceptions:

1. Glass-block assemblies having a fire protection rating of not less than 3/4 hour shall be permitted as opening protectives in accordance with Section 716 in fire barriers, fire partitions and smoke barriers that have a required fire-resistance rating of 1 hour or less and do not enclose exit stairways and ramps or exit passageways.

2. Glass-block assemblies as permitted in Section 404.6, Exception 2.

SECTION 2111
MASONRY FIREPLACES

2111.1 General. The construction of masonry fireplaces, consisting of concrete or masonry shall be in accordance with this section.

2111.2 Fireplace drawings. The construction documents shall describe in sufficient detail the location, size and construction of masonry fireplaces. The thickness and characteristics of materials and the clearances from walls, partitions and ceilings shall be indicated.

2111.3 Footings and foundations. Footings for masonry fireplaces and their chimneys shall be constructed of concrete or solid masonry at least 12 inches (305 mm) thick and shall extend at least 6 inches (153 mm) beyond the face of the fireplace or foundation wall on all sides. Footings shall be founded on natural undisturbed earth or engineered fill below frost depth. In areas not subjected to freezing, footings shall be at least 12 inches (305 mm) below finished grade.

2111.3.1 Ash dump cleanout. Cleanout openings, located within foundation walls below fireboxes, when provided, shall be equipped with ferrous metal or masonry doors and frames constructed to remain tightly closed, except when in use. Cleanouts shall be accessible and located so that ash removal will not create a hazard to combustible materials.

2111.4 Seismic reinforcement. In structures assigned to Seismic Design Category A or B, seismic reinforcement is not required. In structures assigned to Seismic Design Category C or D, masonry fireplaces shall be reinforced and anchored in accordance with Sections 2111.4.1, 2114.2 and 2111.5. In structures assigned to Seismic Design Category E or F, masonry fireplaces shall be reinforced in accordance with the requirements of Sections 2101 through 2108.

2111.4.1 Vertical reinforcing. For fireplaces with chimneys up to 40 inches (1016 mm) wide, four No. 4 continuous vertical bars, anchored in the foundation, shall be placed in the concrete between wythes of solid masonry or within the cells of hollow unit masonry and grouted in accordance with Section 2103.3. For fireplaces with chimneys greater than 40 inches (1016 mm) wide, two additional No. 4 vertical bars shall be provided for each additional 40 inches (1016 mm) in width or fraction thereof.

2111.4.2 Horizontal reinforcing. Vertical reinforcement shall be placed enclosed within 1/4-inch (6.4 mm) ties or other reinforcing of equivalent net cross-sectional area, spaced not to exceed 18 inches (457 mm) on center in concrete; or placed in the bed joints of unit masonry at a minimum of every 18 inches (457 mm) of vertical height. Two such ties shall be provided at each bend in the vertical bars.

2111.5 Seismic anchorage. Masonry fireplaces and foundations shall be anchored at each floor, ceiling or roof line more than 6 feet (1829 mm) above grade with two 3/16-inch by 1-inch (4.8 mm by 25 mm) straps embedded a minimum of 12 inches (305 mm) into the chimney. Straps shall be hooked around the outer bars and extend 6 inches (152 mm) beyond the bend. Each strap shall be fastened to a minimum of four floor joists with two 1/2-inch (12.7 mm) bolts.

Exception: Seismic anchorage is not required for the following:

1. In structures assigned to Seismic Design Category A or B.

2. Where the masonry fireplace is constructed completely within the exterior walls.

2111.6 Firebox walls. Masonry fireboxes shall be constructed of solid masonry units, hollow masonry units grouted solid, stone or concrete. When a lining of firebrick at least 2 inches (51 mm) in thickness or other approved lining is provided, the minimum thickness of back and sidewalls shall each be 8 inches (203 mm) of solid masonry, including the lining. The width of joints between firebricks shall be not greater than 1/4 inch (6.4 mm). When no lining is provided, the total minimum thickness of back and sidewalls shall be 10 inches (254 mm) of solid masonry. Firebrick shall conform to ASTM C 27 or ASTM C 1261 and shall be laid with medium-duty refractory mortar conforming to ASTM C 199.

2111.6.1 Steel fireplace units. Steel fireplace units are permitted to be installed with solid masonry to form a masonry fireplace provided they are installed according to either the requirements of their listing or the requirements of this section. Steel fireplace units incorporating a steel firebox lining shall be constructed with steel not less than $^{1}/_{4}$ inch (6.4 mm) in thickness, and an air-circulating chamber which is ducted to the interior of the building. The firebox lining shall be encased with solid masonry to provide a total thickness at the back and sides of not less than 8 inches (203 mm), of which not less than 4 inches (102 mm) shall be of solid masonry or concrete. Circulating air ducts employed with steel fireplace units shall be constructed of metal or masonry.

2111.7 Firebox dimensions. The firebox of a concrete or masonry fireplace shall have a minimum depth of 20 inches (508 mm). The throat shall be not less than 8 inches (203 mm) above the fireplace opening. The throat opening shall not be less than 4 inches (102 mm) in depth. The cross-sectional area of the passageway above the firebox, including the throat, damper and smoke chamber, shall be not less than the cross-sectional area of the flue.

Exception: Rumford fireplaces shall be permitted provided that the depth of the fireplace is not less than 12 inches (305 mm) and at least one-third of the width of the fireplace opening, and the throat is not less than 12 inches (305 mm) above the lintel, and at least $^{1}/_{20}$ the cross-sectional area of the fireplace opening.

2111.8 Lintel and throat. Masonry over a fireplace opening shall be supported by a lintel of noncombustible material. The minimum required bearing length on each end of the fireplace opening shall be 4 inches (102 mm). The fireplace throat or damper shall be located not less than 8 inches (203 mm) above the top of the fireplace opening.

2111.8.1 Damper. Masonry fireplaces shall be equipped with a ferrous metal damper located not less than 8 inches (203 mm) above the top of the fireplace opening. Dampers shall be installed in the fireplace or at the top of the flue venting the fireplace, and shall be operable from the room containing the fireplace. Damper controls shall be permitted to be located in the fireplace.

2111.9 Smoke chamber walls. Smoke chamber walls shall be constructed of solid masonry units, hollow masonry units grouted solid, stone or concrete. The total minimum thickness of front, back and sidewalls shall be 8 inches (203 mm) of solid masonry. The inside surface shall be parged smooth with refractory mortar conforming to ASTM C 199. When a lining of firebrick not less than 2 inches (51 mm) thick, or a lining of vitrified clay not less than $^{5}/_{8}$ inch (15.9 mm) thick, is provided, the total minimum thickness of front, back and sidewalls shall be 6 inches (152 mm) of solid masonry, including the lining. Firebrick shall conform to ASTM C 1261 and shall be laid with refractory mortar conforming to ASTM C 199. Vitrified clay linings shall conform to ASTM C 315.

2111.9.1 Smoke chamber dimensions. The inside height of the smoke chamber from the fireplace throat to the beginning of the flue shall be not greater than the inside width of the fireplace opening. The inside surface of the smoke chamber shall not be inclined more than 45 degrees (0.76 rad) from vertical when prefabricated smoke chamber linings are used or when the smoke chamber walls are rolled or sloped rather than corbeled. When the inside surface of the smoke chamber is formed by corbeled masonry, the walls shall not be corbeled more than 30 degrees (0.52 rad) from vertical.

2111.10 Hearth and hearth extension. Masonry fireplace hearths and hearth extensions shall be constructed of concrete or masonry, supported by noncombustible materials, and reinforced to carry their own weight and all imposed loads. No combustible material shall remain against the underside of hearths or hearth extensions after construction.

2111.10.1 Hearth thickness. The minimum thickness of fireplace hearths shall be 4 inches (102 mm).

2111.10.2 Hearth extension thickness. The minimum thickness of hearth extensions shall be 2 inches (51 mm).

Exception: When the bottom of the firebox opening is raised not less than 8 inches (203 mm) above the top of the hearth extension, a hearth extension of not less than $^{3}/_{8}$-inch-thick (9.5 mm) brick, concrete, stone, tile or other approved noncombustible material is permitted.

2111.11 Hearth extension dimensions. Hearth extensions shall extend not less than 16 inches (406 mm) in front of, and not less than 8 inches (203 mm) beyond, each side of the fireplace opening. Where the fireplace opening is 6 square feet (0.557 m^2) or larger, the hearth extension shall extend not less than 20 inches (508 mm) in front of, and not less than 12 inches (305 mm) beyond, each side of the fireplace opening.

2111.12 Fireplace clearance. Any portion of a masonry fireplace located in the interior of a building or within the exterior wall of a building shall have a clearance to combustibles of not less than 2 inches (51 mm) from the front faces and sides of masonry fireplaces and not less than 4 inches (102 mm) from the back faces of masonry fireplaces. The airspace shall not be filled, except to provide fireblocking in accordance with Section 2111.13.

Exceptions:

1. Masonry fireplaces listed and labeled for use in contact with combustibles in accordance with UL 127 and installed in accordance with the manufacturer's instructions are permitted to have combustible material in contact with their exterior surfaces.

2. When masonry fireplaces are constructed as part of masonry or concrete walls, combustible materials shall not be in contact with the masonry or concrete walls less than 12 inches (306 mm) from the inside surface of the nearest firebox lining.

3. Exposed combustible trim and the edges of sheathing materials, such as wood siding, flooring and drywall, are permitted to abut the masonry fireplace sidewalls and hearth extension, in accordance with Figure 2111.12, provided such combustible trim or sheathing is not less than 12 inches (306 mm) from the inside surface of the nearest firebox lining.

4. Exposed combustible mantels or trim is permitted to be placed directly on the masonry fireplace front surrounding the fireplace opening, provided such combustible materials shall not be placed within 6 inches (153 mm) of a fireplace opening. Combustible material directly above and within 12 inches (305 mm) of the fireplace opening shall not project more than $^1/_8$ inch (3.2 mm) for each 1-inch (25 mm) distance from such opening. Combustible materials located along the sides of the fireplace opening that project more than $1^1/_2$ inches (38 mm) from the face of the fireplace shall have an additional clearance equal to the projection.

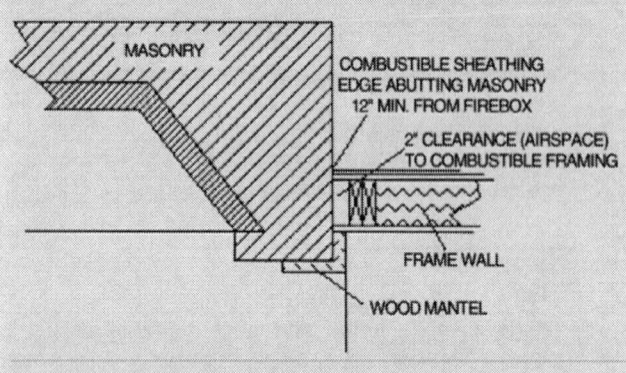

For SI: 1 inch = 25.4 mm

**FIGURE 2111.12
ILLUSTRATION OF EXCEPTION TO
FIREPLACE CLEARANCE PROVISION**

2111.13 Fireplace fireblocking. All spaces between fireplaces and floors and ceilings through which fireplaces pass shall be fireblocked with noncombustible material securely fastened in place. The fireblocking of spaces between wood joists, beams or headers shall be to a depth of 1 inch (25 mm) and shall only be placed on strips of metal or metal lath laid across the spaces between combustible material and the chimney.

2111.14 Exterior air. Factory-built or masonry fireplaces covered in this section shall be equipped with an exterior air supply to ensure proper fuel combustion unless the room is mechanically ventilated and controlled so that the indoor pressure is neutral or positive.

2111.14.1 Factory-built fireplaces. Exterior combustion air ducts for factory-built fireplaces shall be listed components of the fireplace, and installed according to the fireplace manufacturer's instructions.

2111.14.2 Masonry fireplaces. Listed combustion air ducts for masonry fireplaces shall be installed according to the terms of their listing and manufacturer's instructions.

2111.14.3 Exterior air intake. The exterior air intake shall be capable of providing all combustion air from the exterior of the dwelling. The exterior air intake shall not be located within a garage, attic, basement or crawl space of the dwelling nor shall the air intake be located at an elevation higher than the firebox. The exterior air intake shall be covered with a corrosion-resistant screen of $^1/_4$-inch (6.4 mm) mesh.

2111.14.4 Clearance. Unlisted combustion air ducts shall be installed with a minimum 1-inch (25 mm) clearance to combustibles for all parts of the duct within 5 feet (1524 mm) of the duct outlet.

2111.14.5 Passageway. The combustion air passageway shall be not less than 6 square inches (3870 mm^2) and not more than 55 square inches (0.035 m^2), except that combustion air systems for listed fireplaces or for fireplaces tested for emissions shall be constructed according to the fireplace manufacturer's instructions.

2111.14.6 Outlet. The exterior air outlet is permitted to be located in the back or sides of the firebox chamber or within 24 inches (610 mm) of the firebox opening on or near the floor. The outlet shall be closable and designed to prevent burning material from dropping into concealed combustible spaces.

SECTION 2112
MASONRY HEATERS

2112.1 Definition. A masonry heater is a heating appliance constructed of concrete or solid masonry, hereinafter referred to as "masonry," which is designed to absorb and store heat from a solid fuel fire built in the firebox by routing the exhaust gases through internal heat exchange channels in which the flow path downstream of the firebox may include flow in a horizontal or downward direction before entering the chimney and which delivers heat by radiation from the masonry surface of the heater.

2112.2 Installation. Masonry heaters shall be installed in accordance with this section and comply with one of the following:

1. Masonry heaters shall comply with the requirements of ASTM E1602.
2. Masonry heaters shall be listed and labeled in accordance with UL 1482 or EN 15250 and installed in accordance with the manufacturer's instructions.

2112.3 Footings and foundation. The firebox floor of a masonry heater shall be a minimum thickness of 4 inches (102 mm) of noncombustible material and be supported on a noncombustible footing and foundation in accordance with Section 2113.2.

2112.4 Seismic reinforcing. In structures assigned to Seismic Design Category D, E or F, masonry heaters shall be anchored to the masonry foundation in accordance with Section 2113.3. Seismic reinforcing shall not be required within the body of a masonry heater with a height that is equal to or less than 3.5 times its body width and where the masonry chimney serving the heater is not supported by the body of the heater. Where the masonry chimney shares a common wall with the facing of the masonry heater, the chimney portion of the structure shall be reinforced in accordance with Section 2113.

2112.5 Masonry heater clearance. Combustible materials shall not be placed within 36 inches (914 mm) or the distance of the allowed reduction method from the outside surface of a masonry heater in accordance with NFPA 211, Section 12.6, and the required space between the heater and combustible material shall be fully vented to permit the free flow of air around all heater surfaces.

Exceptions:

1. Where the masonry heater wall thickness is at least 8 inches (203 mm) of solid masonry and the wall thickness of the heat exchange channels is not less than 5 inches (127 mm) of solid masonry, combustible materials shall not be placed within 4 inches (102 mm) of the outside surface of a masonry heater. A clearance of not less than 8 inches (203 mm) shall be provided between the gas-tight capping slab of the heater and a combustible ceiling.
2. Masonry heaters listed and labeled in accordance with UL 1482 or EN 15250 and installed in accordance with the manufacturer's instructions.

SECTION 2113
MASONRY CHIMNEYS

2113.1 General. The construction of masonry chimneys consisting of solid masonry units, hollow masonry units grouted solid, stone or concrete shall be in accordance with this section.

2113.2 Footings and foundations. Footings for masonry chimneys shall be constructed of concrete or solid masonry not less than 12 inches (305 mm) thick and shall extend at least 6 inches (152 mm) beyond the face of the foundation or support wall on all sides. Footings shall be founded on natural undisturbed earth or engineered fill below frost depth. In areas not subjected to freezing, footings shall be not less than 12 inches (305 mm) below finished grade.

2113.3 Seismic reinforcement. In structures assigned to Seismic Design Category A or B, seismic reinforcement is not required. In structures assigned to Seismic Design Category C or D, masonry chimneys shall be reinforced and anchored in accordance with Sections 2113.3.1, 2113.3.2 and 2113.4. In structures assigned to Seismic Design Category E or F, masonry chimneys shall be reinforced in accordance with the requirements of Sections 2101 through 2108 and anchored in accordance with Section 2113.4.

2113.3.1 Vertical reinforcement. For chimneys up to 40 inches (1016 mm) wide, four No. 4 continuous vertical bars anchored in the foundation shall be placed in the concrete between wythes of solid masonry or within the cells of hollow unit masonry and grouted in accordance with Section 2103.3. Grout shall be prevented from bonding with the flue liner so that the flue liner is free to move with thermal expansion. For chimneys greater than 40 inches (1016 mm) wide, two additional No. 4 vertical bars shall be provided for each additional 40 inches (1016 mm) in width or fraction thereof.

2113.3.2 Horizontal reinforcement. Vertical reinforcement shall be placed enclosed within $^1/_4$-inch (6.4 mm) ties, or other reinforcing of equivalent net cross-sectional area, spaced not to exceed 18 inches (457 mm) on center in concrete, or placed in the bed joints of unit masonry, at not less than every 18 inches (457 mm) of vertical height. Two such ties shall be provided at each bend in the vertical bars.

2113.4 Seismic anchorage. Masonry chimneys and foundations shall be anchored at each floor, ceiling or roof line more than 6 feet (1829 mm) above grade With two $^3/_{16}$-inch by 1 inch (4.8 mm by 25 mm) straps embedded not less than 12 inches (305 mm) into the chimney. Straps shall be hooked around the outer bars and extend 6 inches (152 mm) beyond the bend. Each strap shall be fastened to not less than four floor joists with two $^1/_2$-inch (12.7 mm) bolts.

Exception: Seismic anchorage is not required for the following:

1. In structures assigned to Seismic Design Category A or B.
2. Where the masonry fireplace is constructed completely within the exterior walls.

Deletion Arrow and Revision Bar

Large arrow refers to text deleted from 2012 IBC.

Thick bar refers to modified text from 2012 IBC.

2113.5 Corbeling. Masonry chimneys shall not be corbeled more than half of the chimney's wall thickness from a wall or foundation, nor shall a chimney be corbeled from a wall or foundation that is less than 12 inches (305 mm) in thickness unless it projects equally on each side of the wall, except that on the second story of a two-story dwelling, corbeling of chimneys on the exterior of the enclosing walls is permitted to equal the wall thickness. The projection of a single course shall not exceed one-half the unit height or one-third of the unit bed depth, whichever is less.

2113.6 Changes in dimension. The chimney wall or chimney flue lining shall not change in size or shape within 6 inches (152 mm) above or below where the chimney passes through floor components, ceiling components or roof components.

2113.7 Offsets. Where a masonry chimney is constructed with a fireclay flue liner surrounded by one wythe of masonry, the maximum offset shall be such that the centerline of the flue above the offset does not extend beyond the center of the chimney wall below the offset. Where the chimney offset is supported by masonry below the offset in an approved manner, the maximum offset limitations shall not apply. Each individual corbeled masonry course of the offset shall not exceed the projection limitations specified in Section 2113.5.

2113.8 Additional load. Chimneys shall not support loads other than their own weight unless they are designed and constructed to support the additional load. Masonry chimneys are permitted to be constructed as part of the masonry walls or concrete walls of the building.

2113.9 Termination. Chimneys shall extend not less than 2 feet (610 mm) higher than any portion of the building within 10 feet (3048 mm), but shall not be less than 3 feet (914 mm) above the highest point where the chimney passes through the roof.

2113.9.1 Chimney caps. Masonry chimneys shall have a concrete, metal or stone cap, sloped to shed water, a drip edge and a caulked bond break around any flue liners in accordance with ASTM C 1283.

2113.9.2 Spark arrestors. Where a spark arrestor is installed on a masonry chimney, the spark arrestor shall meet all of the following requirements:

1. The net free area of the arrestor shall be not less than four times the net free area of the outlet of the chimney flue it serves.
2. The arrestor screen shall have heat and corrosion resistance equivalent to 19-gage galvanized steel or 24-gage stainless steel.
3. Openings shall not permit the passage of spheres having a diameter greater than $^1/_2$ inch (12.7 mm) nor block the passage of spheres having a diameter less than $^3/_8$ inch (9.5 mm).
4. The spark arrestor shall be accessible for cleaning and the screen or chimney cap shall be removable to allow for cleaning of the chimney flue.

2113.9.3 Rain caps. Where a masonry or metal rain cap is installed on a masonry chimney, the net free area under the cap shall be not less than four times the net free area of the outlet of the chimney flue it serves.

2113.10 Wall thickness. Masonry chimney walls shall be constructed of concrete, solid masonry units or hollow masonry units grouted solid with not less than 4 inches (102 mm) nominal thickness.

2113.10.1 Masonry veneer chimneys. Where masonry is used as veneer for a framed chimney, through flashing and weep holes shall be provided as required by Chapter 14.

2113.11 Flue lining (material). Masonry chimneys shall be lined. The lining material shall be appropriate for the type of appliance connected, according to the terms of the appliance listing and the manufacturer's instructions.

2113.11.1 Residential-type appliances (general). Flue lining systems shall comply with one of the following:

1. Clay flue lining complying with the requirements of ASTM C 315.
2. Listed chimney lining systems complying with UL 1777.
3. Factory-built chimneys or chimney units listed for installation within masonry chimneys.
4. Other approved materials that will resist corrosion, erosion, softening or cracking from flue gases and condensate at temperatures up to 1,800°F (982°C).

2113.11.1.1 Flue linings for specific appliances. Flue linings other than those covered in Section 2113.11.1 intended for use with specific appliances shall comply with Sections 2113.11.1.2 through 2113.11.1.4 and Sections 2113.11.2 and 2113.11.3.

2113.11.1.2 Gas appliances. Flue lining systems for gas appliances shall be in accordance with the *International Fuel Gas Code*.

2113.11.1.3 Pellet fuel-burning appliances. Flue lining and vent systems for use in masonry chimneys with pellet fuel-burning appliances shall be limited to flue lining systems complying with Section 2113.11.1 and pellet vents listed for installation within masonry chimneys (see Section 2113.11.1.5 for marking).

2113.11.1.4 Oil-fired appliances approved for use with L-vent. Flue lining and vent systems for use in masonry chimneys with oil-fired appliances approved for use with Type L vent shall be limited to flue lining systems complying with Section 2113.11.1 and listed chimney liners complying with UL 641 (see Section 2113.11.1.5 for marking).

2113.11.1.5 Notice of usage. When a flue is relined with a material not complying with Section 2113.11.1, the chimney shall be plainly and permanently identified by a label attached to a wall, ceiling or other conspicuous location adjacent to where the connector enters the chimney. The label shall include the following message or equivalent language: "This chimney is for use only with (type or category of appliance) that burns (type of fuel). Do not connect other types of appliances."

2113.11.2 Concrete and masonry chimneys for medium-heat appliances.

2113.11.2.1 General. Concrete and masonry chimneys for medium-heat appliances shall comply with Sections 2113.1 through 2113.5.

2113.11.2.2 Construction. Chimneys for medium-heat appliances shall be constructed of solid masonry units or of concrete with walls not less than 8 inches (203 mm) thick, or with stone masonry not less than 12 inches (305 mm) thick.

2113.11.2.3 Lining. Concrete and masonry chimneys shall be lined with an approved medium-duty refractory brick not less than $4^1/_2$ inches (114 mm) thick laid on the $4^1/_2$-inch bed (114 mm) in an approved medium-duty refractory mortar. The lining shall start 2 feet (610 mm) or more below the lowest chimney connector entrance. Chimneys terminating 25 feet (7620 mm) or less above a chimney connector entrance shall be lined to the top.

2113.11.2.4 Multiple passageway. Concrete and masonry chimneys containing more than one passageway shall have the liners separated by a minimum 4-inch-thick (102 mm) concrete or solid masonry wall.

2113.11.2.5 Termination height. Concrete and masonry chimneys for medium-heat appliances shall extend not less than 10 feet (3048 mm) higher than any portion of any building within 25 feet (7620 mm).

2113.11.2.6 Clearance. A minimum clearance of 4 inches (102 mm) shall be provided between the exterior surfaces of a concrete or masonry chimney for medium-heat appliances and combustible material.

2113.11.3 Concrete and masonry chimneys for high-heat appliances.

2113.11.3.1 General. Concrete and masonry chimneys for high-heat appliances shall comply with Sections 2113.1 through 2113.5.

2113.11.3.2 Construction. Chimneys for high-heat appliances shall be constructed with double walls of solid masonry units or of concrete, each wall to be not less than 8 inches (203 mm) thick with a minimum airspace of 2 inches (51 mm) between the walls.

2113.11.3.3 Lining. The inside of the interior wall shall be lined with an approved high-duty refractory brick, not less than $4^1/_2$ inches (114 mm) thick laid on the $4^1/_2$-inch bed (114 mm) in an approved high-duty refractory mortar. The lining shall start at the base of the chimney and extend continuously to the top.

2113.11.3.4 Termination height. Concrete and masonry chimneys for high-heat appliances shall extend not less than 20 feet (6096 mm) higher than any portion of any building within 50 feet (15 240 mm).

2113.11.3.5 Clearance. Concrete and masonry chimneys for high-heat appliances shall have approved clearance from buildings and structures to prevent overheating combustible materials, permit inspection and maintenance operations on the chimney and prevent danger of burns to persons.

2113.12 Clay flue lining (installation). Clay flue liners shall be installed in accordance with ASTM C 1283 and extend from a point not less than 8 inches (203 mm) below the lowest inlet or, in the case of fireplaces, from the top of the smoke chamber to a point above the enclosing walls. The lining shall be carried up vertically, with a maximum slope no greater than 30 degrees (0.52 rad) from the vertical.

Clay flue liners shall be laid in medium-duty nonwatersoluble refractory mortar conforming to ASTM C 199 with tight mortar joints left smooth on the inside and installed to maintain an airspace or insulation not to exceed the thickness of the flue liner separating the flue liners from the interior face of the chimney masonry walls. Flue lining shall be supported on all sides. Only enough mortar shall be placed to make the joint and hold the liners in position.

2113.13 Additional requirements.

2113.13.1 Listed materials. Listed materials used as flue linings shall be installed in accordance with the terms of their listings and the manufacturer's instructions.

2113.13.2 Space around lining. The space surrounding a chimney lining system or vent installed within a masonry chimney shall not be used to vent any other appliance.

Exception: This shall not prevent the installation of a separate flue lining in accordance with the manufacturer's instructions.

2113.14 Multiple flues. When two or more flues are located in the same chimney, masonry wythes shall be built between adjacent flue linings. The masonry wythes shall be at least 4 inches (102 mm) thick and bonded into the walls of the chimney.

Exception: When venting only one appliance, two flues are permitted to adjoin each other in the same chimney with only the flue lining separation between them. The joints of the adjacent flue linings shall be staggered not less than 4 inches (102 mm).

2113.15 Flue area (appliance). Chimney flues shall not be smaller in area than the area of the connector from the appliance. Chimney flues connected to more than one appliance shall be not less than the area of the largest connector plus 50 percent of the areas of additional chimney connectors.

Exceptions:

1. Chimney flues serving oil-fired appliances sized in accordance with NFPA 31.
2. Chimney flues serving gas-fired appliances sized in accordance with the *International Fuel Gas Code*.

2113.16 Flue area (masonry fireplace). Flue sizing for chimneys serving fireplaces shall be in accordance with Section 2113.16.1 or 2113.16.2.

2113.16.1 Minimum area. Round chimney flues shall have a minimum net cross-sectional area not less than $1/12$ of the fireplace opening. Square chimney flues shall have a minimum net cross-sectional area of not less than $1/10$ of the fireplace opening. Rectangular chimney flues with an aspect ratio less than 2 to 1 shall have a minimum net cross-sectional area of not less than $1/10$ of the fireplace opening. Rectangular chimney flues with an aspect ratio of 2 to 1 or more shall have a minimum net cross-sectional area of not less than $1/8$ of the fireplace opening.

2113.16.2 Determination of minimum area. The minimum net cross-sectional area of the flue shall be determined in accordance with Figure 2113.16. A flue size providing not less than the equivalent net cross-sectional area shall be used. Cross-sectional areas of clay flue linings are as provided in Tables 2113.16(1) and 2113.16(2) or as provided by the manufacturer or as measured in the field. The height of the chimney shall be measured from the firebox floor to the top of the chimney flue.

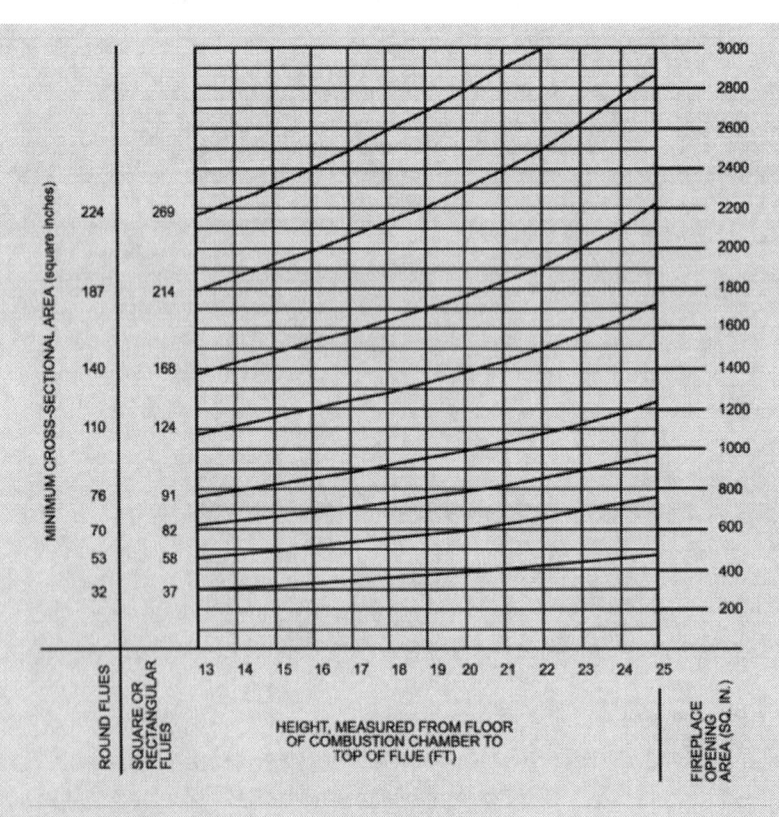

For SI: 1 inch = 25.4 mm, 1 square inch = 645 mm².

FIGURE 2113.16
FLUE SIZES FOR MASONRY CHIMNEYS

Deletion Arrow and Revision Bar

→ Large arrow refers to text deleted from 2012 IBC.

| Thick bar refers to modified text from 2012 IBC.

TABLE 2113.16(1)
NET CROSS-SECTIONAL AREA OF ROUND FLUE SIZES[a]

FLUE SIZE, INSIDE DIAMETER (inches)	CROSS-SECTIONAL AREA (square inches)
6	28
7	38
8	50
10	78
10³/₄	90
12	113
15	176
18	254

For SI: 1 inch = 25.4 mm, 1 square inch = 645.16 mm².
a. Flue sizes are based on ASTM C 315.

TABLE 2113.16(2)
NET CROSS-SECTIONAL AREA OF SQUARE AND RECTANGULAR FLUE SIZES

FLUE SIZE, OUTSIDE NOMINAL DIMENSIONS (inches)	CROSS-SECTIONAL AREA (square inches)
4.5 x 8.5	23
4.5 x 13	34
8 x 8	42
8.5 x 8.5	49
8 x 12	67
8.5 x 13	76
12 x 12	102
8.5 x 18	101
13 x 13	127
12 x 16	131
13 x 18	173
16 x 16	181
16 x 20	222
18 x 18	233
20 x 20	298
20 x 24	335
24 x 24	431

For SI: 1 inch = 25.4 mm, 1 square inch = 645.16 mm².

2113.17 Inlet. Inlets to masonry chimneys shall enter from the side. Inlets shall have a thimble of fireclay, rigid refractory material or metal that will prevent the connector from pulling out of the inlet or from extending beyond the wall of the liner.

2113.18 Masonry chimney cleanout openings. Cleanout openings shall be provided within 6 inches (152 mm) of the base of each flue within every masonry chimney. The upper edge of the cleanout shall be located not less than 6 inches (152 mm) below the lowest chimney inlet opening. The height of the opening shall be not less than 6 inches (152 mm). The cleanout shall be provided with a noncombustible cover.

Exception: Chimney flues serving masonry fireplaces, where cleaning is possible through the fireplace opening.

2113.19 Chimney clearances. Any portion of a masonry chimney located in the interior of the building or within the exterior wall of the building shall have a minimum airspace clearance to combustibles of 2 inches (51 mm). Chimneys located entirely outside the exterior walls of the building, including chimneys that pass through the soffit or cornice, shall have a minimum airspace clearance of 1 inch (25 mm). The airspace shall not be filled, except to provide fireblocking in accordance with Section 2113.20.

Exceptions:

1. Masonry chimneys equipped with a chimney lining system listed and labeled for use in chimneys in contact with combustibles in accordance with UL 1777, and installed in accordance with the manufacturer's instructions, are permitted to have combustible material in contact with their exterior surfaces.

2. Where masonry chimneys are constructed as part of masonry or concrete walls, combustible materials shall not be in contact with the masonry or concrete wall less than 12 inches (305 mm) from the inside surface of the nearest flue lining.

3. Exposed combustible trim and the edges of sheathing materials, such as wood siding, are permitted to abut the masonry chimney sidewalls, in accordance with Figure 2113.19, provided such combustible trim or sheathing is not less than 12 inches (305 mm) from the inside surface of the nearest flue lining. Combustible material and trim shall not overlap the corners of the chimney by more than 1 inch (25 mm).

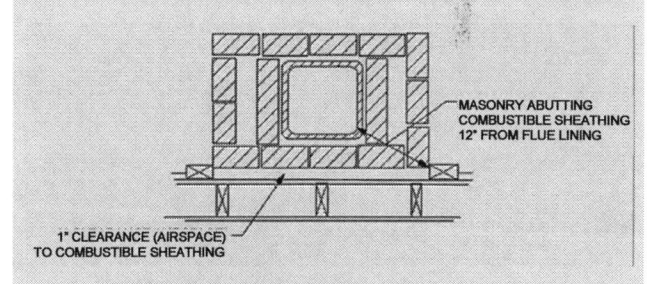

FIGURE 2113.19
ILLUSTRATION OF EXCEPTION THREE CHIMNEY CLEARANCE PROVISION

2113.20 Chimney fireblocking. All spaces between chimneys and floors and ceilings through which chimneys pass shall be fireblocked with noncombustible material securely fastened in place. The fireblocking of spaces between wood joists, beams or headers shall be self-supporting or be placed on strips of metal or metal lath laid across the spaces between combustible material and the chimney.

Deletion Arrow and Revision Bar

➡ Large arrow refers to text deleted from 2012 IBC.

▌ Thick bar refers to modified text from 2012 IBC.

Deletion Arrow and Revision Bar

➡ Large arrow refers to text deleted from 2012 IBC.

| Thick bar refers to modified text from 2012 IBC.

2015 IBC
CHAPTER 23*
WOOD

SECTION 2303
MINIMUM STANDARDS AND QUALITY

2303.1.6 Fiberboard. Fiberboard for its various uses shall conform to ASTM C 208. Fiberboard sheathing, when used structurally, shall be identified by an approved agency as conforming to ASTM C 208.

2303.1.6.3 Wall insulation. Where installed and fireblocked to comply with Chapter 7, fiberboards are permitted as wall insulation in all types of construction. In fire walls and fire barriers, unless treated to comply with Section 803.1 for Class A materials, the boards shall be cemented directly to the concrete, masonry or other noncombustible base and shall be protected with an approved noncombustible veneer anchored to the base without intervening airspaces.

SECTION 2304
GENERAL CONSTRUCTION REQUIREMENTS

2304.11.2 Floor framing. Approved wall plate boxes or hangers shall be provided where wood beams, girders or trusses rest on masonry or concrete walls. Where intermediate beams are used to support a floor, they shall rest on top of girders, or shall be supported by ledgers or blocks securely fastened to the sides of the girders, or they shall be supported by an approved metal hanger into which the ends of the beams shall be closely fitted.

2304.11.4 Floor decks. Floor decks and covering shall not extend closer than $1/2$ inch (12.7 mm) to walls. Such $1/2$-inch (12.7 mm) spaces shall be covered by a molding fastened to the wall either above or below the floor and arranged such that the molding will not obstruct the expansion or contraction movements of the floor. Corbeling of masonry walls under floors is permitted in place of such molding.

2304.12.1.3 Exterior walls below grade. Wood framing members and furring strips in direct contact with the interior of exterior masonry or concrete walls below grade shall be of naturally durable or preservative-treated wood.

2304.12.1.4 Sleepers and sills. Sleepers and sills on a concrete or masonry slab that is in direct contact with earth shall be of naturally durable or preservative-treated wood.

2304.12.2.1 Girder ends. The ends of wood girders entering exterior masonry or concrete walls shall be provided with a $1/2$-inch (12.7 mm) airspace on top, sides and end, unless naturally durable or preservative-treated wood is used.

2304.12.2.2 Posts or columns. Posts or columns supporting permanent structures and supported by a concrete or masonry slab or footing that is in direct contact with the earth shall be of naturally durable or preservative-treated wood.

> **Exceptions:** Posts or columns that are not exposed to the weather, are supported by concrete piers or metal pedestals projected at least 1 inch (25 mm) above the slab or deck and 8 inches (152 mm) above exposed earth and are separated by an impervious moisture barrier.

2304.12.2.5 Supporting members for permeable floors and roofs. Wood structural members that support moisture-permeable floors or roofs that are exposed to the weather, such as concrete or masonry slabs, shall be of naturally durable or preservative-treated wood unless separated from such floors or roofs by an impervious moisture barrier.

2304.13 Long-term loading. Wood members supporting concrete, masonry or similar materials shall be checked for the effects of long-term loading using the provisions of the AWC NDS. The total deflection, including the effects of long-term loading, shall be limited in accordance with Section 1604.3.1 for these supported materials.

> **Exception:** Horizontal wood members supporting masonry or concrete nonstructural floor or roof surfacing not more than 4 inches (102 mm) thick need not be checked for long-term loading.

SECTION 2308
CONVENTIONAL LIGHT-FRAME CONSTRUCTION

2308.2 Limitations. Buildings are permitted to be constructed in accordance with the provisions of conventional light-frame construction, subject to the limitations in Sections 2308.2.1 through 2308.2.6.

---

* Only portions of this section are shown which are particularly applicable to masonry construction. For additional information see the IBC.

Deletion Arrow and Revision Bar

➡ Large arrows refer to text deleted from 2012 IBC.

❙ Thick Bars refer to modified text from 2012 IBC.

2308.2.1 Stories. Structures of conventional light-frame construction shall be limited in story height in accordance with Table 2308.2.1

**TABLE 2308.2.1
ALLOWABLE STORY HEIGHT**

SEISMIC DESIGN CATEGORY	ALLOWABLE STORY ABOVE GRADE PLANE
A and B	Three stories
C	Two stories
D and E[a]	One story

For SI: 1 inch = 25.4 mm.

a. For the purposes of this section, for buildings assigned to Seismic Design Category D or E, cripple walls shall be considered to be a story unless cripple walls are solid blocked and do not exceed 14 inches in height.

2308.2.2 Allowable floor-to-floor height. Maximum floor-to-floor height shall not exceed 11 feet, 7 inches (3531 mm). Exterior bearing wall height and interior braced wall heights shall not exceed a stud height of 10 feet (3048 mm).

2308.2.3 Allowable loads. Loads shall be in accordance with Chapter 16 and shall not exceed the following:

1. Average dead loads shall not exceed 15 psf (718 N/m^2) for combined roof and ceiling, exterior walls, floors and partitions.

   Exceptions:

   1. Subject to the limitations of Section 2308.6.10, stone or masonry veneer up to the lesser of 5 inches (127 mm) thick or 50 psf (2395 N/m^2) and installed in accordance with Chapter 14 is permitted to a height of 30 feet (9144 mm) above a noncombustible foundation, with an additional 8 feet (2438 mm) permitted for gable ends.

   2. Concrete or masonry fireplaces, heaters and chimneys shall be permitted in accordance with the provisions of this code.

2. Live loads shall not exceed 40 psf (1916 N/m^2) for floors.

3. Ground snow loads shall not exceed 50 psf (2395 N/m^2).

2308.2.6 Risk category limitations. The use of the provisions for conventional light-frame construction in this section shall not be permitted for Risk Category IV buildings assigned to Seismic Design Category B, C, D or E.

2308.3 Foundations and footings. Foundations and footings shall be designed and constructed in accordance with Chapter 18. Connections to foundations and footings shall comply with this section.

2308.3.1 Foundation plates or sills. Foundation plates or sills resting on concrete or masonry foundations shall comply with Section 2304.3.1. Foundation plates or sills shall be bolted or anchored to the foundation with not less than $^1/_2$-inch-diameter (12.7 mm) steel bolts or approved anchors spaced to provide equivalent anchorage as the steel bolts. Bolts shall be embedded at least 7 inches (178 mm) into concrete or masonry. Bolts shall be spaced not more than 6 feet (1829 mm) on center and there shall be not less than two bolts or anchor straps per piece with one bolt or anchor strap located not more than 12 inches (305 mm) or less than 4 inches (102 mm) from each end of each piece. A properly sized nut and washer shall be tightened on each bolt to the plate.

Exceptions:

1. Along braced wall lines in structures assigned to Seismic Design Category E, steel bolts with a minimum nominal diameter of $^5/_8$ inch (15.9 mm) or approved anchor straps load-rated in accordance with Section 2304.10.3 and spaced to provide equivalent anchorage shall be used.

2. Bolts in braced wall lines in structures over two stories above grade shall be spaced not more than 4 feet (1219 mm) on center.

2308.4 Floor framing. Floor framing shall comply with this section.

2308.4.1 Girders. Girders for single-story construction or girders supporting loads from a single floor shall be not less than 4 inches by 6 inches (102 mm by 152 mm) for spans 6 feet (1829 mm) or less, provided that girders are spaced not more than 8 feet (2438 mm) on center. Other girders shall be designed to support the loads specified in this code. Girder end joints shall occur over supports.

Where a girder is spliced over a support, an adequate tie shall be provided. The ends of beams or girders supported on masonry or concrete shall not have less than 3 inches (76 mm) of bearing.

2308.4.2 Floor joists. Floor joists shall comply with this section.

2308.4.2.2 Bearing. The ends of each joist shall have not less than $1^1/_2$ inches (38 mm) of bearing on wood or metal, or not less than 3 inches (76 mm) on masonry, except where supported on a 1-inch by 4-inch (25 mm by 102 mm) ribbon strip and nailed to the adjoining stud.

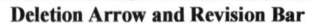

2308.6.7 Connections of braced wall panels. Braced wall panel joints shall occur over studs or blocking Braced wall panels shall be fastened to studs, top and bottom plate and at panel edges. Braced wall panels shall be applied to nominal 2-inch-wide [actual 1 1/2-inch (38 mm)] or larger stud framing.

2308.6.7.3 Sill anchorage. Where foundations are required by Section 2308.6.8, braced wall line sills shall be anchored to concrete or masonry foundations. Such anchorage shall conform to the requirements of Section 2308.3. The anchors shall be distributed along the length of the braced wall line. Other anchorage devices having equivalent capacity are permitted.

2308.6.10 Limitations of concrete or masonry veneer. Concrete or masonry veneer shall comply with Chapter 14 and this section.

2308.6.10.1 Limitations of concrete or masonry in Seismic Design Category B or C. In Seismic Design Categories B and C, concrete or masonry walls and stone or masonry veneer shall not extend above a basement.

Exceptions:

1. In structures assigned to Seismic Design Category B, stone and masonry veneer is permitted to be used in the first two stories above grade plane or the first three stories above grade plane where the lowest story has concrete or masonry walls, provided that wood structural panel wall bracing is used and the length of bracing provided is one and one-half times the required length as specified in Table 2308.6.1.

2. Stone and masonry veneer is permitted to be used in the first story above grade plane or the first two stories above grade plane where the lowest story has concrete or masonry walls.

3. Stone and masonry veneer is permitted to be used in both stories of buildings with two stories above grade plane, provided the following criteria are met:

    3.1. Type of brace in accordance with Section 2308.6.1 shall be WSP and the allowable shear capacity in accordance with Section 2306.3 shall be not less than 350 plf (5108 N/m).

    3.2. Braced wall panels in the second story shall be located in accordance with Section 2308.6.1 and not more than 25 feet (7620 mm) on center, and the total length of braced wall panels shall be not less than 25 percent of the braced wall line length. Braced wall panels in the first story shall be located in accordance with Section 2308.6.1 and not more than 25 feet (7620 mm) on center, and the total length of braced wall panels shall be not less than 45 percent of the braced wall line length.

    3.3. Hold-down connectors with an allowable capacity of 2,000 pounds (8896 N) shall be provided at the ends of each braced wall panel for the second story to the first story connection. Hold-down connectors with an allowable capacity of 3,900 pounds (17 347 N) shall be provided at the ends of each braced wall panel for the first story to the foundation connection. In all cases, the hold-down connector force shall be transferred to the foundation.

    3.4. Cripple walls shall not be permitted.

2308.6.10.2 Limitations of concrete or masonry in Seismic Design Categories D and E. In Seismic Design Categories D and E, concrete or masonry walls and stone or masonry veneer shall not extend above a basement.

Exception: In structures assigned to Seismic Design Category D, stone and masonry veneer is permitted to be used in the first story above grade plane, provided the following criteria are met:

1. Type of brace in accordance with Section 2308.6.1 shall be WSP and the allowable shear capacity in accordance with Section 2306.3 shall be not less than 350 plf (5108 N/m).

2. The braced wall panels in the first story shall be located at each end of the braced wall line and not more than 25 feet (7620 mm) on center, and the total length of braced wall panels shall be not less than 45 percent of the braced wall line length.

3. Hold-down connectors shall be provided at the ends of braced walls for the first floor to foundation with an allowable capacity of 2,100 pounds (9341 N).

4. Cripple walls shall not be permitted.

Deletion Arrow and Revision Bar

➡ Large arrows refer to text deleted from 2012 IBC.

❙ Thick Bars refer to modified text from 2012 IBC.

Deletion Arrow and Revision Bar

➡ Large arrow refers to text deleted from 2012 IBC.

❙ Thick bar refers to modified text from 2012 IBC.

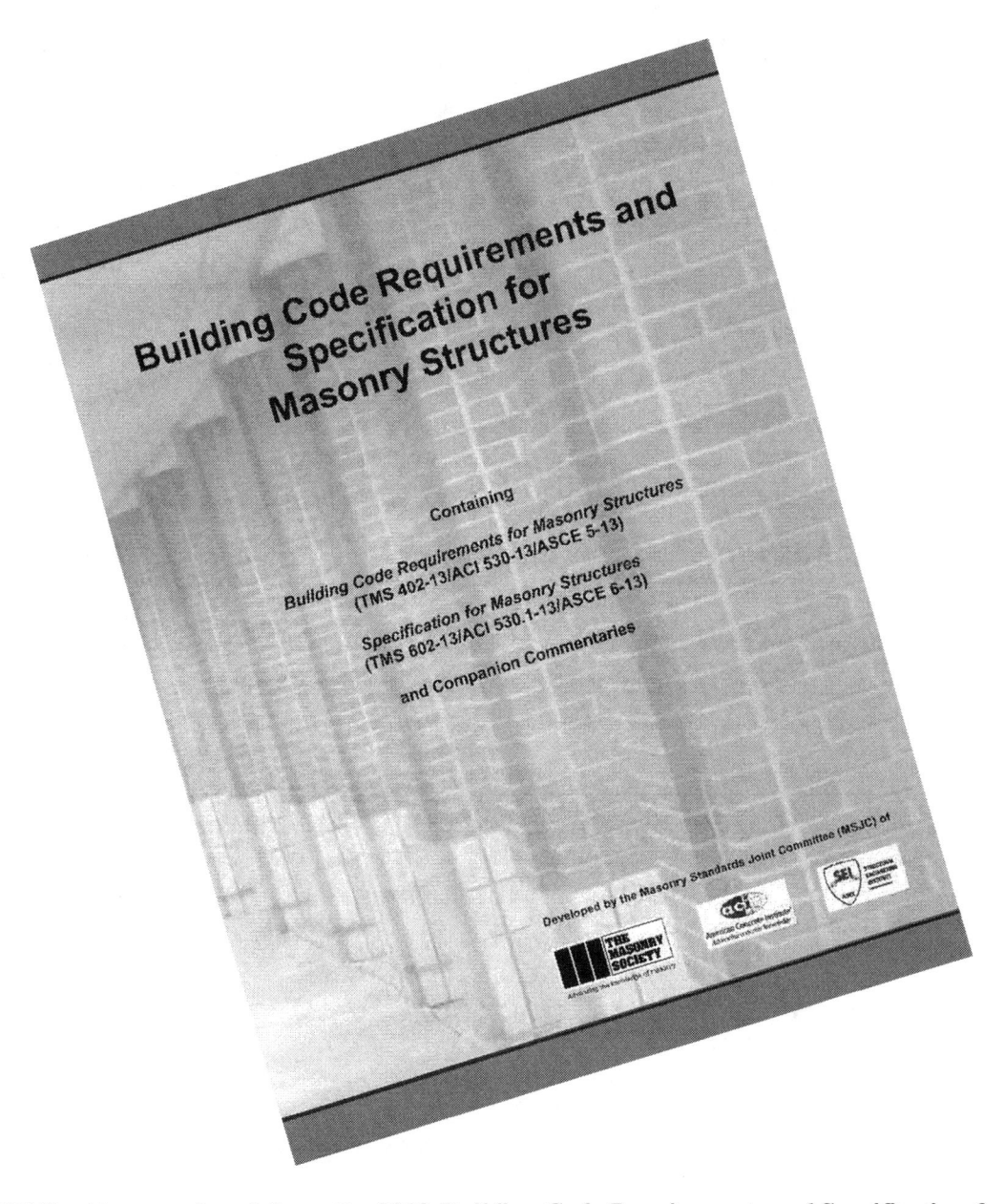

"The MSJC Text is reproduced from the 2013 *Building Code Requirements and Specification for Masonry Structures*, copyright © 2013, with the permission of the publishers, The Masonry Society, the American Concrete Institute, and the Structural Engineering Institute."

The Masonry Society
105 South Sunset Street, Suite Q
Longmont, CO 80501
www.masonrysociety.org

American Concrete Institute
P.O. Box 9094
Farmington Hills, MI 48333
www.concrete.org

Structural Engineering Institute of the American Society of Civil Engineers
1801 Alexander Bell Drive
Reston, VA 20191
www.seinstitute.org

Building Code Requirements and Specification for Masonry Structures

Containing

Building Code Requirements for Masonry Structures
(TMS 402-13/ACI 530-13/ASCE 5-13)

Specification for Masonry Structures
(TMS 602-13/ACI 530.1-13/ASCE 6-13)

and Companion Commentaries

Developed by the Masonry Standards Joint Committee (MSJC)

THE MASONRY SOCIETY
Advancing the knowledge of masonry

The Masonry Society
105 South Sunset Street, Suite Q
Longmont, CO 80501
www.masonrysociety.org

American Concrete Institute®
Advancing concrete knowledge

American Concrete Institute
P.O. Box 9094
Farmington Hills, MI 48333
www.concrete.org

SEI STRUCTURAL ENGINEERING INSTITUTE

Structural Engineering Institute
of the
American Society of Civil Engineers
1801 Alexander Bell Drive
Reston, VA 20191
www.seinstitute.org

ABSTRACT

Building Code Requirements and Specification for Masonry Structures contains two standards and their commentaries: Building Code Requirements for Masonry Structures (TMS 402-13/ACI 530-13/ASCE 5-13) and Specification for Masonry Structures (TMS 602-13/ACI 530.1-13/ASCE 6-13). These standards are produced through the joint efforts of The Masonry Society (TMS), the American Concrete Institute (ACI), and the Structural Engineering Institute of the American Society of Civil Engineers (SEI/ASCE) through the Masonry Standards Joint Committee (MSJC). The Code covers the design and construction of masonry structures while the Specification is concerned with minimum construction requirements for masonry in structures. Some of the topics covered in the Code are: definitions, contract documents; quality assurance; materials; placement of embedded items; analysis and design; strength and serviceability; flexural and axial loads; shear; details and development of reinforcement; walls; columns; pilasters; beams and lintels; seismic design requirements; glass unit masonry; veneers; and autoclaved aerated concrete masonry. An empirical design method and a prescriptive method applicable to buildings meeting specific location and construction criteria are also included. The Specification covers subjects such as quality assurance requirements for materials; the placing, bonding and anchoring of masonry; and the placement of grout and of reinforcement. This Specification is meant to be modified and referenced in the Project Manual. The Code is written as a legal document and the Specification as a master specification required by the Code. The commentaries present background details, committee considerations, and research data used to develop the Code and Specification. The Commentaries are not mandatory and are for information of the user only.

The Masonry Standards Joint Committee, which is sponsored by The Masonry Society, the American Concrete Institute, and the Structural Engineering Institute of the American Society of Civil Engineers, is responsible for these standards and strives to avoid ambiguities, omissions, and errors in these documents. In spite of these efforts, the users of these documents occasionally find information or requirements that may be subject to more than one interpretation or may be incomplete or incorrect. Users who have suggestions for the improvement of these documents are requested to contact TMS.

These documents are intended for the use of individuals who are competent to evaluate the significance and limitations of its content and recommendations and who will accept responsibility for the application of the material it contains. Individuals who use this publication in any way assume all risk and accept total responsibility for the application and use of this information.

All information in this publication is provided "as is" without warranty of any kind, either express or implied, including but not limited to, the implied warranties of merchantability, fitness for a particular purpose or non-infringement.

The sponsoring organizations, TMS, ACI, and SEI/ASCE, and their members disclaim liability for damages of any kind, including any special, indirect, incidental, or consequential damages, including without limitation, lost revenues or lost profits, which may result from the use of this publication.

It is the responsibility of the user of this document to establish health and safety practices appropriate to the specific circumstances involved with its use. The sponsoring organizations do not make any representations with regard to health and safety issues and the use of this document. The user must determine the applicability of all regulatory limitations before applying the document and must comply with all applicable laws and regulations, including but not limited to, United States Occupational Safety and Health Administration (OSHA) health and safety standards.

COPYRIGHT © 2013, The Masonry Society, Longmont, CO, American Concrete Institute, Farmington Hills, MI, Structural Engineering Institute of the American Society of Civil Engineers, Reston, VA. Watch http://www.masonrysociety.org/2013MSJC/Errata.htm for possible additional errata.

ALL RIGHTS RESERVED. This material may not be reproduced or copied, in whole or part, in any printed, mechanical, electronic, film, or other distribution and storage media, without the written consent of TMS.

Adopted as standards of the American Concrete Institute (September 13, 2013), the Structural Engineering Institute of the American Society of Civil Engineers (September 4, 2013), and The Masonry Society (August 27, 2013) to supersede the 2011 edition in accordance with each organization's standardization procedures. These standards were originally adopted by the American Concrete Institute in November, 1988, the American Society of Civil Engineers in August, 1989, and The Masonry Society in July, 1992.

ISBN 978-1-929081-43-1
ISBN 1-929081-43-X
Produced in the United States of America

Masonry Standards Joint Committee

Diane B. Throop - Chair
Richard M. Bennett - Vice Chair
Gerald A. Dalrymple - Secretary

Voting Members on Main Committee[1]

Daniel P. Abrams	Thomas A. Gangel	Richard E. Klingner*	Arturo Ernest Schultz
Jennifer R. Bean Popehn*	David C. Gastgeb	W. Mark McGinley*	Kurtis K. Siggard
Richard M. Bennett*	S. K. Ghosh	David I. McLean	Jennifer E. Tanner
David T. Biggs*	Benchmark H. Harris	Darrell W. McMillian	John G. Tawresey
Robert N. Chittenden	Ronald J. Hunsicker	John M. Melander	Jason J. Thompson
John Chrysler*	Edwin T. Huston	Raymond T. Miller	Margaret L. Thomson
Chukwuma G. Ekwueme	Keith Itzler*	Vilas Mujumdar	Diane B. Throop*
Fernando Fonseca	Rochelle C. Jaffe*	Jerry M. Painter	Charles J. Tucker*
Susan M. Frey+	Eric N. Johnson*	David L. Pierson	Scott W. Walkowicz*
Edward L. Freyermuth	Rashod R. Johnson	Max L. Porter	A. Rhett Whitlock

Voting Members of Subcommittees Only[2]

Bruce Barnes	Mohamed ElGawady	Matthew D. Jackson	Thomas M. Petreshock
Russell H. Brown	James A. Farny	John J. Jacob +	Alan Robinson
Charles B. Clark	James Feagin	Yasser Korany	Paul G. Scott
Thomas M. Corcoran	Sonny J. Fite	James M. LaFave	John J. Smith
George E. Crow III	David Gillick	Walter Laska	Bruce Weems
Terry M. Curtis	Edgar F. Glock Jr.	Nicholas T. Loomis	David B. Woodham
Mark A. Daigle	Dennis W. Graber	Peter J. Loughney	Rick Yelton
Gerald A, Dalrymple	Brian J. Grant	Sunup S. Mathew	Tianyi Yi
Manuel A. Diaz	Charles A. Haynes	James P. Mwangi	
Steve M. Dill	David Chris Hines	Khaled Nahlawi	

Subcommittee Corresponding (C) and Consulting (CN) Members[3]

Sergio M. Alcocer (C)	Richard Filloramo (C)	Shelley Lissel (C)	Donato Pompo (C)
James E. Amrhein (C)+	Hans Rudolf Ganz (C)	John Maloney (C)	Matthew Reiter (C)
Ronald E. Barnett (C)	Claret Heider (C)	John H. Matthys (C)	Drew Rouland (C)
J. Gregg Borchelt (C)	R. Craig Henderson (C)	Scott E. Maxwell (C)	Nigel G. Shrive (CN)
Jim Bryja (C)	Timothy S. Hess (C)	Donald G. McMican (C)	Christopher Sieto (C)
J. Leroy Caldwell (C)	Augusto F. Holmberg (C)	John R. Merk (C)	Dana Smith (C)
Mario J. Catani (CN)	Jason M. Ingham (C)	Ali M. Memari (C)	Gary R. Sturgeon (C)
Angelo Coduto (C)	Mervyn J. Kowalsky (C)	Ehsan Minaie (C)	Christine A. Subasic (C)
Paul Curtis (C)	David G. Kurtanich (C)+	David Mulick (C)	Brian Trimble (C)
Majed A. Dabdoub (C)	James Lai (C)	Mel Oller (C)	Miroslav Vejvoda (C)
Jamie L. Davis (C)	Mark Larsen (C)	Adrian W. Page (CN)	Tyler W. Witthuhn (C)
James Daniel Dolan (C)	Hojin Lee (C)	William D. Palmer Jr. (C)	Thomas C. Young (C)
Dan Eschenasy (C)	Andres Lepage (C)	Guilherme A. Parsekian (C)	Daniel Zechmeister (C)

1 Main Committee Members during the 2013 Revision Cycle. They participated in Committee activities, voted on Main Committee ballots and participated in Subcommittee activities including voting and correspondence.
2 Subcommittee Members during the 2013 Revision Cycle. They participated in Committee activities, voted on Subcommittee ballots and were able to comment on Main Committee ballots.
3 Corresponding and Consulting Members during the 2013 Revision Cycle. They could participate in Subcommittee activities but did not have voting privileges.
* Subcommittee Chair during the 2013 Revision Cycle
+ Deceased

Building Code Requirements for Masonry Structures (TMS 402-13/ACI 530-13/ASCE 5-13)

TABLE OF CONTENTS

SYNOPSIS AND KEYWORDS, pg. C-ix

PART 1 — GENERAL, pg. C-1

CHAPTER 1 — GENERAL REQUIREMENTS, pg. C-1
1.1 — Scope C-1
    1.1.1 Minimum requirements C-1
    1.1.2 Governing building code C-1
    1.1.3 SI information C-1

1.2 — Contract documents and calculations C-2

1.3 — Approval of special systems of design or construction C-4

1.4 — Standards cited in this Code C-4

CHAPTER 2 — NOTATION AND DEFINITIONS, pg. C-7
2.1 — Notation C-7

2.2 — Definitions C-14

CHAPTER 3 — QUALITY AND CONSTRUCTION, pg. C-25
3.1 — Quality Assurance program C-25
    3.1.1 Level A Quality Assurance C-25
    3.1.2 Level B Quality Assurance C-26
    3.1.3 Level C Quality Assurance C-26
    3.1.4 Procedures C-26
    3.1.5 Qualifications C-27
    3.1.6 Acceptance relative to strength requirements C-31

3.2 — Construction considerations C-31
    3.2.1 Grouting, minimum spaces C-31
    3.2.2 Embedded conduits, pipes, and sleeves C-32

PART 2 — DESIGN REQUIREMENTS, pg. C-35

CHAPTER 4 — GENERAL ANALYSIS AND DESIGN CONSIDERATIONS, pg. C-35
4.1 — Loading C-35
    4.1.1 General C-35
    4.1.2 Load provisions C-35
    4.1.3 Lateral load resistance C-35
    4.1.4 Load transfer at horizontal connections C-35
    4.1.5 Other effects C-36
    4.1.6 Lateral load distribution C-36

4.2 — Material properties C-37
    4.2.1 General C-37
    4.2.2 Elastic moduli C-38
    4.2.3 Coefficients of thermal expansion C-40
    4.2.4 Coefficients of moisture expansion for clay masonry C-40

4.2.5 Coefficients of shrinkage	C-40
4.2.6 Coefficients of creep	C-40
4.2.7 Prestressing steel	C-41
4.3 — Section properties	C-41
4.3.1 Stress calculations	C-41
4.3.2 Stiffness	C-42
4.3.3 Radius of gyration	C-43
4.3.4 Bearing area	C-43
4.4 — Connection to structural frames	C-44
4.5 — Masonry not laid in running bond	C-45

CHAPTER 5 — STRUCTURAL ELEMENTS, pg. C-47

5.1 — Masonry assemblies	C-47
5.1.1 Intersecting walls	C-47
5.1.2 Effective compressive width per bar	C-50
5.1.3 Concentrated loads	C-51
5.1.4 Multiwythe masonry elements	C-53
5.2 — Beams	C-57
5.2.1 General beam design	C-57
5.2.2 Deep beams	C-59
5.3 — Columns	C-61
5.3.1 General column design	C-61
5.3.2 Lightly loaded columns	C-62
5.4 — Pilasters	C-63
5.5 — Corbels	C-63
5.5.1 Loadbearing corbels	C-63
5.5.2 Non-loadbearing corbels	C-63

CHAPTER 6 — REINFORCEMENT, METAL ACCESSORIES, AND ANCHOR BOLTS, pg. C-67

6.1 — Details of reinforcement and metal accessories	C-67
6.1.1 Embedment	C-67
6.1.2 Size of reinforcement	C-67
6.1.3 Placement of reinforcement	C-67
6.1.4 Protection of reinforcement and metal accessories	C-68
6.1.5 Standard hooks	C-69
6.1.6 Minimum bend diameter for reinforcing bars	C-69
6.2 — Anchor Bolts	C-71
6.2.1 Placement	C-71
6.2.2 Projected area for axial tension	C-73
6.2.3 Projected area for shear	C-74
6.2.4 Effective embedment length for headed anchor bolts	C-76
6.2.5 Effective embedment length of bent-bar anchor bolts	C-76
6.2.6 Minimum permissible effective embedment length	C-77
6.2.7 Anchor bolt edge distance	C-77

CHAPTER 7 — SEISMIC DESIGN REQUIREMENTS, pg. C-79

7.1 Scope	C-79
7.2 General analysis	C-80
7.2.1 Element interaction	C-80
7.2.2 Load path	C-60

BUILDING CODE REQUIREMENTS FOR MASONRY STRUCTURES

7.2.3 Anchorage design	C-80
7.2.4 Drift limits	C-80
7.3 Element classification	C-82
7.3.1 Nonparticipating elements	C-82
7.3.2 Participating elements	C-82
7.4 Seismic Design Category requirements	C-90
7.4.1 Seismic Design Category A requirements	C-90
7.4.2 Seismic Design Category B requirements	C-90
7.4.3 Seismic Design Category C requirements	C-91
7.4.4 Seismic Design Category D requirements	C-93
7.4.5 Seismic Design Category E and F requirements	C-94

PART 3 — ENGINEERED DESIGN METHODS, pg. C-95

CHAPTER 8 — ALLOWABLE STRESS DESIGN OF MASONRY, pg. C-95

8.1 — General	C-95
8.1.1 Scope	C-95
8.1.2 Design strength	C-95
8.1.3 Anchor bolts embedded in grout	C-96
8.1.4 Shear stress in multiwythe masonry elements	C-98
8.1.5 Bearing stress	C-98
8.1.6 Development of reinforcement embedded in grout	C-99
8.2 — Unreinforced masonry	C-108
8.2.1 Scope	C-108
8.2.2 Design criteria	C-108
8.2.3 Design assumptions	C-108
8.2.4 Axial compression and flexure	C-108
8.2.5 Axial tension	C-114
8.2.6 Shear	C-114
8.3 — Reinforced masonry	C-116
8.3.1 Scope	C-116
8.3.2 Design assumptions	C-116
8.3.3 Steel reinforcement — Allowable stresses	C-116
8.3.4 Axial compression and flexure	C-116
8.3.5 Shear	C-119

CHAPTER 9 — STRENGTH DESIGN OF MASONRY, pg. C-123

9.1 — General	C-123
9.1.1 Scope	C-123
9.1.2 Required strength	C-123
9.1.3 Design strength	C-123
9.1.4 Strength-reduction factors	C-123
9.1.5 Deformation requirements	C-124
9.1.6 Anchor bolts embedded in grout	C-125
9.1.7 Shear strength in multiwythe masonry elements	C-127
9.1.8 Nominal bearing strength	C-127
9.1.9 Material properties	C-127
9.2 — Unreinforced (plain) masonry	C-130
9.2.1 Scope	C-130
9.2.2 Design criteria	C-130
9.2.3 Design assumptions	C-130
9.2.4 Nominal flexural and axial strength	C-130
9.2.5 Axial tension	C-132

9.2.6 Nominal shear strength	C-133

9.3 — Reinforced masonry C-134
    9.3.1 Scope C-134
    9.3.2 Design assumptions C-134
    9.3.3 Reinforcement requirements and details C-135
    9.3.4 Design of beams, piers, and columns C-140
    9.3.5 Wall design for out-of-plane loads C-144
    9.3.6 Wall design for in-plane loads C-147

CHAPTER 10 — PRESTRESSED MASONRY, pg. C-153

    10.1 — General C-153
        10.1.1 Scope C-153

    10.2 — Design methods C-154
        10.2.1 General C-154
        10.2.2 After transfer C-154

    10.3 — Permissible stresses in prestressing tendons C-154
        10.3.1 Jacking force C-154
        10.3.2 Immediately after transfer C-154
        10.3.3 Post-tensioned masonry members C-154
        10.3.4 Effective prestress C-155

    10.4 — Axial compression and flexure C-156
        10.4.1 General C-156
        10.4.2 Service load requirements C-157
        10.4.3 Strength requirements C-158

    10.5 — Axial tension C-159

    10.6 — Shear C-159

    10.7 — Deflection C-160

    10.8 — Prestressing tendon anchorages, couplers, and end blocks C-160
        10.8.1 .... C-160
        10.8.2 .... C-160
        10.8.3 .... C-160
        10.8.4 Bearing stresses C-160

    10.9 — Protection of prestressing tendons and accessories C-161

    10.10 — Development of bonded tendons C-161

CHAPTER 11 — STRENGTH DESIGN OF AUTOCLAVED AERATED CONCRETE (AAC) MASONRY, pg. C-163

    11.1 — General C-163
        11.1.1 Scope C-163
        11.1.2 Required strength C-163
        11.1.3 Design strength C-163
        11.1.4 Strength of joints C-163
        11.1.5 Strength-reduction factors C-164
        11.1.6 Deformation requirements C-164
        11.1.7 Anchor bolts C-165
        11.1.8 Material properties C-165
        11.1.9 Nominal bearing strength C-166

11.1.10 Corbels ... C-167

11.2 — Unreinforced (plain) AAC masonry ... C-168
    11.2.1 Scope ... C-168
    11.2.2 Flexural strength of unreinforced (plain) AAC masonry members ... C-168
    11.2.3 Nominal axial strength of unreinforced (plain) AAC masonry members ... C-169
    11.2.4 Axial tension ... C-169
    11.2.5 Nominal shear strength of unreinforced (plain) AAC masonry members ... C-169
    11.2.6 Flexural cracking ... C-169

11.3 — Reinforced AAC masonry ... C-170
    11.3.1 Scope ... C-170
    11.3.2 Design assumptions ... C-170
    11.3.3 Reinforcement requirements and details ... C-171
    11.3.4 Design of beams, piers, and columns ... C-173
    11.3.5 Wall design for out-of-plane loads ... C-177
    11.3.6 Wall design for in-plane loads ... C-180

PART 4 — PRESCRIPTIVE DESIGN METHODS, pg. C-183

CHAPTER 12 — VENEER, pg. C-183

12.1 — General ... C-183
    12.1.1 Scope ... C-183
    12.1.2 Design of anchored veneer ... C-185
    12.1.3 Design of adhered veneer ... C-187
    12.1.4 Dimension stone ... C-187
    12.1.5 Autoclaved aerated concrete masonry veneer ... C-187
    12.1.6 General design requirements ... C-187

12.2 — Anchored Veneer ... C-188
    12.2.1 Alternative design of anchored masonry veneer ... C-188
    12.2.2 Prescriptive requirements for anchored masonry veneer ... C-188

12.3 — Adhered Veneer ... C-194
    12.3.1 Alternative design of adhered masonry veneer ... C-194
    12.3.2 Prescriptive requirements for adhered masonry veneer ... C-194

CHAPTER 13 — GLASS UNIT MASONRY, pg. C-197

13.1 — General ... C-197
    13.1.1 Scope ... C-197
    13.1.2 General design requirements ... C-197
    13.1.3 Units ... C-197

13.2 — Panel Size ... C-197
    13.2.1 Exterior standard-unit panels ... C-198
    13.2.2 Exterior thin-unit panels ... C-199
    13.2.3 Interior panels ... C-199
    13.2.4 Curved panels ... C-200

13.3 — Support ... C-200
    13.3.1 General requirements ... C-200
    13.3.2 Vertical ... C-200
    13.3.3 Lateral ... C-200

13.4 — Expansion joints ... C-202

13.5 — Base surface treatment C-202

13.6 — Mortar C-202

13.7 — Reinforcement C-202

CHAPTER 14 — MASONRY PARTITION WALLS, pg. C-203

14.1 — General C-203
    14.1.1 Scope C-203
    14.1.2 Design of partition walls C-203

14.2 — Prescriptive design of partition walls C-203
    14.2.1 General C-203
    14.2.2 Thickness limitations C-203
    14.2.3 Limitations C-204

14.3 — Lateral support C-206
    14.3.1 Maximum l/t and h/t C-206
    14.3.2 Openings C-206
    14.3.3 Cantilever walls C-208
    14.3.4 Support elements C-208

14.4 — Anchorage C-208
    14.4.1 General C-208
    14.4.2 Intersecting walls C-208

14.5 — Miscellaneous requirements C-208
    14.5.1 Chases and recesses C-208
    14.5.2 Lintels C-208
    14.5.3 Lap splices C-208

PART 5 — APPENDICES, pg. C-209

APPENDIX A — EMPIRICAL DESIGN OF MASONRY, pg. C-209

A.1 — General C-209
    A.1.1 Scope C-209
    A.1.2 Limitations C-209

A.2 — Height C-213

A.3 — Lateral stability C-213
    A.3.1 Shear walls C-214
    A.3.2 Roofs C-213

A.4 — Compressive stress requirements C-215
    A.4.1 Calculations C-215
    A.4.2 Allowable compressive stresses C-215

A.5 — Lateral support C-218
    A.5.1 Maximum l/t and h/t C-218
    A.5.2 Cantilever walls C-219
    A.5.3 Support elements C-219

A.6 — Thickness of masonry C-219
    A.6.1 General C-219
    A.6.2 Minimum thickness C-219
    A.6.3 Foundation walls C-220
    A.6.4 Foundation piers C-220

A.7 — Bond C-221
    A.7.1 General C-221
    A.7.2 Bonding with masonry headers C-221
    A.7.3 Bonding with wall ties or joint reinforcement C-221
    A.7.4 Natural or cast stone C-222

A.8 — Anchorage C-222
    A.8.1 General C-222
    A.8.2 Intersecting walls C-222
    A.8.3 Floor and roof anchorage C-224
    A.8.4 Walls adjoining structural framing C-224

A.9 — Miscellaneous requirements C-225
    A.9.1 Chases and recesses C-225
    A.9.2 Lintels C-225

APPENDIX B — DESIGN OF MASONRY INFILL, pg. C-227

B.1 — General C-227
    B.1.1 Scope C-227
    B.1.2 Required strength C-228
    B.1.3 Design strength C-228
    B.1.4 Strength-reduction factors C-228
    B.1.5 Limitations C-228

B.2 — Non-Participating Infills C-229
    B.2.1 In-plane isolation joints for non-participating infills C-229
    B.2.2 Design of for non-participating infills for out-of-plane loads C-229

B.3 — Participating Infills C-230
    B.3.1 General C-230
    B.3.2 In-plane connection requirements for participating infills C-230
    B.3.3 Out-of-plane connection requirements for participating infills C-231
    B.3.4 Design of participating infills for in-plane loads C-231
    B.3.5 Design of frame elements with participating infills for in-plane loads C-232
    B.3.6 Design of participating infills for out-of-plane forces C-233

APPENDIX C — LIMIT DESIGN METHOD, pg. C-235
C — General C-235
    C.1 — Yield mechanism C-235

    C.2 — Mechanism strength C-236

    C.3 — Mechanism deformation C-236

EQUATION CONVERSIONS, pg. C-237

CONVERSION OF INCH-POUND UNITS TO SI UNITS, pg. C-249

PREFIXES, pg. C-249

REFERENCE FOR THE CODE COMMENTARY, pg. C-251

Building Code Requirements for Masonry Structures (TMS 402-13/ACI 530-13/ASCE 5-13)

SYNOPSIS

This Code covers the design and construction of masonry structures. It is written in such form that it may be adopted by reference in a legally adopted building code.

Among the subjects covered are: definitions; contract documents; quality assurance; materials; placement of embedded items; analysis and design; strength and serviceability; flexural and axial loads; shear; details and development of reinforcement; walls; columns; pilasters; beams and lintels; seismic design requirements; glass unit masonry; and veneers. An empirical design method applicable to buildings meeting specific location and construction criteria are also included.

The quality, inspection, testing, and placement of materials used in construction are covered by reference to TMS 602-13/ACI 530.1-13/ASCE 6-13 Specification for Masonry Structures and other standards.

Keywords: AAC, masonry, allowable stress design, anchors (fasteners); anchorage (structural); autoclaved aerated concrete masonry, beams; building codes; cements; clay brick; clay tile; columns; compressive strength; concrete block; concrete brick; construction; detailing; empirical design; flexural strength; glass units; grout; grouting; infills; joints; loads (forces); limit design; masonry; masonry cements; masonry load bearing walls; masonry mortars; masonry walls; modulus of elasticity; mortars; pilasters; prestressed masonry, quality assurance; reinforced masonry; reinforcing steel; seismic requirements; shear strength; specifications; splicing; stresses; strength design, structural analysis; structural design; ties; unreinforced masonry; veneers; walls.

This page is intentionally left blank.

PART 1: GENERAL

CHAPTER 1
GENERAL REQUIREMENTS

CODE

1.1 — Scope

1.1.1 *Minimum requirements*
This Code provides minimum requirements for the structural design and construction of masonry elements consisting of masonry units bedded in mortar.

1.1.2 *Governing building code*
This Code supplements the legally adopted building code and shall govern in matters pertaining to structural design and construction of masonry elements. In areas without a legally adopted building code, this Code defines the minimum acceptable standards of design and construction practice.

1.1.3 *SI information*
SI values shown in parentheses are not part of this Code. The equations in this document are for use with the specified inch-pound units only.

COMMENTARY

1.1 — Scope

1.1.1 *Minimum requirements*
This code governs structural design of both structural and non-structural masonry elements. Examples of non-structural elements are masonry veneer, glass unit masonry, and masonry partitions. Structural design aspects of non-structural masonry elements include, but are not limited to, gravity and lateral support, and load transfer to supporting elements.

Masonry structures may be required to have enhanced structural integrity as part of a comprehensive design against progressive collapse due to accident, misuse, sabotage or other causes. General design guidance addressing this issue is available in Commentary Section 1.4 of ASCE 7. Suggestions from that Commentary, of specific application to many masonry structures, include but are not limited to: consideration of plan layout to incorporate returns on walls, both interior and exterior; use of load-bearing interior walls; adequate continuity of walls, ties, and joint rigidity; providing walls capable of beam action; ductile detailing and the use of compartmentalized construction.

1.1.3 *SI information*
The equivalent equations for use with SI units are provided in the Equation Conversions table in Part 5.

CODE

1.2 — Contract documents and calculations

1.2.1 Show all Code-required drawing items on the project drawings, including:

(a) Name and date of issue of Code and supplement to which the design conforms.

(b) Loads used for the design of masonry structures.

(c) Specified compressive strength of masonry at stated ages or stages of construction for which masonry is designed, for each part of the structure, except for masonry designed in accordance with Part 4 or Appendix A.

(d) Size and location of structural elements.

(e) Details of anchorage of masonry to structural members, frames, and other construction, including the type, size, and location of connectors.

(f) Details of reinforcement, including the size, grade, type, lap splice length, and location of reinforcement.

(g) Reinforcing bars to be welded and welding requirements.

(h) Provision for dimensional changes resulting from elastic deformation, creep, shrinkage, temperature, and moisture.

(i) Size and permitted location of conduits, pipes, and sleeves.

1.2.2 Each portion of the structure shall be designed based on the specified compressive strength of masonry for that part of the structure, except for portions designed in accordance with Part 4 or Appendix A.

COMMENTARY

1.2 — Contract documents and calculations

The provisions for preparation of project drawings, project specifications, and issuance of permits are, in general, consistent with those of most legally adopted building codes and are intended as supplements to those codes.

This Code is not intended to be made a part of the contract documents. The contractor should not be required through contract documents to assume responsibility for design (Code) requirements, unless the construction entity is acting in a design-build capacity. A Commentary on TMS 602/ACI 530.1/ASCE 6 accompanies the Specification.

1.2.1 This Code lists some of the more important items of information that must be included in the project drawings or project specifications. This is not an all-inclusive list, and additional items may be required by the building official.

Masonry does not always behave in the same manner as its structural supports or adjacent construction. The designer should consider differential movements and the forces resulting from their restraint. The type of connection chosen should transfer only the loads planned. While some connections transfer loads perpendicular to the wall, other devices transfer loads within the plane of the wall. Figure CC-1.2-1 shows representative wall anchorage details that allow movement within the plane of the wall. While load transfer usually involves masonry attached to structural elements, such as beams or columns, the connection of nonstructural elements, such as door and window frames, should also be addressed.

Connectors are of a variety of sizes, shapes, and uses. In order to perform properly they should be identified on the project drawings.

1.2.2 Masonry design performed in accordance with engineered methods is based on the specified compressive strength of the masonry. For engineered masonry, structural adequacy of masonry construction requires that the compressive strength of masonry equals or exceeds the specified strength. Masonry design by prescriptive approaches relies on rules and masonry compressive strength need not be verified.

CODE

1.2.3 The contract documents shall be consistent with design assumptions.

1.2.4 Contract documents shall specify the minimum level of quality assurance as defined in Section 3.1, or shall include an itemized quality assurance program that equals or exceeds the requirements of Section 3.1.

COMMENTARY

1.2.3 The contract documents must accurately reflect design requirements. For example, joint and opening locations assumed in the design should be coordinated with locations shown on the drawings.

1.2.4 Verification that masonry construction conforms to the contract documents is required by this Code. A program of quality assurance must be included in the contract documents to satisfy this Code requirement.

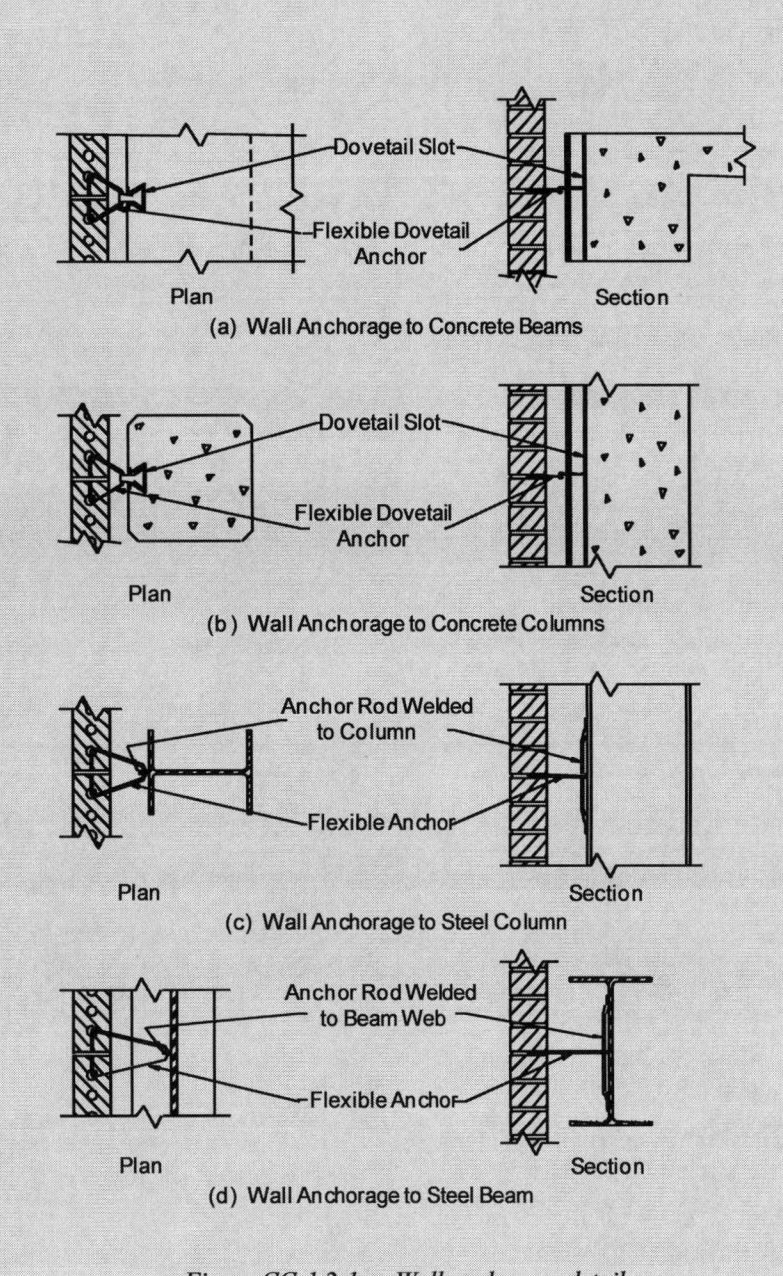

Figure CC-1.2-1 — Wall anchorage details

CODE

1.3 — Approval of special systems of design or construction

Sponsors of any system of design or construction within the scope of this Code, the adequacy of which has been shown by successful use or by analysis or test, but that does not conform to or is not addressed by this Code, shall have the right to present the data on which their design is based to a board of examiners appointed by the building official. The board shall be composed of licensed design professionals and shall have authority to investigate the submitted data, require tests, and formulate rules governing design and construction of such systems to meet the intent of this Code. The rules, when approved and promulgated by the building official, shall be of the same force and effect as the provisions of this Code.

1.4 — Standards cited in this Code

Standards of the American Concrete Institute, the American Society of Civil Engineers, ASTM International, the American Welding Society, and The Masonry Society cited in this Code are listed below with their serial designations, including year of adoption or revision, and are declared to be part of this Code as if fully set forth in this document.

TMS 602-13/ACI 530.1-13/ASCE 6-13 — Specification for Masonry Structures

ASCE 7-10 — Minimum Design Loads for Buildings and Other Structures

ASTM A416/A416M-12 — Standard Specification for Steel Strand, Uncoated Seven-Wire for Prestressed Concrete

ASTM A421/A421M-10 — Standard Specification for Uncoated Stress-Relieved Steel Wire for Prestressed Concrete

ASTM A706/A706M-09b — Standard Specification for Low-Alloy Steel Deformed and Plain Bars for Concrete Reinforcement

ASTM A722/A722M-12 — Standard Specification for Uncoated High-Strength Steel Bars for Prestressing Concrete

ASTM C34-12 — Standard Specification for Structural Clay Load-Bearing Wall Tile

ASTM C140-12a — Standard Test Methods for Sampling and Testing Concrete Masonry Units and Related Units

ASTM C426-10 — Standard Test Method for Linear Drying Shrinkage of Concrete Masonry Units

COMMENTARY

1.3 — Approval of special systems of design or construction

New methods of design, new materials, and new uses of materials must undergo a period of development before being specifically addressed by a code. Hence, valid systems or components might be excluded from use by implication if means were not available to obtain acceptance. This section permits proponents to submit data substantiating the adequacy of their system or component to a board of examiners.

1.4 — Standards cited in this Code

These standards are referenced in this Code. Specific dates are listed here because changes to the standard may result in changes of properties or procedures.

Contact information for these organizations is given below:

American Concrete Institute (ACI)
38800 Country Club Drive
Farmington Hills, MI 48331
www.aci-int.org

American Society of Civil Engineers (ASCE)
1801 Alexander Bell Drive
Reston, VA 20191
www.asce.org

ASTM International
100 Barr Harbor Drive
West Conshohocken, PA 19428-2959
www.astm.org

American Welding Society (AWS)
8669 NW 36th Street, Suite 130
Miami, Florida 33166-6672
www.aws.org

The Masonry Society (TMS)
105 South Sunset Street, Suite Q
Longmont, Colorado 80501-6172
www.masonrysociety.org

CODE

ASTM C476-10 — Standard Specification for Grout for Masonry

ASTM C482-02 (2009) — Standard Test Method for Bond Strength of Ceramic Tile to Portland Cement Paste

ASTM C1006-07 — Standard Test Method for Splitting Tensile Strength of Masonry Units

ASTM C1611/C1611M-09be1 — Standard Test Method for Slump Flow of Self-Consolidating Concrete

ASTM C1693-11 — Standard Specification for Autoclaved Aerated Concrete (AAC)

ASTM E111-04 (2010) — Standard Test Method for Young's Modulus, Tangent Modulus, and Chord Modulus

ASTM E488-96 (2003) — Standard Test Methods for Strength of Anchors in Concrete and Masonry Elements

AWS D 1.4/D1.4M: 2011 — Structural Welding Code — Reinforcing Steel

This page intentionally left blank

CHAPTER 2
NOTATION AND DEFINITIONS

CODE

2.1 — Notation

A_b = cross-sectional area of an anchor bolt, in.^{2} (mm^2)

A_{br} = bearing area, in.^{2} (mm^2)

A_g = gross cross-sectional area of a member, in.^{2} (mm^2)

A_n = net cross-sectional area of a member, in.^{2} (mm^2)

A_{nv} = net shear area, in.^{2} (mm^2)

A_{ps} = area of prestressing steel, in.^{2} (mm^2)

A_{pt} = projected tension area on masonry surface of a right circular cone, in.^{2} (mm^2)

A_{pv} = projected shear area on masonry surface of one-half of a right circular cone, in.^{2} (mm^2)

A_s = area of nonprestressed longitudinal tension reinforcement, in.^{2} (mm^2)

A_{sc} = area of reinforcement placed within the lap, near each end of the lapped reinforcing bars and transverse to them, in.^{2} (mm^2)

A_{st} = total area of laterally tied longitudinal reinforcing steel, in.^{2} (mm^2)

A_v = cross-sectional area of shear reinforcement, in.^{2} (mm^2)

A_1 = loaded area, in.^{2} (mm^2)

A_2 = supporting bearing area, in.^{2} (mm^2)

a = depth of an equivalent compression stress block at nominal strength, in. (mm)

B_a = allowable axial load on an anchor bolt, lb (N)

B_{ab} = allowable axial tensile load on an anchor bolt when governed by masonry breakout, lb (N)

B_{an} = nominal axial strength of an anchor bolt, lb (N)

B_{anb} = nominal axial tensile strength of an anchor bolt when governed by masonry breakout, lb (N)

B_{anp} = nominal axial tensile strength of an anchor bolt when governed by anchor pullout, lb (N)

B_{ans} = nominal axial tensile strength of an anchor bolt when governed by steel yielding, lb (N)

B_{ap} = allowable axial tensile load on an anchor bolt when governed by anchor pullout, lb (N)

B_{as} = allowable axial tensile load on an anchor bolt when governed by steel yielding, lb (N)

COMMENTARY

2.1 — Notation

Notations used in this Code are summarized here.

CODE

B_v = allowable shear load on an anchor bolt, lb (N)

B_{vb} = allowable shear load on an anchor bolt when governed by masonry breakout, lb (N)

B_{vc} = allowable shear load on an anchor bolt when governed by masonry crushing, lb (N)

B_{vn} = nominal shear strength of an anchor bolt, lb (N)

B_{vnb} = nominal shear strength of an anchor bolt when governed by masonry breakout, lb (N)

B_{vnc} = nominal shear strength of an anchor bolt when governed by masonry crushing, lb (N)

B_{vnpry} = nominal shear strength of an anchor bolt when governed by anchor pryout, lb (N)

B_{vns} = nominal shear strength of an anchor bolt when governed by steel yielding, lb (N)

B_{vpry} = allowable shear load on an anchor bolt when governed by anchor pryout, lb (N)

B_{vs} = allowable shear load on an anchor bolt when governed by steel yielding, lb (N)

b = width of section, in. (mm)

b_a = total applied design axial force on an anchor bolt, lb (N)

b_{af} = factored axial force in an anchor bolt, lb (N)

b_v = total applied design shear force on an anchor bolt, lb (N)

b_{vf} = factored shear force in an anchor bolt, lb (N)

b_w = width of wall beam, in. (mm)

C_d = deflection amplification factor

c = distance from the fiber of maximum compressive strain to the neutral axis, in. (mm)

D = dead load or related internal moments and forces

d = distance from extreme compression fiber to centroid of tension reinforcement, in. (mm)

d_b = nominal diameter of reinforcement or anchor bolt, in. (mm)

d_v = actual depth of a member in direction of shear considered, in. (mm)

E = load effects of earthquake or related internal moments and forces

E_{AAC} = modulus of elasticity of AAC masonry in compression, psi (MPa)

E_{bb} = modulus of elasticity of bounding beams, psi (MPa)

COMMENTARY

CODE

E_{bc} = modulus of elasticity of bounding columns, psi (MPa)

E_m = modulus of elasticity of masonry in compression, psi (MPa)

E_{ps} = modulus of elasticity of prestressing steel, psi (MPa)

E_s = modulus of elasticity of steel, psi (MPa)

E_v = modulus of rigidity (shear modulus) of masonry, psi (MPa)

e = eccentricity of axial load, in. (mm)

e_b = projected leg extension of bent-bar anchor, measured from inside edge of anchor at bend to farthest point of anchor in the plane of the hook, in. (mm)

e_u = eccentricity of P_{uf}, in. (mm)

F_a = allowable compressive stress available to resist axial load only, psi (MPa)

F_b = allowable compressive stress available to resist flexure only, psi (MPa)

F_s = allowable tensile or compressive stress in reinforcement, psi (MPa)

F_v = allowable shear stress, psi (MPa)

F_{vm} = allowable shear stress resisted by the masonry, psi (MPa)

F_{vs} = allowable shear stress resisted by the shear reinforcement, psi (MPa)

f_a = calculated compressive stress in masonry due to axial load only, psi (MPa)

f_b = calculated compressive stress in masonry due to flexure only, psi (MPa)

f'_{AAC} = specified compressive strength of AAC masonry, psi (MPa)

f'_g = specified compressive strength of grout, psi (MPa)

f'_m = specified compressive strength of clay masonry or concrete masonry, psi (MPa)

f'_{mi} = specified compressive strength of clay masonry or concrete masonry at the time of prestress transfer, psi (MPa)

f_{ps} = stress in prestressing tendon at nominal strength, psi (MPa)

f_{pu} = specified tensile strength of prestressing tendon, psi (MPa)

f_{py} = specified yield strength of prestressing tendon, psi (MPa)

f_r = modulus of rupture, psi (MPa)

COMMENTARY

CODE

f_{rAAC} = modulus of rupture of AAC, psi (MPa)

f_s = calculated tensile or compressive stress in reinforcement, psi (MPa)

f_{se} = effective stress in prestressing tendon after all prestress losses have occurred, psi (MPa)

f_{tAAC} = splitting tensile strength of AAC as determined in accordance with ASTM C1006, psi (MPa)

f_v = calculated shear stress in masonry, psi (MPa)

f_y = specified yield strength of steel for reinforcement and anchors, psi (MPa)

h = effective height of column, wall, or pilaster, in. (mm)

h_{inf} = vertical dimension of infill, in. (mm)

h_w = height of entire wall or of the segment of wall considered, in. (mm)

I_{bb} = moment of inertia of bounding beam for bending in the plane of the infill, in.^{4} (mm^4)

I_{bc} = moment of inertia of bounding column for bending in the plane of the infill, in.^{4} (mm^4)

I_{cr} = moment of inertia of cracked cross-sectional area of a member, in.^{4} (mm^4)

I_{eff} = effective moment of inertia, in.^{4} (mm^4)

I_g = moment of inertia of gross cross-sectional area of a member, in.^{4} (mm^4)

I_n = moment of inertia of net cross-sectional area of a member, in.^{4} (mm^4)

j = ratio of distance between centroid of flexural compressive forces and centroid of tensile forces to depth, d

K = dimension used to calculate reinforcement development, in. (mm)

K_{AAC} = dimension used to calculate reinforcement development for AAC masonry, in. (mm)

k_c = coefficient of creep of masonry, per psi (per MPa)

k_e = coefficient of irreversible moisture expansion of clay masonry

k_m = coefficient of shrinkage of concrete masonry

k_t = coefficient of thermal expansion of masonry per degree Fahrenheit (degree Celsius)

L = live load or related internal moments and forces

l = clear span between supports, in. (mm)

l_b = effective embedment length of headed or bent anchor bolts, in. (mm)

CODE

l_{be} = anchor bolt edge distance, in. (mm)

l_d = development length or lap length of straight reinforcement, in. (mm)

l_e = equivalent embedment length provided by standard hooks measured from the start of the hook (point of tangency), in. (mm)

l_{eff} = effective span length for a deep beam, in. (mm)

l_{inf} = plan length of infill, in. (mm)

l_p = clear span of the prestressed member in the direction of the prestressing tendon, in. (mm)

l_w = length of entire wall or of the segment of wall considered in direction of shear force, in. (mm)

M = maximum moment at the section under consideration, in.-lb (N-mm)

M_a = maximum moment in member due to the applied unfactored loading for which deflection is calculated, in.-lb (N-mm)

M_{cr} = nominal cracking moment strength, in.-lb (N-mm)

M_n = nominal moment strength, in.-lb (N-mm)

M_{ser} = service moment at midheight of a member, including P-delta effects, in.-lb (N-mm)

M_u = factored moment, magnified by second-order effects where required by the code, in.-lb (N-mm)

$M_{u,0}$ = factored moment from first-order analysis, in.-lb (N-mm)

n = modular ratio, E_s/E_m

N_u = factored compressive force acting normal to shear surface that is associated with the V_u loading combination case under consideration, lb (N)

N_v = compressive force acting normal to shear surface, lb (N)

P = axial load, lb (N)

P_a = allowable axial compressive force in a reinforced member, lb (N)

P_e = Euler buckling load, lb (N)

P_n = nominal axial strength, lb (N)

P_{ps} = prestressing tendon force at time and location relevant for design, lb (N)

P_u = factored axial load, lb (N)

P_{uf} = factored load from tributary floor or roof areas, lb (N)

COMMENTARY

CODE

P_{uw} = factored weight of wall area tributary to wall section under consideration, lb (N)

Q = first moment about the neutral axis of an area between the extreme fiber and the plane at which the shear stress is being calculated, in.^{3} (mm^3)

Q_E = the effect of horizontal seismic (earthquake-induced) forces

$q_{n\,inf}$ = nominal out-of-plane flexural capacity of infill per unit area, psf (Pa)

q_z = velocity pressure determined in accordance with ASCE 7, psf (kPa)

R = response modification coefficient

r = radius of gyration, in. (mm)

S = snow load or related internal moments and forces

S_n = section modulus of the net cross-sectional area of a member, in.^{3} (mm^3)

s = spacing of reinforcement, in. (mm)

s_l = total linear drying shrinkage of concrete masonry units determined in accordance with ASTM C426

t = nominal thickness of member, in. (mm)

t_{inf} = specified thickness of infill, in. (mm)

$t_{net\,inf}$ = net thickness of infill, in. (mm)

t_{sp} = specified thickness of member, in. (mm)

v = shear stress, psi (MPa)

V = shear force, lb (N)

V_{lim} = limiting base-shear strength, lb (N)

V_{nAAC} = nominal shear strength provided by AAC masonry, lb (N)

V_n = nominal shear strength, lb (N)

$V_{n\,inf}$ = nominal horizontal in-plane shear strength of infill, lb (N)

V_{nm} = nominal shear strength provided by masonry, lb (N)

V_{ns} = nominal shear strength provided by shear reinforcement, lb (N)

V_u = factored shear force, lb (N)

V_{ub} = base-shear demand, lb (N)

W = wind load or related internal moments and forces

W_S = dimension of the structural wall strip defined in Sections 14.3.2 and A.5.1 and shown in Figures 14.3.1-1 and A.5.1-1.

COMMENTARY

CODE

W_T = dimension of the tributary length of wall, defined in Sections 14.3.2 and A.5.1 and shown in Figures 14.3.1-1 and A.5.1-1.

w_{inf} = width of equivalent strut, in. (mm)

w_{strut} = horizontal projection of the width of the diagonal strut, in. (mm)

w_u = out-of-plane factored uniformly distributed load, lb/in. (N/mm)

z = internal lever arm between compressive and tensile forces in a deep beam, in. (mm)

α_{arch} = horizontal arching parameter for infill, $lb^{0.25}$ ($N^{0.25}$)

β_{arch} = vertical arching parameter for infill, $lb^{0.25}$ ($N^{0.25}$)

β_b = ratio of area of reinforcement cut off to total area of tension reinforcement at a section

γ = reinforcement size factor

γ_g = grouted shear wall factor

Δ = calculated story drift, in. (mm)

Δ_a = allowable story drift, in. (mm)

δ = moment magnification factor

δ_{ne} = displacements calculated using code-prescribed seismic forces and assuming elastic behavior, in. (mm)

δ_s = horizontal deflection at midheight under allowable stress design load combinations, in. (mm)

δ_u = deflection due to factored loads, in. (mm)

ε_{cs} = drying shrinkage of AAC

ε_{mu} = maximum usable compressive strain of masonry

ξ = lap splice confinement reinforcement factor

θ_{strut} = angle of infill diagonal with respect to the horizontal, degrees

λ_{strut} = characteristic stiffness parameter for infill, $in.^{-1}$ (mm^{-1})

μ_{AAC} = coefficient of friction of AAC

ρ = reinforcement ratio

ρ_{max} = maximum flexural tension reinforcement ratio

ϕ = strength-reduction factor

ψ = magnification factor for second-order effects

CODE

2.2 — Definitions

Anchor — Metal rod, wire, or strap that secures masonry to its structural support.

Anchor pullout — Anchor failure defined by the anchor sliding out of the material in which it is embedded without breaking out a substantial portion of the surrounding material.

Area, gross cross-sectional — The area delineated by the out-to-out dimensions of masonry in the plane under consideration.

Area, net cross-sectional — The area of masonry units, grout, and mortar crossed by the plane under consideration based on out-to-out dimensions.

Area, net shear — The net area of the web of a shear element.

Autoclaved aerated concrete — Low-density cementitious product of calcium silicate hydrates, whose material specifications are defined in ASTM C1693.

Autoclaved aerated concrete (AAC) masonry — Autoclaved aerated concrete units manufactured without reinforcement, set on a mortar leveling bed, bonded with thin-bed mortar, placed with or without grout, and placed with or without reinforcement.

Backing — Wall or surface to which veneer is attached.

Bed joint — The horizontal layer of mortar on which a masonry unit is laid.

COMMENTARY

2.2 — Definitions

For consistent application of this Code, terms are defined that have particular meanings in this Code. The definitions given are for use in application of this Code only and do not always correspond to ordinary usage. Other terms are defined in referenced documents and those definitions are applicable. If any term is defined in both this Code and in a referenced document, the definition in this Code applies. Referenced documents are listed in Section 1.4 and include ASTM standards. Terminology standards include ASTM C1232 Standard Terminology of Masonry and ASTM C1180 Standard Terminology of Mortar and Grout for Unit Masonry. Glossaries of masonry terminology are available from several sources within the industry (BIA TN 2, 1999; NCMA TEK 1-4, 2004; and IMI, 1981).

Area, net shear — The net shear area for a partially grouted flanged shear wall is shown in Figure CC-2.2-1.

Figure CC-2.2-1 — Net shear area

BUILDING CODE REQUIREMENTS FOR MASONRY STRUCTURES AND COMMENTARY

CODE

Bond beam — A horizontal, sloped, or stepped element that is fully grouted, has longitudinal bar reinforcement, and is constructed within a masonry wall.

COMMENTARY

Bond beam – This reinforced member is usually constructed horizontally, but may be sloped or stepped to match an adjacent roof, for example, as shown in Figure CC-2.2-2.

Notes:

(1) Masonry wall
(2) Fully grouted bond beam with reinforcement
(3) Sloped top of wall
(4) Length of noncontact lap splice
(5) Spacing between bars in noncontact lap splice

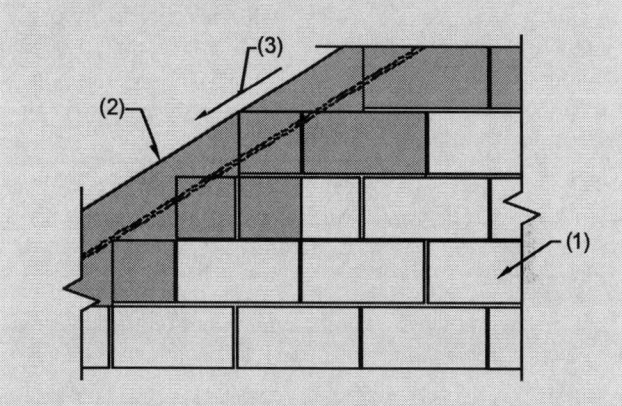

(a) Sloped Bond Beam
(not to scale)

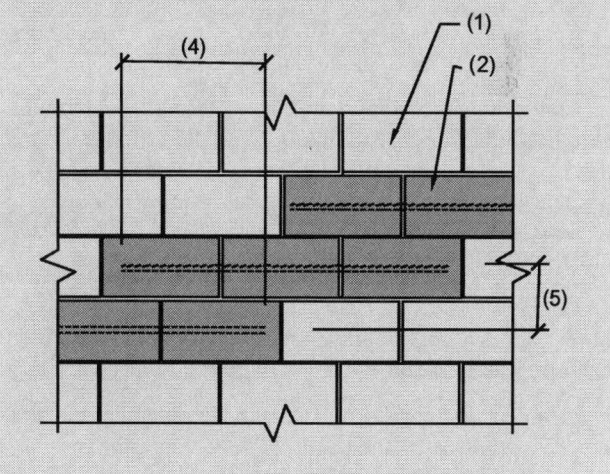

(b) Stepped Bond Beam
(not to scale)

Figure CC-2.2-2 — Sloped and stepped bond beams

CODE

Bonded prestressing tendon — Prestressing tendon encapsulated by prestressing grout in a corrugated duct that is bonded to the surrounding masonry through grouting.

Bounding frame — The columns and upper and lower beams or slabs that surround masonry infill and provide structural support.

Building official — The officer or other designated authority charged with the administration and enforcement of this Code, or the building official's duly authorized representative.

Cavity wall — A masonry wall consisting of two or more wythes, at least two of which are separated by a continuous air space; air space(s) between wythes may contain insulation; and separated wythes must be connected by wall ties.

Collar joint — Vertical longitudinal space between wythes of masonry or between masonry wythe and back-up construction, which is permitted to be filled with mortar or grout.

Column — A structural member, not built integrally into a wall, designed primarily to resist compressive loads parallel to its longitudinal axis and subject to dimensional limitations.

Composite action — Transfer of stress between components of a member designed so that in resisting loads, the combined components act together as a single member.

Composite masonry — Multiwythe masonry members with wythes bonded to produce composite action.

Compressive strength of masonry — Maximum compressive force resisted per unit of net cross-sectional area of masonry, determined by testing masonry prisms or a function of individual masonry units, mortar, and grout, in accordance with the provisions of TMS 602/ACI 530.1/ASCE 6.

Connector — A mechanical device for securing two or more pieces, parts, or members together, including anchors, wall ties, and fasteners.

Contract documents — Documents establishing the required work, and including in particular, the project drawings and project specifications.

Corbel — A projection of successive courses from the face of masonry.

Cover, grout — thickness of grout surrounding the outer surface of embedded reinforcement, anchor, or tie.

Cover, masonry — thickness of masonry units, mortar, and grout surrounding the outer surface of embedded reinforcement, anchor, or tie.

COMMENTARY

Column — Generally, a column spans vertically, though it may have another orientation in space.

CODE

Cover, mortar — thickness of mortar surrounding the outer surface of embedded reinforcement, anchor, or tie.

Deep beam — A beam that has an effective span-to-depth ratio, l_{eff}/d_v, less than 3 for a continuous span and less than 2 for a simple span.

Depth — The dimension of a member measured in the plane of a cross section perpendicular to the neutral axis.

Design story drift — The difference of deflections at the top and bottom of the story under consideration, taking into account the possibility of inelastic deformations as defined in ASCE 7. In the equivalent lateral force method, the story drift is calculated by multiplying the deflections determined from an elastic analysis by the appropriate deflection amplification factor, C_d, from ASCE 7.

Design strength — The nominal strength of an element multiplied by the appropriate strength-reduction factor.

Diaphragm — A roof or floor system designed to transmit lateral forces to shear walls or other lateral-force-resisting elements.

Dimension, nominal — The specified dimension plus an allowance for the joints with which the units are to be laid. Nominal dimensions are usually stated in whole numbers nearest to the specified dimensions.

Dimensions, specified — Dimensions specified for the manufacture or construction of a unit, joint, or element.

Effective height — Clear height of a member between lines of support or points of support and used for calculating the slenderness ratio of a member. Effective height for unbraced members shall be calculated.

Effective prestress — Stress remaining in prestressing tendons after all losses have occurred.

Foundation pier — A vertical foundation member, not built integrally into a foundation wall, empirically designed to support gravity loads and subject to dimensional limitations.

Glass unit masonry — Masonry composed of glass units bonded by mortar.

Grout — (1) A plastic mixture of cementitious materials, aggregates, and water, with or without admixtures, initially produced to pouring consistency without segregation of the constituents during placement. (2) The hardened equivalent of such mixtures.

COMMENTARY

Dimension, nominal — Nominal dimensions are usually used to identify the size of a masonry unit. The thickness or width is given first, followed by height and length. The permitted tolerances for units are given in the appropriate material standards. Permitted tolerances for joints and masonry construction are given in the Specification.

Dimensions, specified — Specified dimensions are most often used for design calculations.

CODE

Grout, self-consolidating — A highly fluid and stable grout typically with admixtures, that remains homogeneous when placed and does not require puddling or vibration for consolidation.

Head joint — Vertical mortar joint placed between masonry units within the wythe at the time the masonry units are laid.

Header (bonder) — A masonry unit that connects two or more adjacent wythes of masonry.

Infill — Masonry constructed within the plane of, and bounded by, a structural frame.

Infill, net thickness — Minimum total thickness of the net cross-sectional area of an infill.

Infill, non-participating — Infill designed so that in-plane loads are not imparted to it from the bounding frame.

Infill, participating — Infill designed to resist in-plane loads imparted to it by the bounding frame.

Inspection, continuous — The Inspection Agency's full-time observation of work by being present in the area where the work is being performed.

Inspection, periodic — The Inspection Agency's part-time or intermittent observation of work during construction by being present in the area where the work has been or is being performed, and observation upon completion of the work.

Laterally restrained prestressing tendon — Prestressing tendon that is not free to move laterally within the cross section of the member.

Laterally unrestrained prestressing tendon — Prestressing tendon that is free to move laterally within the cross section of the member.

COMMENTARY

Infill, net thickness – The net thickness is shown in Figure CC-2.2-3

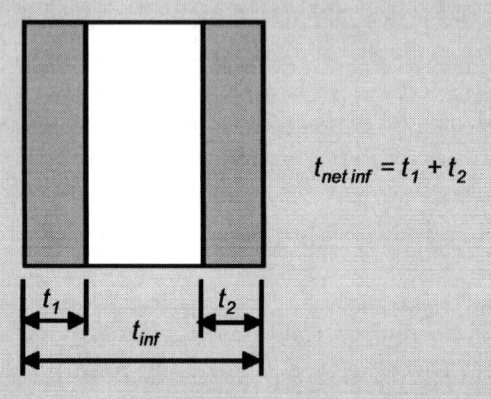

$t_{net\ inf} = t_1 + t_2$

Vertical Section through Hollow Unit in Infill Wall

Figure CC-2.2-3 — Thickness and net thickness of an infill

Inspection, continuous — The Inspection Agency is required to be on the project site whenever masonry tasks requiring continuous inspection are in progress.

Inspection, periodic — During construction requiring periodic inspection, the Inspection Agency is only required to be on the project site intermittently, and is required to observe completed work. The frequency of periodic inspections should be defined by the Architect/Engineer as part of the quality assurance plan, and should be consistent with the complexity and size of the project.

CODE

Licensed design professional — An individual who is licensed to practice design as defined by the statutory requirements of the professional licensing laws of the state or jurisdiction in which the project is to be constructed and who is in responsible charge of the design; in other documents, also referred to as *registered design professional*.

Load, dead — Dead weight supported by a member, as defined by the legally adopted building code.

Load, live — Live load specified by the legally adopted building code.

Load, service — Load specified by the legally adopted building code.

Longitudinal reinforcement — Reinforcement placed parallel to the longitudinal axis of the member.

Masonry breakout — Anchor failure defined by the separation of a volume of masonry, approximately conical in shape, from the member.

Masonry, partially grouted — Construction in which designated cells or spaces are filled with grout, while other cells or spaces are ungrouted.

Masonry unit, hollow — A masonry unit with net cross-sectional area of less than 75 percent of its gross cross-sectional area when measured in any plane parallel to the surface containing voids.

Masonry unit, solid — A masonry unit with net cross-sectional area of 75 percent or more of its gross cross-sectional area when measured in every plane parallel to the surface containing voids.

Modulus of elasticity — Ratio of normal stress to corresponding strain for tensile or compressive stresses below proportional limit of material.

Modulus of rigidity — Ratio of unit shear stress to unit shear strain for unit shear stress below the proportional limit of the material.

Nominal strength — The strength of an element or cross section calculated in accordance with the requirements and assumptions of the strength design methods of these provisions before application of strength-reduction factors.

Partition wall — An interior wall without structural function.

Pier — A reinforced, vertically spanning portion of a wall next to an opening, designed using strength design, and subject to dimensional limitations.

COMMENTARY

Licensed design professional — For convenience, the Commentary uses the term "designer" when referring to the licensed design professional.

Pier — The term "Pier" is used for convenience to define a portion of a wall, and only has meaning for certain reinforced members designed using strength design. The reinforcement requirements for piers are less severe than for columns because piers are part of a wall, have less slender geometry and more restrictive loading limits. A strength-designed member, not meeting the dimensional limits and requirements for a pier, should be designed as a wall or, if not built integrally with a wall, as a column.

CODE

Post-tensioning — Method of prestressing in which a prestressing tendon is tensioned after the masonry has been placed.

Prestressed masonry — Masonry in which internal compressive stresses have been introduced by prestressed tendons to counteract potential tensile stresses resulting from applied loads.

Prestressing grout — A cementitious mixture used to encapsulate bonded prestressing tendons.

Prestressing tendon — Steel elements such as wire, bar, or strand, used to impart prestress to masonry.

Pretensioning — Method of prestressing in which a prestressing tendon is tensioned before the transfer of stress into the masonry.

Prism — An assemblage of masonry units and mortar, with or without grout, used as a test specimen for determining properties of the masonry.

Project drawings — The drawings that, along with the project specifications, complete the descriptive information for constructing the work required by the contract documents.

Project specifications — The written documents that specify requirements for a project in accordance with the service parameters and other specific criteria established by the owner or the owner's agent.

Quality assurance — The administrative and procedural requirements established by the contract documents to assure that constructed masonry is in compliance with the contract documents.

Reinforcement — Nonprestressed steel reinforcement.

Required strength — The strength needed to resist factored loads.

Running bond — The placement of masonry units so that head joints in successive courses are horizontally offset at least one-quarter the unit length.

COMMENTARY

Running bond — This Code concerns itself only with the structural effect of the masonry bond pattern. Therefore, the only distinction made by this Code is between masonry laid in running bond and masonry that is not laid in running bond. For purposes of this Code, architectural bond patterns that do not satisfy the Code definition of running bond are classified as not running bond. Masonry laid in other bond patterns must be reinforced to provide continuity across the heads joints. Stack bond, which is commonly interpreted as a pattern with aligned heads joints, is one bond pattern that is required to be reinforced horizontally.

CODE

Shear wall — A wall, load-bearing or non-load-bearing, designed to resist lateral forces acting in the plane of the wall (sometimes referred to as a vertical diaphragm).

Shear wall, detailed plain (unreinforced) AAC masonry — An AAC masonry shear wall designed to resist lateral forces while neglecting stresses in reinforcement, although provided with minimum reinforcement and connections.

Shear wall, detailed plain (unreinforced) masonry — A masonry shear wall designed to resist lateral forces while neglecting stresses in reinforcement, although provided with minimum reinforcement and connections.

Shear wall, intermediate reinforced masonry — A masonry shear wall designed to resist lateral forces while considering stresses in reinforcement and to satisfy specific minimum reinforcement and connection requirements.

Shear wall, intermediate reinforced prestressed masonry — A prestressed masonry shear wall designed to resist lateral forces while considering stresses in reinforcement and to satisfy specific minimum reinforcement and connection requirements.

Shear wall, ordinary plain (unreinforced) AAC masonry — An AAC masonry shear wall designed to resist lateral forces while neglecting stresses in reinforcement, if present.

Shear wall, ordinary plain (unreinforced) masonry — A masonry shear wall designed to resist lateral forces while neglecting stresses in reinforcement, if present.

Shear wall, ordinary plain (unreinforced) prestressed masonry — A prestressed masonry shear wall designed to resist lateral forces while neglecting stresses in reinforcement, if present.

Shear wall, ordinary reinforced AAC masonry — An AAC masonry shear wall designed to resist lateral forces while considering stresses in reinforcement and satisfying prescriptive reinforcement and connection requirements.

Shear wall, ordinary reinforced masonry — A masonry shear wall designed to resist lateral forces while considering stresses in reinforcement and satisfying prescriptive reinforcement and connection requirements.

Shear wall, special reinforced masonry — A masonry shear wall designed to resist lateral forces while considering stresses in reinforcement and to satisfy special reinforcement and connection requirements.

Shear wall, special reinforced prestressed masonry — A prestressed masonry shear wall designed to resist lateral forces while considering stresses in reinforcement and to satisfy special reinforcement and connection requirements.

CODE

Slump flow — The circular spread of plastic self-consolidating grout, which is evaluated in accordance with ASTM C1611/C1611M.

Special boundary elements — In walls that are designed to resist in-plane load, end regions that are strengthened by reinforcement and are detailed to meet specific requirements, and may or may not be thicker than the wall.

Specified compressive strength of AAC masonry, f'_{AAC} — Minimum compressive strength, expressed as force per unit of net cross-sectional area, required of the AAC masonry used in construction by the contract documents, and upon which the project design is based. Whenever the quantity f'_{AAC} is under the radical sign, the square root of numerical value only is intended and the result has units of psi (MPa).

Specified compressive strength of masonry, f'_m — Minimum compressive strength, expressed as force per unit of net cross-sectional area, required of the masonry used in construction by the contract documents, and upon which the project design is based. Whenever the quantity f'_m is under the radical sign, the square root of numerical value only is intended and the result has units of psi (MPa).

Stirrup — Reinforcement used to resist shear in a flexural member.

Stone masonry — Masonry composed of field, quarried, or cast stone units bonded by mortar.

Stone masonry, ashlar — Stone masonry composed of rectangular units having sawed, dressed, or squared bed surfaces and bonded by mortar.

Stone masonry, rubble — Stone masonry composed of irregular-shaped units bonded by mortar.

Strength-reduction factor, ϕ — The factor by which the nominal strength is multiplied to obtain the design strength.

Tendon anchorage — In post-tensioning, a device used to anchor the prestressing tendon to the masonry or concrete member; in pretensioning, a device used to anchor the prestressing tendon during hardening of masonry mortar, grout, prestressing grout, or concrete.

Tendon coupler — A device for connecting two tendon ends, thereby transferring the prestressing force from end to end.

Tendon jacking force — Temporary force exerted by a device that introduces tension into prestressing tendons.

COMMENTARY

Special boundary elements — Requirements for longitudinal and transverse reinforcement have not been established in general and must be verified by testing. Research in this area is ongoing.

CODE

Thin-bed mortar — Mortar for use in construction of AAC unit masonry whose joints shall not be less than 1/16 in. (1.5 mm).

Tie, lateral — Loop of reinforcing bar or wire enclosing longitudinal reinforcement.

Tie, wall — Metal connector that connects wythes of masonry walls together.

Transfer — Act of applying to the masonry member the force in the prestressing tendons.

Transverse reinforcement — Reinforcement placed perpendicular to the longitudinal axis of the member.

Unbonded prestressing tendon — Prestressing tendon that is not bonded to masonry.

Unreinforced (plain) masonry — Masonry in which the tensile resistance of masonry is taken into consideration and the resistance of reinforcing steel, if present, is neglected.

Veneer, adhered — Masonry veneer secured to and supported by the backing through adhesion.

Veneer, anchored — Masonry veneer secured to and supported laterally by the backing through anchors and supported vertically by the foundation or other structural elements.

Veneer, masonry — A masonry wythe that provides the exterior finish of a wall system and transfers out-of-plane load directly to a backing, but is not considered to add strength or stiffness to the wall system.

Visual stability index (VSI) — An index, defined in ASTM C1611/C1611M, that qualitatively indicates the stability of self-consolidating grout

Wall — A vertical element with a horizontal length to thickness ratio greater than 3, used to enclose space.

Wall, load-bearing — Wall supporting vertical loads greater than 200 lb/linear ft (2919 N/m) in addition to its own weight.

Wall, masonry bonded hollow — A multiwythe wall built with masonry units arranged to provide an air space between the wythes and with the wythes bonded together with masonry units.

Width — The dimension of a member measured in the plane of a cross section parallel to the neutral axis.

Wythe — Each continuous vertical section of a wall, one masonry unit in thickness.

COMMENTARY

This page intentionally left blank

CHAPTER 3
QUALITY AND CONSTRUCTION

CODE

3.1 — Quality Assurance program

The quality assurance program shall comply with the requirements of this section, depending on the Risk Category, as defined in ASCE 7 or the legally adopted building code. The quality assurance program shall itemize the requirements for verifying conformance of material composition, quality, storage, handling, preparation, and placement with the requirements of TMS 602/ACI 530.1/ASCE 6.

3.1.1 *Level A Quality Assurance*
The minimum quality assurance program for masonry in Risk Category I, II, or III structures and designed in accordance with Part 4 or Appendix A shall comply with Table 3.1.1.

COMMENTARY

3.1 — Quality Assurance program

Masonry design provisions in this Code are valid when the quality of masonry construction meets or exceeds that described in the Specification. Therefore, in order to design masonry by this Code, verification of good quality construction is required. The means by which the quality of construction is monitored is the quality assurance program.

A quality assurance program must be defined in the contract documents, to answer questions such as "how to", "what method", "how often", and "who determines acceptance". This information is part of the administrative and procedural requirements. Typical requirements of a quality assurance program include review of material certifications, field inspection, and testing. The acts of providing submittals, inspecting, and testing are part of the quality assurance program.

Because the design and the complexity of masonry construction vary from project to project, so must the extent of the quality assurance program. The contract documents must indicate the testing, Special Inspection, and other measures that are required to assure that the Work is in conformance with the project requirements.

Section 3.1 establishes the minimum criteria required to assure that the quality of masonry construction conforms to the quality upon which the Code-permissible values are based. The scope of the quality assurance program depends on whether the structure is a Risk Category IV structure or not, as defined by ASCE 7 or the legally adopted building code. Because of their importance, Risk Category IV structures are subjected to more extensive quality assurance measures.

The level of required quality assurance depends on whether the masonry was designed in accordance with Part 3, Appendix B, or Appendix C (engineered) or in accordance with Part 4 or Appendix A (empirical or prescriptive).

CODE

3.1.2 *Level B Quality Assurance*

3.1.2.1 The minimum quality assurance program for masonry in Risk Category IV structures and designed in accordance with Chapter 12 or 13 shall comply with Table 3.1.2.

3.1.2.2 The minimum quality assurance program for masonry in Risk Category I, II, or III structures and designed in accordance with chapters other than those in Part 4 or Appendix A shall comply with Table 3.1.2.

3.1.3 *Level C Quality Assurance*

The minimum quality assurance program for masonry in Risk Category IV structures and designed in accordance with chapters other than those in Part 4 or Appendix A shall comply with Table 3.1.3.

3.1.4 *Procedures*

The quality assurance program shall set forth the procedures for reporting and review. The quality assurance program shall also include procedures for resolution of noncompliances.

COMMENTARY

3.1.2 *Level B Quality Assurance*

Implementation of testing and inspection requirements contained in Table 3.1.2 requires detailed knowledge of the appropriate procedures. Comprehensive testing and inspection procedures are available from recognized industry sources (Chrysler, 2010; NCMA, 2008; BIA, 2001; BIA 1988), which may be referenced for assistance in developing and implementing a Quality Assurance program. Certain applications, such as Masonry Veneer (Chapter 12), Masonry Partition Walls (Chapter 14) and Empirical Design of Masonry (Appendix A), do not require compressive strength verification of masonry as indicated in Table 3.1.2.

Installation techniques for AAC masonry and thin-bed mortar differ from concrete and clay masonry. Once it has been demonstrated in the field that compliance is attained for the installation of AAC masonry and thin-bed mortar, the frequency of Special Inspection may be revised from continuous to periodic. However, the frequency of Special Inspection should revert to continuous for the prescribed period whenever new AAC masonry installers work on the project.

3.1.3 *Level C Quality Assurance*

Premixed mortars and grouts are delivered to the project site as "trowel ready" or "pourable" materials, respectively. Preblended mortars and grouts are dry combined materials that are mixed with water at the project site. Verification of proportions of premixed or preblended mortars and grouts can be accomplished by review of manufacture's batch tickets (if applicable), a combination of preconstruction and construction testing, or other acceptable documentation.

3.1.4 *Procedures*

In addition to specifying testing and Special Inspection requirements, the quality assurance program must define the procedures for submitting the testing and inspection reports (that is, how many copies and to whom) and define the process by which those reports are to be reviewed.

Testing and evaluation should be addressed in the quality assurance program. The program should allow for the selection and approval of a testing agency, which agency should be provided with prequalification test information and the rights for sampling and testing of specific masonry construction materials in accordance with referenced standards. The evaluation of test results by the testing agency should indicate compliance or noncompliance with a referenced standard.

Further quality assurance evaluation should allow an appraisal of the testing program and the handling of nonconformance. Acceptable values for all test methods should be given in the contract documents.

Identification and resolution of noncomplying

CODE

3.1.5 *Qualifications*

The quality assurance program shall define the qualifications for testing laboratories and for inspection agencies.

COMMENTARY

conditions should be addressed in the contract documents. A responsible person should be identified to allow resolution of nonconformances. In agreement with others in the design/construct team, the resolutions should be repaired, reworked, accepted as is, or rejected. Repaired and reworked conditions should initiate a reinspection.

Records control should be addressed in the contract documents. The distribution of documents during and after construction should be delineated. The review of documents should persist throughout the construction period so that each party is informed and that records for documenting construction occurrences are available and correct after construction has been completed.

3.1.5 *Qualifications*

The entities verifying compliance must be competent and knowledgeable of masonry construction and the requirements of this Code. Therefore, minimum qualifications for those individuals must also be established by the quality assurance program in the contract documents.

The responsible party performing the quality control measures should document the organizational representatives who will be a part of the quality control segment, their qualifications, and their precise conduct during the performance of the quality assurance phase.

Laboratories that comply with the requirements of ASTM C1093 are more likely to be familiar with masonry materials and testing. Specifying that the testing agencies comply with the requirements of ASTM C1093 should improve the quality of the resulting masonry.

Table 3.1.1 — Level A Quality Assurance

MINIMUM VERIFICATION
Prior to construction, verify certificates of compliance used in masonry construction

Table 3.1.2 — Level B Quality Assurance

MINIMUM TESTS
Verification of Slump flow and Visual Stability Index (VSI) as delivered to the project site in accordance with Specification Article 1.5 B.1.b.3 for self-consolidating grout
Verification of f'_m and f'_{AAC} in accordance with Specification Article 1.4 B prior to construction, except where specifically exempted by this Code

MINIMUM SPECIAL INSPECTION

Inspection Task	Frequency [a] Continuous	Frequency [a] Periodic	Reference for Criteria TMS 402/ ACI 530/ ASCE 5	Reference for Criteria TMS 602/ ACI 530.1/ ASCE 6
1. Verify compliance with the approved submittals		X		Art. 1.5
2. As masonry construction begins, verify that the following are in compliance:				
a. Proportions of site-prepared mortar		X		Art. 2.1, 2.6 A
b. Construction of mortar joints		X		Art. 3.3 B
c. Grade and size of prestressing tendons and anchorages		X		Art. 2.4 B, 2.4 H
d. Location of reinforcement, connectors, and prestressing tendons and anchorages		X		Art. 3.4, 3.6 A
e. Prestressing technique		X		Art. 3.6 B
f. Properties of thin-bed mortar for AAC masonry	X[b]	X[c]		Art. 2.1 C
3. Prior to grouting, verify that the following are in compliance:				
a. Grout space		X		Art. 3.2 D, 3.2 F
b. Grade, type, and size of reinforcement and anchor bolts, and prestressing tendons and anchorages		X	Sec. 6.1	Art. 2.4, 3.4
c. Placement of reinforcement, connectors, and prestressing tendons and anchorages		X	Sec. 6.1, 6.2.1, 6.2.6, 6.2.7	Art. 3.2 E, 3.4, 3.6 A
d. Proportions of site-prepared grout and prestressing grout for bonded tendons		X		Art. 2.6 B, 2.4 G.1.b
e. Construction of mortar joints		X		Art. 3.3 B

Continued on next page

Table 3.1.2 — Level B Quality Assurance (Continued)

MINIMUM SPECIAL INSPECTION				
Inspection Task	Frequency [a]		Reference for Criteria	
	Continuous	Periodic	TMS 402/ ACI 530/ ASCE 5	TMS 602/ ACI 530.1/ ASCE 6
4. Verify during construction:				
a. Size and location of structural elements		X		Art. 3.3 F
b. Type, size, and location of anchors, including other details of anchorage of masonry to structural members, frames, or other construction		X	Sec. 1.2.1(e), 6.1.4.3, 6.2.1	
c. Welding of reinforcement	X		Sec. 8.1.6.7.2, 9.3.3.4 (c), 11.3.3.4(b)	
d. Preparation, construction, and protection of masonry during cold weather (temperature below 40°F (4.4°C)) or hot weather (temperature above 90°F (32.2°C))		X		Art. 1.8 C, 1.8 D
e. Application and measurement of prestressing force	X			Art. 3.6 B
f. Placement of grout and prestressing grout for bonded tendons is in compliance	X			Art. 3.5, 3.6 C
g. Placement of AAC masonry units and construction of thin-bed mortar joints	X[b]	X[c]		Art. 3.3 B.9, 3.3 F.1.b
5. Observe preparation of grout specimens, mortar specimens, and/or prisms		X		Art. 1.4 B.2.a.3, 1.4 B.2.b.3, 1.4 B.2.c.3, 1.4 B.3, 1.4 B.4

(a) Frequency refers to the frequency of Special Inspection, which may be continuous during the task listed or periodic during the listed task, as defined in the table.
(b) Required for the first 5000 square feet (465 square meters) of AAC masonry.
(c) Required after the first 5000 square feet (465 square meters) of AAC masonry.

Table 3.1.3 — Level C Quality Assurance

MINIMUM TESTS
Verification of f'_m and f'_{AAC} in accordance with Specification Article 1.4 B prior to construction and for every 5,000 sq. ft (465 sq. m) during construction
Verification of proportions of materials in premixed or preblended mortar, prestressing grout, and grout other than self-consolidating grout, as delivered to the project site
Verification of Slump flow and Visual Stability Index (VSI) as delivered to the project site in accordance with Specification Article 1.5 B.1.b.3 for self-consolidating grout

MINIMUM SPECIAL INSPECTION				
Inspection Task	**Frequency** [a]		**Reference for Criteria**	
	Continuous	Periodic	TMS 402/ ACI 530/ ASCE 5	TMS 602/ ACI 530.1/ ASCE 6
1. Verify compliance with the approved submittals		X		Art. 1.5
2. Verify that the following are in compliance:				
a. Proportions of site-mixed mortar, grout and prestressing grout for bonded tendons		X		Art. 2.1, 2.6 A, 2.6 B, 2.6 C, 2.4 G.1.b
b. Grade, type, and size of reinforcement and anchor bolts, and prestressing tendons and anchorages		X	Sec. 6.1	Art. 2.4, 3.4
c. Placement of masonry units and construction of mortar joints		X		Art. 3.3 B
d. Placement of reinforcement, connectors, and prestressing tendons and anchorages	X		Sec. 6.1, 6.2.1, 6.2.6, 6.2.7	Art. 3.2 E, 3.4, 3.6 A
e. Grout space prior to grouting	X			Art. 3.2 D, 3.2 F
f. Placement of grout and prestressing grout for bonded tendons	X			Art. 3.5, 3.6 C
g. Size and location of structural elements		X		Art. 3.3 F
h. Type, size, and location of anchors including other details of anchorage of masonry to structural members, frames, or other construction	X		Sec. 1.2.1(e), 6.1.4.3, 6.2.1	
i. Welding of reinforcement	X		Sec. 8.1.6.7.2, 9.3.3.4 (c), 11.3.3.4(b)	
j. Preparation, construction, and protection of masonry during cold weather (temperature below 40°F (4.4°C)) or hot weather (temperature above 90°F (32.2°C))		X		Art. 1.8 C, 1.8 D
k. Application and measurement of prestressing force	X			Art. 3.6 B
l. Placement of AAC masonry units and construction of thin-bed mortar joints	X			Art. 3.3 B.9, 3.3 F.1.b
m. Properties of thin-bed mortar for AAC masonry	X			Art. 2.1 C.1
3. Observe preparation of grout specimens, mortar specimens, and/or prisms	X			Art. 1.4 B.2.a.3, 1.4 B.2.b.3, 1.4 B.2.c.3, 1.4 B.3, 1.4 B.4

(a) Frequency refers to the frequency of Special Inspection, which may be continuous during the task listed or periodic during the listed task, as defined in the table.

CODE

3.1.6 *Acceptance relative to strength requirements*

3.1.6.1 *Compliance with f'_m* — Compressive strength of masonry shall be considered satisfactory if the compressive strength of each masonry wythe and grouted collar joint equals or exceeds the value of f'_m.

3.1.6.2 *Determination of compressive strength* — Compressive strength of masonry shall be determined in accordance with the provisions of TMS 602/ACI 530.1/ASCE 6.

3.2 — Construction considerations

3.2.1 *Grouting, minimum spaces*

The minimum dimensions of spaces provided for the placement of grout shall be in accordance with Table 3.2.1. Grout pours with heights exceeding those shown in Table 3.2.1, cavity widths, or cell sizes smaller than those permitted in Table 3.2.1 or grout lift heights exceeding those permitted by Article 3.5 D of TMS 602/ACI 530.1/ASCE 6 are permitted if the results of a grout demonstration panel show that the grout spaces are filled and adequately consolidated. In that case, the procedures used in constructing the grout demonstration panel shall be the minimum acceptable standard for grouting, and the quality assurance program shall include inspection during construction to verify grout placement.

COMMENTARY

3.1.6 *Acceptance relative to strength requirements*
Fundamental to the structural adequacy of masonry construction is the necessity that the compressive strength of masonry equals or exceeds the specified strength. Rather than mandating design based on different values of f'_m for each wythe of a multiwythe wall construction made of differing material, this Code requires the strength of each wythe and of grouted collar joints to equal or exceed f'_m for the portion of the structure considered. If a multiwythe wall is designed as a composite wall, the compressive strength of each wythe or grouted collar joint should equal or exceed f'_m.

3.2 — Construction considerations

The TMS 602/ACI 530.1/ASCE 6 Specification addresses material and construction requirements. It is an integral part of the Code in terms of minimum requirements relative to the composition, quality, storage, handling, and placement of materials for masonry structures. The Specification also includes provisions requiring verification that construction achieves the quality specified. The construction must conform to these requirements in order for the Code provisions to be valid.

3.2.1 *Grouting, minimum spaces*
Code Table 3.2.1 contains the least clear dimension for grouting between wythes and the minimum cell dimensions when grouting hollow units. Selection of units and bonding pattern should be coordinated to achieve these requirements. Vertical alignment of cells must also be considered. Projections or obstructions into the grout space and the diameter of horizontal reinforcement must be considered when calculating the minimum dimensions. See Figure CC-3.2-1.

Coarse grout and fine grout are differentiated by aggregate size in ASTM C476.

The grout space requirements of Code Table 3.2.1 are based on coarse and fine grouts as defined by ASTM C476, and cleaning practice to permit the complete filling of grout spaces and adequate consolidation using typical methods of construction. Grout spaces smaller than specified in Table 3.2.1 have been used successfully in some areas. When the designer is requested to accept a grouting procedure that does not comply with the limits in Table 3.2.1, construction of a grout demonstration panel is required. Destructive or non-destructive evaluation can confirm that filling and adequate consolidation have been achieved. The designer should establish criteria for the grout demonstration panel to assure that critical masonry elements included in the construction will be represented in the demonstration panel. Because a single grout demonstration panel erected prior to masonry construction cannot account for all conditions that may be encountered during construction, the designer should establish inspection

CODE

3.2.2 *Embedded conduits, pipes, and sleeves*

Conduits, pipes, and sleeves of any material to be embedded in masonry shall be compatible with masonry and shall comply with the following requirements.

3.2.2.1 Conduits, pipes, and sleeves shall not be considered to be structural replacements for the displaced masonry. The masonry design shall consider the structural effects of this displaced masonry.

3.2.2.2 Conduits, pipes, and sleeves in masonry shall be no closer than 3 diameters on center. Minimum spacing of conduits, pipes or sleeves of different diameters shall be determined using the larger diameter.

3.2.2.3 Vertical conduits, pipes, or sleeves placed in masonry columns or pilasters shall not displace more than 2 percent of the net cross section.

3.2.2.4 Pipes shall not be embedded in masonry, unless properly isolated from the masonry, when:

(a) Containing liquid, gas, or vapors at temperature higher than 150° F (66°C).

(b) Under pressure in excess of 55 psi (379 kPa).

(c) Containing water or other liquids subject to freezing.

COMMENTARY

procedures to verify grout placement during construction. These inspection procedures should include destructive or non-destructive evaluation to confirm that filling and adequate consolidation have been achieved.

3.2.2 *Embedded conduits, pipes, and sleeves*

3.2.2.1 Conduits, pipes, and sleeves not harmful to mortar and grout may be embedded within the masonry, but the masonry member strength should not be less than that required by design. Effects of reduction in section properties in the areas of conduit, pipe, or sleeve embedment should be considered.

For the integrity of the structure, conduit and pipe fittings within the masonry should be carefully positioned and assembled. The coupling size should be considered when determining sleeve size.

Aluminum should not be used in masonry unless it is effectively coated or otherwise isolated. Aluminum reacts with ions, and may also react electrolytically with steel, causing cracking, spalling of the masonry, or both. Aluminum electrical conduits present a special problem because stray electric current accelerates the adverse reaction.

Pipes and conduits placed in masonry, whether surrounded by mortar or grout or placed in unfilled spaces, need to allow unrestrained movement.

BUILDING CODE REQUIREMENTS FOR MASONRY STRUCTURES AND COMMENTARY

Table 3.2.1 — Grout space requirements

Grout type[1]	Maximum grout pour height, ft (m)	Minimum clear width of grout space,[2,3] in. (mm)	Minimum clear grout space dimensions for grouting cells of hollow units,[3,4,5] in. x in. (mm x mm)
Fine	1 (0.30)	3/4 (19.1)	1 1/2 x 2 (38.1 x 50.8)
Fine	5.33 (1.63)	2 (50.8)	2 x 3 (50.8 x 76.2)
Fine	12.67 (3.86)	2 1/2 (63.5)	2 1/2 x 3 (63.5 x 76.2)
Fine	24 (7.32)	3 (76.2)	3 x 3 (76.2 x 76.2)
Coarse	1 (0.30)	1 1/2 (38.1)	1 1/2 x 3 (38.1 x 76.2)
Coarse	5.33 (1.63)	2 (50.8)	2 1/2 x 3 (63.5 x 76.2)
Coarse	12.67 (3.86)	2 1/2 (63.5)	3 x 3 (76.2 x 76.2)
Coarse	24 (7.32)	3 (76.2)	3 x 4 (76.2 x 102)

[1] Fine and coarse grouts are defined in ASTM C476.

[2] For grouting between masonry wythes.

[3] Minimum clear width of grout space and minimum clear grout space dimension are the net dimension of the space determined by subtracting masonry protrusions and the diameters of horizontal bars from the as-designed cross-section of the grout space. Grout type and maximum grout pour height shall be specified based on the minimum clear space.

[4] Area of vertical reinforcement shall not exceed 6 percent of the area of the grout space.

[5] Minimum grout space dimension for AAC masonry units shall be 3 in. (76.2 mm) x 3 in. (76.2 mm) or a 3-in. (76.2 mm) diameter cell.

COMMENTARY

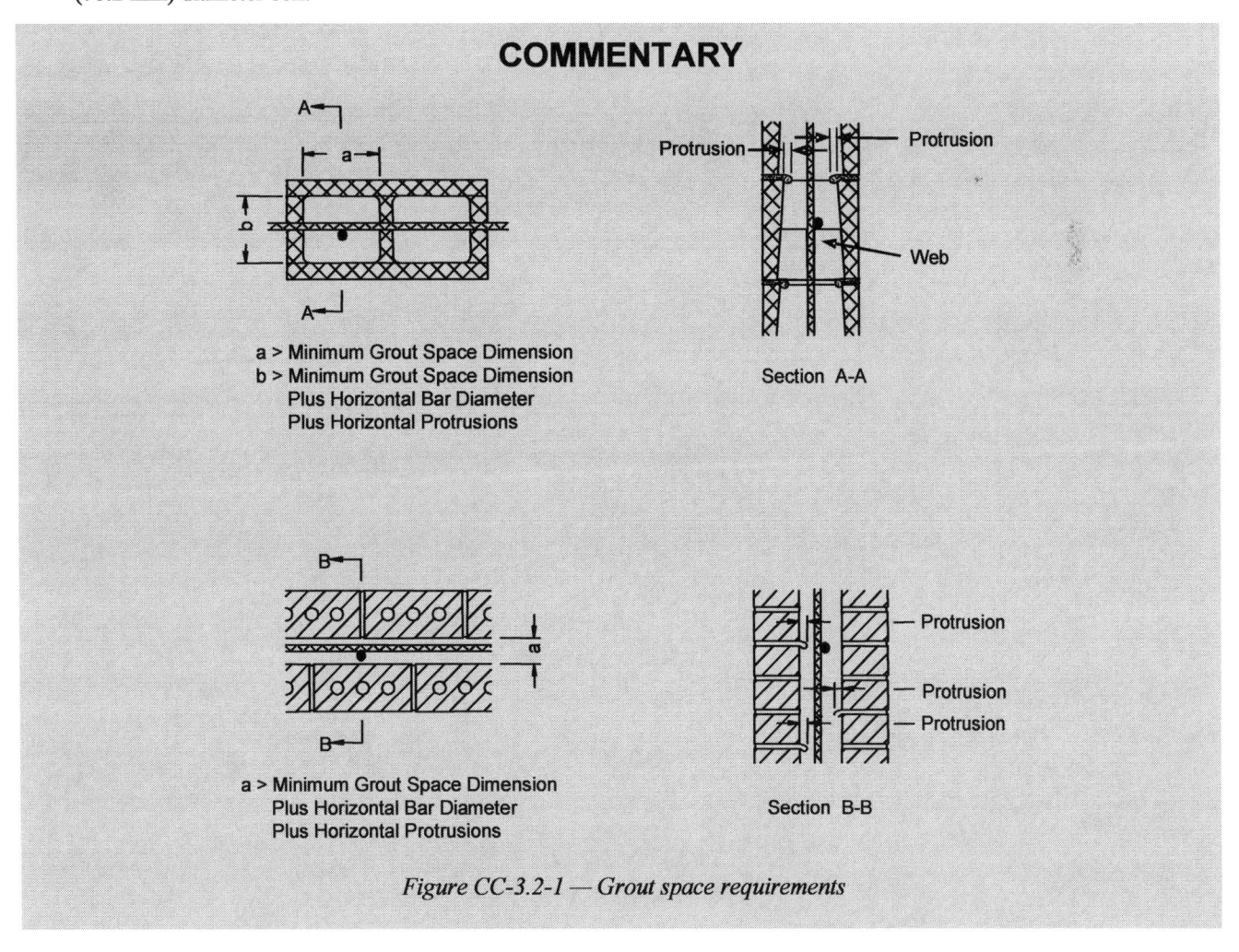

a > Minimum Grout Space Dimension
b > Minimum Grout Space Dimension Plus Horizontal Bar Diameter Plus Horizontal Protrusions

Section A-A

a > Minimum Grout Space Dimension Plus Horizontal Bar Diameter Plus Horizontal Protrusions

Section B-B

Figure CC-3.2-1 — Grout space requirements

This page intentionally left blank

PART 2: DESIGN REQUIREMENTS

CHAPTER 4
GENERAL ANALYSIS AND DESIGN CONSIDERATIONS

CODE

4.1 — Loading

4.1.1 *General*
Masonry shall be designed to resist applicable loads. A continuous load path or paths, with adequate strength and stiffness, shall be provided to transfer forces from the point of application to the final point of resistance.

4.1.2 *Load provisions*
Design loads shall be in accordance with the legally adopted building code of which this Code forms a part, with such live load reductions as are permitted in the legally adopted building code. In the absence of a legally adopted building code, or in the absence of design loads in the legally adopted building code, the load provisions of ASCE 7 shall be used, except as noted in this Code.

4.1.3 *Lateral load resistance*
Buildings shall be provided with a structural system designed to resist wind and earthquake loads and to accommodate the effect of the resulting deformations.

4.1.4 *Load transfer at horizontal connections*
4.1.4.1 Walls, columns, and pilasters shall be designed to resist loads, moments, and shears applied at intersections with horizontal members.

4.1.4.2 Effect of lateral deflection and translation of members providing lateral support shall be considered.

4.1.4.3 Devices used for transferring lateral support from members that intersect walls, columns, or pilasters shall be designed to resist the forces involved.

COMMENTARY

4.1 — Loading
These provisions establish design load requirements. If the design loads specified by the legally adopted building code differ from those of ASCE 7, the legally adopted building code governs. The designer may decide to use the more stringent requirements.

4.1.3 *Lateral load resistance*
Lateral load resistance must be provided by a braced structural system. Interior walls, infill panels, and similar elements may not be a part of the lateral-force-resisting system if isolated. However, when they resist lateral forces due to their rigidity, they should be considered in analysis.

4.1.4 *Load transfer at horizontal connections*
Masonry walls, pilasters, and columns may be connected to horizontal elements of the structure and may rely on the latter for lateral support and stability. The mechanism through which the interconnecting forces are transmitted may involve bond, mechanical anchorage, friction, bearing, or a combination thereof. The designer must assure that, regardless of the type of connection, the interacting forces are safely resisted.

In flexible frame construction, the relative movement (drift) between floors may generate forces within the members and the connections. This Code requires the effects of these movements to be considered in design.

CODE

4.1.5 *Other effects*

Consideration shall be given to effects of forces and deformations due to prestressing, vibrations, impact, shrinkage, expansion, temperature changes, creep, unequal settlement of supports, and differential movement.

4.1.6 *Lateral load distribution*

Lateral loads shall be distributed to the structural system in accordance with member stiffnesses and shall comply with the requirements of this section.

4.1.6.1 Flanges of intersecting walls designed in accordance with Section 5.1.1.2 shall be included in stiffness determination.

4.1.6.2 Distribution of load shall be consistent with the forces resisted by foundations.

4.1.6.3 Distribution of load shall include the effect of horizontal torsion of the structure due to eccentricity of wind or seismic loads resulting from the non-uniform distribution of mass.

COMMENTARY

4.1.5 *Other effects*

Service loads are not the sole source of stresses. The structure may also resist forces from the sources listed. The nature and extent of some of these forces may be greatly influenced by the choice of materials, structural connections, and geometric configuration.

4.1.6 *Lateral load distribution*

The design assumptions for masonry buildings include the use of a lateral-force-resisting system. The distribution of lateral loads to the members of the lateral-force-resisting system is a function of the rigidities of the structural system and of the horizontal diaphragms. The method of connection at intersecting walls and between walls and floor and roof diaphragms determines if the wall participates in the lateral-force-resisting system. Lateral loads from wind and seismic forces are normally considered to act in the direction of the principal axes of the structure. Lateral loads may cause forces in walls both perpendicular and parallel to the direction of the load. Horizontal torsion can be developed due to eccentricity of the applied load with respect to the center of rigidity.

The analysis of lateral load distribution should be in accordance with accepted engineering procedures. The analysis should rationally consider the effects of openings in shear walls and whether the masonry above the openings allows them to act as coupled shear walls. See Figure CC-4.1-1. The interaction of coupled shear walls is complex and further information may be obtained from ASCE, 1978.

Calculation of the stiffness of shear walls should consider shearing and flexural deformations. A guide for solid shear walls (that is, with no openings) is given in Figure CC-4.1-2. For nongrouted hollow unit shear walls, the use of equivalent solid thickness of wall in calculating web stiffness is acceptable.

BUILDING CODE REQUIREMENTS FOR MASONRY STRUCTURES AND COMMENTARY

COMMENTARY

Elevation of Coupled Shear Wall

Elevation of Noncoupled Shear Wall

Figure CC-4.1-1 — Coupled and noncoupled shear walls

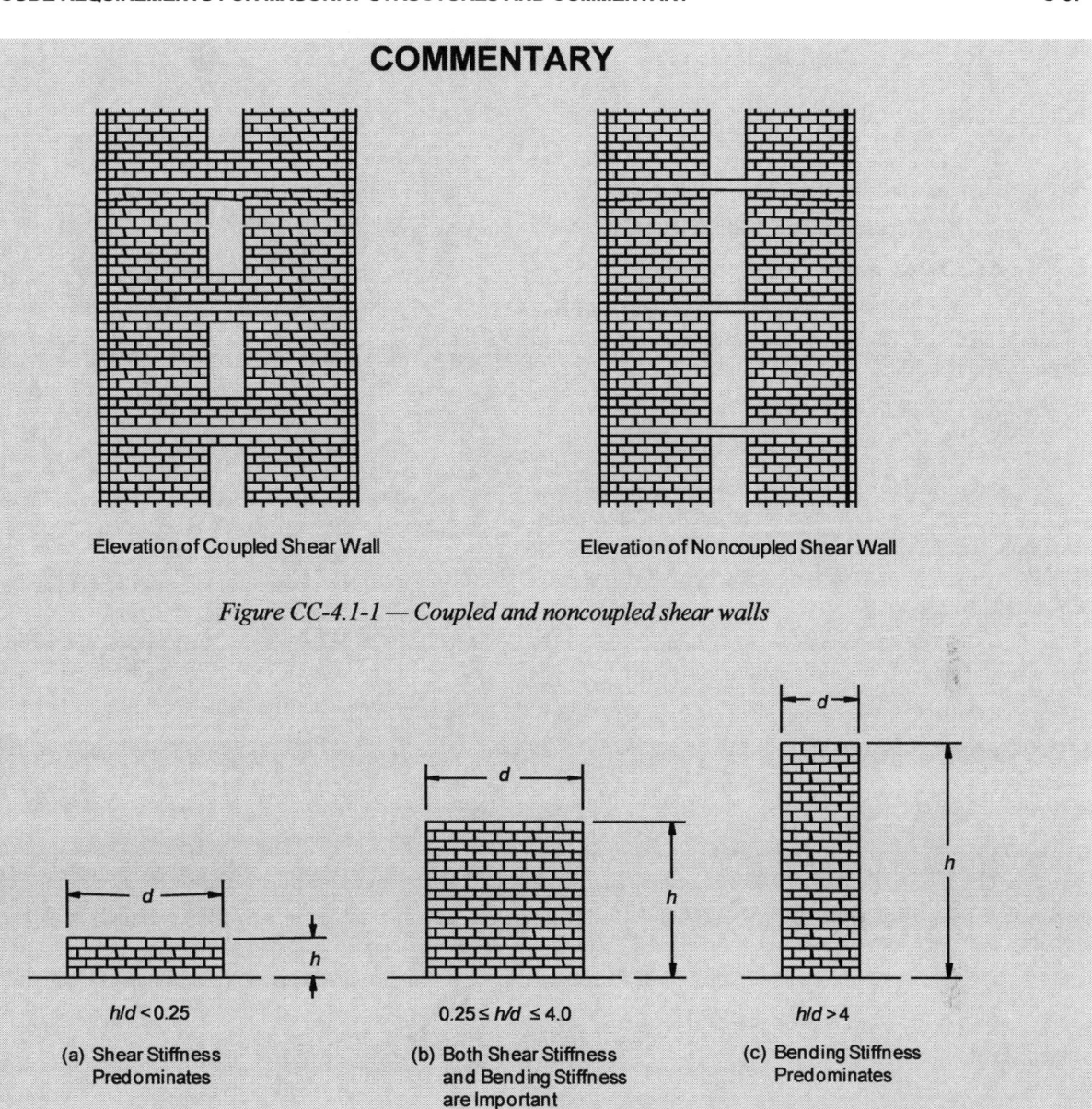

(a) Shear Stiffness Predominates — $h/d < 0.25$

(b) Both Shear Stiffness and Bending Stiffness are Important — $0.25 \leq h/d \leq 4.0$

(c) Bending Stiffness Predominates — $h/d > 4$

Figure CC-4.1-2 — Shear wall stiffness

CODE

4.2 — Material properties

4.2.1 *General*

Unless otherwise determined by test, the following moduli and coefficients shall be used in determining the effects of elasticity, temperature, moisture expansion, shrinkage, and creep.

COMMENTARY

4.2 — Material properties

4.2.1 *General*

Proper evaluation of the building material movement from all sources is an important element of masonry design. Clay masonry and concrete masonry may behave quite differently under normal loading and weather conditions. The Committee has extensively studied available research information in the development of these material properties. The designer is encouraged to review industry standards for further design information and movement joint locations. Material properties can be determined by appropriate tests of the materials to be used.

CODE

4.2.2 *Elastic moduli*

4.2.2.1 *Steel reinforcement* — Modulus of elasticity of steel reinforcement shall be taken as:

$E_s = 29{,}000{,}000$ psi (200,000 MPa)

4.2.2.2 *Clay and concrete masonry*

4.2.2.2.1 The design of clay and concrete masonry shall be based on the following modulus of elasticity values:

$E_m = 700 f'_m$ for clay masonry;

$E_m = 900 f'_m$ for concrete masonry;

or the chord modulus of elasticity taken between 0.05 and 0.33 of the maximum compressive strength of each prism determined by test in accordance with the prism test method, Article 1.4 B.3 of TMS 602/ACI 530.1/ASCE 6, and ASTM E111.

4.2.2.2.2 Modulus of rigidity of clay masonry and concrete masonry shall be taken as:

$E_v = 0.4 E_m$

4.2.2.3 *AAC masonry*

4.2.2.3.1 Modulus of elasticity of AAC masonry shall be taken as:

$E_{AAC} = 6500 (f'_{AAC})^{0.6}$

4.2.2.3.2 Modulus of rigidity of AAC masonry shall be taken as:

$E_v = 0.4 E_{AAC}$

4.2.2.4 *Grout* — Modulus of elasticity of grout shall be taken as $500 f'_g$.

COMMENTARY

4.2.2 *Elastic moduli*

Modulus of elasticity for clay and concrete masonry has traditionally been taken as $1000 f'_m$ in previous masonry codes. Research (Wolde-Tinsae et al, 1993 and Colville et al, 1993) has indicated, however, that there is a large variation in the relationship of elastic modulus versus compressive strength of masonry, and that lower values may be more typical. However, differences in procedures between one research investigation and another may account for much of the indicated variation. Furthermore, the type of elastic moduli being reported (for example, secant modulus, tangent modulus, or chord modulus) is not always identified. The committee decided the most appropriate elastic modulus for allowable-stress design purposes is the slope of the stress-strain curve below a stress value of $0.33 f'_m$. The value of $0.33 f'_m$ was originally chosen because it was the allowable compressive stress prior to the 2011 Code. The committee did not see the need to change the modulus with the increase in allowable compressive stress to $0.45 f'_m$ in the 2011 Code because previous code editions also allowed the allowable compressive stress to be increased by one-third for load combinations including wind or seismic loads and the allowable moment capacity using allowable stress design is not significantly affected by the value of the masonry modulus of elasticity. Data at the bottom of the stress strain curve may be questionable due to the seating effect of the specimen during the initial loading phase if measurements are made on the testing machine platens. The committee therefore decided that the most appropriate elastic modulus for design purposes is the chord modulus from a stress value of 5 to 33 percent of the compressive strength of masonry (see Figure CC-4.2-1). The terms chord modulus and secant modulus have been used interchangeably in the past. The chord modulus, as used here, is defined as the slope of a line intersecting the stress-strain curve at two points, neither of which is the origin of the curve.

For clay and concrete masonry, the elastic modulus is determined as a function of masonry compressive strength using the relations developed from an extensive survey of modulus data by Wolde-Tinsae et al (1993) and results of a test program by Colville et al (1993). Code values for E_m are higher than indicated by a best fit of data relating E_m to the compressive strength of masonry. The higher Code values are based on the fact that actual compressive strength significantly exceeds the specified compressive strength of masonry, f'_m, particularly for clay masonry.

By using the Code values, the contribution of each wythe to composite action is more accurately accounted for in design calculations than would be the case if the elastic modulus of each part of a composite wall were based on one specified compressive strength of masonry.

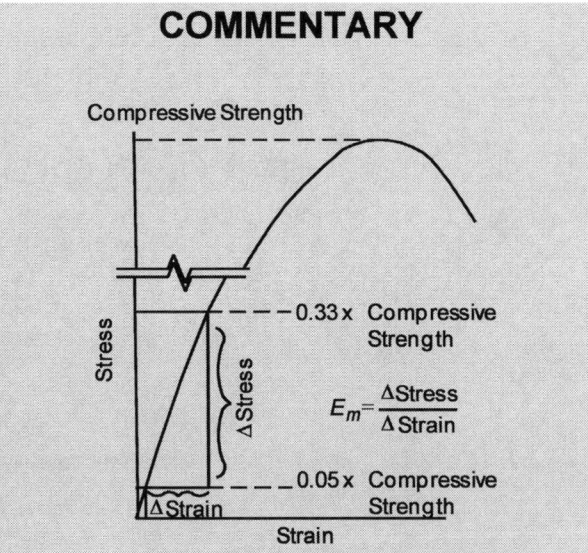

BUILDING CODE REQUIREMENTS FOR MASONRY STRUCTURES AND COMMENTARY

COMMENTARY

Figure CC-4.2-1 — Chord modulus of elasticity

The modulus of elasticity of autoclaved aerated concrete (AAC) masonry depends almost entirely on the modulus of elasticity of the AAC material itself. The relationship between modulus of elasticity and compressive strength is given in Tanner et al, 2005(a) and Argudo, 2003.

The modulus of elasticity of a grouted assemblage of clay or concrete masonry can usually be taken as a factor multiplied by the specified compressive strength, regardless of the extent of grouting, because the modulus of elasticity of the grout is usually close to that of the clay or concrete masonry. However, grout is usually much stiffer than the AAC material. While it is permissible and conservative to calculate the modulus of elasticity of a grouted assemblage of AAC masonry assuming that the modulus of elasticity of the grout is the same as that of the AAC material, it is also possible to recognize the greater modulus of elasticity of the grout by transforming the cross-sectional area of grout into an equivalent cross-sectional area of AAC, using the modular ratio between the two materials.

Because the inelastic stress-strain behavior of grout is generally similar to that of clay or concrete masonry, calculations of element resistance (whether based on allowable-stress or strength design) usually neglect possible differences in strength between grout and the surrounding masonry. For the same reasons noted above, the stress-strain behavior of grout usually differs considerably from that of the surrounding AAC material. It is possible that these differences in stress-strain behavior could also be considered in calculating element resistances. Research is ongoing to resolve this issue.

The relationship between the modulus of rigidity and the modulus of elasticity has historically been given as $0.4\ E_m$. No experimental evidence exists to support this relationship.

CODE

4.2.3 *Coefficients of thermal expansion*

4.2.3.1 *Clay masonry*
$k_t = 4 \times 10^{-6}$ in./in./°F (7.2×10^{-6} mm/mm/°C)

4.2.3.2 *Concrete masonry*
$k_t = 4.5 \times 10^{-6}$ in./in./°F (8.1×10^{-6} mm/mm/°C)

4.2.3.3 *AAC masonry*
$k_t = 4.5 \times 10^{-6}$ in./in./°F (8.1×10^{-6} mm/mm/°C)

4.2.4 *Coefficient of moisture expansion for clay masonry*
$k_e = 3 \times 10^{-4}$ in./in. (3×10^{-4} mm/mm)

4.2.5 *Coefficients of shrinkage*

4.2.5.1 *Concrete masonry*
$k_m = 0.5 \, s_l$

4.2.5.2 *AAC masonry*
$k_m = 0.8 \, \varepsilon_{cs}/100$

where ε_{cs} is determined in accordance with ASTM C1693.

4.2.6 *Coefficients of creep*

4.2.6.1 *Clay masonry*
$k_c = 0.7 \times 10^{-7}$, per psi (0.1×10^{-4}, per MPa)

4.2.6.2 *Concrete masonry*
$k_c = 2.5 \times 10^{-7}$, per psi (0.36×10^{-4}, per MPa)

4.2.6.3 *AAC masonry*
$k_c = 5.0 \times 10^{-7}$, per psi (0.72×10^{-4}, per MPa)

COMMENTARY

4.2.3 *Coefficients of thermal expansion*
Temperature changes cause material expansion and contraction. This material movement is theoretically reversible. These thermal expansion coefficients are slightly higher than mean values for the assemblage (Copeland, 1957; Plummer, 1962; Grimm, 1986).

Thermal expansion for concrete masonry varies with aggregate type (Copeland, 1957; Kalouseb, 1954).

Thermal expansion coefficients are given for AAC masonry in RILEM (1993).

4.2.4 *Coefficient of moisture expansion for clay masonry*
Fired clay products expand upon contact with moisture and the material does not return to its original size upon drying (Plummer, 1962; Grimm, 1986). This is a long-term expansion as clay particles react with atmospheric moisture. Continued moisture expansion of clay masonry units has been reported for 7½ years (Smith, 1973). Moisture expansion is not a design consideration for concrete masonry.

4.2.5 *Coefficients of shrinkage*

4.2.5.1 *Concrete masonry* — Concrete masonry is a cement-based material that shrinks due to moisture loss and carbonation (Kalouseb, 1954). The total linear drying shrinkage is determined in accordance with ASTM C426. The maximum shrinkage allowed by ASTM specifications for concrete masonry units (for example, ASTM C90), other than calcium silicate units, is 0.065%. Further design guidance for estimating the shrinkage due to moisture loss and carbonation is available (NCMA TEK 10-1A, 2005; NCMA TEK 10-2C, 2010; NCMA TEK 10-3, 2003, NCMA TEK18-02B, 2012). The shrinkage of clay masonry is negligible.

4.2.5.2 *AAC masonry* — At time of production, AAC masonry typically has a moisture content of about 30%. That value typically decreases to 15% or less within two to three months, regardless of ambient relative humidity. This process can take place during construction or prior to delivery. ASTM C1693 evaluates AAC material characteristics at moisture contents between 5% and 15%, a range that typifies AAC in service. The shrinkage coefficient of this section reflects the change in strain likely to be encountered within the extremes of moisture content typically encountered in service.

4.2.6 *Coefficients of creep*
When continuously stressed, these materials gradually deform in the direction of stress application. This movement is referred to as creep and is load and time dependent (Kalouseb, 1954; Lenczner and Salahuddin, 1976; RILEM, 1993). The values given are maximum values.

CODE

4.2.7 Prestressing steel

Modulus of elasticity of prestressing steel shall be determined by tests. For prestressing steels not specifically listed in ASTM A416/A416M, A421/A421M, or A722/A722M, tensile strength and relaxation losses shall be determined by tests.

4.3 — Section properties

4.3.1 Stress calculations

4.3.1.1 Members shall be designed using section properties based on the minimum net cross-sectional area of the member under consideration. Section properties shall be based on specified dimensions.

4.3.1.2 In members designed for composite action, stresses shall be calculated using section properties based on the minimum transformed net cross-sectional area of the composite member. The transformed area concept for elastic analysis, in which areas of dissimilar materials are transformed in accordance with relative elastic moduli ratios, shall apply.

COMMENTARY

4.2.7 Prestressing steel

The material and section properties of prestressing steels may vary with each manufacturer. Most significant for design are the prestressing tendon's cross section, modulus of elasticity, tensile strength, and stress-relaxation properties. Values for these properties for various manufacturers' wire, strand, and bar systems are given elsewhere (PTI, 2006). The modulus of elasticity of prestressing steel is often taken equal to 28,000 ksi (193,000 MPa) for design, but can vary and should be verified by the manufacturer. Stress-strain characteristics and stress-relaxation properties of prestressing steels must be determined by test, because these properties may vary between different steel forms (bar, wire, or strand) and types (mild, high strength, or stainless).

4.3 — Section properties

4.3.1 Stress calculations

Minimum net section is often difficult to establish in hollow unit masonry. The designer may choose to use the minimum thickness of the face shells of the units as the minimum net section. The minimum net section may not be the same in the vertical and horizontal directions.

For masonry of hollow units, the minimum cross-sectional area in both directions may conservatively be based on the minimum face-shell thickness (NCMA TEK 14-1B, 2007).

Solid clay masonry units are permitted to have coring up to a maximum of 25 percent of their gross cross-sectional area. For such units, the net cross-sectional area may be taken as equal to the gross cross-sectional area, except as provided in Section 8.1.4.2(c) for masonry headers. Several conditions of net area are shown in Figure CC-4.3-1.

Because the elastic properties of the materials used in members designed for composite action differ, equal strains produce different levels of stresses in the components. To calculate these stresses, a convenient transformed section with respect to the axis of resistance is considered. The resulting stresses developed in each fiber are related to the actual stresses by the ratio E_1/E_x between the modulus of elasticity, E_1, of the most deformable material in the member and the modulus of elasticity, E_x, of the materials in the fiber considered. Thus, to obtain the transformed section, fibers of the actual section are conceptually widened by the ratio E_x/E_1. Stresses calculated based on the section properties of the transformed section, with respect to the axis of resistance considered, are then multiplied by E_x/E_1 to obtain actual stresses.

COMMENTARY

Brick More than 75% Solid
Net Area Equals Gross Area

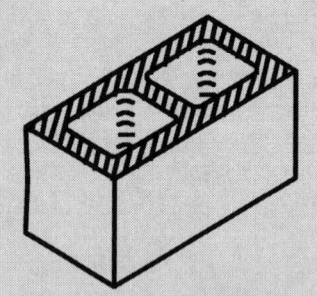

Hollow Unit Full Mortar Bedding
(Requires Alignment of Crosswebs)

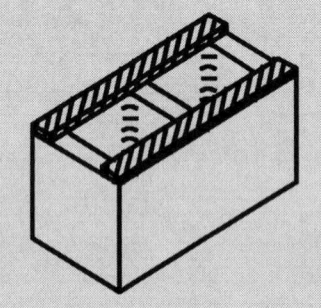

Hollow Unit Face Shell Mortar Bedding

Figure CC-4.3-1 — Net cross-sectional areas

CODE

4.3.2 *Stiffness*

Calculation of stiffness based on uncracked section is permissible. Use of the average net cross-sectional area of the member considered in stiffness calculations is permitted.

COMMENTARY

4.3.2 *Stiffness*

Stiffness is a function of the extent of cracking. Because unreinforced masonry is designed assuming it is uncracked, Code equations for design of unreinforced masonry are based on the member's uncracked moment of inertia and ignoring the effects of reinforcement, if present. Also, because the extent of tension cracking in shear walls is not known in advance, this Code allows the determination of stiffness to be based on uncracked section properties. For reinforced masonry, more accurate estimates may result if stiffness approximations are based on the cracked section.

The section properties of masonry members may vary from point to point. For example, in a single-wythe concrete masonry wall made of hollow ungrouted units, the cross-sectional area varies through the unit height. Also, the distribution of material varies along the length of the wall or unit. For stiffness calculations, an average value of the appropriate section property (cross-sectional area or moment of inertia) is considered adequate for design. The average net cross-sectional area of the member would in turn be based on average net cross-sectional area values of the masonry units and the mortar joints composing the member.

CODE

4.3.3 *Radius of gyration*
Radius of gyration shall be calculated using the average net cross-sectional area of the member considered.

4.3.4 *Bearing area*
The bearing area, A_{br}, for concentrated loads shall not exceed the following:

(a) $A_1\sqrt{A_2/A_1}$

(b) $2A_1$

The area, A_2, is the area of the lower base of the largest frustum of a right pyramid or cone that has the loaded area, A_1, as its upper base, slopes at 45 degrees from the horizontal, and is wholly contained within the support. For walls not laid in running bond, area A_2 shall terminate at head joints.

COMMENTARY

4.3.3 *Radius of gyration*
The radius of gyration is the square root of the ratio of bending moment of inertia to cross-sectional area. Because stiffness is based on the average net cross-sectional area of the member considered, this same area should be used in the calculation of radius of gyration.

4.3.4 *Bearing area*
When the supporting masonry area, A_2, is larger on all sides than the loaded area, A_1, this Code allows distribution of concentrated loads over a bearing area A_{br} larger than A_1. The area A_2 is determined as illustrated in Figure CC-4.3-2. This is permissible because the confinement of the bearing area by surrounding masonry increases the bearing capacity of the masonry under the concentrated loads. When the edge of the loaded area, A_1, coincides with the face or edge of the masonry, the area A_2 is equal to the loaded area A_1.

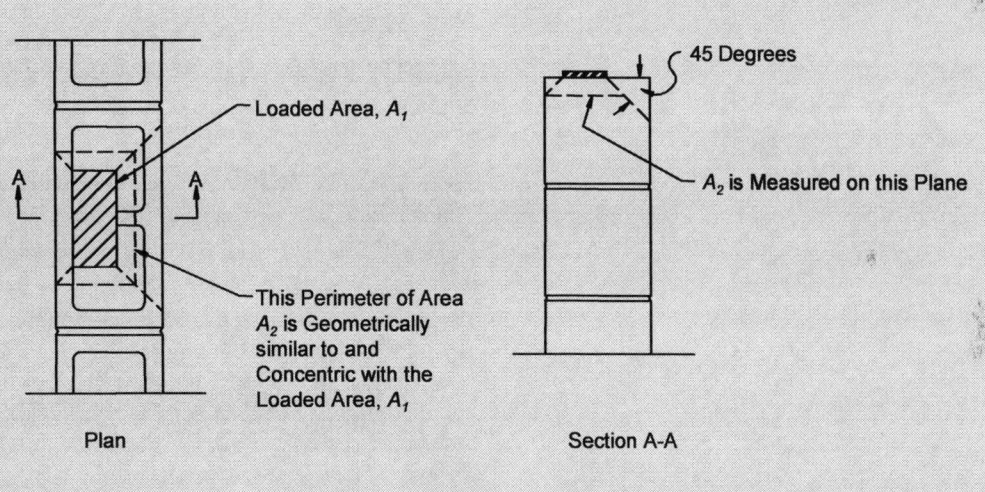

Figure CC-4.3-2 — Bearing areas

CODE

4.4 — Connection to structural frames

Masonry walls shall not be connected to structural frames unless the connections and walls are designed to resist design interconnecting forces and to accommodate calculated deflections.

COMMENTARY

4.4 — Connection to structural frames

Exterior masonry walls connected to structural frames are used primarily as non-load-bearing curtain walls. Regardless of the structural system used for support, there are differential movements between the structure and the wall. These differential movements may occur separately or in combination and may be due to the following:

1) Temperature increase or decrease of either the structural frame or the masonry wall.

2) Moisture and freezing expansion of brick or shrinkage of concrete block walls.

3) Elastic shortening of columns from axial loads, shrinkage, or creep.

4) Deflection of supporting beams.

5) Sidesway in multiple-story buildings.

6) Foundation movement.

Because the tensile strength of masonry is low, these differential movements must be accommodated by sufficient clearance between the frame and masonry and flexible or slip-type connections.

Structural frames and bracing should not be infilled with masonry to increase resistance to in-plane lateral forces without considering the differential movements listed above.

Wood, steel, or concrete columns may be surrounded by masonry serving as a decorative element. Masonry walls may be subject to forces as a result of their interaction with other structural components. Because the masonry element is often much stiffer, the load will be resisted primarily by the masonry. These forces, if transmitted to the surrounding masonry, should not exceed the allowable stresses of the masonry. Alternately, there should be sufficient clearance between the frame and masonry. Flexible ties should be used to allow for the deformations.

Beams or trusses supporting masonry walls are essentially embedded, and their deflections should be limited to the allowable deflections for the masonry being supported. See Section 5.2.1.4 for requirements.

BUILDING CODE REQUIREMENTS FOR MASONRY STRUCTURES AND COMMENTARY

CODE

4.5 — Masonry not laid in running bond

For masonry not laid in running bond, the minimum area of horizontal reinforcement shall be 0.00028 multiplied by the gross vertical cross-sectional area of the wall using specified dimensions. Horizontal reinforcement shall be placed at a maximum spacing of 48 in. (1219 mm) on center in horizontal mortar joints or in bond beams.

COMMENTARY

4.5 — Masonry not laid in running bond

The requirements for masonry laid in running bond are shown in Figure CC-4.5-1. The amount of horizontal reinforcement required in masonry not laid in running bond is a prescriptive amount to provide continuity across the head joints. Because lateral loads are reversible, reinforcement should either be centered in the element thickness by placement in the center of a bond beam, or should be symmetrically located by placing multiple bars in a bond beam or by using joint reinforcement in the mortar bed along each face shell. This reinforcement can be also used to resist load.

Although continuity across head joints in masonry not laid in running bond is a concern for AAC masonry as well as masonry of clay or concrete, the use of horizontal reinforcement to enhance continuity in AAC masonry is generally practical only by the use of bond beams.

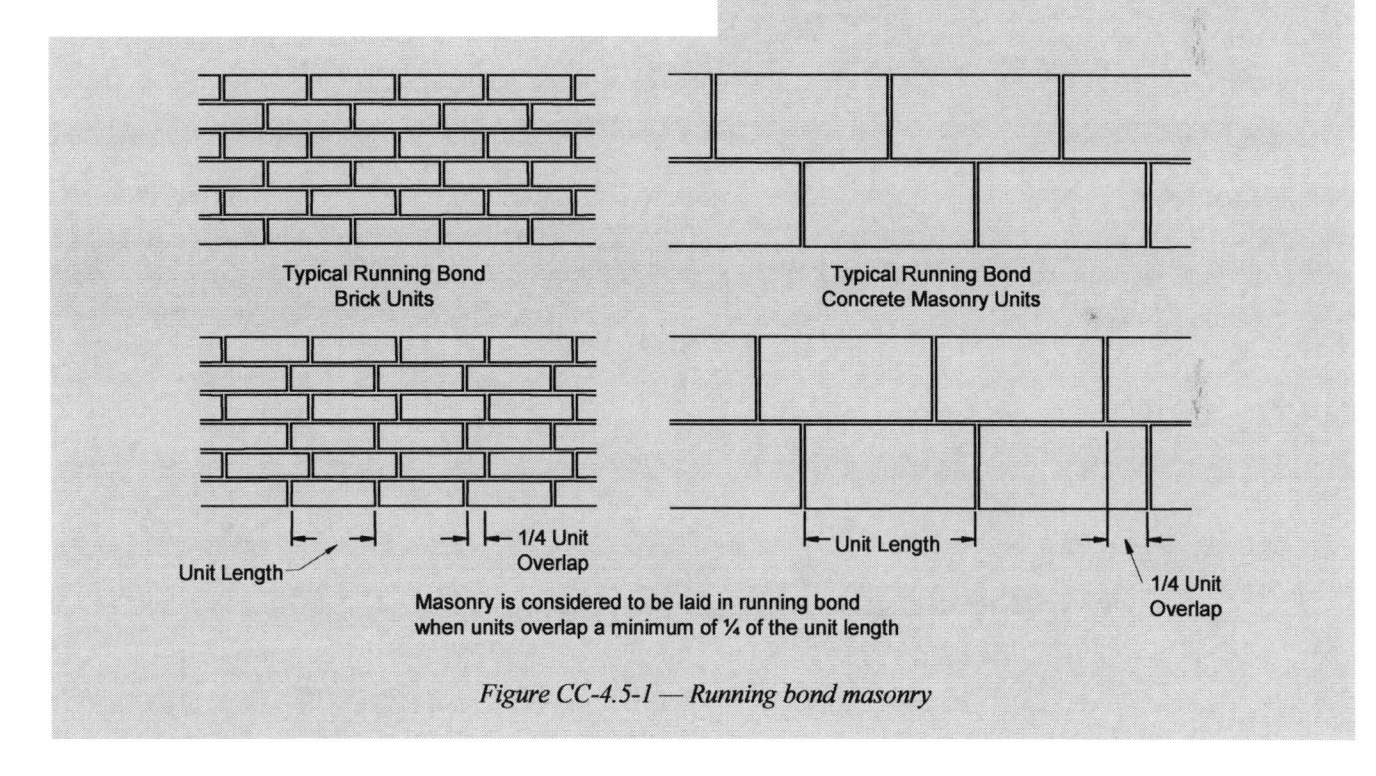

Figure CC-4.5-1 — Running bond masonry

This page intentionally left blank

CHAPTER 5
STRUCTURAL ELEMENTS

CODE

5.1 — Masonry assemblies

5.1.1 *Intersecting walls*

5.1.1.1 Wall intersections shall meet one of the following requirements:

(a) Design shall conform to the provisions of Section 5.1.1.2.

(b) Transfer of shear between walls shall be prevented.

5.1.1.2 *Design of wall intersection*

5.1.1.2.1 Masonry shall be in running bond.

5.1.1.2.2 Flanges shall be considered effective in resisting applied loads.

5.1.1.2.3 The width of flange considered effective on each side of the web shall be the smaller of the actual flange on either side of the web wall or the following:

(a) 6 multiplied by the nominal flange thickness for unreinforced and reinforced masonry, when the flange is in compression

(b) 6 multiplied by the nominal flange thickness for unreinforced masonry, when the flange is in flexural tension

(c) 0.75 multiplied by the floor-to-floor wall height for reinforced masonry, when the flange is in flexural tension.

The effective flange width shall not extend past a movement joint.

5.1.1.2.4 Design for shear, including the transfer of shear at interfaces, shall conform to the requirements of Section 8.2.6; or Section 8.3.5; or Section 9.2.6; or Section 9.3.4.1.2; or Section 10.6; or Section 11.3.4.1.2.

5.1.1.2.5 The connection of intersecting walls shall conform to one of the following requirements:

(a) At least fifty percent of the masonry units at the interface shall interlock.

(b) Walls shall be anchored by steel connectors grouted into the wall and meeting the following requirements:

(1) Minimum size: $1/4$ in. x $1^1/_2$ in. x 28 in. (6.4 mm x 38.1 mm x 711 mm) including 2-in. (50.8-mm) long, 90-degree bend at each end to form a U or Z shape.

(2) Maximum spacing: 48 in. (1219 mm).

(c) Intersecting reinforced bond beams shall be provided at a maximum spacing of 48 in. (1219 mm) on center. The area of reinforcement in each bond beam

COMMENTARY

5.1 — Masonry assemblies

5.1.1 *Intersecting walls*

Connections of webs to flanges of walls may be accomplished by running bond, metal connectors, or bond beams. Achieving stress transfer at a T intersection with running bond only is difficult. A running bond connection is shown in Figure CC-5.1-1 with a "T" geometry over their intersection.

The alternate method, using metal strap connectors, is shown in Figure CC-5.1-2. Bond beams, shown in Figure CC-5.1-3, are the third means of connecting webs to flanges.

When the flanges are connected at the intersection, they are required to be included in the design.

The effective width of the flange for compression and unreinforced masonry in flexural tension is based on shear-lag effects and is a traditional requirement. The effective width of the flange for reinforced masonry in flexural tension is based on the experimental and analytical work of He and Priestley (1992). They showed that the shear-lag effects are significant for uncracked walls, but become less severe after cracking. He and Priestley (1992) proposed that the effective width of the flange be determined as:

$$l_{ef} = \begin{cases} l_f & l_f/h \leq 1.5 \\ 0.75h + 0.5l_f & 1.5 < l_f/h \leq 3.5 \\ 2.5h & l_f/h > 3.5 \end{cases}$$

where l_{ef} is the effective flange width, l_f is the width of the flange, and h is height of the wall. These equations can result in effective flange widths greater than 1.5 times the height of the wall. However, a limit of the effective flange width of 1.5 times the wall height, or ¾ of the wall height on either side of the web, is provided in the code. This limit was chosen because the testing by He and Priestley (1992) was limited to a flange width of 1.4 times the wall height. Designers are cautioned that longitudinal reinforcement just outside the effective flange width specified by the code can affect the ductility and behavior of the wall. Any participation by the reinforcement in resisting the load can lead to other, more brittle, failure modes such as shear or crushing of the compression toe.

CODE

shall not be less than 0.1 in.² per ft (211 mm²/m) multiplied by the vertical spacing of the bond beams in feet (meters). Reinforcement shall be developed on each side of the intersection.

COMMENTARY

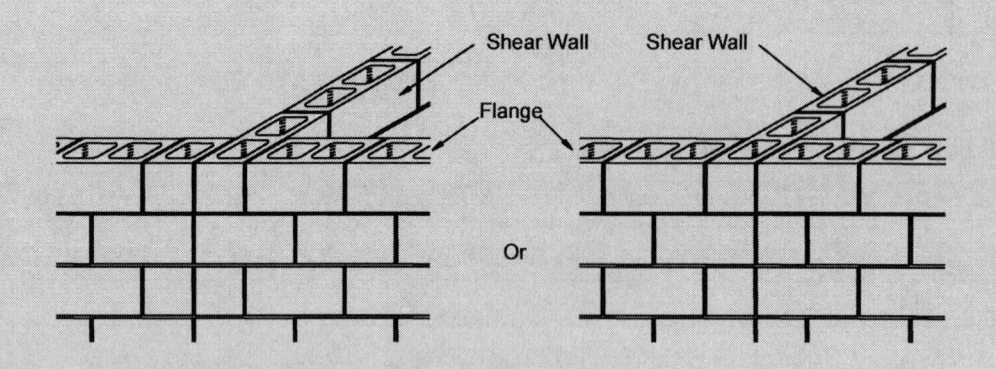

Figure CC-5.1-1 — Running bond lap at intersection

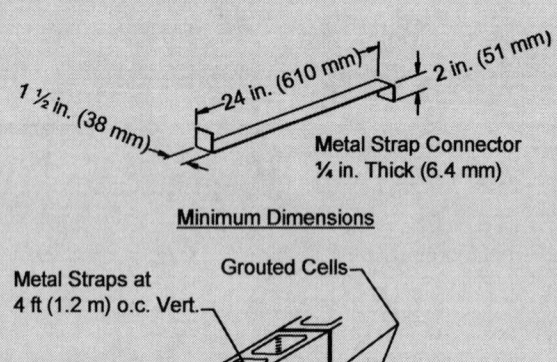

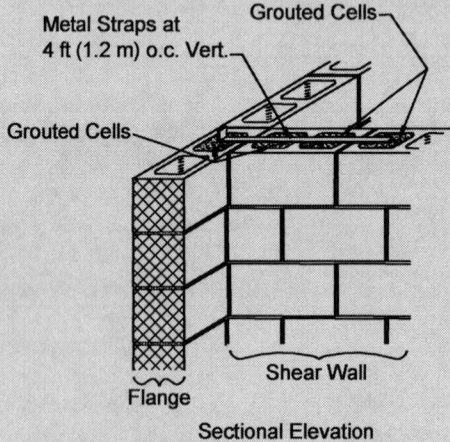

Figure CC-5.1-2 — Metal straps and grouting at wall intersections

COMMENTARY

Figure CC-5.1-3 — Bond beam at wall intersection

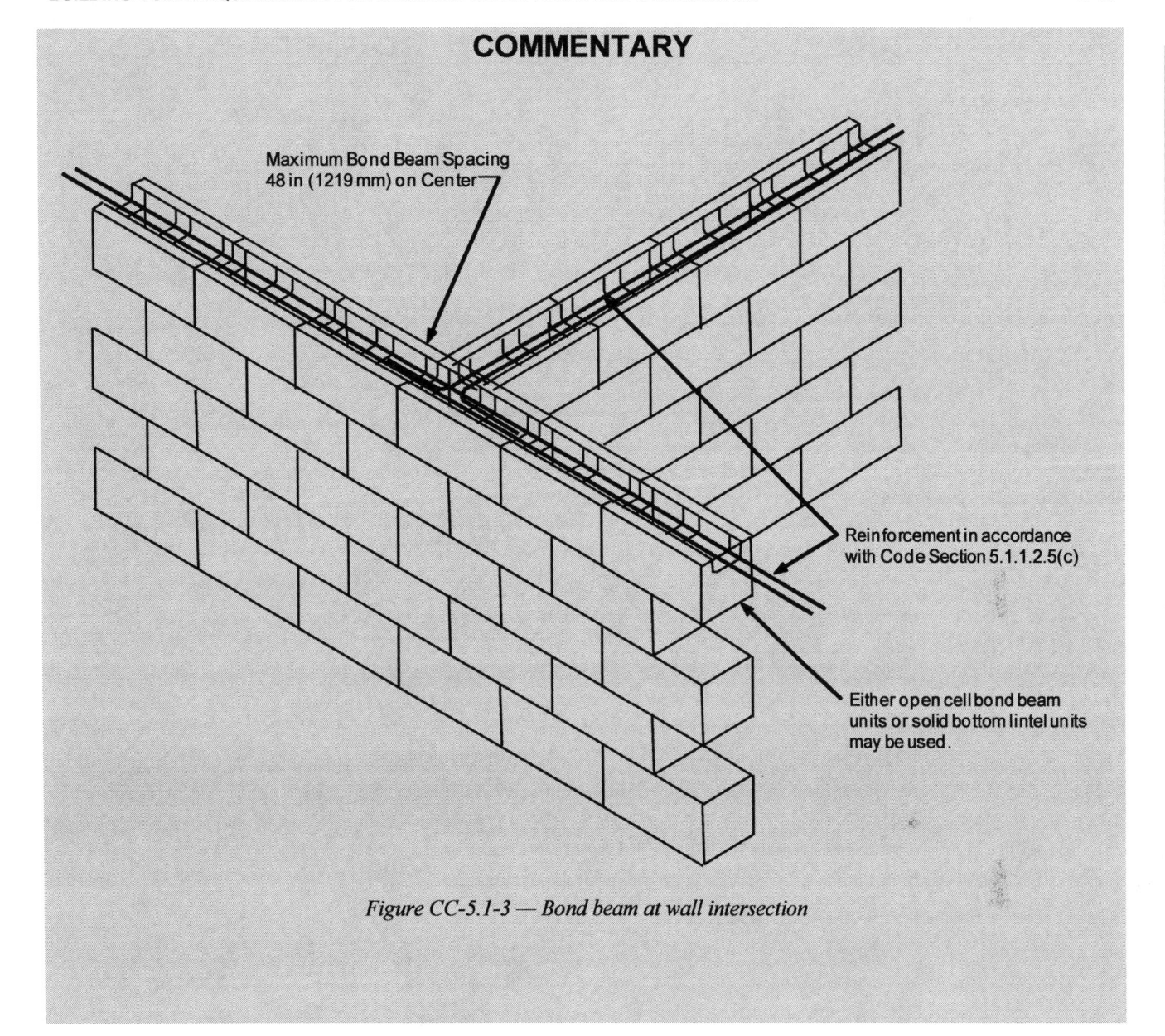

CODE

5.1.2 *Effective compressive width per bar*

5.1.2.1 For masonry not laid in running bond and having bond beams spaced not more than 48 in. (1219 mm) center-to-center, and for masonry laid in running bond, the width of the compression area used to calculate element capacity shall not exceed the least of:

(a) Center-to-center bar spacing.

(b) Six multiplied by the nominal wall thickness.

(c) 72 in. (1829 mm).

5.1.2.2 For masonry not laid in running bond and having bond beams spaced more than 48 in. (1219 mm) center-to-center, the width of the compression area used to calculate element capacity shall not exceed the length of the masonry unit.

COMMENTARY

5.1.2 *Effective compressive width per bar*

The effective width of the compressive area for each reinforcing bar must be established. Figure CC-5.1-4 depicts the limits for the conditions stated. Limited research (Dickey and MacIntosh, 1971) is available on this subject.

The limited ability of head joints to transfer stress when masonry is not laid in running bond is recognized by the requirements for bond beams. Open end masonry units that are fully grouted are assumed to transfer stress as indicated in Section 8.2.6.2(d), as for running bond.

The center-to-center bar spacing maximum is a limit to keep from overlapping areas of compressive stress. The 72-in. (1829-mm) maximum is an empirical choice of the committee.

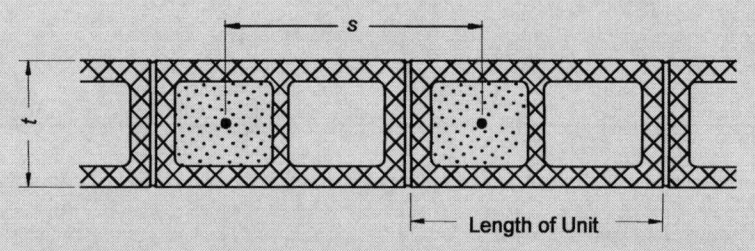

For masonry not laid in running bond with bond beams spaced less than or equal to 48 in. (1219 mm) and running bond masonry, *b* equals the lesser of:
  b = *s*
  b = 6*t*
  b = 72 in. (1829 mm)

For masonry not laid in running bond with bond beams spaced greater than 48 in. (1219 mm), *b* equals the lesser of:
  b = *s*
  b = length of unit

Figure CC-5.1-4 — Width of compression area

CODE

5.1.3 *Concentrated loads*

5.1.3.1 Concentrated loads shall not be distributed over a length greater than the minimum of the following:

(a) The length of bearing area plus the length determined by considering the concentrated load to be dispersed along a 2 vertical: 1 horizontal line. The dispersion shall terminate at half the wall height, a movement joint, the end of the wall, or an opening, whichever provides the smallest length.

(b) The center-to-center distance between concentrated loads.

5.1.3.2 For walls not laid in running bond, concentrated loads shall not be distributed across head joints. Where concentrated loads acting on such walls are applied to a bond beam, the concentrated load is permitted to be distributed through the bond beam, but shall not be distributed across head joints below the bond beams.

COMMENTARY

5.1.3 *Concentrated loads*

Arora (1988) reports the results of tests of a wide variety of specimens under concentrated loads, including AAC masonry, concrete block masonry, and clay brick masonry specimens. Arora (1988) suggests that a concentrated load can be distributed at a 2:1 slope, terminating at half the wall height, where the wall height is from the point of application of the load to the foundation. Tests on the load dispersion through a bond beam on top of hollow masonry reported in Page and Shrive (1987) resulted in an angle from the horizontal of 59° for a 1-course CMU bond beam, 65° for a 2-course CMU bond beam, and 58° for a 2-course clay bond beam, or approximately a 2:1 slope. For simplicity in design, a 2:1 slope is used for all cases of load dispersion of a concentrated load.

Code provisions are illustrated in Figure CC-5.1-5. Figure CC-5.1-5a illustrates the dispersion of a concentrated load through a bond beam. A hollow wall would be checked for bearing under the bond beam using the effective length. Figure CC-5.1-5b illustrates the dispersion of a concentrated load in the wall. The effective length would be used for checking the wall under the axial force. A wall may have to be checked at several locations, such as under a bond beam and at midheight.

COMMENTARY

(a) Distribution of concentrated load through bond beam

(b) Distribution of concentrated load in wall

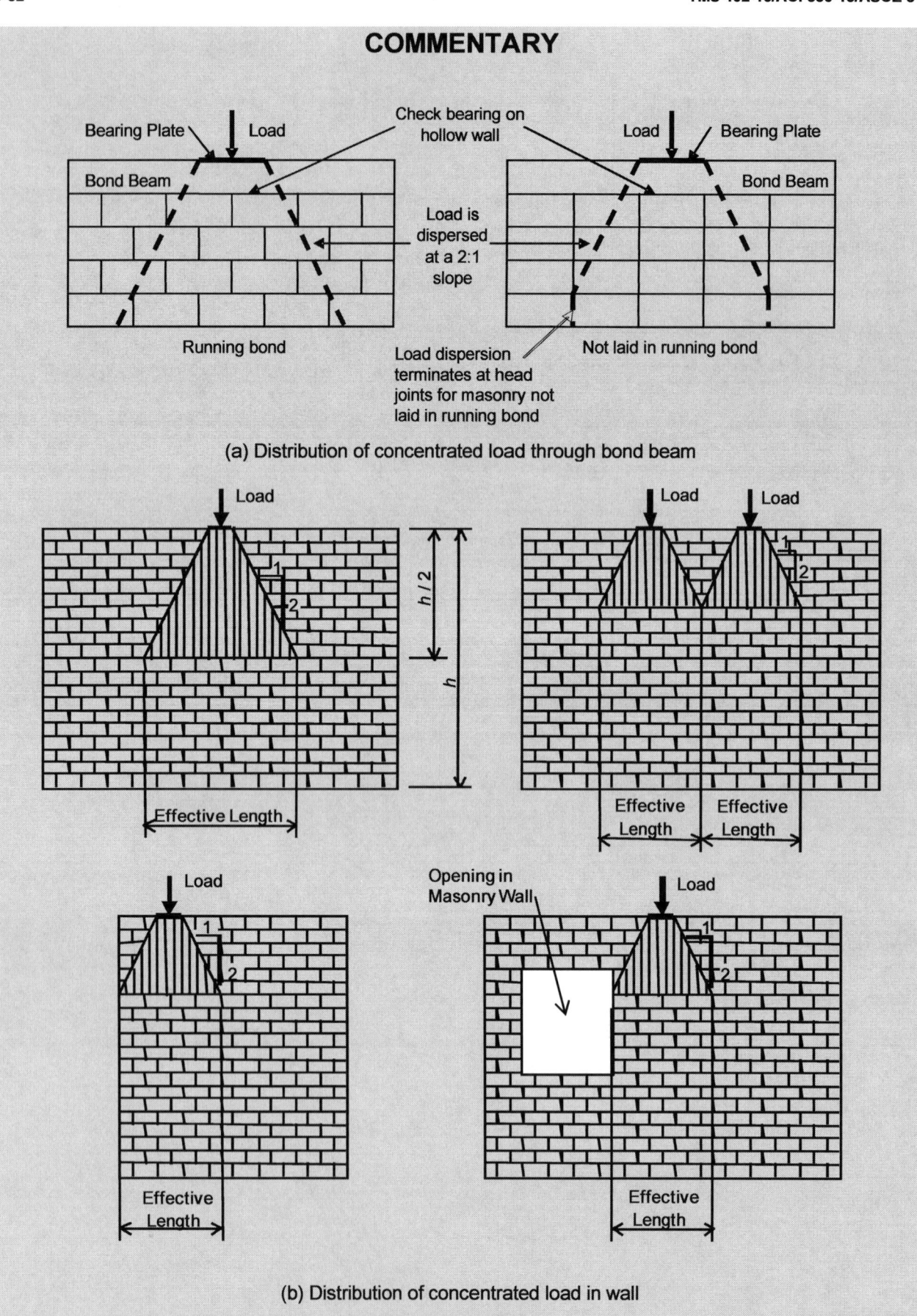

Figure CC-5.1-5. Distribution of concentrated loads

CODE

5.1.4 *Multiwythe masonry elements*
Design of masonry composed of more than one wythe shall comply with the provisions of Section 5.1.4.1, and either 5.1.4.2 or 5.1.4.3.

5.1.4.1 The provisions of Sections 5.1.4.2, and 5.1.4.3 shall not apply to AAC masonry units and glass masonry units.

5.1.4.2 *Composite action*
5.1.4.2.1 Multiwythe masonry designed for composite action shall have collar joints either:

(a) crossed by connecting headers, or

(b) filled with mortar or grout and connected by wall ties.

5.1.4.2.2 Headers used to bond adjacent wythes shall meet the requirements of either Section 8.1.4.2 or Section 9.1.7.2 and shall be provided as follows:

(a) Headers shall be uniformly distributed and the sum of their cross-sectional areas shall be at least 4 percent of the wall surface area.

(b) Headers connecting adjacent wythes shall be embedded a minimum of 3 in. (76.2 mm) in each wythe.

5.1.4.2.3 Wythes not bonded by headers shall meet the requirements of either Section 8.1.4.2 or Section 9.1.7.2 and shall be bonded by non-adjustable ties provided as follows:

Wire size	*Minimum number of ties required*
W1.7 (MW11)	one per $2^2/_3$ ft² (0.25 m²) of masonry surface area
W2.8 (MW18)	one per $4^1/_2$ ft² (0.42 m²) of masonry surface area

The maximum spacing between ties shall be 36 in. (914 mm) horizontally and 24 in. (610 mm) vertically.

The use of rectangular ties to connect masonry wythes of any type of masonry unit shall be permitted. The use of Z ties to connect to a masonry wythe of hollow masonry units shall not be permitted. Cross wires of joint reinforcement shall be permitted to be used instead of ties.

COMMENTARY

5.1.4.2 *Composite action* — Multiwythe masonry acts monolithically if sufficient shear transfer can occur across the interface between the wythes. See Figure CC-5.1-6. Shear transfer is achieved with headers crossing the collar joint or with mortar- or grout-filled collar joints. When mortar- or grout-filled collar joints are relied upon to transfer shear, ties are required to ensure structural integrity of the collar joint. Composite action requires that the shear stresses occurring at the interfaces are within the e limits prescribed.

Composite masonry walls generally consist of brick-to-brick, block-to-block, or brick-to-block wythes. The collar joint thickness ranges from $^3/_8$ to 4 in. (9.5 to 102 mm). The joint may contain either vertical or horizontal reinforcement, or reinforcement may be placed in either the brick or block wythe. Composite masonry is particularly advantageous for resisting high loads, both in-plane and out-of-plane.

5.1.4.2.2 Requirements for masonry headers (Figure CC-A.7-1) are empirical and taken from prior codes. The net area of the header should be used in calculating the stress even if a solid unit, which is allowed to have up to 25 percent coring, is used. Headers do not provide as much ductility as metal tied wythes with filled collar joints. The influence of differential movement is especially critical when headers are used. The committee does not encourage the use of headers.

5.1.4.2.3 The required size, number, and spacing of ties in composite masonry, shown in Figure CC-5.1-7, has been determined from past experience. The limitation of Z-ties to masonry of other than hollow units is also based on past experience.

COMMENTARY

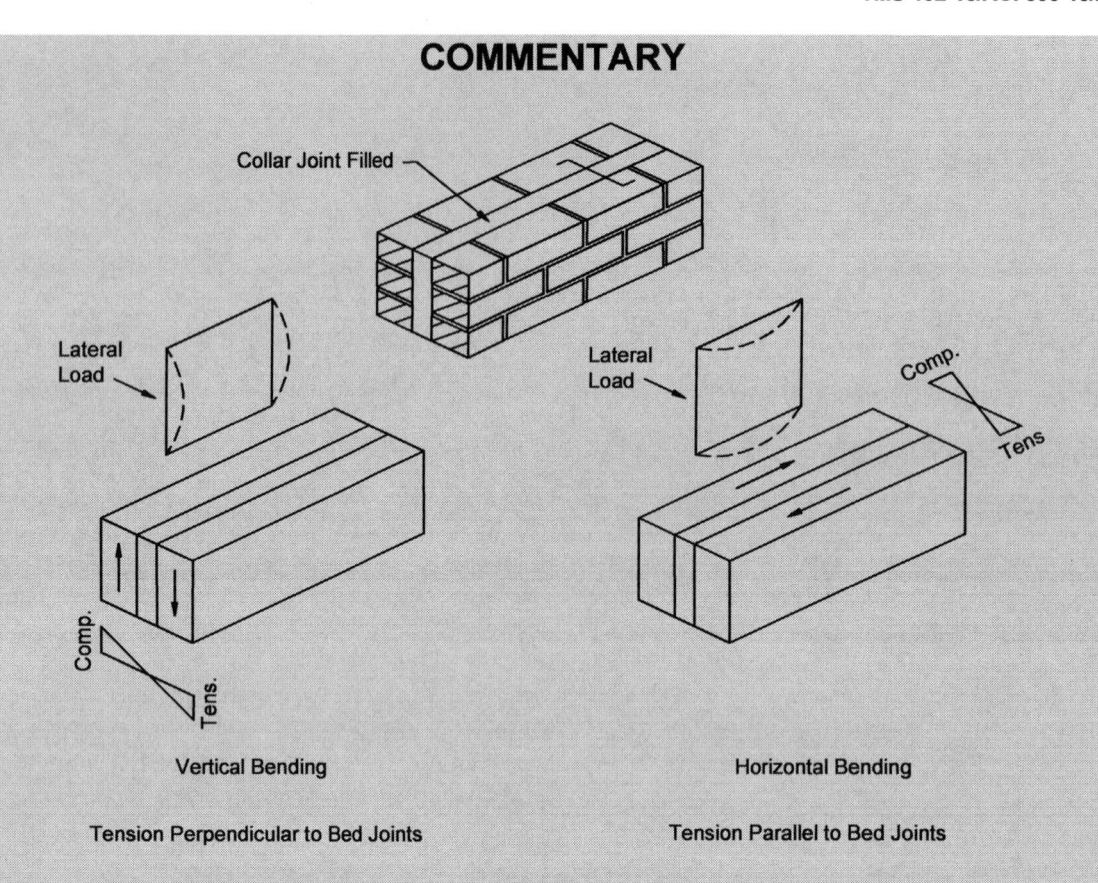

Figure CC-5.1-6 — Stress distribution in composite multiwythe masonry

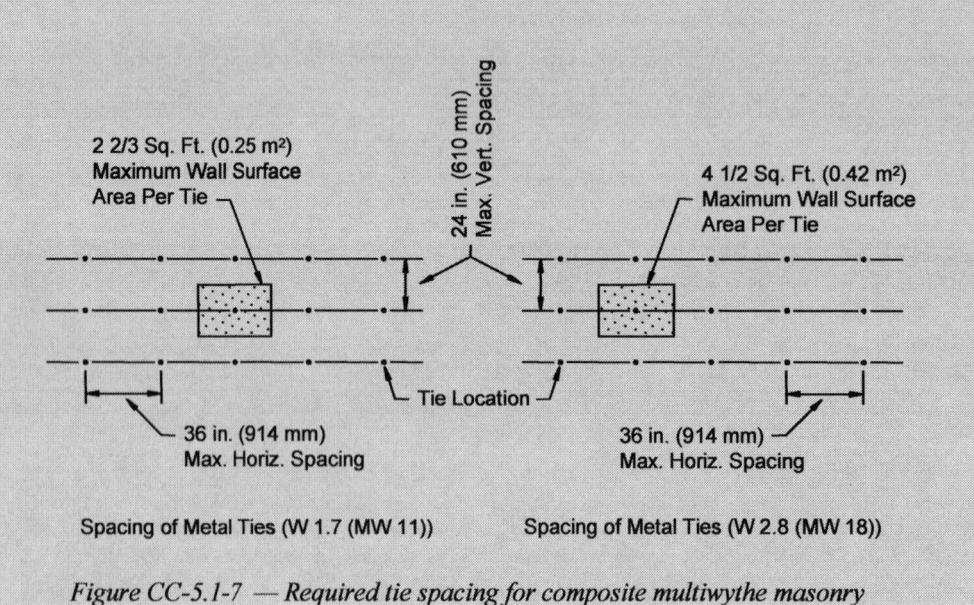

Figure CC-5.1-7 — Required tie spacing for composite multiwythe masonry

CODE

5.1.4.3 *Non-composite action* — The design of multiwythe masonry for non-composite action shall comply with Sections 5.1.4.3.1 and 5.1.4.3.2:

5.1.4.3.1 Each wythe shall be designed to resist individually the effects of loads imposed on it.

Unless a more detailed analysis is performed, the following requirements shall be satisfied:

(a) Collar joints shall not contain headers, grout, or mortar.

(b) Gravity loads from supported horizontal members shall be resisted by the wythe nearest to the center of span of the supported member. Any resulting bending moment about the weak axis of the masonry element shall be distributed to each wythe in proportion to its relative stiffness.

(c) Lateral loads acting parallel to the plane of the masonry element shall be resisted only by the wythe on which they are applied. Transfer of stresses from such loads between wythes shall be neglected.

(d) Lateral loads acting transverse to the plane of the masonry element shall be resisted by all wythes in proportion to their relative flexural stiffnesses.

(e) Specified distances between wythes shall not exceed 4.5 in. (114 mm) unless a detailed tie analysis is performed.

5.1.4.3.2 Wythes of masonry designed for non-composite action shall be connected by ties meeting the requirements of Section 5.1.4.2.3 or by adjustable ties. Where the cross wires of joint reinforcement are used as ties, the joint reinforcement shall be ladder-type or tab-type. Ties shall be without cavity drips.

Adjustable ties shall meet the following requirements:

(a) One tie shall be provided for each 1.77 ft^2 (0.16 m^2) of masonry surface area.

(b) Horizontal and vertical spacing shall not exceed 16 in. (406 mm).

(c) Adjustable ties shall not be used when the misalignment of bed joints from one wythe to the other exceeds $1\frac{1}{4}$ in. (31.8 mm).

(d) Maximum clearance between connecting parts of the tie shall be $\frac{1}{16}$ in. (1.6 mm).

(e) Pintle ties shall have at least two pintle legs of wire size W2.8 (MW18).

COMMENTARY

5.1.4.3 *Non-composite action* — Multiwythe masonry may be constructed so that each wythe is separated from the others by a space that may be crossed only by ties. The ties force compatible lateral deflection, but no composite action exists in the design.

5.1.4.3.1 Weak axis bending moments caused by either gravity loads or lateral loads are assumed to be distributed to each wythe in proportion to its relative stiffness. See Figure CC-5.1-8 for stress distribution in non-composite masonry. In non-composite masonry, the plane of the element is the plane of the space between wythes. Loads due to supported horizontal members are to be resisted by the wythe closest to center of span as a result of the deflection of the horizontal member.

In non-composite masonry, this Code limits the thickness of the cavity to 4½ in. (114 mm) to assure adequate performance. If cavity width exceeds 4½ in. (114 mm), the ties must be designed to resist the loads imposed upon them based on a rational analysis that takes into account buckling, tension, pullout, and load distribution.

The NCMA and Canadian Standards Association (NCMA TEK 12-1B, 2011; CSA, 1984) have recommendations for use in the design of ties for masonry with wide cavities.

5.1.4.3.2 The required size, number, and spacing of metal ties in non-composite masonry (Figure CC-5.1-7) have been determined from past experience. Requirements for adjustable ties are shown in Figure CC-5.1-9. They are based on the results in IIT (1963). Ladder-type or tab-type joint reinforcement is required because truss-type joint reinforcement restricts in-plane differential movement between wythes. However, the use of ties with drips (bends in ties to prevent moisture migration) has been eliminated because of their reduced strength.

COMMENTARY

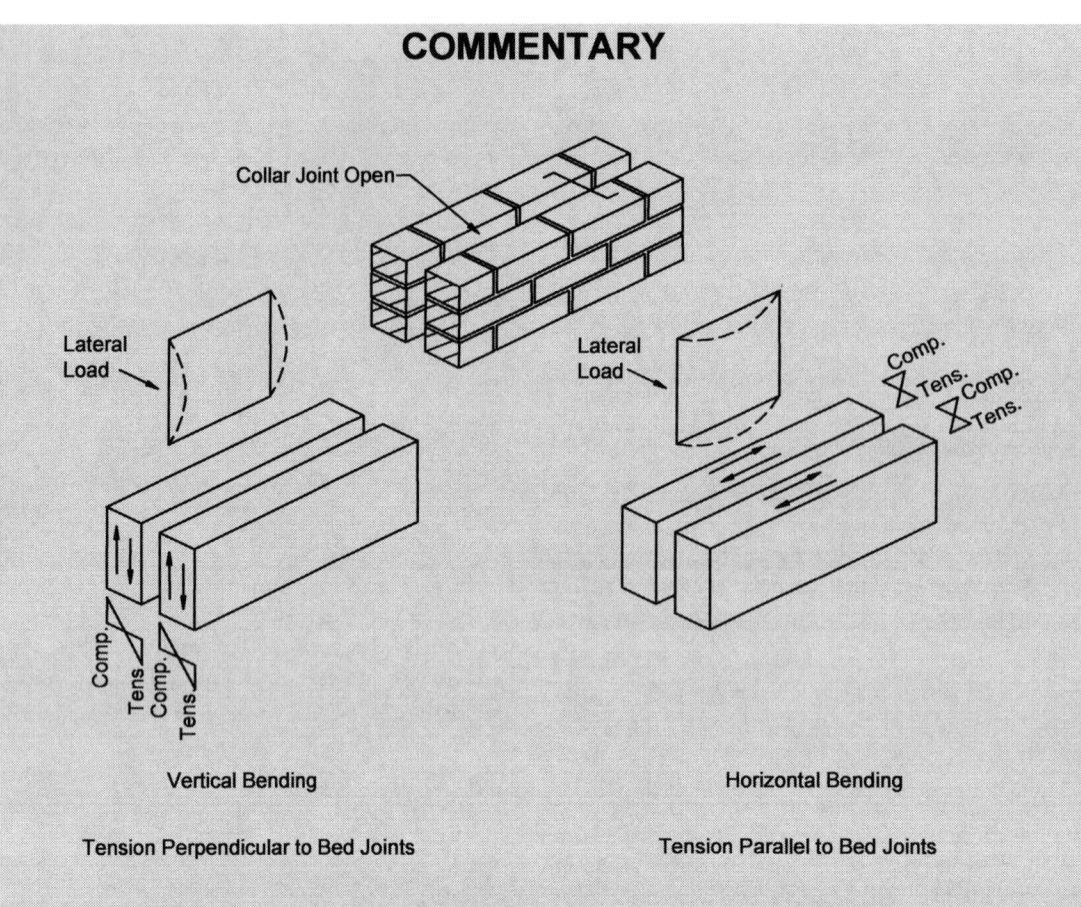

Figure CC-5.1-8 — Stress distribution in non-composite multiwythe masonry

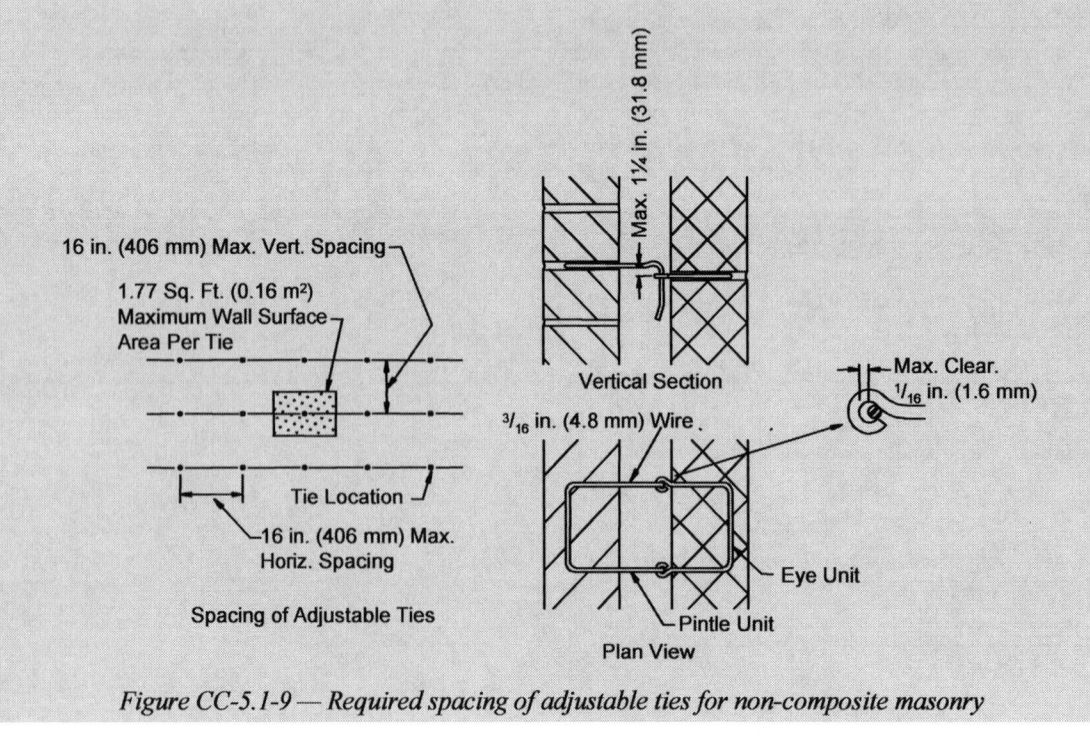

Figure CC-5.1-9 — Required spacing of adjustable ties for non-composite masonry

CODE

5.2 — Beams

Design of beams shall meet the requirements of Section 5.2.1 or Section 5.2.2. Design of beams shall also meet the requirements of Section 8.3, Section 9.3 or Section 11.3. Design requirements for masonry beams shall apply to masonry lintels.

5.2.1 *General beam design*

5.2.1.1 *Span length* — Span length shall be in accordance with the following:

5.2.1.1.1 Span length of beams not built integrally with supports shall be taken as the clear span plus depth of beam, but need not exceed the distance between centers of supports.

5.2.1.1.2 For determination of moments in beams that are continuous over supports, span length shall be taken as the distance between centers of supports.

5.2.1.2 *Lateral support* — The compression face of beams shall be laterally supported at a maximum spacing based on the smaller of:

(a) $32b$

(b) $120b^2/d$

5.2.1.3 *Bearing length* — Length of bearing of beams on their supports shall be a minimum of 4 in. (102 mm) in the direction of span.

5.2.1.4 *Deflections* — Masonry beams shall be designed to have adequate stiffness to limit deflections that adversely affect strength or serviceability.

5.2.1.4.1 The calculated deflection of beams providing vertical support to masonry designed in accordance with Section 8.2, Section 9.2, Section 11.2, Chapter 14, or Appendix A shall not exceed $l/600$ under unfactored dead plus live loads.

COMMENTARY

5.2.1 *General beam design*
5.2.1.1 *Span length*

5.2.1.2 *Lateral support* — To minimize lateral torsional buckling, the Code requires lateral bracing of the compression face. Hansell and Winter (1959) suggest that the slenderness ratios should be given in terms of ld/b^2. Revathi and Menon (2006) report on tests of seven under-reinforced slender concrete beams. In Figure CC-5.2-1, a straight line is fitted to the W_{test}/W_u ratio vs. ld/b^2, where W_{test} is the experimental capacity and W_u is the calculated capacity based on the full cross-sectional moment strength. W_{test}/W_u equals 1 where ld/b^2 equals 146. Based on this, the Code limit of 120 for ld/b^2 is reasonable and slightly conservative.

5.2.1.3 *Bearing length* — The minimum bearing length of 4 in. (102 mm) in the direction of span is considered a reasonable minimum to reduce concentrated compressive stresses at the edge of the support.

5.2.1.4 *Deflections* — The provisions of Section 5.2.1.4 address deflections that may occur at service load levels.

5.2.1.4.1 The deflection limits apply to beams and lintels of all materials that support unreinforced masonry. The deflection requirements may also be applicable to supported reinforced masonry that has vertical reinforcement only.

The deflection limit of $l/600$ should prevent long-term visible deflections and serviceability problems. In most cases, deflections of approximately twice this amount, or $l/300$, are required before the deflection becomes visible (Galambos and Ellingwood, 1986). This deflection limit is for immediate deflections. Creep will cause additional long-term deflections. A larger deflection limit of $l/480$ has been used when considering long-term deflections (CSA, 2004).

CODE

5.2.1.4.2 Deflection of masonry beams shall be calculated using the appropriate load-deflection relationship considering the actual end conditions. Unless stiffness values are obtained by a more comprehensive analysis, immediate deflections shall be calculated with an effective moment of inertia, I_{eff}, as follows.

$$I_{eff} = I_n \left(\frac{M_{cr}}{M_a}\right)^3 + I_{cr}\left[1 - \left(\frac{M_{cr}}{M_a}\right)^3\right] \leq I_n$$

(Equation 5-1)

For continuous beams, I_{eff} shall be permitted to be taken as the average of values obtained from Equation 5-1 for the critical positive and negative moment regions.

For beams of uniform cross-section, I_{eff} shall be permitted to be taken as the value obtained from Equation 5-1 at midspan for simple spans and at the support for cantilevers. For masonry designed in accordance with Chapter 8, the cracking moment, M_{cr}, shall be calculated using the allowable flexural tensile stress taken from Table 8.2.4.2 multiplied by a factor of 2.5. For masonry designed in accordance with Chapter 9, the cracking moment, M_{cr}, shall be calculated using the value for the modulus of rupture, f_r, taken from Table 9.1.9.2. For masonry designed in accordance with Chapter 11, the cracking moment, M_{cr}, shall be calculated using the value for the modulus of rupture, f_{rAAC}, as given by Section 11.1.8.3.

5.2.1.4.3 Deflections of reinforced masonry beams need not be checked when the span length does not exceed 8 multiplied by the effective depth to the reinforcement, d, in the masonry beam.

COMMENTARY

5.2.1.4.2 The effective moment of inertia was developed to provide a transition between the upper and lower bounds of I_g and I_{cr} as a function of the ratio M_{cr}/M_a (Branson, 1965). This procedure was selected as being sufficiently accurate for use to control deflections (Horton and Tadros, 1990). Calculating a more accurate effective moment of inertia using a moment-curvature analysis may be desirable for some circumstances.

Most masonry beams have some end restraint due to being built integrally with a wall. Tests have shown that the end restraint from beams being built integrally with walls reduces the deflections from 20 to 45 percent of those of the simply supported specimens (Longworth and Warwaruk, 1983).

5.2.1.4.3 Reinforced masonry beams and lintels with span lengths of 8 times d have immediate deflections of approximately 1/600 of the span length (Bennett et al, 2007). Masonry beams and lintels with shorter spans should have sufficient stiffness to prevent serviceability problems and, therefore, deflections do not need to be checked.

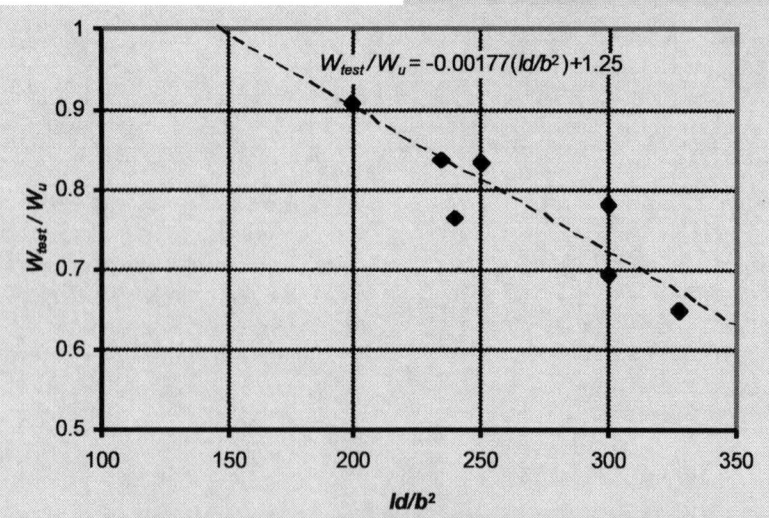

Figure CC-5.2-1 Beam capacity vs. beam slenderness (after Revathi and Menon (2006))

CODE

5.2.2 Deep beams
Design of deep beams shall meet the requirements of Section 5.2.1.2 and 5.2.1.3 in addition to the requirements of Sections 5.2.2.1 through 5.2.2.5.

5.2.2.1 Effective span length — The effective span length, l_{eff}, shall be taken as the center-to-center distance between supports or 1.15 multiplied by the clear span, whichever is smaller.

COMMENTARY

5.2.2 Deep beams
Shear warping of the deep beam cross section and a combination of diagonal tension stress and flexural tension stress in the body of the deep beam require that these members be designed using deep beam theory when the span-to-depth ratio is within the limits given in the definition of deep beams. Background on the development of the deep beam provisions is given in Fonseca et al, 2011.

As per the definition in Section 2.2, a deep beam has an effective span-to-depth ratio, l_{eff}/d_v, less than 3 for a continuous span and less than 2 for a simple span. Sections of masonry over openings may be designed as deep beams if the span-to-depth ratio meets these limits. However, the depth of the beam need not be taken as the entire height of masonry above the opening. A shallower beam can be designed to support the remaining portion of the masonry above in addition to applied loads. This beam can be designed conventionally and need not meet the deep beam provisions if sufficiently shallow. (see Figure CC-5.2-2)

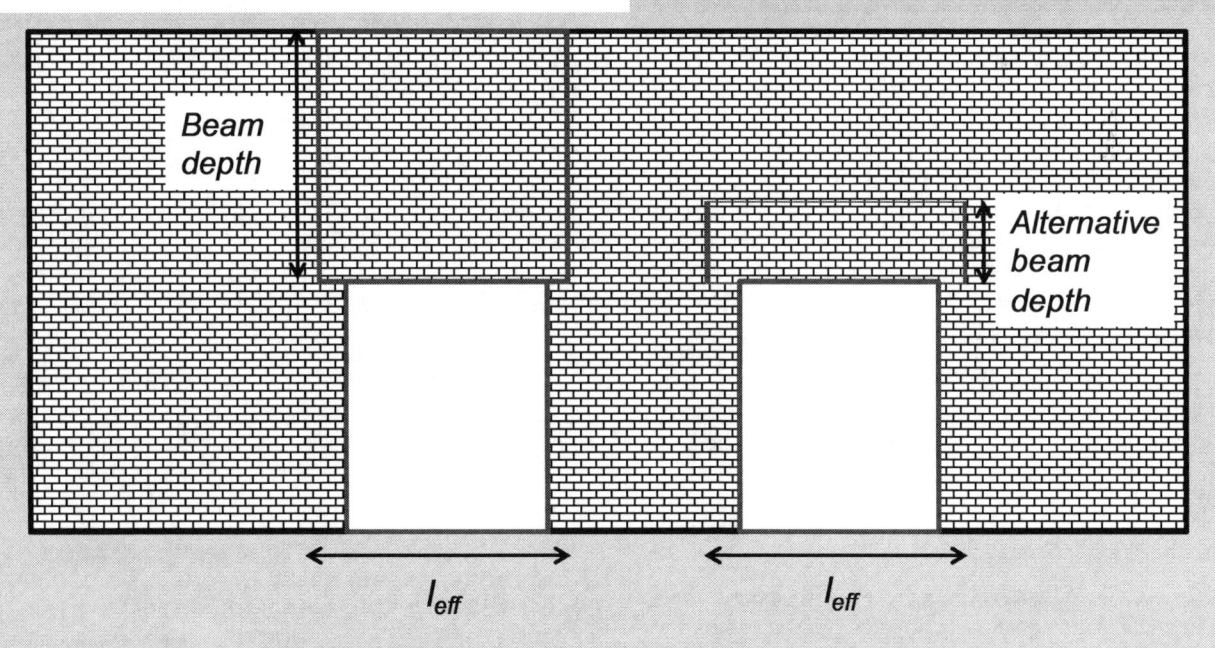

Figure CC-5.2-2 Possible depth of beams over openings in masonry walls

CODE

5.2.2.2 *Internal lever arm* — Unless determined by a more comprehensive analysis, the internal lever arm, z, shall be taken as:

(a) For simply supported spans.

(1) When $1 \leq \dfrac{l_{eff}}{d_v} < 2$

$$z = 0.2(l_{eff} + 2d_v) \quad \text{(Equation 5-2a)}$$

(2) When $\dfrac{l_{eff}}{d_v} < 1$

$$z = 0.6 l_{eff} \quad \text{(Equation 5-2b)}$$

(b) For continuous spans

(1) When $1 \leq \dfrac{l_{eff}}{d_v} < 3$

$$z = 0.2(l_{eff} + 1.5 d_v) \quad \text{(Equation 5-3a)}$$

(2) When $\dfrac{l_{eff}}{d_v} < 1$

$$z = 0.5 l_{eff} \quad \text{(Equation 5-3b)}$$

5.2.2.3 *Flexural reinforcement* — Distributed horizontal flexural reinforcement shall be provided in the tension zone of the beam for a depth equal to half of the beam depth, d_v. The maximum spacing of distributed horizontal flexural reinforcement shall not exceed one-fifth of the beam depth, d_v, nor 16 in. (406 mm). Joint reinforcement shall be permitted to be used as distributed horizontal flexural reinforcement in deep beams. Horizontal flexural reinforcement shall be anchored to develop the yield strength of the reinforcement at the face of supports.

5.2.2.4 *Minimum shear reinforcement* — The following provisions shall apply when shear reinforcement is required in accordance with Section 8.3.5, Section 9.3.4.1.2, or Section 11.3.4.1.2.

(a) The minimum area of vertical shear reinforcement shall be $0.0007\, bd_v$.

(b) Horizontal shear reinforcement shall have cross-sectional area equal to or greater than one half the area of the vertical shear reinforcement. Such reinforcement shall be equally distributed on both side faces of the beam when the nominal width of the beam is greater than 8 in. (203 mm).

(c) The maximum spacing of shear reinforcement shall not exceed one-fifth the beam depth, d_v, nor 16 in. (406 mm).

COMMENTARY

5.2.2.2 *Internal lever arm* — The theory used for design of beams has limited applicability to deep beams. Specifically, there will be a nonlinear distribution of strain in deep beams. The internal lever arm, z, between the centroid of the internal compressive forces and the internal tensile forces will be less than that calculated assuming a linear strain distribution. The Code equations for internal lever arm, z, can be used with either allowable stress design or strength design. For allowable stress design, z is commonly known as jd, and for strength design, z is commonly known as $d-(a/2)$. The internal lever arm provisions in the Code are based on CEB-FIP (1990).

5.2.2.3 *Flexural reinforcement* — The distribution of tensile stress in a deep beam is generally such that the lower one-half of the beam is required to have distributed flexural reinforcement. However, other loading conditions, such as uplift, and support conditions, such as continuous and fixed ends, should be considered in determining the portion of the deep beam that is subjected to tension. Distributed horizontal reinforcement resists tensile stress caused by shear as well as by flexure.

5.2.2.4 *Minimum shear reinforcement* – Distributed flexural reinforcement may be included as part of the provided shear reinforcement to meet the minimum distributed shear reinforcement ratio. The spacing of shear reinforcement is limited to restrain the width of the cracks.

Load applied along the top surface of a deep beam is transferred to supports mainly by arch action. Typically, deep beams do not need transverse reinforcement and it is sufficient to provide distributed flexural reinforcement (Park et al, 1975).

CODE

5.2.2.5 *Total reinforcement* — The sum of the cross-sectional areas of horizontal and vertical reinforcement shall be at least 0.001 multiplied by the gross cross-sectional area, bd_v, of the deep beam, using specified dimensions.

5.3 — Columns

Design of columns shall meet the requirements of Section 5.3.1 or Section 5.3.2. Design of columns shall also meet the requirements of Section 8.3, or Section 9.3, or Section 11.3.

5.3.1 *General column design*

5.3.1.1 *Dimensional limits* — Dimensions shall be in accordance with the following:

(a) The distance between lateral supports of a column shall not exceed 99 multiplied by the least radius of gyration, r.

(b) Minimum side dimension shall be 8 in. (203 mm) nominal.

5.3.1.2 *Construction* — Columns shall be fully grouted.

5.3.1.3 *Vertical reinforcement* — Vertical reinforcement in columns shall not be less than $0.0025A_n$ nor exceed $0.04A_n$. The minimum number of bars shall be four.

COMMENTARY

5.3 — Columns

Columns are defined in Section 2.2. They are isolated members usually under axial compressive loads and flexure. If damaged, columns may cause the collapse of other members; sometimes of an entire structure. These critical structural elements warrant the special requirements of this section.

5.3.1 *General column design*

5.3.1.1 *Dimensional limits* — The limit of 99 for the slenderness ratio, h/r, is judgment based. See Figure CC-5.3-1 for effective height determination. The minimum nominal side dimension of 8 in. (203 mm) results from practical considerations.

5.3.1.3 *Vertical reinforcement* — Minimum vertical reinforcement is required in masonry columns to prevent brittle failure. The maximum percentage limit in column vertical reinforcement was established based on the committee's experience. Four bars are required so ties can be used to provide a confined core of masonry.

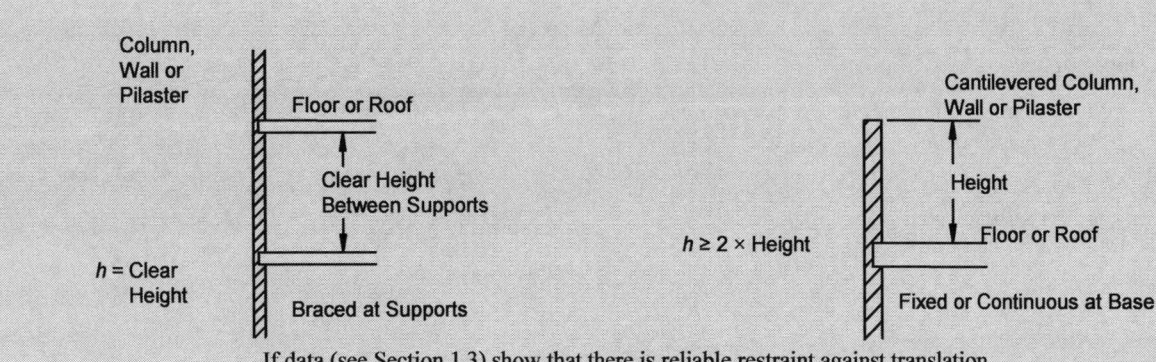

If data (see Section 1.3) show that there is reliable restraint against translation and rotation at the supports, the "effective height" may be taken as low as the distance between points of inflection for the loading case under consideration.

Figure CC-5.3-1 — Effective height, h, of column, wall, or pilaster

CODE

5.3.1.4 *Lateral ties* — Lateral ties shall conform to the following:

(a) Vertical reinforcement shall be enclosed by lateral ties at least $^1/_4$ in. (6.4 mm) in diameter.

(b) Vertical spacing of lateral ties shall not exceed 16 longitudinal bar diameters, 48 lateral tie bar or wire diameters, or least cross-sectional dimension of the member.

(c) Lateral ties shall be arranged so that every corner and alternate longitudinal bar shall have lateral support provided by the corner of a lateral tie with an included angle of not more than 135 degrees. No bar shall be farther than 6 in. (152 mm) clear on each side along the lateral tie from such a laterally supported bar. Lateral ties shall be placed in either a mortar joint or in grout. Where longitudinal bars are located around the perimeter of a circle, a complete circular lateral tie is permitted. Lap length for circular ties shall be 48 tie diameters.

(d) Lateral ties shall be located vertically not more than one-half lateral tie spacing above the top of footing or slab in any story, and shall be spaced not more than one-half a lateral tie spacing below the lowest horizontal reinforcement in beam, girder, slab, or drop panel above

5.3.2 *Lightly loaded columns*

Masonry columns used only to support light frame roofs of carports, porches, sheds or similar structures assigned to Seismic Design Category A, B, or C, which are subject to unfactored gravity loads not exceeding 2,000 lbs (8,900 N) acting within the cross-sectional dimensions of the column are permitted to be constructed as follows:

(a) Minimum side dimension shall be 8 in. (203 mm) nominal.

(b) Height shall not exceed 12 ft (3.66 m).

(c) Cross-sectional area of longitudinal reinforcement shall not be less than 0.2 in.^{2} (129 mm^2) centered in the column.

(d) Columns shall be fully grouted.

COMMENTARY

5.3.1.4 *Lateral ties* — Lateral reinforcement in columns performs two functions. It provides the required support to prevent buckling of longitudinal column reinforcing bars acting in compression and provides resistance to diagonal tension for columns acting in shear (Pfister, 1964). Ties may be located in the mortar joint, when the tie diameter does not exceed ½ the specified mortar joint thickness. For example, ¼ in. (6.4 mm) diameter ties may be placed in ½ in. (12.7 mm) thick mortar joints.

The requirements of this Code are modeled on those for reinforced concrete columns. Except for permitting ¼-in. (6.4-mm) ties in Seismic Design Category A, B, and C, they reflect the applicable provisions of the reinforced concrete code.

5.3.2 *Lightly loaded columns*

Masonry columns are often used to support roofs of carports, porches, sheds or similar light structures. These columns do not need to meet the detailing requirements of Section 5.3.1. The axial load limit of 2,000 pounds (8,900 N) was developed based on the flexural strength of a nominal 8 in. (203 mm) by 8 in. (203 mm) by 12 ft high (3.66 m) column with one No. 4 (M#13) reinforcing bar in the center and f'_m of 1350 psi (9.31 MPa). An axial load of 2,000 pounds (8,900 N) at the edge of the member will result in a moment that is approximately equal to the nominal flexural strength of this member.

CODE

5.4 — Pilasters

Walls interfacing with pilasters shall not be considered as flanges, unless the construction requirements of Sections 5.1.1.2.1 and 5.1.1.2.5 are met. When these construction requirements are met, the pilaster's flanges shall be designed in accordance with Sections 5.1.1.2.2 through 5.1.1.2.4.

5.5 — Corbels

5.5.1 *Load-bearing corbels*

Load-bearing corbels shall be designed in accordance with Chapter 8, 9 or 10.

5.5.2 *Non-load-bearing corbels*

Non-load-bearing corbels shall be designed in accordance with Chapter 8, 9 or 10 or detailed as follows:

(a) Solid masonry units or hollow units filled with mortar or grout shall be used.

(b) The maximum projection beyond the face of the wall shall not exceed:

  (1) one-half the wall thickness for multiwythe walls bonded by mortar or grout and wall ties or masonry headers, or

  (2) one-half the wythe thickness for single wythe walls, masonry bonded hollow walls, multiwythe walls with open collar joints, and veneer walls.

(c) The maximum projection of one unit shall not exceed:

  (1) one-half the nominal unit height.

  (2) one-third the nominal thickness of the unit or wythe.

(d) The back surface of the corbelled section shall remain within 1 in. (25.4 mm) of plane.

COMMENTARY

5.4 — Pilasters

Pilasters are masonry members that can serve several purposes. They may project from one or both sides of the wall, as shown in Figure CC-5.4-1. Pilasters contribute to the lateral load resistance of masonry walls and may resist vertical loads.

5.5 — Corbels

The provision for corbelling up to one-half of the wall or wythe thickness is theoretically valid only if the opposite side of the wall remains in its same plane. The addition of the 1-in. (25.4-mm) intrusion into the plane recognizes the impracticality of keeping the back surface plane. See Figure CC-5.5-1 and CC-5.5-2 for maximum permissible unit projection.

COMMENTARY

Brick Pilasters

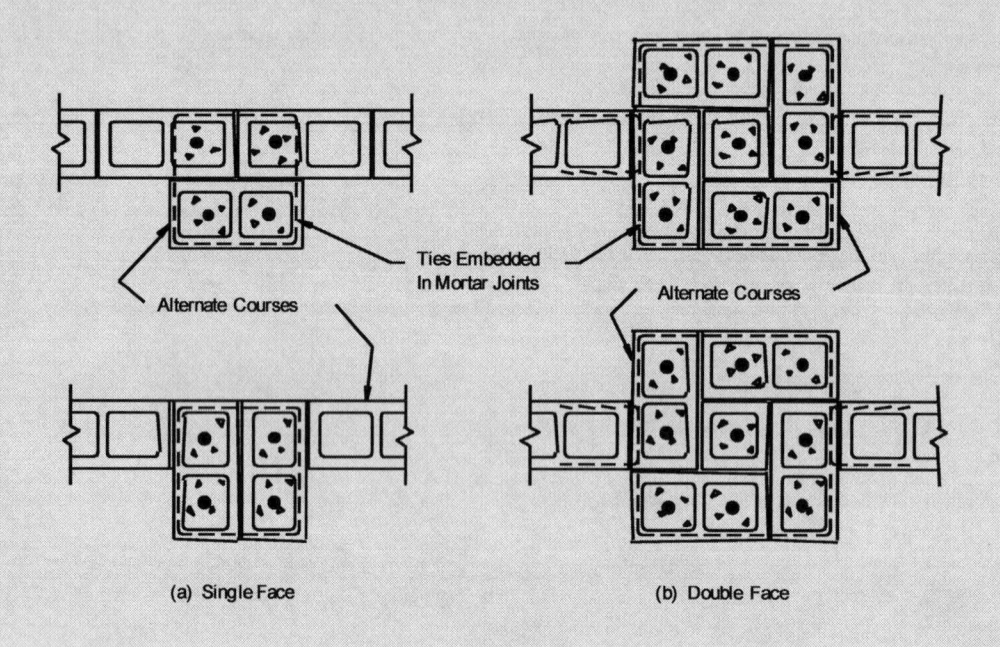

(a) Single Face (b) Double Face

Block Pilasters

(a) Single Face (b) Double Face

Figure CC-5.4-1 — Typical pilasters

COMMENTARY

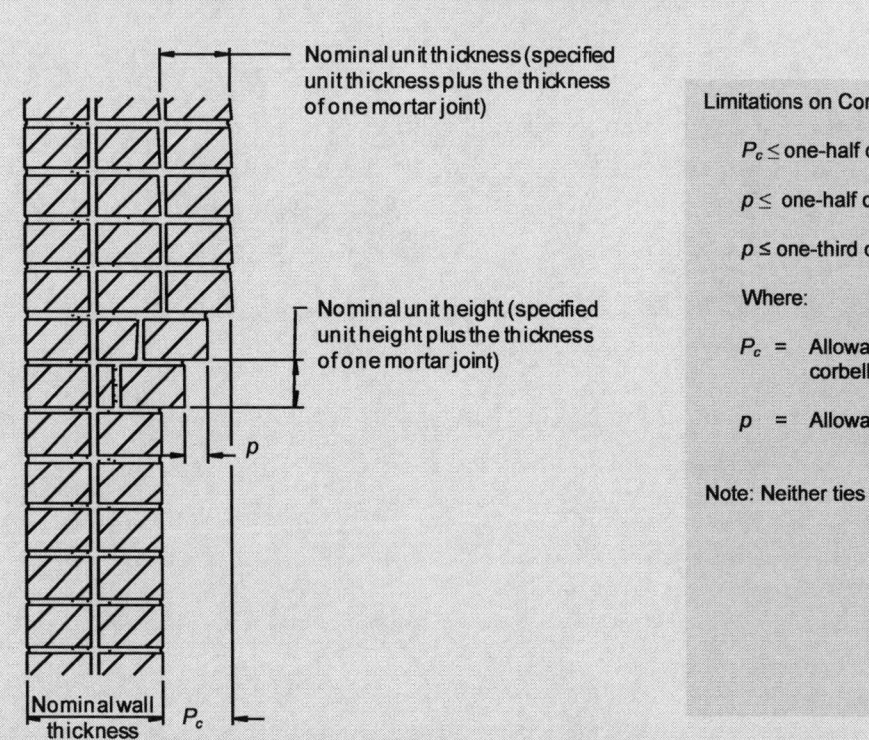

Limitations on Corbelling:

$P_c \leq$ one-half of nominal unit thickness

$p \leq$ one-half of nominal unit height

$p \leq$ one-third of nominal unit thickness

Where:

P_c = Allowable total horizontal projection of corbelling

p = Allowable projection of one unit

Note: Neither ties nor headers shown.

Figure CC-5.5-1 — Limits on corbelling in solid walls

Limitations on Corbelling:

$P_c \leq$ one-half of nominal unit thickness

$p \leq$ one-half of nominal unit height

$p \leq$ one-third of nominal unit thickness

Where:

P_c = Allowable total horizontal projection of corbelling

p = Allowable projection of one unit

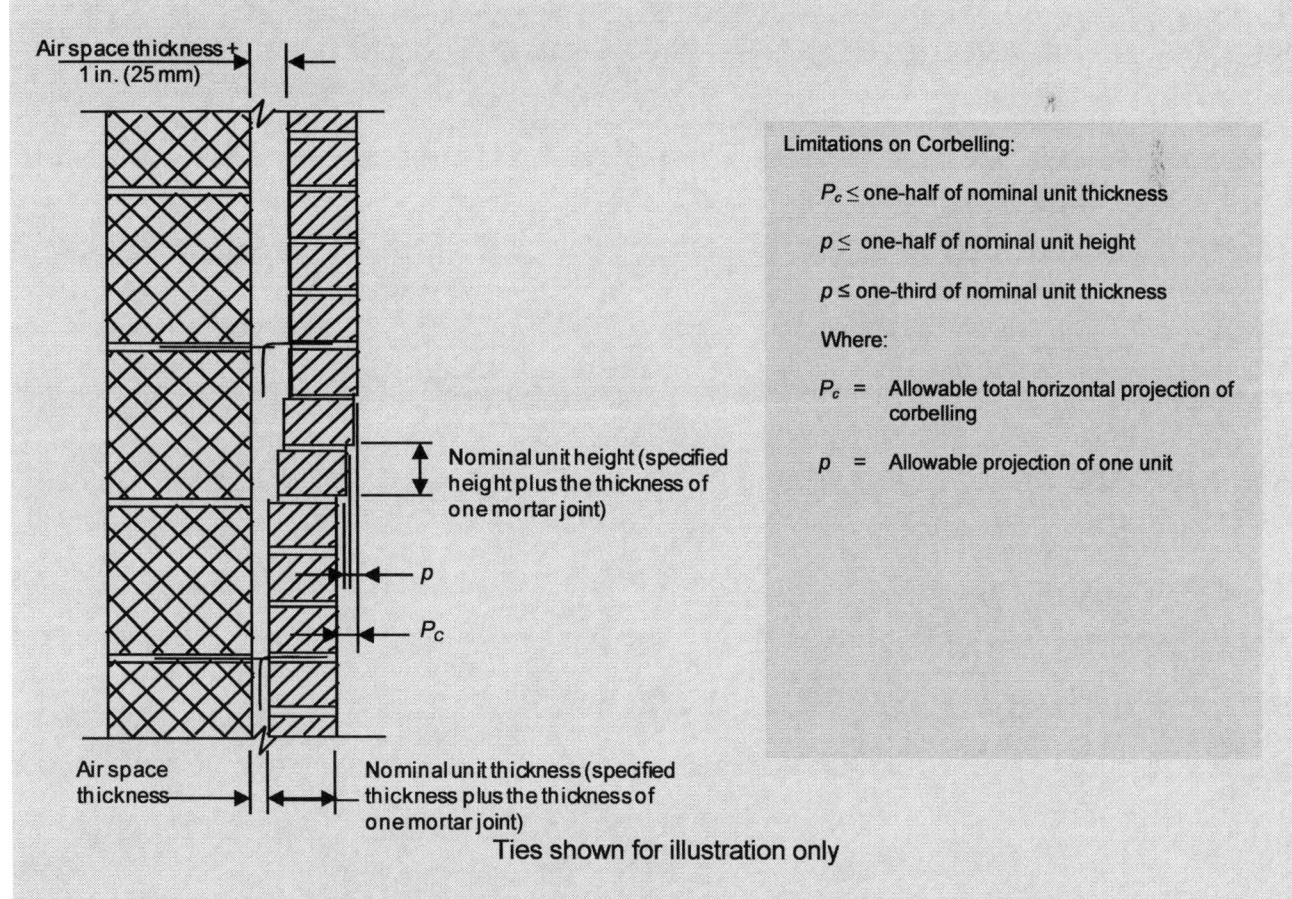

Ties shown for illustration only

Figure CC-5.5-2 – Limits on corbelling in walls with air space

This page intentionally left blank

CHAPTER 6
REINFORCEMENT, METAL ACCESSORIES, AND ANCHOR BOLTS

CODE

6.1 — Details of reinforcement and metal accessories

6.1.1 *Embedment*
Reinforcing bars shall be embedded in grout.

6.1.2 *Size of reinforcement*
    6.1.2.1 The maximum size of reinforcement used in masonry shall be No. 11 (M #36).

    6.1.2.2 The diameter of reinforcement shall not exceed one-half the least clear dimension of the cell, bond beam, or collar joint in which it is placed.

    6.1.2.3 Longitudinal and cross wires of joint reinforcement shall have a minimum wire size of W1.1 (MW7) and a maximum wire size of one-half the joint thickness.

6.1.3 *Placement of reinforcement*
    6.1.3.1 The clear distance between parallel bars shall not be less than the nominal diameter of the bars, nor less than 1 in. (25.4 mm).

    6.1.3.2 In columns and pilasters, the clear distance between vertical bars shall not be less than one and one-half multiplied by the nominal bar diameter, nor less than $1\frac{1}{2}$ in. (38.1 mm).

    6.1.3.3 The clear distance limitations between bars required in Sections 6.1.3.1 and 6.1.3.2 shall also apply to the clear distance between a contact lap splice and adjacent splices or bars.

COMMENTARY

6.1 — Details of reinforcement and metal accessories

When the provisions of this section were originally developed in the late 1980s, the Committee used the 1983 edition of the ACI 318 Code as a guide. Some of the requirements were simplified and others dropped, depending on their suitability for application to masonry.

6.1.2 *Size of reinforcement*
    6.1.2.1 Limits on size of reinforcement are based on accepted practice and successful performance in construction. The No. 11 (M#36) limit is arbitrary, but Priestley and Bridgeman (1974) shows that distributed small bars provide better performance than fewer large bars. Properties of reinforcement are given in Table CC-6.1.2.

    6.1.2.2 Adequate flow of grout necessary for good bond is achieved with this limitation. It also limits the size of reinforcement when combined with Section 3.2.1.

    6.1.2.3 The function of joint reinforcement is to control the size and spacing of cracks caused by volume changes in masonry as well as to resist tension (Dickey, 1982). Joint reinforcement is commonly used in concrete masonry to minimize shrinkage cracking. The restriction on wire size ensures adequate performance. The maximum wire size of one-half the joint thickness allows free flow of mortar around joint reinforcement. Thus, a $\frac{3}{16}$-in. (4.8-mm) diameter wire can be placed in a $\frac{3}{8}$-in. (9.5-mm) joint.

6.1.3 *Placement of reinforcement*
Placement limits for reinforcement are based on successful construction practice over many years. The limits are intended to facilitate the flow of grout between bars. A minimum spacing between bars in a layer prevents longitudinal splitting of the masonry in the plane of the bars. Use of bundled bars in masonry construction is rarely required. Two bars per bundle is considered a practical maximum. It is important that bars be placed accurately. Reinforcing bar positioners are available to control bar position.

CODE

6.1.3.4 Groups of parallel reinforcing bars bundled in contact to act as a unit shall be limited to two in any one bundle. Individual bars in a bundle cut off within the span of a member shall terminate at points at least 40 bar diameters apart.

6.1.3.5 Reinforcement embedded in grout shall have a thickness of grout between the reinforcement and masonry units not less than $1/4$ in. (6.4 mm) for fine grout or $1/2$ in. (12.7 mm) for coarse grout.

6.1.4 *Protection of reinforcement and metal accessories*

6.1.4.1 Reinforcing bars shall have a masonry cover not less than the following:

(a) Masonry face exposed to earth or weather: 2 in. (50.8 mm) for bars larger than No. 5 (M #16); $1^{1}/_{2}$ in. (38.1 mm) for No. 5 (M #16) bars or smaller.

(b) Masonry not exposed to earth or weather: $1^{1}/_{2}$ in. (38.1 mm).

COMMENTARY

6.1.4 *Protection of reinforcement and metal accessories*

6.1.4.1 Reinforcing bars are traditionally not coated for corrosion resistance. The masonry cover retards corrosion of the steel. Cover is measured from the exterior masonry surface to the outermost surface of the reinforcement to which the cover requirement applies. It is measured to the outer edge of stirrups or ties, if transverse reinforcement encloses main bars. Masonry cover includes the thickness of masonry units, mortar, and grout. At bed joints, the protection for reinforcement is the total thickness of mortar and grout from the exterior of the mortar joint surface to outer-most surface of the reinforcement or metal accessory.

The condition "masonry face exposed to earth or weather" refers to direct exposure to moisture changes (alternate wetting and drying) and not just temperature changes.

Table CC-6.1.2 — Physical properties of steel reinforcing wire and bars

Designation	Diameter, in. (mm)	Area, in.^{2} (mm^2)	Perimeter, in. (mm)
Wire			
W1.1 (11 gage) (MW7)	0.121 (3.1)	0.011 (7.1)	0.380 (9.7)
W1.7 (9 gage) (MW11)	0.148 (3.8)	0.017 (11.0)	0.465 (11.8)
W2.1 (8 gage) (MW13)	0.162 (4.1)	0.020 (12.9)	0.509 (12.9)
W2.8 (3/16 in. wire) (MW18)	0.187 (4.8)	0.027 (17.4)	0.587 (14.9)
W4.9 (1/4 in. wire) (MW32)	0.250 (6.4)	0.049 (31.6)	0.785 (19.9)
Bars			
No. 3 (M#10)	0.375 (9.5)	0.11 (71.0)	1.178 (29.9)
No. 4 (M#13)	0.500 (12.7)	0.20 (129)	1.571 (39.9)
No. 5 (M#16)	0.625 (15.9)	0.31 (200)	1.963 (49.9)
No. 6 (M#19)	0.750 (19.1)	0.44 (284)	2.356 (59.8)
No. 7 (M#22)	0.875 (22.2)	0.60 (387)	2.749 (69.8)
No. 8 (M#25)	1.000 (25.4)	0.79 (510)	3.142 (79.8)
No. 9 (M#29)	1.128 (28.7)	1.00 (645)	3.544 (90.0)
No. 10 (M#32)	1.270 (32.3)	1.27 (819)	3.990 (101)
No. 11 (M#36)	1.410 (35.8)	1.56 (1006)	4.430 (113)

CODE

6.1.4.2 Longitudinal wires of joint reinforcement shall be fully embedded in mortar or grout with a minimum cover of $^5/_8$ in. (15.9 mm) when exposed to earth or weather and $^1/_2$ in. (12.7 mm) when not exposed to earth or weather. Joint reinforcement shall be stainless steel or protected from corrosion by hot-dipped galvanized coating or epoxy coating when used in masonry exposed to earth or weather and in interior walls exposed to a mean relative humidity exceeding 75 percent. All other joint reinforcement shall be mill galvanized, hot-dip galvanized, or stainless steel.

6.1.4.3 Wall ties, sheet-metal anchors, steel plates and bars, and inserts exposed to earth or weather, or exposed to a mean relative humidity exceeding 75 percent shall be stainless steel or protected from corrosion by hot-dip galvanized coating or epoxy coating. Wall ties, anchors, and inserts shall be mill galvanized, hot-dip galvanized, or stainless steel for all other cases. Anchor bolts, steel plates, and bars not exposed to earth, weather, nor exposed to a mean relative humidity exceeding 75 percent, need not be coated.

6.1.5 *Standard hooks*
Standard hooks shall consist of the following:

(a) 180-degree bend plus a minimum $4d_b$ extension, but not less than 2-1/2 in. (64 mm), at free end of bar;

(b) 90-degree bend plus a minimum $12d_b$ extension at free end of bar; or

(c) for stirrup and tie hooks for a No. 5 bar and smaller, either a 90-degree or 135-degree bend plus a minimum $6d_b$ extension, but not less than 2-1/2 in. (64 mm), at free end of bar.

6.1.6 *Minimum bend diameter for reinforcing bars*
The diameter of bend measured on the inside of reinforcing bars, other than for stirrups and ties, shall not be less than values specified in Table 6.1.6.

Table 6.1.6 — Minimum diameters of bend

Bar size and type	Minimum diameter
No. 3 through No. 7 (M #10 through #22) Grade 40 (Grade 280)	5 bar diameters
No. 3 through No. 8 (M #10 through #25) Grade 50 or 60 (Grade 350 or 420)	6 bar diameters
No. 9, No. 10, and No. 11 (M #29, #32, and #36) Grade 50 or 60 (Grade 350 or 420)	8 bar diameters

COMMENTARY

6.1.4.2 Because masonry cover protection for joint reinforcement is minimal, the protection of joint reinforcement in masonry is required in accordance with the Specification. Examples of interior walls exposed to a mean relative humidity exceeding 75 percent are natatoria and food processing plants.

6.1.4.3 Corrosion resistance requirements are included because masonry cover varies considerably for these items. The exception for anchor bolts is based on current industry practice.

6.1.5 *Standard hooks*
Standard hooks are shown in Figure CC-6.1-1.

6.1.6 *Minimum bend diameter for reinforcing bars*
Standard bends in reinforcing bars are described in terms of the inside diameter of bend because this is easier to measure than the radius of bend.

A broad survey of bending practices, a study of ASTM bend test requirements, and a pilot study of and experience with bending Grade 60 (Grade 420) bars were considered in establishing the minimum diameter of bend. The primary consideration was feasibility of bending without breakage. Experience has since established that these minimum bend diameters are satisfactory for general use without detrimental crushing of grout.

COMMENTARY

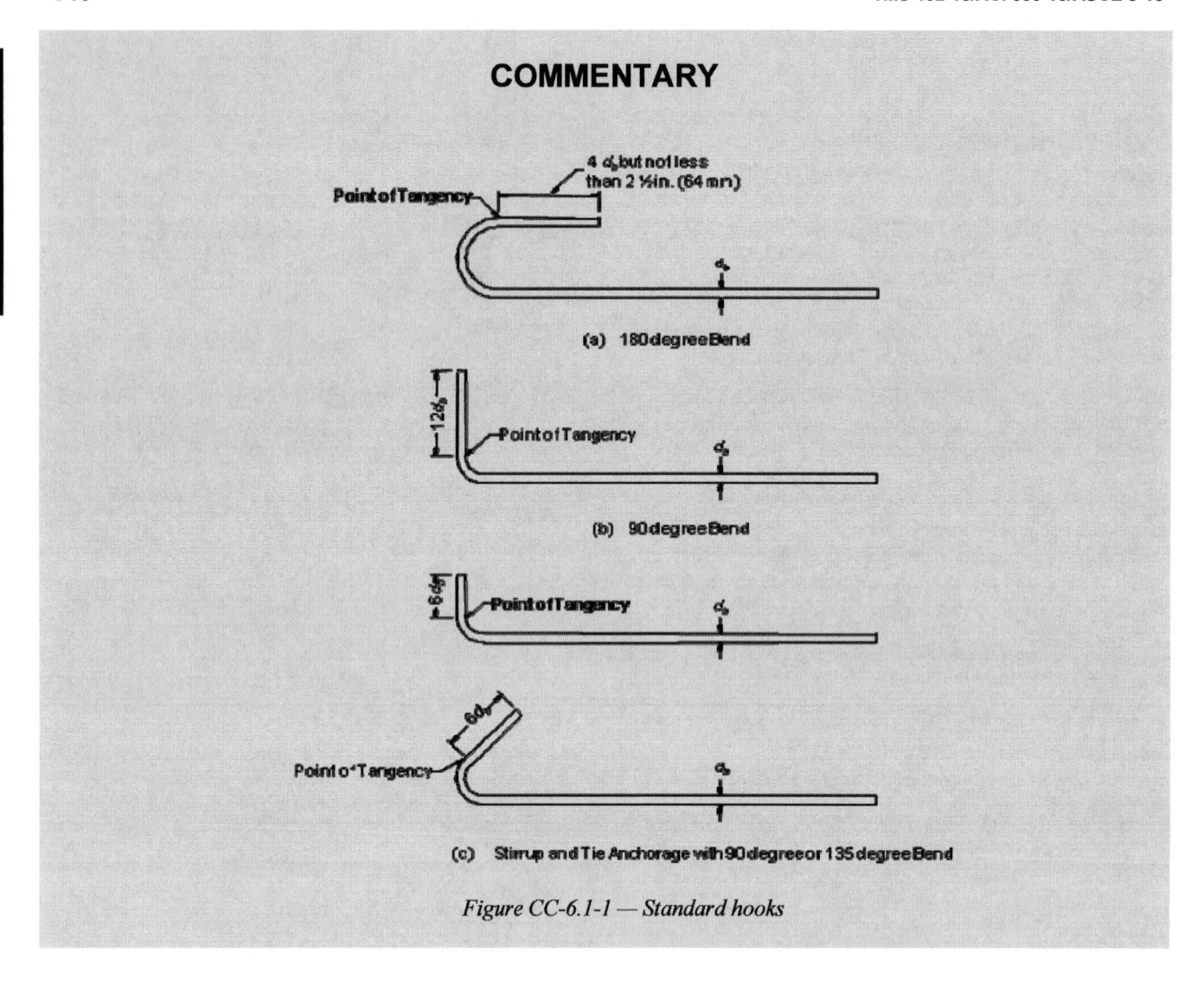

Figure CC-6.1-1 — Standard hooks

CODE

6.2 — Anchor bolts

Headed and bent-bar anchor bolts shall conform to the provisions of Sections 6.2.1 through 6.2.7.

6.2.1 *Placement*

Headed and bent-bar anchor bolts shall be embedded in grout. Anchor bolts of ¼ in. (6.4 mm) diameter are permitted to be placed in mortar bed joints that are at least ½ in. (12.7 mm) in thickness and, for purposes of application of the provisions of Sections 6.2, 8.1.3 and 9.1.6, are permitted to be considered as if they are embedded in grout.

Anchor bolts placed in the top of grouted cells and bond beams shall be positioned to maintain a minimum of ¼ in. (6.4 mm) of fine grout between the bolts and the masonry unit or ½ in. (12.7 mm) of coarse grout between the bolts and the masonry unit. Anchor bolts placed in drilled holes in the face shells of hollow masonry units shall be permitted to contact the masonry unit where the bolt passes through the face shell, but the portion of the bolt that is within the grouted cell shall be positioned to maintain a minimum of ¼ in. (6.4 mm) of fine grout between the head or bent leg of each bolt and the masonry unit or ½ in. (12.7 mm) of coarse grout between the head or bent leg of each bolt and the masonry unit.

The clear distance between parallel anchor bolts shall not be less than the nominal diameter of the anchor bolt, nor less than 1 in. (25.4 mm).

COMMENTARY

6.2 — Anchor bolts

These design values apply only to the specific types of bolts mentioned. These bolts are readily available and are depicted in Figure CC-6.2-1.

6.2.1 *Placement*

Most tests on anchor bolts in masonry have been performed on anchor bolts embedded in grout. Placement limits for anchor bolts are based on successful construction practice over many years. The limits are intended to facilitate the flow of grout between bolts and between bolts and the masonry unit.

Research at Portland State University (Rad and Mueller, 1998) and at Washington State University (Tubbs et al, 2000) has established that there is no difference in the performance of an anchor bolt installed through a tight-fitting hole in the face shell of a grouted hollow masonry unit and in an over-sized hole in the face shell of a grouted hollow masonry unit. Therefore, the clear distance requirement for grout to surround an anchor bolt is not needed where the bolt passes through the face shell. See Figure CC-6.2-2.

Quality/assurance/control (QA) procedures should ensure that there is sufficient clearance around the bolts prior to grout placement. These procedures should also require observation during grout placement to ensure that grout completely surrounds the bolts, as required by the QA Tables in Section 3.1.

Prior to the 2008 MSJC Code, provisions for the allowable shear load of anchors included explicit consideration of bolt edge distances. For edge distance less than 12 bolt diameters, the allowable load in shear was reduced by linear interpolation to zero at an edge distance of 1 in. (25 mm). Since publication of the 2008 MSJC Code, edge distance is considered in provisions for both allowable shear load and shear strength in the calculation of the projected area for shear, A_{pv} (Code Equation 6-2; also see Code and Commentary Section 6.2.3). The projected area is based on an assumed failure cone that originates at the bearing point of the anchor and radiates at 45° in the direction of the shear force towards the free edge of the masonry, thereby accounting for bolt edge distance. The portion of projected area overlapping an open cell, or open head joint, or that lies outside the masonry is deducted from the value of A_{pv}. No minimum edge distance is provided for the placement of anchor bolts. Placement of all anchors, including anchor bolts, must meet the minimum thickness of grout between the anchor and masonry units given in Code Section 6.2.1.

COMMENTARY

Figure CC-6.2-1 — Anchor bolts

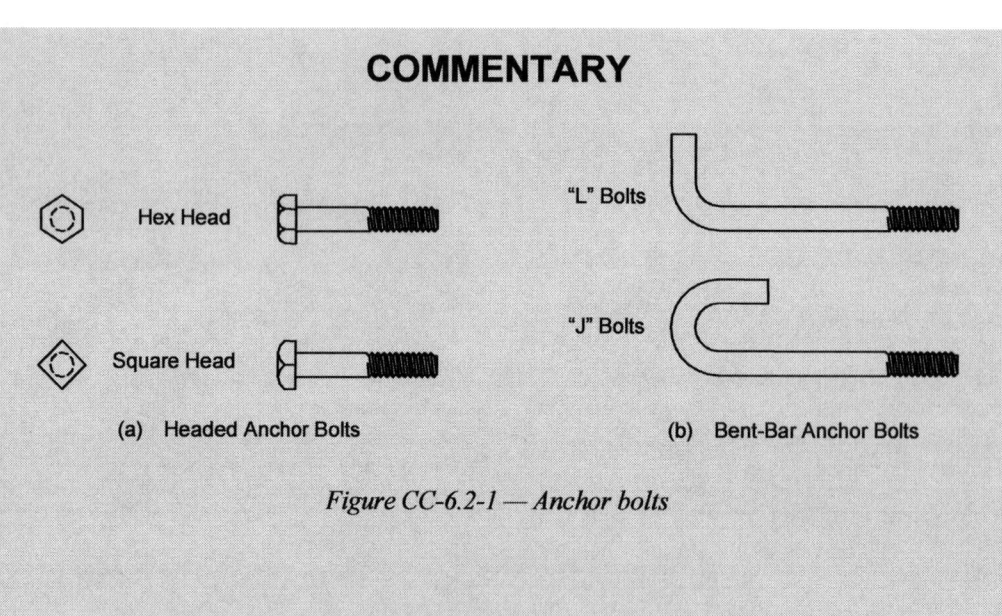

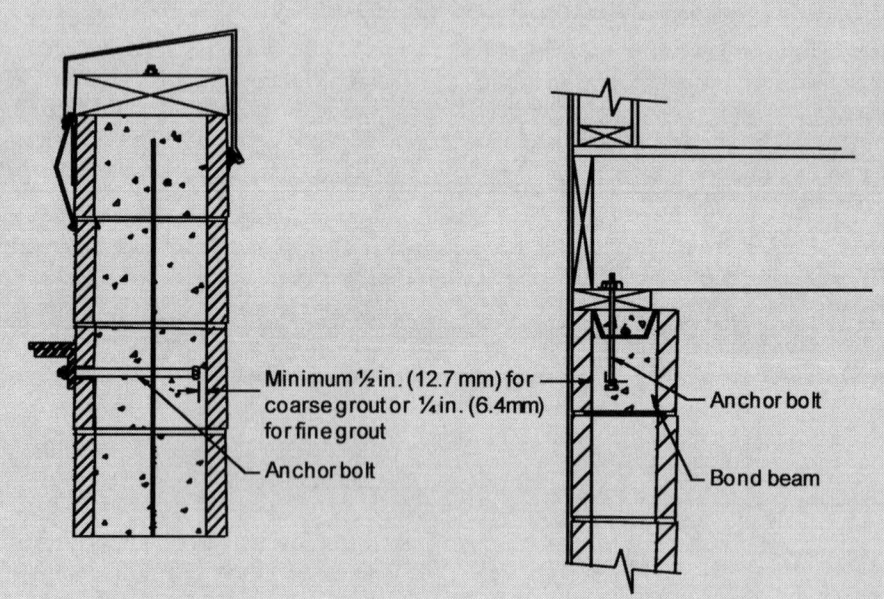

Figure CC-6.2-2 — Anchor bolt clearance requirements for headed anchor bolts – bent-bars are similar

CODE

6.2.2 Projected area for axial tension

The projected area of headed and bent-bar anchor bolts loaded in axial tension, A_{pt}, shall be determined by Equation 6-1.

$$A_{pt} = \pi l_b^2 \quad \text{(Equation 6-1)}$$

The portion of projected area overlapping an open cell, or open head joint, or that lies outside the masonry shall be deducted from the value of A_{pt} calculated using Equation 6-1. Where the projected areas of anchor bolts overlap, the value of A_{pt} calculated using Equation 6-1 shall be adjusted so that no portion of masonry is included more than once.

COMMENTARY

6.2.2 Projected area for axial tension

Results of tests (Brown and Whitlock, 1983; Allen et al, 2000) on headed anchor bolts in tension showed that anchor bolts often failed by breakout of a conically shaped section of masonry. The area, A_{pt}, is the projected area of the assumed failure cone. The cone originates at the compression bearing point of the embedment and radiates at 45° in the direction of the pull (See Figure CC-6.2-3). Other modes of tensile failure are possible. These modes include pullout (straightening of J- or L-bolts) and yield / fracture of the anchor steel.

When anchor bolts are closely spaced, stresses within the masonry begin to become additive, as shown in Figure CC-6.2-4. The Code requires that when projected areas of anchor bolts overlap, an adjustment be made so that the masonry is not overloaded. When the projected areas of two or more anchors overlap, the anchors with overlapping projected areas should be treated as an anchor group. The projected areas of the anchors in the group are summed, this area is adjusted for overlapping areas, and the capacity of the anchor group is calculated using the adjusted area in place of A_{pt}. See Figure CC-6.2-5 for examples of calculating adjusted values of A_{pt}. The equations given in Figure CC-6.2-5 are valid only when the projected areas of the bolts overlap.

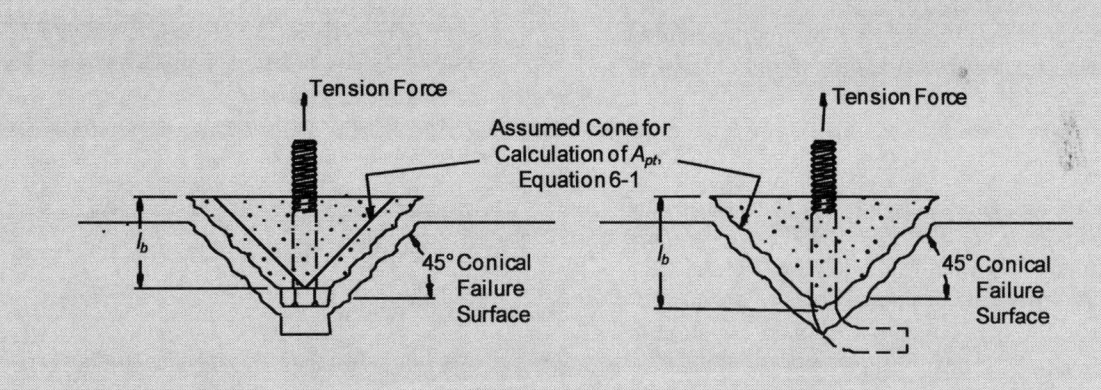

Figure CC-6.2-3 — Anchor bolt tensile breakout cone

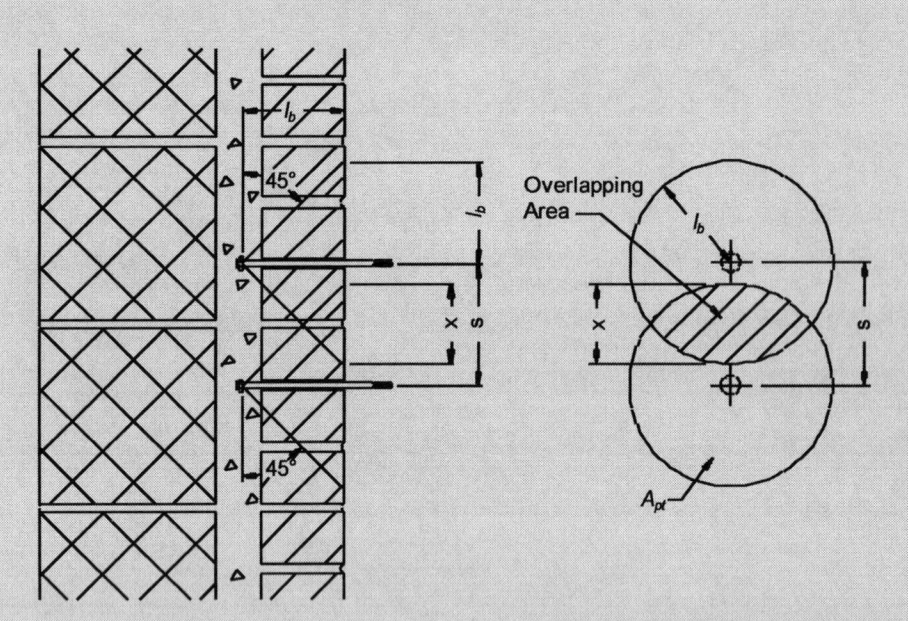

Figure CC-6.2-4 — Overlapping anchor bolt breakout cones

CODE

6.2.3 *Projected area for shear*

The projected area of headed and bent-bar anchor bolts loaded in shear, A_{pv}, shall be determined from Equation 6-2.

$$A_{pv} = \frac{\pi l_{be}^2}{2} \qquad \text{(Equation 6-2)}$$

The portion of projected area overlapping an open cell, or open head joint, or that lies outside the masonry shall be deducted from the value of A_{pv} calculated using Equation 6-2. Where the projected areas of anchor bolts overlap, the value of A_{pv} calculated using Equation 6-2 shall be adjusted so that no portion of masonry is included more than once.

COMMENTARY

6.2.3 *Projected area for shear*

Results of tests (Brown and Whitlock, 1983; Allen et al, 2000) on anchor bolts in shear showed that anchor bolts often failed by breakout of a conically shaped section of masonry. The area A_{pv} is the projected area of the assumed failure cone. The cone originates at the compression bearing point of the embedment and radiates at 45° in the direction of the shear force towards the free edge of the masonry (See Figure CC-6.2-6). Pryout (See Figure CC-6.2-7), masonry crushing, and yielding / fracture of the anchor steel are other possible failure modes.

When the projected areas of two or more anchors overlap, the shear design of these anchors should follow the same procedure as for the tension design of overlapping anchors. See Commentary Section 6.2.2.

COMMENTARY

A_{pt} at Top of Wall for Uplift

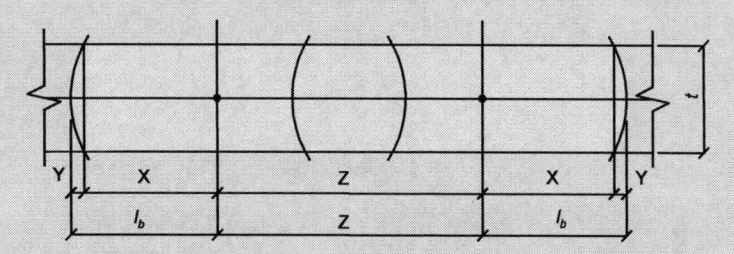

For $l_b \leq z \leq 2X$

$$X = \frac{1}{2}\sqrt{4(l_b)^2 - t^2}$$

$$Y = l_b - X = l_b - \frac{1}{2}\sqrt{4(l_b)^2 - t^2}$$

$$\therefore A_{pt} = (2X+Z)t + l_b^2\left(\frac{\pi\theta}{180} - \sin\theta\right) \quad \text{where } \theta = 2\arcsin\left(\frac{t/2}{l_b}\right) \text{ in degrees}$$

For $0 \leq z \leq l_b$

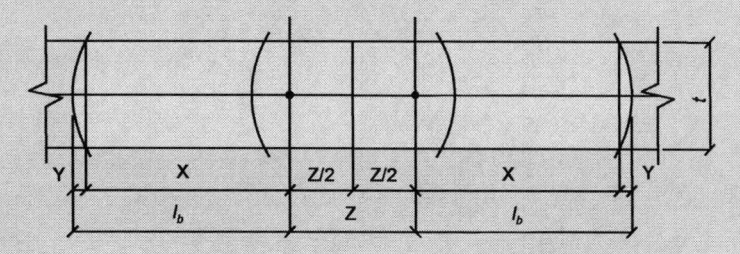

$$\therefore A_{pt} = (2X+Z)t + l_b^2\left(\frac{\pi\theta}{180} - \sin\theta\right) \quad \text{where } \theta = 2\arcsin\left(\frac{t/2}{l_b}\right) \text{ in degrees}$$

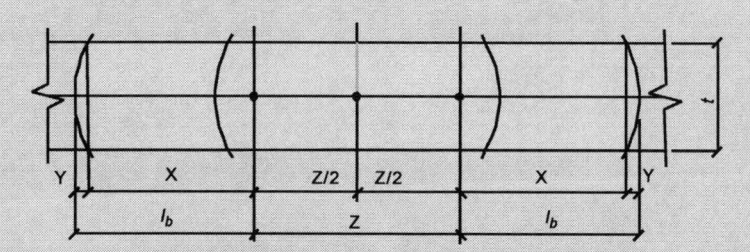

$$\therefore A_{pt} = (2X+Z)t + l_b^2\left(\frac{\pi\theta}{180} - \sin\theta\right) \quad \text{where } \theta = 2\arcsin\left(\frac{t/2}{l_b}\right) \text{ in degrees}$$

Figure CC-6.2-5 — Calculation of Adjusted Values of A_{pt} (Plan Views)

COMMENTARY

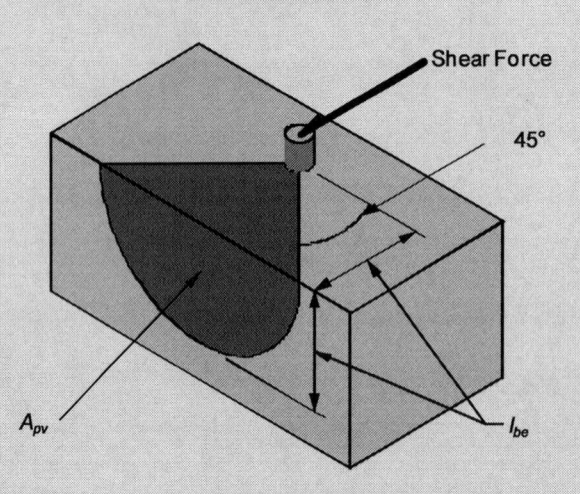

Figure CC-6.2-6 — Anchor bolt shear breakout

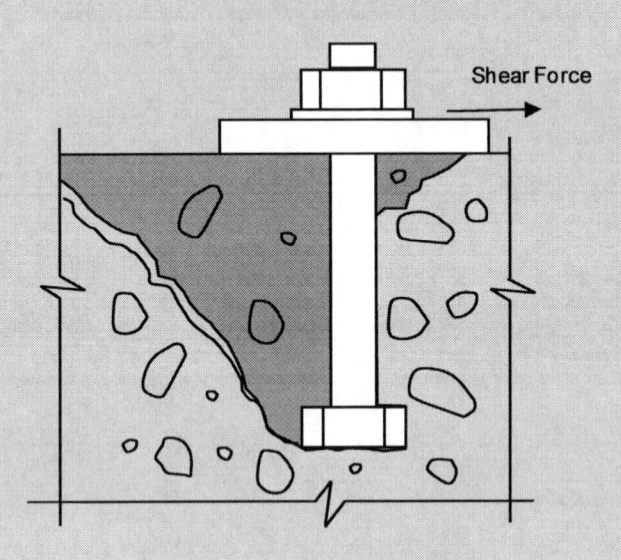

Figure CC-6.2-7 — Anchor bolt shear pryout

CODE

6.2.4 *Effective embedment length for headed anchor bolts*

The effective embedment length for a headed anchor bolt, l_b, shall be the length of the embedment measured perpendicular from the masonry surface to the compression bearing surface of the anchor head.

6.2.5 *Effective embedment length for bent-bar anchor bolts*

The effective embedment for a bent-bar anchor bolt, l_b, shall be the length of embedment measured perpendicular from the masonry surface to the compression bearing surface of the bent end, minus one anchor bolt diameter.

COMMENTARY

6.2.5 *Effective embedment length for bent-bar anchor bolts*

Tests (Brown and Whitlock, 1983) have shown that the pullout strength of bent-bar anchor bolts correlated best with a reduced embedment length. This may be explained with reference to Figure CC-6.2-8. Due to the radius of the bend, stresses are concentrated at a point less than the full embedment length.

BUILDING CODE REQUIREMENTS FOR MASONRY STRUCTURES AND COMMENTARY

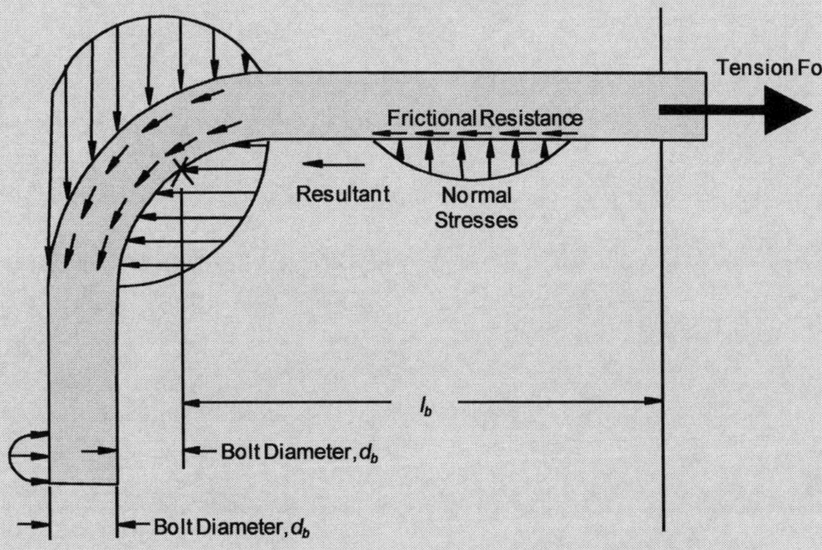

COMMENTARY

Figure CC-6.2-8 — Stress distribution on bent-bar anchor bars

CODE

6.2.6 *Minimum permissible effective embedment length*

The minimum permissible effective embedment length for headed and bent-bar anchor bolts shall be the greater of 4 bolt diameters or 2 in. (50.8 mm).

6.2.7 *Anchor bolt edge distance*

Anchor bolt edge distance, l_{be}, shall be measured in the direction of load from the edge of masonry to center of the cross section of anchor bolt.

COMMENTARY

6.2.6 *Minimum permissible effective embedment length*

The minimum embedment length requirement is considered a practical minimum based on typical construction methods for embedding anchor bolts in masonry. The validity of Code equations for shear and tension capacities of anchor bolts have not been verified by testing of anchor bolts with embedment lengths less than four bolt diameters.

This page intentionally left blank

CHAPTER 7
SEISMIC DESIGN REQUIREMENTS

CODE

7.1 — Scope

The seismic design requirements of Chapter 7 shall apply to the design and construction of masonry, except glass unit masonry and masonry veneer.

COMMENTARY

7.1 — Scope

The requirements in this section have been devised to improve performance of masonry construction when subjected to earthquake loads. Minimum seismic loading requirements are drawn from the legally adopted building code. In the event that the legally adopted building code does not contain appropriate criteria for the determination of seismic forces, the Code requires the use of ASCE 7, which represented the state-of-the-art in seismic design at the time these requirements were developed. Obviously, the seismic design provisions of this section may not be compatible with every edition of every building code that could be used in conjunction with these requirements. As with other aspects of structural design, the designer should understand the implications and limits of combining the minimum loading requirements of other documents with the resistance provisions of this Code.

Seismic design is not optional regardless of the assigned Seismic Design Category, the absolute value of the design seismic loads, or the relative difference between the design seismic loads and other design lateral forces such as wind. Unlike other design loads, seismic design of reinforced masonry elements permits inelastic response of the system, which in turn reduces the seismic design load. This reduction in load presumes an inherent level of inelastic ductility that may not otherwise be present if seismic design was neglected. When nonlinear response is assumed by reducing the seismic loading by an R factor greater than 1.5, the resulting seismic design load may be less than other loading conditions that assume a linear elastic model of the system. This is often misinterpreted by some to mean that the seismic loads do not 'control' the design and can be neglected. For the masonry system to be capable of achieving the ductility-related lower seismic loads, however, the minimum seismic design and detailing requirements of this section must be met.

The seismic design requirements are presented in a cumulative format. Thus, the provisions for Seismic Design Categories E and F include provisions for Seismic Design Category D, which include provisions for Seismic Design Category C, and so on.

This section does not apply to the design or detailing of masonry veneers or glass unit masonry systems. Seismic requirements for masonry veneers are provided in Chapter 12, Veneers. Glass unit masonry systems, by definition and design, are isolated, non-load-bearing elements and therefore cannot be used to resist seismic loads other than those induced by their own mass.

CODE

7.2 — General analysis

7.2.1 *Element interaction* — The interaction of structural and nonstructural elements that affect the linear and nonlinear response of the structure to earthquake motions shall be considered in the analysis.

7.2.2 *Load path* — Structural masonry elements that transmit forces resulting from earthquakes to the foundation shall comply with the requirements of Chapter 7.

7.2.3 *Anchorage design* — Load path connections and minimum anchorage forces shall comply with the requirements of the legally adopted building code. When the legally adopted building code does not prescribe minimum load path connection requirements and anchorage design forces, the requirements of ASCE 7 shall be used.

7.2.4 *Drift limits* — Under loading combinations that include earthquake, masonry structures shall be designed so the calculated story drift, Δ, does not exceed the allowable story drift, Δ_a, obtained from the legally adopted building code. When the legally adopted building code does not prescribe allowable story drifts, structures shall be designed so the calculated story drift, Δ, does not exceed the allowable story drift, Δ_a, obtained from ASCE 7.

It shall be permitted to assume that the following shear wall types comply with the story drift limits of ASCE 7: empirical, ordinary plain (unreinforced), detailed plain (unreinforced), ordinary reinforced, intermediate reinforced, ordinary plain (unreinforced) AAC masonry shear walls, and detailed plain (unreinforced) AAC masonry shear walls.

COMMENTARY

7.2 — General analysis

The designer is permitted to use any of the structural design methods presented in this Code to design to resist seismic loads. There are, however, limitations on some of the design methods and systems based upon the structure's assigned Seismic Design Category. For instance, empirical design (Appendix A) procedures are not permitted to be used in structures assigned to Seismic Design Categories D, E, or F. Further, empirically designed masonry elements can only be used as part of the seismic-force-resisting system in Seismic Design Category A.

7.2.1 *Element interaction* — Even if a nonstructural element is not part of the seismic-force-resisting system, it is possible for it to influence the structural response of the system during a seismic event. This may be particularly apparent due to the interaction of structural and nonstructural elements at displacements larger than those determined by linear elastic analysis.

7.2.2 *Load path* — This section clarifies load path requirements and alerts the designer that the base of the structure as defined in analysis may not necessarily correspond to the ground level.

7.2.3 *Anchorage design* — Previous editions of the Code contained minimum anchorage and connection design forces based upon antiquated service-level earthquake loads and velocity-related acceleration parameters. As these are minimum design loads, their values should be determined using load standards.

Experience has demonstrated that one of the chief causes of failure of masonry construction during earthquakes is inadequate anchorage of masonry walls to floors and roofs. For this reason, an arbitrary minimum anchorage based upon previously established practice has been set as noted in the referenced documents. When anchorage is between masonry walls and wood framed floors or roofs, the designer should avoid the use of wood ledgers in cross-grain bending.

7.2.4 *Drift limits* — Excessive deformation, particularly resulting from inelastic displacements, can potentially result in instability of the seismic-force-resisting system. This section provides procedures for the limitation of story drift. The term "drift" has two connotations:

1. "Story drift" is the maximum calculated lateral displacement within a story (the calculated displacement of one level relative to the level below caused by the effects of design seismic loads).

2. The calculated lateral displacement or deflection due to design seismic loads is the absolute displacement of any point in the structure relative to the base. This is not "story drift" and is not to be used for drift control or stability considerations because it may give a false impression of the effects in critical stories. However, it is important when considering seismic separation requirements.

BUILDING CODE REQUIREMENTS FOR MASONRY STRUCTURES AND COMMENTARY   C-81

CODE

COMMENTARY

Overall or total drift is the lateral displacement of the top of a building relative to the base. The overall drift ratio is the total drift divided by the building height. Story drift is the lateral displacement of one story relative to an adjacent story. The story drift ratio is the story drift divided by the corresponding story height. The overall drift ratio is usually an indication of moments in a structure and is also related to seismic separation demands. The story drift ratio is an indication of local seismic deformation, which relates to seismic separation demands within a story. The maximum story drift ratio could exceed the overall drift ratio.

There are many reasons for controlling drift in seismic design:

(a) To control the inelastic strain within the affected elements. Although the relationship between lateral drift and maximum nonlinear strain is imprecise, so is the current state of knowledge of what strain limitations should be.

(b) Under small lateral deformations, secondary stresses are normally within tolerable limits. However, larger deformations with heavy vertical loads can lead to significant secondary moments from P-delta effects in the design. The drift limits indirectly provide upper bounds for these effects.

(c) Buildings subjected to earthquakes need drift control to restrict damage to partitions, shaft and stair enclosures, glass, and other fragile nonstructural elements and, more importantly, to minimize differential movement demands on the seismic-force-resisting elements.

The designer must keep in mind that the allowable drift limits, Δ_a, correspond to story drifts and, therefore, are applicable to each story. They must not be exceeded in any story even though the drift in other stories may be well below the limit.

Although the provisions of this Code do not give equations for calculating building separations, the distance should be sufficient to avoid damaging contact under total calculated deflection for the design loading in order to avoid interference and possible destructive hammering between buildings. The distance should be equal to the total of the lateral deflections of the two units assumed deflecting toward each other (this involves increasing the separation with height). If the effects of hammering can be shown not to be detrimental, these distances may be reduced. For very rigid shear wall structures with rigid diaphragms whose lateral deflections are difficult to estimate, older code requirements for structural separations of at least 1 in. (25.4 mm) plus ½ in. (12.7 mm) for each 10 ft (3.1 m) of height above 20 ft (6.1 m) could be used as a guide.

Empirical, ordinary plain (unreinforced), detailed plain (unreinforced), ordinary reinforced, intermediate reinforced, ordinary plain (unreinforced) AAC, and detailed plain (unreinforced) AAC masonry shear walls are inherently

CODE

7.3 — Element classification

Masonry elements shall be classified in accordance with Section 7.3.1 and 7.3.2 as either participating or nonparticipating elements of the seismic-force-resisting system.

7.3.1 *Nonparticipating elements* — Masonry elements that are not part of the seismic-force-resisting system shall be classified as nonparticipating elements and shall be isolated in their own plane from the seismic-force-resisting system except as required for gravity support. Isolation joints and connectors shall be designed to accommodate the design story drift.

7.3.2 *Participating elements* — Masonry walls that are part of the seismic-force-resisting system shall be classified as participating elements and shall comply with the requirements of Section 7.3.2.1, 7.3.2.2, 7.3.2.3, 7.3.2.4, 7.3.2.5, 7.3.2.6, 7.3.2.7, 7.3.2.8, 7.3.2.9, 7.3.2.10, 7.3.2.11 or 7.3.2.12.

COMMENTARY

designed to have relatively low inelastic deformations under seismic loads. As such, the Committee felt that requiring designers to check story drifts for these systems of low and moderate ductility was superfluous.

7.3 — Element classification

Classifying masonry elements as either participating or nonparticipating in the seismic-force-resisting system is largely a function of design intent. Participating elements are those that are designed and detailed to actively resist seismic forces, including such elements as shear walls, columns, piers, pilasters, beams, and coupling elements. Nonparticipating elements can be any masonry assembly, but are not designed to collect and resist earthquake loads from other portions of the structure.

7.3.1 *Nonparticipating elements* — In previous editions of the Code, isolation of elements that were not part of the seismic-force-resisting system was not required in Seismic Design Categories A and B, rationalized, in part, due to the low hazard associated with these Seismic Design Categories. Non-isolated, nonparticipating elements, however, can influence a structure's strength and stiffness, and as a result the distribution of lateral loads. In considering the influence nonparticipating elements can inadvertently have on the performance of a structural system, the Committee opted to require that all nonparticipating elements be isolated from the seismic-force-resisting system. The Committee is continuing to discuss alternative design options that would allow non-isolated, nonparticipating elements with corresponding checks for strength, stiffness, and compatibility.

7.3.2 *Participating elements* — A seismic-force-resisting system must be defined for every structure. Most masonry buildings use masonry shear walls to serve as the seismic-force-resisting system, although other systems are sometimes used (such as concrete or steel frames with masonry infill). Such shear walls must be designed by the engineered methods in Part 3, unless the structure is assigned to Seismic Design Category A, in which case empirical provisions of Appendix A may be used.

Twelve shear wall types are defined by the Code. Depending upon the masonry material and detailing method used to design the shear wall, each wall type is intended to have a different capacity for inelastic response and energy dissipation in the event of a seismic event. These twelve shear wall types are assigned system design parameters such as response modification factors, R, based on their expected performance and ductility. Certain shear wall types are permitted in each seismic design category, and unreinforced shear wall types are not permitted in regions of intermediate and high seismic risk. Table CC-7.3.2-1 summarizes the requirements of each of the twelve types of masonry shear walls.

TABLE CC-7.3.2-1 Requirements for Masonry Shear Walls Based on Shear Wall Designation[1]

Shear Wall Designation	Design Methods	Reinforcement Requirements	Permitted In
Empirical Design of Masonry Shear Walls	Section A.3	None	SDC A
Ordinary Plain (Unreinforced) Masonry Shear Walls	Section 8.2 or Section 9.2	None	SDC A and B
Detailed Plain (Unreinforced) Masonry Shear Walls	Section 8.2 or Section 9.2	Section 7.3.2.3.1	SDC A and B
Ordinary Reinforced Masonry Shear Walls	Section 8.3 or Section 9.3	Section 7.3.2.3.1	SDC A, B, and C
Intermediate Reinforced Masonry Shear Walls	Section 8.3 or Section 9.3	Section 7.3.2.5	SDC A, B, and C
Special Reinforced Masonry Shear Walls	Section 8.3 or Section 9.3	Section 7.3.2.6	SDC A, B, C, D, E, and F
Ordinary Plain (Unreinforced) AAC Masonry Shear Walls	Section 11.2	Section 7.3.2.7.1	SDC A and B
Detailed Plain (Unreinforced) AAC Masonry Shear Walls	Section 11.2	Section 7.3.2.8.1	SDC A and B
Ordinary Reinforced AAC Masonry Shear Walls	Section 11.3	Section 7.3.2.9	SDC A, B, C, D, E, and F
Ordinary Plain (Unreinforced) Prestressed Masonry Shear Walls	Chapter 10	None	SDC A and B
Intermediate Reinforced Prestressed Masonry Shear Walls	Chapter 10	Section 7.3.2.11	SDC A, B, and C
Special Reinforced Prestressed Masonry Shear Walls	Chapter 10	Section 7.3.2.12	SDC A, B, C, D, E, and F

[1] Section and Chapter references in this table refer to Code Sections and Chapters.

CODE

7.3.2.1 *Empirical design of masonry shear walls* — Empirical design of shear walls shall comply with the requirements of Section A.3.

7.3.2.2 *Ordinary plain (unreinforced) masonry shear walls* — Design of ordinary plain (unreinforced) masonry shear walls shall comply with the requirements of Section 8.2 or Section 9.2.

7.3.2.3 *Detailed plain (unreinforced) masonry shear walls* — Design of detailed plain (unreinforced) masonry shear walls shall comply with the requirements of Section 8.2 or Section 9.2, and shall comply with the requirements of Section 7.3.2.3.1.

7.3.2.3.1 *Minimum reinforcement requirements* — Vertical reinforcement of at least 0.2 in.^{2} (129 mm^2) in cross-sectional area shall be provided at corners, within 16 in. (406 mm) of each side of openings, within 8 in. (203 mm) of each side of movement joints, within 8 in. (203 mm) of the ends of walls, and at a maximum spacing of 120 in. (3048 mm) on center.

Vertical reinforcement adjacent to openings need not be provided for openings smaller than 16 in. (406 mm), unless the distributed reinforcement is interrupted by such openings.

Horizontal reinforcement shall consist of at least two longitudinal wires of W1.7 (MW11) joint reinforcement spaced not more than 16 in. (406 mm) on center, or at least 0.2 in.^{2} (129 mm^2) in cross-sectional area of bond beam reinforcement spaced not more than 120 in. (3048 mm) on center. Horizontal reinforcement shall also be provided: at the bottom and top of wall openings and shall extend at least 24 in. (610 mm) but not less than 40 bar diameters past the opening; continuously at structurally connected roof and floor levels; and within 16 in. (406 mm) of the top of walls.

Horizontal reinforcement adjacent to openings need not be provided for openings smaller than 16 in. (406 mm), unless the distributed reinforcement is interrupted by such openings.

COMMENTARY

7.3.2.1 *Empirical design of masonry shear walls* — These shear walls are permitted to be used only in Seismic Design Category A. Empirical masonry shear walls are not designed or required to contain reinforcement.

7.3.2.2 *Ordinary plain (unreinforced) masonry shear walls* — These shear walls are permitted to be used only in Seismic Design Categories A and B. Plain masonry walls are designed as unreinforced masonry, although they may in fact contain reinforcement.

7.3.2.3 *Detailed plain (unreinforced) masonry shear walls* — These shear walls are designed as plain (unreinforced) masonry in accordance with the sections noted, but contain minimum reinforcement in the horizontal and vertical directions. Walls that are designed as unreinforced, but that contain minimum prescriptive reinforcement, have more favorable seismic design parameters, including higher response modification coefficients, *R*, than ordinary plain (unreinforced) masonry shear walls.

7.3.2.3.1 *Minimum reinforcement requirements* — The provisions of this section require a judgment-based minimum amount of reinforcement to be included in reinforced masonry wall construction. Tests reported in Gulkan et al (1979) have confirmed that masonry construction, reinforced as indicated, performs adequately considering the highest Seismic Design Category permitted for this shear wall type. This minimum required reinforcement may also be used to resist design loads.

CODE

7.3.2.4 *Ordinary reinforced masonry shear walls* — Design of ordinary reinforced masonry shear walls shall comply with the requirements of Section 8.3 or Section 9.3, and shall comply with the requirements of Section 7.3.2.3.1.

7.3.2.5 *Intermediate reinforced masonry shear walls* — Design of intermediate reinforced masonry shear walls shall comply with the requirements of Section 8.3 or Section 9.3. Reinforcement detailing shall also comply with the requirements of Section 7.3.2.3.1, except that the spacing of vertical reinforcement shall not exceed 48 in. (1219 mm).

7.3.2.6 *Special reinforced masonry shear walls* — Design of special reinforced masonry shear walls shall comply with the requirements of Section 8.3, Section 9.3, or Appendix C. Reinforcement detailing shall also comply with the requirements of Section 7.3.2.3.1 and the following:

(a) The maximum spacing of vertical reinforcement shall be the smallest of one-third the length of the shear wall, one-third the height of the shear wall, and 48 in. (1219 mm) for masonry laid in running bond and 24 in. (610 mm) for masonry not laid in running bond.

(b) The maximum spacing of horizontal reinforcement required to resist in-plane shear shall be uniformly distributed, shall be the smaller of one-third the length of the shear wall and one-third the height of the shear wall, and shall be embedded in grout. The maximum spacing of horizontal reinforcement shall not exceed 48 in. (1219 mm) for masonry laid in running bond and 24 in. (610 mm) for masonry not laid in running bond.

(c) The minimum cross-sectional area of vertical reinforcement shall be one-third of the required shear reinforcement. The sum of the cross-sectional area of horizontal and vertical reinforcement shall be at least 0.002 multiplied by the gross cross-sectional area of the wall, using specified dimensions.

COMMENTARY

7.3.2.4 *Ordinary reinforced masonry shear walls* — These shear walls are required to meet minimum requirements for reinforced masonry as noted in the referenced sections. Because they contain reinforcement, these walls can generally accommodate larger deformations and exhibit higher capacities than similarly configured plain (unreinforced) masonry walls. Hence, they are permitted in both areas of low and moderate seismic risk. Additionally, these walls have more favorable seismic design parameters, including higher response modification factors, R, than plain (unreinforced) masonry shear walls. To provide the minimum level of assumed inelastic ductility, however, minimum reinforcement is required as noted in Section 7.3.2.3.1.

7.3.2.5 *Intermediate reinforced masonry shear walls* — These shear walls are designed as reinforced masonry as noted in the referenced sections, and are also required to contain a minimum amount of prescriptive reinforcement. Because they contain reinforcement, their seismic performance is better than that of plain (unreinforced) masonry shear walls, and they are accordingly permitted in both areas of low and moderate seismic risk. Additionally, these walls have more favorable seismic design parameters including higher response modification factors, R, than plain (unreinforced) masonry shear walls and ordinary reinforced masonry shear walls.

7.3.2.6 *Special reinforced masonry shear walls* — These shear walls are designed as reinforced masonry as noted in the referenced sections and are also required to meet restrictive reinforcement and material requirements. Accordingly, they are permitted to be used as part of the seismic-force-resisting system in any Seismic Design Category. Additionally, these walls have the most favorable seismic design parameters, including the highest response modification factor, R, of any of the masonry shear wall types. The intent of Sections 7.3.2.6(a) through 7.3.2.6(f) is to provide a minimum level of in-plane shear reinforcement to improve ductility.

CODE

1. For masonry laid in running bond, the minimum cross-sectional area of reinforcement in each direction shall be at least 0.0007 multiplied by the gross cross-sectional area of the wall, using specified dimensions.

2. For masonry not laid in running bond, the minimum cross-sectional area of vertical reinforcement shall be at least 0.0007 multiplied by the gross cross-sectional area of the wall, using specified dimensions. The minimum cross-sectional area of horizontal reinforcement shall be at least 0.0015 multiplied by the gross cross-sectional area of the wall, using specified dimensions.

(d) Shear reinforcement shall be anchored around vertical reinforcing bars with a standard hook.

(e) Mechanical splices in flexural reinforcement in plastic hinge zones shall develop the specified tensile strength of the spliced bar.

(f) Masonry not laid in running bond shall be fully grouted and shall be constructed of hollow open-end units or two wythes of solid units.

7.3.2.6.1 *Shear capacity design*

7.3.2.6.1.1 When designing special reinforced masonry shear walls to resist in-plane forces in accordance with Section 9.3, the design shear strength, ϕV_n, shall exceed the shear corresponding to the development of 1.25 times the nominal flexural strength, M_n, of the element, except that the nominal shear strength, V_n, need not exceed 2.5 times required shear strength, V_u.

COMMENTARY

(e) In a structure undergoing inelastic deformations during an earthquake, the tensile stresses in flexural reinforcement in plastic hinge zones may approach the tensile strength of the reinforcement. This requirement is intended to avoid a splice failure in such reinforcement.

7.3.2.6.1 *Shear capacity design* — While different concepts and applications, the requirements of Code Section 7.3.2.6.1.1 and 7.3.2.6.1.2 are different methods of attempting to limit shear failures prior to nonlinear flexural behavior – or if one prefers – increase element ductility. The MSJC recognizes the slight discrepancy between the 2.5 design cap in Code Section 7.3.2.6.1.1 and the 1.5 load factor in Code Section 7.3.2.6.1.2. Given the historical precedence of each of these values, the Committee opted to maintain the two distinct values. When all factors and requirements for special reinforced masonry shear walls are considered, the resulting difference between the two requirements is small.

7.3.2.6.1.1 In previous editions of the Code, this design requirement was applied to all masonry elements designed by the strength design method (elements participating in the seismic-force-resisting system as well as those not participating in the seismic-force-resisting system, reinforced masonry elements, and unreinforced masonry elements) as well as all loading conditions. Upon further review, this design check was considered by the Committee to be related to inelastic ductility demand for seismic resistance and was therefore specifically applied to the seismic design requirements. Further, because unreinforced masonry systems by nature exhibit limited ductility, this check is required only for special reinforced masonry shear walls.

CODE

7.3.2.6.1.2 When designing special reinforced masonry shear walls in accordance with Section 8.3, the shear or diagonal tension stress resulting from in-plane seismic forces shall be increased by a factor of 1.5. The 1.5 multiplier need not be applied to the overturning moment.

7.3.2.7 *Ordinary plain (unreinforced) AAC masonry shear walls* — Design of ordinary plain (unreinforced) AAC masonry shear walls shall comply with the requirements of Section 11.2 and Section 7.3.2.7.1.

7.3.2.7.1 *Anchorage of floor and roof diaphragms in AAC masonry structures* — Floor and roof diaphragms in AAC masonry structures shall be anchored to a continuous grouted bond beam reinforced with at least two longitudinal reinforcing bars, having a total cross-sectional area of at least 0.4 in.^{2} (260 mm^2).

7.3.2.8 *Detailed plain (unreinforced) AAC masonry shear walls* — Design of detailed plain (unreinforced) AAC masonry shear walls shall comply with the requirements of Section 11.2 and Sections 7.3.2.7.1 and 7.3.2.8.1.

7.3.2.8.1 *Minimum reinforcement requirements* — Vertical reinforcement of at least 0.2 in.^{2} (129 mm^2) shall be provided within 24 in. (610 mm) of each side of openings, within 8 in. (203 mm) of movement joints, and within 24 in. (610 mm) of the ends of walls. Vertical reinforcement adjacent to openings need not be provided for openings smaller than 16 in. (406 mm), unless the distributed reinforcement is interrupted by such openings. Horizontal reinforcement shall be provided at the bottom and top of wall openings and shall extend at least 24 in. (610 mm) but not less than 40 bar diameters past the opening. Horizontal reinforcement adjacent to openings need not be provided for openings smaller than 16 in. (406 mm), unless the distributed reinforcement is interrupted by such openings.

COMMENTARY

7.3.2.6.1.2 The 1.5 load factor for reinforced masonry shear walls that are part of the seismic-force-resisting system designed by allowable stress design procedures is applied only to in-plane shear forces. It is not intended to be used for the design of in-plane overturning moments or out-of-plane overturning moments or shear. Increasing the design seismic load is intended to make the flexure mode of failure more dominant, resulting in better ductile performance.

7.3.2.7 *Ordinary plain (unreinforced) AAC masonry shear walls* – These shear walls are philosophically similar in concept to ordinary plain (unreinforced) masonry shear walls. As such, prescriptive mild reinforcement is not required, but may actually be present.

7.3.2.8 *Detailed plain (unreinforced) AAC masonry shear walls* — Prescriptive seismic requirements for AAC masonry shear walls are less severe than for conventional masonry shear walls, and are counterbalanced by more restrictive Code requirements for bond beams and additional requirements for floor diaphragms, contained in evaluation service reports and other documents dealing with floor diaphragms of various materials. AAC masonry shear walls and a full-scale, two-story assemblage specimen with prescriptive reinforcement meeting the requirements of this section have performed satisfactorily under reversed cyclic loads representing seismic excitation (Varela et al, 2006; Tanner, 2005(a)The maximum distance from the edge of an opening or end of a wall to the vertical reinforcement is set at 24 in. (610 mm) because the typical length of an AAC unit is 24 in. (610 mm).

CODE

7.3.2.9 *Ordinary reinforced AAC masonry shear walls* — Design of ordinary reinforced AAC masonry shear walls shall comply with the requirements of Section 11.3 and Sections 7.3.2.7.1 and 7.3.2.8.1.

7.3.2.9.1 *Shear capacity design* — The design shear strength, ϕV_n, shall exceed the shear corresponding to the development of 1.25 times the nominal flexural strength, M_n, of the element, except that the nominal shear strength, V_n, need not exceed 2.5 times required shear strength, V_u.

7.3.2.10 *Ordinary plain (unreinforced) prestressed masonry shear walls* — Design of ordinary plain (unreinforced) prestressed masonry shear walls shall comply with the requirements of Chapter 10.

7.3.2.11 *Intermediate reinforced prestressed masonry shear walls* — Intermediate reinforced prestressed masonry shear walls shall comply with the requirements of Chapter 10, the reinforcement detailing requirements of Section 7.3.2.3.1, and the following:

(a) Reinforcement shall be provided in accordance with Sections 7.3.2.6(a) and 7.3.2.6(b).

(b) The minimum area of horizontal reinforcement shall be $0.0007bd_v$.

(c) Shear walls subjected to load reversals shall be symmetrically reinforced.

(d) The nominal moment strength at any section along the shear wall shall not be less than one-fourth the maximum moment strength.

(e) The cross-sectional area of bonded tendons shall be considered to contribute to the minimum reinforcement in Sections 7.3.2.3.1, 7.3.2.6(a), and 7.3.2.6(b).

(f) Tendons shall be located in cells that are grouted the full height of the wall.

COMMENTARY

7.3.2.9 *Ordinary reinforced AAC masonry shear walls*

7.3.2.10 *Ordinary plain (unreinforced) prestressed masonry shear walls* — These shear walls are philosophically similar in concept to ordinary plain (unreinforced) masonry shear walls. As such, prescriptive mild reinforcement is not required, but may actually be present.

7.3.2.11 *Intermediate reinforced prestressed masonry shear walls* — These shear walls are philosophically similar in concept to intermediate reinforced masonry shear walls. To provide the intended level of inelastic ductility, prescriptive mild reinforcement is required. For consistency with 2003 IBC, intermediate reinforced prestressed masonry shear walls should include the detailing requirements from Section 7.3.2.6 (a) as well as Sections 3.2.3.5 and 3.2.4.3.2 (c) from the 2002 MSJC.

ASCE 7, Tables 12.2-1 and 12.14-1 conservatively combine all prestressed masonry shear walls into one category for seismic coefficients and structural system limitations on seismic design categories and height. The design limitations included in those tables are representative of ordinary plain (unreinforced) prestressed masonry shear walls. The criteria specific to intermediate reinforced prestressed shear walls have not yet been included from IBC 2003, Table 1617.6.2. To utilize the seismic criteria from IBC 2003, the structure would have to be accepted under Section 1.3, Approval of special systems of design and construction.

The seismic coefficients from IBC 2003, Table 1617.6.2 and the building height limitations based upon seismic design category are shown in Table CC-7.3.2-2.

CODE

7.3.2.12 *Special reinforced prestressed masonry shear walls* — Special reinforced prestressed masonry shear walls shall comply with the requirements of Chapter 10, the reinforcement detailing requirements of Sections 7.3.2.3.1 and 7.3.2.11 and the following:

(a) The cross-sectional area of bonded tendons shall be considered to contribute to the minimum reinforcement in Sections 7.3.2.3.1 and 7.3.2.11.

(b) Prestressing tendons shall consist of bars conforming to ASTM A722/A722M.

(c) All cells of the masonry wall shall be grouted.

(d) The requirements of Section 9.3.3.5 or 9.3.6.5 shall be met. Dead load axial forces shall include the effective prestress force, $A_{ps}f_{se}$.

(e) The design shear strength, ϕV_n, shall exceed the shear corresponding to the development of 1.25 times the nominal flexural strength, M_n, of the element, except that the nominal shear strength, V_n, need not exceed 2.5 times required shear strength, V_u.

COMMENTARY

7.3.2.12 *Special reinforced prestressed masonry shear* walls — These shear walls are philosophically similar in concept to special reinforced masonry shear walls. To provide the intended level of inelastic ductility, prescriptive mild reinforcement is required. For consistency with 2003 IBC, special reinforced prestressed masonry shear walls should include the detailing requirements from Sections 3.2.3.5 and 3.2.4.3.2 (c) from the 2002 MSJC.

ASCE 7, Table 12.2-1 and ASCE 7, Table 12.14-1 conservatively combine all prestressed masonry shear walls into one category for seismic coefficients and structural system limitations on seismic design categories and height. The design limitations included in those tables are representative of ordinary plain (unreinforced) prestressed masonry shear walls. The criteria specific to special reinforced prestressed shear walls have not yet been included from IBC 2003, Table 1617.6.2. To utilize the seismic criteria from IBC 2003, the structure would have to be accepted under Section 1.3, Approval of special systems of design and construction.

See Table CC-7.3.2-2. The data in this table is similar to ASCE 7, Table 12.2-1. Users that prefer to use the Simplified Design Procedure in ASCE 7 should interpret the table for use in lieu of ASCE 7, Table 12.14-1.

TABLE CC-7.3.2-2 2003 IBC Seismic Coefficients for Prestressed Masonry Shear Walls

	Response Modification Coefficient, R	System Overstrength Factor, Ω_o	Deflection Amplification Factor, C_d	A or B	C	D	E	F
				\multicolumn{5}{c	}{SYSTEM LIMITATIONS AND BUILDING HEIGHT LIMITATIONS (FEET) BY SEISMIC DESIGN CATEGORY}
Ordinary Plain Prestressed	1½	2½	1¼	NL	NP	NP	NP	NP
Intermediate Reinforced Prestressed	3 for Building Frame System and 2½ for Bearing Wall System	2½	2½	NL	35	NP	NP	NP
Special Reinforced Prestressed	4½	2½	4 for Building Frame System and 3½ for Bearing Wall System	NL	35	35	35	35

NL = no limit   NP = not permitted

CODE

7.4 — Seismic Design Category requirements

The design of masonry elements shall comply with the requirements of Sections 7.4.1 through 7.4.5 based on the Seismic Design Category as defined in the legally adopted building code. When the legally adopted building code does not define Seismic Design Categories, the provisions of ASCE 7 shall be used.

7.4.1 *Seismic Design Category A requirements* — Masonry elements in structures assigned to Seismic Design Category A shall comply with the requirements of Sections 7.1, 7.2, 7.4.1.1, and 7.4.1.2.

7.4.1.1 *Design of nonparticipating elements* — Nonparticipating masonry elements shall comply with the requirements of Section 7.3.1 and Chapter 8, 9, 10, 11, 14, Appendix A, or Appendix B.

7.4.1.2 *Design of participating elements* — Participating masonry elements shall be designed to comply with the requirements of Chapter 8, 9, 10, 11, Appendix A, or Appendix B. Masonry shear walls shall be designed to comply with the requirements of Section 7.3.2.1, 7.3.2.2, 7.3.2.3, 7.3.2.4, 7.3.2.5, 7.3.2.6, 7.3.2.7, 7.3.2.8, 7.3.2.9, 7.3.2.10, 7.3.2.11, or 7.3.2.12.

7.4.2 *Seismic Design Category B requirements* — Masonry elements in structures assigned to Seismic Design Category B shall comply with the requirements of Section 7.4.1 and with the additional requirements of Section 7.4.2.1.

7.4.2.1 *Design of participating elements* — Participating masonry elements shall be designed to comply with the requirements of Chapter 8, 9, 10, 11, or Appendix B. Masonry shear walls shall be designed to comply with the requirements of Section 7.3.2.2, 7.3.2.3, 7.3.2.4, 7.3.2.5, 7.3.2.6, 7.3.2.7, 7.3.2.8, 7.3.2.9, 7.3.2.10, 7.3.2.11, or 7.3.2.12.

COMMENTARY

7.4 — Seismic Design Category requirements

Every structure is assigned to a Seismic Design Category (SDC) in accordance with the legally adopted building code or per the requirements of ASCE 7, whichever govern for the specific project under consideration. Previous editions of the Code included requirements for Seismic Performance Categories and Seismic Zones, each of which is different than a Seismic Design Category.

7.4.1 *Seismic Design Category A requirements* — The general requirements of this Code provide for adequate performance of masonry construction assigned to Seismic Design Category A structures.

7.4.2 *Seismic Design Category B requirements* — Although masonry may be designed by the provisions of Chapter 8, Allowable Stress Design of Masonry; Chapter 9, Strength Design of Masonry; Chapter 10, Prestressed Masonry; Chapter 11, Strength Design of Autoclave Aerated Concrete (AAC) Masonry; or Appendix A, Empirical Design of Masonry; or Appendix B, Design of Masonry Infill, the seismic-force-resisting system for structures assigned to Seismic Design Category B must be designed based on a structural analysis in accordance with Chapter 8, 9, 10 or 11, or Appendix B. The provisions of Appendix A cannot be used to design the seismic-force-resisting system of buildings assigned to Seismic Design Category B or higher.

CODE

7.4.3 *Seismic Design Category C requirements* — Masonry elements in structures assigned to Seismic Design Category C shall comply with the requirements of Section 7.4.2 and with the additional requirements of Section 7.4.3.1 and 7.4.3.2.

7.4.3.1 *Design of nonparticipating elements* — Nonparticipating masonry elements shall comply with the requirements of Section 7.3.1 and Chapter 8, 9, 10, 11, 14, Appendix A, or Appendix B. Nonparticipating masonry elements, except those constructed of AAC masonry, shall be reinforced in either the horizontal or vertical direction in accordance with the following:

(a) *Horizontal reinforcement* — Horizontal reinforcement shall consist of at least two longitudinal wires of W1.7 (MW11) bed joint reinforcement spaced not more than 16 in. (406 mm) on center for walls greater than 4 in. (102 mm) in width and at least one longitudinal W1.7 (MW11) wire spaced not more than 16 in. (406 mm) on center for walls not exceeding 4 in. (102 mm) in width or at least one No. 4 (M #13) bar spaced not more than 48 in. (1219 mm) on center. Where two longitudinal wires of joint reinforcement are used, the space between these wires shall be the widest that the mortar joint will accommodate. Horizontal reinforcement shall be provided within 16 in. (406 mm) of the top and bottom of these masonry walls.

(b) *Vertical reinforcement* — Vertical reinforcement shall consist of at least one No. 4 (M #13) bar spaced not more than 120 in. (3048 mm). Vertical reinforcement shall be located within 16 in. (406 mm) of the ends of masonry walls.

7.4.3.2 *Design of participating elements* — Participating masonry elements shall be designed to comply with the requirements of Section 8.3, 9.3, 11.3, or Appendix B. Masonry shear walls shall be designed to comply with the requirements of Section 7.3.2.4, 7.3.2.5, 7.3.2.6, 7.3.2.9, 7.3.2.11, or 7.3.2.12.

COMMENTARY

7.4.3 *Seismic Design Category C requirements* — In addition to the requirements of Seismic Design Category B, minimum levels of reinforcement and detailing are required. The minimum provisions for improved performance of masonry construction in Seismic Design Category C must be met regardless of the method of design. Shear walls designed as part of the seismic-force-resisting system in Seismic Design Category C and higher must be designed using reinforced masonry methods because of the increased risk and expected intensity of seismic activity. Ordinary reinforced masonry shear walls, ordinary reinforced AAC masonry shear walls, intermediate reinforced masonry shear walls, special reinforced masonry shear walls, or masonry infills are required to be used.

7.4.3.1 *Design of nonparticipating elements* — Reinforcement requirements of Section 7.4.3.1 are traditional for conventional concrete and clay masonry. They are prescriptive in nature. The intent of this requirement is to provide structural integrity for nonparticipating masonry walls. AAC masonry walls differ from concrete masonry walls and clay masonry walls in that the thin-bed mortar strength and associated bond strength is typically greater than that of the AAC units. Also, the unit weight of AAC masonry is typically less than one-third of the unit weight of clay or concrete masonry, reducing seismic inertial forces. This reduced load, combined with a tensile bond strength that is higher than the strength of the AAC material itself, provides a minimum level of structural integrity. Therefore, prescriptive reinforcement is not required. All masonry walls, including non-participating AAC masonry walls, are required to be designed to resist out-of-plane forces. If reinforcement is required, it must be provided in the direction of the span.

CODE

7.4.3.2.1 *Connections to masonry columns* — Where anchor bolts are used to connect horizontal elements to the tops of columns, anchor bolts shall be placed within lateral ties. Lateral ties shall enclose both the vertical bars in the column and the anchor bolts. There shall be a minimum of two No. 4 (M #13) lateral ties provided in the top 5 in. (127 mm) of the column.

7.4.3.2.2 *Anchorage of floor and roof diaphragms in AAC masonry structures* — Seismic load between floor and roof diaphragms and AAC masonry shear walls shall be transferred through connectors embedded in grout and designed in accordance with Section 4.1.4.

7.4.3.2.3 *Material requirements* — ASTM C34, structural clay load-bearing wall tiles, shall not be used as part of the seismic-force-resisting system.

7.4.3.2.4 *Lateral stiffness* — At each story level, at least 80 percent of the lateral stiffness shall be provided by seismic-force-resisting walls. Along each line of lateral resistance at a particular story level, at least 80 percent of the lateral stiffness shall be provided by seismic-force-resisting walls. Where seismic loads are determined based on a seismic response modification factor, R, not greater than 1.5, piers and columns shall be permitted to be used to provide seismic load resistance.

7.4.3.2.5 *Design of columns, pilasters, and beams supporting discontinuous elements* — Columns and pilasters that are part of the seismic-force-resisting system and that support reactions from discontinuous stiff elements shall be provided with transverse reinforcement spaced at no more than one-fourth of the least nominal dimension of the column or pilaster. The minimum transverse reinforcement ratio shall be 0.0015. Beams supporting reactions from discontinuous walls shall be provided with transverse reinforcement spaced at no more than one-half of the nominal depth of the beam. The minimum transverse reinforcement ratio shall be 0.0015.

COMMENTARY

7.4.3.2.1 *Connections to masonry columns* — Connections must be designed to transfer forces between masonry columns and horizontal elements in accordance with the requirements of Section 4.1.4. Experience has demonstrated that connections of structural members to masonry columns are vulnerable to damage during earthquakes unless properly anchored. Requirements are adapted from previously established practice developed as a result of the 1971 San Fernando earthquake.

7.4.3.2.2 *Anchorage of floor and roof diaphragms in AAC masonry structures* — Connectors are required to be placed in grout because of the relatively low strength of connectors embedded in AAC. Different detailing options are available, but often the connectors are placed in bond beams near the top of the wall.

7.4.3.2.3 *Material requirements* — The limitation on the use of ASTM C34 structural clay tile units in the seismic-force-resisting system is based on these units' limited ability to provide inelastic strength.

7.4.3.2.4 *Lateral stiffness* — In order to accurately distribute loads in a structure subjected to lateral loading, the lateral stiffness of all structural members should be considered. Although structures may be designed to use shear walls for lateral-load resistance, columns may also be incorporated for vertical capacity. The stipulation that seismic-force-resisting elements provide at least 80 percent of the lateral stiffness helps ensure that additional elements do not significantly contribute to the lateral stiffness. Based on typical design assumptions, the lateral stiffness of structural elements should be based on cracked section properties for reinforced masonry and uncracked section properties for unreinforced masonry.

The designer may opt to increase the percentage of lateral stiffness provided by piers and columns if the structure is designed to perform elastically under seismic loads.

7.4.3.2.5 *Design of columns, pilasters, and beams supporting discontinuous elements* — Discontinuous stiff members such as shear walls have global overturning forces at their edges that may be supported by columns, pilasters and beams. These vertical support elements are required to have a minimum level of confinement and shear detailing at the discontinuity level. The minimum detailing requirements in this section may be in excess of those requirements that are based on calculations using full-height relative stiffnesses of the elements of the seismic-force-resisting system.

A common example is a building with internal shear walls, such as interior corridor walls, that are discontinuous at the first story above grade or in a basement level. If this structure has a rigid diaphragm at all floor and roof levels; the global (full height) relative stiffnesses of the discontinuous elements is minor in

CODE

7.4.4 *Seismic Design Category D requirements* — Masonry elements in structures assigned to Seismic Design Category D shall comply with the requirements of Section 7.4.3 and with the additional requirements of Sections 7.4.4.1 and 7.4.4.2.

Exception: Design of participating elements of AAC masonry shall comply with the requirements of Section 7.4.3.

7.4.4.1 *Design of nonparticipating elements* — Nonparticipating masonry elements shall comply with the requirements of Chapter 8, 9, 10, 11, or Appendix B. Nonparticipating masonry elements, except those constructed of AAC masonry, shall be reinforced in either the horizontal or vertical direction in accordance with the following:

(a) *Horizontal reinforcement* — Horizontal reinforcement shall comply with Section 7.4.3.1(a).

(b) *Vertical reinforcement* — Vertical reinforcement shall consist of at least one No. 4 (M #13) bar spaced not more than 48 in. (1219 mm). Vertical reinforcement shall be located within 16 in. (406 mm) of the ends of masonry walls.

7.4.4.2 *Design of participating elements* — Masonry shear walls shall be designed to comply with the requirements of Section 7.3.2.6, 7.3.2.9, or 7.3.2.12.

COMMENTARY

comparison to the relative stiffnesses of the continuous elements at the perimeter of the structure. All shear walls above the discontinuity, however, have a forced common interstory displacement. This forced interstory displacement induces overturning forces in the discontinuous shear walls at all levels having this forced story displacement. The accumulated overturning forces at the ends of the walls above the discontinuity in turn are likely to be supported by columns and pilasters in the discontinuous levels and the beams at the level above the discontinuity. This section specifies minimum detailing requirements for these columns, pilasters, and beams. The stiffness of the discontinuous element should be determined based on the relative stiffness of the discontinuous members above and below the discontinuity. If the interstory stiffness of the discontinuous wall below the discontinuity is less than 20% of the interstory stiffness above the discontinuity, the discontinuous element should be considered stiff.

7.4.4 *Seismic Design Category D requirements* — Masonry shear walls for structures assigned to Seismic Design Category D are required to meet the requirements of special reinforced masonry shear walls or ordinary reinforced AAC masonry shear walls because of the increased risk and expected intensity of seismic activity. The minimum amount of wall reinforcement for special reinforced masonry shear walls has been a long-standing, standard empirical requirement in areas of high seismic loading. It is expressed as a percentage of gross cross-sectional area of the wall. It is intended to improve the ductile behavior of the wall under earthquake loading and assist in crack control. Because the minimum required reinforcement may be used to satisfy design requirements, at least $^1/_3$ of the minimum amount is reserved for the lesser stressed direction in order to ensure an appropriate distribution of loads in both directions.

CODE

7.4.4.2.1 Minimum reinforcement for masonry columns — Lateral ties in masonry columns shall be spaced not more than 8 in. (203 mm) on center and shall be at least 3/8 in. (9.5 mm) diameter. Lateral ties shall be embedded in grout.

7.4.4.2.2 Material requirements — Fully grouted participating elements shall be designed and specified with Type S or Type M cement-lime mortar, masonry cement mortar, or mortar cement mortar. Partially grouted participating elements shall be designed and specified with Type S or Type M cement-lime mortar or mortar cement mortar.

7.4.4.2.3 Lateral tie anchorage — Standard hooks for lateral tie anchorage shall be either a 135-degree standard hook or a 180-degree standard hook.

7.4.5 Seismic Design Categories E and F requirements — Masonry elements in structures assigned to Seismic Design Category E or F shall comply with the requirements of Section 7.4.4 and with the additional requirements of Section 7.4.5.1.

7.4.5.1 Minimum reinforcement for nonparticipating masonry elements not laid in running bond — Masonry not laid in running bond in nonparticipating elements shall have a cross-sectional area of horizontal reinforcement of at least 0.0015 multiplied by the gross cross-sectional area of masonry, using specified dimensions. The maximum spacing of horizontal reinforcement shall be 24 in. (610 mm). These elements shall be fully grouted and shall be constructed of hollow open-end units or two wythes of solid units.

COMMENTARY

7.4.4.2.1 Minimum reinforcement for masonry columns — Adequate lateral restraint is important for column reinforcement subjected to overturning forces due to earthquakes. Many column failures during earthquakes have been attributed to inadequate lateral tying. For this reason, closer spacing of ties than might otherwise be required is prudent. An arbitrary minimum spacing has been established through experience. Columns not involved in the seismic-force-resisting system should also be more heavily tied at the tops and bottoms for more ductile performance and better resistance to shear.

7.4.4.2.2 Material requirements — Based on numerous tests by several researchers, (Brown and Melander, 1999, Hamid et al, 1979, Minaie et al, 2009, Klingner et al, 2010) the behavior of fully grouted walls subjected to out-of-plane flexural and in-plane shear loads is dominated by grout and unaffected by mortar formulation. In tests by Minaie et al (2009) and Klingner et al (2010), fully grouted concrete masonry walls exhibited good in-plane response when subjected to seismic loads. For fully grouted participating elements in buildings assigned to Seismic Categories D or higher, no mortar material restrictions are necessary. Historical provisions requiring use of Type S or M cement-lime or mortar cement mortar are retained for partially grouted participating elements in buildings assigned to Seismic Design Categories D or higher.

7.4.5 Seismic Design Categories E and F requirements — See Commentary Sections 7.3.2.3.1 and 7.4.4. The ratio of minimum horizontal reinforcement is increased to reflect the possibility of higher seismic loads. Where fully grouted open end hollow units are used, part of the need for horizontal reinforcement is satisfied by the mechanical continuity provided by the grout core.

PART 3: ENGINEERED DESIGN METHODS

CHAPTER 8
ALLOWABLE STRESS DESIGN OF MASONRY

CODE

8.1 — General

8.1.1 *Scope*

This chapter provides requirements for allowable stress design of masonry. Masonry designed in accordance with this chapter shall comply with the requirements of Part 1, Part 2, Sections 8.1.2 through 8.1.6, and either Section 8.2 or 8.3.

8.1.2 *Design strength*

Calculated stresses shall not exceed the allowable stress requirements of this Chapter.

COMMENTARY

8.1 — General

8.1.1 *Scope*

Chapter 8 design procedures follow allowable stress design methodology, in which the calculated stresses resulting from nominal loads must not exceed permissible masonry and steel stresses.

For allowable stress design, linear elastic materials following Hooke's Law are assumed, that is, deformations (strains) are linearly proportional to the loads (stresses). All materials are assumed to be homogeneous and isotropic, and sections that are plane before bending remain plane after bending. These assumptions are adequate within the low range of working stresses under consideration. The allowable stresses are fractions of the specified compressive strength, resulting in conservative factors of safety.

Service load is the load that is assumed by the legally adopted building code to actually occur when the structure is in service. The stresses allowed under the action of service loads are limited to values within the elastic range of the materials.

Historically, a one-third increase in allowable stress had been permitted for load combinations that included wind or seismic loads. The origin and the reason for the one-third stress increase are unclear (Ellifritt, 1977). From a structural reliability standpoint, the one-third stress increase was a poor way to handle load combination effects. Therefore, the one-third stress increase was removed from this Code beginning with the 2011 edition. The allowable stresses of this Chapter should not be increased by one-third for wind and seismic load combinations.

8.1.2 *Design strength*

Calculated stresses designated by 'f' with subscript indicating stress type are required to be less than allowable stresses designated by 'F' with subscript indicating the same stress type.

CODE

8.1.3 *Anchor bolts embedded in grout*

8.1.3.1 *Design requirements* — Anchor bolts shall be designed using either the provisions of Section 8.1.3.2 or, for headed and bent-bar anchor bolts, by the provisions of Section 8.1.3.3.

8.1.3.2 *Allowable loads determined by test*

8.1.3.2.1 Anchor bolts shall be tested in accordance with ASTM E488, except that a minimum of five tests shall be performed. Loading conditions of the test shall be representative of intended use of the anchor bolt.

8.1.3.2.2 Anchor bolt allowable loads used for design shall not exceed 20 percent of the average failure load from the tests.

8.1.3.3 *Allowable loads determined by calculation for headed and bent-bar anchor bolts* — Allowable loads for headed and bent-bar anchor bolts embedded in grout shall be determined in accordance with the provisions of Sections 8.1.3.3.1 through 8.1.3.3.3.

8.1.3.3.1 *Allowable axial tensile load of headed and bent-bar anchor bolts* — The allowable axial tensile load of headed anchor bolts shall be calculated using the provisions of Sections 8.1.3.3.1.1. The allowable axial tensile load of bent-bar anchor bolts shall be calculated using the provisions of Section 8.1.3.3.1.2.

8.1.3.3.1.1 *Allowable axial tensile load of headed anchor bolts* — The allowable axial tensile load, B_a, of headed anchor bolts embedded in grout shall be the smaller of the values determined by Equation 8-1 and Equation 8-2.

$$B_{ab} = 1.25 A_{pt} \sqrt{f'_m} \quad \text{(Equation 8-1)}$$

$$B_{as} = 0.6 A_b f_y \quad \text{(Equation 8-2)}$$

COMMENTARY

8.1.3 *Anchor bolts embedded in grout*
Significant changes in the anchor bolt design provisions were incorporated into the 2008 MSJC Code. The changes included revising the safety factors for the Allowable Stress Design anchor bolt provisions to correspond to those used for the Strength Design anchor bolt provisions, and requiring consideration of additional failure modes in both the Allowable Stress Design and Strength Design provisions. The result of these changes in the 2008 Code is that the resulting anchor bolt designs produced using Allowable Stress Design or Strength Design should be approximately the same.

See Code Commentary 9.1.6 for additional discussion on the background and application of the anchor bolt design provisions.

8.1.3.3 *Allowable loads determined by calculation for headed and bent-bar anchor bolts* — The anchor provisions in this Code define bolt shear and tension capacities based on the bolt's specified yield strength. Anchors conforming to A307, Grade A specifications are allowed by the Code, but the ASTM A307, Grade A specification does not specify a yield strength. Use of a yield strength of 37 ksi in the Code design equations for A307 anchors will result in anchor capacities similar to those obtained using the American Institute of Steel Construction provisions.

8.1.3.3.1.1 *Allowable axial tensile load of headed anchor bolts* — Equation 8-1 defines the allowable axial tensile load governed by masonry breakout. Equation 8-2 defines the allowable axial tensile load governed by steel yielding. The lower of these loads is the allowable axial tensile load on the anchor.

BUILDING CODE REQUIREMENTS FOR MASONRY STRUCTURES AND COMMENTARY

CODE

8.1.3.3.1.2 *Allowable axial tensile load of bent-bar anchor bolts* — The allowable axial tensile load, B_a, for bent-bar anchor bolts embedded in grout shall be the smallest of the values determined by Equation 8-3, Equation 8-4, and Equation 8-5.

$$B_{ab} = 1.25 A_{pt} \sqrt{f'_m} \quad \text{(Equation 8-3)}$$

$$B_{ap} = 0.6 f'_m e_b d_b + 120\pi(l_b + e_b + d_b)d_b \quad \text{(Equation 8-4)}$$

$$B_{as} = 0.6 A_b f_y \quad \text{(Equation 8-5)}$$

8.1.3.3.2 *Allowable shear load of headed and bent-bar anchor bolts* — The allowable shear load, B_v, of headed and bent-bar anchor bolts embedded in grout shall be the smallest of the values determined by Equation 8-6, Equation 8-7, Equation 8-8, and Equation 8-9.

$$B_{vb} = 1.25 A_{pv} \sqrt{f'_m} \quad \text{(Equation 8-6)}$$

$$B_{vc} = 350 \sqrt[4]{f'_m A_b} \quad \text{(Equation 8-7)}$$

$$B_{vpry} = 2.0 B_{ab} = 2.5 A_{pt} \sqrt{f'_m} \quad \text{(Equation 8-8)}$$

$$B_{vs} = 0.36 A_b f_y \quad \text{(Equation 8-9)}$$

8.1.3.3.3 *Combined axial tension and shear* — Anchor bolts subjected to axial tension in combination with shear shall satisfy Equation 8-10.

$$\frac{b_a}{B_a} + \frac{b_v}{B_v} \leq 1 \quad \text{(Equation 8-10)}$$

COMMENTARY

8.1.3.3.1.2 *Allowable axial tensile load of bent-bar anchor bolts* — Equation 8-3 defines the allowable axial tensile load governed by masonry breakout. Equation 8-4 defines the allowable axial tensile load governed by anchor pullout. Equation 8-5 defines the allowable axial tensile load governed by steel yielding. The lower of these loads is the allowable axial tensile load on the anchor.

8.1.3.3.2 *Allowable shear load of headed and bent-bar anchor bolts* — Equation 8-6 defines the allowable shear load governed by masonry breakout. Equation 8-7 defines the allowable shear load governed by masonry crushing. Equation 8-8 defines the allowable shear load governed by anchor pryout. Equation 8-9 defines the allowable shear load governed by steel yielding. The lower of these loads is the allowable shear load on the anchor.

CODE

8.1.4 *Shear stress in multiwythe masonry elements*

8.1.4.1 Design of multiwythe masonry for composite action shall meet the requirements of Section 5.1.4.2 and Section 8.1.4.2.

8.1.4.2 Shear stresses developed at the interfaces between wythes and collar joints or within headers shall not exceed the following:
(a) mortared collar joints, 7 psi (48.3 kPa).
(b) grouted collar joints, 13 psi (89.6 kPa).
(c) headers,
$1.3\sqrt{\text{specified unit compressive strength of header}}$ psi (MPa) (over net area of header).

8.1.5 *Bearing stress*

Bearing stresses on masonry shall not exceed $0.33 f'_m$ and shall be calculated over the bearing area, A_{br}, as defined in Section 4.3.4.

COMMENTARY

8.1.4 *Shear stress in multiwythe masonry elements*

Limited test data (McCarthy et al, 1985; Williams and Geschwinder, 1982; Colville et al, 1987) are available to document shear strength of collar joints in masonry.

Test results (McCarthy et al, 1985; Williams and Geschwinder, 1982) show that shear bond strength of collar joints could vary from as low as 5 psi (34.5 kPa) to as high as 100 psi (690 kPa), depending on type and condition of the interface, consolidation of the joint, and type of loading. McCarthy et al (1985) reported an average value of 52 psi (359 kPa) with a coefficient of variation of 21.6 percent. An allowable shear stress value of 7 psi (48.3 kPa), which is four standard deviations below the average, is considered to account for the expected high variability of the interface bond. With some units, Type S mortar slushed collar joints may have better shear bond characteristics than Type N mortar. Results show that thickness of joints, unit absorption, and reinforcement have a negligible effect on shear bond strength. Grouted collar joints have higher allowable shear bond stress than the mortared collar joints (Williams and Geschwinder, 1982).

A strength analysis has been demonstrated by Porter et al (1986 and 1987) for composite masonry walls subjected to combined in-plane shear and gravity loads. In addition, these authors have shown adequate behavioral characteristics for both brick-to-brick and brick-to-block composite walls with a grouted collar joint (Wolde-Tinsae et al, 1985(a); Wolde-Tinsae et al, 1985(b); Ahmed et al, 1983(a); Ahmed et al, 1983(b)). Finite element models for analyzing the interlaminar shearing stresses in collar joints of composite walls have been investigated (Anand and Young, 1982; Anand, 1985; Stevens and Anand, 1985; Anand and Rahman, 1986). They found that the shear stresses were principally transferred in the upper portion of the wall near the point of load application for the in-plane loads. Thus, below a certain distance, the overall strength of the composite masonry is controlled by the global strength of the wall, providing that the wythes are acting compositely.

CODE

8.1.6 *Development of reinforcement embedded in grout*

8.1.6.1 *General* — The calculated tension or compression in the reinforcement at each section shall be developed on each side of the section by development length, hook, mechanical device, or combination thereof. Hooks shall not be used to develop bars in compression.

8.1.6.2 *Development of wires in tension* — The development length of wire shall be determined by Equation 8-11, but shall not be less than 6 in. (152 mm).

$$l_d = 0.0015 d_b F_s \qquad \text{(Equation 8-11)}$$

Development length of epoxy-coated wire shall be taken as 150 percent of the length determined by Equation 8-11.

8.1.6.3 *Development of bars in tension or compression* — The required development length of reinforcing bars shall be determined by Equation 8-12, but shall not be less than 12 in. (305 mm).

$$l_d = \frac{0.13 d_b^2 f_y \gamma}{K \sqrt{f_m'}} \qquad \text{(Equation 8-12)}$$

K shall not exceed the smallest of the following: the minimum masonry cover, the clear spacing between adjacent reinforcement splices, and $9d_b$.

γ = 1.0 for No. 3 (M#10) through No. 5 (M#16) bars;

γ = 1.3 for No. 6 (M#19) through No. 7 (M#22) bars;

and

γ = 1.5 for No. 8 (M#25) through No. 11 (M#36) bars.

Development length of epoxy-coated bars shall be taken as 150 percent of the length determined by Equation 8-12.

COMMENTARY

8.1.6 *Development of reinforcement embedded in grout*

8.1.6.1 *General* — From a point of peak stress in reinforcement, some length of reinforcement or anchorage is necessary through which to develop the stress. This development length or anchorage is necessary on both sides of such peak stress points, on one side to transfer stress into and on the other to transfer stress out of the reinforcement. Often the reinforcement continues for a considerable distance on one side of a critical stress point so that calculations need involve only the other side; for example, the negative moment reinforcement continuing through a support to the middle of the next span.

Bars and longitudinal wires must be deformed.

8.1.6.2 *Development of wires in tension* — Equation 8-11 can be derived from the basic development length expression and an allowable bond stress u for deformed bars in grout of 160 psi (1103 kPa) (Gallagher, 1935; Richart, 1949). Research (Treece and Jirsa, 1989) has shown that epoxy-coated reinforcing bars require longer development length than uncoated reinforcing bars. The 50 percent increase in development length does not apply to the 6 in. (152 mm) minimum.

$$l_d = d_b F_s / 4u = d_b F_s / 4(160) = 0.0015 d_b F_s$$

($l_d = 0.22 d_b F_s$ in SI units)

8.1.6.3 *Development of bars in tension or compression* — The 50 percent increase in development length does not apply to the 12 in. (305 mm) minimum.

CODE

8.1.6.4 *Embedment of flexural reinforcement*

8.1.6.4.1 *General*

8.1.6.4.1.1 Tension reinforcement is permitted to be developed by bending across the neutral axis of the member to be anchored or made continuous with reinforcement on the opposite face of the member.

8.1.6.4.1.2 Critical sections for development of reinforcement in flexural members are at points of maximum steel stress and at points within the span where adjacent reinforcement terminates or is bent.

8.1.6.4.1.3 Reinforcement shall extend beyond the point at which it is no longer required to resist flexure for a distance equal to the effective depth of the member or $12d_b$, whichever is greater, except at supports of simple spans and at the free end of cantilevers.

COMMENTARY

8.1.6.4 *Embedment of flexural reinforcement* — Figure CC-8.1-1 illustrates the embedment requirements of flexural reinforcement in a typical continuous beam. Figure CC-8.1-2 illustrates the embedment requirements in a typical continuous wall that is not part of the lateral-force-resisting system.

8.1.6.4.1 *General*

8.1.6.4.1.2 Critical sections for a typical continuous beam are indicated with a "c" or an "x" in Figure CC-8.1-1. Critical sections for a typical continuous wall are indicated with a "c" in Figure CC-8.1-2.

8.1.6.4.1.3 The moment diagrams customarily used in design are approximate. Some shifting of the location of maximum moments may occur due to changes in loading, settlement of supports, lateral loads, or other causes. A diagonal tension crack in a flexural member without stirrups may shift the location of the calculated tensile stress approximately a distance d toward a point of zero moment. When stirrups are provided, this effect is less severe, although still present.

To provide for shifts in the location of maximum moments, this Code requires the extension of reinforcement a distance d or $12d_b$ beyond the point at which it is theoretically no longer required to resist flexure, except as noted.

Cutoff points of bars to meet this requirement are illustrated in Figure CC-8.1-1.

When bars of different sizes are used, the extension should be in accordance with the diameter of bar being terminated. A bar bent to the far face of a beam and continued there may logically be considered effective in satisfying this section, to the point where the bar crosses the middepth of the member.

COMMENTARY

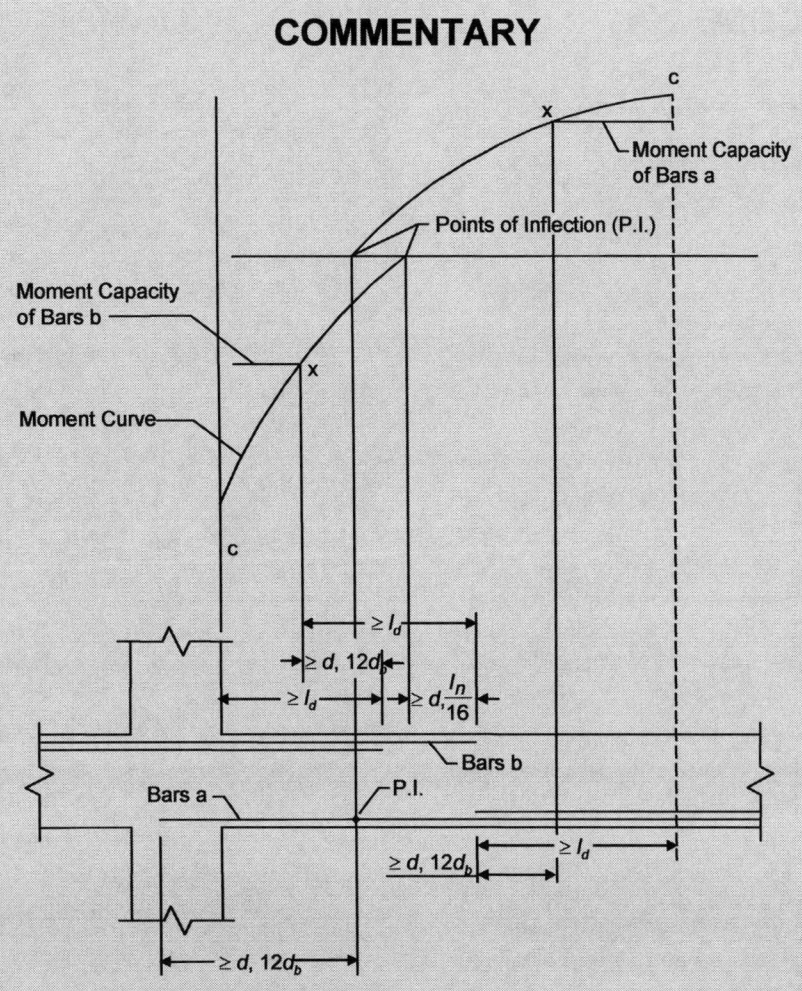

Figure CC-8.1-1 — Development of flexural reinforcement in a typical continuous beam

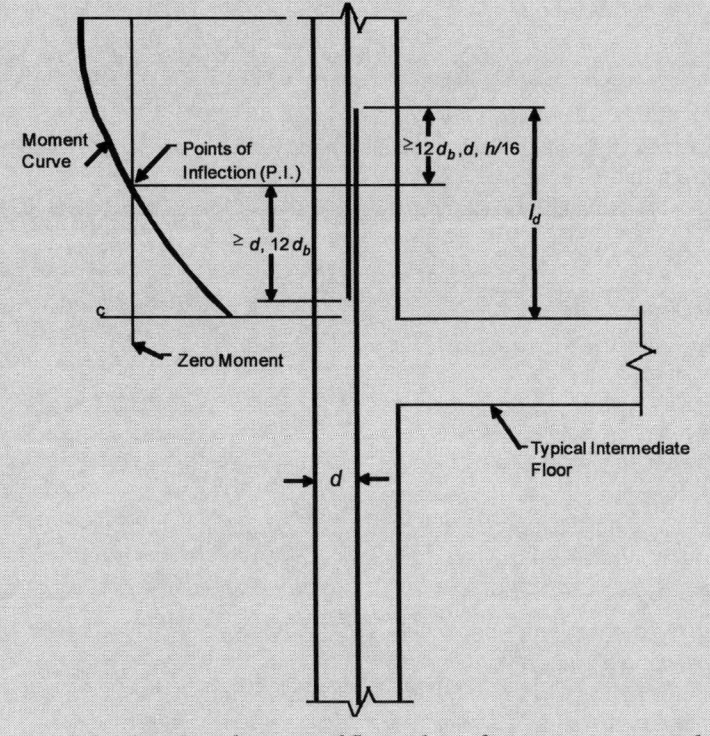

Figure CC-8.1-2 — Development of flexural reinforcement in a typical wall

CODE

8.1.6.4.1.4 Continuing reinforcement shall extend a distance l_d beyond the point where bent or terminated tension reinforcement is no longer required to resist flexure as required by Section 8.1.6.2 or 8.1.6.3.

8.1.6.4.1.5 Flexural reinforcement shall not be terminated in a tension zone unless one of the following conditions is satisfied:

(a) Shear at the cutoff point does not exceed two-thirds of the allowable shear at the section considered.

(b) Stirrup area in excess of that required for shear is provided along each terminated bar or wire over a distance from the termination point equal to three-fourths the effective depth of the member. Excess stirrup area, A_v, shall not be less than $60\, b_w s/f_y$. Spacing s shall not exceed $d/(8\, \beta_b)$.

(c) Continuous reinforcement provides double the area required for flexure at the cutoff point and shear does not exceed three-fourths the allowable shear at the section considered.

8.1.6.4.1.6 Anchorage complying with Section 8.1.6.2 or 8.1.6.3 shall be provided for tension reinforcement in corbels, deep flexural members, variable-depth arches, members where flexural reinforcement is not parallel with the compression face, and in other cases where the stress in flexural reinforcement does not vary linearly through the depth of the section.

8.1.6.4.2 *Development of positive moment reinforcement* — When a wall or other flexural member is part of the lateral-force-resisting system, at least 25 percent of the positive moment reinforcement shall extend into the support and be anchored to develop F_s in tension.

8.1.6.4.3 *Development of negative moment reinforcement*

8.1.6.4.3.1 Negative moment reinforcement in a continuous, restrained, or cantilever member shall be anchored in or through the supporting member in accordance with the provisions of Section 8.1.6.1.

8.1.6.4.3.2 At least one-third of the total reinforcement provided for moment at a support shall extend beyond the point of inflection the greater distance of the effective depth of the member or one-sixteenth of the span.

COMMENTARY

8.1.6.4.1.4 Peak stresses exist in the remaining bars wherever adjacent bars are cut off or bent in tension regions. In Figure CC-8.1-1 an "x" is used to indicate the peak stress points remaining in continuing bars after part of the bars have been cut off. If bars are cut off as short as the moment diagrams allow, these stresses become the full F_s, which requires a full embedment length as indicated. This extension may exceed the length required for flexure.

8.1.6.4.1.5 Evidence of reduced shear strength and loss of ductility when bars are cut off in a tension zone has been reported in Ferguson and Matloob (1959). As a result, the Code does not permit flexural reinforcement to be terminated in a tension zone unless special conditions are satisfied. Flexure cracks tend to open early wherever any reinforcement is terminated in a tension zone. If the stress in the continuing reinforcement and the shear strength are each near their limiting values, diagonal tension cracking tends to develop prematurely from these flexure cracks. Diagonal cracks are less likely to form where shear stress is low. A lower steel stress reduces the probability of such diagonal cracking.

8.1.6.4.1.6 In corbels, deep flexural members, variable-depth arches, members where the tension reinforcement is not parallel with the compression face, or other instances where the steel stress, f_s, in flexural reinforcement does not vary linearly in proportion to the moment, special means of analysis should be used to determine the peak stress for proper development of the flexural reinforcement.

8.1.6.4.2 *Development of positive moment reinforcement* — When a flexural member is part of the lateral-force-resisting system, loads greater than those anticipated in design may cause reversal of moment at supports. As a consequence, some positive reinforcement is required to be anchored into the support. This anchorage assures ductility of response in the event of serious overstress, such as from blast or earthquake. The use of more reinforcement at lower stresses is not sufficient. The full anchorage requirement need not be satisfied for reinforcement exceeding 25 percent of the total that is provided at the support.

8.1.6.4.3 *Development of negative moment reinforcement* — Negative reinforcement must be properly anchored beyond the support faces by extending the reinforcement l_d into the support or by anchoring of the reinforcement with a standard hook or suitable mechanical device.

Section 8.1.6.4.3.2 provides for possible shifting of the moment diagram at a point of inflection, as discussed under Commentary Section 8.1.6.4.1.3. This requirement may exceed that of Section 8.1.6.4.1.3 and the more restrictive governs.

CODE

8.1.6.5 *Hooks*

8.1.6.5.1 Standard hooks in tension shall be considered to develop an equivalent embedment length, l_e, equal to 13 d_b.

8.1.6.5.2 The effect of hooks for bars in compression shall be neglected in design calculations.

8.1.6.6 *Development of shear reinforcement*

8.1.6.6.1 *Bar and wire reinforcement*

8.1.6.6.1.1 Shear reinforcement shall extend to a distance d from the extreme compression face and shall be carried as close to the compression and tension surfaces of the member as cover requirements and the proximity of other reinforcement permit. Shear reinforcement shall be anchored at both ends for its calculated stress.

8.1.6.6.1.2 The ends of single-leg or U-stirrups shall be anchored by one of the following means:

(a) A standard hook plus an effective embedment of 0.5 l_d. The effective embedment of a stirrup leg shall be taken as the distance between the middepth of the member, $d/2$, and the start of the hook (point of tangency).

(b) For No. 5 bar (M #16) and D31 (MD200) wire and smaller, bending around longitudinal reinforcement through at least 135 degrees plus an embedment of 0.33 l_d. The 0.33 l_d embedment of a stirrup leg shall be taken as the distance between middepth of member, $d/2$, and start of hook (point of tangency).

8.1.6.6.1.3 Between the anchored ends, each bend in the continuous portion of a transverse U-stirrup shall enclose a longitudinal bar.

8.1.6.6.1.4 Longitudinal bars bent to act as shear reinforcement, where extended into a region of tension, shall be continuous with longitudinal reinforcement and, where extended into a region of compression, shall be developed beyond middepth of the member, $d/2$.

8.1.6.6.1.5 Pairs of U-stirrups or ties placed to form a closed unit shall be considered properly spliced when length of laps are 1.7 l_d. In grout at least 18 in. (457 mm) deep, such splices with $A_v f_y$ not more than 9,000 lb (40030 N) per leg shall be permitted to be considered adequate if legs extend the full available depth of grout.

COMMENTARY

8.1.6.5 *Hooks*

8.1.6.5.1 Refer to Commentary Section 6.1.5 for more information on hooks.

8.1.6.5.2 In compression, hooks are ineffective and cannot be used as anchorage.

8.1.6.6 *Development of shear reinforcement*

Design and detailing of shear reinforcement locations and anchorage in masonry requires consideration of the masonry module and reinforcement cover and clearance requirements.

8.1.6.6.1 *Bar and wire reinforcement*

8.1.6.6.1.1 Stirrups must be carried as close to the compression face of the member as possible because near ultimate load, flexural tension cracks penetrate deeply.

8.1.6.6.1.2 The requirements for anchorage of U-stirrups for deformed reinforcing bars and deformed wire are illustrated in Figure CC-8.1-3.

(a) When a standard hook is used, 0.5 l_d must be provided between $d/2$ and the point of tangency of the hook.

This provision may require a reduction in size and spacing of web reinforcement, or an increase in the effective depth of the beam, for web reinforcement to be fully effective.

8.1.6.6.1.3 and 8.1.6.6.1.5 U-stirrups that enclose a longitudinal bar have sufficient pullout resistance in the tension zone of the masonry.

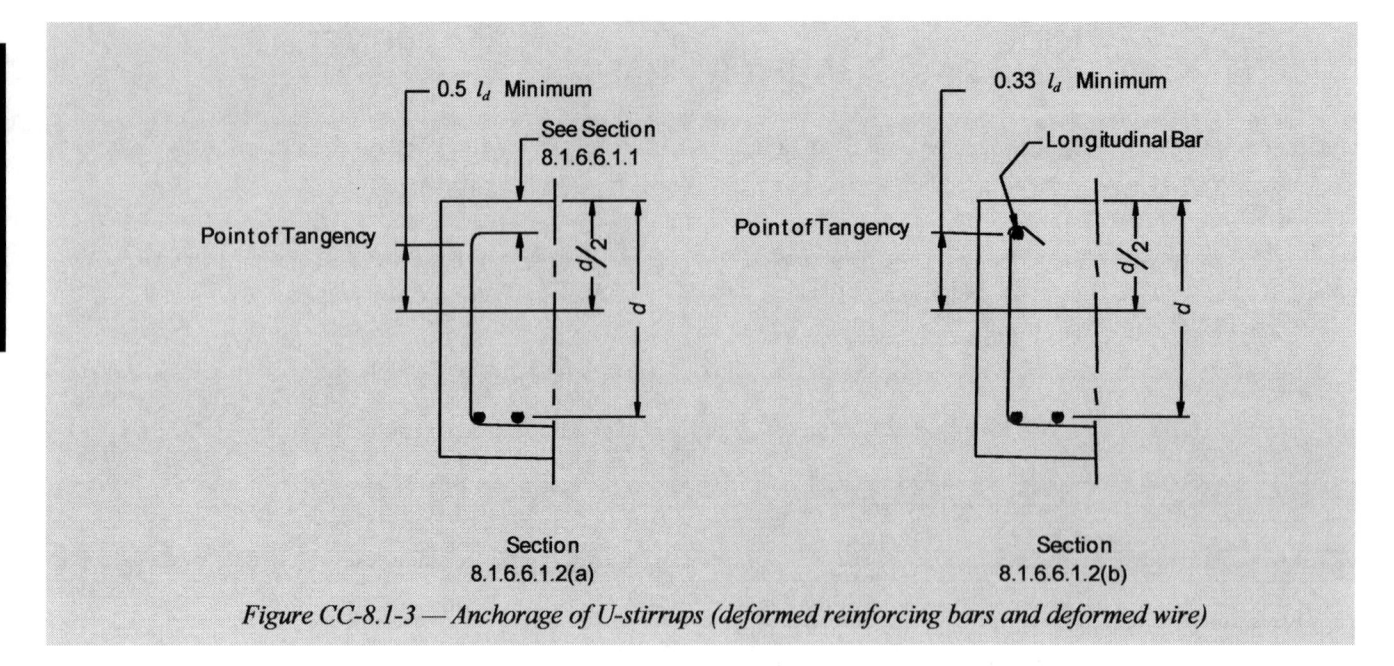

Figure CC-8.1-3 — Anchorage of U-stirrups (deformed reinforcing bars and deformed wire)

CODE

8.1.6.6.2 *Welded wire reinforcement*

8.1.6.6.2.1 For each leg of welded wire reinforcement forming simple U-stirrups, there shall be either:

(a) Two longitudinal wires at a 2-in. (50.8-mm) spacing along the member at the top of the U, or

(b) One longitudinal wire located not more than $d/4$ from the compression face and a second wire closer to the compression face and spaced at least 2 in. (50.8 mm) from the first wire. The second wire shall be located on the stirrup leg beyond a bend, or on a bend with an inside diameter of bend at least $8d_b$.

8.1.6.6.2.2 For each end of a single-leg stirrup of plain or deformed welded wire reinforcement, there shall be two longitudinal wires spaced a minimum of 2 in. (50.8 mm) with the inner wire placed at a distance at least $d/4$ or 2 in. (50.8 mm) from middepth of member, $d/2$. Outer longitudinal wire at tension face shall not be farther from the face than the portion of primary flexural reinforcement closest to the face.

8.1.6.7 *Splices of reinforcement* — Lap splices, welded splices, or mechanical splices are permitted in accordance with the provisions of this section.

COMMENTARY

8.1.6.6.2 *Welded wire reinforcement* — Although not often used in masonry construction, welded wire reinforcement provides a convenient means of placing reinforcement in a filled collar joint. See PCI (1980) for more information.

8.1.6.7 *Splices of reinforcement* — Continuity of reinforcement through proper splicing is necessary to provide force transfer. Effective splices can be provided through various forms: lap splices, welded splices or mechanical splices.

CODE

8.1.6.7.1 *Lap splices*

8.1.6.7.1.1 The minimum length of lap for bars in tension or compression shall be determined by Equation 8-12, but not less than 12 in. (305 mm).

8.1.6.7.1.2 Where reinforcement consisting of No. 3 (M#10) or larger bars is placed transversely within the lap, with at least one bar 8 in. (203 mm) or less from each end of the lap, the minimum length of lap for bars in tension or compression determined by Equation 8-12 shall be permitted to be reduced by multiplying by the confinement factor, ξ, determined in accordance with Equation 8-13. The clear space between the transverse bars and the lapped bars shall not exceed 1.5 in. (38 mm) and the transverse bars shall be fully developed in grouted masonry. The reduced lap splice length shall not be less than $36d_b$.

$$\xi = 1.0 - \frac{2.3 A_{sc}}{d_b^{2.5}} \qquad \text{(Equation 8-13)}$$

Where: $\frac{2.3 A_{sc}}{d_b^{2.5}} \leq 1.0$

A_{sc} is the area of the transverse bars at each end of the lap splice and shall not be taken greater than 0.35 in.^{2} (226 mm^2).

8.1.6.7.1.3 Bars spliced by noncontact lap splices shall not be spaced transversely farther apart than one-fifth the required length of lap nor more than 8 in. (203 mm).

8.1.6.7.2 *Welded splices* — Welded splices shall have the bars butted and welded to develop in tension at least 125 percent of the specified yield strength of the bar. Welding shall conform to AWS D1.4/D1.4M. Reinforcement to be welded shall conform to ASTM A706, or shall be accompanied by a submittal showing its chemical analysis and carbon equivalent as required by AWS D1.4/D1.4M. Existing reinforcement to be welded shall conform to ASTM A706, or shall be analyzed chemically and its carbon equivalent determined as required by AWS D1.4/D1.4M.

COMMENTARY

8.1.6.7.1 *Lap splices* — The required length of the lap splice is based on developing a minimum reinforcing bar stress of $1.25 f_y$. This requirement provides adequate strength while maintaining consistent requirements between lap, mechanical, and welded splices. Historically, the length of lap has been based on the bond stress that is capable of being developed between the reinforcing bar and the surrounding grout. Testing has shown that bond stress failure (or pull-out of the reinforcing bar) is only one possible mode of failure for lap splices. Other failure modes include rupture of the reinforcing bar and longitudinal splitting of masonry along the length of the lap. Experimental results of several independent research programs were combined and analyzed to provide insight into predicting the necessary lap lengths for reinforcing bar splices in masonry construction (Hogan et al, 1997). Equation 8-12 was fitted to the data and has a coefficient of determination, r^2, value of 0.93.

8.1.6.7.1.2 An extensive testing program conducted by the National Concrete Masonry Association (NCMA, 2009) and additional testing done by Washington State University (Mjelde et al, 2009) show that reinforcement provided transverse to lapped bars controls longitudinal tensile splitting of the masonry assembly. These tranverse bars increase the lap performance significantly, as long as there is at least one No. 3 (M#10) transverse reinforcing bar placed within 8 in. (203 mm) of each end of the splice. These bars must be fully developed and have a clear spacing between the transverse bars and the lapped bars not exceeding 1.5 in. (38 mm). Testing also indicated that the lap length must be at least $36d_b$ or the effect of the transverse reinforcement is minimal. As a result, this limit was applied to the lap length. The testing also showed that even when more transverse reinforcement area is provided, it becomes significantly less effective in quantities above 0.35 in.^{2} (226 mm^2). Thus, the transervse reinforcement area at each end of the lap, A_{sc}, is limited to 0.35 in.^{2} (226 mm^2), even if more is provided.

8.1.6.7.1.3 If individual bars in noncontact lap splices are too widely spaced, an unreinforced section is created, which forces a potential crack to follow a diagonal line. Lap splices may occur with the bars in adjacent grouted cells if the requirements of this section are met.

8.1.6.7.2 *Welded splices* — A full welded splice is primarily intended for large bars (No. 6 [M#19] and larger) in main members. The tensile strength requirement of 125 percent of specified yield strength is intended to ensure sound welding, adequate also for compression. It is desirable that splices be capable of developing the ultimate tensile strength of the bars spliced, but practical limitations make this ideal condition difficult to attain. The maximum reinforcement stress used in design under this Code is based upon yield strength. To ensure sufficient strength in splices so that brittle failure can be

CODE

COMMENTARY

avoided, the 25 percent increase above the specified yield strength was selected as both an adequate minimum for safety and a practicable maximum for economy.

When welding of reinforcing bars is required, the weldability of the steel and compatible welding procedures need to be considered. The provisions in AWS D1.4/D1.4M Welding Code cover aspects of welding reinforcing bars, including criteria to qualify welding procedures. Weldability of the steel is based on its chemical composition or carbon equivalent (CE). The Welding Code establishes preheat and interpass temperatures for a range of carbon equivalents and reinforcing bar sizes. Carbon equivalent is calculated from the chemical composition of the reinforcing bars. The Welding Code has two expressions for calculating carbon equivalent. A relatively short expression, considering only the elements carbon and manganese, is to be used for bars other than ASTM A706 material. A more comprehensive expression is given for ASTM A706 bars. The CE formula in the Welding Code for ASTM A706 bars is identical to the CE formula in ASTM A706.

The chemical analysis, for bars other than ASTM A706, required to calculate the carbon equivalent is not routinely provided by the producer of the reinforcing bars. For welding reinforcing bars other than ASTM A706 bars, the design drawings or project specifications should specifically require results of the chemical analysis to be furnished.

ASTM A706 covers low-alloy steel reinforcing bars intended for applications requiring controlled tensile properties or welding. Weldability is accomplished in ASTM A706 by limits or controls on chemical composition and on carbon equivalent (Gustafson and Felder, 1991). The producer is required by ASTM A706 to report the chemical composition and carbon equivalent.

The AWS D1.4/D1.4M Welding Code requires the contractor to prepare written welding procedure specifications conforming to the requirements of the Welding Code. Appendix A of the Welding Code contains a suggested form that shows the information required for such a specification for each joint welding procedure.

Welding to existing reinforcing bars is often necessary even though no mill test report of the existing reinforcement is available. This condition is particularly common in alterations or building expansions. AWS D1.4/D1.4M states for such bars that a chemical analysis may be performed on representative bars. If the chemical composition is not known or obtained, the Welding Code requires a minimum preheat. Welding of the particular bars should be performed in accordance with AWS D1.4/D1.4M, including their preheat. The designer should also determine if additional precautions are in order, based on other considerations such as stress level in the bars, consequences of failure, and heat damage to existing masonry due to welding operations.

CODE

8.1.6.7.3 *Mechanical splices* — Mechanical splices shall have the bars connected to develop in tension or compression, as required, at least 125 percent of the specified yield strength of the bar.

8.1.6.7.4 *End-bearing splices*

8.1.6.7.4.1 In bars required for compression only, the transmission of compressive stress by bearing of square cut ends held in concentric contact by a suitable device is permitted.

8.1.6.7.4.2 Bar ends shall terminate in flat surfaces within $1\frac{1}{2}$ degree of a right angle to the axis of the bars and shall be fitted within 3 degrees of full bearing after assembly.

8.1.6.7.4.3 End-bearing splices shall be used only in members containing closed ties, closed stirrups, or spirals.

8.1.6.7.5 *Splicing of wires in tension*

8.1.6.7.5.1 *Lap splices* — The minimum length of lap for wires in tension shall be determined by Equation 8-11, but shall not be less than 6 in. (152 mm).

8.1.6.7.5.2 *Welded splices* — Welded splices shall have the wires welded to develop at least 125 percent of the specified yield strength of the wire in tension.

8.1.6.7.5.3 *Mechanical splices* — Mechanical splices shall have the wires connected to develop at least 125 percent of the specified yield strength of the wire in tension.

COMMENTARY

Welding of wire to wire, and of wire or welded wire reinforcement to reinforcing bars or structural steel elements is not covered by AWS D1.4/D1.4M. If welding of this type is required on a project, the contract documents should specify requirements or performance criteria for this welding. If cold drawn wires are to be welded, the welding procedures should address the potential loss of yield strength and ductility achieved by the cold working process (during manufacture) when such wires are heated by welding. Machine and resistance welding as used in the manufacture of welded plain and deformed wire reinforcement is covered by ASTM A185 and ASTM A497, respectively, and is not part of this concern.

8.1.6.7.3 *Mechanical splices* — Full mechanical splices are also required to develop 125 percent of the specified yield strength in tension or compression as required, for the same reasons discussed for full welded splices.

8.1.6.7.4 *End-bearing splices* — Experience with end-bearing splices has been almost exclusively with vertical bars in columns. If bars are significantly inclined from the vertical, special attention is required to ensure that adequate end-bearing contact can be achieved and maintained. The lateral tie requirements prevent end-bearing splices from sliding.

8.1.6.7.5.2 *Welded splices* — If welded splices are required on a project, the contract documents should specify requirements or performance criteria for the welding. If cold drawn wires are to be welded, the welding procedures should address the potential loss of yield strength and ductility achieved by the cold working process (manufacturing) when such wires are heated by welding. Machine and resistance welding, as used in the manufacture of welded plain and deformed wire reinforcement, is covered by ASTM A185 and ASTM A497, respectively, and by ASTM A951 for joint reinforcement.

CODE

8.2 — Unreinforced masonry

8.2.1 Scope
This section provides requirements for the design of unreinforced masonry as defined in Section 2.2. Design of unreinforced masonry by the allowable stress method shall comply with the requirements of Part 1, Part 2, Section 8.1, and Section 8.2.

8.2.2 Design criteria
Unreinforced masonry members shall be designed in accordance with the principles of engineering mechanics and shall be designed to remain uncracked.

8.2.3 Design assumptions
The following assumptions shall be used in the design of unreinforced masonry members:

(a) Strain in masonry is directly proportional to the distance from the neutral axis.

(b) Flexural tensile stress in masonry is directly proportional to strain.

(c) Flexural compressive stress in combination with axial compressive stress in masonry is directly proportional to strain.

(d) Stresses in reinforcement, if present, are neglected when determining the resistance of masonry to design loads.

8.2.4 Axial compression and flexure

8.2.4.1 Axial and flexural compression — Members subjected to axial compression, flexure, or to combined axial compression and flexure shall be designed to satisfy Equation 8-14 and Equation 8-15.

$$\frac{f_a}{F_a} + \frac{f_b}{F_b} \leq 1 \quad \text{(Equation 8-14)}$$

$$P \leq (1/4) P_e \quad \text{(Equation 8-15)}$$

where:

(a) For members having an h/r ratio not greater than 99:

$$F_a = (1/4) f'_m \left[1 - \left(\frac{h}{140r}\right)^2\right] \quad \text{(Equation 8-16)}$$

(b) For members having an h/r ratio greater than 99:

$$F_a = (1/4) f'_m \left(\frac{70r}{h}\right)^2 \quad \text{(Equation 8-17)}$$

(c) $F_b = (1/3) f'_m \quad \text{(Equation 8-18)}$

COMMENTARY

8.2 — Unreinforced masonry

8.2.1 Scope
This section provides for the design of masonry members in which tensile stresses, not exceeding allowable limits, are resisted by the masonry. This has previously been referred to as unreinforced or plain masonry. Flexural tensile stresses may result from bending moments, from eccentric vertical loads, or from lateral loads.

8.2.2 Design criteria
A fundamental premise is that under the effects of design loads, masonry remains uncracked. Stresses due to restraint against differential movement, temperature change, moisture expansion, and shrinkage combine with the design load stresses. Stresses due to restraint should be controlled by joints or other construction techniques to ensure that the combined stresses do not exceed the allowable.

8.2.3 Design assumptions
Reinforcement may be placed in masonry walls to control the effects of movements from temperature changes or shrinkage, or as prescriptive seismic reinforcement. This reinforcement is not considered in calculating strength when using unreinforced masonry design.

8.2.4 Axial compression and flexure
8.2.4.1 Axial and flexural compression — Equation (8-14) is a unity interaction equation that is a simple proportioning of the available allowable stresses to the applied loads, and the equation is used to design masonry for combined axial and flexural compressive stresses. The unity equation can be expanded when biaxial bending is present by adding a third term for the bending stress quotient about the second axis of bending.

In this unity interaction equation, secondary bending effects resulting from the axial load are ignored. A more accurate equation would include the use of a moment magnifier applied to the flexure term, f_b/F_b. Although avoidance of a moment magnifier term can produce unconservative results in some cases, the committee decided not to include this term in Equation 8-14 for the following reasons:

- At larger h/r values, where moment magnification is more critical, the allowable axial load on the member is limited by Code Equation 8-15.

- For the practical range of h/r values, errors induced by ignoring the moment magnifier is relatively small, less than 15 percent.

CODE

(d) $P_e = \dfrac{\pi^2 E_m I_n}{h^2}\left(1 - 0.577\dfrac{e}{r}\right)^3$ (Equation 8-19)

COMMENTARY

- The overall safety factor of 4 included in the allowable stress equations is sufficiently large to allow this simplification in the design procedure.

The purpose of Equation 8-15 is to safeguard against a premature stability failure caused by eccentrically applied axial load. The equation is not intended to be used to check adequacy for combined axial compression and flexure. Therefore, in Equation 8-19, the value of the eccentricity "e" that is to be used to calculate P_e is the actual eccentricity of the applied compressive load. The value of "e" is not to be calculated as M_{max} divided by P where M_{max} is a moment caused by other than eccentric load.

Equation 8-15 is an essential check because the allowable compressive stress for members with an h/r ratio in excess of 99 has been developed assuming only a nominal eccentricity of the compressive load. Thus, when the eccentricity of the compressive load exceeds the minimum eccentricity of $0.1t$, Equation 8-17 will overestimate the allowable compressive stress and Equation 8-15 may control.

The allowable stress values for F_a presented in Equations 8-16 and 8-17 are based on an analysis of the results of axial load tests performed on clay and concrete masonry elements. A fit of an empirical curve to this test data, Figure CC-8.2-1, indicates that members having an h/r ratio not exceeding 99 fail under loads below the Euler buckling load at a stress level equal to:

$$f'_m\left[1 - (h/140r)^2\right] \qquad \text{(same with SI units)}$$

Thus, for members having an h/r ratio not exceeding 99, this Code allows axial load stresses not exceeding $1/4$ of the aforementioned failure stress.

Applying the Euler theory of buckling to members having resistance in compression but not in tension, (Colville, 1978; Colville, 1979; Yokel, 1971) show that for a solid section, the critical compressive load for these members can be expressed by the formula

$$P_e = (\pi^2 E_m I_n / h^2)(1 - 2e/t)^3 \qquad \text{(same with SI units)}$$

in which

I_n = uncracked moment of inertia

e = eccentricity of axial compressive load with respect to the member longitudinal centroidal axis.

In the derivation of this buckling load equation, tension cracking is assumed to occur prior to failure.

For h/r values in excess of 99, the limited test data is approximated by the buckling load.

For a solid rectangular section, $r = \sqrt{t^2/12}$. Making this substitution into the buckling load equation gives

COMMENTARY

$$P_e = \frac{\pi^2 E_m I_n}{h^2}\left(1 - 0.577\frac{e}{r}\right)^3 \quad \text{(Equation 8-19)}$$

Transforming the buckling equation using a minimum eccentricity of $0.1t$ (from Section 8.3.4.3) and an elastic modulus equal to $1000 f'_m$, the axial compressive stress at buckling failure amounts approximately to $[70(r/h)]^2 f'_m$. At the time of the development of this equation, the committee had not developed a relationship between E_m and f'_m so the traditional relationship of $E_m = 1000 f'_m$ was used (Colville, 1992). The same equation can be developed using $E_m = 667 f'_m$ and an eccentricity of $0.05t$. Thus, for members having an h/r ratio in excess of 99, this Code allows an axial load compressive stress not exceeding $1/4$ of this failure stress (Equation 8-17).

Tests of masonry have shown (Hatzinikolas et al, 1978; Fattal and Cattaneo, 1976; Yokel and Dikkers, 1971; Yokel and Dikkers, 1973) that the maximum compressive stress at failure under flexural load is higher than the maximum compressive stress at failure under axial load. The higher stress under flexural load is attributed to the restraining effect of less highly strained compressive fibers on the fibers of maximum compressive strain. This effect is less pronounced in hollow masonry than solid masonry; however, the test data indicate that, calculated by the straight-line theory, the compressive stress at failure in hollow masonry subjected to flexure exceeds by $1/3$ that of the masonry under axial load. Thus, to maintain a factor of safety of 4 in design, the committee considered it conservative to establish the allowable compressive stress in flexure as:

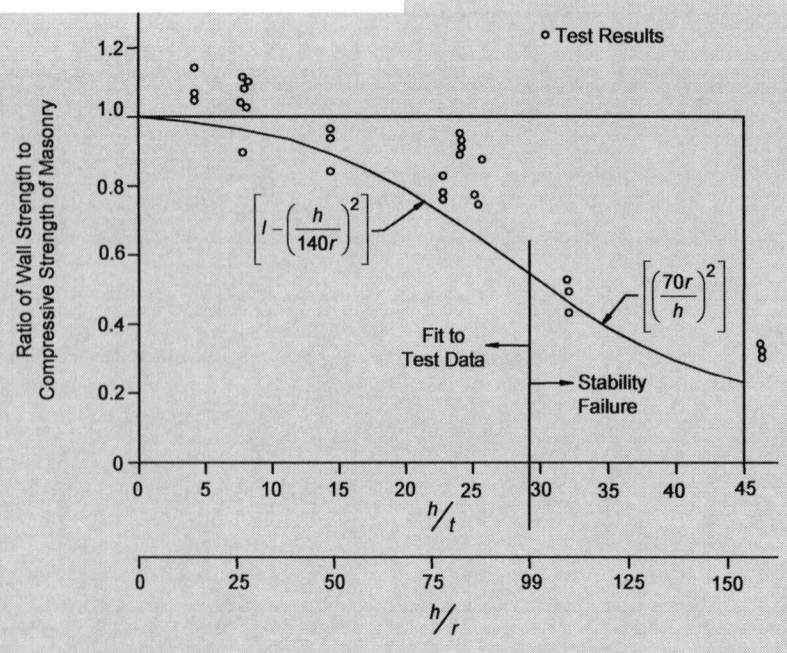

$$F_b = \tfrac{4}{3} \times \left(\tfrac{1}{4}\right) f'_m = \left(\tfrac{1}{3}\right) f'_m$$

Figure CC-8.2-1 — Slenderness effects on axial compressive strength

CODE

8.2.4.2 *Flexural tension* — Allowable tensile stresses for masonry elements subjected to out-of-plane or in-plane bending shall be in accordance with the values in Table 8.2.4.2. For grouted masonry not laid in running bond, tension parallel to the bed joints shall be assumed to be resisted only by the minimum cross-sectional area of continuous grout that is parallel to the bed joints.

COMMENTARY

8.2.4.2 *Flexural tension* — Prior to the 2011 edition of the Code, allowable stresses were permitted to be increased by one-third when considering load combinations including wind or seismic loads. Unreinforced masonry walls designed under codes that permitted the one-third stress increase have had acceptable performance. However, rather than arbitrarily increasing the allowable flexural tensile stresses by one-third, the Committee assessed the allowable flexural tensile stresses using a reliability-based approach to see if an increase in allowable stresses is justified. Kim and Bennett (2002) performed a reliability analysis in which the flexural tensile stress was assumed to follow a lognormal distribution. They used a mean flexural tensile strength of the allowable flexural tensile stress in the 2008 Code multiplied by 5.1 based on the examination of 327 full-scale tests reported in the literature. Coefficients of variations for different data sets (e.g specific mortar type and direction of loading) ranged from 0.10 to 0.51, with a weighted average of 0.42. The coefficient of variation of 0.50 used by Kim and Bennett (2002) is greater than used in previous studies. For example, Ellingwood et al (1980) used a coefficient of variation of 0.24 and Stewart and Lawrence (2000) used a coefficient of variation of 0.30. Kim and Bennett felt, though, that a coefficient of variation of 0.50 is more representative of field conditions. The lognormal distribution was determined by comparing the Anderson-Darling statistic for normal, lognormal, and Weibull probability distributions. For unreinforced masonry walls subjected to wind loading and designed using the one-third stress increase, the reliability index was determined to be 2.66. This is slightly greater than the value of 2.5 that is typical for the design of other materials (Ellingwood et al, 1980). The reliability analysis by Kim and Bennett (2002) assumed the axial load was zero, which is the worst case. With increasing axial load (which has a lower coefficient of variation than 0.50), the reliability index would increase. Based on this reliability analysis, the Code committee felt justified in increasing the allowable flexural tensile stresses by a factor of 4/3 to compensate for the elimination of the previously permitted one-third stress increase.

The allowable tensile strength values are a function of the type of mortar being used. Mortar cement is a product that has bond strength requirements that have been established to provide comparable flexural bond strength to that achieved using portland cement-lime mortar (Melander and Ghosh, 1996; Hedstrom et al, 1991; Borchelt and Tann, 1996).

For masonry cement and air entrained portland-cement lime mortar, there are no conclusive research data and, hence, flexural tensile stresses are based on existing requirements in other codes.

Table 8.2.4.2 — Allowable flexural tensile stresses for clay and concrete masonry, psi (kPa)

Direction of flexural tensile stress and masonry type	Portland cement/lime or mortar cement M or S	Portland cement/lime or mortar cement N	Masonry cement or air entrained portland cement/lime M or S	Masonry cement or air entrained portland cement/lime N
Normal to bed joints				
Solid units	53 (366)	40 (276)	32 (221)	20 (138)
Hollow units[1]				
Ungrouted	33 (228)	25 (172)	20 (138)	12 (83)
Fully grouted	65 (448)	63 (434)	61 (420)	58 (400)
Parallel to bed joints in running bond				
Solid units	106 (731)	80 (552)	64 (441)	40 (276)
Hollow units				
Ungrouted and partially grouted	66 (455)	50 (345)	40 (276)	25 (172)
Fully grouted	106 (731)	80 (552)	64 (441)	40 (276)
Parallel to bed joints in masonry not laid in running bond				
Continuous grout section parallel to bed joints	133 (917)	133 (917)	133 (917)	133 (917)
Other	0 (0)	0 (0)	0 (0)	0 (0)

1   For partially grouted masonry, allowable stresses shall be determined on the basis of linear interpolation between fully grouted hollow units and ungrouted hollow units based on amount (percentage) of grouting.

CODE

COMMENTARY

The allowable tensile stresses are for tension stresses due to flexure under either out-of-plane or in-plane loading. While it is recognized that in-plane and out-of-plane strain gradients are different, for these low stress levels the effect due to any difference is small. Flexural tensile stress can be offset by axial compressive stress, but the net tensile stress due to combined bending and axial compression cannot exceed the allowable flexural tensile stress.

Variables affecting tensile bond strength of brick masonry normal to bed joints include mortar properties, unit initial rate of absorption, surface condition, workmanship, and curing condition. For tension parallel to bed joints, the strength and geometry of the units also affect tensile strength.

Historically, masonry not laid in running bond has been assumed to have no flexural bond strength across mortared head joints; thus the grout area alone is used to resist bending. Examples of continuous grout parallel to the bed joints are shown in Figure CC-8.2-2.

CODE

COMMENTARY

Test data using a bond wrench (Brown and Palm, 1982; Hamid, 1985) revealed tensile bond strength normal to bed joints ranging from 30 psi (207 kPa) to 190 psi (1,310 kPa). This wide range is attributed to the multitude of parameters affecting tensile bond strength.

Test results (Hamid, 1985; Ribar, 1982) show that masonry cement mortars and mortars with high air content generally have lower bond strength than portland cement-lime mortars.

Tests conducted by Hamid (1981) show the significant effect of the aspect ratio (height to least dimension) of the brick unit on the flexural tensile strength. The increase in the aspect ratio of the unit results in an increase in strength parallel to bed joints and a decrease in strength normal to bed joints.

Research work (Drysdale and Hamid, 1984) on flexural strength of concrete masonry has shown that grouting has a significant effect in increasing tensile strength over ungrouted masonry. A three-fold increase in tensile strength normal to bed joints was achieved using fine grout as compared to ungrouted masonry. The results also show that, within a practical range of strength, the actual strength of grout is not of major importance. For tension parallel to bed joints, a 133 percent increase in flexural strength was achieved by grouting the cells. Grout cores change the failure mode from stepped-wise cracking along the bed and head joints for hollow walls to a straight line path along the head joints and unit for grouted walls.

Research (Brown and Melander, 1999) has shown that flexural strength of unreinforced, grouted concrete and clay masonry is largely independent of mortar type or cementitious materials.

For partial grouting, the footnote to Table 8.2.4.2 permits interpolation between the fully grouted value and the hollow unit value based on the percentage of grouting. A concrete masonry wall with Type S portland cement-lime mortar grouted 50 percent and stressed normal to the bed joints would have an allowable stress midway between 86 psi (593 kPa) and 33 psi (228 kPa), hence an allowable stress of 59.5 psi (410 kPa).

The presence of flashing and other conditions at the base of the wall can significantly reduce the flexural bond. The values in Table 8.2.4.2 apply only to the flexural tensile stresses developed between masonry units, mortar, and grout.

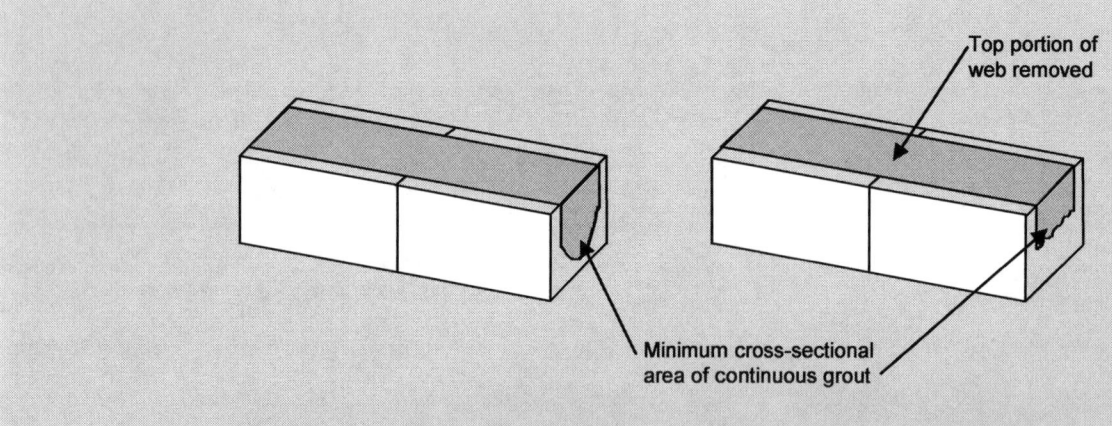

Figure CC-8.2-2 — Continuous grout sections parallel to the bed joints

CODE

8.2.5 *Axial tension*

Axial tension resistance of unreinforced masonry shall be neglected in design.

8.2.6 *Shear*

8.2.6.1 Shear stresses due to forces acting in the direction considered shall be calculated in accordance with Section 4.3.1 and determined by Equation 8-20.

$$f_v = \frac{VQ}{I_n b} \quad \text{(Equation 8-20)}$$

COMMENTARY

8.2.5 *Axial tension*

Net axial tension in unreinforced masonry walls due to axially applied load is not permitted. If axial tension develops in walls due to uplift of connected roofs or floors, the walls must be reinforced to resist the tension. Compressive stress from dead load can be used to offset axial tension.

8.2.6 *Shear*

Three modes of shear failure in unreinforced masonry are possible:

(a) Diagonal tension cracks form through the mortar and masonry units.

(b) Sliding occurs along a straight crack at horizontal bed joints.

(c) Cracks form, that stair-step from head joint to bed joint.

In the absence of suitable research data, the committee recommends that the allowable shear stress values given in Code Section 8.2.6.2 be used for limiting out-of-plane shear stresses.

8.2.6.1 A theoretical parabolic stress distribution is used to define shear stress through the depth. Some other codes use average shear stress so direct comparison of allowable values is not valid. Effective area requirements are given in Section 4.3.1. For rectangular sections, Equation (8-20) produces a maximum shear stress at mid-depth, that is equal to $^3/_2 \times V/A$. Equation (8-20) is also used to calculate shear stresses for composite action and shear stresses, resulting from out-of-plane loading, in the connections between face shells of hollow units.

CODE

8.2.6.2 In-plane shear stresses shall not exceed any of:

(a) $1.5\sqrt{f'_m}$

(b) 120 psi (0.827 MPa)

(c) For running bond masonry not fully grouted;

   37 psi + 0.45 N_v/A_n

(d) For masonry not laid in running bond, constructed of open end units, and fully grouted;

   37 psi + 0.45 N_v/A_n

(e) For running bond masonry fully grouted;

   60 psi + 0.45 N_v/A_n

(f) For masonry not laid in running bond, constructed of other than open end units, and fully grouted;

   15 psi (103 kPa)

8.2.6.3 The minimum normalized web area of concrete masonry units, determined in accordance with ASTM C140, shall not be less than 27 in.²/ft² (187,500 mm²/m²) or the calculated shear stresses in the webs shall not exceed the value given in Section 8.2.6.2(a).

COMMENTARY

8.2.6.2 Shear stress allowable values are applicable to shear walls without reinforcement. The values given are based on research (Woodward and Ranking, 1984; Pook, 1986; Nuss et al, 1978; Hamid et al, 1979). N_v is normally based on dead load.

8.2.6.3 Out-of-plane flexure causes horizontal and vertical shear stresses. Vertical shear stresses are resisted by the connection between the web and face shell of the unit. A normalized web area of 27 in.²/ft² (187,500 mm²/m²) provides sufficient web area so that shear stresses between the web and face shell of a unit, resulting from out-of-plane loading, will not be critical.

CODE

8.3 — Reinforced masonry

8.3.1 *Scope*

This section provides requirements for the design of structures in which reinforcement is used to resist tensile forces in accordance with the principles of engineering mechanics and the contribution of the tensile strength of masonry is neglected, except as provided in Section 8.3.5. Design of reinforced masonry by the allowable stress method shall comply with the requirements of Part 1, Part 2, Section 8.1, and Section 8.3

8.3.2 *Design assumptions*

The following assumptions shall be used in the design of reinforced masonry:

(a) Strain compatibility exists between the reinforcement, grout, and masonry.

(b) Strains in reinforcement and masonry are directly proportional to the distances from the neutral axis.

(c) Stress is linearly proportional to the strain.

(d) The compressive resistance of steel reinforcement does not contribute to the axial and flexural strengths unless lateral reinforcement is provided in compliance with the requirements of Section 5.3.1.4.

(e) Stresses remain in the elastic range.

(f) Masonry in tension does not contribute to axial and flexural resistances. Axial and flexural tension stresses are resisted entirely by steel reinforcement.

8.3.3 *Steel reinforcement — Allowable stresses*

8.3.3.1 Tensile stress in bar reinforcement shall not exceed the following:

(a) Grade 40 or Grade 50 reinforcement: 20,000 psi (137.9 MPa)

(b) Grade 60 reinforcement: 32,000 psi (220.7 MPa)

8.3.3.2 Tensile stress in wire joint reinforcement shall not exceed 30,000 psi (206.9 MPa).

8.3.3.3 When lateral reinforcement is provided in compliance with the requirements of Section 5.3.1.4, the compressive stress in bar reinforcement shall not exceed the values given in Section 8.3.3.1.

8.3.4 *Axial compression and flexure*

8.3.4.1 Members subjected to axial compression, flexure, or combined axial compression and flexure shall be designed in compliance with Sections 8.3.4.2 through 8.3.4.4.

COMMENTARY

8.3 — Reinforced masonry

8.3.1 *Scope*

The requirements in this section pertain to the design of masonry in which flexural tension is assumed to be resisted by reinforcement alone, and the flexural tensile strength of masonry is neglected. Tension still develops in the masonry, but it is not considered to be effective in resisting design loads.

8.3.2 *Design assumptions*

The design assumptions listed have traditionally been used for allowable stress design of reinforced masonry members.

Although tension may develop in the masonry of a reinforced element, it is not considered effective in resisting axial and flexural design loads.

8.3.3 *Steel reinforcement — Allowable stresses —* The allowable steel stresses have a sufficiently large factor of safety that second-order effects do not need to be considered in allowable stress design.

8.3.4 *Axial compression and flexure*
See Commentary for 8.2.4.1.

CODE

8.3.4.2 *Allowable forces and stresses*

8.3.4.2.1 The compressive force in reinforced masonry due to axial load only shall not exceed that given by Equation 8-21 or Equation 8-22:

(a) For members having an *h/r* ratio not greater than 99:

$$P_a = \left(0.25 f'_m A_n + 0.65 A_{st} F_s\right)\left[1 - \left(\frac{h}{140r}\right)^2\right]$$

(Equation 8-21)

(b) For members having an *h/r* ratio greater than 99:

$$P_a = \left(0.25 f'_m A_n + 0.65 A_{st} F_s\right)\left(\frac{70r}{h}\right)^2 \quad \text{(Equation 8-22)}$$

8.3.4.2.2 The compressive stress in masonry due to flexure or due to flexure in combination with axial load shall not exceed $0.45 f'_m$ provided that the calculated compressive stress due to the axial load component, f_a, does not exceed the allowable stress, F_a, in Section 8.2.4.1.

COMMENTARY

8.3.4.2 *Allowable forces and stresses* — This Code limits the compressive stress in masonry members based on the type of load acting on the member. The compressive force at the section resulting from axial loads or from the axial component of combined loads is calculated separately, and is limited to the values permitted in Section 8.3.4.2.1. Equation 8-21 or 8-22 controls the capacity of columns with large axial loads. The coefficient of 0.25 provides a factor of safety of about 4.0 against crushing of masonry. The coefficient of 0.65 was determined from tests of reinforced masonry columns and is taken from previous masonry codes (ACI 531, 1983; BIA, 1969). A second compressive stress calculation must be performed considering the combined effects of the axial load component and flexure at the section and should be limited to the values permitted in Section 8.3.4.2.2. (See Commentary for Section 8.2.4.)

8.3.4.2.2 Figure CC-8.3-1 shows the allowable moment (independent of member size and material strength) versus the ratio of steel reinforcement (Grade 60) multiplied by the steel yield strength and divided by the specified compressive strength of masonry (modified steel reinforcement ratio) for both clay and concrete masonry members subjected to pure flexure. When the masonry compressive stress controls the design, there is little increase in moment capacity with increasing steel reinforcement. This creates a limit on the amount of reinforcement that is practical to use in allowable stress design of masonry. Even when the masonry allowable compressive stress controls the design, the failure of the member will still be ductile. For clay masonry, the masonry stress begins to control the design at $0.39\rho_{bal}$ and for concrete masonry, the masonry stress begins to control the design at $0.38\rho_{bal}$, where ρ_{bal} is the reinforcement ratio at which the masonry would crush simultaneously with yielding of the reinforcement. The reinforcement ratio as a fraction of the balanced reinforcement ratio, ρ_{bal}, is also shown in Figure CC-8.3-1.

The interaction equation used in Section 8.2.4 is not applicable for reinforced masonry and is therefore not included in Section 8.3.

CODE

8.3.4.3 *Columns* — Design axial loads shall be assumed to act at an eccentricity at least equal to 0.1 multiplied by each side dimension. Each axis shall be considered independently.

COMMENTARY

8.3.4.3 *Columns* — The minimum eccentricity of axial load (Figure CC-8.3-2) results from construction imperfections not otherwise anticipated by analysis.

In the event that actual eccentricity exceeds the minimum eccentricity required by this Code, the actual eccentricity should be used. This Code requires that stresses be checked independently about each principal axis of the member (Figure CC-8.3-2).

Additional column design and detailing requirements are given in Section 5.3.

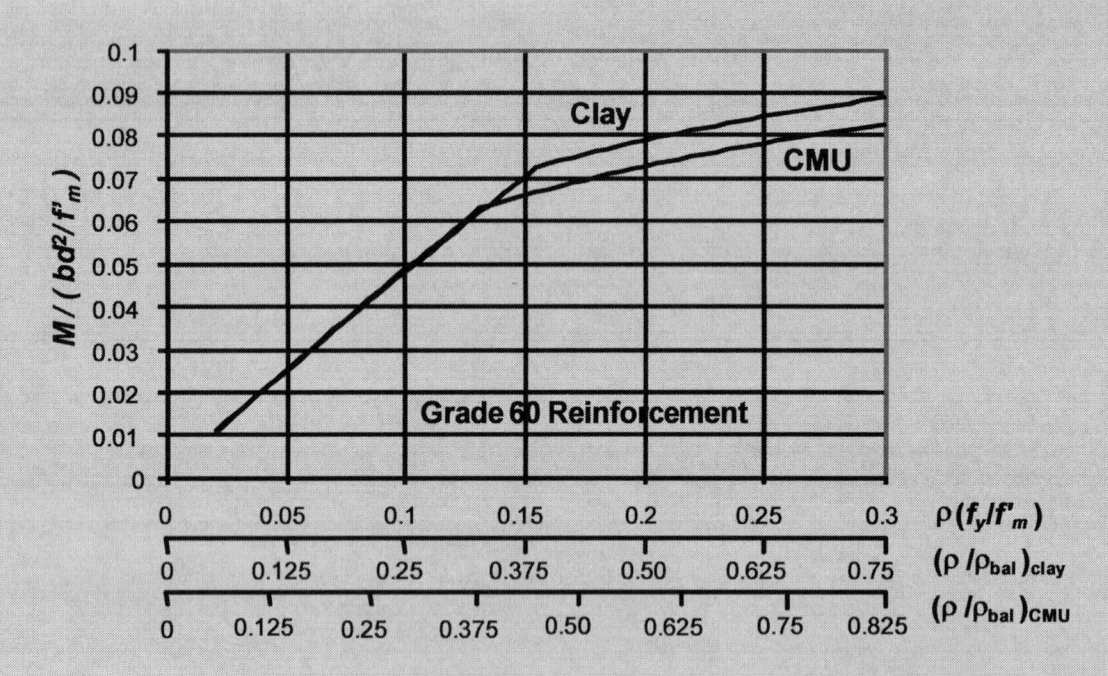

Figure CC-8.3-1 Allowable moment vs. modified steel reinforcement ratio

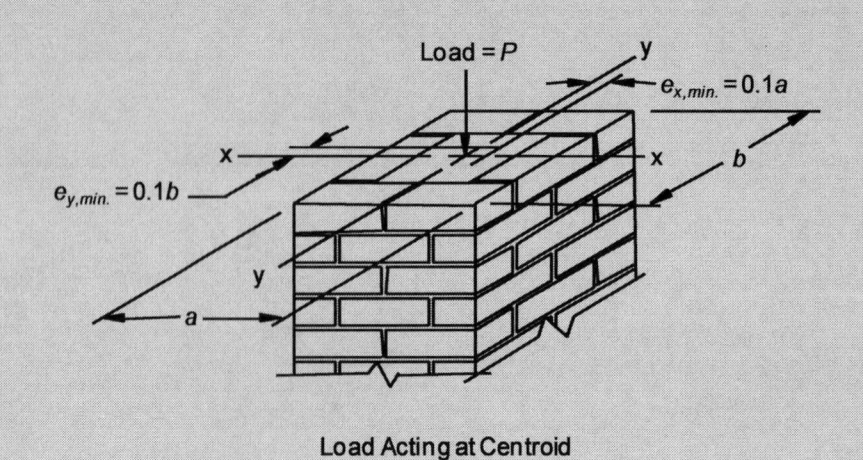

Figure CC-8.3-2 — Minimum design eccentricity in columns

CODE

8.3.4.4 *Walls* — Special reinforced masonry shear walls having a shear span ratio, $M/(Vd_v)$, equal to or greater than 1.0 and having an axial load, P, greater than $0.05f'_m A_n$, which are subjected to in-plane forces, shall have a maximum ratio of flexural tensile reinforcement, ρ_{max}, not greater than that calculated as follows:

$$\rho_{max} = \frac{nf'_m}{2f_y\left(n + \dfrac{f_y}{f'_m}\right)} \quad \text{(Equation 8-23)}$$

The maximum reinforcement ratio does not apply in the out-of-plane direction.

8.3.5 *Shear*

8.3.5.1 Members shall be designed in accordance with Sections 8.3.5.1.1 through 8.3.5.1.4.

8.3.5.1.1 Calculated shear stress in the masonry shall be determined by the relationship:

$$f_v = \frac{V}{A_{nv}} \quad \text{(Equation 8-24)}$$

COMMENTARY

8.3.4.4 *Walls* — The balanced reinforcement ratio for a masonry element with a single layer of reinforcement designed by allowable stress design can be derived by applying principles of engineering mechanics to a cracked, transformed section. The resulting equation is:

$$\rho_b = \frac{nF_b}{2F_s\left(n + \dfrac{F_s}{F_b}\right)}$$

where ρ_b is the balanced reinforcement ratio resulting in a condition in which the reinforcement and the masonry simultaneously reach their specified allowable stresses. However, the ratio of allowable steel tensile stress to the specified yield strength of the reinforcement, and the ratio of allowable masonry compressive stress to the specified compressive strength of the masonry are not consistent (F_s can range from 40 percent to 53 percent of f_y, while F_b is taken equal to $0.45f'_m$). Therefore, allowable stresses in the equation above are replaced with the corresponding specified strengths, as shown in Code Equation 8-23.

The equation is directly applicable for reinforcement concentrated at the end of the shear wall. For distributed reinforcement, the reinforcement ratio is obtained as the total area of tension reinforcement divided by bd.

8.3.5 *Shear*

Prior to the 2011 edition of the Code, the shear resistance provided by the masonry was not added to the shear resistance provided by the shear reinforcement (in allowable stress design). A recent study (Davis et al, 2010) examined eight different methods for predicting the in-plane shear capacity of masonry walls. The design provisions of Chapter 9 (strength design) of this Code were found to be the best predictor of shear strength. The 2008 MSJC's allowable stress design provisions had a greater amount of scatter. Therefore, the provisions of Chapter 9, which allow for the shear resistance provided by the masonry to be added to the shear resistance provided by the shear reinforcement, were appropriately modified and adopted for Chapter 8. See the flow chart for design of masonry members resisting shear shown in Figure CC-8.3-3.

CODE

8.3.5.1.2 The calculated shear stress, f_v, shall not exceed the allowable shear stress, F_v, where F_v shall be calculated using Equation 8-25 and shall not be taken greater than the limits given by Section 8.3.5.1.2 (a) through (c).

$$F_v = (F_{vm} + F_{vs})\gamma_g \quad \text{(Equation 8-25)}$$

(a) Where $M/(Vd_v) \leq 0.25$:

$$F_v \leq \left(3\sqrt{f'_m}\right)\gamma_g \quad \text{(Equation 8-26)}$$

(b) Where $M/(Vd_v) \geq 1.0$

$$F_v \leq \left(2\sqrt{f'_m}\right)\gamma_g \quad \text{(Equation 8-27)}$$

γ_g = 0.75 for partially grouted shear walls and 1.0 otherwise.

(c) The maximum value of F_v for $M/(Vd_v)$ between 0.25 and 1.0 shall be permitted to be linearly interpolated.

8.3.5.1.3 The allowable shear stress resisted by the masonry, F_{vm}, shall be calculated using Equation 8-28 for special reinforced masonry shear walls and using Equation 8-29 for other masonry:

$$F_{vm} = \frac{1}{4}\left[\left(4.0 - 1.75\left(\frac{M}{Vd_v}\right)\right)\sqrt{f'_m}\right] + 0.25\frac{P}{A_n}$$

(Equation 8-28)

$$F_{vm} = \frac{1}{2}\left[\left(4.0 - 1.75\left(\frac{M}{Vd_v}\right)\right)\sqrt{f'_m}\right] + 0.25\frac{P}{A_n}$$

(Equation 8-29)

$M/(Vd_v)$ shall be taken as a positive number and need not be taken greater than 1.0.

8.3.5.1.4 The allowable shear stress resisted by the steel reinforcement, F_{vs}, shall be calculated using Equation 8-30:

$$F_{vs} = 0.5\left(\frac{A_v F_s d_v}{A_{nv} s}\right) \quad \text{(Equation 8-30)}$$

8.3.5.2 Shear reinforcement shall be provided when f_v exceeds F_{vm}. When shear reinforcement is required, the provisions of Section 8.3.5.2.1 and 8.3.5.2.2 shall apply.

COMMENTARY

8.3.5.1.2 Allowable shear stress Equations 8-25 through 8-27 are based on strength design provisions, but reduced by a factor of safety of 2 to obtain allowable stress values. The provisions of this Section were developed through the study of and calibrated to cantilevered shear walls. The ratio $M/(Vd_v)$, can be difficult to interpret or apply consistently for other conditions such as for a uniformly loaded, simply supported beam. Concurrent values of M and Vd_v must be considered at appropriate locations along shear members, such as beams, to determine the critical $M/(Vd_v)$ ratio. To simplify the analytical process, designers are permitted to use $M/(Vd_v) = 1$. Commentary Section 9.3.4.1.2 provides additional information. Partially grouted shear walls can have lower strengths than predicted by the shear capacity equations using just the reduction of net area (Minaie et al, 2010; Nolph and ElGawady, 2011; Schultz, 1996a; Schultz, 1996b; Schultz and Hutchinson, 2001). The grouted shear wall factor, γ_g, is used to compensate for this reduced capacity until methods can be developed to more accurately predict the performance of these elements.

8.3.5.1.3 Equation 8-29 is based on strength design provisions with the masonry shear strength reduced by a factor of safety of 2 and service loads used instead of factored loads.

A reduced value is used for the allowable masonry shear stress in special reinforced masonry shear walls to account for degradation of masonry shear strength in plastic hinging regions. Davis et al (2010) proposed a factor with a value of 1.0 for wall ductility ratios of 2.0 or less, and a linear decrease to zero as the ductility ratio increases from 2.0 to 4.0. The committee chose a constant value of 0.5, resulting in the allowable stress being reduced by a factor of 2, for design convenience.

8.3.5.1.4 Commentary Section 9.3.4.1.2.2 provides additional information.

CODE

8.3.5.2.1 Shear reinforcement shall be provided parallel to the direction of applied shear force. Spacing of shear reinforcement shall not exceed the lesser of $d/2$ or 48 in. (1219 mm).

8.3.5.2.2 Reinforcement shall be provided perpendicular to the shear reinforcement and shall be at least equal to one-third A_v. The reinforcement shall be uniformly distributed and shall not exceed a spacing of 8 ft (2.44 m).

8.3.5.3 In composite masonry walls, shear stresses developed in the planes of interfaces between wythes and filled collar joints or between wythes and headers shall meet the requirements of Section 8.1.4.2.

8.3.5.4 In cantilever beams, the maximum shear shall be used. In noncantilever beams, the maximum shear shall be used except that sections located within a distance $d/2$ from the face of support shall be designed for the same shear as that calculated at a distance $d/2$ from the face of support when the following conditions are met:

(a) support reaction, in direction of applied shear force, introduces compression into the end regions of the beam, and

(b) no concentrated load occurs between face of support and a distance $d/2$ from face.

COMMENTARY

8.3.5.2.1 The assumed shear crack is at 45 degrees to the longitudinal reinforcement. Thus, a maximum spacing of $d/2$ is specified to assure that each crack is crossed by at least one bar. The 48-in. (1219-mm) maximum spacing is an arbitrary choice that has been in codes for many years.

8.3.5.3 Shear across collar joints in composite masonry walls is transferred by the mortar or grout in the collar joint. Shear stress in the collar joint or at the interface between the wythe and the collar joint is limited to the allowable stresses in Section 8.1.4.2. Shear transfer by wall ties or other reinforcement across the collar joint is not considered.

8.3.5.4 The beam or wall loading within $d/2$ of the support is assumed to be transferred in direct compression or tension to the support without increasing the shear force, provided no concentrated load occurs within the $d/2$ distance.

COMMENTARY

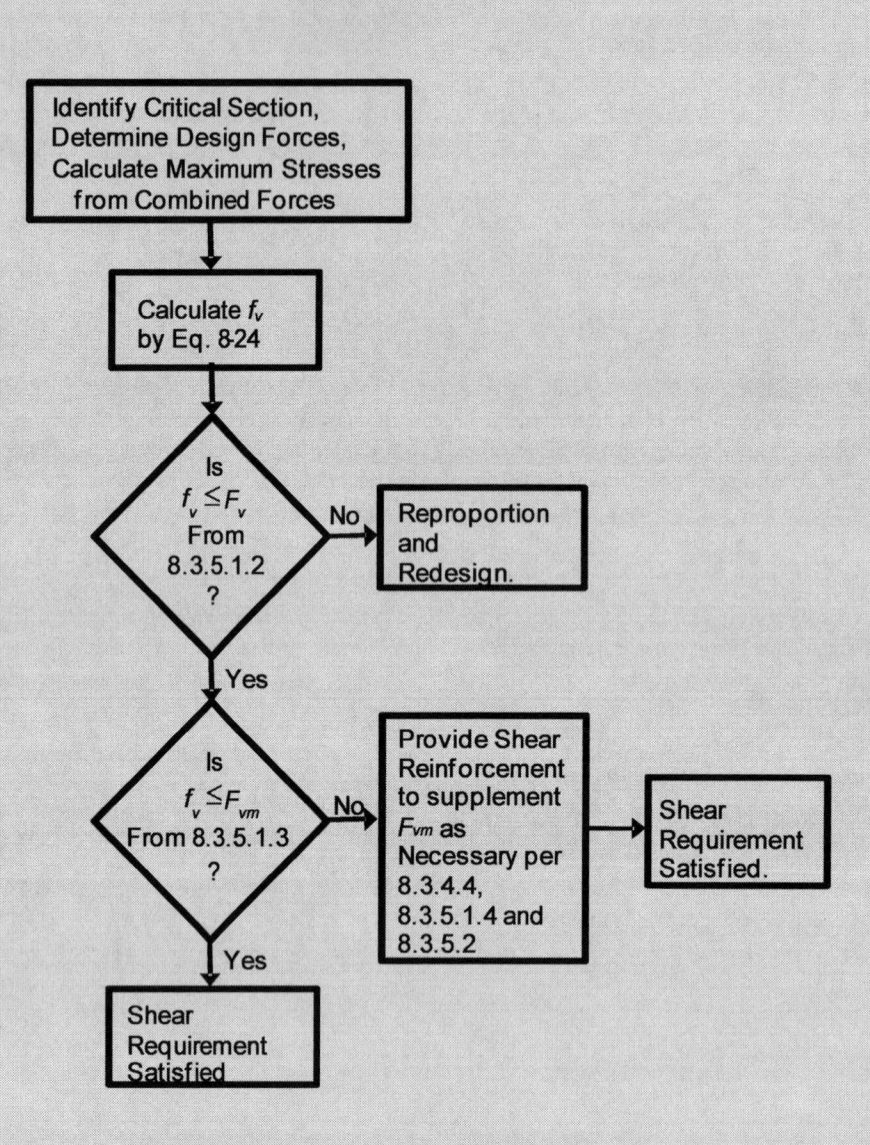

Figure CC-8.3-3 — Flow chart for shear design

BUILDING CODE REQUIREMENTS FOR MASONRY STRUCTURES AND COMMENTARY

CHAPTER 9
STRENGTH DESIGN OF MASONRY

CODE

9.1 — General

9.1.1 *Scope*

This Chapter provides minimum requirements for strength design of masonry. Masonry design by the strength design method shall comply with the requirements of Part 1, Part 2, Sections 9.1.2 through 9.1.9, and either Section 9.2 or 9.3.

9.1.2 *Required strength*

Required strength shall be determined in accordance with the strength design load combinations of the legally adopted building code. Members subject to compressive axial load shall be designed for the factored moment accompanying the factored axial load. The factored moment, M_u, shall include the moment induced by relative lateral displacement.

9.1.3 *Design strength*

Masonry members shall be proportioned so that the design strength equals or exceeds the required strength. Design strength is the nominal strength multiplied by the strength-reduction factor, ϕ, as specified in Section 9.1.4.

9.1.4 *Strength-reduction factors*

COMMENTARY

9.1 — General

9.1.1 *Scope*

Chapter 9 design procedures follow strength design methodology, in which internal forces resulting from application of factored loads must not exceed design strength (nominal member strength reduced by a strength-reduction factor ϕ).

Materials are assumed to be homogenous, isotropic, and exhibit nonlinear behavior. Under loads that exceed service levels, nonlinear material behavior, cracking, and reinforcing bar slip invalidate the assumption regarding the linearity of the stress-strain relation for masonry, grout, and reinforcing steel. If nonlinear behavior is modeled, however, nominal strength can be accurately predicted.

Much of the substantiating data for the strength design criteria in this Chapter was provided by research conducted by the Technical Coordinating Committee for Masonry Research (TCCMaR). This research program resulted in 63 research reports from 1985-1992. These reports are available from The Masonry Society, Longmont, CO. A summary of the TCCMaR program is found in Noland and Kingsley, 1995.

9.1.2 *Required strength*

9.1.3 *Design strength*

Nominal member strengths are typically calculated using minimum specified material strengths.

9.1.4 *Strength-reduction factors*

The strength-reduction factor accounts for the uncertainties in construction, material properties, calculated versus actual member strengths, as well as anticipated mode of failure. Strength-reduction (ϕ) factors are assigned values based on limiting the probability of failure to an acceptably small value, with some adjustment based on judgment and experience.

CODE

9.1.4.1 *Anchor bolts* — For cases where the nominal strength of an anchor bolt is controlled by masonry breakout, by masonry crushing, or by anchor bolt pryout, ϕ shall be taken as 0.50. For cases where the nominal strength of an anchor bolt is controlled by anchor bolt steel, ϕ shall be taken as 0.90. For cases where the nominal strength of an anchor bolt is controlled by anchor pullout, ϕ shall be taken as 0.65.

9.1.4.2 *Bearing* — For cases involving bearing on masonry, ϕ shall be taken as 0.60.

9.1.4.3 *Combinations of flexure and axial load in unreinforced masonry* — The value of ϕ shall be taken as 0.60 for unreinforced masonry subjected to flexure, axial load, or combinations thereof.

9.1.4.4 *Combinations of flexure and axial load in reinforced masonry* — The value of ϕ shall be taken as 0.90 for reinforced masonry subjected to flexure, axial load, or combinations thereof.

9.1.4.5 *Shear* — The value of ϕ shall be taken as 0.80 for masonry subjected to shear.

9.1.5 *Deformation requirements*

9.1.5.1 *Deflection of unreinforced (plain) masonry* — Deflection calculations for unreinforced (plain) masonry members shall be based on uncracked section properties.

9.1.5.2 *Deflection of reinforced masonry* — Deflection calculations for reinforced masonry members shall consider the effects of cracking and reinforcement on member stiffness. The flexural and shear stiffness properties assumed for deflection calculations shall not exceed one-half of the gross section properties, unless a cracked-section analysis is performed.

COMMENTARY

9.1.4.1 *Anchor bolts* — Because of the similarity between the behavior of anchor bolts embedded in grout and in concrete, and because available research data for anchor bolts in grout indicate similarity, the strength-reduction values associated with various controlling anchor bolt failures are derived from expressions based on research into the performance of anchor bolts embedded in concrete.

9.1.4.2 *Bearing* — The value of the strength-reduction factor used in bearing assumes that some degradation has occurred within the masonry material.

9.1.4.3 *Combinations of flexure and axial load in unreinforced masonry* — The same strength-reduction factor is used for the axial load and the flexural tension or compression induced by bending moment in unreinforced masonry elements. The lower strength-reduction factor associated with unreinforced elements (in comparison to reinforced elements) reflects an increase in the coefficient of variation of the measured strengths of unreinforced elements when compared to similarly configured reinforced elements.

9.1.4.4 *Combinations of flexure and axial load in reinforced masonry* — The same strength-reduction factor is used for the axial load and the flexural tension or compression induced by bending moment in reinforced masonry elements. The higher strength-reduction factor associated with reinforced elements (in comparison to unreinforced elements) reflects a decrease in the coefficient of variation of the measured strengths of reinforced elements when compared to similarly configured unreinforced elements.

9.1.4.5 *Shear* — The strength-reduction factor for calculating the design shear strength recognizes the greater uncertainty in calculating nominal shear strength than in calculating nominal flexural strength.

9.1.5 *Deformation requirements*

9.1.5.1 *Deflection of unreinforced (plain) masonry* — The deflection calculations of unreinforced masonry are based on elastic performance of the masonry assemblage as outlined in the design criteria of Section 9.2.2.

9.1.5.2 *Deflection of reinforced masonry* — Values of I_{eff} are typically about one-half of I_g for common configurations of elements that are fully grouted. Calculating a more accurate value using the cracked transformed section may be desirable for some circumstances (Abboud et al, 1993; Hamid et al, 1990).

CODE

9.1.6 *Anchor bolts embedded in grout*

9.1.6.1 *Design requirements* — Anchor bolts shall be designed using either the provisions of 9.1.6.2 or, for headed and bent-bar anchor bolts, by the provisions of Section 9.1.6.3.

9.1.6.2 *Nominal strengths determined by test*

9.1.6.2.1 Anchor bolts shall be tested in accordance with ASTM E488, except that a minimum of five tests shall be performed. Loading conditions of the test shall be representative of intended use of the anchor bolt.

9.1.6.2.2 Anchor bolt nominal strengths used for design shall not exceed 65 percent of the average failure load from the tests.

9.1.6.3 *Nominal strengths determined by calculation for headed and bent-bar anchor bolts* — Nominal strengths of headed and bent-bar anchor bolts embedded in grout shall be determined in accordance with the provisions of Sections 9.1.6.3.1 through 9.1.6.3.3.

9.1.6.3.1 *Nominal tensile strength of headed and bent-bar anchor bolts* — The nominal axial tensile strength of headed anchor bolts shall be calculated using the provisions of Sections 9.1.6.3.1.1. The nominal axial tensile strength of bent-bar anchor bolts shall be calculated using the provisions of Section 9.1.6.3.1.2.

COMMENTARY

9.1.6 *Anchor bolts embedded in grout*
Design of anchor bolts embedded in grout may be based on physical testing or, for headed and bent-bar anchor bolts, by calculation. Due to the wide variation in configurations of post-installed anchors, designers are referred to product literature published by manufacturers for these anchors.

9.1.6.1 *Design requirements*

9.1.6.2 *Nominal strengths determined by test* — Many types of anchor bolts, such as expansion anchors, toggle bolts, sleeve anchors, etc., are not addressed by Code Section 9.1.6.3 and, therefore, such anchors must be designed using test data. Testing may also be used to establish higher strengths than those calculated by Code Section 9.1.6.3. ASTM E488 requires only three tests. The variability of anchor bolt strength in masonry and the possibility that anchor bolts may be used in a non-redundant manner warrants an increase to the minimum of five tests stipulated by the Code. Assuming a normal probability distribution and a coefficient of variation of 20 percent for the test data, a fifth-percentile value for nominal strength is 67 percent, which is rounded to 65 percent of the average strength value. Failure modes obtained from testing should be reported and the associated ϕ factors should be used when establishing design strengths.

9.1.6.3 *Nominal strength determined by calculation for headed and bent-bar anchor bolts* — Design equations provided in the Code stem from research (Brown and Whitlock, 1983; Hatzinikolos et al, 1980; Rad et al, 1998; Tubbs et al, 1999; Allen et al, 2000; Brown et al, 2001; Weigel et al, 2002) conducted on headed anchor bolts and bent-bar anchor bolts (J- or L-bolts) embedded in grout.

The anchor provisions in this Code define bolt shear and tension capacities based on the bolt's specified yield strength. Anchors conforming to A307, Grade A specifications are allowed by the Code, but the ASTM A307, Grade A specification does not specify a yield strength. Use of a yield strength of 37 ksi in the Code design equations for A307 anchors will result in anchor capacities similar to those obtained using the American Institute of Steel Construction provisions.

9.1.6.3.1 *Nominal tensile strength of headed and bent-bar anchor bolts*

CODE

9.1.6.3.1.1 *Axial tensile strength of headed anchor bolts* — The nominal axial tensile strength, B_{an}, of headed anchor bolts embedded in grout shall be determined by Equation 9-1 (nominal axial tensile strength governed by masonry breakout) or Equation 9-2 (nominal axial tensile strength governed by steel yielding). The design axial tensile strength, ϕB_{an}, shall be the smaller of the values obtained from Equations 9-1 and 9-2 multiplied by the applicable ϕ value.

$$B_{anb} = 4 A_{pt} \sqrt{f'_m} \quad \text{(Equation 9-1)}$$

$$B_{ans} = A_b f_y \quad \text{(Equation 9-2)}$$

9.1.6.3.1.2 *Axial tensile strength of bent-bar anchor bolts* – The nominal axial tensile strength, B_{an}, for bent-bar anchor bolts embedded in grout shall be determined by Equation 9-3 (nominal axial tensile strength governed by masonry breakout), Equation 9-4 (nominal axial tensile strength governed by anchor bolt pullout), or Equation 9-5 (nominal axial tensile strength governed by steel yielding). The design axial tensile strength, ϕB_{an}, shall be the smallest of the values obtained from Equations 9-3, 9-4 and 9-5 multiplied by the applicable ϕ value.

$$B_{anb} = 4 A_{pt} \sqrt{f'_m} \quad \text{(Equation 9-3)}$$

$$B_{anp} = 1.5 f'_m e_b d_b + 300\pi (l_b + e_b + d_b) d_b$$

$$\text{(Equation 9-4)}$$

$$B_{ans} = A_b f_y \quad \text{(Equation 9-5)}$$

9.1.6.3.2 *Shear strength of headed and bent-bar anchor bolts* — The nominal shear strength, B_{vn}, of headed and bent-bar anchor bolts shall be determined by Equation 9-6 (nominal shear strength governed by masonry breakout), Equation 9-7 (nominal shear strength governed by masonry crushing), Equation 9-8 (nominal shear strength governed by anchor bolt pryout) or Equation 9-9 (nominal shear strength governed by steel yielding). The design shear strength ϕB_{vn}, shall be the smallest of the values obtained from Equations 9-6, 9-7, 9-8 and 9-9 multiplied by the applicable ϕ value.

$$B_{vnb} = 4 A_{pv} \sqrt{f'_m} \quad \text{(Equation 9-6)}$$

$$B_{vnc} = 1050 \sqrt[4]{f'_m A_b} \quad \text{(Equation 9-7)}$$

$$B_{vnpry} = 2.0 B_{anb} = 8 A_{pt} \sqrt{f'_m} \quad \text{(Equation 9-8)}$$

$$B_{vns} = 0.6 A_b f_y \quad \text{(Equation 9-9)}$$

COMMENTARY

9.1.6.3.1.1 *Axial tensile strength of headed anchor bolts* — Tensile strength of a headed anchor bolt is governed by yield of the anchor steel, Equation 9-2, or by breakout of an approximately conical volume of masonry starting at the anchor head and having a fracture surface oriented at approximately 45 degrees to the masonry surface, Equation 9-1. Steel strength is calculated using the effective tensile stress area of the anchor (that is, including the reduction in area of the anchor shank due to threads).

9.1.6.3.1.2 *Axial tensile strength of bent-bar anchor bolts* — The tensile strength of a bent-bar anchor bolt (J- or L-bolt) is governed by yield of the anchor steel, Equation 9-5, by tensile cone breakout of the masonry, Equation 9-3, or by straightening and pullout of the anchor bolt from the masonry, Equation 9-4. Capacities corresponding to the first two failure modes are calculated as for headed anchor bolts. Code Equation 9-4 corresponds to anchor bolt pullout. The second term in Equation 9-4 is the portion of the anchor bolt capacity due to bond between bolt and grout. Accordingly, Specification Article 3.2A requires that precautions be taken to ensure that the shanks of the bent-bar anchor bolts are clean and free of debris that would otherwise interfere with the bond between anchor bolt and grout.

9.1.6.3.2 *Shear strength of headed and bent-bar anchor bolts* -- Shear strength of a headed or bent-bar anchor bolt is governed by yielding of the anchor steel, Equation 9-9, by masonry crushing, Equation 9-7, or by masonry shear breakout, Equation 9-6. Steel strength is calculated using the effective tensile stress area (that is, threads are conservatively assumed to lie in the critical shear plane). Pryout (see Figure CC-6.2-7) is also a possible failure mode. The pryout equation (Equation 9-8) is adapted from concrete research (Fuchs et al, 1995).

Under static shear loading, bent-bar anchor bolts do not exhibit straightening and pullout. Under reversed cyclic shear, however, available research (Malik et al, 1982) suggests that straightening and pullout may occur.

CODE

9.1.6.3.3 *Combined axial tension and shear* — Anchor bolts subjected to axial tension in combination with shear shall satisfy Equation 9-10.

$$\frac{b_{af}}{\phi B_{an}} + \frac{b_{vf}}{\phi B_{vn}} \leq 1 \qquad \text{(Equation 9-10)}$$

9.1.7 *Shear strength in multiwythe masonry elements*

9.1.7.1 Design of multiwythe masonry for composite action shall meet the requirements of Sections 5.1.4.2 and 9.1.7.2.

9.1.7.2 The nominal shear strength at the interfaces between wythes and collar joints or within headers shall be determined so that shear stresses shall not exceed the following:

(a) mortared collar joints, 14 psi (96.5 kPa).

(b) grouted collar joints, 26 psi (179.3 kPa).

(c) headers,

$2.6\sqrt{\text{specified unit compressive strength of header}}$ psi (MPa) (over net area of header).

9.1.8 *Nominal bearing strength*

The nominal bearing strength of masonry shall be calculated as $0.8 f'_m$ multiplied by the bearing area, A_{br}, as defined in Section 4.3.4.

9.1.9 *Material properties*

9.1.9.1 *Compressive strength*

9.1.9.1.1 *Masonry compressive strength* — The specified compressive strength of masonry, f'_m, shall equal or exceed 1,500 psi (10.34 MPa). The value of f'_m used to determine nominal strength values in this chapter shall not exceed 4,000 psi (27.58 MPa) for concrete masonry and shall not exceed 6,000 psi (41.37 MPa) for clay masonry.

9.1.9.1.2 *Grout compressive strength* — For concrete masonry, the specified compressive strength of grout, f'_g, shall equal or exceed the specified compressive strength of masonry, f'_m, but shall not exceed 5,000 psi (34.47 MPa). For clay masonry, the specified compressive strength of grout, f'_g, shall not exceed 6,000 psi (41.37 MPa).

COMMENTARY

9.1.7 *Shear strength in multiwythe masonry elements*

The nominal shear strength is based on shear stresses that are twice the allowable shear stresses in allowable stress design. Commentary Section 8.1.4 provides additional information.

9.1.8 *Nominal bearing strength*

Commentary Section 4.3.4 provides further information.

9.1.9 *Material properties*

Commentary Section 4.2 provides additional information.

9.1.9.1 *Compressive strength*

9.1.9.1.1 *Masonry compressive strength* — Design criteria are based on TCCMaR research (Noland and Kingsley, 1995) conducted on structural masonry components having compressive strength in the range of 1,500 to 4,000 psi (10.34 to 27.58 MPa) for concrete masonry and 1,500 to 6,000 psi (10.34 to 41.37 MPa) for clay masonry. Thus, the upper limits given represent the upper values that were tested in the research.

9.1.9.1.2 *Grout compressive strength* — Because most empirically derived design equations calculate nominal strength as a function of the specified compressive strength of the masonry, the specified compressive strength of the grout is required to be at least equal to the specified compressive strength for concrete masonry. This requirement is an attempt to ensure that where the grout compressive strength controls the design (such as anchors embedded in grout), the nominal strength will not be affected. The limitation on the maximum grout compressive strength is due to the lack of available research using higher material strengths.

CODE

9.1.9.2 *Masonry modulus of rupture* — The modulus of rupture, f_r, for masonry elements subjected to out-of-plane or in-plane bending shall be in accordance with the values in Table 9.1.9.2. For grouted masonry not laid in running bond, tension parallel to the bed joints shall be assumed to be resisted only by the minimum cross-sectional area of continuous grout that is parallel to the bed joints.

COMMENTARY

9.1.9.2 *Masonry modulus of rupture* — The modulus of rupture values provided in Code Table 9.1.9.2 are directly proportional to the allowable stress values for flexural tension. In-plane and out-of-plane strain gradients are recognized as being different, but at these low stress levels this effect should be small.

Historically, masonry not laid in running bond has been assumed to have no flexural bond strength across mortared head joints; thus, the grout area alone is used to resist bending. Examples of a continuous grout section parallel to the bed joints are shown in Figure CC-8.2-2.

The presence of flashing and other conditions at the base of the wall can significantly reduce the flexural bond. The values in Table 9.1.9.2 apply only to the flexural tensile stresses developed between masonry units, mortar, and grout.

Table 9.1.9.2 — Modulus of rupture, f_r, psi (kPa)

Direction of flexural tensile stress and masonry type	Mortar types			
	Portland cement/lime or mortar cement		Masonry cement or air entrained portland cement/lime	
	M or S	N	M or S	N
Normal to bed joints				
Solid units	133 (919)	100 (690)	80 (552)	51 (349)
Hollow units[1]				
Ungrouted	84 (579)	64 (441)	51 (349)	31 (211)
Fully grouted	163 (1124)	158 (1089)	153 (1055)	145 (1000)
Parallel to bed joints in running bond				
Solid units	267 (1839)	200 (1379)	160 (1103)	100 (689)
Hollow units				
Ungrouted and partially grouted	167 (1149)	127 (873)	100 (689)	64 (441)
Fully grouted	267 (1839)	200 (1379)	160 (1103)	100 (689)
Parallel to bed joints in masonry not laid in running bond				
Continuous grout section parallel to bed joints	335 (2310)	335 (2310)	335 (2310)	335 (2310)
Other	0 (0)	0 (0)	0 (0)	0 (0)

[1] For partially grouted masonry, modulus of rupture values shall be determined on the basis of linear interpolation between fully grouted hollow units and ungrouted hollow units based on amount (percentage) of grouting.

CODE

9.1.9.3 *Reinforcement strengths*

9.1.9.3.1 *Reinforcement for in-plane flexural tension and flexural tension perpendicular to bed joints* — Masonry design shall be based on a reinforcement strength equal to the specified yield strength of reinforcement, f_y, which shall not exceed 60,000 psi (413.7 MPa). The actual yield strength shall not exceed 1.3 multiplied by the specified yield strength.

9.1.9.3.2 *Reinforcement for in-plane shear and flexural tension parallel to bed joints* — Masonry design shall be based on a specified yield strength, f_y, which shall not exceed 60,000 psi (413.7 MPa) for reinforcing bars and which shall not exceed 85,000 psi (586 MPa) for reinforcing wire.

COMMENTARY

9.1.9.3 *Reinforcement strengths*

9.1.9.3.1 *Reinforcement for in-plane flexural tension and flexural tension perpendicular to bed joints* — TCCMaR Research (Noland and Kingsley, 1995) conducted on reinforced masonry components used Grade 60 reinforcement. To be consistent with laboratory documented investigations, design is based on a nominal steel yield strength of 60,000 psi (413.7 MPa). The limitation on the flexural steel yield strength of 130 percent of the nominal yield strength is to minimize the over-strength unintentionally incorporated into a design.

9.1.9.3.2 *Reinforcement for in-plane shear and flexural tension parallel to bed joints* — Studies of minimum shear reinforcement requirements (Schultz, 1996a, Baenziger and Porter, 2011, Porter and Baenziger, 2007, Sveinsson et al, 1985, and Schultz and Hutchinson, 2001a) have shown that when sufficient area, strength, and strain elongation properties of reinforcement are provided to resist the load transferred from the masonry after cracking, then the reinforcement does not rupture upon cracking of the masonry. Equivalent performance of shear walls with bond beams and shear walls with bed joint reinforcement under simulated seismic loading was observed in the laboratory tests (Baenziger and Porter, 2011 and Schultz and Hutchinson, 2001a). Minimum Code requirements have been provided (Schultz, 1996a) to satisfy both strength and energy criteria. The limitation on steel yield strength of 130 percent of the nominal yield strength, in Section 9.1.9.3.1, does not apply to shear reinforcement because the risk of brittle shear failures is reduced with higher yield strength.

Joint reinforcement of at least 3/16 in. (4.8 mm) diameter longitudinal wire is deemed to have sufficient strain elongation and, thus, was selected as the minimum size when joint reinforcement is used as the primary shear and flexural reinforcement. The research (Baenziger and Porter, 2011) was for walls that contained a minimum of two 3/16 in. (4.8 mm) diameter longitudinal wires in a bed joint. Other research (Schultz and Hutchinson, 2001a) contained two No. 9 gage (0.148 in. (3.76 mm)) diameter longitudinal wires or two No. 5 gage (0.207 in. (5.26 mm)) diameter longitudinal wires in a bed joint. The No. 5 gage longitudinal wires exhibited similar ductility to the joint reinforcement in the Baenziger/Porter research.

CODE

9.2 — Unreinforced (plain) masonry

9.2.1 *Scope*

Design of unreinforced masonry by the strength design method shall comply with the requirements of Part 1, Part 2, Section 9.1, and Section 9.2.

9.2.2 *Design criteria*

Unreinforced masonry members shall be designed in accordance with the principles of engineering mechanics and shall be designed to remain uncracked.

9.2.3 *Design assumptions*

The following assumptions shall be used in the design of unreinforced masonry members:

(a) Strain in masonry shall be directly proportional to the distance from the neutral axis.

(b) Flexural tension in masonry shall be assumed to be directly proportional to strain.

(c) Flexural compressive stress in combination with axial compressive stress in masonry shall be assumed to be directly proportional to strain.

(d) Stresses in the reinforcement are not accounted for in determining the resistance to design loads.

9.2.4 *Nominal flexural and axial strength*

9.2.4.1 *Nominal strength* — The nominal strength of unreinforced (plain) masonry cross-sections for combined flexure and axial loads shall be determined so that:

(a) the compressive stress does not exceed $0.80 f'_m$.

(b) the tensile stress does not exceed the modulus of rupture determined from Section 9.1.9.2.

9.2.4.2 *Nominal axial strength* — The nominal axial strength, P_n, shall not be taken greater than the following:

(a) For members having an h/r ratio not greater than 99:

$$P_n = 0.80 \left\{ 0.80 A_n f'_m \left[1 - \left(\frac{h}{140r} \right)^2 \right] \right\}$$

(Equation 9-11)

COMMENTARY

9.2 — Unreinforced (plain) masonry

9.2.1 *Scope*

Commentary Section 8.2.1 provides further information.

9.2.2 *Design criteria*

The design of unreinforced masonry requires that the structure performs elastically under design loads. The system response factors used in the design of unreinforced masonry assume an elastic response. Commentary Section 8.2.2 provides further information.

9.2.3 *Design assumptions*

Commentary Section 8.2.3 provides further information.

9.2.4 *Nominal flexural and axial strength*

9.2.4.1 *Nominal strength* — This section gives requirements for constructing an interaction diagram for unreinforced masonry members subjected to combined flexure and axial loads. The requirements are illustrated in Figure CC-9.2-1. Also shown in Figure CC-9.2-1 are the requirements of Section 9.2.4.2, which give a maximum axial force.

9.2.4.2 *Nominal axial strength* — Commentary Section 9.3.4.1.1 gives additional information.

BUILDING CODE REQUIREMENTS FOR MASONRY STRUCTURES AND COMMENTARY C-131

COMMENTARY

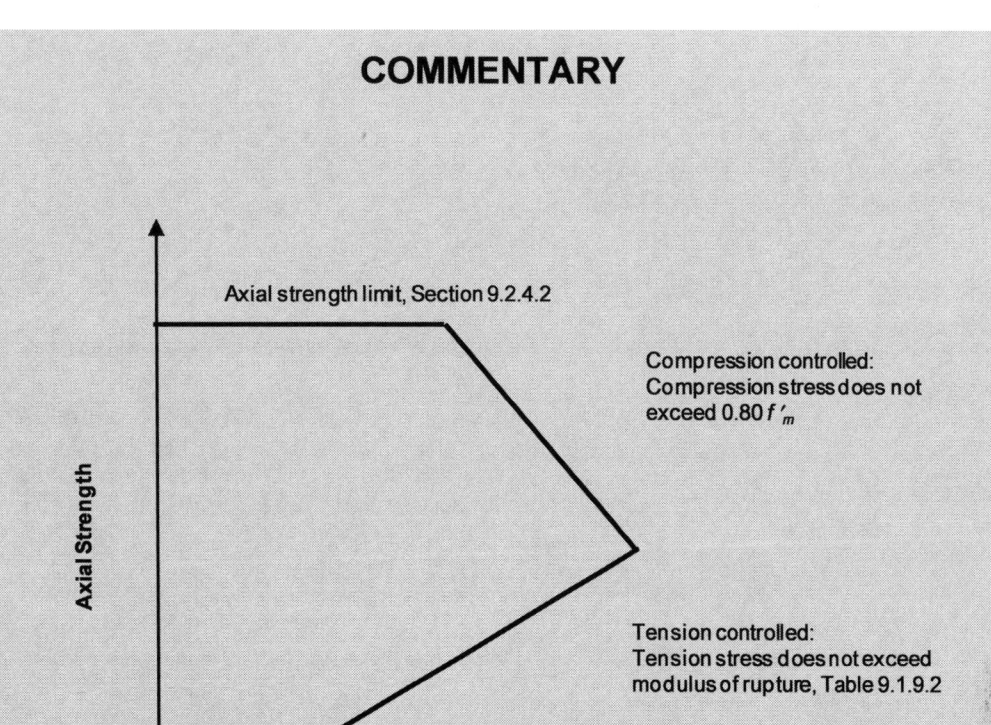

Figure CC-9.2-1 Interaction diagram for unreinforced masonry members

CODE

(b) For members having an h/r ratio greater than 99:

$$P_n = 0.80\left[0.80 A_n f'_m \left(\frac{70r}{h}\right)^2\right] \quad \text{(Equation 9-12)}$$

9.2.4.3 P-Delta effects

9.2.4.3.1 Members shall be designed for the factored axial load, P_u, and the moment magnified for the effects of member curvature, M_u.

9.2.4.3.2 The magnified moment, M_u, shall be determined either by a second-order analysis, or by a first-order analysis and Equations 9-13 and 9-14.

$$M_u = \psi M_{u,0} \quad \text{(Equation 9-13)}$$

$$\psi = \cfrac{1}{1 - \cfrac{P_u}{A_n f'_m \left(\cfrac{70r}{h}\right)^2}} \quad \text{(Equation 9-14)}$$

9.2.4.3.3 A value of $\psi = 1$ shall be permitted for members in which $h/r \le 45$.

COMMENTARY

9.2.4.3 P-delta effects — P-delta effects are either determined by a second-order analysis, which includes P-delta effects, or a first-order analysis, which excludes P-delta effects and the use of moment magnifier. The moment magnifier is determined as:

$$\psi = \cfrac{C_m}{1 - \cfrac{P_u}{\phi_k P_{euler}}}$$

where ϕ_k is a stiffness reduction factor or a resistance factor to account for variability in stiffness, C_m is a factor relating the actual moment diagram to an equivalent uniform moment diagram, and P_{euler} is Euler's buckling load. For reinforced concrete design, a value of $\phi_k = 0.75$ is used (Mirza et al, 1987).

Euler's buckling load is obtained as $P_{euler} = \pi^2 E_m A_n r^2 / h^2$. Using $E_m = 700 f'_m$, which is the lower value of clay and concrete masonry, Euler's buckling load becomes:

CODE

9.2.4.3.4 A value of $\psi = 1$ shall be permitted for members in which $45 < h/r \leq 60$, provided that the nominal strength defined in Section 9.2.4.1 is reduced by 10 percent.

COMMENTARY

$$P_{euler} = \frac{\pi^2 E_m A_n r^2}{h^2}$$

$$= \frac{\pi^2 700 f'_m A_n r^2}{h^2} = A_n f'_m \left(\frac{83.1r}{h}\right)^2$$

Current design provisions calculate the axial strength of walls with $h/r > 99$ as $A_n f'_m (70r/h)^2$. Section 8.2.4.1 of the Commentary gives the background of this equation. It is based on using $E_m = 1000 f'_m$, neglecting the tensile strength of the masonry, and considering an accidental eccentricity of $0.10t$. In spite of the fact that this equation was developed using a higher modulus than in the current Code, the equation gives a strength of $(70/83.1)^2 = 0.71$ of Euler's buckling load for clay masonry. The value of 0.71 is approximately the value of ϕ_k that has been used as a stiffness reduction factor. For ease of use and because of designer's familiarity, a value of $(70 r / h)$ is used for Euler's buckling load instead of an explicit stiffness reduction factor. For most walls, $C_m = 1$. The moment magnifier can thus be determined as:

$$\psi = \frac{1}{1 - \frac{P_u}{A_n f'_m \left(\frac{70r}{h}\right)^2}}$$

Figure CC-9.2-2 shows the ratio of the second-order stress, $\frac{P_u}{A_n} + \frac{\delta M_u}{S_n}$, divided by the first-order stress, $\frac{P_u}{A_n} + \frac{M_u}{S_n}$, when the second-order stress is at the strength design limit $\phi(0.8 f'_m)$. Typically slenderness effects are ignored if they contribute less than 5 percent (MacGregor et al, 1970). From Figure CC-9.2-2, slenderness effects contribute less than 5 percent for values of $h/r \leq 45$. An intermediate wall is one with a slenderness h/r greater than 45 but not greater than 60. Slenderness effects contribute about 10 percent to the design at $h/r = 60$. Intermediate walls can be designed using either the moment magnifier approach or a simplified method in which the nominal stresses are reduced by 10 percent. The Code requires walls with $h/r > 60$ to be designed using the moment magnifier approach.

9.2.5 *Axial tension* — Axial tension resistance of unreinforced masonry shall be neglected in design.

9.2.5 *Axial tension*
Commentary Section 8.2.5 provides further information.

CODE

9.2.6 Nominal shear strength

9.2.6.1 Nominal shear strength, V_n, shall be the smallest of (a), (b) and the applicable condition of (c) through (f):

(a) $3.8 A_{nv} \sqrt{f'_m}$

(b) $300 A_{nv}$

(c) For running bond masonry not fully grouted;
$56 A_{nv} + 0.45 N_u$

(d) For masonry not laid in running bond, constructed of open end units, and fully grouted;
$56 A_{nv} + 0.45 N_u$

(e) For running bond masonry fully grouted;
$90 A_{nv} + 0.45 N_u$

(f) For masonry not laid in running bond, constructed of other than open end units, and fully grouted;
$23 A_{nv}$

9.2.6.2 The minimum normalized web area of concrete masonry units, determined in accordance with ASTM C140, shall not be less than 27 in.^{2}/ft^2 (187,500 mm^2/m^2) or the nominal shear strength of the web shall not exceed $3.8 A_{nv} \sqrt{f'_m} I_n b / Q$.

COMMENTARY

9.2.6 Nominal shear strength

9.2.6.1 For a rectangular cross-section, the shear stress is assumed to follow a parabolic distribution. The Code is based on an average shear stress, which is two-thirds of the maximum shear stress for a parabolic shear stress distribution. Commentary Section 8.2.6 provides further information.

9.2.6.2 Out-of-plane flexure causes horizontal and vertical shear stresses. Vertical shear stresses are resisted by the connection between the web and face shell of the unit. A normalized web area of 27 in.^{2}/ft^2 (187,500 mm^2/m^2) provides sufficient web area so that shear stresses between the web and face shell of a unit, resulting from out-of-plane loading, will not be critical. For simplicity, the same nominal out-of-plane shear strength as for in-plane shear is conservatively used, although peak shear stresses instead of average shear stresses are being checked.

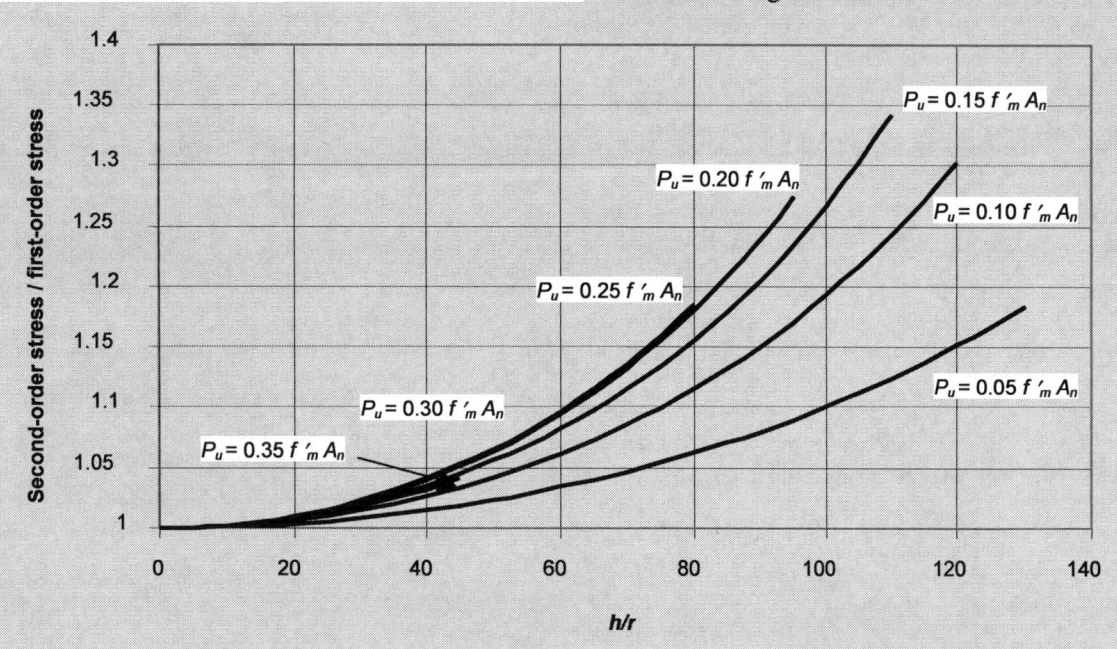

Figure CC-9.2-2 Ratio of second-order stress to first-order stress

CODE

9.3 — Reinforced masonry

9.3.1 *Scope*

This section provides requirements for the design of structures in which reinforcement is used to resist tensile forces in accordance with the principles of engineering mechanics and the contribution of the tensile resistance of the masonry is neglected except as provided in Section 9.3.4.1.2. Design of reinforced masonry by the strength design method shall comply with the requirements of Part 1, Part 2, Section 9.1, and Section 9.3.

9.3.2 *Design assumptions*

The following assumptions shall be used in the design of reinforced masonry:

(a) Strain compatibility exists between the reinforcement, grout, and masonry.

(b) The nominal strength of reinforced masonry cross-sections for combined flexure and axial load is based on applicable conditions of equilibrium.

(c) The maximum usable strain, ε_{mu}, at the extreme masonry compression fiber is 0.0035 for clay masonry and 0.0025 for concrete masonry.

(d) Strains in reinforcement and masonry are directly proportional to the distance from the neutral axis.

(e) Compression and tension stress in reinforcement is E_s multiplied by the steel strain, but not greater than f_y. Except as permitted in Section 9.3.3.5.1 (e) for determination of maximum area of flexural reinforcement, the compressive stress of steel reinforcement does not contribute to the axial and flexural resistance unless lateral restraining reinforcement is provided in compliance with the requirements of Section 5.3.1.4.

(f) Masonry in tension does not contribute to axial and flexural strengths. Axial and flexural tension stresses are resisted entirely by steel reinforcement.

(g) The relationship between masonry compressive stress and masonry strain is defined by the following:

Masonry stress of $0.80 f'_m$ is uniformly distributed over an equivalent compression stress block bounded by edges of the cross section and a straight line located parallel to the neutral axis and located at a distance $a = 0.80 c$ from the fiber of maximum compressive strain. The distance c from the fiber of maximum strain to the neutral axis shall be measured perpendicular to the neutral axis.

COMMENTARY

9.3 — Reinforced masonry

9.3.1 *Scope*

The high tensile strength of reinforcement complements the high compressive strength of masonry. Increased strength and greater ductility result from the use of reinforcement in masonry structures as compared with unreinforced masonry.

9.3.2 *Design assumptions*

The design assumptions listed have traditionally been used for strength design of reinforced masonry members.

The values for the maximum usable strain are based on research on masonry materials (Assis and Hamid, 1990; Brown, 1987). Concern has been raised as to the implied precision of the values. However, the Committee agrees that the reported values for the maximum usable strain reasonably represent those observed during testing.

Although tension may develop in the masonry of a reinforced element, it is not considered in calculating axial and flexural strengths.

CODE

9.3.3 *Reinforcement requirements and details*
9.3.3.1 *Reinforcement size limitations*

(a) Reinforcing bars used in masonry shall not be larger than No. 9 (M#29). The nominal bar diameter shall not exceed one-eighth of the nominal member thickness and shall not exceed one-quarter of the least clear dimension of the cell, course, or collar joint in which the bar is placed. The area of reinforcing bars placed in a cell or in a course of hollow unit construction shall not exceed 4 percent of the cell area.

(b) Joint reinforcement longitudinal wire used in masonry as shear reinforcement shall be at least 3/16 in. (4.8 mm) diameter.

9.3.3.2 *Standard hooks* — Standard hooks in tension shall be considered to develop an equivalent embedment length, l_e, as determined by Equation 9-15:

$$l_e = 13 d_b \quad \text{(Equation 9-15)}$$

9.3.3.3 *Development* — The required tension or compression reinforcement shall be developed in accordance with the following provisions:

The required development length of reinforcement shall be determined by Equation 9-16, but shall not be less than 12 in. (305 mm).

$$l_d = \frac{0.13 d_b^2 f_y \gamma}{K \sqrt{f_m'}} \quad \text{(Equation 9-16)}$$

K shall not exceed the smallest of the following: the minimum masonry cover, the clear spacing between adjacent reinforcement splices, and $9d_b$.

γ = 1.0 for No. 3 (M#10) through No. 5 (M#16) bars;

γ = 1.3 for No. 6 (M#19) through No. 7 (M#22) bars; and

γ = 1.5 for No. 8 (M#25) through No. 9 (M#29) bars.

Development length of epoxy-coated reinforcing bars shall be taken as 150 percent of the length determined by Equation 9-16.

COMMENTARY

9.3.3 *Reinforcement requirements and details*
9.3.3.1 *Reinforcement size limitations*

(a) The limit of using a No. 9 (M #29) bar is motivated by the goal of having a larger number of smaller diameter bars to transfer stresses rather than a fewer number of larger diameter bars. Some TCCMaR research investigations (Noland and Kingsley, 1995) have concluded that in certain applications masonry reinforced with more uniformly distributed smaller diameter bars performs better than similarly configured masonry elements using fewer larger diameter bars. While not every investigation is conclusive, the Committee does agree that incorporating larger diameter reinforcement may dictate unreasonable cover distances or development lengths. The limitations on clear spacing and percentage of cell area are indirect methods of preventing problems associated with over-reinforcing and grout consolidation. At sections containing lap splices, the maximum area of reinforcement should not exceed 8 percent of the cell area.

(b) The limit of using at least 3/16 in. longitudinal wire in joint reinforcement used as shear reinforcement is to provide sufficient strain capacity to avoid rupture. The minimum wire size does not apply to wire reinforcement used to satisfy prescriptive seismic reinforcement.

9.3.3.2 *Standard hooks* — Refer to Commentary Section 6.1.5 for further information.

9.3.3.3 *Development* — The clear spacing between adjacent reinforcement does not apply to the reinforcing bars being spliced together. Refer to Commentary 8.1.6.7.1 for further information.

Schultz (2005) studied the performance of the 2005 MSJC formula for splice lengths in masonry relative to a database of splice tests. Schultz (2004, 2005) found that for clear cover in excess of $5d_b$, the 2005 MSJC lap splice formula gains accuracy, relative to the experimental database, when a $5d_b$ limit is not imposed on the coefficient. Additional testing and subsequent analysis by the National Concrete Masonry Association (2009) also found the $5d_b$ overly conservative and recommended that the limit on K be increased to 8.8 which is rounded to the current $9d_b$ limit.

The 50 percent increase in development length for epoxy-coated bars does not apply to the 12 in. (305 mm) minimum.

CODE

9.3.3.3.1 Reinforcement spliced by noncontact lap splices shall not be spaced farther apart than one-fifth the required length of lap nor more than 8 in. (203 mm).

9.3.3.3.2 Shear reinforcement shall extend the depth of the member less cover distances.

9.3.3.3.2.1 Except at wall intersections, the end of a horizontal reinforcing bar needed to satisfy shear strength requirements of Section 9.3.4.1.2 shall be bent around the edge vertical reinforcing bar with a 180-degree hook. The ends of single-leg or U-stirrups shall be anchored by one of the following means:

(a) A standard hook plus an effective embedment of $l_d/2$. The effective embedment of a stirrup leg shall be taken as the distance between the mid-depth of the member, $d/2$, and the start of the hook (point of tangency).

(b) For No. 5 (M #16) bars and smaller, bending around longitudinal reinforcement through at least 135 degrees plus an embedment of $l_d/3$. The $l_d/3$ embedment of a stirrup leg shall be taken as the distance between mid-depth of the member, $d/2$, and the start of the hook (point of tangency).

(c) Between the anchored ends, each bend in the continuous portion of a transverse U-stirrup shall enclose a longitudinal bar.

9.3.3.3.2.2 At wall intersections, horizontal reinforcing bars needed to satisfy shear strength requirements of Section 9.3.4.1.2 shall be bent around the edge vertical reinforcing bar with a 90-degree standard hook and shall extend horizontally into the intersecting wall a minimum distance at least equal to the development length.

9.3.3.3.2.3 Joint reinforcement used as shear reinforcement and needed to satisfy the shear strength requirements of Section 9.3.4.1.2 shall be anchored around the edge reinforcing bar in the edge cell, either by bar placement between adjacent cross-wires or with a 90-degree bend in longitudinal wires bent around the edge cell and with at least 3-in. (76-mm) bend extensions in mortar or grout.

9.3.3.3.3 *Development of wires in tension* — The development length of wire shall be determined by Equation 9-17, but shall not be less than 6 in. (152 mm).

$$l_d = 48 d_b \qquad \text{(Equation 9-17)}$$

Development length of epoxy-coated wire shall be taken as 150 percent of the length determined by Equation 9-17.

COMMENTARY

9.3.3.3.1 If individual bars in noncontact lap splices are too widely spaced, an unreinforced section is created, which forces a potential crack to follow a zigzag line. Lap splices may occur with the bars in adjacent grouted cells if the requirements of this section are met.

9.3.3.3.2.1 In a wall without an intersecting wall at its end, the edge vertical bar is the bar closest to the end of the wall. When the wall has an intersecting wall at its end, the edge vertical bar is the bar at the intersection of walls. Hooking the horizontal reinforcement around a vertical bar located within the wall running parallel to the horizontal reinforcement would cause the reinforcement to protrude from the wall.

9.3.3.3.2.3 Wire reinforcement should be anchored around or beyond the edge reinforcing bar. Joint reinforcement longitudinal wires and wire bends are placed over masonry unit face shells in mortar and wire extensions can be placed in edge cell mortar or can extend into edge cell grout. Both joint reinforcement longitudinal wires and cross wires can be used to confine vertical reinforcing bars and grouted cells because wires are developed within a short length.

9.3.3.3.3 *Development of wires in tension* — Commentary 8.1.6.2. explains the development of wires as being $0.0015 F_s d_b$. The term $F_s d_b$ is equivalent to 45 d_b since $F_s = 30,000$ psi. The value was rounded up to 48 d_b to be consistent with other sections of the Code.

CODE

9.3.3.4 *Splices* — Reinforcement splices shall comply with one of the following:

(a) The minimum length of lap for bars shall be 12 in. (305 mm) or the development length determined by Equation 9-16, whichever is greater.

(b) Where reinforcement consisting of No. 3 (M#10) or larger bars is placed within the lap, with at least one bar 8 in. (203 mm) or less from each end of the lap, the minimum length of lap for bars in tension or compression determined by Equation 9-16 shall be permitted to be reduced by multiplying the confinement reinforcement factor, ξ. The clear space between the transverse bars and the lapped bars shall not exceed 1.5 in. (38 mm) and the transverse bars shall be fully developed in grouted masonry. The reduced lap splice length shall not be less than $36d_b$.

$$\xi = 1.0 - \frac{2.3 A_{sc}}{d_b^{2.5}} \qquad \text{(Equation 9-18)}$$

Where : $\dfrac{2.3 A_{sc}}{d_b^{2.5}} \leq 1.0$

A_{sc} is the area of the transverse bars at each end of the lap splice and shall not be taken greater than 0.35 in^2 (226 mm^2).

(c) A welded splice shall have the bars butted and welded to develop at least 125 percent of the yield strength, f_y, of the bar in tension or compression, as required. Welding shall conform to AWS D1.4/D1.4M. Reinforcement to be welded shall conform to ASTM A706, or shall be accompanied by a submittal showing its chemical analysis and carbon equivalent as required by AWS D1.4/D1.4M. Existing reinforcement to be welded shall conform to ASTM A706, or shall be analyzed chemically and its carbon equivalent determined as required by AWS D1.4/D1.4M.

(d) Mechanical splices shall have the bars connected to develop at least 125 percent of the yield strength, f_y, of the bar in tension or compression, as required.

(e) Where joint reinforcement is used as shear reinforcement, the splice length of the longitudinal wires shall be a minimum of $48d_b$.

COMMENTARY

9.3.3.4 *Splices* — Refer to Code Commentary Section 8.1.6.7.1 for information on splices.

(c) See Code Commentary Section 8.1.6.7.2 for additional information on welded splices.

(e) Research studies (Porter and Braun, 1998 and Porter and Braun, 1999) of gage length, embedment, and anchorage of joint reinforcement supports wire development lengths consistent with the traditional Code formula of 48 d_b. To avoid expanding the thickness of the bed joint, joint reinforcement wires should be lapped in the same bed joint without stacking longitudinal wires and cross wires. Where cells are fully grouted, interior longitudinal wires may be angled into the grout to accomplish the lap.

CODE

9.3.3.5 *Maximum area of flexural tensile reinforcement*

9.3.3.5.1 For masonry members where $M_u/(V_u d_v) \geq 1$, the cross-sectional area of flexural tensile reinforcement shall not exceed the area required to maintain axial equilibrium under the following conditions:

(a) A strain gradient shall be assumed, corresponding to a strain in the extreme tensile reinforcement equal to 1.5 multiplied by the yield strain and a maximum strain in the masonry as given by Section 9.3.2(c).

(b) The design assumptions of Section 9.3.2 shall apply.

(c) The stress in the tension reinforcement shall be taken as the product of the modulus of elasticity of the steel and the strain in the reinforcement, and need not be taken greater than f_y.

(d) Axial forces shall be taken from the loading combination given by $D + 0.75L + 0.525Q_E$.

(e) The effect of compression reinforcement, with or without lateral restraining reinforcement, shall be permitted to be included for purposes of calculating maximum flexural tensile reinforcement.

9.3.3.5.2 For intermediate reinforced masonry shear walls subject to in-plane loads where $M_u/(V_u d_v) \geq 1$, a strain gradient corresponding to a strain in the extreme tensile reinforcement equal to 3 multiplied by the yield strain and a maximum strain in the masonry as given by Section 9.3.2(c) shall be used. For intermediate reinforced masonry shear walls subject to out-of-plane loads, the provisions of Section 9.3.3.5.1 shall apply.

9.3.3.5.3 For special reinforced masonry shear walls subject to in-plane loads where $M_u/(V_u d_v) \geq 1$, a strain gradient corresponding to a strain in the extreme tensile reinforcement equal to 4 multiplied by the yield strain and a maximum strain in the masonry as given by Section 9.3.2(c) shall be used. For special reinforced masonry shear walls subject to out-of-plane loads, the provisions of Section 9.3.3.5.1 shall apply.

COMMENTARY

9.3.3.5 *Maximum area of flexural tensile reinforcement* — Longitudinal reinforcement in flexural members is limited to a maximum amount to ensure that masonry compressive strains will not exceed ultimate values. In other words, the compressive zone of the member will not crush before the tensile reinforcement develops the inelastic strain consistent with the curvature ductility implied by the R value used in design.

For masonry components that are part of the lateral-force-resisting system, maximum reinforcement is limited in accordance with a prescribed strain distribution based on a tensile strain equal to a factor times the yield strain for the reinforcing bar closest to the edge of the member, and a maximum masonry compressive strain equal to 0.0025 for concrete masonry or 0.0035 for clay-unit masonry. By limiting longitudinal reinforcement in this manner, inelastic curvature capacity is directly related to the strain gradient.

The tensile strain factor varies in accordance with the amount of curvature ductility expected, and ranges from 1.5 to 4 for specially reinforced masonry shear walls. Expected curvature ductility, controlled by the factor on tensile yield strain, is assumed to be associated directly with the displacement ductility, or the value of C_d as given for the type of component. For example, a strain factor of 3 for intermediate reinforced masonry shear walls corresponds to the slightly smaller C_d factor of 2.5, and a strain factor of 4 for specially reinforced walls corresponds to the slightly smaller C_d factor of 3.5.

The maximum reinforcement is determined by considering the prescribed strain distribution, determining the corresponding stress and force distribution, and using statics to sum axial forces.

For a fully grouted shear wall subjected to in-plane loads with uniformly distributed reinforcement, the maximum area of reinforcement per unit length of wall is determined as:

$$\frac{A_s}{d_v} = \frac{0.64 f'_m b \left(\dfrac{\varepsilon_{mu}}{\varepsilon_{mu} + \alpha \varepsilon_y} \right) - \dfrac{P}{d_v}}{f_y \left(\dfrac{\alpha \varepsilon_y - \varepsilon_{mu}}{\varepsilon_{mu} + \alpha \varepsilon_y} \right)}$$

CODE

9.3.3.5.4 For masonry members where $M_u/(V_u d_v) \leq 1$ and when designed using $R \leq 1.5$, there is no upper limit to the maximum flexural tensile reinforcement. For masonry members where $M_u/(V_u d_v) \leq 1$ and when designed using $R \geq 1.5$, the provisions of Section 9.3.3.5.1 shall apply.

COMMENTARY

For a fully grouted member with only concentrated tension reinforcement, the maximum reinforcement is:

$$\rho = \frac{A_s}{bd} = \frac{0.64 f'_m \left(\dfrac{\varepsilon_{mu}}{\varepsilon_{mu} + \alpha \varepsilon_y}\right) - \dfrac{P}{bd}}{f_y}$$

If there is concentrated compression reinforcement with an area equal to the concentrated tension reinforcement, A_s, the maximum reinforcement is:

$$\rho = \frac{A_s}{bd} = \frac{0.64 f'_m \left(\dfrac{\varepsilon_{mu}}{\varepsilon_{mu} + \alpha \varepsilon_y}\right) - \dfrac{P}{bd}}{f_y - \min\left\{\varepsilon_{mu} - \dfrac{d'}{d}(\varepsilon_{mu} + \alpha \varepsilon_y), \varepsilon_y\right\} E_s}$$

where d' is the distance from the extreme compression fiber to the centroid of the compression reinforcement.

For partially grouted cross-sections subjected to out-of-plane loads, the maximum reinforcement is determined based on a fully grouted member with tension reinforcement only, provided that the neutral axis is in the flange. If the neutral axis is in the web, the maximum reinforcement is determined as:

$$\rho = \frac{A_s}{bd}$$

$$\rho = \frac{0.64 f'_m \left(\dfrac{\varepsilon_{mu}}{\varepsilon_{mu} + \alpha \varepsilon_y}\right)\left(\dfrac{b_w}{b}\right) + 0.80 f'_m \, t_{fs}\left(\dfrac{b - b_w}{bd}\right) - \dfrac{P}{bd}}{f_y}$$

where b_w is the width of the compression section minus the sum of the length of ungrouted cells, and t_{fs} is the specified face-shell thickness for hollow masonry units.

Because axial force is implicitly considered in the determination of maximum longitudinal reinforcement, inelastic curvature capacity can be relied on no matter what the level of axial compressive force. Thus, the strength-reduction factors, ϕ, for axial load and flexure can be the same as for flexure alone. Also, confinement reinforcement is not required because the maximum masonry compressive strain will be less than ultimate values.

The axial force is the expected load at the time of the design earthquake. It is derived from ASCE 7 Allowable Stress Load Combination 6 and consideration of the horizontal component of the seismic loading. The vertical component of the earthquake load, E_v, should not be included in calculating the axial force for purposes of determining maximum area of flexural tensile reinforcement.

CODE

9.3.3.6 *Bundling of reinforcing bars* — Reinforcing bars shall not be bundled.

9.3.3.7 *Joint reinforcement used as shear reinforcement* — Joint reinforcement used as shear reinforcement shall consist of at least two 3/16 in. (4.8 mm) diameter longitudinal wires located within a bed joint and placed over the masonry unit face shells. The maximum spacing of joint reinforcement used as shear reinforcement shall not exceed 16 in. (406 mm) for Seismic Design Categories (SDC) A and B and shall not exceed 8 in. (203 mm) in partially grouted walls for SDC C, D, E, and F. Joint reinforcement used as shear reinforcement in fully grouted walls for SDC C, D, E and F shall consist of four 3/16 in. (4.8 mm) diameter longitudinal wires at a spacing not to exceed 8 in. (203 mm).

9.3.4 *Design of beams, piers, and columns*

Member design forces shall be based on an analysis that considers the relative stiffness of structural members. The calculation of lateral stiffness shall include the contribution of all beams, piers, and columns. The effects of cracking on member stiffness shall be considered.

9.3.4.1 *Nominal strength*

9.3.4.1.1 *Nominal axial and flexural strength* — The nominal axial strength, P_n, and the nominal flexural strength, M_n, of a cross section shall be determined in accordance with the design assumptions of Section 9.3.2 and the provisions of this Section. The nominal flexural strength at any section along a member shall not be less than one-fourth of the maximum nominal flexural strength at the critical section.

The nominal axial compressive strength shall not exceed Equation 9-19 or Equation 9-20, as appropriate.

(a) For members having an *h/r* ratio not greater than 99:

$$P_n = 0.80\left[0.80 f'_m (A_n - A_{st}) + f_y A_{st}\right]\left[1 - \left(\frac{h}{140r}\right)^2\right]$$

(Equation 9-19)

(b) For members having an *h/r* ratio greater than 99:

$$P_n = 0.80\left[0.80 f'_m (A_n - A_{st}) + f_y A_{st}\right]\left(\frac{70r}{h}\right)^2$$

(Equation 9-20)

COMMENTARY

For structures intended to undergo inelastic deformation, Sections 9.3.3.5.1, 9.3.3.5.2 and 9.3.3.5.3 are technically sound ways of achieving the design objective of inelastic deformation capacity. These provisions are, however, unnecessarily restrictive for those structures not required to undergo inelastic deformation under the design earthquake. Section 9.3.3.5.4 addresses a relaxation of the maximum reinforcement limits.

9.3.3.6 *Bundling of reinforcing bars* — This requirement stems from the lack of research on masonry with bundled bars.

9.3.3.7 *Joint reinforcement used as shear reinforcement* — The quantities of joint reinforcement indicated are minimums and the designer should evaluate whether additional reinforcement is required to satisfy specific seismic conditions. Research (Schultz, 1997) provides additional guidelines as to the strength and energy requirements of shear reinforcement. Other research (Schultz and Hutchinson, 2001a and Baenziger and Porter, 2011) provides additional perspective on the behavior of joint reinforcement under cyclic loading conditions.

9.3.4 *Design of beams, piers, and columns*

9.3.4.1 *Nominal strength*

9.3.4.1.1 *Nominal axial and flexural strength* — The nominal flexural strength of a member may be calculated using the assumption of an equivalent rectangular stress block as outlined in Section 9.3.2. Commentary Section 8.2.4 gives further information regarding slenderness effects on axial load strength as taken into account with the use of Equation 9-19 and Equation 9-20. Equation 9-19 and Equation 9-20 apply to simply supported end conditions, with or without transverse loading, which result in a symmetric deflection (curvature) about the midheight of the element. Where other support conditions or loading scenarios are known to exist, Equation 9-19 and Equation 9-20 should be modified accordingly to account for the effective height of the element or shape of the bending moment diagram over the clear span of the element. The weak-axis radius of gyration should be used in calculating slenderness-dependent reduction factors. The first coefficient, 0.80, in Equation 9-19 and Equation 9-20 accounts for unavoidable minimum eccentricity in the axial load.

CODE

9.3.4.1.2 *Nominal shear strength* — Nominal shear strength, V_n, shall be calculated using Equation 9-21, and shall not be taken greater than the limits given by 9.3.4.1.2 (a) through (c).

$$V_n = (V_{nm} + V_{ns})\gamma_g \qquad \text{(Equation 9-21)}$$

(a) Where $M_u/(V_u d_v) \leq 0.25$:

$$V_n \leq \left(6 A_{nv} \sqrt{f'_m}\right)\gamma_g \qquad \text{(Equation 9-22)}$$

(b) Where $M_u/(V_u d_v) \geq 1.0$

$$V_n \leq \left(4 A_{nv} \sqrt{f'_m}\right)\gamma_g \qquad \text{(Equation 9-23)}$$

γ_g = 0.75 for partially grouted shear walls and 1.0 otherwise.

(c) The maximum value of V_n for $M_u/(V_u d_v)$ between 0.25 and 1.0 shall be permitted to be linearly interpolated.

9.3.4.1.2.1 *Nominal masonry shear strength* — Shear strength provided by the masonry, V_{nm}, shall be calculated using Equation 9-24:

$$V_{nm} = \left[4.0 - 1.75\left(\frac{M_u}{V_u d_v}\right)\right] A_{nv}\sqrt{f'_m} + 0.25 P_u$$

(Equation 9-24)

$M_u/(V_u d_v)$ shall be taken as a positive number and need not be taken greater than 1.0

9.3.4.1.2.2 *Nominal shear strength provided by reinforcement* — Nominal shear strength provided by shear reinforcement, V_{ns}, shall be calculated as follows:

$$V_{ns} = 0.5\left(\frac{A_v}{s}\right) f_y d_v \qquad \text{(Equation 9-25)}$$

COMMENTARY

9.3.4.1.2 *Nominal shear strength* — The shear strength equations in Section 9.3.4.1.2 are derived from research (Shing et al, 1990a; Shing et al, 1990b). The equations have been compared with results from fifty-six tests of masonry walls failing in in-plane shear (Davis et al, 2010). The test data encompassed both concrete masonry walls and clay masonry walls, all of which were fully grouted. The average ratio of the test strength to the calculated strength was 1.17 with a coefficient of variation of 0.15.

The limitations on maximum nominal shear strength are included to preclude critical (brittle) shear-related failures.

The provisions of this Section were developed through the study of and calibrated to cantilevered shear walls. The ratio $M_u/(V_u d_v)$ can be difficult to interpret or apply consistently for other conditions such as for a uniformly loaded, simply supported beam. Concurrent values of M_u and $V_u d_v$ must be considered at appropriate locations along shear members, such as beams, to determine the critical $M_u/(V_u d_v)$ ratio. To simplify the analytical process, designers are permitted to use $M_u/(V_u d_v) = 1$.

Partially grouted walls can produce lower strengths than predicted by the shear strength equations using just the reduction of net area (Minaie et al, 2010; Nolph and ElGawady, 2011; Schultz, 1996b; Schultz, 1996c; Schultz and Hutchinson, 2001b). The grouted shear wall factor is used to compensate for this reduced strength until methods can be developed to more accurately predict the performance of these elements.

9.3.4.1.2.1 *Nominal masonry shear strength* — Equation 9-24 is empirically derived from research (Shing et al, 1990a; Shing et al, 1990b).

9.3.4.1.2.2 *Nominal shear strength provided by reinforcement* — Equation 9-25 is empirically derived from research (Shing et al, 1990a; Shing et al, 1990b). The nominal shear strength provided by shear reinforcement, Equation 9-25, represents half the theoretical contribution. In other words, the nominal shear strength is determined as the full masonry contribution plus one-half the contribution from the shear reinforcement. Other coefficients were evaluated (0.6, 0.8, and 1.0), but the best fit to the experimental data was obtained using the 0.5 factor (Davis et al, 2010).

CODE

9.3.4.2 *Beams* — Design of beams shall meet the requirements of Section 5.2 and the additional requirements of Sections 9.3.4.2.1 through 9.3.4.2.4.

9.3.4.2.1 The factored axial compressive force on a beam shall not exceed $0.05 A_n f'_m$.

9.3.4.2.2 *Longitudinal reinforcement*

9.3.4.2.2.1 The variation in longitudinal reinforcing bars in a beam shall not be greater than one bar size. Not more than two bar sizes shall be used in a beam.

9.3.4.2.2.2 The nominal flexural strength of a beam shall not be less than 1.3 multiplied by the nominal cracking moment of the beam, M_{cr}. The modulus of rupture, f_r, for this calculation shall be determined in accordance with Section 9.1.9.2.

9.3.4.2.2.3 The requirements of Section 9.3.4.2.2.2 need not be applied if at every section the area of tensile reinforcement provided is at least one-third greater than that required by analysis.

9.3.4.2.3 *Transverse reinforcement* — Transverse reinforcement shall be provided where V_u exceeds ϕV_{nm}. The factored shear, V_u, shall include the effects of lateral load. When transverse reinforcement is required, the following provisions shall apply:

(a) Transverse reinforcement shall be a single bar with a 180-degree hook at each end.

(b) Transverse reinforcement shall be hooked around the longitudinal reinforcement.

(c) The minimum area of transverse reinforcement shall be $0.0007 bd_v$.

(d) The first transverse bar shall not be located more than one-fourth of the beam depth, d_v, from the end of the beam.

(e) The maximum spacing shall not exceed one-half the depth of the beam nor 48 in. (1219 mm).

COMMENTARY

9.3.4.2 *Beams* — This section applies to the design of lintels and beams.

9.3.4.2.2 *Longitudinal reinforcement*

9.3.4.2.2.1 Restricting the variation of bar sizes in a beam is included to increase the depth of the member compression zone and to increase member ductility. When incorporating two bars of significantly different sizes in a single beam, the larger bar requires a much higher force to reach yield strain, in effect "stiffening" the beam.

9.3.4.2.2.2 The requirement that the nominal flexural strength of a beam not be less than 1.3 multiplied by the nominal cracking moment is imposed to prevent brittle failures. This situation may occur where a beam is so lightly reinforced that the bending moment required to cause yielding of the reinforcement is less than the bending moment required to cause cracking.

9.3.4.2.2.3 This exception provides sufficient additional reinforcement in members in which the amount of reinforcement required by Section 9.3.4.2.2.2 would be excessive.

9.3.4.2.3 *Transverse reinforcement* — Beams recognized in this section of the Code are often designed to resist only shear forces due to gravity loads. Beams that are controlled by high seismic forces and lateral drift should be designed as ductile elements.

(a) Although some concerns have been raised regarding the difficulty in constructing beams containing a single bar stirrup, the Committee feels such spacing limitations within beams inhibits the construction of necessary lap lengths required for two-bar stirrups. Furthermore, the added volume of reinforcement as a result of lap splicing stirrups may prevent adequate consolidation of the grout.

(b) The requirement that shear reinforcement be hooked around the longitudinal reinforcement not only facilitates construction but also confines the longitudinal reinforcement and contributes to the development of the shear reinforcement.

(c) A minimum area of transverse reinforcement is established to prevent brittle shear failures.

(d) Although different codes contain different spacing requirements for the placement of transverse reinforcement, the Committee has conservatively established this requirement.

(e) The requirements of this section establish limitations on the spacing and placement of reinforcement in order to increase member ductility.

CODE

9.3.4.2.4 *Construction* — Beams shall be fully grouted.

9.3.4.3 *Piers*

9.3.4.3.1 The factored axial compression force on piers shall not exceed $0.3\, A_n f'_m$.

9.3.4.3.2 *Longitudinal reinforcement* — A pier subjected to in-plane stress reversals shall be reinforced symmetrically about the neutral axis of the pier. Longitudinal reinforcement of piers shall comply with the following:

(a) At least one bar shall be provided in each end cell.

(b) The minimum area of longitudinal reinforcement shall be 0.0007 bd.

9.3.4.3.3 *Dimensional limits* — Dimensions shall be in accordance with the following:

(a) The nominal thickness of a pier shall not exceed 16 in. (406 mm).

(b) The distance between lateral supports of a pier shall not exceed 25 multiplied by the nominal thickness of a pier except as provided for in Section 9.3.4.3.3(c).

(c) When the distance between lateral supports of a pier exceeds 25 multiplied by the nominal thickness of the pier, design shall be based on the provisions of Section 9.3.5.

(d) The nominal length of a pier shall not be less than three multiplied by its nominal thickness nor greater than six multiplied by its nominal thickness. The clear height of a pier shall not exceed five multiplied by its nominal length.

Exception: When the factored axial force at the location of maximum moment is less than $0.05 f'_m A_g$, the length of a pier shall be permitted to be equal to the thickness of the pier.

COMMENTARY

9.3.4.2.4 *Construction* — Although beams can physically be constructed of partially grouted masonry, the lack of research supporting the performance of partially grouted beams combined with the increased probability of brittle failure dictates this requirement.

9.3.4.3 *Piers*

9.3.4.3.1 Due to the less severe requirements imposed for the design of piers with respect to similar requirements for columns, the maximum axial force is arbitrarily limited to a relatively lower value.

9.3.4.3.2 *Longitudinal reinforcement* — These provisions are predominantly earthquake-related and are intended to provide ductility. Piers not subject to in-plane stress reversals are not required to comply with this section.

9.3.4.3.3 *Dimensional limits* — Judgment-based dimensional limits are established for piers to distinguish their design from walls and to prevent local instability or buckling modes.

CODE

9.3.5 *Wall design for out-of-plane loads*

9.3.5.1 *Scope* — The requirements of Section 9.3.5 shall apply to the design of walls for out-of-plane loads.

9.3.5.2 *Nominal axial and flexural strength* — The nominal axial strength, P_n, and the nominal flexural strength, M_n, of a cross-section shall be determined in accordance with the design assumptions of Section 9.3.2. The nominal axial compressive strength shall not exceed that determined by Equation 9-19 or Equation 9-20, as appropriate.

9.3.5.3 *Nominal shear strength* — The nominal shear strength shall be determined by Section 9.3.4.1.2.

9.3.5.4 *P-delta effects*

9.3.5.4.1 Members shall be designed for the factored axial load, P_u, and the moment magnified for the effects of member curvature, M_u. The magnified moment shall be determined either by Section 9.3.5.4.2 or Section 9.3.5.4.3.

9.3.5.4.2 Moment and deflection calculations in this section are based on simple support conditions top and bottom. For other support and fixity conditions, moments and deflections shall be calculated using established principles of mechanics.

The procedures set forth in this Section shall be used when the factored axial load stress at the location of maximum moment satisfies the requirement calculated by Equation 9-26.

COMMENTARY

9.3.5 *Wall design for out-of-plane loads*

9.3.5.1 *Scope*

9.3.5.2 *Nominal axial and flexural strength* — When the depth of the equivalent stress block is in the face shell of a wall that is fully or partially grouted, the nominal moment may be found from:

$$M_n = \left(P_u/\phi + A_s f_y\right)\left(\frac{t_{sp} - a}{2}\right) + A_s f_y \left(d - \frac{t_{sp}}{2}\right)$$

$$a = \frac{A_s f_y + P_u/\phi}{0.80 f'_m b}$$

The above equations are valid for both centered and noncentered flexural reinforcement. For centered flexural reinforcement, $d = t_{sp}/2$ and the nominal moment, M_n, is obtained as:

$$M_n = \left(P_u/\phi + A_s f_y\right)\left(d - \frac{a}{2}\right)$$

These equations take into account the effect of compressive vertical loads increasing the flexural strength of the section. In the case of axial tension, the flexural strength is decreased.

9.3.5.4.2 The provisions of this section are derived from results of tests on simply supported specimens. Because the maximum bending moment and deflection occur near the mid-height of those specimens, this section includes only design equations for that condition. When actual conditions are not simple supports, the curvature of a wall under out-of-plane lateral loading will be different than that assumed by these equations. Using the principles of mechanics, the points of inflection can be determined and actual moments and deflections can be calculated under different support conditions. The designer should examine all moment and deflection conditions to locate the critical section using the assumptions outlined in Section 9.3.5.

The criterion to limit vertical load on a cross section was included because the slender wall design method was based on data from testing with typical roof loads. For h/t ratios greater than 30, there is an additional limitation on the axial stress.

BUILDING CODE REQUIREMENTS FOR MASONRY STRUCTURES AND COMMENTARY

CODE

$$\left(\frac{P_u}{A_g}\right) \leq 0.20 f'_m \quad \text{(Equation 9-26)}$$

When the ratio of effective height to nominal thickness, h/t, exceeds 30, the factored axial stress shall not exceed $0.05f'_m$.

Factored moment and axial force shall be determined at the midheight of the wall and shall be used for design. The factored moment, M_u, at the midheight of the wall shall be calculated using Equation 9-27.

$$M_u = \frac{w_u h^2}{8} + P_{uf}\frac{e_u}{2} + P_u \delta_u \quad \text{(Equation 9-27)}$$

Where:

$$P_u = P_{uw} + P_{uf} \quad \text{(Equation 9-28)}$$

The deflection due to factored loads (δ_u) shall be obtained using Equations 9-29 and 9-30.

(a) Where $M_u < M_{cr}$

$$\delta_u = \frac{5M_u h^2}{48 E_m I_n} \quad \text{(Equation 9-29)}$$

(b) Where $M_{cr} \leq M_u \leq M_n$

$$\delta_u = \frac{5M_{cr} h^2}{48 E_m I_n} + \frac{5(M_u - M_{cr})h^2}{48 E_m I_{cr}}$$

$$\text{(Equation 9-30)}$$

9.3.5.4.3 The factored moment, M_u, shall be determined either by a second-order analysis, or by a first-order analysis and Equations 9-31 through 9-33.

$$M_u = \psi M_{u,0} \quad \text{(Equation 9-31)}$$

Where $M_{u,0}$ is the factored moment from first-order analysis.

$$\psi = \frac{1}{1 - \frac{P_u}{P_e}} \quad \text{(Equation 9-32)}$$

Where:

$$P_e = \frac{\pi^2 E_m I_{eff}}{h^2} \quad \text{(Equation 9-33)}$$

For $M_u < M_{cr}$, I_{eff} shall be taken as $0.75 I_n$. For $M_u \geq M_{cr}$, I_{eff} shall be taken as I_{cr}. P_u/P_e cannot exceed 1.0.

COMMENTARY

The required moment due to lateral loads, eccentricity of axial load, and lateral deformations is assumed maximum at mid-height of the wall. In certain design conditions, such as large eccentricities acting simultaneously with small lateral loads, the design maximum moment may occur elsewhere. When this occurs, the designer should use the maximum moment at the critical section rather than the moment determined from Equation 9-27.

9.3.5.4.3 The moment magnifier provisions in this section were developed to provide an alternative to the traditional P-delta methods of Section 9.3.5.4.2. These provisions also allow other second-order analyses to be used.

The proposed moment magnification equation is very similar to that used for slender wall design for reinforced concrete. Concrete design provisions use a factor of 0.75 in the denominator of the moment magnifier to account for uncertainties in the wall stiffness. This factor is retained for uncracked walls. It is not used for cracked walls. Instead, the cracked moment of inertia is conservatively used for the entire wall height. Trial designs indicated that using this approach matches design using Section 9.3.5.4.2. If a 0.75 factor were included along with using the cracked moment of inertia for the entire height would result in design moments approximately 7% greater than using Section 9.3.5.4.2. The committee did not see any reason for the additional conservatism.

CODE

9.3.5.4.4 The cracking moment of the wall shall be calculated using the modulus of rupture, f_r, taken from Table 9.1.9.2.

9.3.5.4.5 The neutral axis for determining the cracked moment of inertia, I_{cr}, shall be determined in accordance with the design assumptions of Section 9.3.2. The effects of axial load shall be permitted to be included when calculating I_{cr}.

Unless stiffness values are obtained by a more comprehensive analysis, the cracked moment of inertia for a wall that is partially or fully grouted and whose neutral axis is in the face shell shall be obtained from Equation 9-34 and Equation 9-35.

$$I_{cr} = n\left(A_s + \frac{P_u}{f_y}\frac{t_{sp}}{2d}\right)(d-c)^2 + \frac{bc^3}{3} \quad \text{(Equation 9-34)}$$

$$c = \frac{A_s f_y + P_u}{0.64 f'_m b} \quad \text{(Equation 9-35)}$$

9.3.5.5 *Deflections* — The horizontal midheight deflection, δ_s, under allowable stress design load combinations shall be limited by the relation:

$$\delta_s \leq 0.007\, h \quad \text{(Equation 9-36)}$$

P-delta effects shall be included in deflection calculation using either Section 9.3.5.5.1 or Section 9.3.5.5.2.

9.3.5.5.1 For simple support conditions top and bottom, the midheight deflection, δ_s, shall be calculated using either Equation 9-29 or Equation 9-30, as applicable, and replacing M_u with M_{ser} and δ_u with δ_s.

COMMENTARY

9.3.5.4.4 The cracking moment, M_{cr}, is the calculated moment corresponding to first cracking. The Code permits the applied axial force to be included in the calculation of the cracking moment.

9.3.5.4.5 The Code requires that the neutral axis used to calculate the cracked moment of inertia be determined using the strain distribution at nominal strength. Amrhein and Lee (1984) used this condition to develop the original slender wall design provisions.

Equations 9-34 and 9-35 are valid for both centered and non-centered vertical reinforcement. The modification term of ($t_{sp}/2d$) in Equation 9-34 accounts for a reduction in the contribution of the axial load to the cracked moment of inertia when the reinforcement is near the face of the wall.

9.3.5.5 *Deflections* — Historically, the recommendation has been to limit the deflection under allowable stress load combinations to $0.01h$. The committee has chosen a more stringent value of $0.007h$.

The Code limits the lateral deflection under allowable stress load combinations. A wall loaded in this range returns to its original vertical position when the lateral load is removed, because the stress in the reinforcement is within its elastic limit.

9.3.5.5.1 Equation 9-30 is for mid-height deflection of a simply supported wall for an uncracked section, and Equation 9-30 is for mid-height deflection for a cracked section. A wall is assumed to deflect as an uncracked section until the modulus of rupture is reached, after which it is assumed to deflect as a cracked section. The cracked moment of inertia is conservatively assumed to apply over the entire height of the wall. The cracked moment of inertia, I_{cr}, for a fully grouted or partially grouted cross section is usually the same as that for a hollow section because the compression stress block is generally within the thickness of the face shell.

These equations represent good approximations to test results, assuming that the wall is simply supported top and bottom, and is subjected to a uniformly distributed lateral load. If the wall is fixed at top, bottom, or both, other formulas should be developed considering the support conditions at the top or bottom and considering the possible deflection or rotation of the foundation, roof, or floor diaphragm.

CODE

9.3.5.5.2 The deflection, δ_s, shall be determined by a second-order analysis that includes the effects of cracking, or by a first-order analysis with the calculated deflections magnified by a factor of $1/(1-P/P_e)$, where P_e is determined from Equation (9-33).

9.3.6 Wall design for in-plane loads

9.3.6.1 *Scope* — The requirements of Section 9.3.6 shall apply to the design of walls to resist in-plane loads.

9.3.6.2 *Reinforcement* — Reinforcement shall be provided perpendicular to the shear reinforcement and shall be at least equal to one-third A_v. The reinforcement shall be uniformly distributed and shall not exceed a spacing of 8 ft (2.44 m).

9.3.6.3 *Flexural and axial strength* — The nominal flexural and axial strength shall be determined in accordance with Section 9.3.4.1.1.

9.3.6.4 *Shear strength* — The nominal shear strength shall be calculated in accordance with Section 9.3.4.1.2.

9.3.6.5 The maximum reinforcement requirements of Section 9.3.3.5 shall not apply if a shear wall is designed to satisfy the requirements of Sections 9.3.6.5.1 through 9.3.6.5.5.

COMMENTARY

The Code requires that the neutral axis used to calculate the cracked moment of inertia be determined using the strain distribution at nominal strength. Amrhein and Lee (1984) used this condition to develop the original slender wall design provisions.

Equation 9-34 and 9-35 are valid for both centered and non-centered vertical reinforcement. The modification term of $(t_{sp}/2d)$ in Equation 9-34 accounts for a reduction in the contribution of the axial load to the cracked moment of inertia when the reinforcement is near the face of the wall.

9.3.5.5.2 This section allows other second-order analyses to be used to predict wall deflections, including first-order deflections amplified using a moment magnification factor.

Less conservative estimation for first-order wall deformation can be obtained using an effective I value that accounts for partial cracking of the sections, such as that described in Section 5.2.1.4.2.

9.3.6 Wall design for in-plane loads

9.3.6.5 The maximum reinforcement requirements of Section 9.3.3.5 are intended to ensure that an intermediate or a special reinforced masonry shear wall has sufficient inelastic deformation capacity under the design-basis earthquake of ASCE 7 or the model building codes. Inelastic deformability is the ability of a structure or structural element to continue to sustain gravity loads as it deforms laterally under earthquake (or some other type of) excitation beyond the stage where the response of the structure or the structural element to that excitation is elastic (that is, associated with no residual displacement or damage). In the alternative shear wall design approach given in Sections 9.3.6.5.1 through 9.3.6.5.5, such inelastic deformability is sought to be ensured by means of specially confined boundary elements, making compliance with the maximum reinforcement requirements unnecessary. These requirements are therefore waived.

CODE

9.3.6.5.1 Special boundary elements need not be provided in shear walls meeting the following conditions:

1. $P_u \leq 0.10 A_g f'_m$ for geometrically symmetrical wall sections

   $P_u \leq 0.05 A_g f'_m$ for geometrically unsymmetrical wall sections; and either

2. $\dfrac{M_u}{V_u d_v} \leq 1.0$

   or

3. $V_u \leq 3 A_{nv} \sqrt{f'_m}$ and $\dfrac{M_u}{V_u d_v} \leq 3.0$

9.3.6.5.2 The need for special boundary elements at the edges of shear walls shall be evaluated in accordance with Section 9.3.6.5.3 or 9.3.6.5.4. The requirements of Section 9.3.6.5.5 shall also be satisfied.

9.3.6.5.3 This Section applies to walls bending in single curvature in which the flexural limit state response is governed by yielding at the base of the wall. Walls not satisfying those requirements shall be designed in accordance with Section 9.3.6.5.4

(a) Special boundary elements shall be provided over portions of compression zones where:

$$c \geq \dfrac{l_w}{600 \left(C_d \delta_{ne} / h_w \right)}$$

and c is calculated for the P_u given by ASCE 7 Strength Design Load Combination 5 ($1.2D + 1.0E + L + 0.2S$) or the corresponding strength design load combination of the legally adopted building code, and the corresponding nominal moment strength, M_n, at the base critical section. The load factor on L in Combination 5 is reducible to 0.5, as per exceptions to Section 2.3.2 of ASCE 7.

COMMENTARY

9.3.6.5.1 This subsection sets up some "screens" with the expectation that many, if not most, shear walls will go through the screens, in which case no special boundary elements would be required. This situation will be the case when a shear wall is lightly axially loaded and it is either short or is moderate in height and is subject to only moderate shear stresses.

The threshold values are adapted from the design procedure for special reinforced concrete shear walls in the 1997 Uniform Building Code (UBC). In the early 1990s, when this procedure of the 1997 UBC was first being developed, an ad hoc subcommittee within the Seismology Committee of the Structural Engineers Association of California had limited, unpublished parametric studies done, showing that a reinforced concrete shear wall passing through the "screens" could not develop sufficiently high compressive strains in the concrete to warrant special confinement. In the case of masonry, strains requiring special confinement would be values exceeding the maximum usable strains of Section 9.3.2 (c).

9.3.6.5.2 Two approaches for evaluating detailing requirements at wall boundaries are included in Section 9.3.6.5.2. Section 9.3.6.5.3 allows the use of displacement-based design of walls, in which the structural details are determined directly on the basis of the expected lateral displacements of the wall under the design-basis earthquake. The provisions of Section 9.3.6.5.4 are conservative for assessing required transverse reinforcement at wall boundaries for many walls. The requirements of Section 9.3.6.5.5 apply to shear walls designed by either Section 9.3.6.5.3 or 9.3.6.5.4.

9.3.6.5.3 Section 9.3.6.5.3 is based on the assumption that inelastic response of the wall is dominated by flexural action at a critical, yielding section – typically at the base. The wall should be proportioned so that the critical section occurs where intended (at the base).

(a) The following explanation, including Figure CC-9.3-3, is adapted from a paper by Wallace and Orakcal (2002). The relationship between the wall top displacement and wall curvature for a wall of uniform cross-section with a single critical section at the base is presented in Figure CC-9.3-3. The provisions of this Code are based on a simplified version of the model presented in Figure CC-9.3-3(a). The simplified model, shown in Figure CC-9.3-3(b), neglects the contribution of elastic deformations to the top displacement, and moves the center of the plastic hinge to the base of the wall. Based on the model of Figure CC-9.3-3, the relationship between the top displacement and the curvature at the base of the wall is:

$$C_d \delta_{ne} = \theta_p h_w = (\phi_u \ell_p) h_w = \left(\phi_u \dfrac{\ell_w}{2} \right) h_w$$

(Equation 1)

CODE

COMMENTARY

assuming that $\ell_p = \ell_w/2$, as is permitted to be assumed by the 1997 UBC,

where ϕ_u = ultimate curvature, and

θ_p = plastic rotation at the base of the wall.

If at the stage where the top deflection of the wall is δ_{ne}, the extreme fiber compressive strain at the critical section at the base does not exceed ε_{mu}, no special confinement would be required anywhere in the wall. Figure CC-9.3-4 illustrates such a strain distribution at the critical section. The neutral axis depth corresponding to this strain distribution is c_{cr}, and the corresponding ultimate curvature is $\phi_u = \varepsilon_{mu}/c_{cr}$. From Equation 1,

$$C_d \delta_{ne} = \left(\frac{\varepsilon_{mu}}{c_{cr}} \frac{\ell_w}{2}\right) h_w \qquad \text{(Equation 2a)}$$

$$\text{or, } c_{cr} = \frac{\varepsilon_{mu}}{2} \frac{\ell_w}{(C_d \delta_{ne}/h_w)} \qquad \text{(Equation 2b)}$$

From the equations above (see Figure CC-9.3-4), special detailing would be required if:

$$c \geq \frac{\varepsilon_{mu}}{2} \frac{\ell_w}{(C_d \delta_{ne}/h_w)} = \frac{0.003}{2} \frac{\ell_w}{(C_d \delta_{ne}/h_w)}$$

$$= \frac{\ell_w}{667(C_d \delta_{ne}/h_w)} \approx \frac{\ell_w}{600(C_d \delta_{ne}/h_w)}$$

because if the neutral axis depth exceeded the critical value, the extreme fiber compressive strain would exceed the maximum usable strain ε_{mu}. For purposes of this derivation, and to avoid having separate sets of drift-related requirements for clay and concrete masonry, a single useful strain of 0.003 is used, representing an average of the design values of 0.0025 for concrete masonry and 0.0035 for clay masonry.

(b) Where special boundary elements are required by Section 9.3.6.5.3 (a), the special boundary element reinforcement shall extend vertically from the critical section a distance not less than the larger of l_w or $M_u/4V_u$.

(b) These special extensions are intended to be an upper-bound estimate of the plastic hinge length for special reinforced masonry shear walls.

COMMENTARY

Figure CC-9.3-3 – Wall curvature and displacement

Figure CC-9.3-4 – Strain distribution at critical section

CODE

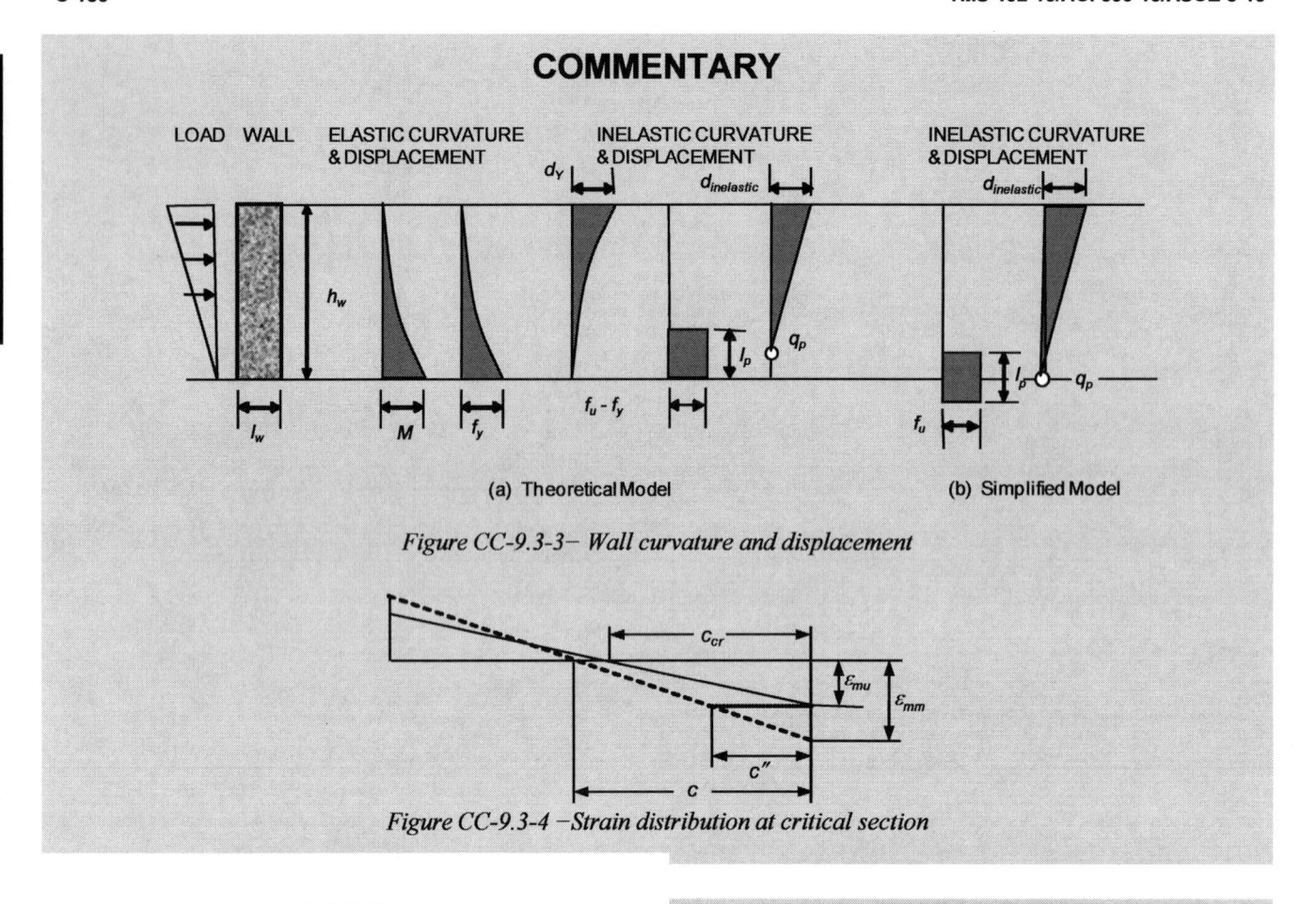

9.3.6.5.4 Shear walls not designed by Section 9.3.6.5.3 shall have special boundary elements at boundaries and edges around openings in shear walls where the maximum extreme fiber compressive stress, corresponding to factored forces including earthquake effect, exceeds $0.2 f'_m$. The special boundary element shall be permitted to be discontinued where the calculated compressive stress is less than $0.15 f'_m$. Stresses shall be calculated for the factored forces using a linearly elastic model and gross section properties. For walls with flanges, an effective flange width as defined in Section 5.1.1.2.3 shall be used.

9.3.6.5.5 Where special boundary elements are required by Section 9.3.6.5.3 or 9.3.6.5.4, requirements (a) through (d) in this section shall be satisfied and tests shall be performed to verify the strain capacity of the element:

COMMENTARY

9.3.6.5.4 A stress-based approach is included to address wall configurations to which the application of displacement-based approach is not appropriate (for example, walls with openings, walls with setbacks, walls not controlled by flexure).

The Code has adopted the stress-based triggers of ACI 318-99 for cases where the displacement-based approach is not applicable, simply changing the threshold values of $0.2 f'_c$ and $0.15 f'_c$ for reinforced concrete walls to $0.2 f'_m$ and $0.15 f'_m$, respectively, for reinforced masonry walls. Other aspects of the ACI 318-99 approach are retained. Design for flexure and axial loads does not change depending on whether the neutral axis-based trigger or the stress-based trigger is used.

9.3.6.5.5 This Code requires that testing be done to verify that the detailing provided is capable of developing a strain capacity in the boundary element that would be in excess of the maximum imposed strain. Reasonably extensive tests need to be conducted to develop prescriptive detailing requirements for specially confined boundary elements of intermediate as well as special reinforced masonry shear walls.

BUILDING CODE REQUIREMENTS FOR MASONRY STRUCTURES AND COMMENTARY   C-151

CODE

(a) The special boundary element shall extend horizontally from the extreme compression fiber a distance not less than the larger of $(c - 0.1 l_w)$ and $c/2$.

COMMENTARY

(a) Figure CC-9.3-4 shows that when the neutral axis depth c exceeds the critical neutral axis depth c_{cr}, the extreme compression fiber strain in the masonry reaches a value ε_{mm} in excess of the maximum usable strain ε_{mu}. The corresponding ultimate curvature ϕ is ε_{mu}/c. Based on the model of Figure CC-9.3-3(b),

$$C_d \delta_{ne} = \theta_p h_w = (\phi_u \ell_p) h_w = \left(\frac{\varepsilon_{mm}}{c} \frac{\ell_w}{2}\right) h_w \qquad \text{(Equation 3)}$$

From Equation 3:

$$\varepsilon_{mm} = 2\left(\frac{C_d \delta_{ne}}{h_w}\right)\left(\frac{c}{\ell_w}\right) \qquad \text{(Equation 4)}$$

The wall length over which the strains exceed the limiting value of ε_{mu}, denoted as c'', can be determined using similar triangles from Figure CC-9.3-4:

$$c'' = c\left(1 - \frac{\varepsilon_{mu}}{\varepsilon_{mm}}\right) \qquad \text{(Equation 5)}$$

An expression for the required length of confinement can be developed by combining Equations 2 and 3:

$$\frac{c''}{\ell_w} = \frac{c}{\ell_w} - \frac{(\varepsilon_{mu}/2)}{(C_d \delta_{ne}/h_w)} \qquad \text{(Equation 6)}$$

The term c/ℓ_w in Equation 4 accounts for the influence of material properties (f'_m, f_y), axial load, geometry, and quantities and distribution of reinforcement, whereas the term $(\varepsilon_{mu}/2)/(C_d \delta_{ne}/h_w)$ accounts for the influence of system response (roof displacement) and the maximum usable strain of masonry.

The wall length over which special transverse reinforcement is to be provided is based on Equation 6, with a value of $C_d \delta_{ne}/h_w = 0.015$:

$$\frac{c''}{\ell_w} = \frac{c}{\ell_w} - \frac{(0.003/2)}{0.015} = \frac{c}{\ell_w} - 0.1 \geq \frac{c}{2} \qquad \text{(Equation 7)}$$

The value of $C_d \delta_{ne}/h_w$ was selected to provide an upper-bound estimate of the mean drift ratio of typical shear wall buildings constructed in the United States of America (Wallace and Moehle, 1992). Thus, the length of the wall that must be confined is conservative for many buildings. The value of $c/2$ represents a minimum length of confinement, is adopted from ACI 318-99, and is arbitrary.

CODE

(b) In flanged sections, the special boundary element shall include the effective flange width in compression and shall extend at least 12 in. (305 mm) into the web.

(c) Special boundary element transverse reinforcement at the wall base shall extend into the support a minimum of the development length of the largest longitudinal reinforcement in the boundary element unless the special boundary element terminates on a footing or mat, where special boundary element transverse reinforcement shall extend at least 12 in. (305 mm) into the footing or mat.

(d) Horizontal shear reinforcement in the wall web shall be anchored to develop the specified yield strength, f_y, within the confined core of the boundary element.

COMMENTARY

(b) This requirement originated in the 1997 UBC. Where flanges are highly stressed in compression, the web-to-flange interface is likely to be highly stressed and may sustain local crushing failure unless special boundary element reinforcement extends into the web.

(c) The same extension is required for special boundary element transverse reinforcement in special reinforced concrete shear walls and for special transverse reinforcement in reinforced concrete columns supporting reactions from discontinued stiff members in buildings assigned to high seismic design categories.

(d) Because horizontal reinforcement is likely to act as web reinforcement in walls requiring boundary elements, it needs to be fully anchored in boundary elements that act as flanges. Achievement of this anchorage is difficult when large transverse cracks occur in the boundary elements. Standard 90-degree hooks or mechanical anchorage schemes, instead of straight bar development are recommended.

CHAPTER 10
PRESTRESSED MASONRY

CODE

10.1 — General

10.1.1 *Scope*

This chapter provides requirements for design of masonry walls that are prestressed with bonded or unbonded prestressing tendons.

10.1.2 Walls shall be designed for strength requirements and checked for service load requirements.

10.1.3 The wall provisions of Part 1, Part 2, and Section 8.1 shall apply to prestressed masonry walls.

10.1.4 The provisions of Section 10.4.3 shall apply for the calculation of nominal moment strength.

10.1.5 Masonry shall be laid in running bond unless a bond beam or other technique is used to distribute anchorage forces.

COMMENTARY

10.1 — General

10.1.1 *Scope*

Prestressing forces are used in masonry walls to reduce or eliminate tensile stresses due to externally applied loads by using controlled precompression. The precompression is generated by prestressing tendons, either bars, wires, or strands, that are contained in openings in the masonry, which may be grouted. The prestressing tendons can be pre-tensioned (stressed against external abutments prior to placing the masonry), or post-tensioned (stressed against the masonry after it has been placed). Because most research and applications to date have focused on walls, the chapter applies only to walls, not columns, beams, nor lintels. (Provisions for columns, beams, and lintels will be developed in future editions of the Code.)

Most construction applications to date have involved post-tensioned, ungrouted masonry for its ease of construction and overall economy. Consequently, these code provisions primarily focus on post-tensioned masonry. Although not very common, pre-tensioning has been used to construct prefabricated masonry panels. A more detailed review of prestressed masonry systems and applications is given elsewhere (Schultz and Scolforo, 1991).

Throughout this Code and Specification, references to "reinforcement" apply to non-prestressed reinforcement. These references do not apply to prestressing tendons, except as explicitly noted in Chapter 10. Requirements for prestressing tendons use the terms "prestressing tendon" or "tendon." The provisions of Chapter 10 do not require a mandatory quantity of reinforcement or bonded prestressing tendons for prestressed masonry walls.

Anchorage forces are distributed within a wall similar to the way in which concentrated loads are distributed (as described in Section 5.1.3; see Figure CC-5.1-5). However, research (Woodham and Hamilton, 2003) has indicated that prestress losses can distribute to adjacent tendons as far laterally from the anchorage as the height of the wall.

CODE

10.1.6 For prestressed masonry members, the prestressing force shall be added to load combinations, except as modified by Section 10.4.2.

10.2 — Design methods

10.2.1 *General*

Prestressed masonry members shall be designed by elastic analysis using loading and load combinations in accordance with the provisions of Sections 4.1.2, except as noted in Section 10.4.3.

10.2.2 *After transfer*

Immediately after the transfer of prestressing force to the masonry, limitations on masonry stresses given in this chapter shall be based upon f'_{mi}.

10.3 — Permissible stresses in prestressing tendons

10.3.1 *Jacking force*

The stress in prestressing tendons due to the jacking force shall not exceed $0.94 f_{py}$, nor $0.80 f_{pu}$, nor the maximum value recommended by the manufacturer of the prestressing tendons or anchorages.

10.3.2 *Immediately after transfer*

The stress in the prestressing tendons immediately after transfer of the prestressing force to the masonry shall not exceed $0.82 f_{py}$ nor $0.74 f_{pu}$.

10.3.3 *Post-tensioned masonry members*

At the time of application of prestress, the stress in prestressing tendons at anchorages and couplers shall not exceed $0.78 f_{py}$ nor $0.70 f_{pu}$.

COMMENTARY

10.2 — Design methods

Originally, prestressed masonry was designed using allowable stress design with a moment strength check for walls with laterally restrained tendons. The British code for prestressed masonry (BSI, 1985; Phipps, 1992) and extensive research on the behavior of prestressed masonry were considered. Summaries of prestressed masonry research and proposed design criteria are available in the literature (Schultz and Scolforo, 1992(a and b); VSL, 1990; Curtin et al, 1988; Phipps and Montague, 1976). Design methods are now based upon strength provisions with serviceability checks.

A masonry wall is typically prestressed prior to 28 days after construction, sometimes within 24 hours after construction. The specified compressive strength of the masonry at the time of prestressing (f'_{mi}) is used to determine allowable prestressing levels. This strength will likely be a fraction of the 28-day specified compressive strength. Assessment of masonry compressive strength immediately before the transfer of prestress should be by testing of masonry prisms or by a record of strength gain over time of masonry prisms constructed of similar masonry units, mortar, and grout, when subjected to similar curing conditions.

10.3 — Permissible stresses in prestressing tendons

Allowable prestressing-tendon stresses are based on criteria established for prestressed concrete (ACI 318, 2011). Allowable prestressing-tendon stresses are for jacking forces and for the state of stress in the prestressing tendon immediately after the prestressing has been applied, or transferred, to the masonry. When calculating the prestressing-tendon stress immediately after transfer of prestress, consider all sources of short term prestress losses. These sources include such items as anchorage seating loss, elastic shortening of masonry, and friction losses.

CODE

10.3.4 *Effective prestress*

The calculated effective stress in the prestressing tendons under service loads, f_{se}, shall include the effects of the following:

(a) anchorage seating losses,

(b) elastic shortening of masonry,

(c) creep of masonry,

(d) shrinkage of concrete masonry,

(e) relaxation of prestressing tendon stress,

(f) friction losses,

(g) irreversible moisture expansion of clay masonry, and

(h) thermal effects.

COMMENTARY

10.3.4 *Effective prestress*

The state of stress in a prestressed masonry wall must be checked for each stage of loading. For each loading condition, the effective level of prestress should be used in the calculation of stresses and wall strength. Effective prestress is not a fixed quantity over time. Research on the loss and gain of prestress in prestressed masonry is extensive and includes testing of time-dependent phenomena such as creep, shrinkage, moisture expansion, and prestressing-tendon stress relaxation (PCI, 1975; Lenczner, 1985; Lenczner, 1987; Shrive, 1988).

Instantaneous deformation of masonry due to the application of prestress may be calculated by the modulus of elasticity of masonry given in Section 4.2.2. Creep, shrinkage, and moisture expansion of masonry may be calculated by the coefficients given in Section 4.2. Change in effective prestress due to elastic deformation, creep, shrinkage, and moisture expansion should be based on relative modulus of elasticity of masonry and prestressing steel.

The stressing operation and relative placement of prestressing tendons should be considered in calculating losses. Elastic shortening during post-tensioning can reduce the stress in adjacent tendons that have already been stressed. Consequently, elastic shortening of the wall should be calculated considering the incremental application of post-tensioning. That elastic shortening should then be used to estimate the total loss of prestress. Alternatively, post-tensioning tendons can be prestressed to compensate for the elastic shortening caused by the incremental stressing operation.

Prestressing steel that is stressed to a large fraction of its yield strength and held at a constant strain will relax, requiring less stress to maintain a constant strain. The phenomenon of stress relaxation is associated with plastic deformation and its magnitude increases with steel stress as a fraction of steel strength. ASTM A416 (2006), A421 (2005), and A722 (2007) prestressing steels are stabilized for low relaxation losses during production. Other steel types that do not have this stabilization treatment may exhibit considerably higher relaxation losses. Their relaxation losses must be carefully assessed by testing. The loss of effective prestress due to stress relaxation of the prestressing tendon is dependent upon the level of prestress, which changes with time-dependent phenomenon such as creep, shrinkage, and moisture expansion of the masonry. An appropriate formula for predicting prestress loss due to relaxation has been developed (Lenczner, 1985; Lenczner, 1987; Shrive, 1988). Alternately, direct addition of the steel stress-relaxation value provided by the manufacturer can be used to calculate prestress losses and gains.

Friction losses are minimal or nonexistent for most post-tensioned masonry applications, because prestressing tendons are usually straight and contained in cavities. For anchorage losses, manufacturers' information should be used to calculate prestress losses. Changes in prestress due

CODE

COMMENTARY

to thermal fluctuations may be neglected if masonry is prestressed with high-strength prestressing steels. Loss of prestressing should be calculated for each design to determine effective prestress. Calculations should be based on the particular construction materials and methods as well as the climate and environmental conditions. Committee experience, research, and field experience with post-tensioned wall designs from Switzerland, Great Britain, Australia, and New Zealand has indicated that prestress losses are expected to be in the following ranges (Woodham and Hamilton, 2003; Hamilton and Badger, 2000; Biggs and Ganz, 1998; NCMA TEK 14-20A, 2002):

(a) Initial loss after jacking – 5% to 10%

(b) Total losses after long-term service for concrete masonry – 30% to 35%

(c) Total losses after long-term service for clay masonry – 20% to 25%

The values in (b) and (c) include both the short-term and long-term losses expected for post-tensioning. The Committee believes these ranges provide reasonable estimates for typical wall applications, unless calculations, experience, or construction techniques indicate different losses are expected.

10.4 — Axial compression and flexure

10.4.1 *General*

10.4.1.1 Walls subjected to axial compression, flexure, or to combined axial compression and flexure shall be designed according to the provisions of Section 8.2.4, except as noted in Section 10.4.1.2, 10.4.1.3, 10.4.2, and 10.4.3.

10.4.1.2 The allowable compressive stresses due to axial loads, F_a, and flexure, F_b, and the allowable axial force in Equation 8-15 shall be permitted to be increased by 20 percent for the stress condition immediately after transfer of prestress.

10.4.1.3 Masonry shall not be subjected to flexural tensile stress from the combination of prestressing force and dead load.

10.4 — Axial compression and flexure

10.4.1 *General*

The requirements for prestressed masonry walls subjected to axial compression and flexure are separated into those with laterally unrestrained prestressing tendons and those with laterally restrained prestressing tendons. This separation was necessary because the flexural behavior of a prestressed masonry wall significantly depends upon the lateral restraint of the prestressing tendon. Lateral restraint of a prestressing tendon is typically provided by grouting the cell or void containing the tendon before or after transfer of prestressing force to the masonry. Lateral restraint may be provided by placing the masonry in contact with the tendon or the protective sheathing of the tendon at periodic intervals along the length of the prestressing tendon.

Allowable compressive stresses for prestressed masonry address two distinct loading stages; stresses immediately after transfer of prestressing force to the masonry wall and stresses after all prestress losses and gains have taken place. The magnitude of allowable axial compressive stress and bending compressive stress after all prestress losses and gains are consistent with those for unreinforced masonry in Section 8.2. Immediately after transfer of prestressing, allowable compressive stresses and applied axial load should be based upon f'_{mi} and may be increased by 20 percent. This means that the factors of safety at the time of the transfer of prestress may be lower than those after prestress losses and gains occur. The first reason for this is that the effective precompression stress at

CODE

COMMENTARY

the time of transfer of prestressing almost certainly decreases over time and masonry compressive strength most likely increases over time. Second, loads at the time of transfer of prestressing, namely prestress force and dead loads, are known more precisely than loads throughout the remainder of service life.

Cracking of prestressed masonry under permanent loads is to be avoided. The prestressing force and the dead weight of the wall are permanent loads. Cracking under permanent loading conditions is not desirable due to the potential for significant water penetration, which may precipitate corrosion of the prestressing tendons and accessories and damage to interior finishes. Masonry provides a significant flexural tensile resistance to cracking, as reflected by the allowable flexural tensile stress values stated in Section 8.2. Consequently, elimination of tensile stress under prestressing force and dead loads alone is a conservative measure, but one the committee deemed reasonable and reflective of current practice for prestressed masonry members.

10.4.2 *Service load requirements*

10.4.2.1 For walls with laterally unrestrained prestressing tendons, the prestressing force, P_{ps}, shall be included in the calculation of the axial load, P, in Equation 8-15 and in the calculation of the eccentricity of the axial load, e, in Equation 8-19.

10.4.2 *Service load requirements*

10.4.2.1 Because masonry walls with laterally unrestrained prestressing tendons are equivalent to masonry walls subjected to applied axial loads, the design approach for unreinforced masonry in Section 8.2 has been adopted for convenience and consistency. Buckling of masonry walls under prestressing force must be avoided for walls with laterally unrestrained prestressing tendons. The prestressing force, P_{ps}, is to be added to the design axial load, P, for stress and load calculations and in the calculation of the eccentricity of the axial resultant, e.

10.4.2.2 For walls with laterally restrained prestressing tendons, the prestressing force, P_{ps}, shall not be considered for the calculation of the axial load, P, in Equation 8-15. The prestressing force, P_{ps}, shall be considered for the calculation of the eccentricity of the axial resultant load, e, in Equation 8-19.

10.4.2.2 Lateral restraint of a prestressing tendon is typically provided by grouting the cell or void containing the tendon before or after transfer of prestressing force to the masonry. Lateral restraint may also be provided by placing the masonry in contact with the tendon or the tendon's protective sheath at periodic intervals along the length of the prestressing tendon (Stierwalt and Hamilton, 2000). In general, three intermediate contacts within a laterally unsupported wall length or height can be considered to provide full lateral support of the tendon but the analysis and decision are the responsibility of the designer.

Prestressed masonry walls with laterally restrained prestressing tendons require a modified design approach from the criteria in Section 8.2. If the prestressing tendon is laterally restrained, the wall cannot buckle under its own prestressing force. Any tendency to buckle under prestressing force induces a lateral deformation that is resisted by an equal and opposite restraining force provided by the prestressing tendon. Such walls are susceptible to buckling under axial loads other than prestressing, however, and this loading condition must be checked (Scolforo and Borchelt, 1992). For this condition, with both concentrically and eccentrically prestressed masonry walls, the prestressing force must be considered in the calculation of the eccentricity of this axial resultant, e, in Equation 8-19 of

CODE

10.4.3 Strength requirements

10.4.3.1 Required strength shall be determined in accordance with the factored load combinations of the legally adopted building code. When the legally adopted building code does not provide factored load combinations, structures and members shall be designed to resist the combination of loads specified in ASCE 7 for strength design. Walls subject to compressive axial load shall be designed for the factored design moment and the accompanying factored axial load. The factored moment, M_u, shall include the moment induced by relative lateral displacement.

10.4.3.2 Values of the response modification coefficient (R) and the deflection amplification factor (C_d), indicated in ASCE 7 Table 12.2-1 for ordinary plain (unreinforced) masonry shear walls shall be used in determining base shear and design story drift.

10.4.3.3 The design moment strength shall be taken as the nominal moment strength, M_n, multiplied by a strength-reduction factor (ϕ) of 0.8.

10.4.3.4 For cross sections with uniform width, b, over the depth of the compression zone, the depth of the equivalent compression stress block, a, shall be determined by the following equation:

$$a = \frac{f_{ps} A_{ps} + f_y A_s + P_u/\phi}{0.80 \, f'_m \, b} \quad \text{(Equation 10-1)}$$

For other cross sections, Equation 10-1 shall be modified to consider the variable width of compression zone.

10.4.3.5 For walls with (a) uniform width, b, (b) concentric reinforcement and prestressing tendons, and (c) concentric axial load, the nominal moment strength, M_n, shall be calculated by the following equation:

$$M_n = \left(f_{ps} A_{ps} + f_y A_s + P_u/\phi\right)\left(d - \frac{a}{2}\right) \quad \text{(Equation 10-2)}$$

10.4.3.5.1 The quantity a shall be calculated according to Section 10.4.3.4 and f_{ps} shall be calculated according to Section 10.4.3.7.

10.4.3.5.2 The nominal moment strength for other conditions shall be based on static moment equilibrium principles.

10.4.3.5.3 The distance d shall be calculated as the actual distance from the centerline of the tendon to the compression face of the member. For walls with laterally unrestrained prestressing tendons and loaded out of plane, d shall not exceed the face-shell thickness plus one-half the tendon diameter plus 0.375 in. (9.5 mm).

COMMENTARY

the Code. The flexural stress induced by eccentric prestressing causes an increase or decrease in the axial buckling load, depending upon the location and magnitude of the applied axial load relative to the prestressing force.

10.4.3 Strength requirements

Calculation of the moment strength of prestressed masonry walls is similar to the method for prestressed concrete (ACI 318, 2011). For bonded tendons, the simplification of taking the tendon stress at nominal moment strength equal to the yield strength can be more conservative for bars than for strands because the yield strength of a prestressing bar is a smaller percentage of the ultimate strength of the tendon.

The equation for the unbonded prestressing tendon stress, f_{ps}, at the moment strength condition (Equation 10-3) is based on tests of prestressed masonry walls, which were loaded out-of-plane. Equation 10-3 is used for calculating unbonded tendon stress at nominal moment strength for members loaded out-of-plane containing either laterally restrained or laterally unrestrained tendons. This equation provides improved estimates of the tendon stresses over previous equations in the Code (Schultz et al, 2003; Bean Popehn and Schultz, 2003; Bean Popehn and Schultz, 2010; Bean Popehn, 2007). Equation 10-3 can be solved iteratively for f_{ps}. For the first iteration, f_{ps} in the parenthetical term can be taken equal to f_{se}.

The equation for the nominal moment strength, M_n, is for the general case of a masonry wall with concentrically applied axial load and concentric tendons and reinforcement. This is representative of most prestressed masonry applications to date. For other conditions, the designer should refer to first principles of structural mechanics to determine the nominal moment strength of the wall.

The depth of the equivalent compression stress block must be determined with consideration of the cross section of the wall, the tensile resistance of tendons and reinforcement, and the factored design axial load, P_u. P_u is an additive quantity in Code Equations 10-1 and 10-2. Prestressing adds to the resistance for ultimate strength evaluations and is used with a load factor of 1.0. Equation 10-1 defining the depth of the equivalent compression stress block, a, is modified to match the value for the equivalent uniform stress parameter specified in Chapter 9 (Strength Design of Masonry) of the Code ($0.80 f'_m$). A review of existing tests of post-tensioned masonry walls indicates that the flexural strength of the walls is more accurately calculated using uniform stresses smaller than the value specified in previous editions of the Code ($0.85 f'_m$) (Schultz et al, 2003; Bean Popehn and Schultz, 2003).

The ratio, a/d, is limited to assure ductile performance in flexure when using tendons fabricated from steel with yield strengths between 60 ksi (420 MPa) and 270 ksi (1865 MPa). As with reinforced masonry designed in accordance with Chapters 8 and 9, the calculated depth in

CODE

10.4.3.5.4 When tendons are not placed in the center of the wall, d shall be calculated in each direction for out-of-plane bending.

10.4.3.6 The ratio a/d shall not exceed 0.38.

10.4.3.7 *Calculation of f_{ps} for out-of-plane bending*

10.4.3.7.1 For walls with bonded prestressing tendons, f_{ps} shall be calculated based on strain compatibility or shall be taken equal to f_{py}.

10.4.3.7.2 For walls with laterally restrained or laterally unrestrained unbonded prestressing tendons, the following equation shall be permitted to be used instead of a more accurate determination of f_{ps}:

$$f_{ps} = f_{se} + 0.03\left(\frac{E_{ps}d}{l_p}\right)\left(1 - 1.56\frac{A_{ps}f_{ps} + P}{f'_m bd}\right)$$

(Equation 10-3)

10.4.3.7.3 In Equation 10-3, the value of f_{ps} shall be not less than f_{se}, and not larger than f_{py}.

10.4.3.8 *Calculation of f_{ps} for shear walls* — For walls with bonded prestressing tendons, f_{ps} shall be calculated based on strain compatibility or shall be taken equal to f_{py}. Instead of a more accurate determination, f_{ps} for members with unbonded prestressing tendons shall be f_{se}.

10.5 — Axial tension

Axial tension shall be resisted by reinforcement, prestressing tendons, or both.

10.6 — Shear

10.6.1 For walls without bonded mild reinforcement, nominal shear strength, V_n, shall be calculated in accordance with Sections 9.2.6a, 9.2.6b, 9.2.6c, and 9.2.6e. N_u shall include the effective prestress force, $A_{ps}f_{se}$.

10.6.2 For walls with bonded mild reinforcement, nominal shear strength, V_n, shall be calculated in accordance with Section 9.3.4.1.2.

10.6.2.1 Nominal masonry shear strength, V_{nm}, shall be calculated in accordance with Section 9.3.4.1.2.1. P_u shall include the effective prestress force, $A_{ps}f_{se}$.

10.6.2.2 Nominal shear strength provided by reinforcement, V_{ns}, shall be calculated in accordance with Section 9.3.4.1.2.

COMMENTARY

compression should be compared to the depth available to resist compressive stresses. For sections with uniform width, the value of the compression block depth, a, should be compared to the solid bearing depth available to resist compressive stresses. For hollow sections that are ungrouted or partially-grouted, the available depth may be limited to the face shell thickness of the masonry units, particularly if the webs are not mortared. The a/d limitation is intended to ensure significant yielding of the prestressing tendons prior to masonry compression failure. In such a situation, the nominal moment strength is determined by the strength of the prestressing tendon, which is the basis for a strength-reduction factor equal to 0.8. This ductility limit was determined for sections with bonded tendons, and when more experimental and field data are available on the ductility of both unbonded and bonded systems, this limit will again be reviewed.

The calculation of this limit assumes that the effective prestressing stress is equivalent to 0.65 f_y. If the magnitude of the initial effective prestress (i.e., f_{se}) is less than 0.65 f_y, then the strain in the steel at ultimate strength ε_s should be compared to the yield strain (i.e., $\varepsilon_y = f_y / E_s$). The steel strain at ultimate strength ε_s can be approximated by assuming the strain in the steel is equal to an initial strain due to the effective prestressing ($\varepsilon_{s,i} = f_{se}/E_s$) plus additional strain due to flexure ($\varepsilon_{s,flex} = 0.003 \times ((d - 1.25a)/1.25a)$.

10.5 — Axial tension

The axial tensile strength of masonry in a prestressed masonry wall is to be neglected, which is a conservative measure. This requirement is consistent with that of Section 8.3. If axial tension develops, for example due to wind uplift on the roof structure, the axial tension must be resisted by reinforcement, tendons, or both.

10.6 — Shear

This section applies to both in-plane and out-of-plane shear.

The shear strength of prestressed walls is calculated using the provisions of the Chapter 9. Calculation of shear strength is dictated by the presence or absence of bonded mild reinforcement. While the MSJC acknowledges that prestressed masonry walls are reinforced, for walls without bonded mild reinforcement, the unreinforced (plain) masonry shear provisions of Chapter 9 are used to calculate shear strength. When bonded mild reinforcement is provided, then the reinforced masonry shear provisions of Chapter 9 are used to calculate shear strength.

No shear strength enhancement due to arching action of the masonry is recognized in this Code for prestressed masonry walls. The formation of compression struts and tension ties in prestressed masonry is possible, but this phenomenon has not been considered.

CODE

10.7 — Deflection

Calculation of member deflection shall include camber, the effects of time-dependent phenomena, and P-delta effects.

10.8 — Prestressing tendon anchorages, couplers, and end blocks

10.8.1 Prestressing tendons in masonry construction shall be anchored by either:

(a) mechanical anchorage devices bearing directly on masonry or placed inside an end block of concrete or fully grouted masonry, or

(b) bond in reinforced concrete end blocks or members.

10.8.2 Anchorages and couplers for prestressing tendons shall develop at least 95 percent of the specified tensile strength of the prestressing tendons when tested in an unbonded condition, without exceeding anticipated set.

10.8.3 Reinforcement shall be provided in masonry members near anchorages if tensile stresses created by bursting, splitting, and spalling forces induced by the prestressing tendon exceed the capacity of the masonry.

10.8.4 *Bearing stresses*

10.8.4.1 In prestressing tendon anchorage zones, local bearing stress on the masonry shall be calculated based on the contact surface between masonry and the mechanical anchorage device or between masonry and the end block.

10.8.4.2 Bearing stresses on masonry due to maximum jacking force of the prestressing tendon shall not exceed $0.50 f'_{mi}$.

COMMENTARY

10.7 — Deflection

In accordance with Section 4.3.2, prestressed masonry wall deflection should be calculated based on uncracked section properties. Calculation of wall deflection must include the effect of time-dependent phenomenon such as creep and shrinkage of masonry and relaxation of prestressing tendons. There are no limits for the out-of-plane deflection of prestressed masonry walls. This is because appropriate out-of-plane deflection limits are project-specific. The designer should consider the potential for damage to interior finishes, and should limit deflections accordingly.

10.8 — Prestressing tendon anchorages, couplers, and end blocks

The provisions of this section of the Code are used to design the tendon anchorages, couplers, and end blocks to withstand the prestressing operation and effectively transfer prestress force to the masonry wall without distress to the masonry or the prestressing accessories. Anchorages are designed for adequate pull-out strength from their foundations.

Because the actual stresses are quite complicated around post-tensioning anchorages, experimental data, or a refined analysis should be used whenever possible. Appropriate formulas from the references (PTI, 1990) should be used as a guide to size prestressing tendon anchorages when experimental data or more refined analysis are not available. Additional guidance on design and details for post-tensioning anchorage zones is given in the references (Sanders et al, 1987).

In most cases, f'_{mi} is equal to or greater than $0.75 f'_m$ for prestressed masonry. At $0.75 f'_m$, the prestressed bearing stress of $0.50 f'_{mi}$ is equivalent to $0.375 f'_m$. If f'_{mi} is specified as equal to f'_m, the maximum permitted bearing stress would be the equivalent of $0.50 f'_m$.

CODE

10.9 — Protection of prestressing tendons and accessories

10.9.1 Prestressing tendons, anchorages, couplers, and end fittings in exterior walls exposed to earth or weather, or walls exposed to a mean relative humidity exceeding 75 percent, shall be corrosion-protected.

10.9.2 Corrosion protection of prestressing tendons shall not rely solely on masonry cover.

10.9.3 Parts of prestressing tendons not embedded in masonry shall be provided with mechanical and fire protection equivalent to that of the embedded parts of the tendon.

10.10 — Development of bonded tendons

Development of bonded prestressing tendons in grouted corrugated ducts, anchored in accordance with Section 10.8.1, does not need to be calculated.

COMMENTARY

10.9 — Protection of prestressing tendons and accessories

Corrosion protection of the prestressing tendon and accessories is required in masonry walls subject to a moist and corrosive environment. Methods of corrosion protection are addressed in the Specification. Masonry and grout cover is not considered adequate protection due to variable permeability and the sensitivity of prestressing tendons to corrosion. The methods of corrosion protection given in the Specification provide a minimum level of corrosion protection. The designer may wish to impose more substantial corrosion protection requirements, especially in highly corrosive environments.

10.10 — Development of bonded tendons

Consistent with design practice in prestressed concrete, development of post-tensioned tendons away from the anchorage does not need to be calculated.

This page intentionally left blank

CHAPTER 11
STRENGTH DESIGN OF AUTOCLAVED AERATED CONCRETE (AAC) MASONRY

CODE

11.1 — General

11.1.1 *Scope*

This Chapter provides minimum requirements for design of AAC masonry.

11.1.1.1 Except as stated elsewhere in this Chapter, design of AAC masonry shall comply with the requirements of Part 1 and Part 2, excluding Sections 5.5.1, 5.5.2(d) and 5.3.2.

11.1.1.2 Design of AAC masonry shall comply with Sections 11.1.2 through 11.1.9, and either Section 11.2 or 11.3.

11.1.2 *Required strength*

Required strength shall be determined in accordance with the strength design load combinations of the legally adopted building code. Members subject to compressive axial load shall be designed for the maximum design moment accompanying the axial load. The factored moment, M_u, shall include the moment induced by relative lateral displacement.

11.1.3 *Design strength*

AAC masonry members shall be proportioned so that the design strength equals or exceeds the required strength. Design strength is the nominal strength multiplied by the strength-reduction factor, ϕ, as specified in Section 11.1.5.

11.1.4 *Strength of joints*

AAC masonry members shall be made of AAC masonry units. The tensile bond strength of AAC masonry joints shall not be taken greater than the limits of Section 11.1.8.3. When AAC masonry units with a maximum height of 8 in. (203 mm) (nominal) are used, head joints shall be permitted to be left unfilled between AAC masonry units laid in running bond, provided that shear capacity is calculated using the formulas of this Code corresponding to that condition. Open head joints shall not be permitted in AAC masonry not laid in running bond.

COMMENTARY

11.1 — General

11.1.1 *Scope*

Design procedures in Chapter 11 are strength design methods in which internal forces resulting from application of factored loads must not exceed design strength (nominal member strength reduced by a strength-reduction factor ϕ).

Refer to Section 11.1.10 for requirements for corbels constructed of AAC masonry.

11.1.4 *Strength of joints*

Design provisions of Chapter 11 and prescriptive seismic reinforcement requirements of Chapter 7 are based on monolithic behavior of AAC masonry. The reduction in shear strength of AAC masonry shear walls laid in running bond with unfilled head joints is accounted for in Equation 11-13b. AAC masonry walls constructed with AAC masonry units greater in height than 8 in. (203 mm) (nominal) with unfilled head joints and AAC masonry walls not laid in running bond with unfilled head joints do not have sufficient test data to develop design provisions and thus are not permitted at this time.

CODE

11.1.5 *Strength-reduction factors*

11.1.5.1 *Anchor bolts* — For cases where the nominal strength of an anchor bolt is controlled by AAC masonry breakout, ϕ shall be taken as 0.50. For cases where the nominal strength of an anchor bolt is controlled by anchor bolt steel, ϕ shall be taken as 0.90. For cases where the nominal strength of an anchor bolt is controlled by anchor pullout, ϕ shall be taken as 0.65.

11.1.5.2 *Bearing* — For cases involving bearing on AAC masonry, ϕ shall be taken as 0.60.

11.1.5.3 *Combinations of flexure and axial load in unreinforced AAC masonry* — The value of ϕ shall be taken as 0.60 for unreinforced AAC masonry designed to resist flexure, axial load, or combinations thereof.

11.1.5.4 *Combinations of flexure and axial load in reinforced AAC masonry* — The value of ϕ shall be taken as 0.90 for reinforced AAC masonry designed to resist flexure, axial load, or combinations thereof.

11.1.5.5 *Shear* — The value of ϕ shall be taken as 0.80 for AAC masonry designed to resist shear.

11.1.6 *Deformation requirements*

11.1.6.1 *Deflection of unreinforced (plain) AAC masonry* — Deflection calculations for unreinforced (plain) AAC masonry members shall be based on uncracked section properties.

COMMENTARY

11.1.5 *Strength-reduction factors*

The strength-reduction factor incorporates the difference between the nominal strength provided in accordance with the provisions of Chapter 11 and the expected strength of the as-built AAC masonry. The strength-reduction factor also accounts for the uncertainties in construction, material properties, calculated versus actual member strengths, and anticipated mode of failure.

11.1.5.1 *Anchor bolts* — Anchor bolts embedded in grout in AAC masonry behave like those addressed in Chapter 9 and are designed identically. Anchors for use in AAC masonry units are available from a variety of manufacturers, and nominal resistance should be based on tested capacities.

11.1.5.2 *Bearing* — The value of the strength-reduction factor used in bearing assumes that some degradation has occurred within the masonry material.

11.1.5.3 *Combinations of flexure and axial load in unreinforced AAC masonry* — The same strength-reduction factor is used for the axial load and the flexural tension or compression induced by bending moment in unreinforced masonry elements. The lower strength-reduction factor associated with unreinforced elements (in comparison to reinforced elements) reflects an increase in the coefficient of variation of the measured strengths of unreinforced elements when compared to similarly configured reinforced elements.

11.1.5.4 *Combinations of flexure and axial load in reinforced AAC masonry* — The same strength-reduction factor is used for the axial load and the flexural tension or compression induced by bending moment in reinforced AAC masonry elements. The higher strength-reduction factor associated with reinforced elements (in comparison to unreinforced elements) reflects a decrease in the coefficient of variation of the measured strengths of reinforced elements when compared to similarly configured unreinforced elements.

11.1.5.5 *Shear* — Strength-reduction factors for calculating the design shear strength are commonly more conservative than those associated with the design flexural strength. However, the capacity design provisions of Chapter 11 require that shear capacity significantly exceed flexural capacity. Hence, the strength-reduction factor for shear is taken as 0.80, a value 33 percent larger than the historical value.

11.1.6 *Deformation requirements*

11.1.6.1 *Deflection of unreinforced (plain) AAC masonry* — The deflection calculations of unreinforced masonry are based on elastic performance of the masonry assemblage as outlined in the design criteria of Section 9.2.2.

CODE

11.1.6.2 *Deflection of reinforced AAC masonry* — Deflection calculations for reinforced AAC masonry members shall be based on cracked section properties including the reinforcement and grout. The flexural and shear stiffness properties assumed for deflection calculations shall not exceed one-half of the gross section properties unless a cracked-section analysis is performed.

11.1.7 *Anchor bolts*

Headed and bent-bar anchor bolts shall be embedded in grout, and shall be designed in accordance with Section 9.1.6 using f'_g instead of f'_m and neglecting the contribution of AAC to the edge distance and embedment depth. Anchors embedded in AAC without grout shall be designed using nominal capacities provided by the anchor manufacturer and verified by an independent testing agency.

11.1.8 *Material properties*

11.1.8.1 *Compressive strength*

11.1.8.1.1 *Masonry compressive strength* — The specified compressive strength of AAC masonry, f'_{AAC}, shall equal or exceed 290 psi (2.0 MPa).

11.1.8.1.2 *Grout compressive strength* — The specified compressive strength of grout, f'_g, shall equal or exceed 2,000 psi (13.8 MPa) and shall not exceed 5,000 psi (34.5 MPa).

11.1.8.2 *Masonry splitting tensile strength* — The splitting tensile strength f_{tAAC} shall be determined by Equation 11-1.

$$f_{tAAC} = 2.4\sqrt{f'_{AAC}} \qquad \text{(Equation 11-1)}$$

COMMENTARY

11.1.6.2 *Deflection of reinforced AAC masonry* — Values of I_{eff} are typically about one-half of I_g for common configurations of elements that are fully grouted. Calculating a more accurate effective moment of inertia using a moment curvature analysis may be desirable for some circumstances. Historically, an effective moment of inertia has been calculated using net cross-sectional area properties and the ratio of the cracking moment strength based on appropriate modulus of rupture values to the applied moment resulting from unfactored loads as shown in the following equation. This equation has successfully been used for estimating the post-cracking flexural stiffness of both concrete and masonry.

$$I_{eff} = I_n \left(\frac{M_{cr}}{M_a}\right)^3 + I_{cr}\left[1 - \left(\frac{M_{cr}}{M_a}\right)^3\right] \leq I_n \leq 0.5 I_g$$

11.1.7 *Anchor bolts*

Headed and bent-bar anchor bolts embedded in grout in AAC masonry behave like those addressed in Chapter 9 and are designed identically. Anchors for use in AAC masonry units are available from a variety of manufacturers.

11.1.8 *Material properties*

11.1.8.1 *Compressive strength*

11.1.8.1.1 *Masonry compressive strength* — Research (Varela et al, 2006; Tanner et al, 2005(a), Tanner et al, 2005(b); Argudo, 2003) has been conducted on structural components of AAC masonry with a compressive strength of 290 to 1,500 psi (2.0 to 10.3 MPa). Design criteria are based on these research results.

11.1.8.1.2 *Grout compressive strength* — Because most empirically derived design equations relate the calculated nominal strength as a function of the specified compressive strength of the masonry, the specified compressive strength of the grout is required to be at least equal to the specified compressive strength. Additionally, due to the hydrophilic nature of AAC masonry, care should be taken to control grout shrinkage by pre-wetting cells to be grouted or by using other means, such as non-shrink admixtures. Bond between grout and AAC units is equivalent to bond between grout and other masonry units (Tanner et al, 2005(a), Tanner et al, 2005(b); Argudo, 2003).

11.1.8.2 *Masonry splitting tensile strength* — The equation for splitting tensile strength is based on ASTM C1006 tests (Tanner et al, 2005(b); Argudo, 2003).

CODE

11.1.8.3 *Masonry modulus of rupture* — The modulus of rupture, f_{rAAC}, for AAC masonry elements shall be taken as twice the masonry splitting tensile strength, f_{tAAC}. If a section of AAC masonry contains a Type M or Type S horizontal leveling bed of mortar, the value of f_{rAAC} shall not exceed 50 psi (345 kPa) at that section. If a section of AAC masonry contains a horizontal bed joint of thin-bed mortar and AAC, the value of f_{rAAC} shall not exceed 80 psi (552 kPa) at that section.

11.1.8.4 *Masonry direct shear strength* — The direct shear strength, f_v, across an interface of AAC material shall be determined by Equation 11-2, and shall be taken as 50 psi (345 kPa) across an interface between grout and AAC material.

$$f_v = 0.15 f'_{AAC} \qquad \text{(Equation 11-2)}$$

11.1.8.5 *Coefficient of friction* — The coefficient of friction between AAC and AAC shall be 0.75. The coefficient of friction between AAC and thin-bed mortar or between AAC and leveling-bed mortar shall be 1.0.

11.1.8.6 *Reinforcement strength* — Masonry design shall be based on a reinforcement strength equal to the specified yield strength of reinforcement, f_y, which shall not exceed 60,000 psi (413.7 MPa). The actual yield strength shall not exceed 1.3 multiplied by the specified yield strength.

11.1.9 *Nominal bearing strength*

11.1.9.1 The nominal bearing strength of AAC masonry shall be calculated as f'_{AAC} multiplied by the bearing area, A_{br}, as defined in Section 4.3.4.

11.1.9.2 *Bearing for simply supported precast floor and roof members on AAC masonry shear walls* — The following minimum requirements shall apply so that after the consideration of tolerances, the distance from the edge of the supporting wall to the end of the precast member in the direction of the span is at least:

For AAC floor panels	2 in. (51 mm)
For solid or hollow-core slabs	2 in. (51 mm)
For beams or stemmed members	3 in. (76 mm)

COMMENTARY

11.1.8.3 *Masonry modulus of rupture* — The modulus of rupture is based on tests conducted in accordance with ASTM C78 (2002) on AAC masonry with different compressive strengths (Tanner et al, 2005(b); Argudo, 2003; Fouad, 2002). Modulus of rupture tests show that a thin-bed mortar joint can fail before the AAC material indicating that the tensile-bond strength of the thin-bed mortar is less than the modulus of rupture of the AAC. This critical value is 80 psi (552 kPa). The data are consistent with the formation of cracks in thin-bed mortar joints observed in AAC shear wall tests (Tanner et al, 2005(b); Argudo, 2003). Shear wall tests (Tanner et al, 2005(b)) show that when a leveling bed is present, flexural cracking capacity may be controlled by the tensile bond strength across the interface between the AAC and the leveling mortar, which is usually less than the modulus of rupture of the AAC material itself.

11.1.8.4 *Masonry direct shear strength* — The equation for direct shear strength is based on shear tests (Tanner et al, 2005(b); Argudo, 2003). Based on tests by Kingsley et al (1985), interface shear strength between grout and conventional masonry units varies from 100 to 250 psi (689 to 1,723 kPa). Based on tests by Forero and Klingner (2011), interface shear strength between grout and AAC material had a 5% fractile (lower characteristic) value of 50 psi (345 kPa).

11.1.8.5 *Coefficient of friction* — The coefficient of friction between AAC and AAC is based on direct shear tests performed at The University of Texas at Austin and. the coefficient of friction between AAC and leveling mortar is based on tests on shear walls at the same institution.

11.1.8.6 *Reinforcement strength* — Research[3.11] conducted on reinforced masonry components used Grade 60 steel. To be consistent with laboratory documented investigations, design is based on a nominal steel yield strength of 60,000 psi (413.7 MPa). The limitation on the steel yield strength of 130 percent of the nominal yield strength limits the over-strength that may be present in the construction.

11.1.9 *Nominal bearing strength*

11.1.9.1 Commentary Section 4.3.4 gives further information.

11.1.9.2 *Bearing for simply supported precast floor and roof members on AAC shear walls* — Bearing should be checked wherever floor or roof elements rest on AAC walls. The critical edge distance for bearing and the critical section for shear to be used in this calculation are shown in Figure CC-11.1-1.

CODE

11.1.10 *Corbels* — Load-bearing corbels of AAC masonry shall not be permitted. Non-load-bearing corbels of AAC masonry shall conform to the requirements of Section 5.5.2(a) through 5.5.2(c). The back section of the corbelled section shall remain within ¼ in. (6.4 mm) of plane.

COMMENTARY

11.1.10 *Corbels* — Load-bearing corbels of AAC masonry are not permitted due to the possibility of a brittle shear failure. Non-load-bearing corbels of AAC masonry are permitted, provided that the back section of the corbelled wall remains plane within the code limits. The relative ease in which AAC masonry can be cut and shaped makes this requirement practical.

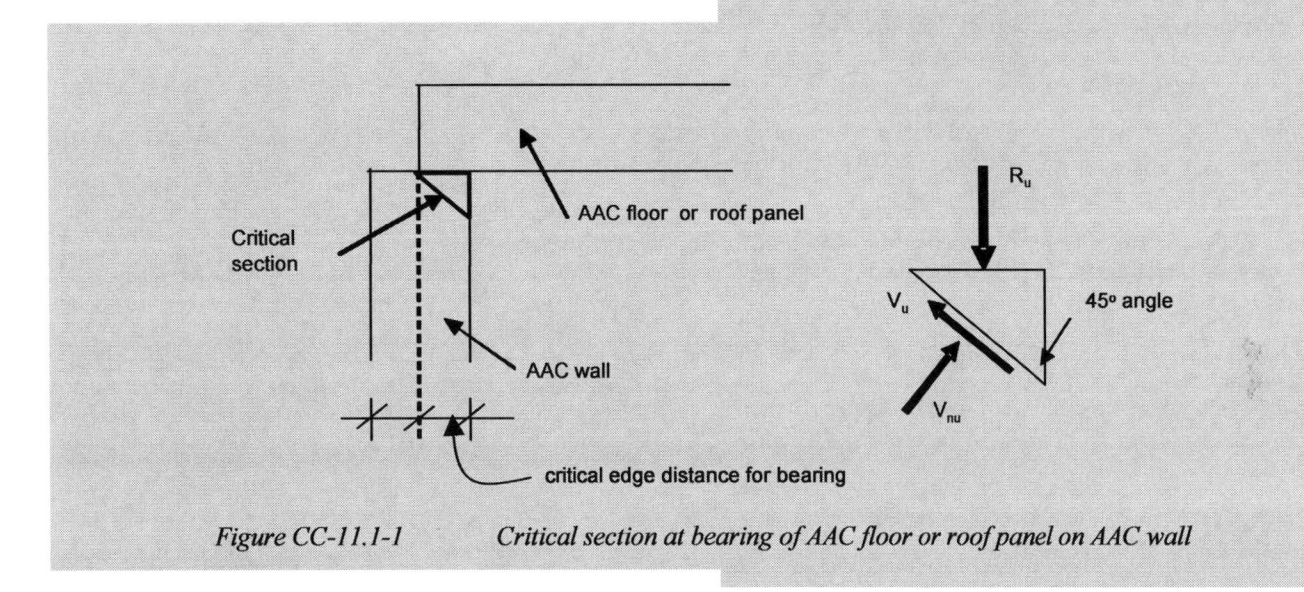

Figure CC-11.1-1   Critical section at bearing of AAC floor or roof panel on AAC wall

CODE

11.2 — Unreinforced (plain) AAC masonry

11.2.1 *Scope*
The requirements of Section 11.2 are in addition to the requirements of Part 1, Part 2, and Section 11.1, and govern masonry design in which AAC masonry is used to resist tensile forces.

11.2.1.1 *Strength for resisting loads* — Unreinforced (plain) AAC masonry members shall be designed using the strength of masonry units, mortar, and grout in resisting design loads.

11.2.1.2 *Strength contribution from reinforcement* — Stresses in reinforcement shall not be considered effective in resisting design loads.

11.2.1.3 *Design criteria* — Unreinforced (plain) AAC masonry members shall be designed to remain uncracked.

11.2.2 *Flexural strength of unreinforced (plain) AAC masonry members*
The following assumptions shall apply when determining the flexural strength of unreinforced (plain) AAC masonry members:

(a) Strength design of members for factored flexure and axial load shall be in accordance with principles of engineering mechanics.

(b) Strain in masonry shall be directly proportional to the distance from the neutral axis.

(c) Flexural tension in masonry shall be assumed to be directly proportional to strain.

(d) Flexural compressive stress in combination with axial compressive stress in masonry shall be assumed to be directly proportional to strain. Nominal compressive strength shall not exceed a stress corresponding to $0.85 f'_{AAC}$.

(e) The nominal flexural tensile strength of AAC masonry shall be determined from Section 11.1.8.3.

CODE

11.2.3 *Nominal axial strength of unreinforced (plain) AAC masonry members*

Nominal axial strength, P_n, shall be calculated using Equation 11-3 or Equation 11-4.

(a) For members having an *h/r* ratio not greater than 99:

$$P_n = 0.80\left\{0.85 A_n f'_{AAC}\left[1-\left(\frac{h}{140r}\right)^2\right]\right\}$$

(Equation 11-3)

(b) For members having an *h/r* ratio greater than 99:

$$P_n = 0.80\left[0.85 A_n f'_{AAC}\left(\frac{70r}{h}\right)^2\right] \quad \text{(Equation 11-4)}$$

11.2.4 *Axial tension*

The tensile strength of unreinforced AAC masonry shall be neglected in design when the masonry is subjected to axial tension forces.

11.2.5 *Nominal shear strength of unreinforced (plain) AAC masonry members*

The nominal shear strength of AAC masonry, V_{nAAC}, shall be the least of the values calculated by Sections 11.3.4.1.2.1 through 11.3.4.1.2.3. In evaluating nominal shear strength by Section 11.3.4.1.2.3, effects of reinforcement shall be neglected. The provisions of 11.3.4.1.2 shall apply to AAC shear walls not laid in running bond. The provisions of Section 11.3.4.1.2.4 shall apply to AAC walls loaded out-of-plane.

11.2.6 *Flexural cracking*

The flexural cracking strength shall be calculated in accordance with Section 11.3.6.5.

COMMENTARY

11.2.4 *Axial tension*

Commentary Section 8.2.5 provides further information.

CODE

11.3 — Reinforced AAC masonry

11.3.1 *Scope*

The requirements of this section are in addition to the requirements of Part 1, Part 2, and Section 11.1 and govern AAC masonry design in which reinforcement is used to resist tensile forces.

11.3.2 *Design assumptions*

The following assumptions apply to the design of reinforced AAC masonry:

(a) There is strain compatibility between the reinforcement, grout, and AAC masonry.

(b) The nominal strength of reinforced AAC masonry cross sections for combined flexure and axial load shall be based on applicable conditions of equilibrium.

(c) The maximum usable strain, ε_{mu}, at the extreme AAC masonry compression fiber shall be assumed to be 0.0012 for Class 2 AAC masonry and 0.003 for Class 4 AAC masonry and higher.

(d) Strain in reinforcement and AAC masonry shall be assumed to be directly proportional to the distance from the neutral axis.

(e) Tension and compression stresses in reinforcement shall be calculated as the product of steel modulus of elasticity, E_s, and steel strain, ε_s, but shall not be greater than f_y. Except as permitted in Section 11.3.3.5 for determination of maximum area of flexural reinforcement, the compressive stress of steel reinforcement shall be neglected unless lateral restraining reinforcement is provided in compliance with the requirements of Section 5.3.1.4.

(f) The tensile strength of AAC masonry shall be neglected in calculating axial and flexural strength.

(g) The relationship between AAC masonry compressive stress and masonry strain shall be assumed to be defined by the following: AAC masonry stress of $0.85 f'_{AAC}$ shall be assumed uniformly distributed over an equivalent compression stress block bounded by edges of the cross section and a straight line parallel to the neutral axis and located at a distance $a = 0.67 c$ from the fiber of maximum compressive strain. The distance c from the fiber of maximum strain to the neutral axis shall be measured perpendicular to the neutral axis.

COMMENTARY

11.3 — Reinforced AAC masonry

Provisions are identical to those of concrete or clay masonry, with a few exceptions. Only those exceptions are addressed in this Commentary.

11.3.2 *Design assumptions*

For AAC, test results indicate that ε_{mu} for Class 4 AAC masonry and higher is 0.003 and the value of the stress in the equivalent rectangular stress block is $0.85 f'_{AAC}$ with $a = 0.67c$ (Argudo, 2003 and Tanner et al, 2005a). Additional testing has indicated a ε_{mu} of 0.0012 for Class 2 AAC masonry (Cancino, 2003 and Tanner et al, 2011).

CODE

11.3.3 Reinforcement requirements and details

11.3.3.1 *Reinforcing bar size limitations* — Reinforcing bars used in AAC masonry shall not be larger than No. 9 (M#29). The nominal bar diameter shall not exceed one-eighth of the nominal member thickness and shall not exceed one-quarter of the least clear dimension of the grout space in which it is placed. In plastic hinge zones, the area of reinforcing bars placed in a grout space shall not exceed 3 percent of the grout space area. In other than plastic hinge zones, the area of reinforcing bars placed in a grout space shall not exceed 4.5 percent of the grout space area.

11.3.3.2 *Standard hooks* — The equivalent embedment length to develop standard hooks in tension, l_e, shall be determined by Equation 11-5:

$$l_e = 13 d_b \qquad \text{(Equation 11-5)}$$

11.3.3.3 *Development*

11.3.3.3.1 *Development of tension and compression reinforcement* — The required tension or compression reinforcement shall be developed in accordance with the following provisions:

The required development length of reinforcement shall be determined by Equation 11-6, but shall not be less than 12 in. (305 mm).

$$l_d = \frac{0.13 \, d_b^2 \, f_y \, \gamma}{K_{AAC} \sqrt{f'_g}} \qquad \text{(Equation 11-6)}$$

K_{AAC} shall not exceed the smallest of the following: the minimum grout cover, the clear spacing between adjacent reinforcement splices, and $9 d_b$.

$\gamma = 1.0$ for No. 3 (M#10) through No. 5 (M#16) bars;

$\gamma = 1.3$ for No. 6 (M#19) through No. 7 (M#22) bars;

and

$\gamma = 1.5$ for No. 8 (M#25) through No. 9 (M#29) bars.

11.3.3.3.2 *Development of shear reinforcement* — Shear reinforcement shall extend the depth of the member less cover distances.

11.3.3.3.2.1 Except at wall intersections, the end of a horizontal reinforcing bar needed to satisfy shear strength requirements of Section 11.3.4.1.2, shall be bent around the edge vertical reinforcing bar with a 180-degree hook. The ends of single-leg or U-stirrups shall be anchored by one of the following means:

(a) A standard hook plus an effective embedment of $l_d/2$. The effective embedment of a stirrup leg shall be taken as the distance between the mid-depth of the member, $d/2$, and the start of the hook (point of tangency).

COMMENTARY

11.3.3 Reinforcement requirements and details

11.3.3.1 *Reinforcing bar size limitations* — Grout spaces may include, but are not limited to, cores, bond beams, and collar joints. At sections containing lap splices, the maximum area of reinforcement specified in the Code may be doubled.

11.3.3.3.1 *Development of tension and compression reinforcement* — Development and lap splice detailing provisions for conventional masonry are calibrated to the masonry assembly strength, f'_m, which includes the contribution of each constituent material (unit, grout, and mortar). Due to the low compressive strength of AAC, however, the AAC masonry component is ignored and the calibration is based on f'_g.

CODE

(b) For No. 5 (M #16) bars and smaller, bending around longitudinal reinforcement through at least 135 degrees plus an embedment of $l_d/3$. The $l_d/3$ embedment of a stirrup leg shall be taken as the distance between mid-depth of the member, $d/2$, and the start of the hook (point of tangency).

(c) Between the anchored ends, each bend in the continuous portion of a transverse U-stirrup shall enclose a longitudinal bar.

11.3.3.3.2.2 At wall intersections, horizontal reinforcing bars needed to satisfy shear strength requirements of Section 11.3.4.1.2 shall be bent around the edge vertical reinforcing bar with a 90-degree standard hook and shall extend horizontally into the intersecting wall a minimum distance at least equal to the development length.

11.3.3.4 *Splices* — Reinforcement splices shall comply with one of the following:

(a) The minimum length of lap for bars shall be 12 in. (305 mm) or the development length determined by Equation 11-6, whichever is greater.

(b) A welded splice shall have the bars butted and welded to develop at least 125 percent of the yield strength, f_y, of the bar in tension or compression, as required. Welding shall conform to AWS D1.4. Reinforcement to be welded shall conform to ASTM A706, or shall be accompanied by a submittal showing its chemical analysis and carbon equivalent as required by AWS D1.4. Existing reinforcement to be welded shall conform to ASTM A706, or shall be analyzed chemically and its carbon equivalent determined as required by AWS D1.4.

(c) Mechanical splices shall have the bars connected to develop at least 125 percent of the yield strength, f_y, of the bar in tension or compression, as required.

11.3.3.5 *Maximum reinforcement percentages* — The ratio of reinforcement, ρ, shall be calculated in accordance with Section 9.3.3.5 with the following exceptions:

The maximum usable strain, ε_{mu}, at the extreme masonry compression fiber shall be in accordance with Section 11.3.2.c.

The strength of the compression zone shall be calculated as 85 percent of f'_{AAC} multiplied by 67 percent of the area of the compression zone.

11.3.3.6 *Bundling of reinforcing bars* — Reinforcing bars shall not be bundled.

COMMENTARY

11.3.3.4 *Splices* — See Code Commentary Section 8.1.6.7.2 for additional information on welded splices.

CODE

11.3.4 *Design of beams, piers, and columns*

Member design forces shall be based on an analysis that considers the relative stiffness of structural members. The calculation of lateral stiffness shall include the contribution of beams, piers, and columns. The effects of cracking on member stiffness shall be considered.

11.3.4.1 *Nominal strength*

11.3.4.1.1 *Nominal axial and flexural strength* — The nominal axial strength, P_n, and the nominal flexural strength, M_n, of a cross section shall be determined in accordance with the design assumptions of Section 11.3.2 and the provisions of Section 11.3.4.1. For any value of nominal flexural strength, the corresponding calculated nominal axial strength shall be modified for the effects of slenderness in accordance with Equation 11-7 or 11-8. The nominal flexural strength at any section along a member shall not be less than one-fourth of the maximum nominal flexural strength at the critical section.

The nominal axial compressive strength shall not exceed Equation 11-7 or Equation 11-8, as appropriate.

(a) For members having an h/r ratio not greater than 99:

$$P_n = 0.80 \left[0.85 f'_{AAC} (A_n - A_{st}) + f_y A_{st} \right] \left[1 - \left(\frac{h}{140r} \right)^2 \right]$$

(Equation 11-7)

(b) For members having an h/r ratio greater than 99:

$$P_n = 0.80 \left[0.85 f'_{AAC} (A_n - A_{st}) + f_y A_{st} \right] \left(\frac{70r}{h} \right)^2$$

(Equation 11-8)

11.3.4.1.2 *Nominal shear strength* — Nominal shear strength, V_n, shall be calculated using Equation 11-9 through Equation 11-12, as appropriate.

$$V_n = V_{nAAC} + V_{ns} \quad \text{(Equation 11-9)}$$

where V_n shall not exceed the following:

(a) $V_n = \mu_{AAC} P_u$ (Equation 11-10)

At an interface of AAC and thin-bed mortar or leveling-bed mortar, the nominal sliding shear strength shall be calculated using Equation 11-10 and using the coefficient of friction from Section 11.1.8.5.

(b) Where $M_u/(V_u d_v) \leq 0.25$:

$$V_n \leq 6 A_{nv} \sqrt{f'_{AAC}} \quad \text{(Equation 11-11)}$$

(c) Where $M_u/(V_u d_v) \geq 1.0$

$$V_n \leq 4 A_{nv} \sqrt{f'_{AAC}} \quad \text{(Equation 11-12)}$$

(d) The maximum value of V_n for $M_u/(V_u d_v)$ between 0.25 and 1.0 shall be permitted to be linearly interpolated.

COMMENTARY

11.3.4.1.2 *Nominal shear strength* — The nominal shear strength of AAC walls is based on testing at UT Austin (Tanner et al, 2005(b); Argudo, 2003). Test results show that factory-installed, welded-wire reinforcement is developed primarily by bearing of the cross-wires on the AAC material, which normally crushes before the longitudinal wires develop significant stress. Therefore, the additional shear strength provided by the horizontal reinforcement should be neglected. Joint-type reinforcement will probably behave similarly and is not recommended. In contrast, deformed reinforcement placed in grouted bond beams is effective and should be included in calculating V_{ns}.

The upper limit on V_n, defined by Equation 11-10, is based on sliding shear. Flexural cracking can result in an unbonded interface, which typically occurs at a horizontal joint in a shear wall. For this reason, the shear capacity of an AAC bed joint is conservatively limited to the frictional resistance, without considering initial adhesion. The sliding shear capacity should be based on the frictional capacity consistent with the perpendicular force on the compressive stress block, including the compressive force required to equilibrate the tensile force

CODE

The nominal masonry shear strength shall be taken as the least of the values calculated using Section 11.3.4.1.2.1 and 11.3.4.1.2.2.

11.3.4.1.2.1 *Nominal masonry shear strength as governed by web-shear cracking* — Nominal masonry shear strength as governed by web-shear cracking, V_{nAAC}, shall be calculated using Equation 11-13a for AAC masonry with mortared head joints, and Equation 11-13b for masonry with unmortared head joints:

$$V_{nAAC} = 0.95 \, l_w \, t \, \sqrt{f'_{AAC}} \, \sqrt{1 + \frac{P_u}{2.4\sqrt{f'_{AAC}} \, l_w \, t}}$$

(Equation 11-13a)

$$V_{nAAC} = 0.66 \, l_w \, t \, \sqrt{f'_{AAC}} \, \sqrt{1 + \frac{P_u}{2.4\sqrt{f'_{AAC}} \, l_w \, t}}$$

(Equation 11-13b)

For AAC masonry not laid in running bond, nominal masonry shear strength as governed by web-shear cracking, V_{nAAC}, shall be calculated using Equation 11-13c:

$$V_{nAAC} = 0.9 \sqrt{f'_{AAC}} \, A_{nv} + 0.5 P_u \quad \text{(Equation 11-13c)}$$

11.3.4.1.2.2 *Nominal shear strength as governed by crushing of diagonal compressive strut* — For walls with $M_u/(V_u d_v) < 1.5$, nominal shear strength, V_{nAAC}, as governed by crushing of a diagonal strut, shall be calculated as follows:

$$V_{nAAC} = 0.17 f'_{AAC} t \frac{h \cdot l_w^2}{h^2 + (\tfrac{3}{4} l_w)^2} \quad \text{(Equation 11-14)}$$

For walls with $M_u/(V_u d_v)$ equal to or exceeding 1.5, capacity as governed by crushing of the diagonal compressive strut need not be calculated.

11.3.4.1.2.3 *Nominal shear strength provided by shear reinforcement* — Nominal shear strength provided by reinforcement, V_{ns}, shall be calculated as follows:

$$V_{ns} = 0.5 \left(\frac{A_v}{s}\right) f_y d_v \quad \text{(Equation 11-15)}$$

Nominal shear strength provided by reinforcement, V_{ns}, shall include only deformed reinforcement embedded in grout for AAC shear walls.

COMMENTARY

in the flexural reinforcement. Dowel action should not be included.

11.3.4.1.2.1 *Nominal masonry shear strength as governed by web-shear cracking* — Equations 11-13a and 11-13b were developed based on observed web shear cracking in shear walls tested at the University of Texas at Austin (Tanner et al, 2005(b); Argudo, 2003) and Hebel AG (Vratsanou and Langer, 2001) in Germany. Independent testing has validated these equations (Costa et al, 2011; Tanner et al, 2011). During testing at the University of Texas at Austin, flexur-shear cracking of AAC shear walls was observed, as predicted, in 6 shear wall tests (Varela et al, 2006; Tanner et al, 2005(a); Tanner et al, 2005(b)). The presence of flexur-shear cracks did not reduce the strength or stiffness of tested AAC shear walls. Another AAC shear wall tested by Cancino (2003) performed in a similar manner. The results in both testing efforts indicate the hysteretic behavior was not changed after the formation of flexure-shear cracks. Thus, flexure-shear cracking does not constitute a limit state in AAC masonry and design equations are not provided.

Masonry units not laid in running bond may exhibit discontinuities at head joints. The nominal masonry shear strength calculation for AAC masonry not laid in running bond considers the likelihood of vertical discontinuities at head joints and is based on test results for AAC walls made of vertical panels with open vertical joints between some panels.

11.3.4.1.2.2 *Nominal shear strength as governed by crushing of diagonal compressive strut* — This mechanism limits the shear strength at large levels of axial load. It was based on test results (Tanner et al, 2005(b)), using a diagonal strut width of $0.25 l_w$ based on test observations.

11.3.4.1.2.3 *Nominal shear strength provided by shear reinforcement* — Equation 11-15 is based on Equation 9-24. Equation 9-24 was developed based on results of reversed cyclic load tests on masonry wall segments with horizontal reinforcement distributed over their heights. The reason for the 0.5 efficiency factor is the non-uniform distribution of tensile strain in the horizontal reinforcement over the height of the element. The formation of an inclined diagonal compressive strut from one corner of the wall segment to the diagonally opposite corner creates a strain field in which the horizontal shear reinforcement at the top and bottom of the segment may not yield. For that reason, not all of the horizontal shear reinforcement in the wall may be fully

CODE

11.3.4.1.2.4 Nominal shear strength for beams and for out-of-plane loading of other members shall be calculated as follows:

$$V_{nAAC} = 0.8 \sqrt{f'_{AAC}}\, bd \qquad \text{(Equation 11-16)}$$

11.3.4.2 *Beams* — Design of beams shall meet the requirements of Section 5.2 and the additional requirements of Sections 11.3.4.2.1 through 11.3.4.2.5.

11.3.4.2.1 The factored axial compressive force on a beam shall not exceed $0.05 A_n f'_{AAC}$.

11.3.4.2.2 *Longitudinal reinforcement*
11.3.4.2.2.1 The variation in longitudinal reinforcing bars shall not be greater than one bar size. Not more than two bar sizes shall be used in a beam.

11.3.4.2.2.2 The nominal flexural strength of a beam shall not be less than 1.3 multiplied by the nominal cracking moment of the beam, M_{cr}. The modulus of rupture, f_{rAAC}, for this calculation shall be determined in accordance with Section 11.1.8.3.

COMMENTARY

effective or efficient in resisting shear forces.

AAC masonry walls differ from concrete masonry walls and clay masonry walls in that horizontal joint reinforcement is not used for horizontal shear reinforcement. For reasons of constructability, AAC walls are traditionally reinforced horizontally with deformed steel reinforcing bars in grout-filled bond beams. In addition, the strength of the thin set AAC mortar exceeds the strength of the AAC masonry units, which would suggest that AAC walls will behave in a manner similar to reinforced concrete. Assemblage testing conducted on AAC masonry walls also suggested that horizontal joint reinforcement provided in concrete bond beams could be fully effective in resisting shear. For this reason, earlier additions of the Code presented Equation 11-15 without the 0.5 efficiency factor, mimicking the reinforced concrete design equation for strength provided by shear reinforcement.

Although this appeared reasonable in the original judgment of the committee, no tests have been performed with AAC masonry walls having deformed horizontal reinforcement in concrete bond beams Until such testing is performed, the 0.5 efficiency factor is being included in Equation 11-15 to be consistent with design procedures associated with concrete masonry and clay masonry, and to provide a conservative design approach.

11.3.4.2.2.2 Section 9.3.4.2.2.3 permits reducing the minimum tensile reinforcement requirement of 1.3 multiplied by the nominal cracking moment of the beam, M_{cr} to one-third greater than that required by analysis. Because AAC masonry beams tend to be lightly reinforced, this reduction is not appropriate in AAC masonry design.

CODE

11.3.4.2.3 *Transverse reinforcement* — Transverse reinforcement shall be provided where V_u exceeds ϕV_{nAAC}. The factored shear, V_u, shall include the effects of lateral load. When transverse reinforcement is required, the following provisions shall apply:

(a) Transverse reinforcement shall be a single bar with a 180-degree hook at each end.

(b) Transverse reinforcement shall be hooked around the longitudinal reinforcement.

(c) The minimum area of transverse reinforcement shall be $0.0007 bd_v$.

(d) The first transverse bar shall not be located more than one-fourth of the beam depth, d_v, from the end of the beam.

(e) The maximum spacing shall not exceed the lesser of one-half the depth of the beam or 48 in. (1219 mm).

11.3.4.2.4 *Construction* — Beams shall be fully grouted.

11.3.4.2.5 *Dimensional limits* — The nominal depth of a beam shall not be less than 8 in. (203 mm).

11.3.4.3 *Piers*

11.3.4.3.1 The factored axial compression force on the piers shall not exceed $0.3 A_n f'_{AAC}$.

11.3.4.3.2 *Longitudinal reinforcement* — A pier subjected to in-plane stress reversals shall be reinforced symmetrically about the geometric center of the pier. The longitudinal reinforcement of piers shall comply with the following:

(a) At least one bar shall be provided in each end cell.

(b) The minimum area of longitudinal reinforcement shall be $0.0007 bd$.

11.3.4.3.3 *Dimensional limits* — Dimensions shall be in accordance with the following:

(a) The nominal thickness of a pier shall not be less than 6 in. (152 mm) and shall not exceed 16 in. (406 mm).

(b) The distance between lateral supports of a pier shall not exceed 25 multiplied by the nominal thickness of a pier except as provided for in Section 11.3.4.3.3(c).

(c) When the distance between lateral supports of a pier exceeds 25 multiplied by the nominal thickness of the pier, design shall be based on the provisions of Section 11.3.5.

CODE

(d) The nominal length of a pier shall not be less than three multiplied by its nominal thickness nor greater than six multiplied by its nominal thickness. The clear height of a pier shall not exceed five multiplied by its nominal length.

Exception: When the factored axial force at the location of maximum moment is less than $0.05 f'_{AAC} A_g$, the length of a pier shall be permitted to be taken equal to the thickness of the pier.

11.3.5 Wall design for out-of-plane loads

11.3.5.1 *Scope* — The requirements of Section 11.3.5 shall apply to the design of walls for out-of-plane loads.

11.3.5.2 *Maximum reinforcement* — The maximum reinforcement ratio shall be determined by Section 11.3.3.5.

11.3.5.3 *Nominal axial and flexural strength* — The nominal axial strength, P_n, and the nominal flexural strength, M_n, of a cross-section shall be determined in accordance with the design assumptions of Section 11.3.2. The nominal axial compressive strength shall not exceed that determined by Equation 11-7 or Equation 11-8, as appropriate.

11.3.5.4 *Nominal shear strength* — The nominal shear strength shall be determined by Section 11.3.4.1.2.

11.3.5.5 *P-delta effects*

11.3.5.5.1 Members shall be designed for the factored axial load, P_u, and the moment magnified for the effects of member curvature, M_u. The magnified moment shall be determined either by Section 11.3.5.5.2 or Section 11.3.5.5.3.

11.3.5.5.2 Moment and deflection calculations in this Section are based on simple support conditions top and bottom. For other support and fixity conditions, moments, and deflections shall be calculated using established principles of mechanics.

The procedures set forth in this section shall be used when the factored axial load stress at the location of maximum moment satisfies the requirement calculated by Equation 11-17.

$$\left(\frac{P_u}{A_g}\right) \leq 0.20 f'_{AAC} \qquad \text{(Equation 11-17)}$$

When the ratio of effective height to nominal thickness, h/t, exceeds 30, the factored axial stress shall not exceed $0.05 f'_{AAC}$

Factored moment and axial force shall be determined at the midheight of the wall and shall be used for design. The factored moment, M_u, at the midheight of the wall shall be calculated using Equation 11-18.

COMMENTARY

11.3.5.5.2 This section only includes design equations based on walls having simple support conditions at the top and bottom of the walls. In actual design and construction, there may be varying support conditions, thus changing the curvature of the wall under lateral loading. Through proper calculation and using the principles of mechanics, the points of inflection can be determined and actual moments and deflection can be calculated under different support conditions. The designer should examine moment and deflection conditions to locate the critical section using the assumptions outlined in Section 11.3.5.

The required moment due to lateral loads, eccentricity of axial load, and lateral deformations is assumed maximum at mid-height of the wall. In certain design conditions, such as large eccentricities acting simultaneously with small lateral loads, the design maximum moment may occur elsewhere. When this occurs, the designer should use the maximum moment at the critical section rather than the moment determined from Equation 11-18.

CODE

$$M_u = \frac{w_u h^2}{8} + P_{uf}\frac{e_u}{2} + P_u \delta_u \quad \text{(Equation 11-18)}$$

Where:

$$P_u = P_{uw} + P_{uf} \quad \text{(Equation 11-19)}$$

The deflection due to factored loads (δ_u) shall be obtained using Equations (11-20) and (11-21)

a) Where $M_u < M_{cr}$

$$\delta_u = \frac{5 M_u h^2}{48 E_{AAC} I_n} \quad \text{(Equation 11-20)}$$

b) Where $M_{cr} \le M_u \le M_n$

$$\delta_u = \frac{5 M_{cr} h^2}{48 E_{AAC} I_n} + \frac{5(M_u - M_{cr}) h^2}{48 E_{AAC} I_{cr}}$$

$$\text{(Equation 11-21)}$$

11.3.5.5.3 The factored moment, M_u, shall be determined either by a second-order analysis, or by a first-order analysis and Equations 11-22 through 11-24.

$$M_u = \psi M_{u,0} \quad \text{(Equation 11-22)}$$

Where $M_{u,0}$ is the factored moment from first-order analysis.

$$\psi = \frac{1}{1 - \frac{P_u}{P_e}} \quad \text{(Equation 11-23)}$$

Where:

$$P_e = \frac{\pi^2 E_{AAC} I_{eff}}{h^2} \quad \text{(Equation 11-24)}$$

For $M_u < M_{cr}$, I_{eff} shall be taken as $0.75 I_n$. For $M_u \ge M_{cr}$, I_{eff} shall be taken as I_{cr}. P_u/P_e cannot exceed 1.0.

11.3.5.5.4 The cracking moment of the wall shall be calculated using Equation 11-25, where f_{rAAC} is given by Section 11.1.8.3:

$$M_{cr} = S_n \left(f_{rAAC} + \frac{P}{A_n} \right) \quad \text{(Equation 11-25)}$$

If the section of AAC masonry contains a horizontal leveling bed, the value of f_{rAAC} shall not exceed 50 psi (345 kPa).

11.3.5.5.5 The neutral axis for determining the cracked moment of inertia, I_{cr}, shall be determined in accordance with the design assumptions of Section 11.3.2. The effects of axial load shall be permitted to be included when calculating I_{cr}.

COMMENTARY

11.3.5.5.3 The moment magnifier provisions in this section were developed to provide an alternative to the traditional P-delta methods of Section 11.3.5.5.2. These provisions also allow other second-order analyses to be used.

The proposed moment magnification equation is very similar to that used for slender wall design for reinforced concrete. Concrete design provisions use a factor of 0.75 in the denominator of the moment magnifier to account for uncertainties in the wall stiffness. This factor is retained for uncracked walls. It is not used for cracked walls. Instead, the cracked moment of inertia is conservatively used for the entire wall height. Trial designs indicated that using this approach matches design using Section 11.3.5.5.2. If a 0.75 factor were included along with using the cracked moment of inertia for the entire height would result in design moments approximately 7% greater than using Section 11.3.5.5.2. The committee did not see any reason for the additional conservatism.

CODE

Unless stiffness values are obtained by a more comprehensive analysis, the cracked moment of inertia for a solidly grouted wall or a partially grouted wall with the neutral axis in the face shell shall be obtained from Equation 11-26 and Equation 11-27.

$$I_{cr} = n\left(A_s + \frac{P_u}{f_y}\frac{t_{sp}}{2d}\right)(d-c)^2 + \frac{b(c)^3}{3}$$

(Equation 11-26)

$$c = \frac{A_s f_y + P_u}{0.57 f'_{AAC} b}$$

(Equation 11-27)

11.3.5.5.6 The design strength for out-of-plane wall loading shall be in accordance with Equation 11-28.

$$M_u \leq \phi M_n$$

(Equation 11-28)

The nominal moment shall be calculated using Equations 11-29 and 11-30 if the reinforcing steel is placed in the center of the wall.

$$M_n = (A_s f_y + P_u)\left(d - \frac{a}{2}\right)$$

(Equation 11-29)

$$a = \frac{(P_u + A_s f_y)}{0.85 f'_{AAC} b}$$

(Equation 11-30)

11.3.5.6 *Deflections* — The horizontal midheight deflection, δ_s, under allowable stress design load combinations shall be limited by the relation:

$$\delta_s \leq 0.007 h$$

(Equation 11-31)

P-delta effects shall be included in deflection calculation using either Section 11.3.5.6.1 or Section 11.3.5.6.2.

11.3.5.6.1 For simple support condition top and bottom, the midheight deflection, δ_s, shall be calculated using either Equation 11-20 or Equation 11-21, as applicable, and replacing M_u with M_{ser} and δ_u with δ_s.

11.3.5.6.2 The deflection, δ_s, shall be determined by a second-order analysis that includes the effects of cracking, or by a first-order analysis with the calculated deflections magnified by a factor of $1/(1-P/P_e)$, where P_e is determined from Equation 11-24.

CODE

11.3.6 *Wall design for in-plane loads*

11.3.6.1 *Scope* — The requirements of Section 11.3.6 shall apply to the design of walls to resist in-plane loads.

11.3.6.2 *Reinforcement* — Reinforcement shall be in accordance with the following:

(a) Reinforcement shall be provided perpendicular to the shear reinforcement and shall be at least equal to one-third A_v. The reinforcement shall be uniformly distributed and shall not exceed a spacing of 8 ft (2.44 m).

(b) The maximum reinforcement ratio shall be determined in accordance with Section 11.3.3.5.

11.3.6.3 *Flexural and axial strength* — The nominal flexural and axial strength shall be determined in accordance with Section 11.3.4.1.1.

11.3.6.4 *Shear strength* — The nominal shear strength shall be calculated in accordance with Section 11.3.4.1.2.

11.3.6.5 *Flexural cracking strength* — The flexural cracking strength shall be calculated in accordance with Equation 11-32, where f_{rAAC} is given by Section 11.1.8.3:

$$V_{cr} = \frac{S_n}{h}\left(f_{rAAC} + \frac{P}{A_n}\right) \qquad \text{(Equation 11-32)}$$

If the section of AAC masonry contains a horizontal leveling bed, the value of f_{rAAC} shall not exceed 50 psi (345 kPa).

11.3.6.6 The maximum reinforcement requirements of Section 11.3.3.5 shall not apply if a shear wall is designed to satisfy the requirements of Sections 11.3.6.6.1 through 11.3.6.6.4.

11.3.6.6.1 The need for special boundary elements at the edges of shear walls shall be evaluated in accordance with Section 11.3.6.6.2 or 11.3.6.6.3. The requirements of Section 11.3.6.6.4 shall also be satisfied.

11.3.6.6.2 This Section applies to walls bending in single curvature in which the flexural limit state response is governed by yielding at the base of the wall. Walls not satisfying those requirements shall be designed in accordance with Section 11.3.6.6.3.

(a) Special boundary elements shall be provided over portions of compression zones where:

$$c \geq \frac{l_w}{600\left(C_d \delta_{ne}/h_w\right)}$$

COMMENTARY

11.3.6.6 While requirements for confined boundary elements have not been developed for AAC shear walls, they have not been developed for conventional masonry shear walls either, and the monolithic nature of AAC shear walls favors possible applications involving boundary elements. Also see Commentary Section 9.3.6.5.

11.3.6.6.1 See Commentary Section 9.3.6.5.2.

11.3.6.6.2 See Commentary Section 9.3.6.5.3.

BUILDING CODE REQUIREMENTS FOR MASONRY STRUCTURES AND COMMENTARY

CODE

and c is calculated for the P_u given by ASCE 7 Load Combination 5 $(1.2D + 1.0E + L + 0.2S)$ or the corresponding strength design load combination of the legally adopted building code, and the corresponding nominal moment strength, M_n, at the base critical section. The load factor on L in Load Combination 5 is reducible to 0.5, as per exceptions to Section 2.3.2 of ASCE 7.

(b) Where special boundary elements are required by Section 11.3.6.6.2 (a), the special boundary element reinforcement shall extend vertically from the critical section a distance not less than the larger of l_w or $M_u/4V_u$.

11.3.6.6.3 Shear walls not designed to the provisions of Section 11.3.6.6.2 shall have special boundary elements at boundaries and edges around openings in shear walls where the maximum extreme fiber compressive stress, corresponding to factored forces including earthquake effect, exceeds $0.2f'_{AAC}$. The special boundary element shall be permitted to be discontinued where the calculated compressive stress is less than $0.15f'_{AAC}$. Stresses shall be calculated for the factored forces using a linearly elastic model and gross section properties. For walls with flanges, an effective flange width as defined in Section 5.1.1.2.3 shall be used.

11.3.6.6.4 Where special boundary elements are required by Section 11.3.6.6.2 or 11.3.6.6.3, (a) through (d) shall be satisfied and tests shall be performed to verify the strain capacity of the element:

(a) The special boundary element shall extend horizontally from the extreme compression fiber a distance not less than the larger of $(c - 0.1l_w)$ and $c/2$.

(b) In flanged sections, the special boundary element shall include the effective flange width in compression and shall extend at least 12 in. (305 mm) into the web.

(c) Special boundary element transverse reinforcement at the wall base shall extend into the support at least the development length of the largest longitudinal reinforcement in the boundary element unless the special boundary element terminates on a footing or mat, where special boundary element transverse reinforcement shall extend at least 12 in. (305 mm) into the footing or mat.

(d) Horizontal shear reinforcement in the wall web shall be anchored to develop the specified yield strength, f_y, within the confined core of the boundary element.

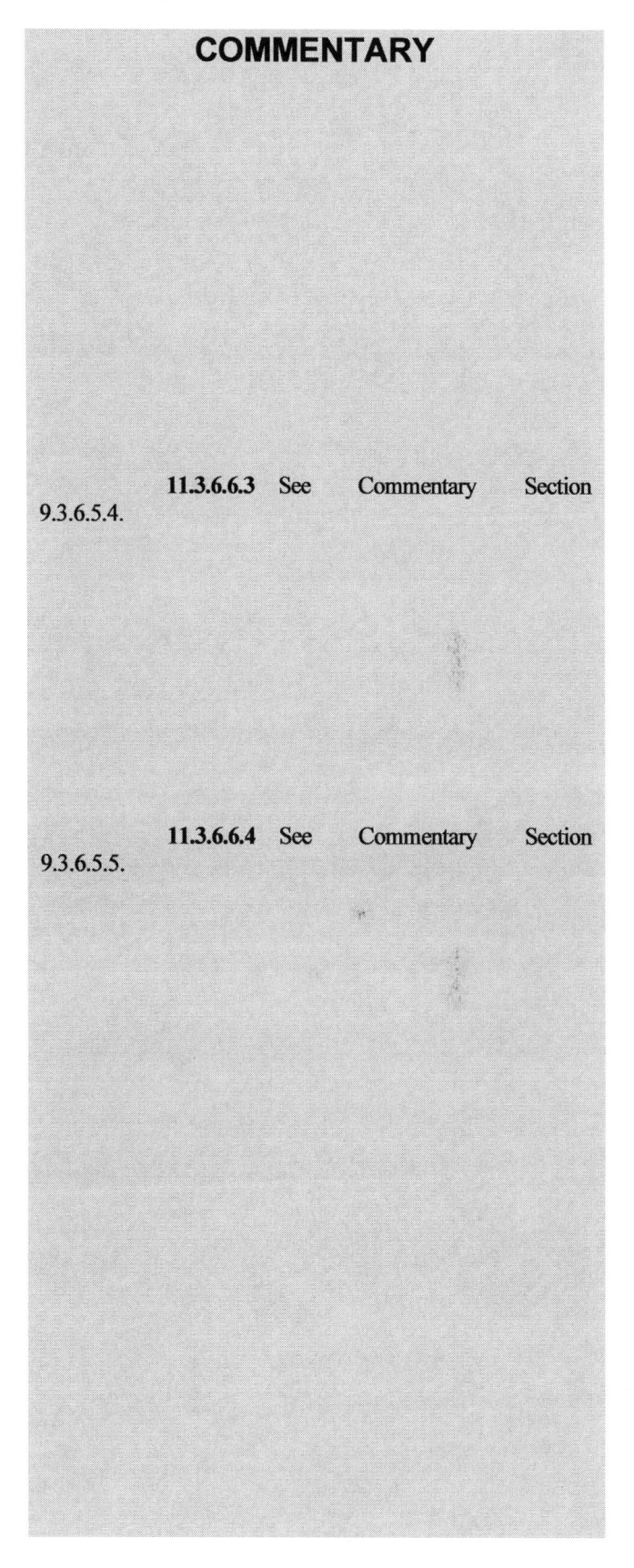

COMMENTARY

11.3.6.6.3 See Commentary Section 9.3.6.5.4.

11.3.6.6.4 See Commentary Section 9.3.6.5.5.

This page intentionally left blank

PART 4: PRESCRIPTIVE DESIGN METHODS

CHAPTER 12
VENEER

CODE

12.1 — General

12.1.1 *Scope*

This chapter provides requirements for design and detailing of anchored masonry veneer and adhered masonry veneer.

COMMENTARY

12.1 — General

12.1.1 *Scope*

Adhered and anchored veneer definitions given in Section 2.2 are variations of those used in model building codes. Modifications have been made to the definitions to clearly state how the veneer is handled in design. Veneer is an element that is not considered to add strength or stiffness to the wall. The design of the veneer backing should be in compliance with the appropriate standard for the material. See Figures CC-12.1-1 and CC-12.1-2 for typical examples of anchored and adhered veneer, respectively.

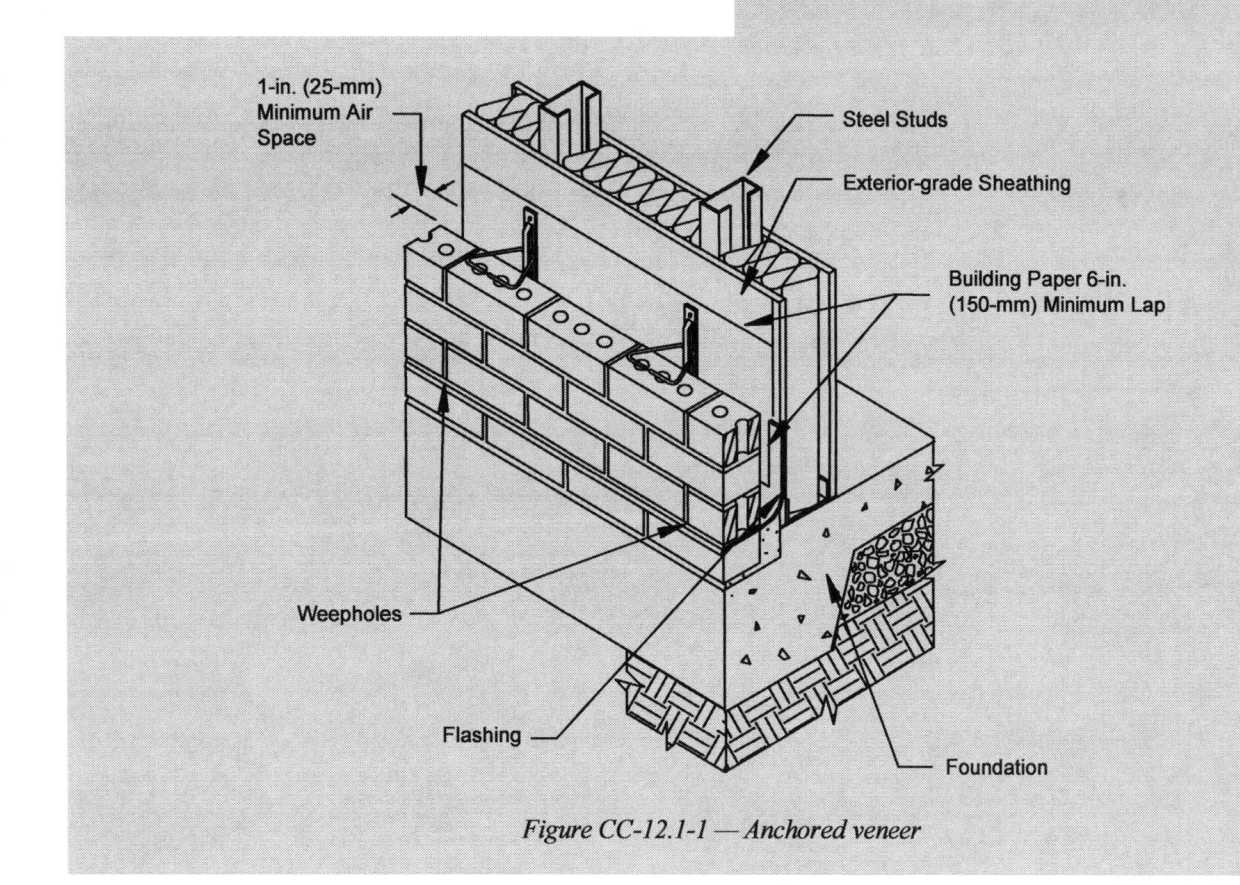

Figure CC-12.1-1 — Anchored veneer

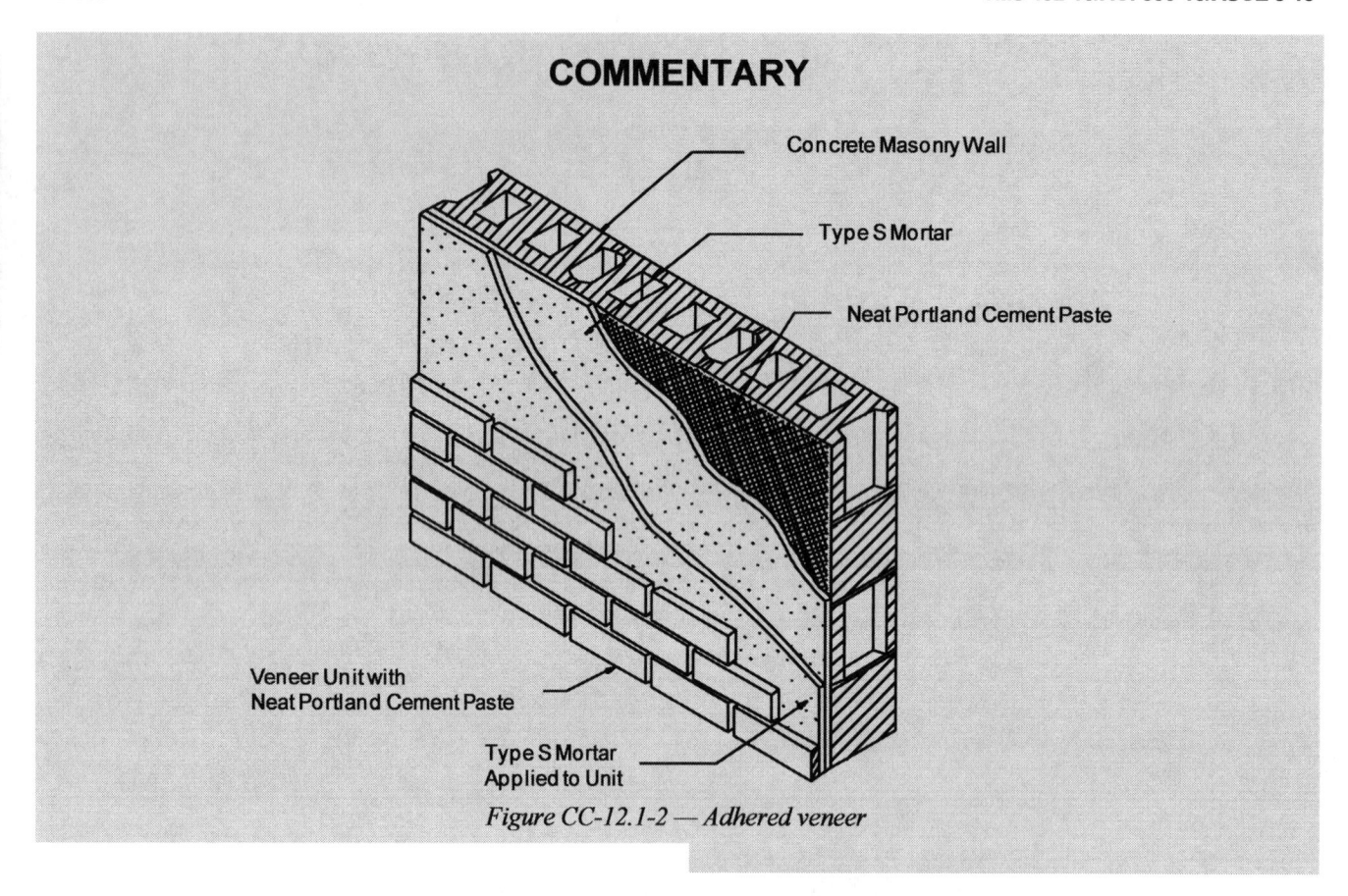

COMMENTARY

Figure CC-12.1-2 — Adhered veneer

CODE

12.1.1.1 The provisions of Part 1, excluding Sections 1.2.1(c) and 1.2.2; Chapter 4, excluding Sections 4.1 and 4.3; and Chapter 6 shall apply to design of anchored and adhered veneer except as specifically stated in this Chapter.

12.1.1.2 Section 4.5 shall not apply to adhered veneer.

12.1.1.3 Articles 1.4 A and B and 3.4 C of TMS 602/ACI 530.1/ASCE 6 shall not apply to any veneer. Articles 3.4 B and F shall not apply to anchored veneer. Articles 3.3 B and 3.4 A, B, E and F shall not apply to adhered veneer.

12.1.2 *Design of anchored veneer*

Anchored veneer shall meet the requirements of Section 12.1.6 and shall be designed rationally by Section 12.2.1 or detailed by the prescriptive requirements of Section 12.2.2.

COMMENTARY

12.1.1.1 Because there is no consideration of stress in the veneer, there is no need to specify the compressive strength of masonry.

12.1.1.3 The Specification was written for construction of masonry subjected to design stresses in accordance with the other chapters of this Code. Masonry veneer, as defined by this Code, is not subject to those design provisions. The Specification articles that are excluded address materials and requirements that are not applicable to veneer construction or are items addressed by specific requirements in this Chapter and are put here to be inclusive.

12.1.2 *Design of anchored veneer*

Implicit within these requirements is the knowledge that the veneer transfers out-of-plane loads through the veneer anchors to the backing. The backing accepts and resists the anchor loads and is designed to resist the out-of-plane loads.

When utilizing anchored masonry veneer, the designer should consider the following conditions and assumptions:

a) The veneer may crack in flexure under service load.

b) Deflection of the backing should be limited to control crack width in the veneer and to provide veneer stability.

c) Connections of the anchor to the veneer and to the backing should be sufficient to transfer applied loads.

d) Differential movement should be considered in the design, detailing, and construction.

e) Water will penetrate the veneer, and the wall system should be designed, detailed, and constructed to prevent water penetration into the building.

f) Requirements for corrosion protection and fire resistance must be included.

If the backing is masonry and the exterior masonry wythe is not considered to add to the strength of the wall in resisting out-of-plane load, the exterior wythe is masonry veneer. However, if the exterior wythe is considered to add to the strength of the wall in resisting out-of-plane load, the wall is properly termed either a multiwythe, non-composite or composite wall rather than a veneer wall.

CODE

COMMENTARY

Manufacturers of steel studs and sheathing materials have published literature on the design of steel stud backing for anchored masonry veneer. Some recommendations have included composite action between the stud and the sheathing and load carrying participation by the veneer. The Metal Lath/Steel Framing Association has promoted a deflection limit of stud span length divided by 360 (Brown and Arumula, 1982). The Brick Industry Association has held that an appropriate deflection limit should be in the range of stud span length divided by 600 to 720. The deflection is calculated assuming that all of the load is resisted by the studs (BIA TN 28B, 2005). Neither set of assumptions will necessarily ensure that the veneer remains uncracked at service load. In fact, the probability of cracking may be high (Grimm and Klingner, 1990). However, post-cracking performance is satisfactory if the wall is properly designed, constructed and maintained with appropriate materials (Kelly et al, 1990). Plane frame computer programs are available for the rational structural design of anchored masonry veneer (Grimm and Klingner, 1990).

A deflection limit of stud span length divided by 200 multiplied by the specified veneer thickness provides a maximum uniform crack width for various heights and various veneer thicknesses. Deflection limits do not reflect the actual distribution of load. They are simply a means of obtaining a minimum backing stiffness. The National Concrete Masonry Association provides a design methodology by which the stiffness properties of the masonry veneer and its backing are proportioned to achieve compatibility (NCMA TEK 16-3A, 1995).

Masonry veneer with wood frame backing has been used successfully on one- and two-family residential construction for many years. Most of these applications are installed without a deflection analysis.

CODE

12.1.3 Design of adhered veneer

Adhered veneer shall meet the requirements of Section 12.1.6, and shall be designed rationally by Section 12.3.1 or detailed by the prescriptive requirements of Section 12.3.2.

12.1.4 Dimension stone

The provisions of Sections 12.1.1, 12.1.3 and 12.3 shall apply to design of adhered dimension stone veneer. Anchored dimension stone veneer is not addressed by this Code. Such a veneer system shall be considered a Special System, and consideration for approval of its use shall be submitted to the Building Official.

12.1.5 Autoclaved aerated concrete masonry veneer

Autoclaved aerated concrete masonry as a veneer wythe is not addressed by this Chapter. Such a veneer system shall be considered a Special System, and consideration for approval of its use shall be submitted to the Building Official.

12.1.6 General design requirements

12.1.6.1 Design and detail the backing system of exterior veneer to resist water penetration. Exterior sheathing shall be covered with a water-resistant membrane, unless the sheathing is water resistant and the joints are sealed.

12.1.6.2 Design and detail flashing and weep holes in exterior veneer wall systems to resist water penetration into the building interior. Weepholes shall be at least $3/16$ in. (4.8 mm) in diameter and spaced less than 33 in. (838 mm) on center.

12.1.6.3 Design and detail the veneer to accommodate differential movement.

COMMENTARY

12.1.3 Design of adhered veneer

Adhered veneer differs from anchored veneer in its means of attachment. The designer should consider conditions and assumptions given in Code Section 12.3.1 when designing adhered veneer.

12.1.4 Dimension stone

Anchored dimension stone veneer should be considered as a Special System of Construction, under Code Section 1.3.

12.1.5 Autoclaved aerated concrete masonry veneer

Veneer anchors described in Chapter 12 are not suitable for use in AAC masonry because of the narrow joints. No testing of such anchors has been performed for AAC masonry. Therefore AAC masonry anchored veneer must be considered a Special System. The method of adhering veneer, as described in Specification Article 3.3 C, has not been evaluated with AAC masonry and shear strength requirements for adhesion of AAC masonry veneer have not been established. Therefore, AAC masonry adhered veneer must be considered a Special System.

12.1.6 General design requirements

Water penetration through the exterior veneer is expected. The wall system must be designed and constructed to prevent water from entering the building.

The requirements given here and the minimum air space dimensions of Sections 12.2.2.6.3, 12.2.2.7.4, and 12.2.2.8.2 are those required for a drainage wall system. Proper drainage requires weep holes and a clear air space. It may be difficult to keep a 1-in. (25-mm) air space free from mortar bridging. Other options are to provide a wider air space, a vented air space, or to use the rain screen principle.

Masonry veneer can be designed with horizontal and vertical bands of different materials. The dissimilar physical properties of the materials should be considered when deciding how to accommodate differential movement.

Industry recommendations are available regarding horizontal bands of clay and concrete masonry, and address such items as joint reinforcement, slip joints, and sealant joints (NCMA TEK 5-2A, 2002; BIA, 2000; BIA TN 18A, 2006). Vertical movement joints can be used to accommodate differential movement between vertical bands of dissimilar materials.

CODE

12.2 — Anchored veneer

12.2.1 *Alternative design of anchored masonry veneer*
The alternative design of anchored veneer, which is permitted under Section 1.3, shall satisfy the following conditions:

(a) Loads shall be distributed through the veneer to the anchors and the backing using principles of mechanics.

(b) Out-of-plane deflection of the backing shall be limited to maintain veneer stability.

(c) The veneer is not subject to the flexural tensile stress provisions of Section 8.2 or the nominal flexural tensile strength provisions of Section 9.1.9.2.

(d) The provisions of Section 12.1, Section 12.2.2.9, and Section 12.2.2.10 shall apply.

12.2.2 *Prescriptive requirements for anchored masonry veneer*

12.2.2.1 Except as provided in Section 12.2.2.11, prescriptive requirements for anchored masonry veneer shall not be used in areas where the velocity pressure, q_z, exceeds 40 psf (1.92 kPa) as given in ASCE 7.

12.2.2.2 Connect anchored veneer to the backing with anchors that comply with Section 12.2.2.5 and Article 2.4 of TMS 602/ACI 530.1/ASCE 6.

12.2.2.3 *Vertical support of anchored masonry veneer*

12.2.2.3.1 The weight of anchored veneer shall be supported vertically on concrete or masonry foundations or other noncombustible structural construction, except as permitted in Sections 12.2.2.3.1.1, 12.2.2.3.1.4, and 12.2.2.3.1.5.

COMMENTARY

12.2 — Anchored veneer

12.2.1 *Alternative design of anchored masonry veneer*
There are no rational design provisions for anchored veneer in any code or standard. The intent of Section 12.2.1 is to permit the designer to use alternative means of supporting and anchoring masonry veneer. See Commentary Section 12.1.1 for conditions and assumptions to consider. The designer may choose to not consider stresses in the veneer or may limit them to a selected value, such as the allowable stresses of Section 8.2, the anticipated cracking stress, or some other limiting condition. The rational analysis used to distribute the loads must be consistent with the assumptions made. See Commentary Section 12.2.2.5 for information on anchors.

The designer should provide support of the veneer; control deflection of the backing; consider anchor loads, stiffness, strength and corrosion; water penetration; and air and vapor transmission.

12.2.2 *Prescriptive requirements for anchored masonry veneer*
The provisions are based on the successful performance of anchored masonry veneer. These have been collected from a variety of sources and reflect current industry practices. Changes result from logical conclusions based on engineering consideration of the backing, anchor, and veneer performance.

12.2.2.1 The wind speed triggers used in the 2008 MSJC were replaced with strength level velocity pressures in the 2011 edition. These velocity pressure triggers were based on the 25 psf (1.20 kPa) working stress velocity pressure that had been used in previous editions of this Code multiplied by 1.6 to convert to strength levels.

12.2.2.3 *Vertical support of anchored masonry veneer* — These requirements are based on current industry practice and current model building codes. Support does not need to occur at the floor level; it can occur at a window head or other convenient location.

12.2.2.3.1 There are no restrictions on the height limit of veneer backed by masonry or concrete, nor are there any requirements that the veneer weight be carried by intermediate supports. The designer should consider the effects of differential movement on the anchors and connection of the veneer to other building components.

CODE

12.2.2.3.1.1 Anchored veneer shall be permitted to be supported vertically by preservative-treated wood foundations. The height of veneer supported by wood foundations shall not exceed 18 ft (5.49 m) above the support.

12.2.2.3.1.2 Anchored veneer with a backing of wood framing shall not exceed 30 ft (9.14 m), or 38 ft (11.58 m) at a gable, in height above the location where the veneer is supported.

12.2.2.3.1.3 If anchored veneer with a backing of cold-formed steel framing exceeds 30 ft (9.14 m), or 38 ft (11.58 m) at a gable, in height above the location where the veneer is supported, the weight of the veneer shall be supported by noncombustible construction at each story above 30 ft (9.14 m) in height.

12.2.2.3.1.4 When anchored veneer is used as an interior finish on wood framing, it shall have a weight of 40 psf (195 kg/m^2) or less and be installed in conformance with the provisions of this Chapter.

12.2.2.3.1.5 Exterior masonry veneer having an installed weight of 40 psf (195 kg/m^2) or less and height of no more than 12 ft (3.7 m) shall be permitted to be supported on wood construction. A vertical movement joint in the masonry veneer shall be used to isolate the veneer supported by wood construction from that supported by the foundation. Masonry shall be designed and constructed so that masonry is not in direct contact with wood. The horizontally spanning element supporting the masonry veneer shall be designed so that deflection due to dead plus live loads does not exceed $l/600$ or 0.3 in. (7.6 mm).

12.2.2.3.2 When anchored veneer is supported by floor construction, the floor shall be designed to limit deflection as required in Section 5.2.1.4.1.

12.2.2.3.3 Provide noncombustible lintels or supports attached to noncombustible framing over openings where the anchored veneer is not self-supporting. Lintels shall have a length of bearing not less than 4 in. (102 mm). The deflection of such lintels or supports shall conform to the requirements of Section 5.2.1.4.1.

12.2.2.4 *Masonry units* — Masonry units shall be at least $2^5/_8$ in. (66.7 mm) in actual thickness.

12.2.2.5 *Anchor requirements*
12.2.2.5.1 *Corrugated sheet-metal anchors*
12.2.2.5.1.1 Corrugated sheet-metal anchors shall be at least $^7/_8$ in. (22.2 mm) wide, have a base metal thickness of at least 0.03 in. (0.8 mm), and shall have corrugations with a wavelength of 0.3 to 0.5 in. (7.6 to 12.7 mm) and an amplitude of 0.06 to 0.10 in. (1.5 to 2.5 mm).

COMMENTARY

12.2.2.3.1.1 The full provisions for preservative-treated wood foundations are given in the National Forest Products Association Technical Report 7 (NFPA TR No. 7, 1987).

12.2.2.3.1.5 Support of anchored veneer on wood is permitted in previous model building codes. The vertical movement joint between the veneer on different supports reduces the possibility of cracking due to differential settlement. The height limit of 12 ft (3.7 m) was considered to be the maximum single story height and is considered to be a reasonable fire safety risk.

12.2.2.5 *Anchor requirements* — It could be argued that the device between the veneer and its backing is not an anchor as defined in the Code. That device is often referred to as a tie. However, the term anchor is used because of the widespread use of anchored veneer in model building codes and industry publications, and the desire to differentiate from tie as used in other chapters.

CODE

12.2.2.5.1.2 Corrugated sheet-metal anchors shall be placed as follows:

(a) With solid units, embed anchors in the mortar joint and extend into the veneer a minimum of $1\frac{1}{2}$ in. (38.1 mm), with at least $\frac{5}{8}$-in. (15.9-mm) mortar cover to the outside face.

(b) With hollow units, embed anchors in mortar or grout and extend into the veneer a minimum of $1\frac{1}{2}$ in. (38.1 mm), with at least $\frac{5}{8}$-in. (15.9-mm) mortar or grout cover to the outside face.

12.2.2.5.2 *Sheet-metal anchors*

12.2.2.5.2.1 Sheet-metal anchors shall be at least $\frac{7}{8}$ in. (22.2 mm) wide, shall have a base metal thickness of at least 0.06 in. (1.5 mm), and shall:

(a) have corrugations as given in Section 12.2.2.5.1.1, or

(b) be bent, notched, or punched to provide equivalent performance in pull-out or push-through.

12.2.2.5.2.2 Sheet-metal anchors shall be placed as follows:

(a) With solid units, embed anchors in the mortar joint and extend into the veneer a minimum of $1\frac{1}{2}$ in. (38.1 mm), with at least $\frac{5}{8}$-in. (15.9-mm) mortar cover to the outside face.

(b) With hollow units, embed anchors in mortar or grout and extend into the veneer a minimum of $1\frac{1}{2}$ in. (38.1 mm), with at least $\frac{5}{8}$-in. (15.9-mm) mortar or grout cover to the outside face.

12.2.2.5.3 *Wire anchors*

12.2.2.5.3.1 Wire anchors shall be at least wire size W1.7 (MW11) and have ends bent to form an extension from the bend at least 2 in. (50.8 mm) long. Wire anchors shall be without drips.

12.2.2.5.3.2 Wire anchors shall be placed as follows:

(a) With solid units, embed anchors in the mortar joint and extend into the veneer a minimum of $1\frac{1}{2}$ in. (38.1 mm), with at least $\frac{5}{8}$-in. (15.9-mm) mortar cover to the outside face.

(b) With hollow units, embed anchors in mortar or grout and extend into the veneer a minimum of $1\frac{1}{2}$ in. (38.1 mm), with at least $\frac{5}{8}$-in. (15.9-mm) mortar or grout cover to the outside face.

12.2.2.5.4 *Joint reinforcement*

12.2.2.5.4.1 Ladder-type or tab-type joint reinforcement is permitted. Cross wires used to anchor masonry veneer shall be at least wire size W1.7 (MW11) and shall be spaced at a maximum of 16 in. (406 mm) on center. Cross wires shall be welded to longitudinal wires, which shall be at least wire size W1.7 (MW11). Cross wires

COMMENTARY

When first introduced in 1995, U.S. industry practice was combined with the requirements of the Canadian Standards Association (CSA, 1984) to produce the requirements given at that time. Each anchor type has physical requirements that must be met. Minimum embedment requirements have been set for each of the anchor types to ensure load resistance against push-through or pull-out of the mortar joint. Maximum air space dimensions are set in Sections 12.2.2.6 through 12.2.2.8.

There are no performance requirements for veneer anchors in previous codes. Indeed, there are none in the industry. Tests on anchors have been reported (Brown and Arumula, 1982; BIA TN 28, 1966). Many anchor manufacturers have strength and stiffness data for their proprietary anchors.

Veneer anchors typically allow for movement in the plane of the wall but resist movement perpendicular to the veneer. The mechanical play in adjustable anchors and the stiffness of the anchor influence load transfer between the veneer and the backing. Stiff anchors with minimal mechanical play provide more uniform transfer of load, increase the stress in the veneer, and reduce veneer deflection.

Veneer anchors of wire with drips are not permitted because of their reduced load capacity. The anchors listed in Section 12.2.2.5.6.1 are thought to have lower strength or stiffness than the more rigid plate-type anchors. Thus fewer plate-type anchors are required. The number of anchors required by this Code is based on the requirements of the 1991 UBC. The number of required anchors is increased in the higher Seismic Design Categories. Anchor spacing is independent of backing type.

Anchor frequency should be calculated independently for the wall surface in each plane. That is, horizontal spacing of veneer anchors should not be continued from one plane of the veneer to another.

In the 1995 edition of the Code, when anchored veneer provisions were first introduced, the use of adjustable single-pintle anchors was not permitted. Based on testing of tie capacity (Klingner and Torrealva, 2005) and Committee consideration, the use of the adjustable single pintle anchor was permitted in the 2011 edition of the Code.

The term "offset" in Code Section 12.2.2.5.5.4 refers to the vertical distance between a wire eye and the horizontal leg of a bent wire tie inserted into that eye, or the vertical distance between functionally similar components of a pintle anchor.

CODE

and tabs shall be without drips.

12.2.2.5.4.2 Embed longitudinal wires of joint reinforcement in the mortar joint with at least $^5/_8$-in. (15.9-mm) mortar cover on each side.

12.2.2.5.5 *Adjustable anchors*

12.2.2.5.5.1 Sheet-metal and wire components of adjustable anchors shall conform to the requirements of Section 12.2.2.5.2 or 12.2.2.5.3. Adjustable anchors with joint reinforcement shall also meet the requirements of Section 12.2.2.5.4.

12.2.2.5.5.2 Maximum clearance between connecting parts of the tie shall be $^1/_{16}$ in. (1.6 mm).

12.2.2.5.5.3 Adjustable anchors shall be detailed to prevent disengagement.

12.2.2.5.5.4 Pintle anchors shall have one or more pintle legs of wire size W2.8 (MW18) and shall have an offset not exceeding $1^1/_4$ in. (31.8 mm).

12.2.2.5.5.5 Adjustable anchors of equivalent strength and stiffness to those specified in Sections 12.2.2.5.5.1 through 12.2.2.5.5.4 are permitted.

12.2.2.5.6 *Anchor spacing*

12.2.2.5.6.1 For adjustable two-piece anchors, anchors of wire size W1.7 (MW11), and 22 gage (0.8 mm) corrugated sheet-metal anchors, provide at least one anchor for each 2.67 ft^2 (0.25 m^2) of wall area.

12.2.2.5.6.2 For other anchors, provide at least one anchor for each 3.5 ft^2 (0.33 m^2) of wall area.

12.2.2.5.6.3 Space anchors at a maximum of 32 in. (813 mm) horizontally and 25 in. (635 mm) vertically, but not to exceed the applicable requirements of Section 12.2.2.5.6.1 or 12.2.2.5.6.2.

12.2.2.5.6.4 Provide additional anchors around openings larger than 16 in. (406 mm) in either dimension. Space anchors around perimeter of opening at a maximum of 3 ft (0.91 m) on center. Place anchors within 12 in. (305 mm) of openings.

12.2.2.5.7 *Joint thickness for anchors* — Mortar bed joint thickness shall be at least twice the thickness of the embedded anchor.

CODE

12.2.2.6 *Masonry veneer anchored to wood backing*

12.2.2.6.1 Veneer shall be attached with any anchor permitted in Section 12.2.2.5.

12.2.2.6.2 Attach each anchor to wood studs or wood framing with a corrosion-resistant 8d common nail, or with a fastener having equivalent or greater pullout strength. For corrugated sheet-metal anchors, locate the nail or fastener within $\frac{1}{2}$ in. (12.7 mm) of the 90-degree bend in the anchor.

12.2.2.6.3 When corrugated sheet metal anchors are used, a maximum distance between the inside face of the veneer and outside face of the solid sheathing of 1 in. (25.4 mm) shall be specified. When other anchors are used, a maximum distance between the inside face of the veneer and the wood stud or wood framing of 4½ in. (114 mm) shall be specified. A 1-in. (25.4-mm) minimum air space shall be specified.

12.2.2.7 *Masonry veneer anchored to steel backing*

12.2.2.7.1 Attach veneer with adjustable anchors.

12.2.2.7.2 Attach each anchor to steel framing with at least a No. 10 corrosion-resistant screw (nominal shank diameter of 0.190 in. (4.8 mm)), or with a fastener having equivalent or greater pullout strength.

12.2.2.7.3 Cold-formed steel framing shall be corrosion resistant and have a minimum base metal thickness of 0.043 in. (1.1 mm).

12.2.2.7.4 A 4½ in. (114-mm) maximum distance between the inside face of the veneer and the steel framing shall be specified. A 1 in. (25.4 mm) minimum air space shall be specified.

12.2.2.8 *Masonry veneer anchored to masonry or concrete backing*

12.2.2.8.1 Attach veneer to masonry backing with wire anchors, adjustable anchors, or joint reinforcement. Attach veneer to concrete backing with adjustable anchors.

12.2.2.8.2 A 4½ in. (114-mm) maximum distance between the inside face of the veneer and the outside face of the masonry or concrete backing shall be specified. A 1 in. (25.4 mm) minimum air space shall be specified.

12.2.2.9 *Veneer not laid in running bond* — Anchored veneer not laid in running bond shall have joint reinforcement of at least one wire, of size W1.7 (MW11), spaced at a maximum of 18 in. (457 mm) on center vertically.

COMMENTARY

12.2.2.6 *Masonry veneer anchored to wood backing* — These requirements are similar to those used by industry and given in model building codes for years. The limitation on fastening corrugated anchors at a maximum distance from the bend is to achieve better performance. The maximum distances between the veneer and the sheathing or wood stud is provided in order to obtain minimum compression capacity of anchors.

12.2.2.7 *Masonry veneer anchored to steel backing* — These requirements generally follow recommendations in current use (BIA TN 28B, 2005; Drysdale and Suter, 1991). The minimum base metal thickness is given to provide sufficient pull-out resistance of screws.

Increasingly energy efficiency building envelopes may require more than 3.5 in. (89 mm) of insulation in the wall cavity. With a code requirement for a minimum specified air gap of 1 in. (25 mm), the system would need to be designed using the alternative procedures.

12.2.2.8 *Masonry veneer anchored to masonry or concrete backing* — These requirements are similar to those used by industry and have been given in model building codes for many years.

Increasingly energy efficiency building envelopes may require more than 3.5 in. (89 mm) of insulation in the wall cavity. With a code requirement for a minimum specified air gap of 1 in. (25 mm), the system would need to be designed using the alternative procedures.

12.2.2.9 *Veneer not laid in running bond* — Masonry not laid in running bond has similar requirements in Section 4.5. The area of joint reinforcement required in Section 12.2.2.9 is equivalent to that in Section 4.5 for a nominal 4-in. (102-mm) wythe.

CODE

12.2.2.10 *Requirements in seismic areas*
12.2.2.10.1 *Seismic Design Category C*
12.2.2.10.1.1 The requirements of this section apply to anchored veneer for buildings in Seismic Design Category C.

12.2.2.10.1.2 Isolate the sides and top of anchored veneer from the structure so that vertical and lateral seismic forces resisted by the structure are not imparted to the veneer.

12.2.2.10.2 *Seismic Design Category D*
12.2.2.10.2.1 The requirements for Seismic Design Category C and the requirements of this section apply to anchored veneer for buildings in Seismic Design Category D.

12.2.2.10.2.2 Reduce the maximum wall area supported by each anchor to 75 percent of that required in Sections 12.2.2.5.6.1 and 12.2.2.5.6.2. Maximum horizontal and vertical spacings are unchanged.

12.2.2.10.2.3 For masonry veneer anchored to wood backing, attach each veneer anchor to wood studs or wood framing with a corrosion-resistant 8d ring-shank nail, a No. 10 corrosion-resistant screw with a minimum nominal shank diameter of 0.190 in. (4.8 mm) or with a fastener having equivalent or greater pullout strength.

12.2.2.10.3 *Seismic Design Categories E and F*
12.2.2.10.3.1 The requirements for Seismic Design Category D and the requirements of this section apply to anchored veneer for buildings in Seismic Design Categories E and F.

12.2.2.10.3.2 Support the weight of anchored veneer for each story independent of other stories.

12.2.2.11 *Requirements in areas of high winds* — The following requirements apply in areas where the velocity pressure, q_z, exceeds 40 psf (1.92 kPa) but does not exceed 55 psf (2.63 kPa) and the building's mean roof height is less than or equal to 60 ft (18.3 m):

(a) Reduce the maximum wall area supported by each anchor to 70 percent of that required in Sections 12.2.2.5.6.1 and 12.2.2.5.6.2.

(b) Space anchors at a maximum 18 in. (457 mm) horizontally and vertically.

(c) Provide additional anchors around openings larger than 16 in. (406 mm) in either direction. Space anchors around perimeter of opening at a maximum of 24 in. (610 mm) on center. Place anchors within 12 in. (305 mm) of openings.

COMMENTARY

12.2.2.10 *Requirements in seismic areas* — These requirements provide several cumulative effects to improve veneer performance under seismic load. Many of them are based on similar requirements given in earlier model building codes (UBC, 1991). The isolation from the structure reduces accidental loading and permits larger building deflections to occur without veneer damage. Support at each floor articulates the veneer and reduces the size of potentially damaged areas. An increased number of anchors increases veneer stability and reduces the possibility of falling debris. Added expansion joints further articulate the veneer, permit greater building deflection without veneer damage and limit stress development in the veneer.

Shake table tests of panel (Klingner et al, 2010(a)) and full-scale wood frame/brick veneer buildings (Reneckis and LaFave, 2009) have demonstrated that 8d common nails are not sufficient to resist seismic loading under certain conditions. 8d ring-shank nails or #10 screws were recommended by the researchers for use in areas of significant seismic loading.

12.2.2.10.3 *Seismic Design Categories E and F* — The 1995 through 2011 editions of the MSJC Code required that masonry veneer in Seismic Design Categories E and F be provided with joint reinforcement, mechanically attached to anchors with clips or hooks. Shaking-table research (Klingner et al, 2010(b)) has shown that the requirement is not necessary or useful so the requirement was removed in the 2013 edition of the MSJC Code.

12.2.2.11 *Requirements in areas of high winds* — The provisions in this section are based on a reduction in tributary area by 30%. The velocity pressure trigger was therefore raised by 1/0.7, and rounded to 55 psf (2.63 kPa).

CODE

12.3 — Adhered veneer

12.3.1 *Alternative design of adhered masonry veneer*
The alternative design of adhered veneer, which is permitted under Section 1.3, shall satisfy the following conditions:

(a) Loads shall be distributed through the veneer to the backing using principles of mechanics.

(b) Out-of-plane curvature shall be limited to prevent veneer unit separation from the backing.

(c) The veneer is not subject to the flexural tensile stress provisions of Section 8.2 or the nominal flexural tensile strength provisions of Section 9.1.9.2.

(d) The provisions of Section 12.1 shall apply.

12.3.2 *Prescriptive requirements for adhered masonry veneer*

12.3.2.1 *Unit sizes* — Adhered veneer units shall not exceed $2^5/_8$ in. (66.7 mm) in specified thickness, 36 in. (914 mm) in any face dimension, nor more than 5 ft^2 (0.46 m^2) in total face area, and shall not weigh more than 15 psf (73 kg/m^2).

12.3.2.2 *Wall area limitations* — The height, length, and area of adhered veneer shall not be limited except as required to control restrained differential movement stresses between veneer and backing.

12.3.2.3 *Backing* — Backing shall provide a continuous, moisture-resistant surface to receive the adhered veneer. Backing is permitted to be masonry, concrete, or metal lath and portland cement plaster applied to masonry, concrete, steel framing, or wood framing.

COMMENTARY

12.3 — Adhered veneer

12.3.1 *Alternative design of adhered masonry veneer*
There are no rational design provisions for adhered veneer in any code or standard. The intent of Section 12.3.1 is to permit the designer to use alternative unit thicknesses and areas for adhered veneer. The designer should provide for adhesion of the units, control curvature of the backing, and consider freeze-thaw cycling, water penetration, and air and vapor transmission. The Tile Council of America limits the deflection of the backing supporting ceramic tiles to span length divided by 360 (TCA, 1996).

12.3.2 *Prescriptive requirements for adhered masonry veneer*
Similar requirements for adhered veneer first appeared in the 1967 Uniform Building Code. The construction requirements for adhered veneer in the Specification have a history of successful performance (Dickey, 1982).

12.3.2.1 *Unit sizes* — The dimension, area, and weight limits are imposed to reduce the difficulties of handling and installing large units and to assure good bond.

12.3.2.2 *Wall area limitations* — Selecting proper location for movement joints involves many variables. These include: changes in moisture content, inherent movement of materials, temperature exposure, temperature differentials, strength of units, and stiffness of the backing.

12.3.2.3 *Backing* — These surfaces have demonstrated the ability to provide the necessary adhesion when using the construction method described in the Specification. Model building codes contain provisions for metal lath and portland cement plaster. For masonry or concrete backing, it may be desirable to apply metal lath and plaster. Also, refer to ACI 524R, "Guide to Portland Cement Plastering" (1993) for metal lath, accessories, and their installation. These publications also contain recommendations for control of cracking.

CODE

12.3.2.4 Adhesion developed between adhered veneer units and backing shall have a shear strength of at least 50 psi (345 kPa) based on gross unit surface area when tested in accordance with ASTM C482, or shall be adhered in compliance with Article 3.3 C of TMS 602/ACI 530.1/ASCE 6.

COMMENTARY

12.3.2.4 The required shear strength of 50 psi (345 kPa) is an empirical value based on judgment derived from historical use of adhered veneer systems similar to those permitted by Article 3.3 C of TMS 602/ACI 530.1/ASCE 6. This value is easily obtained with workmanship complying with the Specification. It is anticipated that the 50 psi (345 kPa) will account for differential shear stress between the veneer and its backing in adhered veneer systems permitted by this Code and Specification.

The test method is used to verify shear strength of adhered veneer systems that do not comply with the construction requirements of the Specification or as a quality assurance test for systems that do comply.

This page intentionally left blank

CHAPTER 13
GLASS UNIT MASONRY

CODE

13.1 — General

13.1.1 *Scope*
This chapter provides requirements for empirical design of glass unit masonry as non-load-bearing elements in exterior or interior walls.

13.1.1.1 The provisions of Part 1 and Part 2, excluding Sections 1.2.1(c), 1.2.2, 4.1, 4.2, and 4.3, shall apply to design of glass unit masonry, except as stated in this Chapter.

13.1.1.2 Article 1.4 of TMS 602/ACI 530.1/ASCE 6 shall not apply to glass unit masonry.

13.1.2 *General design requirements*
Design and detail glass unit masonry to accommodate differential movement.

13.1.3 *Units*
13.1.3.1 Hollow or solid glass block units shall be standard or thin units.

13.1.3.2 The specified thickness of standard units shall be at least $3^{7}/_{8}$ in. (98.4 mm).

13.1.3.3 The specified thickness of thin units shall be $3^{1}/_{8}$ in. (79.4 mm) for hollow units or 3 in. (76.2 mm) for solid units.

13.2 — Panel size

COMMENTARY

13.1 — General

13.1.1 *Scope*
Glass unit masonry is used as a non-load-bearing element in interior and exterior walls, partitions, window openings, and as an architectural feature. Design provisions in the Code are empirical. These provisions are cited in previous codes, are based on successful performance, and are recommended by manufacturers.

13.1.1.1 Because there is no consideration of stress in glass unit masonry, there is no need to specify the compressive strength of masonry.

13.2 — Panel size

The Code limitations on panel size are based on structural and performance considerations. Height limits are more restrictive than length limits based on historical requirements rather than actual field experience or engineering principles. Fire resistance rating tests of assemblies may also establish limitations on panel size. Glass block manufacturers can be contacted for technical data on the fire resistance ratings of panels and local building code should be consulted for required fire resistance ratings for glass unit masonry panels. In addition, fire resistance ratings for glass unit masonry panels may be listed with Underwriters Laboratories, Inc. at www.ul.com.

CODE

13.2.1 *Exterior standard-unit panels*

The maximum area of each individual standard-unit panel shall be based on the design wind pressure, in accordance with Figure 13.2-1. The maximum dimension between structural supports shall be 25 ft (7.62 m) horizontally or 20 ft (6.10 m) vertically.

COMMENTARY

13.2.1 *Exterior standard-unit panels*

The wind load resistance curve (Pittsburgh Corning, 1992; Glashaus, 1992, NCMA, 1992) (Figure CC-13.2-1) is representative of the ultimate load limits for a variety of panel conditions. Historically, a 144-ft^2 (13.37-m^2) area limit has been referenced in building codes as the maximum area permitted in exterior applications, without reference to any safety factor or design wind pressure. The 144-ft^2 (13.37-m^2) area also reflects the size of panels tested by the National Concrete Masonry Association (NCMA, 1992). The 144-ft^2 (13.37-m^2) area limitation provides a safety factor of 2.7 when the design wind pressure is 20 psf (958 Pa) (Smolenski, 1992).

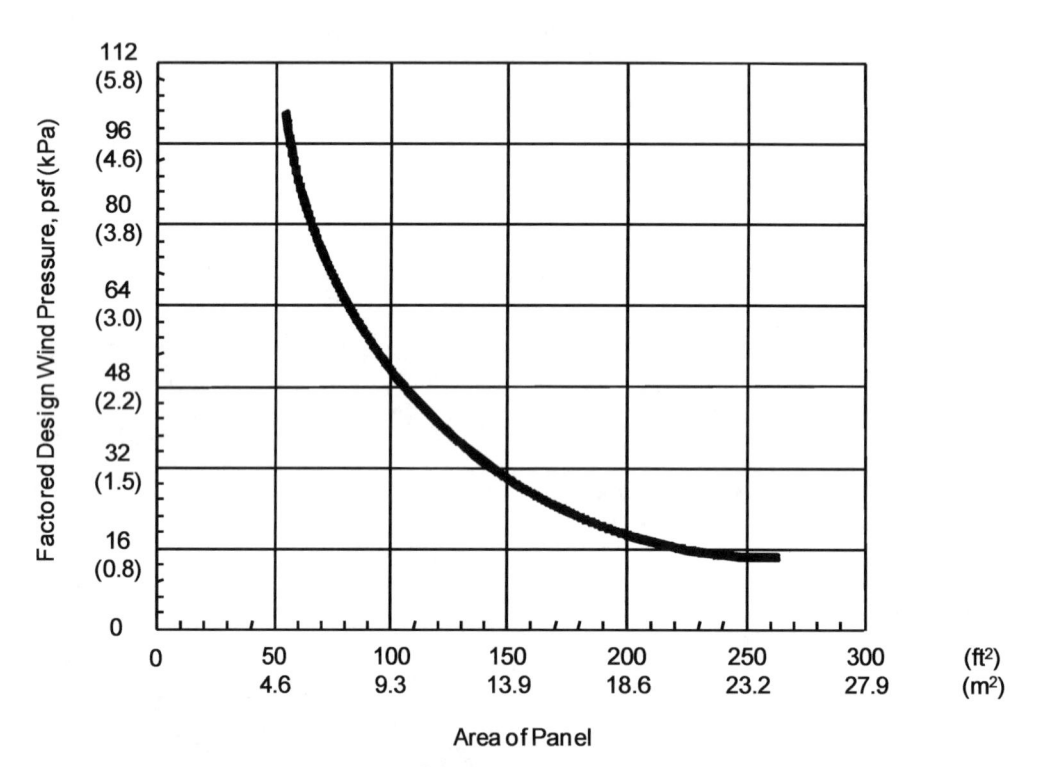

Figure 13.2-1 — Factored design wind pressure for glass unit masonry

COMMENTARY

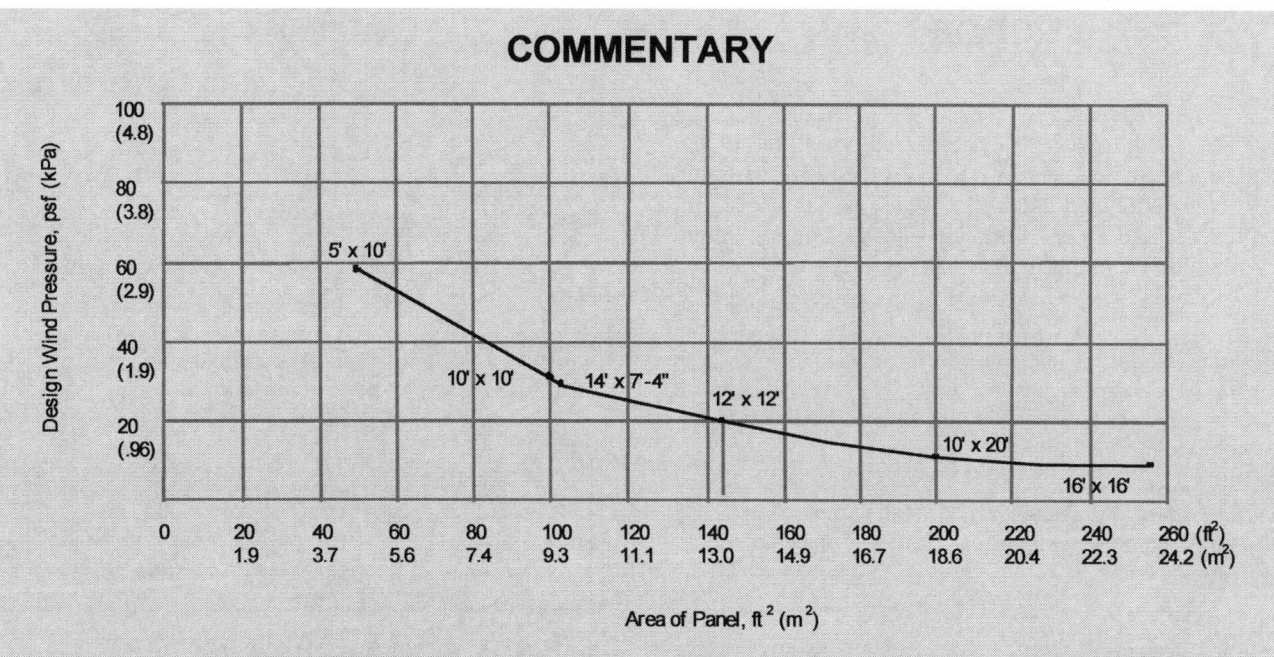

Note: The above historical glass masonry design chart reflects different sizes of glass unit masonry panels tested to ultimate pressures, resulting in a design curve that reflects a 2.7 safety factor of ultimate loading. As an example, the recommended design pressure of 20 psf (958 Pa) for a 144 ft^2 (13.37-m^2) panel, reflects a safety factor of 2.7 of its ultimate strength of 54 psf (2,586 Pa).

Figure CC-13.2-1 — Historical glass masonry design chart

CODE

13.2.2 *Exterior thin-unit panels*
The maximum area of each individual thin-unit panel shall be 100 ft^2 (9.29 m^2). The maximum dimension between structural supports shall be 15 ft (4.57 m) wide or 10 ft (3.05 m) high. Thin units shall not be used in applications where the factored design wind pressure per ASCE 7 exceeds 32 psf (1,532 Pa).

13.2.3 *Interior panels*
13.2.3.1 When the factored wind pressure does not exceed 16 psf (768 Pa), the maximum area of each individual standard-unit panel shall be 250 ft^2 (23.22 m^2) and the maximum area of each thin-unit panel shall be 150 ft^2 (13.94 m^2). The maximum dimension between structural supports shall be 25 ft (7.62 m) wide or 20 ft (6.10 m) high.

13.2.3.2 When the factored wind pressure exceeds 16 psf (768 Pa), standard-unit panels shall be designed in accordance with Section 13.2.1 and thin-unit panels shall be designed in accordance with Section 13.2.2.

COMMENTARY

13.2.2 *Exterior thin-unit panels*
There is limited historical data for developing a curve for thin units. The Committee recommends limiting the exterior use of thin units to areas where the factored design wind pressure does not exceed 32 psf (1,532 Pa).

CODE

13.2.4 *Curved panels*

The width of curved panels shall conform to the requirements of Sections 13.2.1, 13.2.2, and 13.2.3, except additional structural supports shall be provided at locations where a curved section joins a straight section and at inflection points in multi-curved walls.

13.3 — Support

13.3.1 *General requirements*

Glass unit masonry panels shall be isolated so that in-plane loads are not imparted to the panel.

13.3.2 *Vertical*

13.3.2.1 Maximum total deflection of structural members supporting glass unit masonry shall not exceed $l/600$.

13.3.2.2 Glass unit masonry having an installed weight of 40 psf (195 kg/m^2) or less and a maximum height of 12 ft (3.7 m) shall be permitted to be supported on wood construction.

13.3.2.3 A vertical expansion joint in the glass unit masonry shall be used to isolate the glass unit masonry supported by wood construction from that supported by other types of construction.

13.3.3 *Lateral*

13.3.3.1 Glass unit masonry panels, more than one unit wide or one unit high, shall be laterally supported along the top and sides of the panel. Lateral support shall be provided by panel anchors along the top and sides spaced not more than 16 in. (406 mm) on center or by channel-type restraints. Glass unit masonry panels shall be recessed at least 1 in. (25.4 mm) within channels and chases. Channel-type restraints must be oversized to accommodate expansion material in the opening, and packing and sealant between the framing restraints and the glass unit masonry perimeter units. Lateral supports for glass unit masonry panels shall be designed to resist applied loads, or a minimum of 200 lb per linear ft (2919 N/m) of panel, whichever is greater.

13.3.3.2 Glass unit masonry panels that are no more than one unit wide shall conform to the requirements of Section 13.3.3.1, except that lateral support at the top of the panel is not required.

13.3.3.3 Glass unit masonry panels that are no more than one unit high shall conform to the requirements of Section 13.3.3.1, except that lateral support at the sides of the panels is not required.

13.3.3.4 Glass unit masonry panels that are a single glass masonry unit shall conform to the requirements of Section 13.3.3.1, except that lateral support shall not be provided by panel anchors.

COMMENTARY

13.3 — Support

13.3.1 *General requirements*

13.3.2 *Vertical*

Support of glass unit masonry on wood has historically been permitted in model building codes. The Code requirements for expansion joints and for asphalt emulsion at the sill isolate the glass unit masonry within the wood framing. These requirements also reduce the possibility of contact of the glass units and mortar with the wood framing. The height limit of 12 ft. (3.7 m) was considered to be the maximum single story height.

13.3.3 *Lateral*

The Code requires glass unit masonry panels to be laterally supported by panel anchors or channel-type restraints. See Figures CC-13.3-1 and CC-13.3-2 for panel anchor construction and channel-type restraint construction, respectively. Glass unit masonry panels may be laterally supported by either construction type or by a combination of construction types. The channel-type restraint construction can be made of any channel-shaped concrete, masonry, metal, or wood elements so long as they provide the required lateral support.

COMMENTARY

Figure CC-13.3-1 — Panel anchor construction

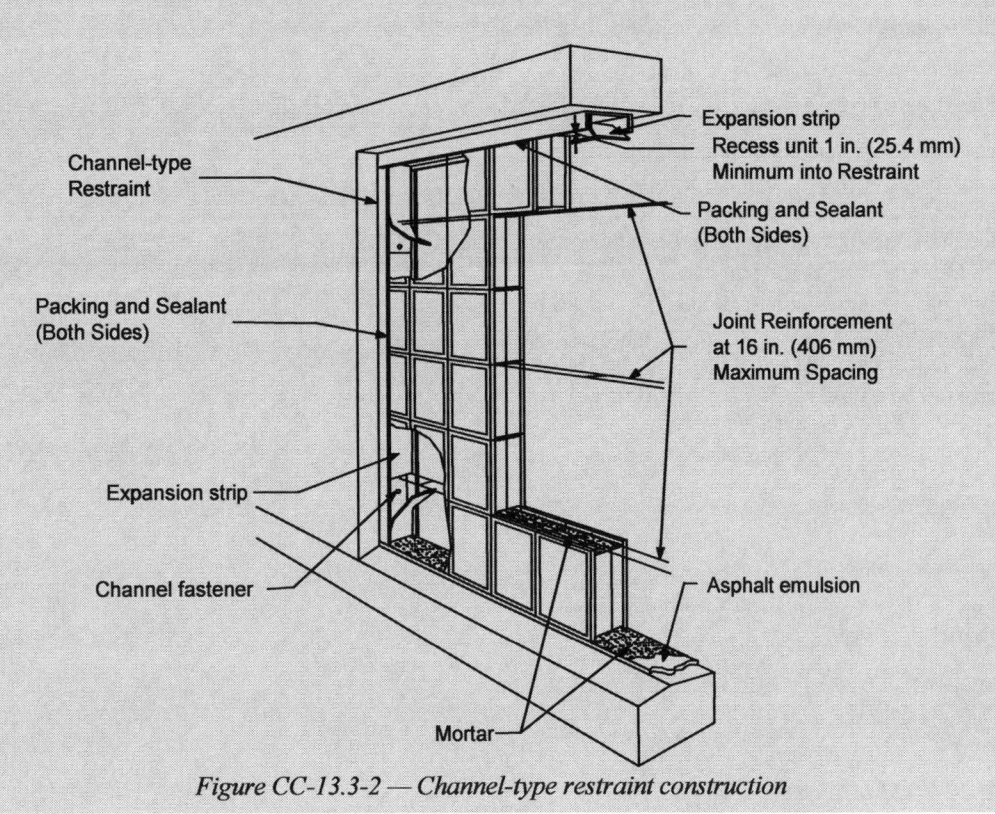

Figure CC-13.3-2 — Channel-type restraint construction

CODE

13.4 — Expansion joints

Glass unit masonry panels shall be provided with expansion joints along the top and sides at structural supports. Expansion joints shall have sufficient thickness to accommodate displacements of the supporting structure, but shall not be less than $3/8$ in. (9.5 mm) in thickness. Expansion joints shall be entirely free of mortar or other debris and shall be filled with resilient material.

13.5 — Base surface treatment

The surface on which glass unit masonry panels are placed shall be coated with a water-based asphaltic emulsion or other elastic waterproofing material prior to laying the first course.

13.6 — Mortar

Glass unit masonry shall be laid with Type S or N mortar.

13.7 — Reinforcement

Glass unit masonry panels shall have horizontal joint reinforcement spaced not more than 16 in. (406 mm) on center, located in the mortar bed joint, and extending the entire length of the panel but not across expansion joints. Longitudinal wires shall be lapped a minimum of 6 in. (152 mm) at splices. Joint reinforcement shall be placed in the bed joint immediately below and above openings in the panel. The reinforcement shall have at least two parallel longitudinal wires of size W1.7 (MW11) and have welded cross wires of size W1.7 (MW11).

COMMENTARY

13.5 — Base surface treatment

Current industry practice and recommendations by glass block manufacturers state that surfaces on which glass unit masonry is placed be coated with an asphalt emulsion Pittsburgh Corning, 1992; Glashau, 1992). The asphalt emulsion provides a slip plane at the panel base. This is in addition to the expansion provisions at head and jamb locations. The asphalt emulsion also waterproofs porous panel bases.

Glass unit masonry panels subjected to structural investigation tests by the National Concrete Masonry Association (1992) to confirm the validity and use of the Glass Unit Masonry Design Wind Load Resistance chart (Figure CC-13.2-1), were constructed on bases coated with asphalt emulsion. Asphalt emulsion on glass unit masonry panel bases is needed to be consistent with these tests.

CHAPTER 14
MASONRY PARTITION WALLS

CODE

14.1 — General

14.1.1 *Scope*
This chapter provides requirements for the design of masonry partition walls.

14.1.2 *Design of partition walls*
Partition walls shall be designed by one of the following:

(a) the requirements of Part 1, Part 2 and the requirements of Chapter 8, Chapter 9, Chapter 10, Chapter 11, or Chapter 13; or

(b) the prescriptive design requirements of Section 14.2 through 14.5.

14.2 — Prescriptive design of partition walls

14.2.1 *General*

14.2.1.1 The provisions of Part 1 and Part 2, excluding Sections 1.2.1(c), 1.2.2, 4.1, 4.2, and 4.3, shall apply to prescriptive design of masonry partition walls

14.2.1.2 Article 1.4 of TMS 602/ACI 530.1/ASCE 6 shall not apply to prescriptively designed masonry partition walls.

14.2.2 *Thickness Limitations*
14.2.2.1 *Minimum thickness* — The minimum nominal thickness of partition walls shall be 4 in. (102 mm).
14.2.2.2 *Maximum thickness* — The maximum nominal thickness of partition walls shall be 12 in. (305 mm).

COMMENTARY

14.2 — Prescriptive design of partition walls

The prescriptive design requirements of this Chapter were originally based on empirical rules and formulas for the design of masonry structures that were developed by experience. Design is based on the condition that vertical loads are reasonably centered on the walls and lateral loads are limited. Walls have minimum and maximum thicknesses and additional limitations as noted in Section 14.2.3. The masonry is laid in running bond for horizontally spanning walls. Specific limitations on building height, seismic, wind, and horizontal loads exist. Buildings are of limited height. Members not participating in the lateral-force-resisting system of a building may be designed by the prescriptive provisions of this Chapter even though the lateral-force-resisting system is designed under another Chapter.

14.2.2 *Thickness Limitations* — The minimum and maximum thicknesses set practical limits on walls to be designed with this simplified prescriptive method. The permitted l/t or h/t values in Table 14.3.1(5) and Table 14.3.1(10) are based on analyses of partition walls ranging from 4 in. (102 mm) to 12 in. (305 mm) in nominal thickness.

CODE

14.2.3 Limitations

14.2.3.1 *Vertical loads* — The prescriptive design requirements of Chapter 14 shall not apply to the design of partition walls that support vertical compressive, service loads of more than 200 lb/linear ft (2919 N/m) in addition to their own weight. The resultant of vertical loads shall be placed within the center third of the wall thickness. The prescriptive design requirements of Chapter 14 shall not apply to the design of partition walls that resist net axial tension.

14.2.3.2 *Lateral loads* — The prescriptive design requirements of Chapter 14 shall not apply to partition walls resisting service level unfactored lateral loads that exceed 5 psf (0.239 kPa) when using Table 14.3.1(5) or 10 psf (0.479 kPa) when using Table 14.3.1(10).

14.2.3.3 *Seismic Design Category* — The prescriptive design requirements of Chapter 14 shall not apply to the design of masonry partition walls in Seismic Design Categories D, E, or F.

COMMENTARY

14.2.3 Limitations

14.2.3.1 *Vertical loads* — This provision allows miscellaneous light loads, such as pictures, emergency lighting, etc., to be applied to interior partition walls, while limiting the load to less than what the Code defines as a load-bearing wall, which is a wall supporting vertical loads greater than 200 lb/linear ft (2919 N/m) in addition to its own weight. The allowable stress analyses performed to establish the permitted span to thickness ratios included a 200 lb/linear ft (2919 N/m) compressive, unfactored load applied at the top of the wall with an eccentricity of $t/6$.

Net axial tension is not permitted in partition walls designed in accordance with this chapter.

14.2.3.2 *Lateral loads* — Out-of-plane loads on the partition walls must not exceed 5 psf (0.239 kPa) service load levels to use Table 14.3.1(5) and must not exceed 10 psf (0.479 kPa) service load levels to use Table 14.3.1(10).

Section 1607.14 "Interior walls and partitions" in the Live Loads section of the 2012 International Building Code (IBC) states, "Interior walls and partitions that exceed 6 feet (1829 mm) in height, including their finish materials, shall have adequate strength to resist the loads to which they are subjected but not less than a horizontal load of 5 psf (0.240 kN/m2)."

Two tables are provided for the use of the designer, one for the code prescribed minimum lateral load of 5 psf (0.239 kPa) as noted above and, one that may be used at the designer's discretion for conditions where the 5 psf (0.239 kPa) minimum is exceeded (but no more than 10 psf (0.479 kPa) maximum) – one example: seismic loading includes out-of-plane lateral loading as a factor of self-weight even on nonparticipating elements such as partition walls and those lateral loads may exceed 5 psf (0.239kPa) in certain conditions especially in Seismic Design Category C.

CODE

14.2.3.4 *Nonparticipating Elements* — Partition walls designed using the prescriptive requirements of Chapter 14 shall be designed as 'nonparticipating elements' in accordance with the requirements of Section 7.3.1.

14.2.3.5 *Enclosed Buildings* — The prescriptive design requirements of Chapter 14 shall only be permitted to be applied to the design of masonry partition walls in Enclosed Buildings as defined by ASCE 7.

14.2.3.6 *Risk Category IV* — The prescriptive design requirements of Chapter 14 shall not apply to the design of masonry partition walls in Risk Category IV as defined in ASCE 7.

14.2.3.7 *Masonry not laid in running bond* — The prescriptive design requirements of Chapter 14 shall not apply to the design of masonry not laid in running bond in horizontally spanning walls.

14.2.3.8 *Glass unit masonry* — The prescriptive design requirements of Chapter 14 shall not apply to the design of glass unit masonry.

14.2.3.9 *AAC masonry* — The prescriptive design requirements of Chapter 14 shall not apply to the design of AAC masonry.

14.2.3.10 *Concrete masonry* — Concrete masonry, designed in accordance with Chapter 14, shall comply with one of the following:

(a) The minimum normalized web area of concrete masonry units, determined in accordance with ASTM C140, shall not be less than 27 in.^{2}/ft^2 (187,500 mm^2/m^2), or

(b) the member shall be grouted solid.

14.2.3.11 *Support* — The provisions of Chapter 14 shall not apply to masonry vertically supported on wood construction.

COMMENTARY

14.2.3.5 *Enclosed Buildings* — Partition walls, as defined by this Code, are interior walls without structural function. Therefore, the requirement that the provisions of this Chapter be limited to Enclosed Buildings as defined by ASCE 7 is appropriate.

14.2.3.7 *Masonry not laid in running bond* — The analyses performed in establishing the permitted span to thickness ratios for the prescriptive design of partition walls were based on the allowable flexural tensile stresses for clay masonry and concrete masonry. This Code does not permit flexural tensile stress parallel to bed joints in unreinforced masonry not laid in running bond unless the masonry has a continuous grout section parallel to the span. Therefore, the prescriptive requirements of Chapter 14 limit the use of masonry that is not laid in running bond to vertically spanning walls that are solidly grouted.

14.2.3.10 *Concrete masonry* — Concrete masonry units are required to have a normalized web area of 27 in.^{2}/ft^2 (187,500 mm^2/m^2) to allow designers to avoid checking shear stress by providing sufficient web area such that web shear stresses do not control a design. This approach is consistent with the goal of keeping the provisions of the Chapter 14 more prescriptive and simplified. If the normalized web area is less than 27 in.^{2}/ft^2 (187,500 mm^2/m^2), solid grouting is required to provide additional shear area.

CODE

14.3 — Lateral support

14.3.1 *Maximum l/t and h/t*
Masonry partition walls without openings shall be laterally supported in either the horizontal or the vertical direction so that *l/t* or *h/t* does not exceed the values given in Table 14.3.1(5) or Table 14.3.1(10). It shall not be permitted to decrease the cross-section of the partition wall between supports unless permitted by Section 14.3.2.

14.3.2 *Openings*
Masonry partition walls with single or multiple openings shall be laterally supported in either the horizontal or vertical direction so that *l/t* or *h/t* does not exceed the values given in Table 14.3.1(5) or Table 14.3.1(10) divided by $\sqrt{W_T/W_S}$.

W_S is the dimension of the structural wall strip measured perpendicular to the span of the wall strip and perpendicular to the thickness as shown in Figure 14.3.1-1. W_S is measured from the edge of the opening. W_S shall be no less than $3t$ on each side of each opening. Therefore, at walls with multiple openings, jambs shall be no less than $6t$ between openings. For design purposes, the effective W_S shall not be assumed to be greater than $6t$. At non-masonry lintels, the edge of the opening shall be considered the edge of the non-masonry lintel. W_S shall occur uninterrupted over the full span of the wall.

W_T is the dimension, parallel to W_S, from the center of the opening to the opposite end of W_S as shown in Figure 14.3.1-1. Where there are multiple openings perpendicular to W_S, W_T shall be measured from the center of a virtual opening that encompasses such openings. Masonry elements within the virtual opening must be designed in accordance with Chapter 8 or 9.

For walls with openings that span no more than 4 feet, parallel to W_S, if W_S is no less than 4 feet, then it shall be permitted to ignore the effect of those openings.

The span of openings, parallel to W_S, shall be limited so that the span divided by t does not exceed the values given in Table 14.3.1(5) or Table 14.3.1(10).

COMMENTARY

14.3 — Lateral support

14.3.1 *Maximum l/t and h/t*
Lateral support requirements are included to limit the flexural tensile stress due to out-of-plane loads. The requirements provide relative out-of-plane resistance that limit the maximum width of opening and provide sufficient masonry sections between the openings.

The permitted span to thickness ratios for prescriptively designed partition walls were established based on Allowable Stress Design and service level (unfactored) loads of no more than 200 lb/ft (2919 N/m) (vertical) and 5 psf (0.239 kPa) (lateral out-of-plane) or 10 psf (0.479 kPa) (lateral out-of-plane), and a conservative wall self-weight. Critical sections were assumed to be at mid-span and the walls were conservatively assumed to be pinned at both supports.

Table 14.3.1(5) and Table 14.3.1(10) provide maximum *l/t* and *h/t* ratios that are a function of mortar type, mortar cementitious materials, unit solidity, and extent of grouting. Second order effects of axial forces combined with progressively larger deflections were not calculated explicitly. However, the combined effects of axial and flexure loads were analyzed using Allowable Stress Design provisions. Secondary bending effects resulting from the axial loads are ignored since axial forces are limited.

Decreases in partition wall cross-section must be accounted for by treating any such decrease as an opening. As one example, a vertical movement joint would need to be accounted for in a horizontal spanning wall.

Table 14.3.1(5) — Maximum l/t^1 or h/t^1 for 5 psf (0.239 kPa) lateral load[2]

Unit and Masonry Type	Mortar types			
	Portland cement/lime or mortar cement		Masonry cement or air entrained portland cement/lime	
	M or S	N	M or S	N
Ungrouted and partially grouted hollow units[3]	26	24	22	18
Solid units and fully grouted hollow units[3]	40	36	33	26

[1] t by definition is the nominal thickness of member

[2] See Section 14.2.3.2.

[3] For non-cantilevered walls laterally supported at both ends. See Section 14.3.3 for cantilevered walls.

Table 14.3.1(10) — Maximum l/t^1 or h/t^1 for 10 psf (0.479 kPa) lateral load[2]

Unit and Masonry Type	Mortar types			
	Portland cement/lime or mortar cement		Masonry cement or air entrained portland cement/lime	
	M or S	N	M or S	N
Ungrouted and partially grouted hollow units[3]	18	16	14	12
Solid units and fully grouted hollow units[3]	28	24	22	18

[1] t by definition is the nominal thickness of member

[2] See Section 14.2.3.2.

[3] For non-cantilevered walls laterally supported at both ends. See Section 14.3.3 for cantilevered walls.

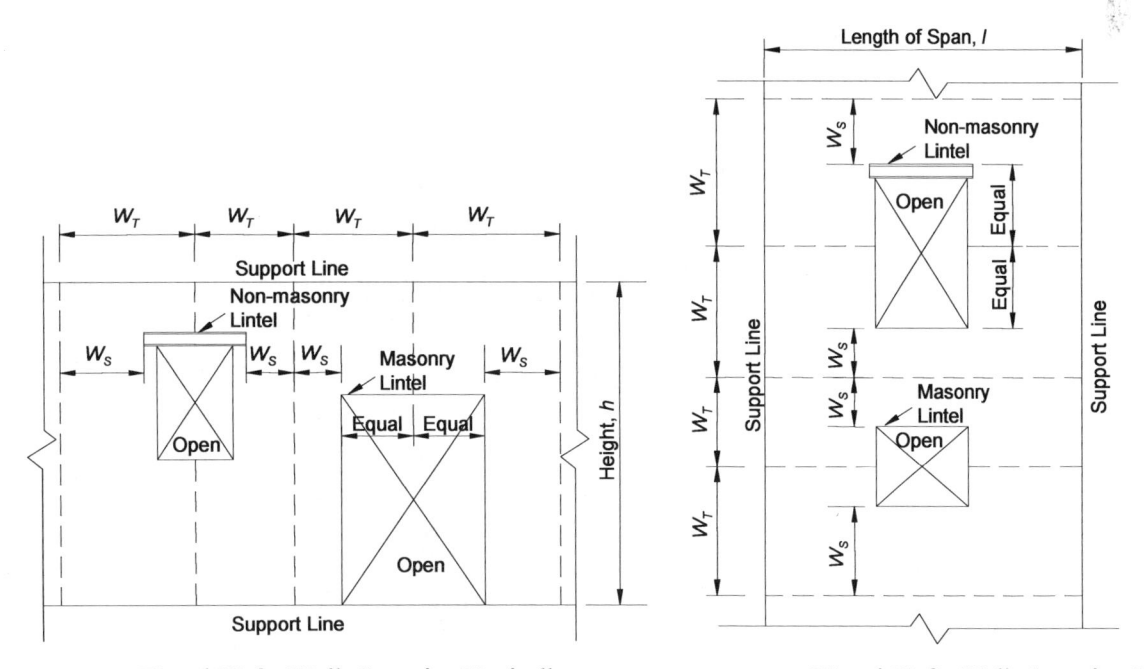

W_S and W_T for Walls Spanning Vertically

W_S and W_T for Walls Spanning Horizontally

Figure 14.3.1-1 — Graphical representation of W_S and W_T

CODE

14.3.3 *Cantilever walls*
The ratio of height-to-nominal-thickness for cantilevered partition walls shall not exceed 6 for solid masonry or 4 for hollow masonry.

14.3.4 *Support elements*
Lateral support shall be provided by cross walls, pilasters, or structural frame members when the limiting distance is taken horizontally; or by floors, roofs acting as diaphragms, or structural frame members when the limiting distance is taken vertically.

14.4 — Anchorage

14.4.1 *General*
Masonry partition walls shall be anchored in accordance with this section.

14.4.2 *Intersecting walls*
Masonry partition walls depending upon one another for lateral support shall be anchored or bonded at locations where they meet or intersect by one of the following methods:

14.4.2.1 Fifty percent of the units at the intersection shall be laid in an overlapping masonry bonding pattern, with alternate units having a bearing of at least 3 in. (76.2 mm) on the unit below.

14.4.2.2 Walls shall be anchored at their intersection at vertical intervals of not more than 16 in. (406 mm) with joint reinforcement or $1/4$ in. (6.4 mm) mesh galvanized hardware cloth.

14.4.2.3 Other metal ties, joint reinforcement or anchors, if used, shall be spaced to provide equivalent area of anchorage to that required by Section 14.4.2.2.

14.5 — Miscellaneous requirements

14.5.1 *Chases and recesses*
Masonry directly above chases or recesses wider than 12 in. (305 mm) shall be supported on lintels.

14.5.2 *Lintels*
The design of masonry lintels shall be in accordance with the provisions of Section 5.2.

14.5.3 *Lap splices*
Lap splices for bar reinforcement or joint reinforcement, required by Section 7.4.3.1 and located in masonry partition walls designed in accordance with this Chapter, shall be a minimum of $48d_b$.

COMMENTARY

14.3.3 *Cantilever walls*
The span to thickness ratios permitted for cantilevered walls are based on historical use and confirming analyses using design assumptions similar to those used to develop Table 14.3.1(5) and Table 14.3.1(10).

14.5.3 *Lap splices*
Because Chapter 14 does not have f'_m requirements, designing lap splice requirements in accordance with Chapters 8 or 9 can be problematic when using Chapter 14. The reinforcement required in Chapter 14 is prescriptive. The minimum lap splice length of $48 d_b$ for reinforcement is based on historical use and the use of a No. 4 (M13) bar or W1.7 (MW1.1) joint reinforcement.

PART 5: APPENDICES

APPENDIX A
EMPIRICAL DESIGN OF MASONRY

CODE

A.1 — General

A.1.1 *Scope*
This appendix provides requirements for empirical design of masonry.

A.1.1.1 The provisions of Part 1 and Part 2, excluding Part 1 Sections 1.2.1(c), 1.2.2, 4.1, 4.2, and 4.3, shall apply to empirical design, except as specifically stated in this Chapter.

A.1.1.2 Article 1.4 of TMS 602/ACI 530.1/ASCE 6 shall not apply to empirically designed masonry.

A.1.2 *Limitations*

A.1.2.1 *Gravity Loads* — The resultant of gravity loads shall be placed within the center third of the wall thickness and within the central area bounded by lines at one-third of each cross-sectional dimension of foundation piers.

A.1.2.2 *Seismic* — Empirical requirements shall not apply to the design of masonry for buildings, parts of buildings or other structures in Seismic Design Categories D, E, or F as defined in ASCE 7, and shall not apply to the design of the seismic-force-resisting system for structures in Seismic Design Categories B or C.

COMMENTARY

A.1 — General

Empirical design procedures of Appendix A are permitted in certain instances. Empirical design is permitted for buildings of limited height and low seismic risk.

Empirical rules and formulas for the design of masonry structures were developed by experience. These are part of the legacy of masonry's long use, predating engineering analysis. Design is based on the condition that gravity loads are reasonably centered on the load-bearing walls and foundation piers. Figure CC-A.1-1 illustrates the location of the resultant of gravity loads on foundation piers. The effect of any steel reinforcement, if used, is neglected. The masonry should be laid in running bond. Specific limitations on building height, seismic, wind, and horizontal loads exist. Buildings are of limited height. Members not participating in the lateral-force-resisting system of a building, other than partition walls, may be empirically designed even though the lateral-force-resisting system is designed under other chapters of this Code.

These procedures have been compiled through the years (Baker, 1909; NBS, 1924; NBS, 1931; ASA, 1944; ANSI, 1953). The most recent of these documents (ANSI, 1953) is the basis for this chapter.

Empirical design is a procedure of sizing and proportioning masonry elements. It is not design analysis. This procedure is conservative for most masonry construction. Empirical design of masonry was developed for buildings of smaller scale, with more masonry interior walls and stiffer floor systems than built today. Thus, the limits imposed are valid.

Because empirically designed masonry is based on the gross compressive strength of the units, there is no need to specify the compressive strength of masonry.

Table CC-A.1.1 is a checklist to assist the Architect/Engineer in designing masonry structures using the empirical design provisions. The checklist identifies the applicable Sections of the Code that need to be considered. There may be additional specific Code requirements that have to be met, depending on the project. The checklist simply serves as a guide.

CODE

A.1.2.3 *Wind* — Empirical requirements shall be permitted to be applied to the design of masonry elements defined by Table A.1.1, based on building height and basic wind speed that are applicable to the building.

A.1.2.4 *Buildings and other structures in Risk Category IV* — Empirical requirements shall not apply to the design of masonry for buildings, parts of buildings or other structures in Risk Category IV as defined in ASCE 7.

A.1.2.5 *Other horizontal loads* — Empirical requirements shall not apply to structures resisting horizontal loads other than permitted wind or seismic loads or foundation walls as provided in Section A.6.3.

A.1.2.6 *Glass unit masonry* — The provisions of Appendix A shall not apply to glass unit masonry.

A.1.2.7 *AAC masonry* — The provisions of Appendix A shall not apply to AAC masonry.

A.1.2.8 *Concrete masonry* — Concrete masonry, designed in accordance with Appendix A, shall comply with one of the following:

(a) The minimum normalized web area of concrete masonry units, determined in accordance with ASTM C140, shall not be less than 27 in.^{2}/ft^2 (187,500 mm^2/m^2), or

(b) the member shall be grouted solid.

A.1.2.9 *Support* — The provisions of Appendix A shall not apply to masonry vertically supported on wood construction.

A.1.2.10 *Partition walls* — The provisions of Appendix A shall not apply to partition walls.

COMMENTARY

A.1.2.3 *Wind* — There is a change in the wind speed values listed in the table from previous versions of the Code. The values listed were adjusted to strength levels for use with ASCE 7-10 wind speed maps and are designed to maintain the strength level velocity pressures below approximately 40 psf (1.92 kPa) for a wide range of building configurations.

A.1.2.8 *Concrete masonry* — When the empirical design provisions for masonry were initially standardized in the early 20th century, only a limited number of concrete masonry unit configurations were available. In contrast, there is a near limitless array of unit shapes and sizes commercially available today. The requirements of this section establish a minimum web area that is consistent with the configuration of concrete masonry units prevalent when the empirical design provisions first began to take shape, thereby maintaining assumptions inherent in the empirical design provisions. Alternatively, the assembly can be solidly grouted, in which case the configuration of the web is structurally irrelevant.

A.1.2.10 *Partition walls* — Partition wall design is not included in this Appendix. Partition wall design is permitted by Chapter 14 of this code.

Table A.1.1 Limitations based on building height and basic wind speed

Element Description	Building Height, ft (m)	Basic Wind Speed, mph (mps)[1]			
		Less than or equal to 115 (51)	Over 115 (51) and less than or equal to 120 (54)	Over 120 (54) and less than or equal to 125 (56)	Over 125 (56)
Masonry elements that are part of the lateral-force-resisting system	35 (11) and less	Permitted			Not Permitted
Interior masonry loadbearing elements that are not part of the lateral-force-resisting system in buildings other than enclosed as defined by ASCE 7	Over 180 (55)	Not Permitted			
	Over 60 (18) and less than or equal to 180 (55)	Permitted	Not Permitted		
	Over 35 (11) and less than or equal to 60 (18)	Permitted		Not Permitted	
	35 (11) and less	Permitted			Not Permitted
Exterior masonry elements that are not part of the lateral-force-resisting system	Over 180 (55)	Not Permitted			
	Over 60 (18) and less than or equal to 180 (55)	Permitted	Not Permitted		
	Over 35 (11) and less than or equal to 60 (18)	Permitted		Not Permitted	
Exterior masonry elements	35 (11) and less	Permitted			Not Permitted

[1] Basic wind speed as given in ASCE 7

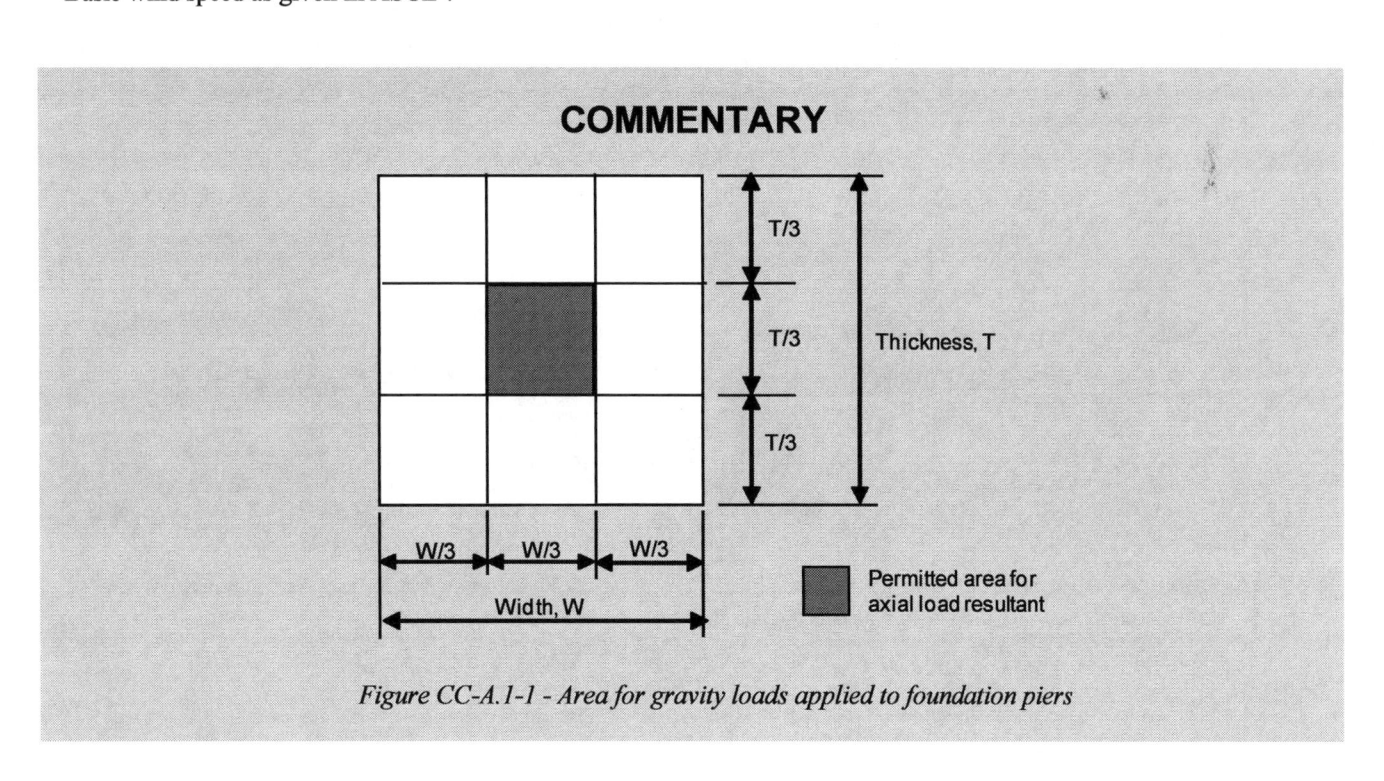

COMMENTARY

Figure CC-A.1-1 - Area for gravity loads applied to foundation piers

COMMENTARY

Table CC-A.1.1 — Checklist for use of Appendix A – Empirical Design of Masonry

1.	Risk Category IV structures are not permitted to be designed using Appendix A.
2.	Partitions are not permitted to be designed using Appendix A.
3.	Use of empirical design is limited based on Seismic Design Category, as described in the following table. | Seismic Design Category	Participating Walls	Non-Participating Walls, except partition walls
---	---	---
A	Allowed by Appendix A	Allowed by Appendix A
B	Not Allowed	Allowed by Appendix A
C	Not Allowed	With prescriptive reinforcement per 7.4.3.1[1]
D, E, and F	Not Allowed	Not Allowed	 [1] Lap splices are required to be designed and detailed in accordance with the requirements of Chapters 8 or 9.
4.	Use of empirical design is limited based on wind speed at the project site, as described in Code A.1.2.3 and Code Table A.1.1.
5.	If wind uplift on roofs result in net tension, empirical design is not permitted (A.8.3.1).
6.	Loads used in the design of masonry must be listed on the design drawings (1.2.1b).
7.	Details of anchorage to structural frames must be included in the design drawings (1.2.1e).
8.	The design is required to include provisions for volume change (1.2.1h). The design drawings are required to include the locations and sizing of expansion, control, and isolation joints.
9.	If walls are connected to structural frames, the connections and walls are required to be designed to resist the interconnecting forces and to accommodate deflections (4.4). This provision requires a lateral load and uplift analysis for exterior walls that receive wind load and are supported by or are supporting a frame or roofing system.
10.	Masonry not laid in running bond (for example, stack bond masonry) is required to have horizontal reinforcement (4.5).
11.	A project quality assurance plan is required (3.1) with minimum requirements given in Table 3.1.1.
12.	The resultant of gravity loads must be determined and assured to be located within certain limitations for walls and piers (A.1.2.1).
13.	Ensure compliance of the design with prescriptive floor, roof, and wall-to-structural framing anchorage requirements, as well as other anchorage requirements (A.8.3 and A.8.4).
14.	Type N mortar is not permitted for foundation walls (A.6.3.1(g)).
15.	Design shear wall lengths, spacings, and orientations to meet the requirements of Code A.3.1.

CODE

A.2 — Height

Buildings relying on masonry walls as part of their lateral-force-resisting system shall not exceed 35 ft (10.67 m) in height.

A.3 — Lateral stability

A.3.1 *Shear walls*

Where the structure depends upon masonry walls for lateral stability, shear walls shall be provided parallel to the direction of the lateral forces resisted.

A.3.1.1 In each direction in which shear walls are required for lateral stability, shear walls shall be positioned in at least two separate planes parallel with the direction of the lateral force. The minimum cumulative length of shear walls provided along each plane shall be 0.2 multiplied by the long dimension of the building. Cumulative length of shear walls shall not include openings or any element whose length is less than one-half its height.

A.3.1.2 Shear walls shall be spaced so that the length-to-width ratio of each diaphragm transferring lateral forces to the shear walls does not exceed values given in Table A.3.1.

A.3.2 *Roofs*

The roof construction shall be designed so as not to impart out-of-plane lateral thrust to the walls under roof gravity load.

COMMENTARY

A.3 — Lateral stability

Lateral stability requirements are a key provision of empirical design. Obviously, shear walls must be in two directions to provide stability. Load-bearing walls can serve as shear walls. The height of a wall refers to the shortest unsupported height in the plane of the wall such as the shorter of a window jamb on one side and a door jamb on the other. See Figure CC-A.3-1 for cumulative length of shear walls. See Figure CC-A.3-2 for diaphragm panel length to width ratio determination.

Table A.3.1 — Diaphragm length-to-width ratios

Floor or roof diaphragm construction	Maximum length-to-width ratio of diaphragm panel
Cast-in-place concrete	5:1
Precast concrete	4:1
Metal deck with concrete fill	3:1
Metal deck with no fill	2:1
Wood	2:1

COMMENTARY

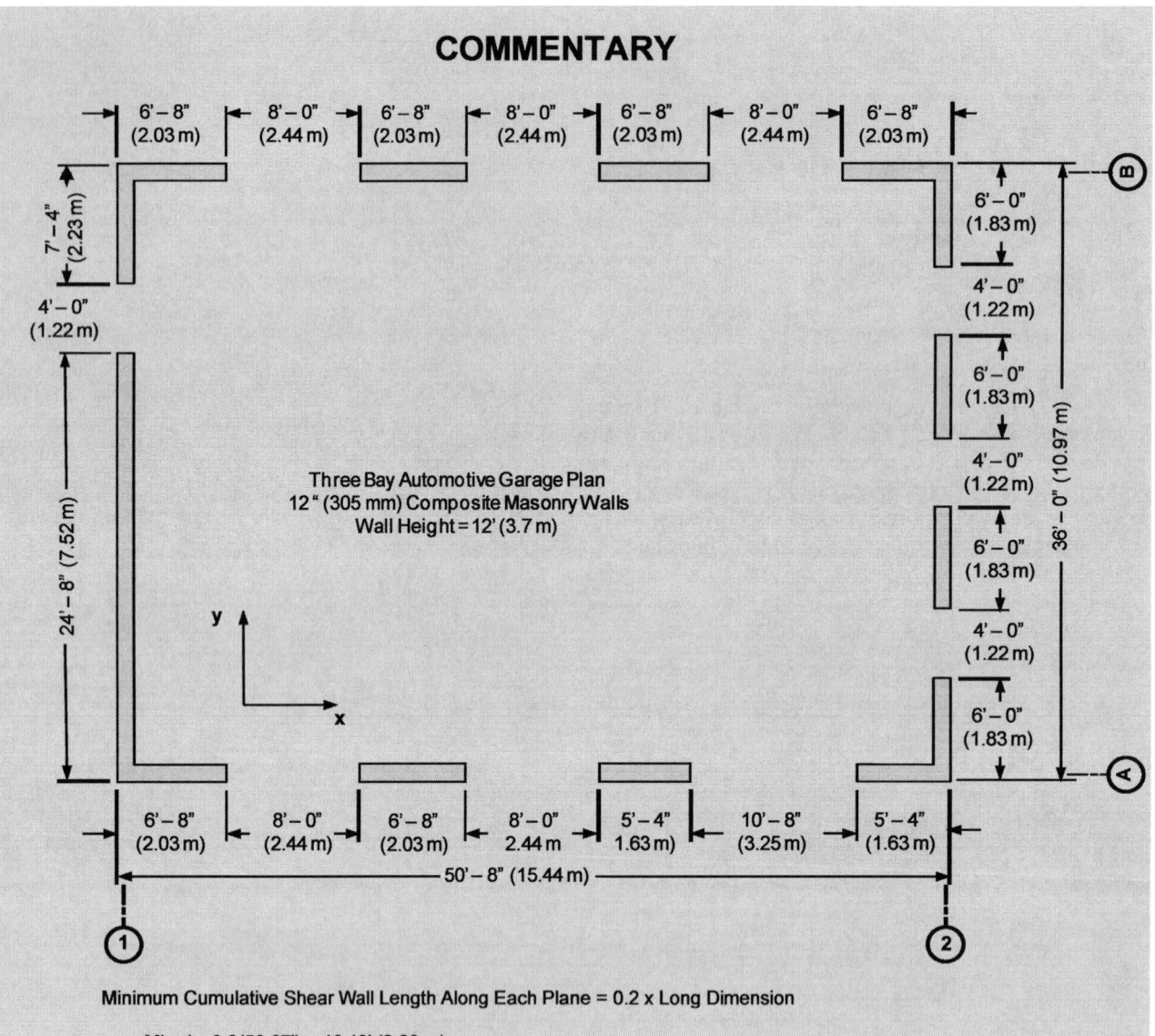

Three Bay Automotive Garage Plan
12" (305 mm) Composite Masonry Walls
Wall Height = 12' (3.7 m)

Minimum Cumulative Shear Wall Length Along Each Plane = 0.2 x Long Dimension

Min. l = 0.2(50.67') = 10.13' (3.09 m)

Wall line 1: l = (24.67 + 7.33) = 32.0' > 10.13' OK
l = (7.52 m + 2.23 m) = 9.75 m > 3.09 m OK

Wall line 2: l = (6.0' + 6.0' + 6.0' + 6.0') = 24.0' > 10.13' OK
l = (1.83 m + 1.83 m + 1.83 m + 1.83 m) = 7.32 m > 3.09 m OK

Wall line A: Note, 5'-4"(1.62 m) wall segments not included as they are less than ½ of 12' (3.66 m) wall height
l = (6.67' + 6.67') = 13.33' > 10.13' OK
l = (2.03 m + 2.03 m) = 4.06 m > 3.09 m OK

Wall line B: l = (6.67' + 6.67' + 6.67' + 6.67') = 26.67' > 10.13' OK
l = (2.03 m + 2.03 m + 2.03 m + 2.03 m) = 8.13 m > 3.09 m OK

Figure CC-A.3-1 — Cumulative length of shear walls

BUILDING CODE REQUIREMENTS FOR MASONRY STRUCTURES AND COMMENTARY

COMMENTARY

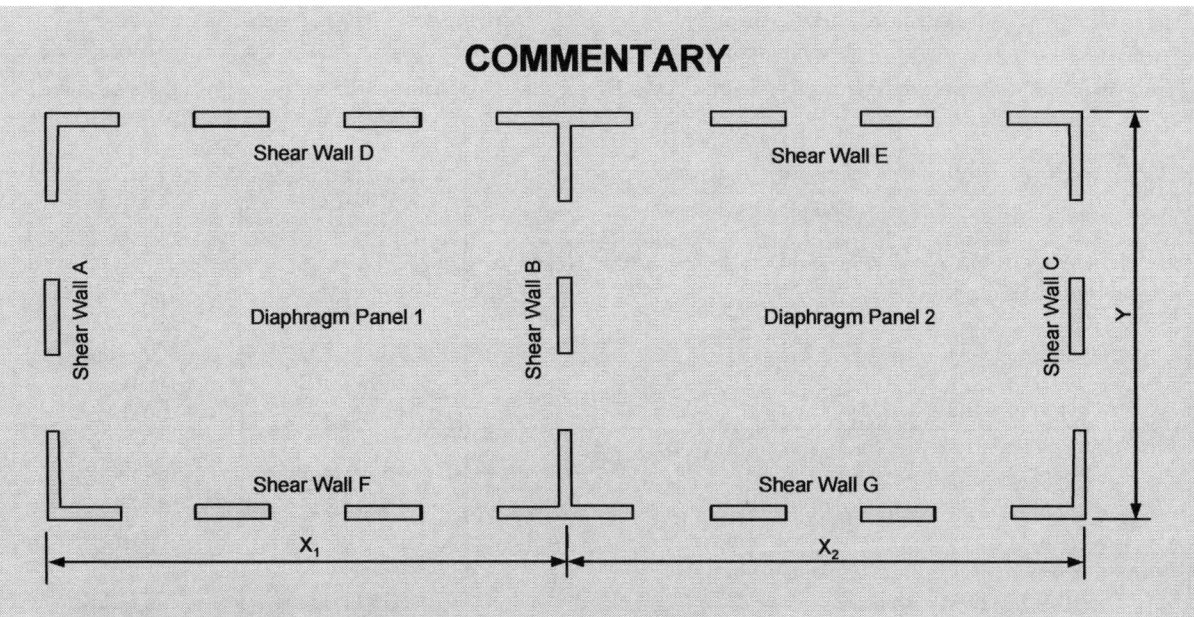

Diaphragm Panel Length = Dimension perpendicular to the resisting shear wall

Diaphragm Panel Width = Dimension parallel to the resisting shear wall

For example:

    For Shear Walls A and B, the diaphragm panel length to width ratio is X_1/Y

    For Shear Walls D and F, the diaphragm panel length to width ratio is Y/X_1

Note: Shear walls should be placed on all four sides of the diaphragm panel or the resulting torsion should be accounted for.

Figure CC-A.3-2 — Diaphragm panel length to width ratio determination for shear wall spacing

CODE

A.4 — Compressive stress requirements

A.4.1 *Calculations*

Dead loads and live loads shall be in accordance with the legally adopted building code of which this Code forms a part, with such live load reductions as are permitted in the legally adopted building code. Compressive stresses in masonry due to vertical dead plus live loads (excluding wind or seismic loads) shall be determined in accordance with the following:

(a) Stresses shall be calculated based on specified dimensions.

(b) Calculated compressive stresses for single wythe walls and for multiwythe composite masonry walls shall be determined by dividing the design load by the gross cross-sectional area of the member. The area of openings, chases, or recesses in walls shall not be included in the gross cross-sectional area of the wall.

A.4.2 *Allowable compressive stresses*

The compressive stresses in masonry shall not exceed the values given in Table A.4.2. In multiwythe walls, the allowable stresses shall be based on the weakest combination of the units and mortar used in each wythe.

COMMENTARY

A.4 — Compressive stress requirements

These are average compressive stresses based on gross area using specified dimensions. The course immediately under the point of bearing should be a solid unit or fully filled with mortar or grout.

Table A.4.2 — Allowable compressive stresses for empirical design of masonry

Construction; compressive strength of masonry unit, gross area, psi (MPa)	Allowable compressive stresses[1] based on gross cross-sectional area, psi (MPa)	
	Type M or S mortar	**Type N mortar**
Solid masonry of brick and other solid units of clay or shale; sand-lime or concrete brick:		
8,000 (55.16) or greater	350 (2.41)	300 (2.07)
4,500 (31.03)	225 (1.55)	200 (1.38)
2,500 (17.23)	160 (1.10)	140 (0.97)
1,500 (10.34)	115 (0.79)	100 (0.69)
Grouted masonry of clay or shale; sand-lime or concrete:		
4,500 (31.03) or greater	225 (1.55)	200 (1.38)
2,500 (17.23)	160 (1.10)	140 (0.97)
1,500 (10.34)	115 (0.79)	100 (0.69)
Solid masonry of solid concrete masonry units:		
3,000 (20.69) or greater	225 (1.55)	200 (1.38)
2,000 (13.79)	160 (1.10)	140 (0.97)
1,200 (8.27)	115 (0.79)	100 (0.69)
Masonry of hollow load-bearing units of clay or shale[2]:		
2,000 (13.79) or greater	140 (0.97)	120 (0.83)
1,500 (10.34)	115 (0.79)	100 (0.69)
1,000 (6.90)	75 (0.52)	70 (0.48)
700 (4.83)	60 (0.41)	55 (0.38)
Masonry of hollow load-bearing concrete masonry units, up to and including 8 in. (203 mm) nominal thickness:		
2,000 (13.79) or greater	140 (0.97)	120 (0.83)
1,500 (10.34)	115 (0.79)	100 (0.69)
1,000 (6.90)	75 (0.52)	70 (0.48)
700 (4.83)	60 (0.41)	55 (0.38)
Masonry of hollow load-bearing concrete masonry units, greater than 8 and up to 12 in. (203 to 305 mm) nominal thickness:		
2,000 (13.79) or greater	125 (0.86)	110 (0.76)
1,500 (10.34)	105 (0.72)	90 (0.62)
1,000 (6.90)	65 (0.45)	60 (0.41)
700 (4.83)	55 (0.38)	50 (0.35)

Table A.4.2 (continued) — Allowable compressive stresses for empirical design of masonry

Construction; compressive strength of masonry unit, gross area, psi (MPa)	Allowable compressive stresses[1] based on gross cross-sectional area, psi (MPa)	
	Type M or S mortar	**Type N mortar**
Masonry of hollow load-bearing concrete masonry units, 12 in. (305 mm) nominal thickness and greater:		
2,000 (13.79) or greater	115 (0.79)	100 (0.69)
1,500 (10.34)	95 (0.66)	85 (0.59)
1,000 (6.90)	60 (0.41)	55 (0.38)
700 (4.83)	50 (0.35)	45 (0.31)
Multiwythe non-composite walls[2]:		
Solid units:		
2500 (17.23) or greater	160 (1.10)	140 (0.97)
1500 (10.34)	115 (0.79)	100 (0.69)
Hollow units of clay or shale	75 (0.52)	70 (0.48)
Hollow units of concrete masonry of nominal thickness,		
up to and including 8 in. (203 mm):	75 (0.52)	70 (0.48)
greater than 8 and up to 12 in. (203-305 mm):	70 (0.48)	65 (0.45)
12 in. (305 mm) and greater:	60 (0.41)	55 (0.38)
Stone ashlar masonry:		
Granite	720 (4.96)	640 (4.41)
Limestone or marble	450 (3.10)	400 (2.76)
Sandstone or cast stone	360 (2.48)	320 (2.21)
Rubble stone masonry:		
Coursed, rough, or random	120 (0.83)	100 (0.69)

1 Linear interpolation shall be permitted for determining allowable stresses for masonry units having compressive strengths which are intermediate between those given in the table.

2 In non-composite walls, where floor and roof loads are carried upon one wythe, the gross cross-sectional area is that of the wythe under load; if both wythes are loaded, the gross cross-sectional area is that of the wall minus the area of the cavity between the wythes.

CODE

A.5 — Lateral support

A.5.1 *Maximum l/t and h/t*

Masonry walls without openings shall be laterally supported in either the horizontal or the vertical direction so that l/t or h/t does not exceed the values given in Table A.5.1.

Masonry walls with single or multiple openings shall be laterally supported in either the horizontal or vertical direction so that l/t or h/t does not exceed the values given in Table A.5.1 divided by $\sqrt{W_T/W_S}$.

W_S is the dimension of the structural wall strip measured perpendicular to the span of the wall strip and perpendicular to the thickness as shown in Figure A.5.1-1. W_S is measured from the edge of the opening. W_S shall be no less than $3t$ on each side of each opening. Therefore, at walls with multiple openings, jambs shall be no less than $6t$ between openings. For design purposes, the effective W_S shall not be assumed to be greater than $6t$. At non-masonry lintels, the edge of the opening shall be considered the edge of the non-masonry lintel. W_S shall occur uninterrupted over the full span of the wall.

W_T is the dimension, parallel to W_S, from the center of the opening to the opposite end of W_S as shown in Figure A.5.1-1. Where there are multiple openings perpendicular to W_S, W_T shall be measured from the center of a virtual opening that encompasses such openings. Masonry elements within the virtual opening must be designed in accordance with Chapter 8 or 9.

For walls with openings that span no more than 4 feet, parallel to W_S, if W_S is no less than 4 feet, then it shall be permitted to ignore the effect of those openings.

The span of openings, parallel to W_S, shall be limited so that the span divided by t does not exceed the values given in Table A.5.1.

In addition to these limitations, lintels shall be designed for gravity loads in accordance with Section A.9.2.

COMMENTARY

A.5 — Lateral support

Lateral support requirements are included to limit the flexural tensile stress due to out-of-plane loads. Masonry headers resist shear stress and permit the entire cross-section to perform as a single element. This is not the case for non-composite walls connected with wall ties. For such non-composite walls, the use of the sum of the thicknesses of the wythes has been used successfully for a long time and is a traditional approach that is acceptable within the limits imposed by Code Table A.5.1. Requirements were added in the 2008 edition to provide relative out-of-plane resistance that limit the maximum width of opening and provide sufficient masonry sections between the openings.

Table A.5.1 — Wall lateral support requirements

Construction	Maximum l/t or h/t
Load-bearing walls	
Solid units or fully grouted	20
Other than solid units or fully grouted	18
Non-load-bearing walls	
Exterior	18

In calculating the ratio for multiwythe walls, use the following thickness:
1. The nominal wall thicknesses for solid walls and for hollow walls bonded with masonry headers (Section A.7.2).
2. The sum of the nominal thicknesses of the wythes for non-composite walls connected with wall ties (Section A.7.3).

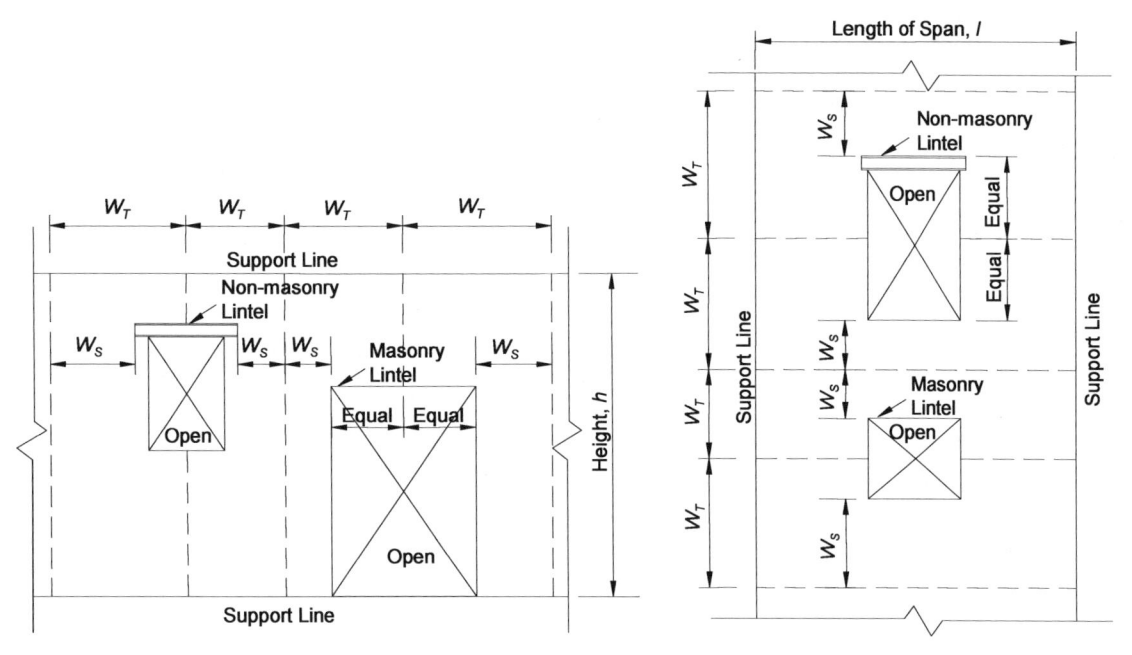

W_S and W_T for Walls Spanning Vertically

W_S and W_T for Walls Spanning Horizontally

Figure A.5.1-1 — Graphical representation of W_S and W_T

CODE

A.5.2 *Cantilever walls*
Except for parapets, the ratio of height-to-nominal-thickness for cantilever walls shall not exceed 6 for solid masonry or 4 for hollow masonry. For parapets see Section A.6.4.

A.5.3 *Support elements*
Lateral support shall be provided by cross walls, pilasters, or structural frame members when the limiting distance is taken horizontally; or by floors, roofs acting as diaphragms, or structural frame members when the limiting distance is taken vertically.

A.6 — Thickness of masonry

A.6.1 *General*
Minimum thickness requirements shall be based on nominal dimensions of masonry.

A.6.2 *Minimum thickness*
A.6.2.1 *Load-bearing walls* — The minimum thickness of load-bearing walls of one story buildings shall be 6 in. (152 mm). The minimum thickness of load-bearing walls of buildings more than one story high shall be 8 in. (203 mm).

A.6.2.2 *Rubble stone walls* — The minimum thickness of rough, random, or coursed rubble stone walls shall be 16 in. (406 mm).

COMMENTARY

A.6 — Thickness of masonry

A.6.1 *General*
Experience of the committee has shown that the present ANSI A 41.1 (1953) thickness ratios are not always conservative. These requirements represent the consensus of the committee for more conservative design.

CODE

A.6.2.3 *Shear walls* — The minimum thickness of masonry shear walls shall be 8 in. (203 mm).

A.6.2.4 *Foundation walls* — The minimum thickness of foundation walls shall be 8 in. (203 mm).

A.6.2.5 *Foundation piers* — The minimum thickness of foundation piers shall be 8 in. (203 mm).

A.6.2.6 *Parapet walls* — The minimum thickness of parapet walls shall be 8 in. (203 mm).

A.6.2.7 *Partition walls* — The minimum thickness of partition walls shall be 4 in. (102 mm).

A.6.2.8 *Change in thickness* — Where walls of masonry of hollow units or masonry bonded hollow walls are decreased in thickness, a course or courses of solid masonry units or fully grouted hollow masonry units shall be interposed between the wall below and the thinner wall above, or special units or construction shall be used to transmit the loads from face shells or wythes above to those below.

A.6.3 Foundation walls

A.6.3.1 Foundation walls shall comply with the requirements of Table A.6.3.1, which are applicable when:

(a) the foundation wall does not exceed 8 ft (2.44 m) in height between lateral supports,

(b) the terrain surrounding foundation walls is graded to drain surface water away from foundation walls,

(c) backfill is drained to remove ground water away from foundation walls,

(d) lateral support is provided at the top of foundation walls prior to backfilling,

(e) the length of foundation walls between perpendicular masonry walls or pilasters is a maximum of 3 multiplied by the basement wall height,

(f) the backfill is granular and soil conditions in the area are non-expansive, and

(g) masonry is laid in running bond using Type M or S mortar.

A.6.3.2 Where the requirements of Section A.6.3.1 are not met, foundation walls shall be designed in accordance with Part 1, Part 2, and Chapter 8, 9, or 10.

A.6.4 Foundation piers

Design of foundation piers shall comply with Appendix A and the following:

(a) Length, measured perpendicular to its thickness, shall not exceed 3 times its thickness.

(b) Height shall be equal to or less than 4 times its thickness

COMMENTARY

A.6.3 Foundation walls

Empirical criteria for masonry foundation wall thickness related to the depth of unbalanced fill have been contained in building codes and federal government standards for many years. The use of Code Table A.6.3.1, which lists the traditional allowable backfill depths, is limited by a number of requirements that were not specified in previous codes and standards. These restrictions are enumerated in Section A.6.3.1. Further precautions are recommended to guard against allowing heavy earth-moving or other equipment close enough to the foundation wall to develop high earth pressures. Experience with local conditions should be used to modify the values in Table A.6.3.1 when appropriate.

A.6.4 Foundation piers

Foundation piers are masonry members that are unique to the Empirical Design method. Use of empirically designed foundation piers has been common practice in many areas of the country for many years. ANSI A 41.1 (1953) provisions for empirically designed piers (Section A.3) include a requirement for a maximum h/t ratio of 4. The length and height requirements provide the basis on which the design requirements were developed and differentiate a foundation pier from other members such as columns or piers.

Table A.6.3.1 — Foundation wall construction

Wall construction	Nominal wall thickness, in. (mm)	Maximum depth of unbalanced backfill, ft (m)
Masonry of hollow units	8 (203)	5 (1.52)
	10 (254)	6 (1.83)
	12 (305)	7 (2.13)
Masonry of solid units	8 (203)	5 (1.52)
	10 (254)	7 (2.13)
	12 (305)	7 (2.13)
Fully grouted masonry	8 (203)	7 (2.13)
	10 (254)	8 (2.44)
	12 (305)	8 (2.44)

CODE

A.7 — Bond

A.7.1 *General*

Wythes of multiple wythe masonry walls shall be bonded in accordance with the requirements of Section A.7.2, Section A.7.3, or Section A.7.4.

A.7.2 *Bonding with masonry headers*

A.7.2.1 *Solid units* — Where adjacent wythes of solid masonry walls are bonded by means of masonry headers, no less than 4 percent of the wall surface area of each face shall be composed of headers extending not less than 3 in. (76.2 mm) into each wythe. The distance between adjacent full-length headers shall not exceed 24 in. (610 mm) either vertically or horizontally. In multi-wythe walls that are thicker than the length of a header, each wythe shall be connected to the adjacent wythe by adjacent headers that overlap a minimum of 3 in. (76.2 mm).

A.7.2.2 *Hollow units* — Where two or more wythes are constructed using hollow units, the stretcher courses shall be bonded at vertical intervals not exceeding 34 in. (864 mm) by lapping at least 3 in. (76.2 mm) over the unit below, or by lapping at vertical intervals not exceeding 17 in. (432 mm) with units which are at least 50 percent greater in thickness than the units below.

A.7.3 *Bonding with wall ties or joint reinforcement*

A.7.3.1 Where adjacent wythes of masonry walls are bonded with wire size W2.8 (MW18) wall ties or metal wire of equivalent stiffness embedded in the horizontal mortar joints, there shall be at least one metal tie for each $4\frac{1}{2}$ ft^2 (0.42 m^2) of wall area. The maximum vertical distance between ties shall not exceed 24 in. (610 mm), and the maximum horizontal distance shall not exceed 36 in. (914 mm). Rods or ties bent to rectangular shape shall be used with hollow masonry units laid with the cells vertical. In other walls, the ends of ties shall be bent to 90-degree angles to provide hooks no less than 2 in. (50.8 mm) long. Wall ties shall be without drips and shall be non-adjustable. Additional bonding ties shall be provided at openings, spaced not more than 3 ft (0.91 m) apart around the perimeter and within 12 in. (305 mm) of the opening.

COMMENTARY

A.7 — Bond

Figure CC-A.7-1 depicts the requirements listed. Wall ties with drips are not permitted because of their reduced load capacity.

CODE

A.7.3.2 Where adjacent wythes of masonry are bonded with prefabricated joint reinforcement, there shall be at least one cross wire serving as a tie for each $2^2/_3$ ft^2 (0.25 m^2) of wall area. The vertical spacing of the joint reinforcement shall not exceed 24 in. (610 mm). Cross wires on prefabricated joint reinforcement shall be not smaller than wire size W1.7 (MW11) and shall be without drips. The longitudinal wires shall be embedded in the mortar.

A.7.4 *Natural or cast stone*

A.7.4.1 *Ashlar masonry* — In ashlar masonry, uniformly distributed bonder units shall be provided to the extent of not less than 10 percent of the wall area. Such bonder units shall extend not less than 4 in. (102 mm) into the backing wall.

A.7.4.2 *Rubble stone masonry* — Rubble stone masonry 24 in. (610 mm) or less in thickness shall have bonder units with a maximum spacing of 3 ft (0.91 m) vertically and 3 ft (0.91 m) horizontally, and if the masonry is of greater thickness than 24 in. (610 mm), shall have one bonder unit for each 6 ft^2 (0.56 m^2) of wall surface on both sides.

A.8 — Anchorage

A.8.1 *General*

Masonry elements shall be anchored in accordance with this section.

A.8.2 *Intersecting walls*

Masonry walls depending upon one another for lateral support shall be anchored or bonded at locations where they meet or intersect by one of the following methods:

A.8.2.1 Fifty percent of the units at the intersection shall be laid in an overlapping masonry bonding pattern, with alternate units having a bearing of not less than 3 in. (76.2 mm) on the unit below.

A.8.2.2 Walls shall be anchored by steel connectors having a minimum section of $^1/_4$ in. (6.4 mm) by $1^1/_2$ in. (38.1 mm) with ends bent up at least 2 in. (50.8 mm), or with cross pins to form anchorage. Such anchors shall be at least 24 in. (610 mm) long and the maximum spacing shall be 4 ft (1.22 m).

A.8.2.3 Walls shall be anchored by joint reinforcement spaced at a maximum distance of 8 in. (203 mm). Longitudinal wires of such reinforcement shall be at least wire size W1.7 (MW11) and shall extend at least 30 in. (762 mm) in each direction at the intersection.

A.8.2.4 Other metal ties, joint reinforcement or anchors, if used, shall be spaced to provide equivalent area of anchorage to that required by Sections A.8.2.2 through A.8.2.4.

COMMENTARY

A.8 — Anchorage

The requirements of Sections A.8.2.2 through A.8.2.4 are less stringent than those of Section 5.1.1.2.5. Anchorage requirements in Section A.8.3.3 are intended to comply with the Steel Joist Institute's Standard Specification (SJI, 2002) for end anchorage of steel joists.

COMMENTARY

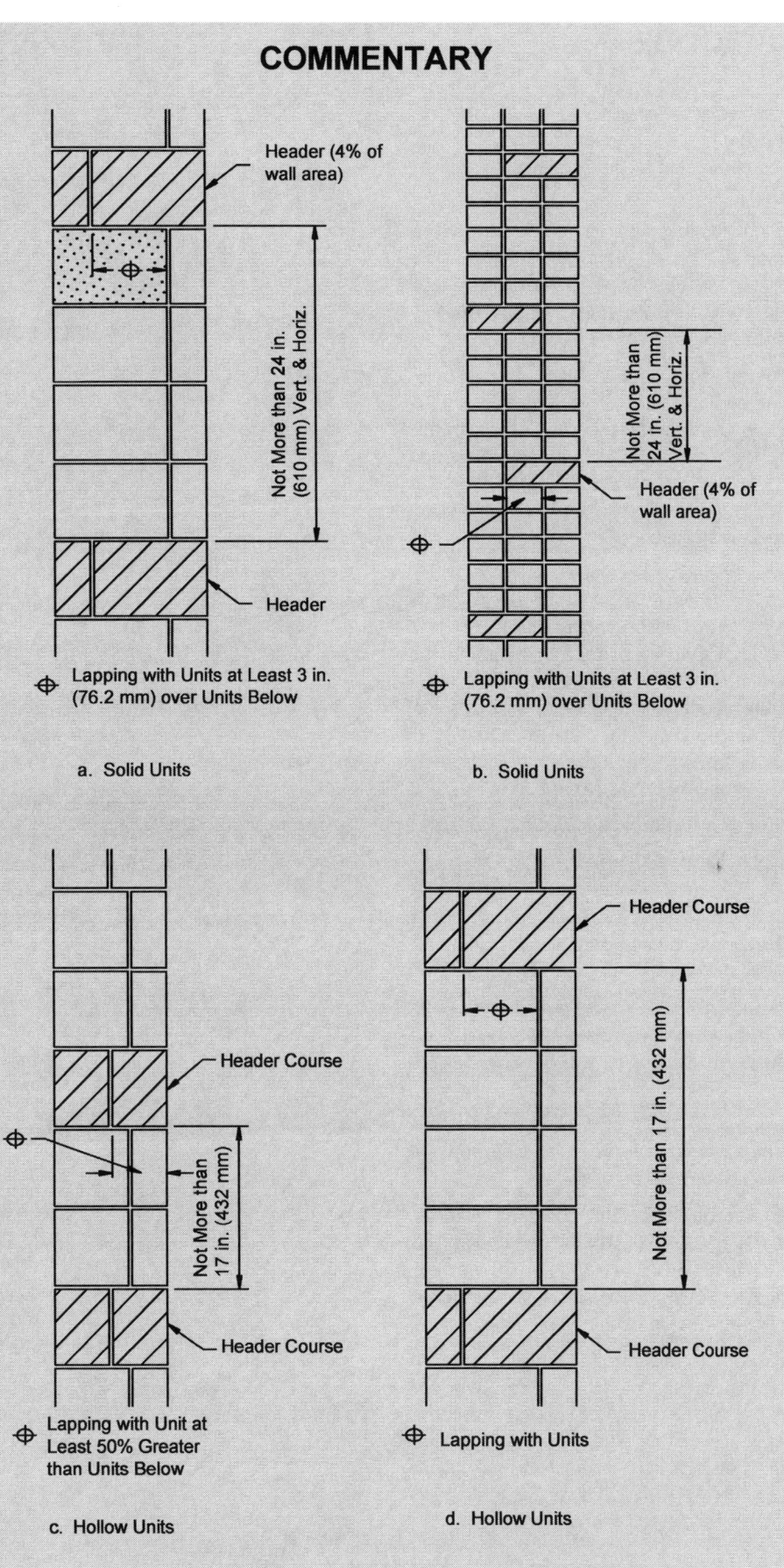

a. Solid Units — Lapping with Units at Least 3 in. (76.2 mm) over Units Below; Header (4% of wall area); Not More than 24 in. (610 mm) Vert. & Horiz.

b. Solid Units — Lapping with Units at Least 3 in. (76.2 mm) over Units Below; Header (4% of wall area); Not More than 24 in. (610 mm) Vert. & Horiz.

c. Hollow Units — Lapping with Unit at Least 50% Greater than Units Below; Header Course; Not More than 17 in. (432 mm)

d. Hollow Units — Lapping with Units; Header Course; Not More than 17 in. (432 mm)

Figure CC-A.7-1 — Cross section of wall elevations

CODE

A.8.3 *Floor and roof anchorage*

Floor and roof diaphragms providing lateral support to masonry shall be connected to the masonry by one of the following methods:

A.8.3.1 Roof loading shall be determined by the provisions of Section 4.1.2 and, where net uplift occurs, uplift shall be resisted entirely by an anchorage system designed in accordance with the provisions of Sections 8.1 and 8.3, Sections 9.1 and 9.3, or Chapter 10.

A.8.3.2 Wood floor joists bearing on masonry walls shall be anchored to the wall at intervals not to exceed 6 ft (1.83 m) by metal strap anchors. Joists parallel to the wall shall be anchored with metal straps spaced not more than 6 ft (1.83 m) on centers extending over or under and secured to at least 3 joists. Blocking shall be provided between joists at each strap anchor.

A.8.3.3 Steel joists that are supported by masonry walls shall bear on and be connected to steel bearing plates. Maximum joist spacing shall be 6 ft (1.83 m) on center. Each bearing plate shall be anchored to the wall with a minimum of two ½ in. (12.7 mm) diameter bolts, or their equivalent. Where steel joists are parallel to the wall, anchors shall be located where joist bridging terminates at the wall and additional anchorage shall be provided to comply with Section A.8.3.4.

A.8.3.4 Roof and floor diaphragms shall be anchored to masonry walls with a minimum of ½ in. (12.7 mm) diameter bolts at a maximum spacing of 6 ft (1.83 m) on center or their equivalent.

A.8.3.5 Bolts and anchors required by Sections A.8.3.3 and A.8.3.4 shall comply with the following:

(a) Bolts and anchors at steel floor joists and floor diaphragms shall be embedded in the masonry at least 6 in. (152 mm) or shall comply with Section A.8.3.5 (c).

(b) Bolts at steel roof joists and roof diaphragms shall be embedded in the masonry at least 15 in. (381 mm) or shall comply with Section A.8.3.5(c).

(c) In lieu of the embedment lengths listed in Sections A.8.3.5(a) and A.8.3.5(b), bolts shall be permitted to be hooked or welded to not less than 0.20 in.^{2} (129 mm^2) of bond beam reinforcement placed not less than 6 in. (152 mm) below joist bearing or bottom of diaphragm.

A.8.4 *Walls adjoining structural framing*

Where walls are dependent upon the structural frame for lateral support, they shall be anchored to the structural members with metal anchors or otherwise keyed to the structural members. Metal anchors shall consist of ½-in. (12.7-mm) bolts spaced at 4 ft (1.22 m) on center embedded 4 in. (102 mm) into the masonry, or their equivalent area.

COMMENTARY

CODE

A.9 — Miscellaneous requirements

A.9.1 *Chases and recesses*

Masonry directly above chases or recesses wider than 12 in. (305 mm) shall be supported on lintels.

A.9.2 *Lintels*

The design of masonry lintels shall be in accordance with the provisions of Section 5.2.

COMMENTARY

This page intentionally left blank

APPENDIX B
DESIGN OF MASONRY INFILL

CODE

B.1 — General

B.1.1 *Scope*

This chapter provides minimum requirements for the structural design of concrete masonry, clay masonry, and AAC masonry infills, either non-participating or participating. Infills shall comply with the requirements of Part 1, Part 2, excluding Sections 5.2, 5.3, 5.4, and 5.5, Section B.1, and either Section B.2 or B.3.

COMMENTARY

B.1 — General

B.1.1 *Scope*

The provisions of Appendix B outline a basic set of design provisions for masonry infill based upon experimental research and anecdotal performance of these masonry assemblies. The provisions address both non-participating infills, which are structurally isolated from the lateral force-resisting system, as well as participating infills, which are used to resist in-plane forces due to wind and earthquake. While masonry infills have been a part of contemporary construction for nearly a century, research investigations into their performance, particularly during seismic events, is still ongoing. A comprehensive review of available research data on the performance of masonry infills is provided by Tucker (2007).

As with masonry systems designed by other chapters of the Code, masonry infill must also be designed per the applicable requirements of Part 1 and Part 2. By reference to Part 1, masonry infill must comply with the prescriptive requirements of Chapter 7 for seismic design and detailing. This includes the prescriptive detailing requirements of Section 7.3.1 for non-participating infills and Section 7.3.2 for participating infills. Properly detailed masonry infills have shown considerable system ductility (Henderson et al, 2006). When participating infills are used to resist in-plane loads as part of a concrete or steel frame structure, a hybrid system is effectively created that may not otherwise be defined in Table 12.2-1 of ASCE 7 for seismic force-resistance. Until further research is completed, the Committee recommends using the smallest R and C_d value for the combination of the frame and masonry infill be used to design the system.

Over time, masonry materials expand and contract due to fluctuations in temperature and moisture content as discussed in Code Commentary Sections 4.2.3, 4.2.4, and 4.2.5. Volumetric changes in the masonry infill will open and close the gap between the infill and the bounding frame, which can have a significant impact on the strength and performance of the infill assembly. Such volumetric changes must be considered as required by Section 4.1.5.

When Appendix B (Design of Masonry Infill) was originally developed, information was not available regarding the performance of infills made of AAC masonry and designed according to the provisions of that Appendix. Information has subsequently become available regarding that performance (Ravichandran, 2009; Ravichandran and Klingner, 2011a; Ravichandran and Klingner, 2011b; Ravichandran and Klingner, 2011c). Infills of AAC masonry can safely be designed

CODE

COMMENTARY

using the provisions of Appendix B, and using f'_{AAC} instead of f'_m.

While Ravichandran's investigation illustrated that the provisions of Appendix B are accurate for stiffness and give conservative (low) values for strength, the user should be aware that underestimating the strength of AAC masonry infill may in turn underestimate the forces that can be transmitted from the infill to the bounding frame. Ravichandran (2009) suggests that this can be addressed by designing the frame for an upper fractile of the calculated infill capacity.

B.1.2 *Required strength*

Required strength shall be determined in accordance with the strength design load combinations of the legally adopted building code. When the legally adopted building code does not provide load combinations, structures and members shall be designed to resist the combination of loads specified in ASCE 7 for strength design.

B.1.3 *Design strength*

Infills shall be proportioned so that the design strength equals or exceeds the required strength. Design strength is the nominal strength multiplied by the strength-reduction factor, ϕ, as specified in Section B.1.4.

B.1.4 *Strength-reduction factors*

The value of ϕ shall be taken as 0.60, and applied to the shear, flexure, and axial strength of a masonry infill panel.

B.1.4 *Strength-reduction factors*

See Code Commentary Section 9.1.4 for additional discussion on strength reduction factors applicable to concrete and clay masonry. See Code Commentary Section 11.1.5 for additional discussion on strength reduction factors applicable to AAC masonry. The strength reduction factor applies only to the design of the masonry infill. The strength reduction factors for the anchorage (Section 9.1.4.1 or 11.1.5.1, as appropriate) and bearing (Section 9.1.4.2 or 11.1.5.2, as appropriate) remain unchanged.

B.1.5 *Limitations*

Partial infills and infills with openings shall not be considered as part of the lateral force-resisting system. Their effect on the bounding frame, however, shall be considered.

B.1.5 *Limitations*

Structures with partial-height infills have generally performed very poorly during seismic events. Partial-height infills create short columns, which attract additional load due to their increased stiffness. This has led to premature column failure. Concrete columns bounding partial-height infills are particularly vulnerable to shear failure (Chiou et al, 1999).

CODE

B.2 — Non-participating infills

Non-participating infills shall comply with the requirements of Sections B.2.1 and B.2.2.

B.2.1 *In-plane isolation joints for non-participating infills*

B.2.1.1 In-plane isolation joints shall be designed between the infill and the sides and top of the bounding frame.

B.2.1.2 In-plane isolation joints shall be specified to be at least 3/8 in. (9.5 mm) wide in the plane of the infill, and shall be sized to accommodate the design displacements of the bounding frame.

B.2.1.3 In-plane isolation joints shall be free of mortar, debris, and other rigid materials, and shall be permitted to contain resilient material, provided that the compressibility of that material is considered in establishing the required size of the joint.

B.2.2 *Design of non-participating infills for out-of-plane loads*

Connectors supporting non-participating infills against out-of-plane loads shall be designed to meet the requirements of Sections B.2.2.1 through B.2.2.4. The infill shall be designed to meet the requirements of Section B.2.2.5.

B.2.2.1 The connectors shall be attached to the bounding frame.

B.2.2.2 The connectors shall not transfer in-plane forces.

B.2.2.3 The connectors shall be designed to satisfy the requirements of ASCE 7.

B.2.2.4 The connectors shall be spaced at a maximum of 4 ft (1.22 m) along the supported perimeter of the infill.

B.2.2.5 The infill shall be designed to resist out-of-plane bending between connectors in accordance with Section 9.2 for unreinforced concrete masonry or clay masonry infill, Section 11.2 for unreinforced AAC masonry infill, Section 9.3 for reinforced concrete masonry or clay masonry infill, or Section 11.3 for reinforced AAC masonry infill.

COMMENTARY

B.2.1 *In-plane isolation joints for non-participating infills*

To preclude the unintentional transfer of in-plane loads from the bounding frame to the non-participating infill, gaps are required between the top and sides of the masonry infill assembly. These gaps must be free of materials that could transfer loads between the infill and bounding frame and must be capable of accommodating frame displacements, including inelastic deformation during seismic events.

B.2.2 *Design of non-participating infills for out-of-plane loads*

Mechanical connection between the infill and bounding frame is required for out-of-plane support of the masonry. Masonry infill can be modeled as spanning vertically, horizontally, or both. Connectors between the infill and the bounding frame must be sized and located to maintain load path continuity.

CODE

B.3 — Participating infills

Participating infills shall comply with the requirements of Sections B.3.1 through B.3.6.

B.3.1 *General*

Infills with in-plane isolation joints not meeting the requirements of Section B.2.1 shall be considered as participating infills. For such infills the displacement shall be taken as the bounding frame displacement minus the specified width of the gap between the bounding column and infill.

B.3.1.1 The maximum ratio of the nominal vertical dimension to nominal thickness of participating infills shall not exceed 30.

B.3.1.2 Participating infills that are not constructed in contact with the bounding beam or slab adjacent to their upper edge shall be designed in accordance with Section B.3.1.2.1 or B.3.1.2.2.

B.3.1.2.1 Where the specified gap between the bounding beam or slab at the top of the infill is less than 3/8 in. (9.5 mm) or the gap is not sized to accommodate design displacements, the infill shall be designed in accordance with Sections B.3.4 and B.3.5, except that the calculated stiffness and strength of the infill shall be multiplied by a factor of 0.5.

B.3.1.2.2 If the gap between the infill and the overlying bounding beam or slab is sized such that in-plane forces cannot be transferred between the bounding beam or slab and the infill, the infill shall be considered a partial infill and shall comply with Section B.1.5.

B.3.2 *In-plane connection requirements for participating infills*

Mechanical connections between the infill and the bounding frame shall be permitted provided that they do not transfer in-plane forces between the infill and the bounding frame.

COMMENTARY

B.3.1 *General*

Flanagan and Bennett (1999a) tested an infilled frame with a 1.0 in. gap between the infill and bounding column. Once the gap was closed, the specimen performed like an infilled frame with no gap.

B.3.1.1 The maximum permitted ratio of height to thickness is based on practical conditions for stability.

B.3.1.2.1 Dawe and Seah (1989a) noted a slight decrease in stiffness and strength when a bond breaker (a polyethylene sheet) was used at the top interface. Riddington (1984) showed an approximate 50% decrease in stiffness but little reduction in peak load with a top gap that was 0.1% of the height of the infill. Dawe and Seah (1989a) showed an approximate 50% reduction in stiffness and a 60% reduction in strength with a top gap that was 0.8% of the height of the infill. A top gap that is in compliance with Section B.2.1.2 is generally less than 0.5% of the infill height. Thus, a 50% reduction in strength and stiffness seems appropriate.

B.3.1.2.2 In cases where the gap at the top of the infill is sufficiently large so that forces cannot be transferred between the bounding frame or beam and the masonry infill, the infill is considered to be partial infill and not permitted to considered part of the lateral force-resisting system.

B.3.2 *In-plane connection requirements for participating infills*

The modeling provisions of Appendix B for participating infills assume that in-plane loads are resisted by the infill by a diagonal compression strut, which does not rely upon mechanical connectors to transfer in-plane load. While mechanical connections, including the use of reinforcement, are permitted, they must be detailed to preclude load transfer between the infill and bounding frame. This is because mechanical connectors between the infill and bounding frame can cause premature damage along the boundaries of the infill under in-plane loading (Dawe and Seah, 1989a). This damage actually reduces the out-of-plane capacity of the infill, as the ability of the infill to have arching action is reduced.

CODE

B.3.3 *Out-of-plane connection requirements for participating infills*

B.3.3.1 Participating infills shall be supported out-of-plane by connectors attached to the bounding frame.

B.3.3.2 Connectors providing out-of-plane support shall be designed to satisfy the requirements of ASCE 7.

B.3.3.3 Connectors providing out-of-plane support shall be spaced at a maximum of 4 ft (1.22 m) along the supported perimeter of the infill.

B.3.4 *Design of participating infills for in-plane forces*

B.3.4.1 Unless the stiffness of the infill is obtained by a more comprehensive analysis, a participating infill shall be analyzed as an equivalent strut, capable of resisting compression only; whose width is calculated using Equation B-1; whose thickness is the specified thickness of the infill; and whose elastic modulus is the elastic modulus of the infill.

$$w_{inf} = \frac{0.3}{\lambda_{strut} \cos\theta_{strut}} \quad \text{(Equation B-1)}$$

where

$$\lambda_{strut} = \sqrt[4]{\frac{E_m \, t_{net\,inf} \, \sin 2\theta_{strut}}{4 \, E_{bc} \, I_{bc} \, h_{inf}}} \quad \text{(Equation B-2a)}$$

for the design of concrete masonry and clay masonry infill; and

$$\lambda_{strut} = \sqrt[4]{\frac{E_{AAC} \, t_{net\,inf} \, \sin 2\theta_{strut}}{4 \, E_{bc} \, I_{bc} \, h_{inf}}} \quad \text{(Equation B-2b)}$$

for the design of AAC masonry infill.

B.3.4.2 Design forces in equivalent struts, as defined in Section B.3.4.1, shall be determined from an elastic analysis of a braced frame including such equivalent struts.

CODE

B.3.4.3 $V_{n\,inf}$ shall be the smallest of (a), (b), and (c) for concrete masonry and clay masonry infill and (b), (d), and (e) for AAC masonry infill:

(a) $(6.0 \text{ in.}) t_{net\,inf} f'_m$ (Equation B-3)

(b) the calculated horizontal component of the force in the equivalent strut at a horizontal racking displacement of 1.0 in. (25 mm)

(c) $\dfrac{V_n}{1.5}$ (Equation B-4)

where V_n is the smallest nominal shear strength from Section 9.2.6, calculated along a bed joint.

(d) $(6.0 \text{ in.}) t_{net\,inf} f'_{AAC}$ (Equation B-5)

(e) $\dfrac{V_{nAAC}}{1.5}$ (Equation B-6)

where V_{nAAC} is the smallest nominal shear strength from Section 11.2.5, calculated along a bed joint.

B.3.5 *Design of frame elements with participating infills for in-plane loads*

B.3.5.1 Design each frame member not in contact with an infill for shear, moment, and axial force not less than the results from the equivalent strut frame analysis.

B.3.5.2 Design each bounding column in contact with an infill for shear and moment equal to not less than 1.1 multiplied by the results from the equivalent strut frame analysis, and for axial force not less than the results from that analysis. In addition, increase the design shear at each end of the column by the horizontal component of the equivalent strut force acting on that end under design loads.

B.3.5.3 Design each beam or slab in contact with an infill for shear and moment equal to at least 1.1 multiplied by the results from the equivalent strut frame analysis, and for an axial force not less than the results from that analysis. In addition, increase the design shear at each end of the beam or slab by the vertical component of the equivalent strut force acting on that end under design loads.

COMMENTARY

B.3.4.3 The capacity of the infill material is often referred to as corner crushing, although the failure may occur elsewhere as well. Flanagan and Bennett (1999a) compared six methods for determining the strength of the infill material to experimental results of structural clay tile infills in steel frames. The method given in the Code is the simplest method, and also quite accurate, with a coefficient of variation of the ratio of the measured strength to the predicted strength of the infill of 24%. Flanagan and Bennett (2001) examined the performance of this method for predicting the strength of 58 infill tests reported in the literature. Clay tile, clay brick, and concrete masonry infills in both steel and concrete bounding frames were examined. For the 58 tests considered, the coefficient of variation of the ratio of measured to predicted strength of the infill was 21%.

Flanagan and Bennett (1999a) determined that in-plane displacement is a better indicator of infill performance than in-plane drift (displacement divided by height). This was based on comparing the results of approximately 8-ft high (2.4 m) infill tests to 24-ft (7.3 m) high infill tests on similar material. Thus, a displacement limit rather than a drift limit is given in the Code. As a general rule, the strength of the infill is reached at smaller displacements for stiffer bounding columns. For more flexible bounding columns, the strength of the infill is controlled by the displacement limit of 1.0 in. (25 mm).

Equation B-4 is intended to address shear failure along a bed joint. The use of a formula from Section 9.2 is not intended to imply that concrete masonry and clay masonry infills are necessarily unreinforced. Shear resistance along a bed joint is similar for the equations of Section 9.2 and Section 9.3, and the former are more clearly related to failure along a bed joint. The same reasoning applies to Equation B-6 for AAC masonry infill.

CODE

B.3.6 *Design of participating infills for out-of-plane forces*

The nominal out-of-plane flexural capacity to resist out-of-plane forces of the infill per unit area shall be determined in accordance with Equation B-7a for concrete masonry and clay masonry and Equation B-7b for AAC masonry:

$$q_{n\,\text{inf}} = 105(f'_m)^{0.75} t_{\text{inf}}^2 \left(\frac{\alpha_{arch}}{l_{\text{inf}}^{2.5}} + \frac{\beta_{arch}}{h_{\text{inf}}^{2.5}} \right)$$

(Equation B-7a)

$$q_{n\,\text{inf}} = 105(f'_{AAC})^{0.75} t_{\text{inf}}^2 \left(\frac{\alpha_{arch}}{l_{\text{inf}}^{2.5}} + \frac{\beta_{arch}}{h_{\text{inf}}^{2.5}} \right)$$

(Equation B-7b)

where:

$$\alpha_{arch} = \frac{1}{h_{\text{inf}}} (E_{bc} I_{bc} h_{\text{inf}}^2)^{0.25} < 35$$

(Equation B-8)

$$\beta_{arch} = \frac{1}{l_{\text{inf}}} (E_{bb} I_{bb} l_{\text{inf}}^2)^{0.25} < 35$$

(Equation B-9)

In Equation B-7, t_{inf} shall not be taken greater than $1/8\ h_{inf}$. When bounding columns of different cross-sectional properties are used on either side of the infill, average properties shall be used to calculate this capacity. When bounding beams of different cross-sectional properties are used above and below the infill, average properties shall be used to calculate this capacity. In the case of a single story frame, the cross-sectional properties of the bounding beam above the infill shall be used to calculate this capacity. When a side gap is present, α_{arch} shall be taken as zero. When a top gap is present, β_{arch} shall be taken as zero.

COMMENTARY

B.3.6 *Design of participating infills for out-of-plane forces*

It is not appropriate to calculate the out-of-plane flexural capacity of unreinforced masonry infills using values for flexural tensile capacity. The predominant out-of-plane resisting mechanism for masonry infills is arching. Even infills with dry-stacked block have been shown to have significant out-of-plane strength (Dawe and Seah, 1989b).

The out-of-plane resistance of masonry infill as calculated by Equation B-7 is based upon an arching model of the infill in the bounding frame and therefore neglects the contribution of any reinforcement that may be present in the infill in determining the out-of-plane flexural strength of participating infill. Masonry infill may require reinforcement, however, to resist out-of-plane flexure between points of connection with the bounding frame, or to meet the prescriptive seismic detailing requirements of Chapter 7.

The thickness used in calculations of out-of-plane flexural resistance is limited because infills with low height-to-thickness ratios are less influenced by membrane compression and more influenced by plate bending.

The out-of-plane flexural capacity of the masonry infill is determined based on the work of Dawe and Seah (1989b). They first developed a computer program based on a modified yield line analysis that included the flexibility of the bounding frame. The program coincided quite well with their experimental results, with an average ratio of observed to predicted capacity of 0.98 and a coefficient of variation of 6%. Dawe and Seah (1989b) then used the program for an extensive parametric study that resulted in the empirical equation given here.

Two other equations are available. The first, proposed by Abrams et al. (1993), is used in ASCE 41 (2006). The second was proposed by Klingner et al. (1997). In Flanagan and Bennett (1999b), each of these three proposed equations is checked against the results of 31 experimental tests from seven different test programs including clay brick infills in concrete frames, clay tile infills in steel frames, clay brick infills in steel frames, and concrete masonry infills in steel frames. Flanagan and Bennett (1999b) determined that Dawe and Seah's (1989b) equation is the best predictor of out-of-plane strength, with an average ratio of observed to predicted strength of 0.92, and a coefficient of variation of 0.28. The coefficient of variation of observed to predicted capacity was 28%. Results are summarized in Figure CC-B.3-1. The experimental tests involved infills with height-to-thickness ratios ranging from 6.8 to 35.3. Some infills had joint reinforcement, but this did not affect the results. Two of the specimens had a top gap. Arching still occurred, but was one-way arching. The code equation is thus quite robust.

COMMENTARY

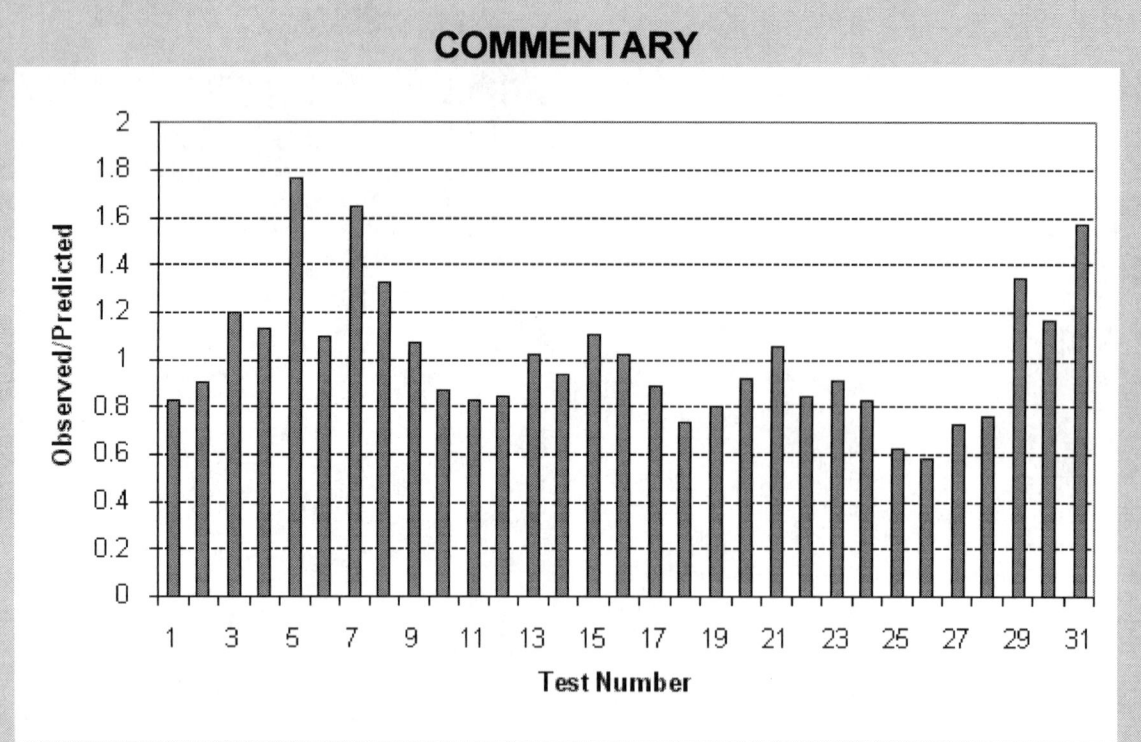

Figure CC-B.3-1: Ratios of observed to predicted strengths for infills loaded out-of-plane (Flanagan and Bennett 1999b)

APPENDIX C
LIMIT DESIGN METHOD

CODE

C. *General* — The limit design method shall be permitted to be applied to a line of lateral load resistance consisting of special reinforced masonry shear walls that are designed per the strength design provisions of Chapter 9, except that the provisions of Section 9.3.3.5 and Section 9.3.6.5 shall not apply.

C.1 *Yield mechanism* — It shall be permitted to use limit analysis to determine the controlling yield mechanism and its corresponding base-shear strength, V_{lim}, for a line of lateral load resistance, provided that (a) through (e) are satisfied:

(a) The relative magnitude of lateral seismic forces applied at each floor level shall correspond to the loading condition producing the maximum base shear at the line of resistance in accordance with analytical procedures permitted in Section 12.6 of ASCE 7.

(b) In the investigation of potential yield mechanisms induced by seismic loading, plastic hinges shall be considered to form at the faces of joints and at the interfaces between masonry components and the foundation.

(c) The axial forces associated with Load Combination 7 of Section 2.3.2 of ASCE 7 shall be used when determining the strength of plastic hinges, except that axial loads due to horizontal seismic forces shall be permitted to be neglected.

(d) The strength assigned to plastic hinges shall be based on the nominal flexural strength, M_n, but shall not exceed the moment associated with one-half of the nominal shear strength, V_n, calculated using MSJC Section 9.3.4.1.2.

(e) At locations other than the plastic hinges identified in C.1(b), moments shall not exceed the strengths assigned in C.1(d) using the assumptions of C.1(c).

COMMENTARY

C. *General* — This section provides alternative design provisions for special reinforced masonry shear walls subjected to in-plane seismic loading. The limit design method is presented as an alternative to the requirements of 9.3.3.5 and 9.3.6.5. All other sections in Chapter 9 are applicable. Limit design is considered to be particularly useful for perforated wall configurations for which a representative yield mechanism can be determined (Lepage et al, 2011).

C.1 *Yield mechanism* — This section defines the basic conditions for allowing the use of limit analysis to determine the base shear strength of a line of resistance subjected to seismic loading.

Item (a) allows the use of conventional methods of analysis permitted in ASCE 7 to determine the distribution of lateral loads. The designer should use the seismic loading condition that produces the maximum base shear demand at the line of resistance.

Item (b) allows the location of yielding regions at the interfaces between wall segments and their supporting members.

Item (c) prescribes the use of the loading condition that induces the lowest axial force due to gravity loads. For wall segments loaded with axial forces below the balanced point, this loading condition gives the lowest flexural strength and therefore leads to lower mechanism strengths. Axial loads from seismic overturning are permitted to be neglected only in the initial process of establishing the plastic capacity of the selected mechanism. Axial loads from seismic overturning are required to be considered subsequently, in determining the deformation capacity of plastic hinges.

Item (d) limits the flexural strength that is assigned to a plastic hinge so that the maximum shear that can be developed does not exceed one-half the shear strength of the wall segment. This stratagem effectively reduces the strength of the controlling yield mechanism involving wall segments vulnerable to shear failure. In addition to a reduction in strength there is a reduction in deformation capacity as indicated in C.3.2.

Item (e) requires the designer to verify that the selected mechanism is the critical one. If yielding is detected away from the selected plastic hinge locations, the designer has the choice of changing the selected plastic hinge location to recognize that yielding, or of placing additional reinforcement at the section where yielding is detected.

CODE

C.2 *Mechanism strength* — The yield mechanism associated with the limiting base-shear strength, V_{lim}, shall satisfy the following:

$$\phi V_{lim} \geq V_{ub} \quad \text{(Equation C-1)}$$

The value of ϕ assigned to the mechanism strength shall be taken as 0.8. The base-shear demand, V_{ub}, shall be determined from analytical procedures permitted in Section 12.6 of ASCE 7.

C.3 *Mechanism deformation* — The rotational deformation demand on plastic hinges shall be determined by imposing the design displacement, δ_u, at the roof level of the yield mechanism. The rotational deformation capacity of plastic hinges shall satisfy C.3.1 to C.3.3.

C.3.1 The rotational deformation capacity of plastic hinges shall be taken as $0.5\, l_w\, \varepsilon_{mu}/c$. The value of c shall be calculated for the P_u corresponding to Load Combination 5 of Section 2.3.2 of ASCE 7.

C.3.2 The angular deformation capacity of masonry components whose plastic hinge strengths are limited by shear as specified in C.1(d), shall be taken as $1/400$. The angular deformation capacity shall be permitted to be taken as $1/200$ for masonry components satisfying the following requirements:

(a) The areas of transverse and longitudinal reinforcement shall each not be less than 0.001 multiplied by the gross cross-sectional area of the component, using specified dimensions;

(b) Spacing of transverse and longitudinal reinforcement shall not exceed the smallest of 24 in. (610 mm), $l_w/2$, and $h_w/2$.

(c) Reinforcement ending at a free edge of masonry shall be anchored around perpendicular reinforcing bars with a standard hook.

C.3.3 The P_u corresponding to load combination 5 of Section 2.3.2 of ASCE 7 shall not exceed a compressive stress of $0.3\, f_m'\, A_g$ at plastic hinges in the controlling mechanism.

COMMENTARY

C.2 *Mechanism strength* — Because the controlling yield mechanism is investigated using nominal strengths, an overall strength reduction factor of $\phi = 0.8$ is applied to the limiting base shear strength. For simplicity, a single value of ϕ is adopted.

C.3 *Mechanism deformation* — This section defines the ductility checks required by the limit design method. The deformation demands at locations of plastic hinges are determined by imposing the calculated design roof displacement to the controlling yield mechanism.

C.3.1 The rotational deformation capacity is calculated assuming an ultimate curvature of ε_{mu}/c over a plastic hinge length of $0.5\, l_w$. The resulting expression is similar to that used in 9.3.6.5.3(a) to determine the need for special boundary elements. In the latter case, it is multiplied by wall height. The value of P_u includes earthquake effects, and may be calculated using a linearly elastic model.

C.3.2 In shear-dominated elements (elements whose hinge strength is assigned a value lower than their nominal flexural strength due to limitations in C.1(d)), the angular deformation capacity is limited to $1/400$ or $1/200$, depending on the percentage and maximum spacing of transverse and longitudinal reinforcement.

C.3.3 The limit of 30% of f_m' is intended to ensure that all yielding components respond below the balanced point of the P-M interaction diagram.

BUILDING CODE REQUIREMENTS FOR MASONRY STRUCTURES AND COMMENTARY

EQUATION CONVERSIONS

The equations in this Code are for use with the specified inch-pound units only. The equivalent equations for use with SI units follow.

Code Equation No. or Section No.	SI Unit Equivalent Equation	Units
4.2.2.2.1	$E_m = 700 f'_m$ for clay masonry $E_m = 900 f'_m$ for concrete masonry	E_m in MPa f'_m in MPa
4.2.2.3.1	$E_{AAC} = 888 (f'_{AAC})^{0.6}$	E_{AAC} in MPa f'_{AAC} in MPa
4.2.2.4	$500 f'_g$	f'_g in MPa
(5-1)	$I_{eff} = I_n \left(\dfrac{M_{cr}}{M_a}\right)^3 + I_{cr}\left[1-\left(\dfrac{M_{cr}}{M_a}\right)^3\right] \le I_n$	I_{eff} in mm^4 I_n in mm^4 I_{cr} in mm^4 M_{cr} in N-mm M_a in N-mm
(5-2a)	(1) When $1 \le \dfrac{l_{eff}}{d_v} < 2$, $z = 0.2(l_{eff} + 2d_v)$	l_{eff} in mm d_v in mm z in mm
(5-2b)	(2) When $\dfrac{l_{eff}}{d_v} < 1$, $z = 0.6 l_{eff}$	l_{eff} in mm d_v in mm z in mm
(5-3a)	(1) When $1 \le \dfrac{l_{eff}}{d_v} < 3$, $z = 0.2(l_{eff} + 1.5 d_v)$	l_{eff} in mm d_v in mm z in mm
(5-3b)	(2) When $\dfrac{l_{eff}}{d_v} < 1$, $z = 0.5 l_{eff}$	l_{eff} in mm d_v in mm z in mm
(6-1)	$A_{pt} = \pi l_b^2$	A_{pt} in mm^2 l_b in mm
(6-2)	$A_{pv} = \dfrac{\pi l_{be}^2}{2}$	A_{pv} in mm^2 l_{be} in mm
(8-1)	$B_{ab} = 0.104 A_{pt} \sqrt{f'_m}$	A_{pt} in mm^2 B_{ab} in N f'_m in MPa $\sqrt{f'_m}$ result in MPa
(8-2)	$B_{as} = 0.6 A_b f_y$	A_b in mm^2 B_{as} in N f_y in MPa
(8-3)	$B_{ab} = 0.104 A_{pt} \sqrt{f'_m}$	A_{pt} in mm^2 B_{ab} in N f'_m in MPa $\sqrt{f'_m}$ result in MPa
(8-4)	$B_{ap} = 0.6 f'_m e_b d_b + 0.83 \pi (l_b + e_b + d_b) d_b$	f'_m in MPa e_b in mm d_b in mm l_b in mm B_{ap} in N
(8-5)	$B_{as} = 0.6 A_b f_y$	A_b in mm^2 B_{as} in N f_y in MPa

Code Equation No. or Section No.	SI Unit Equivalent Equation	Units
(8-6)	$B_{vb} = 0.104 A_{pv} \sqrt{f'_m}$	A_{pv} in mm^2 B_{vb} in N f'_m in MPa $\sqrt{f'_m}$ result in MPa
(8-7)	$B_{vc} = 1072 \sqrt[4]{f'_m A_b}$	A_b in mm^2 B_{vc} in N f'_m in MPa $\sqrt[4]{f'_m A_b}$ result in N
(8-8)	$B_{vpry} = 2.0 B_{ab} = 0.208 A_{pt} \sqrt{f'_m}$	A_{pt} in mm^2 B_{ab} in N B_{vpry} in N f'_m in MPa $\sqrt{f'_m}$ result in MPa
(8-9)	$B_{vs} = 0.36 A_b f_y$	A_b in mm^2 B_{vs} in N f_y in MPa
(8-10)	$\dfrac{b_a}{B_a} + \dfrac{b_v}{B_v} \leq 1$	b_a in N b_v in N B_a in N B_v in N
8.1.4.2(c)	$0.108 \sqrt{\text{specified unit compressive strength of header}}$	in MPa
(8-11)	$l_d = 0.22 d_b F_s$	d_b in mm F_s in MPa l_d in mm
(8-12)	$l_d = \dfrac{1.57 d_b^2 f_y \gamma}{K \sqrt{f'_m}}$	d_b in mm f'_m in MPa $\sqrt{f'_m}$ result in MPa f_y in MPa K in mm l_d in mm
8.1.6.4.1.5(b)	$A_v \geq 0.41 \left(\dfrac{b_w s}{f_y} \right)$ $s \leq \left(\dfrac{d}{8 \beta_b} \right)$	A_v in mm^2 b_w in mm s in mm f_y in MPa d in mm β_b is dimensionless
(8-13)	$\xi = 1.0 - \dfrac{11.6 A_{sc}}{d_b^{2.5}}$ where $\dfrac{11.6 A_{sc}}{d_b^{2.5}} \leq 1.0$	A_{sc} in mm^2 d_b in mm
(8-14)	$\dfrac{f_a}{F_a} + \dfrac{f_b}{F_b} \leq 1$	F_a in MPa F_b in MPa f_a in MPa f_b in MPa
(8-15)	$P \leq (1/4) P_e$	P in N P_e in N
(8-16)	$F_a = (1/4) f'_m \left[1 - \left(\dfrac{h}{140 r} \right)^2 \right]$	F_a in MPa f'_m in MPa h in mm r in mm

BUILDING CODE REQUIREMENTS FOR MASONRY STRUCTURES AND COMMENTARY

Code Equation No. or Section No.	SI Unit Equivalent Equation	Units
(8-17)	$F_a = \left(\frac{1}{4}\right) f'_m \left(\frac{70r}{h}\right)^2$	F_a in MPa f'_m in MPa h in mm r in mm
(8-18)	$F_b = \left(\frac{1}{3}\right) f'_m$	F_b in MPa f'_m in MPa
(8-19)	$P_e = \frac{\pi^2 E_m I_n}{h^2}\left(1 - 0.577\frac{e}{r}\right)^3$	E_m in MPa e in mm h in mm I_n in mm^4 P_e in N r in mm
(8-20)	$f_v = \frac{VQ}{I_n b}$	b in mm f_v in MPa I_n in mm^4 Q in mm^3 V in N
8.2.6.2(a)	$0.125\sqrt{f'_m}$	f'_m in MPa $\sqrt{f'_m}$ result in MPa
8.2.6.2(c)	$0.255 + 0.45\, N_v/A_n$	A_n in mm^2 N_v in N Answer in MPa
8.2.6.2(d)	$0.255 + 0.45\, N_v/A_n$	A_n in mm^2 N_v in N Answer in MPa
8.2.6.2(e)	$0.414 + 0.45\, N_v/A_n$	A_n in mm^2 N_v in N Answer in MPa
(8-21)	$P_a = (0.25 f'_m A_n + 0.65 A_{st} F_s)\left[1 - \left(\frac{h}{140r}\right)^2\right]$	A_n in mm^2 A_{st} in mm^2 F_s in MPa f'_m in MPa h in mm P_a in N r in mm
(8-22)	$P_a = (0.25 f'_m A_n + 0.65 A_{st} F_s)\left(\frac{70r}{h}\right)^2$	A_n in mm^2 A_{st} in mm^2 F_s in MPa f'_m in MPa h in mm P_a in N r in mm
(8-23)	$\rho_{max} = \dfrac{n f'_m}{2 f_y\left(n + \dfrac{f_y}{f'_m}\right)}$	f_y in MPa f'_m in MPa
(8-24)	$f_v = \dfrac{V}{A_{nv}}$	A_{nv} in mm^2 f_v in MPa V in N
(8-25)	$F_v = (F_{vm} + F_{vs})\gamma_g$	F_v in MPa F_{vm} in MPa F_{vs} in MPa

Code Equation No. or Section No.	SI Unit Equivalent Equation	Units
(8-26)	$F_v \leq \left(0.249\sqrt{f'_m}\right)\gamma_g$ For $M/(Vd_v) \leq 0.25$	d_v in mm f'_m in MPa $\sqrt{f'_m}$ result in MPa F_v in MPa M in N-mm V in N
(8-27)	$F_v \leq \left(0.167\sqrt{f'_m}\right)\gamma_g$ For $M/(Vd_v) \geq 1.0$	d_v in mm f'_m in MPa $\sqrt{f'_m}$ result in MPa F_v in MPa M in N-mm V in N
(8-28)	$F_{vm} = 0.021\left[\left(4.0 - 1.75\left(\dfrac{M}{Vd_v}\right)\right)\sqrt{f'_m}\right] + 0.25\dfrac{P}{A_n}$	A_n in mm^2 d_v in mm f'_m in MPa $\sqrt{f'_m}$ result in MPa F_{vm} in MPa M in N-mm P in N V in N
(8-29)	$F_{vm} = 0.042\left[\left(4.0 - 1.75\left(\dfrac{M}{Vd_v}\right)\right)\sqrt{f'_m}\right] + 0.25\dfrac{P}{A_n}$	A_n in mm^2 d_v in mm f'_m in MPa $\sqrt{f'_m}$ result in MPa F_{vm} in MPa M in N-mm P in N V in N
(8-30)	$F_{vs} = 0.5\left(\dfrac{A_v F_s d_v}{A_{nv} s}\right)$	A_{nv} in mm^2 A_v in mm^2 d_v in mm F_s in MPa F_{vs} in MPa s in mm
(9-1)	$B_{anb} = 0.332 A_{pt}\sqrt{f'_m}$	A_{pt} in mm^2 B_{anb} in N f'_m in MPa $\sqrt{f'_m}$ result in MPa
(9-2)	$B_{ans} = A_b f_y$	A_b in mm^2 f_y in MPa B_{ans} in N
(9-3)	$B_{anb} = 0.332 A_{pt}\sqrt{f'_m}$	A_{pt} in mm^2 B_{anb} in N f'_m in MPa $\sqrt{f'_m}$ result in MPa

BUILDING CODE REQUIREMENTS FOR MASONRY STRUCTURES AND COMMENTARY

Code Equation No. or Section No.	SI Unit Equivalent Equation	Units
(9-4)	$B_{anp} = 1.5 f'_m e_b d_b + 2.07\pi(l_b + e_b + d_b)d_b$	f'_m in MPa e_b in mm d_b in mm l_b in mm B_{anp} in N
(9-5)	$B_{ans} = A_b f_y$	A_b in mm^2 f_y in MPa B_{ans} in N
(9-6)	$B_{vnb} = 0.332 A_{pv} \sqrt{f'_m}$	A_{pv} in mm^2 B_{vnb} in N f'_m in MPa $\sqrt{f'_m}$ result in MPa
(9-7)	$B_{vnc} = 3216 \sqrt[4]{f'_m A_b}$	A_b in mm^2 B_{vnc} in N f'_m in MPa $\sqrt[4]{f'_m A_b}$ result in N
(9-8)	$B_{vnpry} = 2.0 B_{anb} = 0.664 A_{pt} \sqrt{f'_m}$	A_{pt} in mm^2 B_{anb} in N B_{vnpry} in N f'_m in MPa $\sqrt{f'_m}$ result in MPa
(9-9)	$B_{vns} = 0.6 A_b f_y$	A_b in mm^2 f_y in MPa B_{vns} in N
(9-10)	$\dfrac{b_{af}}{\phi B_{an}} + \dfrac{b_{vf}}{\phi B_{vn}} \leq 1$	b_{af} in N b_{vf} in N B_{an} in N B_{vn} in N
9.1.7.2(c)	$0.216 \sqrt{\text{specified unit compressive strength of header}}$	in MPa
(9-11)	$P_n = 0.80 \left\{ 0.80 A_n f'_m \left[1 - \left(\dfrac{h}{140r} \right)^2 \right] \right\}$ For $\dfrac{h}{r} \leq 99$	P_n in N A_n in mm^2 f'_m in MPa h in mm r in mm
(9-12)	$P_n = 0.80 \left(0.80 A_n f'_m \left(\dfrac{70r}{h} \right)^2 \right)$ For $\dfrac{h}{r} > 99$	P_n in N A_n in mm^2 f'_m in MPa h in mm r in mm
(9-13)	$M_u = \psi M_{u,0}$	M_u in N-mm $M_{u,0}$ in N-mm
(9-14)	$\psi = \dfrac{1}{1 - \dfrac{P_u}{A_n f'_m \left(\dfrac{70r}{h}\right)^2}}$	A_n in mm^2 f'_m in MPa P_u in N h in mm r in mm
9.2.6.1(a)	$0.316 A_{nv} \sqrt{f'_m}$ in N	A_{nv} in mm^2 f'_m in MPa
9.2.6.1(b)	$2.07 A_{nv}$ in N	A_{nv} in mm^2
9.2.6.1(c)	$0.386 A_{nv} + 0.45 N_u$ in N	A_{nv} in mm^2 N_u in N

Code Equation No. or Section No.	SI Unit Equivalent Equation	Units
9.2.6.1(d)	$0.386 A_{nv} + 0.45 N_u$ in N	A_{nv} in mm^2 N_u in N
9.2.6.1(e)	$0.620 A_{nv} + 0.45 N_u$ in N	A_{nv} in mm^2 N_u in N
9.2.6.1(f)	$0.159 A_{nv}$ in N	A_{nv} in mm^2
(9-15)	$l_e = 13 d_b$	l_e in mm d_b in mm
(9-16)	$l_d = \dfrac{1.57 d_b^2 f_y \gamma}{K \sqrt{f'_m}}$	d_b in mm f'_m in MPa $\sqrt{f'_m}$ result in MPa f_y in MPa K in mm l_d in mm
(9-17)	$l_d = 48 d_b$	l_d in mm d_b in mm
(9-18)	$\xi = 1.0 - \dfrac{11.6 A_{sc}}{d_b^{2.5}}$ where $\dfrac{11.6 A_{sc}}{d_b^{2.5}} \leq 1.0$	A_{sc} in mm^2 d_b in mm
(9-19)	$P_n = 0.80 \left[0.80 f'_m (A_n - A_{st}) + f_y A_{st} \right] \left[1 - \left(\dfrac{h}{140r} \right)^2 \right]$	A_n in mm^2 A_{st} in mm^2 f'_m in MPa f_y in MPa P_n in N h in mm r in mm
(9-20)	$P_n = 0.80 \left[0.80 f'_m (A_n - A_{st}) + f_y A_{st} \right] \left(\dfrac{70r}{h} \right)^2$	A_n in mm^2 A_{st} in mm^2 f'_m in MPa f_y in MPa P_n in N h in mm r in mm
(9-21)	$V_n = (V_{nm} + V_{ns}) \gamma_g$	V_{nm} in N V_{ns} in N V_n in N
(9-22)	$V_n \leq \left(0.498 A_{nv} \sqrt{f'_m} \right) \gamma_g$ For $\dfrac{M_u}{V_u d_v} \leq 0.25$	A_{nv} in mm^2 M_u in N-mm V_u in N d_v in mm f'_m in MPa $\sqrt{f'_m}$ result in MPa V_n in N
(9-23)	$V_n \leq \left(0.332 A_{nv} \sqrt{f'_m} \right) \gamma_g$ For $\dfrac{M_u}{V_u d_v} \geq 1.0$	A_{nv} in mm^2 M_u in N-mm V_u in N f'_m in MPa $\sqrt{f'_m}$ result in MPa d_v in mm V_n in N

BUILDING CODE REQUIREMENTS FOR MASONRY STRUCTURES AND COMMENTARY

Code Equation No. or Section No.	SI Unit Equivalent Equation	Units
(9-24)	$V_{nm} = 0.083\left[4.0 - 1.75\left(\dfrac{M_u}{V_u d_v}\right)\right] A_{nv} \sqrt{f'_m} + 0.25 P_u$	A_{nv} in mm^2 M_u in N-mm V_u in N f'_m in MPa $\sqrt{f'_m}$ result in MPa d_v in mm P_u in N V_{nm} in N
(9-25)	$V_{ns} = 0.5\left(\dfrac{A_v}{s}\right) f_y d_v$	A_v in mm^2 f_y in MPa d_v in mm s in mm V_{ns} in N
(9-26)	$\left(\dfrac{P_u}{A_g}\right) \leq 0.20 f'_m$	P_u in N A_g in mm^2 f'_m in MPa
(9-27)	$M_u = \dfrac{w_u h^2}{8} + P_{uf}\dfrac{e_u}{2} + P_u \delta_u$	h in mm w_u in N/mm P_{uf} in N e_u in mm P_u in N δ_u in mm M_u in N-mm
(9-28)	$P_u = P_{uw} + P_{uf}$	P_u in N P_{uf} in N P_{uw} in N
(9-29)	$\delta_u = \dfrac{5 M_u h^2}{48 E_m I_n}$ For $M_u < M_{cr}$	δ_u in mm h in mm E_m in MPa I_n in mm^4 M_u in N-mm M_{cr} in N-mm
(9-30)	$\delta_u = \dfrac{5 M_{cr} h^2}{48 E_m I_n} + \dfrac{5(M_u - M_{cr})h^2}{48 E_m I_{cr}}$ For $M_{cr} \leq M_u \leq M_n$	δ_u in mm h in mm E_m in MPa I_{cr} in mm^4 I_n in mm^4 M_{cr} in N-mm M_n in N-mm M_u in N-mm
(9-31)	$M_u = \psi M_{u,0}$	
(9-32)	$\psi = \dfrac{1}{1 - \dfrac{P_u}{P_e}}$	
(9-33)	$P_e = \dfrac{\pi^2 E_m I_{eff}}{h^2}$	

Code Equation No. or Section No.	SI Unit Equivalent Equation	Units
(9-34)	$I_{cr} = n\left(A_s + \dfrac{P_u}{f_y}\dfrac{t_{sp}}{2d}\right)(d-c)^2 + \dfrac{bc^3}{3}$	I_{cr} in mm^4 A_s in mm^2 P_u in N t_{sp} in mm f_y in MPa d in mm c in mm b in mm
(9-35)	$c = \dfrac{A_s f_y + P_u}{0.64 f'_m b}$	c in mm A_s in mm^2 f_y in MPa P_u in N f'_m in MPa b in mm
(9-36)	$\delta_s \leq 0.007\, h$	δ_s in mm h in mm
9.3.6.5.1	$P_u \leq 0.10\, A_g f'_m$ $P_u \leq 0.05\, A_g f'_m$	P_u in N A_g in mm^2 f'_m in MPa
9.3.6.5.1	$\dfrac{M_u}{V_u d_v} \leq 1.0$	M_u in N-mm V_u in N l_w in mm
9.3.6.5.1	$V_u \leq 0.25\, A_{nv}\sqrt{f'_m}$ and $\dfrac{M_u}{V_u d_v} \leq 3.0$	A_n in mm^2 f'_m in MPa l_w in mm M_u in N-mm V_u in N
9.3.6.5.3 (a)	$c \geq \dfrac{l_w}{600\,(C_d \delta_{ne}/h_w)}$	c in mm h_w in mm l_w in mm δ_{ne} in mm
(10-1)	$a = \dfrac{f_{ps}A_{ps} + f_y A_s + P_u/\phi}{0.80\, f'_m b}$	a in mm f_{ps} in MPa A_{ps} in mm^2 f_y in MPa A_s in mm^2 P_u in N f'_m in MPa b in mm
(10-2)	$M_n = \left(f_{ps}A_{ps} + f_y A_s + P_u/\phi\right)\left(d - \dfrac{a}{2}\right)$	M_n in N-mm f_{ps} in MPa A_{ps} in mm^2 f_y in MPa A_s in mm^2 P_u in N d in mm a in mm
(10-3)	$f_{ps} = f_{se} + 0.03\left(\dfrac{E_{ps}d}{l_p}\right)\left(1 - 1.56\dfrac{A_{ps}f_{ps}+P}{f'_m b d}\right)$	f_{ps} in MPa f_{se} in MPa d in mm l_p in mm E_{ps} in MPa A_{ps} in mm^2 b in mm f'_m in MPa P in N

BUILDING CODE REQUIREMENTS FOR MASONRY STRUCTURES AND COMMENTARY

Code Equation No. or Section No.	SI Unit Equivalent Equation	Units
(11-1)	$f_{tAAC} = 0.199\sqrt{f'_{AAC}}$	f_{tAAC} in MPa $\sqrt{f'_{AAC}}$ result in MPa
(11-2)	$f_v = 0.15 f'_{AAC}$	f_v in MPa f'_{AAC} in MPa
(11-3)	$P_n = 0.80\left\{0.85 A_n f'_{AAC}\left[1-\left(\dfrac{h}{140r}\right)^2\right]\right\}$	h in mm r in mm A_n in mm² f'_{AAC} in MPa P_n in N
(11-4)	$P_n = 0.80\left[0.85 A_n f'_{AAC}\left(\dfrac{70r}{h}\right)^2\right]$	h in mm r in mm A_n in mm² f'_{AAC} in MPa P_n in N
(11-5)	$l_e = 13 d_b$	l_e in mm d_b in mm
(11-6)	$l_d = \dfrac{1.57 d_b^2 f_y \gamma}{K_{AAC}\sqrt{f'_g}}$	l_d in mm d_b in mm f'_g in MPa $\sqrt{f'_g}$ result in MPa f_y in MPa K_{AAC} in mm
(11-7)	$P_n = 0.80\left[0.85 f'_{AAC}(A_n - A_{st}) + f_y A_{st}\right]\left[1-\left(\dfrac{h}{140r}\right)^2\right]$	h in mm r in mm A_n in mm² A_{st} in mm² f_y in MPa f'_{AAC} in MPa P_n in N
(11-8)	$P_n = 0.80\left[0.85 f'_{AAC}(A_n - A_{st}) + f_y A_{st}\right]\left(\dfrac{70r}{h}\right)^2$	h in mm r in mm A_n in mm² A_{st} in mm² f_y in MPa f'_{AAC} in MPa P_n in N
(11-9)	$V_n = V_{nAAC} + V_{ns}$	V_n in N V_{nAAC} in N V_{ns} in N
(11-10)	$V_n = \mu_{AAC} P_u$	V_n in N P_u in N
(11-11)	$V_n \le 0.498 A_{nv}\sqrt{f'_{AAC}}$	V_n in N f'_{AAC} in MPa $\sqrt{f'_{AAC}}$ result in MPa A_{nv} in mm²
(11-12)	$V_n \le 0.332 A_{nv}\sqrt{f'_{AAC}}$	V_n in N f'_{AAC} in MPa $\sqrt{f'_{AAC}}$ result in MPa A_{nv} in mm²

Code Equation No. or Section No.	SI Unit Equivalent Equation	Units
(11-13a)	$V_{nAAC} = 0.0789\, l_w\, t\, \sqrt{f'_{AAC}}\, \sqrt{1 + \dfrac{P_u}{0.199\sqrt{f'_{AAC}}\, l_w\, t}}$	V_{nAAC} in N P_u in N f'_{AAC} in MPa $\sqrt{f'_{AAC}}$ result in MPa l_w in mm t in mm
(11-13b)	$V_{nAAC} = 0.0548\, l_w\, t\, \sqrt{f'_{AAC}}\, \sqrt{1 + \dfrac{P_u}{0.199\sqrt{f'_{AAC}}\, l_w\, t}}$	V_{nAAC} in N P_u in N f'_{AAC} in MPa $\sqrt{f'_{AAC}}$ result in MPa l_w in mm t in mm
(11-13c)	$V_{nAAC} = 0.0747\sqrt{f'_{AAC}}\, A_{nv} + 0.05 P_u$	V_{nAAC} in N P_u in N f'_{AAC} in MPa $\sqrt{f'_{AAC}}$ result in MPa A_{nv} in mm^2
(11-14)	$V_{nAAC} = 0.17\, f'_{AAC}\, t\, \dfrac{h \cdot l_w^{\,2}}{h^2 + (\tfrac{3}{4} l_w)^2}$	V_{nAAC} in N f'_{AAC} in MPa t in mm h in mm l_w in mm
(11-15)	$V_{ns} = 0.5\left(\dfrac{A_v}{s}\right) f_y d_v$	V_{ns} in N f_y in MPa s in mm d_v in mm A_v in mm^2
(11-16)	$V_{nAAC} = 0.0664\sqrt{f'_{AAC}}\, bd$	V_{nAAC} in N f'_{AAC} in MPa $\sqrt{f'_{AAC}}$ result in MPa b in mm d in mm
(11-17)	$\dfrac{P_u}{A_g} \le 0.2 f'_{AAC}$	P_u in N f'_{AAC} in MPa A_g in mm^2
(11-18)	$M_u = \dfrac{w_u h^2}{8} + P_{uf}\dfrac{e_u}{2} + P_u \delta_u$	P_u in N P_{uf} in N h in mm e_u in mm δ_u in mm w_u in N/mm M_u in N-mm
(11-19)	$P_u = P_{uw} + P_{uf}$	P_u in N P_{uw} in N P_{uf} in N
(11-20)	$\delta_u = \dfrac{5 M_u h^2}{48 E_{AAC} I_n}$	δ_u in mm I_n in mm^4 h in mm E_{AAC} in MPa M_u in N-mm

Code Equation No. or Section No.	SI Unit Equivalent Equation	Units
(11-21)	$\delta_u = \dfrac{5 M_{cr} h^2}{48 E_{AAC} I_n} + \dfrac{5(M_u - M_{cr}) h^2}{48 E_{AAC} I_{cr}}$	δ_u in mm I_n in mm^4 I_{cr} in mm^4 h in mm E_{AAC} in MPa M_{cr} in N-mm M_u in N-mm
(11-22)	$M_u = \psi\, M_{u,0}$	M_u in N-mm $M_{u,0}$ in N-mm
(11-23)	$\psi = \dfrac{1}{1 - \dfrac{P_u}{P_e}}$	P_e in N P_u in N
(11-24)	$P_e = \dfrac{\pi^2 E_{AAC} I_{eff}}{h^2}$	P_e in N E_{AAC} in MPa I_{eff} in mm^4 h in mm
(11-25)	$M_{cr} = S_n \left(f_{rAAC} + \dfrac{P}{A_n} \right)$	S_n in mm^3 A_n in mm^2 f_{rAAC} in MPa P in N M_{cr} in N-mm.
(11-26)	$I_{cr} = n\left(A_s + \dfrac{P_u}{f_y} \dfrac{t_{sp}}{2d} \right)(d-c)^2 + \dfrac{b(c)^3}{3}$	I_{cr} in mm^4 A_s in mm^2 P_u in N t_{sp} in mm f_y in MPa d in mm c in mm b in mm
(11-27)	$c = \dfrac{A_s f_y + P_u}{0.57 f'_{AAC} b}$	c in mm A_s in mm^2 f_y in MPa P_u in N f'_{AAC} in MPa b in mm
(11-28)	$M_u \leq \phi M_n$	M_u in N-mm M_n in N-mm
(11-29)	$M_n = \left(A_s f_y + P_u \right)\left(d - \dfrac{a}{2} \right)$	P_u in N a in mm d in mm A_s in mm^2 f_y in MPa M_n in N-mm
(11-30)	$a = \dfrac{(P_u + A_s f_y)}{0.85 f'_{AAC} b}$	a in mm P_u in N b in mm A_s in mm^2 f'_{AAC} in MPa f_y in MPa
(11-31)	$\delta_s \leq 0.007\, h$	δ_s in mm h in mm

Code Equation No. or Section No.	SI Unit Equivalent Equation	Units
(11-32)	$V_{cr} = \dfrac{S_n}{h}\left(f_{rAAC} + \dfrac{P}{A_n}\right)$	S_n in mm^3 A_n in mm^2 h in mm f_{rAAC} in MPa P in N V_{cr} in N
11.3.6.6.2 (a)	$c \geq \dfrac{l_w}{600\,(C_d \delta_{ne}/h_w)}$	c in mm h_w in mm l_w in mm δ_{ne} in mm
(B-1)	$w_{inf} = \dfrac{0.3}{\lambda_{strut}\cos\theta_{strut}}$	w_{inf} in. mm θ_{strut} in degrees λ_{strut} = mm^{-1}
(B-2a)	$\lambda_{strut} = \sqrt[4]{\dfrac{E_m\, t_{net\,inf}\,\sin 2\theta_{strut}}{4\,E_{bc}\,I_{bc}\,h_{inf}}}$	λ_{strut} = mm^{-1} E_{bc} in MPa E_m in MPa h_{inf} in mm I_{bc} in mm^4 $t_{net\,inf}$ in mm θ_{strut} in degrees
(B-2b)	$\lambda_{strut} = \sqrt[4]{\dfrac{E_{AAC}\, t_{net\,inf}\,\sin 2\theta_{strut}}{4\,E_{bc}\,I_{bc}\,h_{inf}}}$	λ_{strut} = mm^{-1} E_{AAC} in MPa E_{bc} in MPa h_{inf} in mm I_{bc} in mm^4 $t_{net\,inf}$ in mm θ_{strut} in degrees
(B-3)	$(150\text{mm})\, t_{net\,inf}\, f'_m$	f'_m in MPa $t_{net\,inf}$ in mm
(B-4)	$\dfrac{V_n}{1.5}$	V_n in N
(B-5)	$(150\text{mm})\, t_{net\,inf}\, f'_{AAC}$	f'_{AAC} in MPa $t_{net\,inf}$ in mm
(B-6)	$\dfrac{V_{nAAC}}{1.5}$	V_{nAAC} in N
(B-7a)	$q_{n\,inf} = 729000\,(f'_m)^{0.75}\, t_{inf}^2 \left(\dfrac{\alpha_{arch}}{l_{inf}^{2.5}} + \dfrac{\beta_{arch}}{h_{inf}^{2.5}}\right)$	$q_{n\,inf}$ in Pa f'_m in MPa h_{inf} in mm l_{inf} in mm t_{inf} in mm α_{arch} in N$^{0.25}$ β_{arch} in N$^{0.25}$
(B-7b)	$q_{n\,inf} = 729000\,(f'_{AAC})^{0.75}\, t_{inf}^2 \left(\dfrac{\alpha_{arch}}{l_{inf}^{2.5}} + \dfrac{\beta_{arch}}{h_{inf}^{2.5}}\right)$	$q_{n\,inf}$ in Pa f'_{AAC} in MPa h_{inf} in mm l_{inf} in mm t_{inf} in mm α_{arch} in N$^{0.25}$ β_{arch} in N$^{0.25}$
(B-8)	$\alpha_{arch} = \dfrac{1}{h_{inf}}(E_{bc}\,I_{bc}\,h_{inf}^2)^{0.25} < 50$	α_{arch} in N$^{0.25}$ E_{bc} in MPa h_{inf} in mm I_{bc} in mm^4

BUILDING CODE REQUIREMENTS FOR MASONRY STRUCTURES AND COMMENTARY

Code Equation No. or Section No.	SI Unit Equivalent Equation	Units
(B-9)	$\beta_{arch} = \dfrac{1}{l_{inf}}(E_{bb} I_{bb} l_{inf}^2)^{0.25} < 50$	β_{arch} in $N^{0.25}$ E_{bb} in MPa l_{inf} in mm I_{bb} in mm^4
(C-1)	$\phi V_{lim} \geq V_{ub}$	V_{lim} in N V_{ub} in N

CONVERSION OF INCH-POUND UNITS TO SI UNITS

TO CONVERT FROM	TO	MULTIPLY BY
inches (in.)	millimeters (mm)	25.4
square inches (in.^{2})	square millimeters (mm^2)	645.2
cubic inches (in.^{3})	cubic millimeters (mm^3)	16,390
inches to the fourth power (in.^{4})	millimeters to the fourth power (mm^4)	416,200
pound-force (lb)	newton (N)	4.448
pounds per linear foot (plf)	newtons per millimeter (N/mm)	0.01459
pounds per square inch (psi)	megapascal (MPa)	0.006895
pounds per square foot (psf)	kilo pascal (kPa)	0.04788
inch-pounds (in-lb)	newton-millimeters (N-mm)	113.0
$\sqrt{psi}$, result in psi	$\sqrt{MPa}$, result in MPa	0.08304

PREFIXES

POWER	PREFIX	ABBREVIATION
$1{,}000{,}000 = 10^6$	mega	M
$1{,}000 = 10^3$	kilo	k
$0.001 = 10^{-3}$	milli	m

This page is intentionally left blank.

Specification for Masonry Structures (TMS 602-13/ACI 530.1-13/ASCE 6-13)

TABLE OF CONTENTS

SYNOPSIS AND KEYWORDS, pg. S-iii

PREFACE, S-1

PART 1 — GENERAL, pg. S-3
- 1.1 — Summary S-3
- 1.2 — Definitions S-3
- 1.3 — Reference standards S-9
- 1.4 — System description S-14
- 1.5 — Submittals S-22
- 1.6 — Quality assurance S-23
- 1.7 — Delivery, storage, and handling S-28
- 1.8 — Project conditions S-29

PART 2 — PRODUCTS, pg. S-33
- 2.1 — Mortar materials S-33
- 2.2 — Grout materials S-36
- 2.3 — Masonry unit materials S-37
- 2.4 — Reinforcement, prestressing tendons, and metal accessories S-40
- 2.5 — Accessories S-48
- 2.6 — Mixing S-51
- 2.7 — Fabrication S-53

PART 3 — EXECUTION, pg. S-55
- 3.1 — Inspection S-55
- 3.2 — Preparation S-56
- 3.3 — Masonry erection S-57
- 3.4 — Reinforcement, tie, and anchor installation S-64
- 3.5 — Grout placement S-73
- 3.6 — Prestressing tendon installation and stressing procedure S-77
- 3.7 — Field quality control S-78
- 3.8 — Cleaning S-78

FOREWORD TO SPECIFICATION CHECKLISTS, pg. S-79
- Mandatory Requirements Checklist S-80
- Optional Requirements Checklist S-82

REFERENCES FOR THE SPECIFICATION COMMENTARY, pg. S-83

This page is intentionally left blank.

Specification for Masonry Structures (TMS 602-13/ACI 530.1-13/ASCE 6-13)

SYNOPSIS

This Specification for Masonry Structures (TMS 602-13/ACI 530.1-13/ASCE 6-13) is written as a master specification and is required by Building Code Requirements for Masonry Structures (TMS 402-13/ACI 530-13/ASCE 5-13) to control materials, labor, and construction. Thus, this Specification covers minimum construction requirements for masonry in structures. Included are quality assurance requirements for materials; the placing, bonding, and anchoring of masonry; and the placement of grout and of reinforcement. This Specification may be referenced in the Project Manual. Individual project requirements may supplement the provisions of this Specification.

Keywords: AAC masonry, anchors; autoclaved aerated concrete (AAC) masonry, clay brick; clay tile; concrete block; concrete brick; construction; construction materials; curing; grout; grouting; inspection; joints; masonry; materials handling; mortars (material and placement); quality assurance and quality control; reinforcing steel; specifications; ties; tests; tolerances.

This page is intentionally left blank.

SPECIFICATION

PREFACE

P1. This Preface is included for explanatory purposes only; it does not form a part of Specification TMS 602-13/ACI 530.1-13/ASCE 6-13.

P2. TMS 602-13/ACI 530.1-13/ASCE 6-13 is written in the three-part section format of the Construction Specifications Institute. The language is generally imperative and terse.

P3. TMS 602-13/ACI 530.1-13/ASCE 6-13 establishes minimum construction requirements for the materials and workmanship used to construct masonry structures. It is the means by which the designer conveys to the contractor the performance expectations upon which the design is based, in accordance with Code TMS 402-13/ACI 530-13/ASCE 5-13.

P4. The Checklists and the Forward to Specification Checklists are non-mandatory and do not form part of Specification TMS 602-13/ACI 530.1-13/ASCE 6-13.

COMMENTARY

COMMENTARY

Part 1 of the *Building Code Requirements for Masonry Structures* (TMS 402-13/ACI 530-13/ASCE 5-13) makes the *Specification for Masonry Structures* (TMS 602-13/ACI 530.1-13/ASCE 6-13) an integral part of the Code. TMS 602-13/ACI 530.1-13/ASCE 6-13 Specification sets minimum construction requirements regarding the materials used in and the erection of masonry structures. Specifications are written to set minimum acceptable levels of performance for the contractor. This commentary is directed to the Architect/Engineer writing the project specifications.

This Commentary explains some of the topics that the Masonry Standards Joint Committee (MSJC) considered in developing the provisions of this Specification. Comments on specific provisions are made under the corresponding article numbers of this Specification.

Specification TMS 602-13/ACI 530.1-13/ASCE 6-13 is a reference standard that the Architect/Engineer may cite in the contract documents for any project. It establishes the minimum construction requirements to assure compliance of the construction with the Code-based design. Owners, through their representatives (Architect/Engineer), may write requirements into contract documents that are more stringent than those of TMS 602-13/ACI 530.1-13/ASCE 6-13. As an example, requirements to satisfy visual aesthetics may be added in a project specification. This can be accomplished with supplemental specifications to this Specification.

The contractor should not be required through contract documents to comply with the Code or to assume responsibility regarding design (Code) requirements. The Code is not intended to be made a part of the contract documents.

The Preface and the Foreword to Specification Checklists contain information that explains the scope of this Specification. The Checklists are a summary of the Articles that require a decision by the Architect/Engineer preparing the contract documents. Project specifications should include the information that relates to those Checklist items that are pertinent to the project. Each project requires response to the mandatory requirements.

This page is intentionally left blank.

PART 1 — GENERAL

SPECIFICATION

1.1 — Summary

1.1 A. This Specification addresses requirements for materials and construction of masonry structures. SI values shown in parentheses are provided for information only and are not part of this Specification.

1.1 B. The Specification supplements the legally adopted building code and governs the construction of masonry elements designed in accordance with the Code. In areas without a legally adopted building code, this Specification defines the minimum acceptable standards of construction practice.

1.1 C. This article addresses the furnishing and construction of masonry including the following:

1. Furnishing and placing masonry units, grout, mortar, masonry lintels, sills, copings, through-wall flashing, and connectors.
2. Furnishing, erecting and maintaining of bracing, forming, scaffolding, rigging, and shoring.
3. Furnishing and installing other equipment for constructing masonry.
4. Cleaning masonry and removing surplus material and waste.
5. Installing lintels, nailing blocks, inserts, window and door frames, connectors, and construction items to be built into the masonry, and building in vent pipes, conduits and other items furnished and located by other trades.

1.2 — Definitions

A. *Acceptable, accepted* — Acceptable to or accepted by the Architect/Engineer.

B. *Architect/Engineer* — The architect, engineer, architectural firm, engineering firm, or architectural and engineering firm, issuing drawings and specifications, or administering the work under project specifications and project drawings, or both.

C. *Area, gross cross-sectional* — The area delineated by the out-to-out dimensions of masonry in the plane under consideration.

D. *Area, net cross-sectional* — The area of masonry units, grout, and mortar crossed by the plane under consideration based on out-to-out dimensions.

E. *Autoclaved aerated concrete* — Low-density cementitious product of calcium silicate hydrates.

COMMENTARY

1.1 — Summary

1.1 C. The scope of the work is outlined in this article. All of these tasks and materials will not appear in every project.

1.2 — Definitions

For consistent application of this Specification, it is necessary to define terms that have particular meaning in this Specification. The definitions given are for use in application of this Specification only and do not always correspond to ordinary usage. Other terms are defined in referenced documents and those definitions are applicable. If any term is defined in both this Specification and in a referenced document, the definition in this Specification applies. Referenced documents include ASTM standards. Terminology standards include ASTM C1232 Standard Terminology of Masonry and ASTM C1180 Standard Terminology of Mortar and Grout for Unit Masonry. Definitions have been coordinated between the Code and Specification.

SPECIFICATION

1.2 — Definitions (Continued)

F. *Autoclaved aerated concrete (AAC) masonry* — Autoclaved aerated concrete units, manufactured without reinforcement, set on a mortar leveling bed, bonded with thin-bed mortar, placed with or without grout, and placed with or without reinforcement.

G. *Bond beam* — A horizontal, sloped, or stepped element that is fully grouted, has longitudinal bar reinforcement, and is constructed within a masonry wall.

COMMENTARY

G. *Bond beam* — This reinforced member is usually constructed horizontally, but may be sloped or stepped to match an adjacent roof, for example, as shown in Figure CC-2.2-2.

Notes:

(1) Masonry wall
(2) Fully grouted bond beam with reinforcement
(3) Sloped top of wall
(4) Length of noncontact lap splice
(5) Spacing between bars in noncontact lap splice

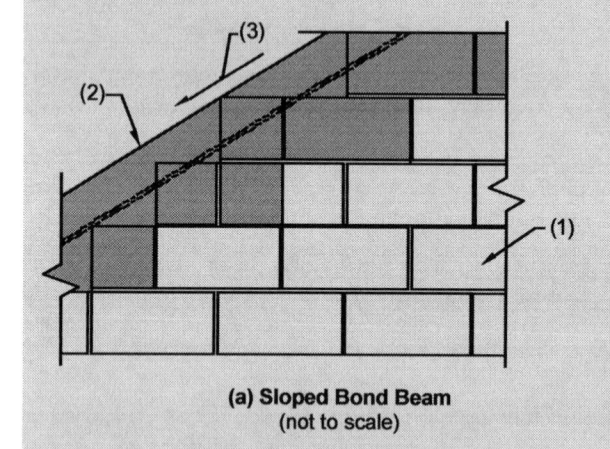

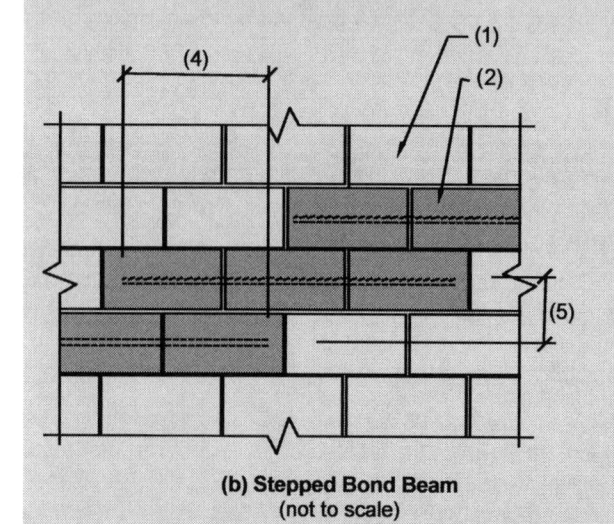

(a) Sloped Bond Beam
(not to scale)

(b) Stepped Bond Beam
(not to scale)

Figure SC-1— Sloped and stepped bond beams

SPECIFICATION

1.2 — Definitions (Continued)

H. *Bonded prestressing tendon* — Prestressing tendon that is encapsulated by prestressing grout in a corrugated duct that is bonded to the surrounding masonry through grouting.

I. *Cleanouts* — Openings that are sized and spaced to allow removal of debris from the bottom of the grout space.

J. *Collar joint* — Vertical longitudinal space between wythes of masonry or between masonry and back up construction, which is permitted to be filled with mortar or grout.

K. *Compressive strength of masonry* — Maximum compressive force resisted per unit of net cross-sectional area of masonry, determined by testing masonry prisms; or a function of individual masonry units, mortar and grout in accordance with the provisions of this Specification.

L. *Contract Documents* — Documents establishing the required Work, and including in particular, the Project Drawings and Project Specifications.

M. *Contractor* — The person, firm, or corporation with whom the Owner enters into an agreement for construction of the Work.

N. *Cover, grout* — Thickness of grout surrounding the outer surface of embedded reinforcement, anchor, or tie.

O. *Cover, masonry* — Thickness of masonry units, mortar, and grout surrounding the outer surface of embedded reinforcement, anchor, or tie.

P. *Cover, mortar* — thickness of mortar surrounding the outer surface of embedded reinforcement, anchor, or tie.

Q. *Dimension, nominal* — The specified dimension plus an allowance for the joints with which the units are to be laid. Nominal dimensions are usually stated in whole numbers. Thickness is given first, followed by height and then length.

R. *Dimensions, specified* — Dimensions specified for the manufacture or construction of a unit, joint, or element.

S. *Glass unit masonry* — Non-load-bearing masonry composed of glass units bonded by mortar.

T. *Grout* — (1) A plastic mixture of cementitious materials, aggregates, and water, with or without admixtures, initially produced to pouring consistency without segregation of the constituents during placement. (2) The hardened equivalent of such mixtures.

U. *Grout, self-consolidating* — A highly fluid and stable grout typically with admixtures, that remains homogeneous when placed and does not require puddling or vibration for consolidation.

COMMENTARY

Q & R. The permitted tolerances for units are given in the appropriate materials standards. Permitted tolerances for joints and masonry construction are given in this Specification. Nominal dimensions are usually used to identify the size of a masonry unit. The thickness or width is given first, followed by height and length. Nominal dimensions are normally given in whole numbers nearest to the specified dimensions. Specified dimensions are most often used for design calculations.

SPECIFICATION

1.2 — Definitions (Continued)

V. *Grout lift* — An increment of grout height within a total grout pour. A grout pour consists of one or more grout lifts.

W. *Grout pour* — The total height of masonry to be grouted prior to erection of additional masonry. A grout pour consists of one or more grout lifts.

X. *Inspection, continuous* — The Inspection Agency's full-time observation of work by being present in the area where the work is being performed.

Y. *Inspection, periodic* — The Inspection Agency's part-time or intermittent observation of work during construction by being present in the area where the work has been or is being performed, and observation upon completion of the work.

Z. *Masonry, partially grouted* — Construction in which designated cells or spaces are filled with grout, while other cells or spaces are ungrouted.

AA. *Masonry unit, hollow* — A masonry unit with net cross-sectional area of less than 75 percent of its gross cross-sectional area when measured in any plane parallel to the surface containing voids.

AB. *Masonry unit, solid* — A masonry unit with net cross-sectional area of 75 percent or more of its gross cross-sectional area when measured in every plane parallel to the surface containing voids.

AC. *Mean daily temperature* — The average daily temperature of temperature extremes predicted by a local weather bureau for the next 24 hours.

AD. *Minimum daily temperature* — The low temperature forecast by a local weather bureau to occur within the next 24 hours.

AE. *Minimum/maximum (not less than . . . not more than)* — Minimum or maximum values given in this Specification are absolute. Do not construe that tolerances allow lowering a minimum or increasing a maximum.

AF. *Otherwise required* — Specified differently in requirements supplemental to this Specification.

AG. *Owner* — The public body or authority, corporation, association, partnership, or individual for whom the Work is provided.

AH. *Partition wall* — An interior wall without structural function.

COMMENTARY

X & Y. The Inspection Agency is required to be on the project site whenever masonry tasks requiring continuous inspection are in progress. During construction requiring periodic inspection, the Inspection Agency is only required to be on the project site intermittently, and is required to observe completed work. The frequency of periodic inspections should be defined by the Architect/Engineer as part of the quality assurance plan, and should be consistent with the complexity and size of the project.

SPECIFICATION

1.2 — Definitions (Continued)

AI. *Post-tensioning* — Method of prestressing in which prestressing tendons are tensioned after the masonry has been placed.

AJ. *Prestressed masonry* — Masonry in which internal compressive stresses have been introduced by prestressed tendons to counteract potential tensile stresses resulting from applied loads.

AK. *Prestressing grout* — A cementitious mixture used to encapsulate bonded prestressing tendons.

AL. *Prestressing tendon* — Steel element such as wire, bar, or strand, or a bundle of such elements, used to impart prestress to masonry.

AM. *Pretensioning* — Method of prestressing in which prestressing tendons are tensioned before the transfer of stress into the masonry.

AN. *Prism* — An assemblage of masonry units and mortar, with or without grout, used as a test specimen for determining properties of the masonry.

AO. *Project Drawings* — The Drawings that, along with the Project Specifications, complete the descriptive information for constructing the Work required or referred to in the Contract Documents.

AP. *Project Specifications* — The written documents that specify requirements for a project in accordance with the service parameters and other specific criteria established by the Owner or his agent.

AQ. *Quality assurance* — The administrative and procedural requirements established by the Contract Documents to assure that constructed masonry is in compliance with the Contract Documents.

AR. *Reinforcement* — Nonprestressed steel reinforcement.

AS. *Running bond* — The placement of masonry units such that head joints in successive courses are horizontally offset at least one-quarter the unit length.

AT. *Slump flow* — The circular spread of plastic self-consolidating grout, which is evaluated in accordance with ASTM C1611/C1611M.

AU. *Specified compressive strength of masonry, f'_m* — Minimum compressive strength, expressed as force per unit of net cross-sectional area, required of the masonry used in construction by the Project Specifications or Project Drawings, and upon which the project design is based.

COMMENTARY

AS. *Running bond* — The Code requires horizontal reinforcement in masonry that is not laid in running bond. Stack bond, which is commonly interpreted as a pattern with aligned head joints, is one bond pattern that is required to be reinforced horizontally

SPECIFICATION

1.2 — Definitions (Continued)

AV. *Stone masonry* — Masonry composed of field, quarried, or cast stone units bonded by mortar.

　1. *Stone masonry, ashlar* — Stone masonry composed of rectangular units having sawed, dressed, or squared bed surfaces and bonded by mortar.

　2. *Stone masonry, rubble* — Stone masonry composed of irregular shaped units bonded by mortar.

AW. *Submit, submitted* — Submit, submitted to the Architect/Engineer for review.

AX. *Tendon anchorage* — In post-tensioning, a device used to anchor the prestressing tendon to the masonry or concrete member; in pretensioning, a device used to anchor the prestressing tendon during hardening of masonry mortar, grout, prestressing grout, or concrete.

AY. *Tendon coupler* — A device for connecting two tendon ends, thereby transferring the prestressing force from end to end.

AZ. *Tendon jacking force* — Temporary force exerted by a device that introduces tension into prestressing tendons.

BA. *Unbonded prestressing tendon* — Prestressing tendon that is not bonded to masonry.

BB. *Veneer, adhered* — Masonry veneer secured to and supported by the backing through adhesion.

BC. *Visual stability index (VSI)* — An index, defined in ASTM C1611/C1611M, that qualitatively indicates the stability of self-consolidating grout

BD. *Wall* — A vertical element with a horizontal length to thickness ratio greater than 3, used to enclose space.

BE. *Wall, load-bearing* — A wall supporting vertical loads greater than 200 lb per linear foot (2919 N/m) in addition to its own weight.

BF. *Wall, masonry bonded hollow* — A multiwythe wall built with masonry units arranged to provide an air space between the wythes and with the wythes bonded together with masonry units.

BG. *When required* — Specified in requirements supplemental to this Specification.

BH. *Work* — The furnishing and performance of equipment, services, labor, and materials required by the Contract Documents for the construction of masonry for the project or part of project under consideration.

BI. *Wythe* — Each continuous vertical section of a wall, one masonry unit in thickness.

SPECIFICATION

1.3 — Reference standards

Standards referred to in this Specification are listed below with their serial designations, including year of adoption or revision, and are declared to be part of this Specification as if fully set forth in this document except as modified here.

American Concrete Institute

A. ACI 117-10 Standard Specifications for Tolerances for Concrete Construction and Materials

American National Standards Institute

B. ANSI A 137.1-08 Standard Specification for Ceramic Tile

ASTM International

C. ASTM A36/A36M-08 Standard Specification for Carbon Structural Steel

D. ASTM A82/A82M-07 Standard Specification for Steel Wire, Plain, for Concrete Reinforcement

E. ASTM A123/A123M-12 Standard Specification for Zinc (Hot-Dip Galvanized) Coatings on Iron and Steel Products

F. ASTM A153/A153M-09 Standard Specification for Zinc Coating (Hot-Dip) on Iron and Steel Hardware

G. ASTM A185/A185M-07 Standard Specification for Steel Welded Wire Reinforcement, Plain, for Concrete

H. ASTM A240/A240M-12a Standard Specification for Chromium and Chromium-Nickel Stainless Steel Plate, Sheet, and Strip for Pressure Vessels and for General Applications

I. ASTM A307-12 Standard Specification for Carbon Steel Bolts, Studs, and Threaded Rod, 60,000 PSI Tensile Strength

J. ASTM A416/A416M-12a Standard Specification for Steel Strand, Uncoated Seven-Wire for Prestressed Concrete

K. ASTM A421/A421M-10 Standard Specification for Uncoated Stress-Relieved Steel Wire for Prestressed Concrete

L. ASTM A480/A480M-12 Standard Specification for General Requirements for Flat-Rolled Stainless and Heat-Resisting Steel Plate, Sheet, and Strip

M. ASTM A496/A496M-07 Standard Specification for Steel Wire, Deformed, for Concrete Reinforcement

COMMENTARY

1.3 — Reference standards

This list of standards includes material specifications, sampling, test methods, detailing requirements, design procedures, and classifications. Standards produced by ASTM International (ASTM) are referenced whenever possible. Material manufacturers and testing laboratories are familiar with ASTM standards that are the result of a consensus process. In the few cases where standards do not exist for materials or methods, the committee developed requirements. Specific dates are given because changes to the standards alter this Specification. Many of these standards require compliance with additional standards.

Contact information for these organizations is given below:

American Concrete Institute
38800 Country Club Drive
Farmington Hills, MI 48331
www.aci-int.org

American National Standards Institute
25 West 43rd Street,
New York, NY 10036
www.ansi.org

ASTM International
100 Barr Harbor Drive
West Conshohocken, PA 19428-2959
www.astm.org

American Welding Society
8669 NW 36th Street, Suite 130
Miami, Florida 33166-6672
www.aws.org

Federal Test Method Standard from:
U.S. Army General Material and Parts Center
Petroleum Field Office (East)
New Cumberland Army Depot
New Cumberland, PA 17070

SPECIFICATION

1.3 — Reference standards (Continued)

N. ASTM A497/A497M-07 Standard Specification for Steel Welded Wire Reinforcement, Deformed, for Concrete

O. ASTM A510/A510M-11 Standard Specification for General Requirements for Wire Rods and Coarse Round Wire, Carbon Steel

P. ASTM A580/A580M-12a Standard Specification for Stainless Steel Wire

Q. ASTM A615/A615M-12 Standard Specification for Deformed and Plain Carbon-Steel Bars for Concrete Reinforcement

R. ASTM A641/A641M-09a Standard Specification for Zinc-Coated (Galvanized) Carbon Steel Wire

S. ASTM A653/A653M-11 Standard Specification for Steel Sheet, Zinc-Coated (Galvanized) or Zinc-Iron Alloy-Coated (Galvannealed) by the Hot-Dip Process

T. ASTM A666-10 Standard Specification for Annealed or Cold-Worked Austenitic Stainless Steel Sheet, Strip, Plate, and Flat Bar

U. ASTM A706/A706M-09b Standard Specification for Low-Alloy Steel Deformed and Plain Bars for Concrete Reinforcement

V. ASTM A722/A722M-12 Standard Specification for Uncoated High-Strength Steel Bars for Prestressing Concrete

W. ASTM A767/A767M-09 Standard Specification for Zinc-Coated (Galvanized) Steel Bars for Concrete Reinforcement

X. ASTM A775/A775M-07b Standard Specification for Epoxy-Coated Steel Reinforcing Bars

Y. ASTM A884/A884M-12 Standard Specification for Epoxy-Coated Steel Wire and Welded Wire Reinforcement

Z. ASTM A899-91(2007) Standard Specification for Steel Wire, Epoxy-Coated

AA. ASTM A951/A951M-11 Standard Specification for Steel Wire for Masonry Joint Reinforcement

AB. ASTM A996/A996M-09b Standard Specification for Rail-Steel and Axle-Steel Deformed Bars for Concrete Reinforcement

AC. ASTM A1008/A1008M-12a Standard Specification for Steel, Sheet, Cold-Rolled, Carbon, Structural, High-Strength Low-Alloy, High-Strength Low-Alloy with Improved Formability, Solution Hardened, and Bake Hardenable

AD. ASTM B117-11 Standard Practice for Operating Salt Spray (Fog) Apparatus

SPECIFICATION

1.3 — Reference standards (Continued)

AE. ASTM C34-12 Standard Specification for Structural Clay Load-Bearing Wall Tile

AF. ASTM C55-11 Standard Specification for Concrete Building Brick

AG. ASTM C56-12 Standard Specification for Structural Clay Nonloadbearing Tile

AH. ASTM C62-12 Standard Specification for Building Brick (Solid Masonry Units Made from Clay or Shale)

AI. ASTM C67-12 Standard Test Methods for Sampling and Testing Brick and Structural Clay Tile

AJ. ASTM C73-10 Standard Specification for Calcium Silicate Brick (Sand-Lime Brick)

AK. ASTM C90-12 Standard Specification for Loadbearing Concrete Masonry Units

AL. ASTM C109/C109M-12 Standard Test Method for Compressive Strength of Hydraulic Cement Mortars (Using 2-in. or [50-mm] Cube Specimens)

AM. ASTM C126-12a Standard Specification for Ceramic Glazed Structural Clay Facing Tile, Facing Brick, and Solid Masonry Units

AN. ASTM C129-11 Standard Specification for Nonloadbearing Concrete Masonry Units

AO. ASTM C143/C143M-12 Standard Test Method for Slump of Hydraulic-Cement Concrete

AP. ASTM C144-11 Standard Specification for Aggregate for Masonry Mortar

AQ. ASTM C150/C150M-12 Standard Specification for Portland Cement

AR. ASTM C212-10 Standard Specification for Structural Clay Facing Tile

AS. ASTM C216-12a Standard Specification for Facing Brick (Solid Masonry Units Made from Clay or Shale)

AT. ASTM C270-12a Standard Specification for Mortar for Unit Masonry

AU. ASTM C476-10 Standard Specification for Grout for Masonry

AV. ASTM C482-02 (2009) Standard Test Method for Bond Strength of Ceramic Tile to Portland Cement Paste

AW. ASTM C503/C503M-10 Standard Specification for Marble Dimension Stone

AX. ASTM C568/C568M-10 Standard Specification for Limestone Dimension Stone

COMMENTARY

SPECIFICATION

1.3 — Reference standards (Continued)

AY. ASTM C615/C615M-11 Standard Specification for Granite Dimension Stone

AZ. ASTM C616/C616M-10 Standard Specification for Quartz-Based Dimension Stone

BA. ASTM C629/C629-10 Standard Specification for Slate Dimension Stone

BB. ASTM C652-12a Standard Specification for Hollow Brick (Hollow Masonry Units Made from Clay or Shale)

BC. ASTM C744-11 Standard Specification for Prefaced Concrete and Calcium Silicate Masonry Units

BD. ASTM C901-10 Standard Specification for Prefabricated Masonry Panels

BE. ASTM C920-11 Standard Specification for Elastomeric Joint Sealants

BF. ASTM C1006-07 Standard Test Method for Splitting Tensile Strength of Masonry Units

BG. ASTM C1019-11 Standard Test Method for Sampling and Testing Grout

BH. ASTM C1072-12 Standard Tests Method for Measurement of Masonry Flexural Bond Strength

BI. ASTM C1088-12 Standard Specification for Thin Veneer Brick Units Made from Clay or Shale

BJ. ASTM C1314-12 Standard Test Method for Compressive Strength of Masonry Prisms

BK. ASTM C1405-12 Standard Specification for Glazed Brick (Single Fired, Brick Units)

BL. ASTM C1532/C1532M-12 Standard Practice for Selection, Removal and Shipment of Manufactured Masonry Units and Masonry Specimens from Existing Construction

BM. ASTM C1611/C1611M-09b[ε1] Standard Test Method for Slump Flow of Self-Consolidating Concrete

BN. ASTM C1634-11 Standard Specification for Concrete Facing Brick

BO. ASTM C 1660-10 Standard Specification for Thin-bed Mortar for Autoclaved Aerated Concrete (AAC) Masonry

BP. ASTM C1691-11 Standard Specification for Unreinforced Autoclaved Aerated Concrete (AAC) Masonry Units

BQ. ASTM C 1692-11 Standard Practice for Construction and Testing of Autoclaved Aerated Concrete (AAC) Masonry.

SPECIFICATION

1.3 — Reference standards (Continued)

BR. ASTM D92-12a Standard Test Method for Flash and Fire Points by Cleveland Open Cup Tester

BS. ASTM D95-05(2010) Standard Test Method for Water in Petroleum Products and Bituminous Materials by Distillation

BT. ASTM D512-12 Standard Test Methods for Chloride Ion in Water

BU. ASTM D566-02(2009) Standard Test Method for Dropping Point of Lubricating Grease

BV. ASTM D610-08(2012) Standard Practice for Evaluating Degree of Rusting on Painted Steel Surfaces

BW. ASTM D638-10 Standard Test Method for Tensile Properties of Plastics

BX. ASTM D994/D994M-11 Standard Specification for Preformed Expansion Joint Filler for Concrete (Bituminous Type)

BY. ASTM D1056-07 Standard Specification for Flexible Cellular Materials — Sponge or Expanded Rubber

BZ. ASTM D1187/D1187M-97 (2011)^{e1} Standard Specification for Asphalt-Base Emulsions for Use as Protective Coatings for Metal

CA. ASTM D1227-95 (2007) Standard Specification for Emulsified Asphalt Used as a Protective Coating for Roofing

CB. ASTM D2000-12 Standard Classification System for Rubber Products in Automotive Applications

CC. ASTM D2265-06 Standard Test Method for Dropping Point of Lubricating Grease Over Wide Temperature Range

CD. ASTM D2287-12 Standard Specification for Nonrigid Vinyl Chloride Polymer and Copolymer Molding and Extrusion Compounds

CE. ASTM D4289-03 (2008) Standard Test Method for Elastomer Compatibility of Lubricating Greases and Fluids

CF. ASTM E72-10 Standard Test Methods of Conducting Strength Tests of Panels for Building Construction

CG. ASTM E328-02 (2008) Standard Test Methods for Stress Relaxation Tests for Materials and Structures

CH. ASTM E518/E518M-10 Standard Test Methods for Flexural Bond Strength of Masonry

CI. ASTM E519/E519M-10 Standard Test Method for Diagonal Tension (Shear) in Masonry Assemblages

COMMENTARY

SPECIFICATION

1.3 — Reference standards (Continued)

CJ. ASTM F959-09 Standard Specification for Compressible-Washer-Type Direct Tension Indicators for Use with Structural Fasteners

CK. ASTM F959M-07 Standard Specification for Compressible-Washer-Type Direct Tension Indicators for Use with Structural Fasteners [Metric]

American Welding Society

CL. AWS D 1.4/D1.4M:2011 Structural Welding Code – Reinforcing Steel

Federal Test Method Standard

CM. FTMS 791B (1974) Oil Separation from Lubricating Grease (Static Technique). Federal Test Method Standard from the U.S. Army General Material and Parts Center, Petroleum Field Office (East), New Cumberland Army Depot, New Cumberland, PA 17070

1.4 — System description

1.4 A. *Compressive strength requirements* — Compressive strength of masonry in each masonry wythe and grouted collar joint shall equal or exceed the applicable f'_m or f'_{AAC}. For partially grouted masonry, the compressive strength of both the grouted and ungrouted masonry shall equal or exceed the applicable f'_m. At the transfer of prestress, the compressive strength of the masonry shall equal or exceed f'_{mi}.

1.4 B. *Compressive strength determination*

1. *Methods for determination of compressive strength* — Determine the compressive strength for each wythe by the unit strength method or by the prism test method as specified here.

COMMENTARY

1.4 — System description

1.4 A. *Compressive strength requirements* — Design is based on a certain f'_m or f'_{AAC} and this compressive strength value must be achieved or exceeded. In a multiwythe wall designed as a composite wall, the compressive strength of masonry for each wythe or grouted collar joint must equal or exceed f'_m or f'_{AAC}.

1.4 B. *Compressive strength determination*

1. *Methods for determination of compressive strength* — Two methods are permitted to verify compliance with the specified compressive strength of masonry during construction: the unit strength method and the prism test method. The unit strength method has several advantages. It is less expensive than the prism test method, and it eliminates the possibility of unrepresentative low values due to errors in the construction, transport and testing of prisms. The prism test method also has advantages. Although it often requires specialized testing equipment that may not be readily available in all areas, it generally provides higher values than the unit strength method, when properly executed. Local practices and jobsite conditions may favor one method over the other.

The Specification permits the contractor to select the method of verifying compliance with the specified compressive strength of masonry, unless a method is stipulated in the Project Specifications or Project Drawings.

SPECIFICATION

2. *Unit strength method*

a. *Clay masonry* — Use Table 1 to determine the compressive strength of clay masonry based on the strength of the units and the type of mortar specified. The following requirements apply to masonry:

1) Units are sampled and tested to verify conformance with ASTM C62, ASTM C216, or ASTM C652.

2) Thickness of bed joints does not exceed $5/8$ in. (15.9 mm).

3) For grouted masonry, the grout conforms to Article 2.2:

COMMENTARY

2. *Unit strength method* — Compliance with the requirement for f'_m, based on the compressive strength of masonry units, grout, and mortar type, is permitted instead of prism testing.

The influence of mortar joint thickness is noted by the maximum joint thickness. Grout strength greater than or equal to f'_m fulfills the requirements of Specification Article 1.4 A and Code Section 3.1.6.1.

a. *Clay masonry* — The original values of net area compressive strength of clay masonry in Table 1 were derived from research conducted by the Structural Clay Products Institute (SCPI, 1969).

The original values were based on testing of solid clay masonry units (SCPI, 1969) and portland cement-lime mortar. Further testing (Brown and Borchelt, 1990) has shown that the values are applicable for hollow and solid clay masonry units with all mortar types. A plot of the data is shown in Figure SC-2.

SCPI (1969) uses a height-to-thickness ratio of five as a basis to establish prism compressive strength. The Code uses a different method to design for axial stress so it was necessary to change the basic prism h/t ratio to two. This corresponds to the h/t ratio used for concrete masonry in the Code and for all masonry in other codes. The net effect is to increase the net area compressive strength of brick masonry as shown in Table 1 by 22 percent over that in Figure SC-2.

Table 1 — Compressive strength of masonry based on the compressive strength of clay masonry units and type of mortar used in construction

Net area compressive strength of clay masonry, psi (MPa)	Net area compressive strength of clay masonry units, psi (MPa)	
	Type M or S mortar	Type N mortar
1,000 (6.90)	1,700 (11.72)	2,100 (14.48)
1,500 (10.34)	3,350 (23.10)	4,150 (28.61)
2,000 (13.79)	4,950 (34.13)	6,200 (42.75)
2,500 (17.24)	6,600 (45.51)	8,250 (56.88)
3,000 (20.69)	8,250 (56.88)	10,300 (71.02)
3,500 (24.13)	9,900 (68.26)	—
4,000 (27.58)	11,500 (79.29)	—

COMMENTARY

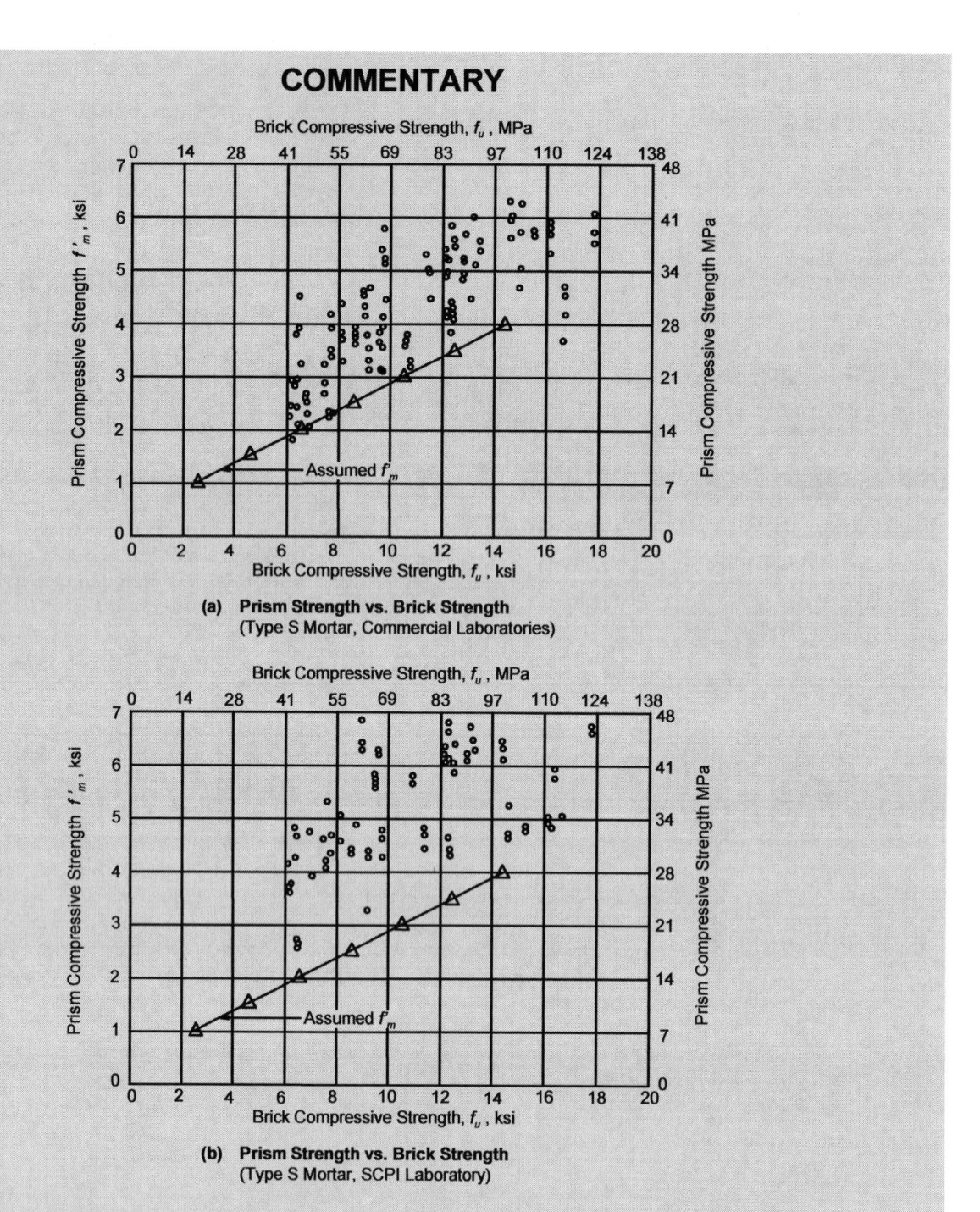

(a) Prism Strength vs. Brick Strength
(Type S Mortar, Commercial Laboratories)

(b) Prism Strength vs. Brick Strength
(Type S Mortar, SCPI Laboratory)

*Figure SC-2 — Compressive strength of masonry versus clay masonry unit strength
(See Commentary Article 1.4 B.2.a)*

SPECIFICATION

1.4 B.2. *Unit strength method* (Continued)

b. *Concrete masonry* — Use Table 2 to determine the compressive strength of concrete masonry based on the strength of the unit and type of mortar specified. The following Articles must be met:

1) Units are sampled and tested to verify conformance with, ASTM C90.

2) Thickness of bed joints does not exceed $5/8$ in. (15.9 mm).

3) For grouted masonry, the grout conforms to Article 2.2.

COMMENTARY

b. *Concrete masonry* — Prior to the 2013 Specification, the standardized correlations between unit compressive strength, mortar type, and resulting assembly compressive strength of concrete masonry were established using prism test results collected from the 1950s through the 1980s. The result was a database of prism compressive strengths with statistically high variability, which when introduced into the Specification, drove the lower bound design values between unit, mortar, and prism to very conservative values. The reasons for the inherent historical conservatism in the unit strength table are twofold: 1) When originally introduced, the testing procedures and equipment used to develop the prism test data were considerably less refined than they are today. Changes introduced into ASTM C1314, particularly requirements for stiffer/thicker bearing platens on testing equipment, produce more consistent, repeatable compressive strength results. 2) Previous testing procedures either did not control the construction, curing, and testing of masonry prisms, or permitted many procedures for doing so. As a result, a single set of materials could produce prism test results that varied significantly depending upon how the prisms were constructed, cured, and tested. Often, a field-constructed and field-cured prism would test to a lower value than a laboratory-constructed and laboratory-cured prism. Consequently, the compressive-strength values for concrete masonry prisms used to develop historical versions of the unit strength tables are not directly comparable to the compressive-strength values that would be obtained today.

Table 2 — Compressive strength of masonry based on the compressive strength of concrete masonry units and type of mortar used in construction

Net area compressive strength of concrete masonry, psi (MPa)	Net area compressive strength of concrete masonry units, psi (MPa)	
	Type M or S mortar	Type N mortar
1,700 (11.72)	---	1,900 (13.10)
1,900 (13.10)	1,900 (13.10)	2,350 (16.20)
2,000 (13.79)	2,000 (13.79)	2,650 (18.27)
2,250 (15.51)	2,600 (17.93)	3,400 (23.44)
2,500 (17.24)	3,250 (22.41)	4,350 (28.96)
2,750 (18.96)	3,900 (26.89)	-----
3,000 (20.69)	4,500 (31.03)	-----

[1]For units of less than 4 in. (102 mm) nominal height, use 85 percent of the values listed.

SPECIFICATION

1.4 B.2.b *Unit strength method* (Continued)

COMMENTARY

In 2010, the National Concrete Masonry Association (NCMA, 2012) began compiling prism test data to create a new database that would permit the development of a new unit strength table for concrete masonry that would better represent results from current prism tests. Concrete brick (ASTM C55 and ASTM C1634) are not included in Table 2 because the NCMA research program did not include these units. Most concrete brick are used in applications not requiring that f'_m be specified (such as veneer). Where f'_m is required for concrete brick applications, prism testing is required to verify the compressive strength.

The unit strength method was generated using prism test data as shown in Figures SC-3 and SC-4. The values in Table 2 are based on a consistent statistical criterion, with slight modifications based on engineering judgment.

For each specified unit strength and mortar type, the resulting masonry assembly compressive strengths were assumed to be normally distributed. Using the NCMA data for each specified unit strength and mortar type, and including the effects of sample size, the 75th percent confidence level on the 10-percentile value was calculated. That is, the value that would be expected to exceed the lower 10% fractile of the entire population 75% of the time. The criterion gives results that are reasonably consistent with other codes and standards (Bennett, 2010). Choosing the 10-percentile value results in an approximately 1% probability that the average of three prism test specimens will be less than the tabulated value.

For a given unit strength and mortar type, the resulting masonry assembly compressive strength also depends on the height of the units. The lateral expansion of the unit due to unit and mortar incompatibility increases with reduced unit height (Drysdale et al, 1999). A reduction factor in the compressive strength of masonry is required for masonry constructed of units less than 4 in. (102 mm) in nominal height, but need not be applied to masonry in which occasional units are cut to fit.

COMMENTARY

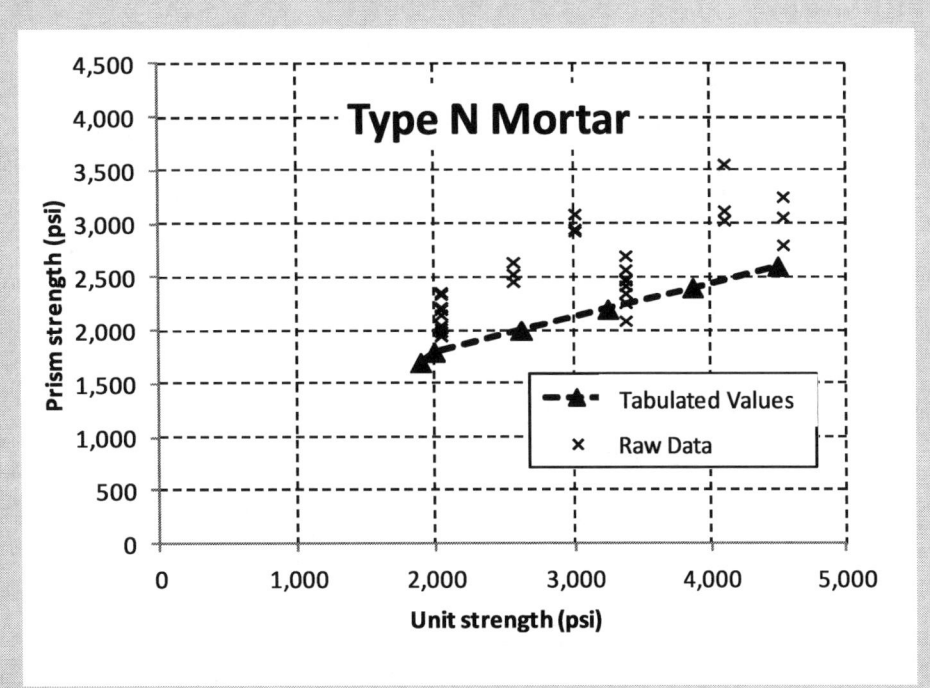

Figure SC-3 — Compressive strength of concrete masonry versus compressive strength of concrete masonry units – Type N Mortar

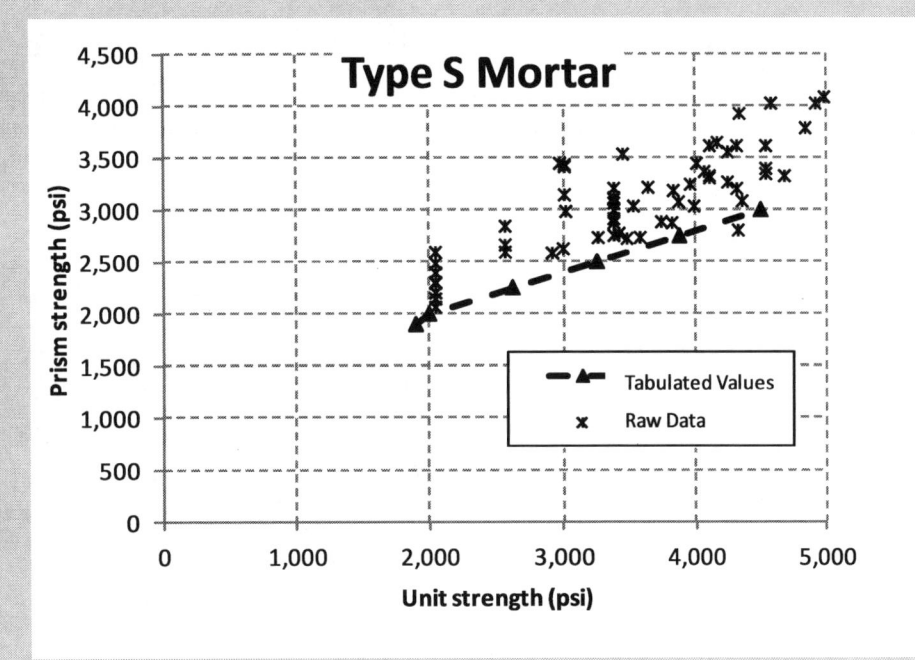

Figure SC-4 — Compressive strength of concrete masonry versus compressive strength of concrete masonry units – Type S Mortar

SPECIFICATION

1.4 B.2. *Unit strength method* (Continued)

  c. *AAC masonry* — Determine the compressive strength of masonry based on the strength of the AAC masonry unit only. The following requirements apply to the masonry:

  1) Units conform to Article 2.3 E.

  2) Thickness of bed joints does not exceed 1/8 in. (3.2 mm).

  3) For grouted masonry, the grout conforms to Article 2.2.

3. *Prism test method* — Determine the compressive strength of clay masonry and concrete masonry by the prism test method in accordance with ASTM C1314.

COMMENTARY

  c. *AAC masonry* — The strength of AAC masonry, f'_{AAC}, is controlled by the strength class of the AAC unit as defined by ASTM C1693. The strength of the thin-bed mortar and its bond in compression and shear will exceed the strength of the unit.

3. *Prism test method* — The prism test method described in ASTM C1314 was selected as a uniform method of testing clay masonry and concrete masonry to determine their compressive strengths. Masonry design is based on the compressive strength established at 28 days. The prism test method is used as an alternative to the unit strength method.

ASTM C1314 provides for testing masonry prisms at 28 days or at any designated test age. Therefore, a shorter time period, such as a 7-day test, could be used to estimate the 28-day strength based on a previously established relationship between the results of tests conducted at the shorter time period and results of the 28 day tests. Materials and workmanship of the previously established relationship must be representative of the prisms being tested.

Compliance with the specified compressive strength of masonry can be determined by the prism method instead of the unit strength method. ASTM C1314 uses the same materials and workmanship to construct the prisms as those to be used in the structure. Atkinson and Kingsley (1985), Priestley and Elder (1983), Miller et al (1979), Noland (1982) and Hegemier et al (1978) discuss prism testing. Many more references on the prism test method parameters and results could be added. The adoption of ASTM C1314 alleviates most of the concerns stated in the above references. ASTM C1314 replaced ASTM E447 (1997), which was referenced in editions of the Specification prior to 1999.

SPECIFICATION

1.4 B. *Compressive strength determination* (Continued)

4. *Testing prisms from constructed masonry* — When approved by the building official, acceptance of masonry that does not meet the requirements of Article 1.4 B.2 or 1.4 B.3 is permitted to be based on tests of prisms cut from the masonry construction.

   a. *Prism sampling and removal* — For each 5,000 square feet (465 m^2) of wall area in question, saw-cut a minimum of three prisms from completed masonry. Select, remove and transport prisms in accordance with ASTM C1532/C1532M. Determine the length, width and height dimensions of the prism and test prisms when at least 28 days old in accordance with ASTM C1314.

   b. *Compressive strength calculations* — Calculate the compressive strength of prisms in accordance with ASTM C1314.

   c. *Compliance* — Strengths determined from saw-cut prisms shall equal or exceed the specified compressive strength of masonry. Additional testing of specimens cut from construction in question is permitted.

1.4 C. *Adhered veneer requirements* — When adhered veneer is not placed in accordance with Article 3.3 C, determine the adhesion of adhered veneer unit to backing in accordance with ASTM C482.

COMMENTARY

4. *Testing prisms from constructed masonry* — While uncommon, there are times when the compressive strength of masonry determined by the unit strength method or prism test method may be questioned or may be lower than the specified strength. Because low strengths could be a result of inappropriate testing procedures or unintentional damage to the test specimens, prisms may be saw-cut from the completed masonry wall and tested. This section prescribes procedures for such tests.

Such testing is difficult, is performed on masonry walls constructed at least 28 days before the test, and requires replacement of the sampled wall area. Therefore, concerted efforts should be taken so that strengths determined by the unit strength method or prism test method are adequate.

   a. *Prism sampling and removal* — Removal of prisms from a constructed wall requires care so that the prism is not damaged and that damage to the wall is minimal. Prisms must be representative of the wall, yet not contain any reinforcing steel, which would bias the results. As with a prism test taken during construction, a prism test from existing masonry requires three prism specimens.

   b. *Compressive strength calculations* — Compressive strength calculations from saw-cut specimens must be based on the net mortar bedded area, or the net mortar bedded area plus the grouted area for grouted prisms. The net area must be determined by the testing agency before the prism is tested.

1.4 C. *Adhered veneer requirements* — Adhesion should be verified if a form release agent, an applied coating, or a smooth surface is present on the backing.

SPECIFICATION

1.5 — Submittals

1.5 A. Obtain written acceptance of submittals prior to the use of the materials or methods requiring acceptance.

1.5 B. Submit the following:

1. Mix designs and test results

    a. One of the following for each mortar mix, excluding thin-bed mortar for AAC:

    1) Mix designs indicating type and proportions of ingredients in compliance with the proportion specification of ASTM C270, or

    2) Mix designs and mortar tests performed in accordance with the property specification of ASTM C270.

    b. One of the following for each grout mix:

    1) Mix designs indicating type and proportions of the ingredients according to the proportion requirements of ASTM C476, or

    2) Mix designs and grout strength test performed in accordance with ASTM C476, or

    3) Compressive strength tests performed in accordance with ASTM C1019, and slump flow and Visual Stability Index (VSI) as determined by ASTM C1611/C1611M.

2. Material certificates — Material certificates for the following, certifying that each material is in compliance.

    a. Reinforcement

    b. Anchors, ties, fasteners, and metal accessories

    c. Masonry units

    d. Mortar, thin-bed mortar for AAC, and grout materials

    e. Self-consolidating grout

3. Construction procedures

    a. Cold weather construction procedures

    b. Hot weather construction procedures

COMMENTARY

1.5 — Submittals

Submittals and their subsequent acceptance or rejection on a timely basis will keep the project moving smoothly. If the specifier wishes to require a higher level of quality assurance than the minimum required by this Specification, submittals may be required for one or more of the following: shop drawings for reinforced masonry and lintels; sample specimens of masonry units, colored mortar, each type of movement joint accessory, anchor, tie, fastener, and metal accessory; and test results for masonry units, mortar, and grout.

SPECIFICATION

1.6 — Quality assurance

1.6 A. *Testing Agency's services and duties*

1. Sample and test in accordance with Table 3, 4, or 5, as specified for the project.

2. Unless otherwise required, report test results to the Architect/Engineer, Inspection Agency, and Contractor promptly after they are performed. Include in test reports a summary of conditions under which test specimens were stored prior to testing and state what portion of the construction is represented by each test.

3. When there is reason to believe that any material furnished or work performed by the Contractor fails to fulfill the requirements of the Contract Documents, report such discrepancy to the Architect/Engineer, Inspection Agency, and Contractor.

4. Unless otherwise required, the Owner will retain the Testing Agency.

COMMENTARY

1.6 — Quality assurance

Quality assurance consists of the actions taken by an owner or owner's representative, including establishing the quality assurance requirements, to provide assurance that materials and workmanship are in accordance with the contract documents. Quality assurance includes quality control measures as well as testing and inspection to verify compliance. The term quality control was not used in the Specification because its meaning varies with the perspective of the parties involved in the project.

The owner and Architect/Engineer may require a testing laboratory to provide some or all of the tests mentioned in Specification Tables 3, 4, and 5.

The quality objectives are met when the building is properly designed, completed using materials complying with product specifications using adequate construction practices, and is adequately maintained. Special Inspection and testing are important components of the quality assurance program, which is used to meet the objective of quality in construction.

Laboratories that comply with the requirements of ASTM C1093 are more likely to be familiar with masonry materials and testing. Specifying that the testing agencies comply with the requirements of ASTM C1093 is suggested.

1.6 A. *Testing Agency's services and duties* — Implementation of testing and inspection requirements contained in the Quality Assurance Tables requires detailed knowledge of the appropriate procedures. Comprehensive (Chrysler, 2010; NCMA, 2008; BIA TN 39, 2001; BIA TN 39B, 1988) and summary (SCI and MIA, 2006(a) SCI and MIA, 2006(a)) testing and inspection procedures are available from recognized industry sources which may be referenced for assistance in complying with the specified Quality Assurance program.

Table 3 — Level A Quality Assurance

MINIMUM VERIFICATION
Prior to construction, verify certificates of compliance used in masonry construction

Table 4 — Level B Quality Assurance

MINIMUM TESTS
Verification of Slump flow and Visual Stability Index (VSI) as delivered to the project site in accordance with Article 1.5 B.1.b.3 for self-consolidating grout
Verification of f'_m and f'_{AAC} in accordance with Article 1.4 B prior to construction, except where specifically exempted by the Code.

MINIMUM SPECIAL INSPECTION

Inspection Task	Frequency [a] Continuous	Frequency [a] Periodic	Reference for Criteria TMS 402/ ACI 530/ ASCE 5	Reference for Criteria TMS 602/ ACI 530.1/ ASCE 6
1. Verify compliance with the approved submittals		X		Art. 1.5
2. As masonry construction begins, verify that the following are in compliance:				
a. Proportions of site-prepared mortar		X		Art. 2.1, 2.6 A
b. Construction of mortar joints		X		Art. 3.3 B
c. Grade and size of prestressing tendons and anchorages		X		Art. 2.4 B, 2.4 H
d. Location of reinforcement, connectors, and prestressing tendons and anchorages		X		Art. 3.4, 3.6 A
e. Prestressing technique		X		Art. 3.6 B
f. Properties of thin-bed mortar for AAC masonry	X[b]	X[c]		Art. 2.1 C
3. Prior to grouting, verify that the following are in compliance:				
a. Grout space		X		Art. 3.2 D, 3.2 F
b. Grade, type, and size of reinforcement and anchor bolts, and prestressing tendons and anchorages		X	Sec. 6.1	Art. 2.4, 3.4
c. Placement of reinforcement, connectors, and prestressing tendons and anchorages		X	Sec. 6.1, 6.2.1, 6.2.6, 6.2.7	Art. 3.2 E, 3.4, 3.6 A
d. Proportions of site-prepared grout and prestressing grout for bonded tendons		X		Art. 2.6 B, 2.4 G.1.b
e. Construction of mortar joints		X		Art. 3.3 B

SPECIFICATION FOR MASONRY STRUCTURES AND COMMENTARY S-25

Table 4 — Level B Quality Assurance (Continued)

<table>
<tr><th colspan="5">MINIMUM SPECIAL INSPECTION</th></tr>
<tr><th rowspan="2">Inspection Task</th><th colspan="2">Frequency [a]</th><th colspan="2">Reference for Criteria</th></tr>
<tr><th>Continuous</th><th>Periodic</th><th>TMS 402/
ACI 530/
ASCE 5</th><th>TMS 602/
ACI 530.1/
ASCE 6</th></tr>
<tr><td>4. Verify during construction:</td><td></td><td></td><td></td><td></td></tr>
<tr><td>   a. Size and location of structural elements</td><td></td><td>X</td><td></td><td>Art. 3.3 F</td></tr>
<tr><td>   b. Type, size, and location of anchors, including other details of anchorage of masonry to structural members, frames, or other construction</td><td></td><td>X</td><td>Sec. 1.2.1(e), 6.1.4.3, 6.2.1</td><td></td></tr>
<tr><td>   c. Welding of reinforcement</td><td>X</td><td></td><td>Sec. 8.1.6.7.2, 9.3.3.4 (c), 11.3.3.4(b)</td><td></td></tr>
<tr><td>   d. Preparation, construction, and protection of masonry during cold weather (temperature below 40°F (4.4°C)) or hot weather (temperature above 90°F (32.2°C))</td><td></td><td>X</td><td></td><td>Art. 1.8 C, 1.8 D</td></tr>
<tr><td>   e. Application and measurement of prestressing force</td><td>X</td><td></td><td></td><td>Art. 3.6 B</td></tr>
<tr><td>   f. Placement of grout and prestressing grout for bonded tendons is in compliance</td><td>X</td><td></td><td></td><td>Art. 3.5, 3.6 C</td></tr>
<tr><td>   g. Placement of AAC masonry units and construction of thin-bed mortar joints</td><td>X[b]</td><td>X[c]</td><td></td><td>Art. 3.3 B.9, 3.3 F.1.b</td></tr>
<tr><td>5. Observe preparation of grout specimens, mortar specimens, and/or prisms</td><td></td><td>X</td><td></td><td>Art. 1.4 B.2.a.3, 1.4 B.2.b.3, 1.4 B.2.c.3, 1.4 B.3, 1.4 B.4</td></tr>
</table>

(a) Frequency refers to the frequency of Special Inspection, which may be continuous during the task listed or periodic during the listed task, as defined in the table.

(b) Required for the first 5000 square feet (465 square meters) of AAC masonry.

(c) Required after the first 5000 square feet (465 square meters) of AAC masonry.

Table 5 — Level C Quality Assurance

MINIMUM TESTS
Verification of f'_m and f'_{AAC} in accordance with Article 1.4 B prior to construction and for every 5,000 sq. ft (465 sq. m) during construction
Verification of proportions of materials in premixed or preblended mortar, prestressing grout, and grout other than self-consolidating grout as delivered to the project site
Verification of Slump flow and Visual Stability Index (VSI) as delivered to the project site in accordance with Article 1.5 B.1.b.3 for self-consolidating grout

MINIMUM SPECIAL INSPECTION				
Inspection Task	Frequency [a]		Reference for Criteria	
	Continuous	Periodic	TMS 402/ ACI 530/ ASCE 5	TMS 602/ ACI 530.1/ ASCE 6
1. Verify compliance with the approved submittals		X		Art. 1.5
2. Verify that the following are in compliance:				
a. Proportions of site-mixed mortar, grout, and prestressing grout for bonded tendons		X		Art. 2.1, 2.6 A, 2.6 B, 2.6 C, 2.4 G.1.b
b. Grade, type, and size of reinforcement and anchor bolts, and prestressing tendons and anchorages		X	Sec. 6.1	Art. 2.4, 3.4
c. Placement of masonry units and construction of mortar joints		X		Art. 3.3 B
d. Placement of reinforcement, connectors, and prestressing tendons and anchorages	X		Sec. 6.1, 6.2.1, 6.2.6, 6.2.7	Art. 3.2 E, 3.4, 3.6 A
e. Grout space prior to grouting	X			Art. 3.2 D, 3.2 F
f. Placement of grout and prestressing grout for bonded tendons	X			Art. 3.5, 3.6 C
g. Size and location of structural elements		X		Art. 3.3 F
h. Type, size, and location of anchors including other details of anchorage of masonry to structural members, frames, or other construction	X		Sec. 1.2.1(e), 6.1.4.3, 6.2.1	
i. Welding of reinforcement	X		Sec. 8.1.6.7.2, 9.3.3.4 (c), 11.3.3.4(b)	
j. Preparation, construction, and protection of masonry during cold weather (temperature below 40°F (4.4°C)) or hot weather (temperature above 90°F (32.2°C))		X		Art. 1.8 C, 1.8 D
k. Application and measurement of prestressing force	X			Art. 3.6 B
l. Placement of AAC masonry units and construction of thin-bed mortar joints	X			Art. 3.3 B.9, 3.3 F.1.b
m. Properties of thin-bed mortar for AAC masonry	X			Art. 2.1 C.1
3. Observe preparation of grout specimens, mortar specimens, and/or prisms	X			Art. 1.4 B.2.a.3, 1.4 B.2.b.3, 1.4 B.2.c.3, 1.4 B.3, 1.4 B.4

(a) Frequency refers to the frequency of Special Inspection, which may be continuous during the task listed or periodic during the listed task, as defined in the table.

SPECIFICATION

1.6 B. *Inspection Agency's services and duties*

1. Inspect and evaluate in accordance with Table 3, 4, or 5, as specified for the project.

2. Unless otherwise required, report inspection results to the Architect/Engineer, and Contractor promptly after they are performed. Include in inspection reports a summary of conditions under which the inspections were made and state what portion of the construction is represented by each inspection.

3. Furnish inspection reports to the Architect/Engineer and Contractor.

4. When there is reason to believe that any material furnished or work performed by the Contractor fails to fulfill the requirements of the Contract Documents, report such discrepancy to the Architect/Engineer and to the Contractor.

5. Submit a final signed report stating whether the Work requiring Special Inspection was, to the best of the Inspection Agency's knowledge, in conformance. Submit the final report to the Architect/Engineer and Contractor.

6. Unless otherwise required, the Owner will retain the Inspection Agency.

1.6 C. *Contractor's services and duties*

1. Permit and facilitate access to the construction sites and the performance of activities for quality assurance by the Testing and Inspection Agencies.

2. The use of testing and inspection services does not relieve the Contractor of the responsibility to furnish materials and construction in full compliance.

3. To facilitate testing and inspection, comply with the following:

   a. Furnish necessary labor to assist the designated testing agency in obtaining and handling samples at the Project.

   b. Advise the designated Testing Agency and Inspection Agency sufficiently in advance of operations to allow for completion of quality assurance measures and for the assignment of personnel.

   c. Provide masonry materials required for preconstruction and construction testing.

4. Provide and maintain adequate facilities for the sole use of the testing agency for safe storage and proper curing of test specimens on the Project Site.

5. In the submittals, include the results of testing performed to qualify the materials and to establish mix designs.

COMMENTARY

1.6 B. *Inspection Agency's services and duties* — The Code and this Specification require that masonry be inspected. The design provisions used in the Code are based on the premise that the work will be inspected, and that quality assurance measures will be implemented. Minimum testing and minimum Special Inspection requirements are given in Specification Tables 3, 4, and 5. The Architect/Engineer may increase the amount of testing and inspection required. Certain applications, such as Masonry Veneer (Chapter 12), Masonry Partition Walls (Chapter 14) and Empirical Design of Masonry (Appendix A), do not require compressive strength verification of masonry as indicated in Table 4. The method of payment for inspection services is usually addressed in general conditions or other contract documents and usually is not governed by this article.

1.6 C. *Contractor's services and duties* — The contractor establishes mix designs, the source for supply of materials, and suggests change orders.

The listing of duties of the inspection agency, testing agency, and contractor provide for a coordination of their tasks and a means of reporting results. The contractor is bound by contract to supply and place the materials required by the contract documents. Perfection is obviously the goal, but factors of safety included in the design method recognize that some deviation from perfection will exist. Engineering judgment must be used to evaluate reported discrepancies. Tolerances listed in Specification Article 3.3 F were established to assure structural performance and were not based on aesthetic criteria.

SPECIFICATION

1.6 D. *Sample panels*

1. For masonry governed by Level B or C Quality Assurance (Table 4 or Table 5), construct sample panels of masonry walls.

    a. Use materials and procedures accepted for the Work.

    b. The minimum sample panel dimensions are 4 ft by 4 ft (1.22 m by 1.22 m).

2. The acceptable standard for the Work is established by the accepted panel.

3. Retain sample panels at the project site until Work has been accepted.

1.6 E. *Grout demonstration panel* — Prior to masonry construction, construct a grout demonstration panel if proposed grouting procedures, construction techniques, or grout space geometry do not conform to the applicable requirements of Articles 3.5 C, 3.5 D, and 3.5 E.

1.7 — Delivery, storage, and handling

1.7 A. Do not use damaged masonry units, damaged components of structure, or damaged packaged material.

1.7 B. Protect cementitious materials for mortar and grout from precipitation and groundwater.

1.7 C. Do not use masonry materials that are contaminated.

1.7 D. Store different aggregates separately.

1.7 E. Protect reinforcement, ties, and metal accessories from permanent distortions and store them off the ground.

COMMENTARY

1.6 D. *Sample panels* — Sample panels should contain the full range of unit and mortar color. Each procedure, including cleaning and application of coatings and sealants, should be demonstrated on the sample panel. The effect of these materials and procedures on the masonry can then be determined before large areas are treated. Because it serves as a comparison of the finished work, the sample panel should be maintained until the work has been accepted. Certain elements of sample panels, such as the type of mortar joint, can have structural implications with the performance of masonry. Construct sample panels within the tolerances of Article 3.3 F. The specifier has the option of permitting a segment of the masonry construction to serve as a sample panel or requiring a separate stand-alone panel.

1.7 — Delivery, storage, and handling

The performance of masonry materials can be reduced by contamination by dirt, water, and other materials during delivery or at the project site.

Reinforcement and metal accessories are less prone than masonry materials to damage from handling.

SPECIFICATION

1.8 — Project conditions

1.8 A. *Construction loads* — Do not apply construction loads that exceed the safe superimposed load capacity of the masonry and shores, if used.

1.8 B. *Masonry protection* — Cover top of unfinished masonry work to protect it from moisture intrusion.

1.8 C. *Cold weather construction* — When ambient air temperature is below 40°F (4.4°C), implement cold weather procedures and comply with the following:

1. Do not lay glass unit masonry.

2. *Preparation* — Comply with the following requirements prior to conducting masonry work:

   a. Do not lay masonry units having either a temperature below 20°F (-6.7°C) or containing frozen moisture, visible ice, or snow on their surface.

   b. Remove visible ice and snow from the top surface of existing foundations and masonry to receive new construction. Heat these surfaces above freezing, using methods that do not result in damage.

COMMENTARY

1.8 — Project conditions

1.8 B. *Masonry protection* — Many geographic areas are subject to unpredictable weather. Masonry under construction needs to be protected from detrimental moisture intrusion, particularly when there is a possibility of freezing temperatures. In areas where dry weather is consistent, covering walls to protect against moisture intrusion during the normal progress of construction may not be required.

1.8 C. *Cold weather construction* — The procedure described in this article represents the committee's consensus of current good construction practice and has been framed to generally agree with masonry industry recommendations (IMI, 1973).

The provisions of Article 1.8 C are mandatory, even if the procedures submitted under Article 1.5 B.3.a are not required. The contractor has several options to achieve the results required in Article 1.8 C. The options are available because of the climatic extremes and their duration. When the air temperature at the project site or unit temperatures fall below 40 F (4.4 C), the cold weather protection plan submitted becomes mandatory. Work stoppage may be justified if a short cold spell is anticipated. Enclosures and heaters can be used as necessary.

Temperature of the masonry mortar may be measured using a metal tip immersion thermometer inserted into a sample of the mortar. The mortar sample may be mortar as contained in the mixer, in hoppers for transfer to the working face of the masonry or as available on mortar boards currently being used. The critical mortar temperatures are the temperatures at the mixer and mortar board locations. The ideal mortar temperature is 60°F to 80°F (15.6°C to 26.7°C).

Temperature of the masonry unit may be measured using a metallic surface contact thermometer. Temperature of the units may be below the ambient temperature if the requirements of Article 1.8 C.2.a are met.

The contractor may choose to enclose the entire area rather than make the sequential materials conditioning and protection modifications. Ambient temperature conditions apply while work is in progress. Minimum daily temperatures apply to the time after grouted masonry is placed. Mean daily temperatures apply to the time after ungrouted masonry is placed.

Grout made with Type III portland cement gains strength more quickly than grout mixed with Type I portland cement. This faster strength gain eliminates the need to protect masonry for the additional 24 hr period.

SPECIFICATION

1.8 C. *Cold weather construction* (Continued)

3. *Construction* — These requirements apply to work in progress and are based on ambient air temperature. Do not heat water or aggregates used in mortar or grout above 140°F (60°C). Comply with the following requirements when the following ambient air temperatures exist:

a. 40°F to 32°F (4.4°C to 0°C):

1) Heat sand or mixing water to produce mortar temperature between 40°F (4.4°C) and 120°F (48.9°C) at the time of mixing.

2) Heat grout materials when the temperature of the materials is below 32°F (0°C).

b. Below 32°F to 25°F (0°C to -3.9°C):

1) Heat sand and mixing water to produce mortar temperature between 40°F (4.4°C) and 120°F (48.9°C) at the time of mixing. Maintain mortar temperature above freezing until used in masonry.

2) Heat grout aggregates and mixing water to produce grout temperature between 70°F (21.1°C) and 120°F (48.9°C) at the time of mixing. Maintain grout temperature above 70°F (21.1°C) at the time of grout placement.

3) Heat AAC units to a minimum temperature of 40°F (4.4°C) before installing thin-bed mortar.

c. Below 25°F to 20°F (-3.9°C to –6.7°C): Comply with Article 1.8 C.3.b and the following:

1) Heat masonry surfaces under construction to a minimum temperature of 40°F (4.4°C).

2) Use wind breaks or enclosures when the wind velocity exceeds 15 mph (24 km/h).

3) Heat masonry to a minimum temperature of 40°F (4.4°C) prior to grouting.

d. Below 20°F (-6.7°C): Comply with Article 1.8 C.3.c and the following: Provide an enclosure and auxiliary heat to maintain air temperature above 32°F (0°C) within the enclosure.

COMMENTARY

Construction experience, though not formally documented, suggests that AAC thin-bed mortar reaches full strength significantly faster than masonry mortar; however, it is more sensitive to cold weather applications. AAC masonry also holds heat considerably longer than concrete masonry. Cold weather requirements are therefore different for thin-bed mortar applications as compared to conventional mortar. Cold weather requirements for leveling course mortar and grout remain the same as for other masonry products.

SPECIFICATION

1.8 C *Cold weather construction* (Continued)

4. *Protection* — These requirements apply after masonry is placed and are based on anticipated minimum daily temperature for grouted masonry and anticipated mean daily temperature for ungrouted masonry. Protect completed masonry in the following manner:

   a. Maintain the temperature of glass unit masonry above 40°F (4.4°C) for the first 48 hr after construction.

   b. Maintain the temperature of AAC masonry above 32°F (0°C) for the first 4 hr after thin-bed mortar application.

   c. 40°F to 25°F (4.4°C to -3.9°C): Protect newly constructed masonry by covering with a weather-resistive membrane for 24 hr after being completed.

   d. Below 25°F to 20°F (-3.9°C to -6.7°C): Cover newly constructed masonry completely with weather-resistive insulating blankets, or equal protection, for 24 hr after completion of work. Extend time period to 48 hr for grouted masonry, unless the only cement in the grout is Type III portland cement.

   e. Below 20°F (-6.7°C): Maintain newly constructed masonry temperature above 32°F (0°C) for at least 24 hr after being completed by using heated enclosures, electric heating blankets, infared lamps, or other acceptable methods. Extend time period to 48 hr for grouted masonry, unless the only cement in the grout is Type III portland cement.

SPECIFICATION

1.8 D. *Hot weather construction* — Implement approved hot weather procedures and comply with the following provisions:

1. *Preparation* — Prior to conducting masonry work:

   a. When the ambient air temperature exceeds 100°F (37.8°C), or exceeds 90°F (32.2°C) with a wind velocity greater than 8 mph (12.9 km/hr):

   1) Maintain sand piles in a damp, loose condition.

   2) Provide necessary conditions and equipment to produce mortar having a temperature below 120°F (48.9°C).

   b. When the ambient temperature exceeds 115°F (46.1°C), or exceeds 105°F (40.6°C) with a wind velocity greater than 8 mph (12.9 km/hr), implement the requirements of Article 1.8 D.1.a and shade materials and mixing equipment from direct sunlight.

2. *Construction* — While masonry work is in progress:

   a. When the ambient air temperature exceeds 100°F (37.8°C), or exceeds 90°F (32.2°C) with a wind velocity greater than 8 mph (12.9 km/hr):

   1) Maintain temperature of mortar and grout below 120°F (48.9°C).

   2) Flush mixer, mortar transport container, and mortar boards with cool water before they come into contact with mortar ingredients or mortar.

   3) Maintain mortar consistency by retempering with cool water.

   4) Use mortar within 2 hr of initial mixing.

   5) Spread thin-bed mortar no more than four feet ahead of AAC masonry units.

   6) Set AAC masonry units within one minute after spreading thin-bed mortar.

   b. When the ambient temperature exceeds 115°F (46.1°C), or exceeds 105°F (40.6°C) with a wind velocity greater than 8 mph (12.9 km/hr), implement the requirements of Article 1.8 D.2.a and use cool mixing water for mortar and grout. Ice is permitted in the mixing water prior to use. Do not permit ice in the mixing water when added to the other mortar or grout materials.

3. *Protection* — When the mean daily temperature exceeds 100°F (37.8°C) or exceeds 90°F (32.2°C) with a wind velocity greater than 8 mph (12.9 km/hr), fog spray newly constructed masonry until damp, at least three times a day until the masonry is three days old.

COMMENTARY

1.8 D. *Hot weather construction* — High temperature and low relative humidity increase the rate of moisture evaporation. These conditions can lead to "dryout" (drying of the mortar or grout before sufficient hydration has taken place) of the mortar and grout (Tomasetti, 1990). Dryout adversely affects the properties of mortar and grout because dryout signals improper curing and associated reduction of masonry strength development. The preparation, construction, and protection requirements in the Specification are minimum requirements to avoid dryout of mortar and grout and to allow for proper curing. They are based on industry practice (BIA, 1992; PCA, 1993; Panarese et al, 1991). More stringent and extensive hot weather practices may be prudent where temperatures are high, winds are strong, and humidity is low.

During hot weather, shading masonry materials and equipment reduces mortar and grout temperatures. Scheduling construction to avoid hotter periods of the day should be considered.

See Specification Commentary Article 2.1 for considerations in selecting mortar materials. The most effective way of reducing mortar and grout batch temperatures is by using cool mixing water. Small batches of mortar are preferred over larger batches to minimize drying time on mortar boards. Mortar should not be used after a maximum of 2 hr after initial mixing in hot weather conditions. Use of cool water to retemper, when tempering is permitted, restores plasticity and reduces the mortar temperature (IMI, 1973; BIA, 1992; PCA, 1993).

Most mason's sand is delivered to the project in a damp, loose condition with a moisture content of about 4 to 6 percent. Sand piles should be kept cool and in a damp, loose condition by sprinkling and by covering with a plastic sheet to limit evaporation.

Research suggests that covering and moist curing of concrete masonry walls dramatically improves flexural bond strength compared to walls not covered or moist cured (NCMA, 1994).

PART 2 — PRODUCTS

SPECIFICATION

2.1 — Mortar materials

2.1 A. Provide mortar of the type and color specified, and conforming with ASTM C270.

COMMENTARY

2.1 — Mortar materials

ASTM C270 contains standards for materials used to make mortar. Thus, component material specifications need not be listed. The Architect/Engineer may wish to include only certain types of materials, or exclude others, to gain better control.

There are two methods of specifying mortar under ASTM C270: proportion and property. The proportion specification directs the contractor to mix the materials in the volumetric proportions given in ASTM C270. These are repeated in Table SC-1. The property specification instructs the contractor to develop a mortar mix that will yield the specified properties under laboratory testing conditions. Table SC-2 contains the required results outlined in ASTM C270. The results are submitted to the Architect/Engineer and the mix proportions developed in the laboratory are maintained in the field. Water added in the field is determined by the mason for both methods of specifying mortar. A mortar mixed in accordance with the proportion requirements of Table SC-1 may have different physical properties than of a mortar of the same type (i.e. Type M, S, N, or O) mixed in accordance with proportions established by laboratory testing to meet the property specification requirements of Table SC-2. Higher lime content increases workability and water retentivity. ASTM C270 has an Appendix with information that can be useful in selecting mortar.

Either proportions or properties, but not both, should be specified. A good rule of thumb is to specify the weakest mortar that will perform adequately, not the strongest. Excessive amounts of pigments used to achieve mortar color may reduce both the compressive and bond strength of the masonry. Conformance to the maximum percentages indicated will limit the loss of strength to acceptable amounts. Due to the fine particle size, the water demand of the mortar increases when coloring pigments are used. Admixtures containing excessive amounts of chloride ions are detrimental to steel items placed in mortar or grout.

ASTM C270 specifies mortar testing under laboratory conditions only for acceptance of mortar mixes under the property specifications. Field sampling and testing of mortar is conducted under ASTM C780 and is used to verify consistency of materials and procedures, not mortar strength. ASTM C1586 provides guidance on appropriate testing of mortar for quality assurance.

COMMENTARY

Table SC-1 — ASTM C270 mortar proportion specification requirements

Mortar	Type	Portland cement or blended cement	Mortar cement M	Mortar cement S	Mortar cement N	Masonry cement M	Masonry cement S	Masonry cement N	Hydrated lime or lime putty	Aggregate ratio (measured in damp, loose conditions)
Cement-lime	M	1	-	-	-	-	-	-	¼	
	S	1	-	-	-	-	-	-	over ¼ to ½	
	N	1	-	-	-	-	-	-	over ½ to 1¼	
	O	1	-	-	-	-	-	-	over 1¼ to 2½	
Mortar cement	M	1	-	-	1	-	-	-	-	Not less than 2 ¼ and not more than 3 times the sum of the separate volumes of cementitious materials.
	M	-	1	-	-	-	-	-	-	
	S	½	-	-	1	-	-	-	-	
	S	-	-	1	-	-	-	-	-	
	N	-	-	-	1	-	-	-	-	
	O	-	-	-	1	-	-	-	-	
Masonry cement	M	1	-	-	-	-	-	1	-	
	M	-	-	-	-	1	-	-	-	
	S	½	-	-	-	-	-	1	-	
	S	-	-	-	-	-	1	-	-	
	N	-	-	-	-	-	-	1	-	
	O	-	-	-	-	-	-	1	-	

Two air entraining materials shall not be combined in mortar.

Table SC-2 — ASTM C270 property specification requirements for laboratory prepared mortar

Mortar	Type	Average compressive strength at 28 days, psi (MPa)	Water retention min, percent	Air content max, percent	Aggregate ratio (measured in damp, loose conditions)
Cement-lime	M	2500 (17.2)	75	12	
	S	1800 (12.4)	75	12	
	N	750 (5.2)	75	14[1]	
	O	350 (2.4)	75	14[1]	
Mortar cement	M	2500 (17.2)	75	12	Not less than 2¼ and not more than 3½ times the sum of the separate volumes of cementitious materials
	S	1800 (12.4)	75	12	
	N	750 (5.2)	75	14[1]	
	O	350 (2.4)	75	14[1]	
Masonry cement	M	2500 (17.2)	75	18	
	S	1800 (12.4)	75	18	
	N	750 (5.2)	75	20[2]	
	O	350 (2.4)	75	20[2]	

1. When structural reinforcement is incorporated in cement-lime or mortar cement mortar, the maximum air content shall be 12 percent.
2. When structural reinforcement is incorporated in masonry cement mortar, the maximum air content shall be 18 percent.

SPECIFICATION

2.1 B. *Glass unit masonry* — For glass unit masonry, provide Type S or N mortar that conforms to Article 2.1A.

2.1 C. *AAC masonry*
1. Provide thin-bed mortar specifically manufactured for use with AAC masonry. Testing to verify mortar properties shall be conducted by the thin-bed mortar manufacturer and confirmed by an independent testing agency.

   a. Provide thin-bed mortar with compressive strength that meets or exceeds the strength of the AAC masonry units. Conduct compressive strength tests in accordance with ASTM C109/C109M.

   b. Provide thin-bed mortar with shear strength that meets or exceeds the strength of the AAC masonry units. Conduct shear strength tests in accordance with ASTM E519. Cure the gypsum capping for at least 6 hours prior to testing.

   c. For each specified strength class, provide thin-bed mortar with flexural tensile strength that is not less than the smaller of: the maximum value specified in the governing building code; and the modulus of rupture of the masonry units. Conduct flexural strength tests in accordance with ASTM E72, ASTM E518 Method A or ASTM C1072.

   1) For conducting flexural strength tests in accordance with ASTM E518, construct at least five test specimens as stack-bonded prisms at least 32 in. (810 mm) high. Use the type of mortar specified by the AAC unit manufacturer.

   2) For flexural strength tests in accordance with ASTM C1072, construct test specimens as stack-bonded prisms comprised of at least 3 bed joints. Test a total of at least 5 joints. Use the type of mortar specified by the AAC unit manufacturer.

COMMENTARY

2.1 B. *Glass unit masonry* — In exterior applications, certain exposure conditions or panel sizes may warrant the use of mortar type with high bond strength. Type S mortar has a higher bond strength than Type N mortar. Portland cement-lime mortars and mortar-cement mortars have a higher bond strength than some masonry cement mortars of the same type. The performance of locally available materials and the size and exposure conditions of the panel should be considered when specifying the type of mortar. Manufacturers of glass units recommend using mortar containing a water-repellent admixture or a cement containing a water-repellent addition (Pittsburgh Corning, 1992; Glashaus, 1992; Beall, 1989) A workable, highly water-retentive mortar is recommended for use when conditions of high heat and low relative humidity exist during construction.

2.1 C. *AAC masonry* — ASTM E72 measures the flexural strength of a full-sized panel, whereas ASTM E518 and ASTM C1072 measure the flexural strength of small scale test specimens. ASTM E72 was developed to provide the most realistic assessment of a wall's performance under flexural loading.

c When tested, flexural tensile strength values of thin-bed mortar must not be less than the criteria contained in Code Section 11.1.8.3.

SPECIFICATION

2.1 C.1 *AAC masonry* (Continued)

d. Perform splitting tensile strength tests in accordance with ASTM C1660. The splitting tensile strength of each specimen tested used to calculate the average splitting tensile strength shall not be less than the minimum splitting tensile strength of the specimen for the corresponding AAC Strength Class specified in Table 1 of ASTM C1660.

2. Mortar for leveling course shall be Type M or S. Conform to the requirements of Article 2.1A.

2.2 — Grout materials

2.2 A. Unless otherwise required, provide grout that conforms to the requirements of ASTM C476.

2.2 B. When f'_m exceeds 2,000 psi (13.79 MPa), provide grout compressive strength that equals or exceeds f'_m. Determine compressive strength of grout in accordance with ASTM C1019.

2.2 C. Do not use admixtures unless acceptable. Field addition of admixtures is not permitted in self-consolidating grout.

COMMENTARY

d. ASTM C1660 requires that the average splitting tensile strength of mortared AAC specimens shall equal or exceed the minimum splitting strength of AAC as it relates to the minimum compressive strength for the AAC Class specified. In any sample of specimens tested to calculate the average there is no minimum value or statistical method specified in ASTM C1660 to allow for under strength specimens in a group of specimens used to calculate the average strength. The Specification clarifies this requirement by requiring that all samples must exceed the minimum value. This requirement is consistent with the specified strength requirements for AAC masonry units and the requirement that thin-bed mortar exceed the strength of the AAC in compression, shear and tensile splitting strength.

2.2 — Grout materials

ASTM C476 contains standards for materials used to make grout. Thus, component material specifications need not be listed.

Admixtures for grout include those to increase flow and to reduce shrinkage. Because self-consolidating grouts include admixtures and are delivered to the project site premixed or preblended and certified by the manufacturer, the addition of admixtures in the field is not permitted.

Self-consolidating grout meets the material requirements in ASTM C476. Because the mix is highly fluid, traditional slump cone tests for masonry grout are not applicable. The material is qualified by measuring its slump flow and determining its Visual Stability Index (VSI) using ASTM C1611/C1611 M.

Because the strength of AAC units never approaches 2,000 psi (13.79 MPa), and f'_{AAC} cannot be designed above the unit material strength, AAC masonry only requires grout that meets the requirements of Article 2.2 A.

This article does not apply to prestressing grout; see Article 2.4 G.1.b.

SPECIFICATION

2.3 — Masonry unit materials

2.3 A. Provide concrete masonry units that conform to ASTM C55, C73, C90, C129, C744, or C1634 as specified.

COMMENTARY

2.3 — Masonry unit materials

2.3 A. Concrete masonry units are made from lightweight and normal weight aggregate, water, and cement. The units are available in a variety of shapes, sizes, colors, and strengths. Because the properties of the concrete vary with the aggregate type and mix proportions, there is a range of physical properties and weights available in concrete masonry units.

Masonry units are selected for the use and appearance desired, with minimum requirements addressed by each respective ASTM standard. When particular features are desired such as surface textures for appearance or bond, finish, color, or particular properties such as weight classification, higher compressive strength, fire resistance, thermal or acoustical performance, these features should be specified separately by the purchaser. Local suppliers should be consulted as to the availability of units having the desired features.

ASTM C73 designates sand-lime brick as either Grade SW or Grade MW. Grade SW brick are intended for use where they will be exposed to freezing temperatures in the presence of moisture. Grade MW brick are limited to applications in which they may be subjected to freezing temperature but in which they are unlikely to be saturated with water.

Table SC-3 summarizes the requirements for various concrete masonry units given in the referenced standards.

ASTM C744 addresses the properties of units with a resin facing. The units must meet the requirements of one of the other referenced standards.

Table SC-3 — Concrete masonry unit requirements

ASTM Specification	Unit	Strength	Weight	Type	Grade
C55	Concrete brick	yes	yes	no	no
C73	Sand-lime brick	yes	no	no	yes
C90	Load-bearing units	yes	yes	no	no
C129	Non-load-bearing units	yes	yes	no	no
C744	Prefaced units	—	—	—	—
C1634	Concrete facing brick	yes	yes	no	no

SPECIFICATION

2.3 B. Provide clay or shale masonry units that conform to ASTM C34, C56, C62, C126, C212, C216, C652, C1088, or C1405 or to ANSI A 137.1, as specified.

COMMENTARY

2.3 B. Clay or shale masonry units are formed from those materials and referred to as brick or tile. Clay masonry units may be molded, pressed, or extruded into the desired shape. Physical properties depend upon the raw materials, the method of forming, and the firing temperature. Incipient fusion, a melting and joining of the clay particles, is necessary to develop the strength and durability of clay masonry units. A wide variety of unit shapes, sizes, colors, and strengths is available.

The intended use determines which standard specification is applicable. Generally, brick units are smaller than tile, tile is always cored, and brick may be solid or cored. Clay brick is normally exposed in use, but clay tile is usually not exposed. Grade or class is determined by exposure condition and has requirements for durability, usually given by compressive strength and absorption. Dimensional variations and allowable chips and cracks are controlled by type.

Table SC-4 summarizes the requirements given in the referenced standards.

Table SC-4 — Clay brick and tile requirements

ASTM Specification	Unit	Minimum % solid	Grade Strength	Grade Weight	Grade Type
C34	Load-bearing wall tile	a	yes	yes	no
C56	Non-load-bearing wall tile	b	no	yes	no
C62	Building brick (solid)	75	yes	yes	no
C126	Ceramic glazed units	c	yes	no	yes
C212	Structural facing tile	b	yes	no	yes
C216	Facing brick (solid)	75	yes	yes	yes
C652	Hollow brick	a	yes	yes	yes

Notes:
a. A minimum percent is given in this specification. The percent solid is a function of the requirements for size and/or number of cells as well as the minimum shell and web thicknesses.
b. No minimum percent solid is given in this specification. The percent solid is a function of the requirements for the number of cells and weights per square foot.
c. Solid masonry units minimum percent solid is 75 percent. Hollow masonry units — no minimum percent solid is given in this specification. Their percent solid is a function of the requirements for number of cells and the minimum shell and web thicknesses.

SPECIFICATION

2.3 C. Provide stone masonry units that conform to ASTM C503, C568, C615, C616, or C629, as specified.

2.3 D. Provide hollow glass units that are partially evacuated and have a minimum average glass face thickness of $^3/_{16}$ in. (4.8 mm). Provide solid glass block units when required. Provide units in which the surfaces intended to be in contact with mortar are treated with polyvinyl butyral coating or latex-based paint. Do not use reclaimed units.

COMMENTARY

2.3 C. Stone masonry units are typically selected by color and appearance. The referenced standards classify building stones by the properties shown in Table SC-5. The values given in the standards serve as minimum requirements. Stone is often ordered by a particular quarry or color rather than the classification method in the standard.

2.3 D. Hollow glass masonry units are formed by fusing two molded halves of glass together to produce a partial vacuum in the resulting cavity. The resulting glass block units are available in a variety of shapes, sizes, and patterns.

The block edges are usually treated in the factory with a coating that can be clear or opaque. The primary purpose of the coating is to provide an expansion/contraction mechanism to reduce stress cracking and to improve the mortar bond.

Table SC-5 — Stone requirements

ASTM Specification	Stone	Absorption	Density	Compressive strength	Modulus of rupture	Abrasion resistance	Acid resistance
C503	Marble	minimum	range	minimum	minimum	minimum	none
C568	Limestone	range	range	range	range	range	none
C615	Granite	minimum	minimum	minimum	minimum	minimum	none
C616	Sandstone	range	range	range	range	range	none
C629	Slate	range	none	none	minimum	minimum	range

SPECIFICATION

2.3 E. Provide AAC masonry units that conform to ASTM C1691 and ASTM C1693 for the strength class specified in the Contract Documents.

2.4 — Reinforcement, prestressing tendons, and metal accessories

2.4 A. *Reinforcing bars* — Provide deformed reinforcing bars that conform to one of the following as specified:

1. ASTM A615/A615M
2. ASTM A706/A706M
3. ASTM A767/A767M
4. ASTM A775/A775M
5. ASTM A996/A996M

COMMENTARY

2.3 E. AAC masonry units are specified by both compressive strength and density. Various density ranges are given in ASTM C1693 for specific compressive strengths. Generally, the density is specified based on consideration of thermal, acoustical, and weight requirements. AAC masonry is structurally designed based on the minimum compressive strength of the AAC material as determined by ASTM C1691. ASTM C1386, the predecessor standard to ASTM C1693, specified average compressive strength values that corresponded to the minimum compressive strength of each grade of AAC specified. Average specified compressive strengths for AAC 2, AAC 4 and AAC 6 were 360 psi, 725 psi and 1090 psi, respectively. ASTM C1691 deletes the requirement for minimum average strength for each grade of AAC masonry and allows the AAC manufacturer to determine required target strengths in the manufacturing process in order to achieve the specified minimum strength.

2.4 — Reinforcement, prestressing tendons, and metal accessories

2.4 A. *Reinforcing bars* — Code Sections 9.1.9.3.1 and 9.1.9.3.2 limit the reinforcing bar's specified and actual yield strengths when the reinforcement is used to resist in-plane flexural tension, flexural tension perpendicular to bed joints, in-plane shear, or flexural tension parallel to bed joints in strength design. Test reports should be reviewed to verify conformance with the Code requirement.

See Table SC-6 for a summary of properties.

Table SC-6 — Reinforcement and metal accessories

ASTM specification	Material	Use	Yield strength, ksi (MPa)	Yield stress, MPa
A36/A36M	Structural steel	Connectors	36 (248.2)	250
A82/A82 M	Steel wire	Joint reinforcement, ties	70 (482.7)	485
A167	Stainless steel	Bolts, reinforcement, ties	30 (206.9)	205
A185/A185 M	Steel welded wire reinforcement	Welded wire reinforcement	75 (517.1)	485
A307	Carbon steel	Connectors	a	—
A366/A366M	Carbon steel	Connectors	—	—
A496/A496 M	Steel wire	Reinforcement	75 (517.1)	485
A497/A497 M	Steel welded wire reinforcement	Reinforcement, welded wire reinforcement	70 (482.7)	485
A615/A615M	Carbon-steel	Reinforcement	40, 60 (275.8, 413.7)	300, 420
A996/A996M	Rail and axle steel	Reinforcement	40, 50, 60 (275.8, 344.8, 413.7)	300, 350, 420
A706/A706M	Low-alloy steel	Reinforcement	60 (413.7)	—

a. ASTM does not define a yield strength value for ASTM A307, Grade A anchor bolts.

SPECIFICATION

2.4 B. *Prestressing tendons*

1. Provide prestressing tendons that conform to one of the following standards, except for those permitted in Articles 2.4 B.2 and 2.4 B.3:

   a. Wire .... ASTM A421/A421M

   b. Low-relaxation wire .... ASTM A421/A421M

   c. Strand .... ASTM A416/A416M

   d. Low-relaxation strand .... ASTM A416/A416M

   e. Bar .... ASTM A722/A722M

2. Wire, strands, and bars not specifically listed in ASTM A416/A416M, A421/A421M, or A722/A722M are permitted, provided that they conform to the minimum requirements in ASTM A416/A416M, A421/A421M, or A722/A722M and are approved by the Architect/Engineer.

3. Bars and wires of less than 150 ksi (1034 MPa) tensile strength and conforming to ASTM A82/A82M, A510/A510M, A615/A615M, A996/A996M, or A706/A706M are permitted to be used as prestressed tendons, provided that the stress relaxation properties have been assessed by tests according to ASTM E328 for the maximum permissible stress in the tendon.

2.4 C. *Joint reinforcement*

1. Provide joint reinforcement that conforms to ASTM A951. Maximum spacing of cross wires in ladder-type joint reinforcement and of points of connection of cross wires to longitudinal wires of truss-type joint reinforcement shall be 16 in. (400 mm).

2. *Deformed reinforcing wire* — Provide deformed reinforcing wire that conforms to ASTM A496/A496M.

3. *Welded wire reinforcement* — Provide welded wire reinforcement that conforms to one of the following specifications:

   a. Plain .... ASTM A185/A185M

   b. Deformed .... ASTM A497/A497M

2.4 D. *Anchors, ties, and accessories* — Provide anchors, ties, and accessories that conform to the following specifications, except as otherwise specified:

1. Plate and bent-bar anchors .... ASTM A36/A36M

2. Sheet-metal anchors and ties .... ASTM A1008/A1008M

3. Wire mesh ties .... ASTM A185/A185M

4. Wire ties and anchors .... ASTM A82/A82M

5. Headed anchor bolts .... ASTM A307, Grade A

COMMENTARY

2.4 B. *Prestressing tendons* — The constructibility aspects of prestressed masonry favor the use of rods or rigid strands with mechanical anchorage in ungrouted construction. Mild strength steel bars have been used in prestressed masonry installations in the United States (Schultz and Scolforo, 1991). The stress-relaxation characteristics of mild strength bars (of less than 150 ksi [1034 MPa]) should be determined by tests and those results should be documented.

2.4 C. *Joint reinforcement* — Code Section 9.1.9.3.2 limits the specified yield strength of joint reinforcement used to resist in-plane shear and flexural tension parallel to bed joints in strength design.

SPECIFICATION

2.4 D. *Anchors, ties, and accessories* (Continued)

6. Panel anchors (for glass unit masonry) — Provide 1 3/4-in. (44.5-mm) wide, 24-in. (610-mm) long, 20-gage steel strips, punched with three staggered rows of elongated holes, galvanized after fabrication.

2.4 E. *Stainless steel* — Stainless steel items shall be AISI Type 304 or Type 316, and shall conform to the following:

1. Joint reinforcement .... ASTM A580/A580M

2. Plate and bent-bar anchors...
 .... ASTM A480/A480M and ASTM A666

3. Sheet-metal anchors and ties ....
 .... ASTM A480/A480M and ASTM A240/A240M

4. Wire ties and anchors .... ASTM A580/A580M

2.4 F. *Coatings for corrosion protection* — Unless otherwise required, protect carbon steel joint reinforcement, ties, anchors, and steel plates and bars from corrosion by galvanizing or epoxy coating in conformance with the following minimums:

1. Galvanized coatings:

 a. Mill galvanized coatings:

 1) Joint reinforcement ....
 ASTM A641/A641M (0.1 oz/ft^2) (0.031 kg/m^2)

 2) Sheet-metal ties and sheet-metal anchors ......
 ASTM A653/A653M Coating Designation G60

 b. Hot-dip galvanized coatings:

 1) Joint reinforcement, wire ties, and wire anchors
 ASTM A153/A153M (1.50 oz/ft^2) (458 g/m^2)

 2) Sheet-metal ties and sheet-metal anchors ....
 .... ASTM A153/A153M Class B

 3) Steel plates and bars (as applicable to size and form indicated)................ASTM A123/A123M
 .... or ASTM A153/A153M, Class B

2. Epoxy coatings:

 a. Joint reinforcement ....
 .... ASTM A884/A884M Class A
 .... Type 1 — 7 mils (175 μm)

 b. Wire ties and anchors..
 ASTM A899/A899M Class C — 20 mils (508 μm)

 c. Sheet-metal ties and anchors..................................
 ....20 mils (508 μm) per surface
 .... or manufacturer's specification

COMMENTARY

2.4 E. *Stainless steel* — Corrosion resistance of stainless steel is greater than that of the other steels listed. Thus, it does not have to be coated for corrosion resistance.

2.4 F. *Coatings for corrosion protection* — Amount of galvanizing required increases with severity of exposure (Grimm, 1985; Catani, 1985; NCMA TEK 12-4D, 2006). Project documents should specify the level of corrosion protection as required by Code Section 6.1.4.

SPECIFICATION

2.4 G. *Corrosion protection for tendons* — Protect tendons from corrosion when they are in exterior walls exposed to earth or weather or walls exposed to a mean relative humidity exceeding 75 percent). Select corrosion protection methods for bonded and unbonded tendons from one of the following:

1. *Bonded tendons* — Encapsulate bonded tendons in corrosion resistant and watertight corrugated ducts complying with Article 2.4 G.1.a. Fill ducts with prestressing grout complying with Article 2.4 G.1.b.

   a. Ducts — High-density polyethylene or polypropylene.

      1) Use ducts that are mortar-tight and non-reactive with masonry, tendons, and grout.

      2) Provide ducts with an inside diameter at least 1/4 in. (6.4 mm) larger than the tendon diameter.

      3) Maintain ducts free of water if members to be grouted are exposed to temperatures below freezing prior to grouting.

      4) Provide openings at both ends of ducts for grout injection.

COMMENTARY

2.4 G. *Corrosion protection for tendons* — The specified methods of corrosion protection for unbonded prestressing tendons are consistent with corrosion protection requirements developed for single-strand prestressing tendons in concrete (PTI, 2006). Masonry cover is not sufficient corrosion protection for bonded prestressing tendons in an environment with relative humidity over 75%. Therefore, complete encapsulation into plastic ducts is required. This requirement is consistent with corrosion protection for unbonded tendons. Alternative methods of corrosion protection, such as the use of stainless steel tendons or galvanized tendons, are permitted. Evidence should be provided that the galvanizing used on the tendons does not cause hydrogen embrittlement of the prestressing tendon.

Protection of prestressing tendons against corrosion is provided by a number of measures. Typically, a proprietary system is used that includes sheathing the prestressing tendon with a waterproof plastic tape or duct. Discussion of the various corrosion-protection systems used for prestressed masonry is available in the literature (Garrity, 1995). One example of a corrosion-protection system for the prestressing tendon is shown in Figure SC-5.

Chlorides, fluorides, sulfites, nitrates, or other chemicals in the prestressing grout may harm prestressing tendons and should not be used in harmful concentrations.

Historically, aggregates have not been used in grouts for bonded, post-tensioned concrete construction.

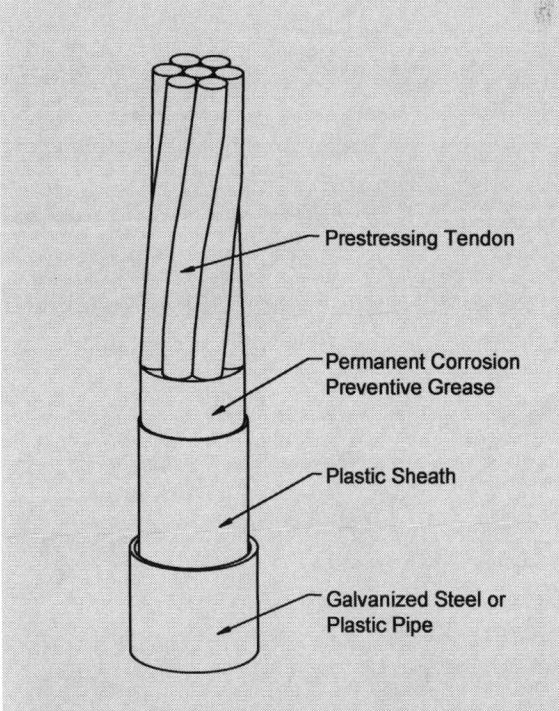

Figure SC-5 — An example of a corrosion-protection system for an unbonded tendon

SPECIFICATION

2.4 G. *Corrosion protection for tendons* (Continued)

b. Prestressing grout

1) Select proportions of materials for prestressing grout using either of the following methods as accepted by the Architect/Engineer:

a) Results of tests on fresh and hardened prestressing grout — prior to beginning grouting operations, or

b) Prior documented experience with similar materials and equipment and under comparable field conditions.

2) Use portland cement conforming to ASTM C150, Type I, II, or III, that corresponds to the type upon which selection of prestressing grout was based.

3) Use the minimum water content necessary for proper pumping of prestressing grout; however, limit the water-cement ratio to a maximum of 0.45 by weight.

4) Discard prestressing grout that has begun to set due to delayed use.

5) Do not use admixtures, unless acceptable to the Architect/Engineer.

6) Use water that is potable and free of materials known to be harmful to masonry materials and reinforcement.

COMMENTARY

b. Prestressing grout is a cementitious mixture, not conforming to ASTM C476, and is unique to bonded tendons.

SPECIFICATION

2.4 G. *Corrosion protection for tendons* (Continued)

2. *Unbonded tendons* — Coat unbonded tendons with a material complying with Article 2.4 G.2b and wrap with a sheathing complying with Article 2.4 G.2a. Acceptable materials include a corrosion-inhibiting coating material with a tendon sheathing.

   a. Provide continuous tendon sheathing over the entire tendon length to prevent loss of coating materials during tendon installation and stressing procedures. Provide a sheathing of medium-density or high-density polyethylene or polypropylene with the following properties:

      1) Sufficient strength to withstand damage during fabrication, transport, installation, and tensioning.

      2) Water-tightness over the entire sheathing length.

      3) Chemical stability without embrittlement or softening over the anticipated exposure temperature range and service life of the structure.

      4) Non-reactive with masonry and the tendon corrosion-inhibiting coating.

      5) In normal (non-corrosive) environments, a sheathing thickness of at least 0.025 in. (0.6 mm). In corrosive environments, a sheathing thickness of at least 0.040 in. (1.0 mm).

      6) An inside diameter at least 0.010 in. (0.3 mm) greater than the maximum diameter of the tendon.

      7) For applications in corrosive environments, connect the sheathing to intermediate and fixed anchorages in a watertight fashion, thus providing a complete encapsulation of the tendon.

   b. Provide a corrosion-inhibiting coating material with the following properties:

      1) Lubrication between the tendon and the sheathing.

      2) Resist flow from the sheathing within the anticipated temperature range of exposure.

      3) A continuous non-brittle film at the lowest anticipated temperature of exposure.

      4) Chemically stable and non-reactive with the tendon, sheathing material, and masonry.

      5) An organic coating with appropriate polar-moisture displacing and corrosion-preventive additives.

COMMENTARY

SPECIFICATION

2.4 G.2.b. *Unbonded tendons* (Continued)

    6) A minimum weight not less than 2.5 lb of coating material per 100 ft (37.2 g of coating material per m) of 0.5-in. (12.7-mm) diameter tendon and 3.0 lb of coating material per 100 ft (44.6 g of coating material per m) of 0.6-in. (15.2-mm) diameter tendon. Use a sufficient amount of coating material to ensure filling of the annular space between tendon and sheathing.

    7) Extend the coating over the entire tendon length.

    8) Provide test results in accordance with Table 6 for the corrosion-inhibiting coating material.

3. Alternative methods of corrosion protection that provide a protection level equivalent to Articles 2.4 G.1 and 2.4 G.2 are permitted. Stainless steel prestressing tendons or tendons galvanized according to ASTM A153/A153M, Class B, are acceptable alternative methods. If galvanized, further evidence must be provided that the coating will not produce hydrogen embrittlement of the steel.

2.4 H. *Prestressing anchorages, couplers, and end blocks*

1. Provide anchorages and couplers that develop at least 95 percent of the specified breaking strength of the tendons or prestressing steel when tested in an unbonded condition, without exceeding anticipated set.

2. Place couplers where accepted by Architect/Engineer. Enclose with housing that permits anticipated movements of the couplers during stressing.

3. Protect anchorages, couplers, and end fittings against corrosion

4. Protect exposed anchorages, couplers, and end fittings to achieve the fire-resistance rating required for the element by the legally adopted building code.

COMMENTARY

2.4 H. *Prestressing anchorages, couplers, and end blocks* — Typical anchorage and coupling devices are shown in Figure SC-6. Strength of anchorage and coupling devices should be provided by the manufacturer.

Protection of anchorage devices typically includes filling the opening of bearing pads with grease, grouting the recess in bearing pads, and providing drainage of cavities housing prestressing tendons with base flashing and weep holes.

When anchorages and end fittings are exposed, additional precautions to achieve the required fire ratings and mechanical protection for these elements must be taken.

Table 6 — Performance specification for corrosion-inhibiting coating

Test	Test Method	Acceptance Criteria
Dropping Point, °F (°C)	ASTM D566 or ASTM D2265	Minimum 300 (148.9)
Oil Separation @ 160°F (71.1°C) % by weight	FTMS 791B Method 321.2	Maximum 0.5
Water, % maximum	ASTM D95	0.1
Flash Point, °F (°C) (Refers to oil component)	ASTM D92	Minimum 300 (148.9)
Corrosion Test 5% Salt Fog @ 100°F (37.8°C) 5 mils (0.13 mm), minimum hours (Q Panel type S)	ASTM B117	For normal environments: Rust Grade 7 or better after 720 hr of exposure according to ASTM D610. For corrosive environments : Rust Grade 7 or better after 1000 hr of exposure according to ASTM D610.[1]
Water Soluble Ions[2] a. Chlorides, ppm maximum b. Nitrates, ppm maximum c. Sulfides, ppm maximum	ASTM D512	10 10 10
Soak Test 5% Salt Fog at 100°F (37.8°C) 5 mils (0.13 mm) coating, Q panels, type S. Immerse panels 50% in a 5% salt solution and expose to salt fog	ASTM B117 (Modified)	No emulsification of the coating after 720 hr of exposure
Compatibility with Sheathing a. Hardness and volume change of polymer after exposure to grease, 40 days @ 150°F (65.6°C). b. Tensile strength change of polymer after exposure to grease, 40 days @ 150°F (65.6°C).	ASTM D4289 ASTM D638	Permissible change in hardness 15% Permissible change in volume 10% Permissible change in tensile strength 30%

[1] Extension of exposure time to 1000 hours for greases used in corrosive environments requires use of more or better corrosion-inhibiting additives.

[2] Procedure: The inside (bottom and sides) of a 33.8 oz (1L) Pyrex beaker, approximate O.D. 4.1 in. (105 mm), height 5.7 in. (145 mm), is thoroughly coated with 35.3 ± 3.5 oz (1000 ± 100 g) corrosion-inhibiting coating material. The coated beaker is filled with approximately 30.4 oz (900 cc) of distilled water and heated in an oven at a controlled temperature of 100°F ± 2°F (37.8°C ± 1°C) for 4 hours. The water extraction is tested by the noted test procedures for the appropriate water soluble ions. Results are reported as ppm in the extracted water.

SPECIFICATION

2.5 — Accessories

2.5 A. Unless otherwise required, provide contraction (shrinkage) joint material that conforms to one of the following standards:

1. ASTM D2000, M2AA-805 Rubber shear keys with a minimum durometer hardness of 80.
2. ASTM D2287, Type PVC 654-4 PVC shear keys with a minimum durometer hardness of 85.
3. ASTM C920.

2.5 B. Unless otherwise required, provide expansion joint material that conforms to one of the following standards:

1. ASTM C920.
2. ASTM D994.
3. ASTM D1056, Class 2A

2.5 C. *Asphalt emulsion* — Provide asphalt emulsion as follows:

1. Metal surfaces .... ASTM D1187, Type II
2. Porous surfaces ... ASTM D1227, Type III, Class 1

2.5 D. *Masonry cleaner*

1. Use potable water and detergents to clean masonry unless otherwise acceptable.
2. Unless otherwise required, do not use acid or caustic solutions.

2.5 E. *Joint fillers* — Use the size and shape of joint fillers specified.

COMMENTARY

2.5 — Accessories

2.5 A. and B. Movement joints are used to allow dimensional changes in masonry, minimize random wall cracks, and other distress. Contraction joints (also called control joints or shrinkage joints) are used in concrete masonry to accommodate shrinkage. These joints are free to open as shrinkage occurs. Expansion joints permit clay brick masonry to expand. Material used in expansion joints must be compressible.

Placement of movement joints is recommended by several publications (Grimm, 1988; BIA TN 19, 2006; BIA TN 18A, 2006; NCMA TEK 10-2C, 2010). Typical movement joints are illustrated in Figure SC-7. Shear keys keep the wall sections on either side of the movement joint from moving out of plane. Proper configuration must be available to fit properly.

ASTM C920 addresses elastomeric joint sealants, either single or multi-component. Sealants that qualify as Grade NS, Class 50 (50% movement capability) or alternatively Class 25 (25% movement capability), Use M are applicable to masonry construction. Expansion joint fillers must be compressible so the anticipated expansion of the masonry can occur without imposing stress.

2.5 D. *Masonry cleaner* — Adverse reactions can occur between certain cleaning agents and masonry units. Hydrochloric acid has been observed to cause corrosion of metal ties. Care should be exercised in its use to minimize this potential problem. Manganese staining, efflorescence, "burning" of the units, white scum removal of the cement paste from the surface of the joints, and damage to metals can occur through improper cleaning. The manufacturers of the masonry units should be consulted for recommended cleaning agents.

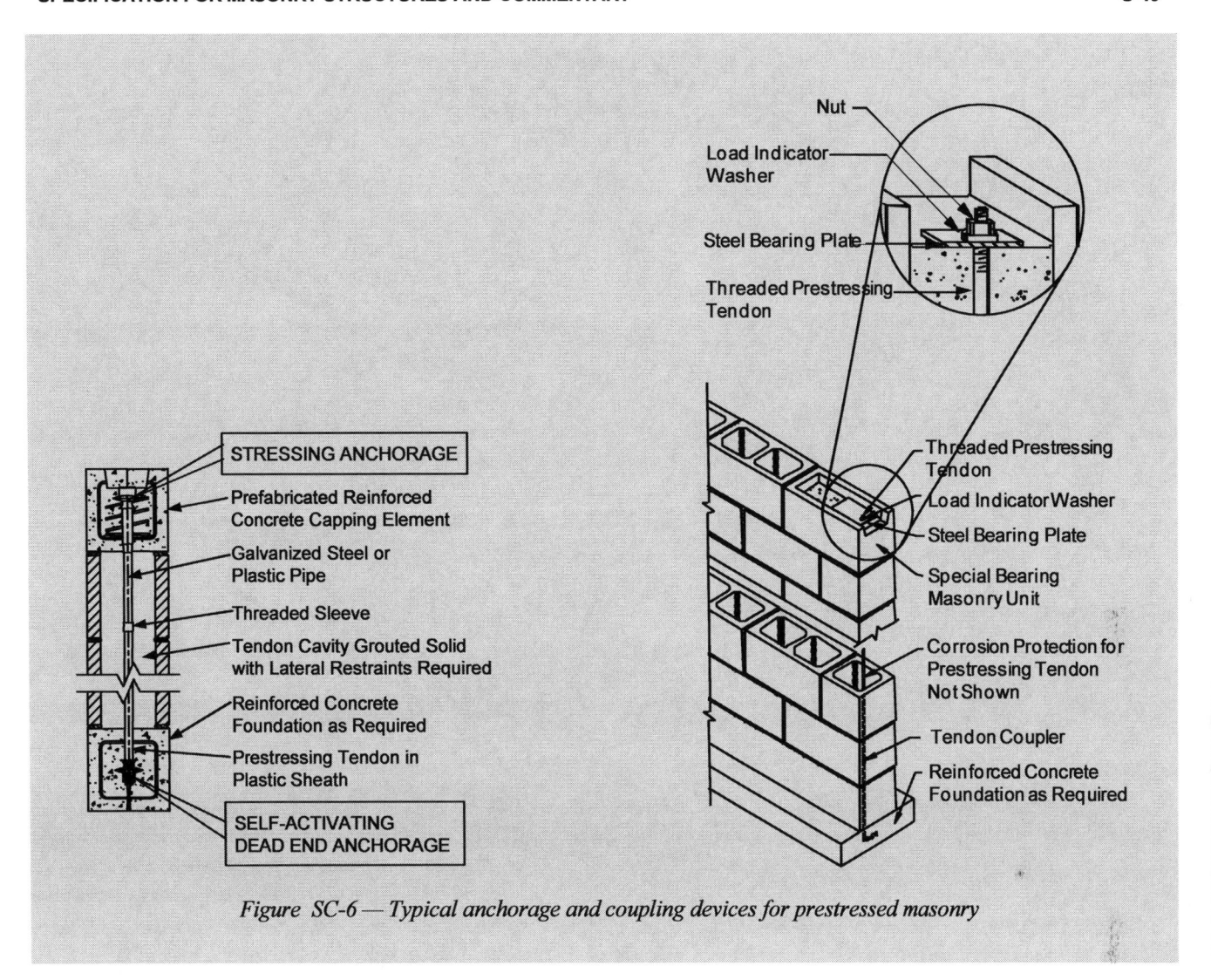

Figure SC-6 — Typical anchorage and coupling devices for prestressed masonry

COMMENTARY

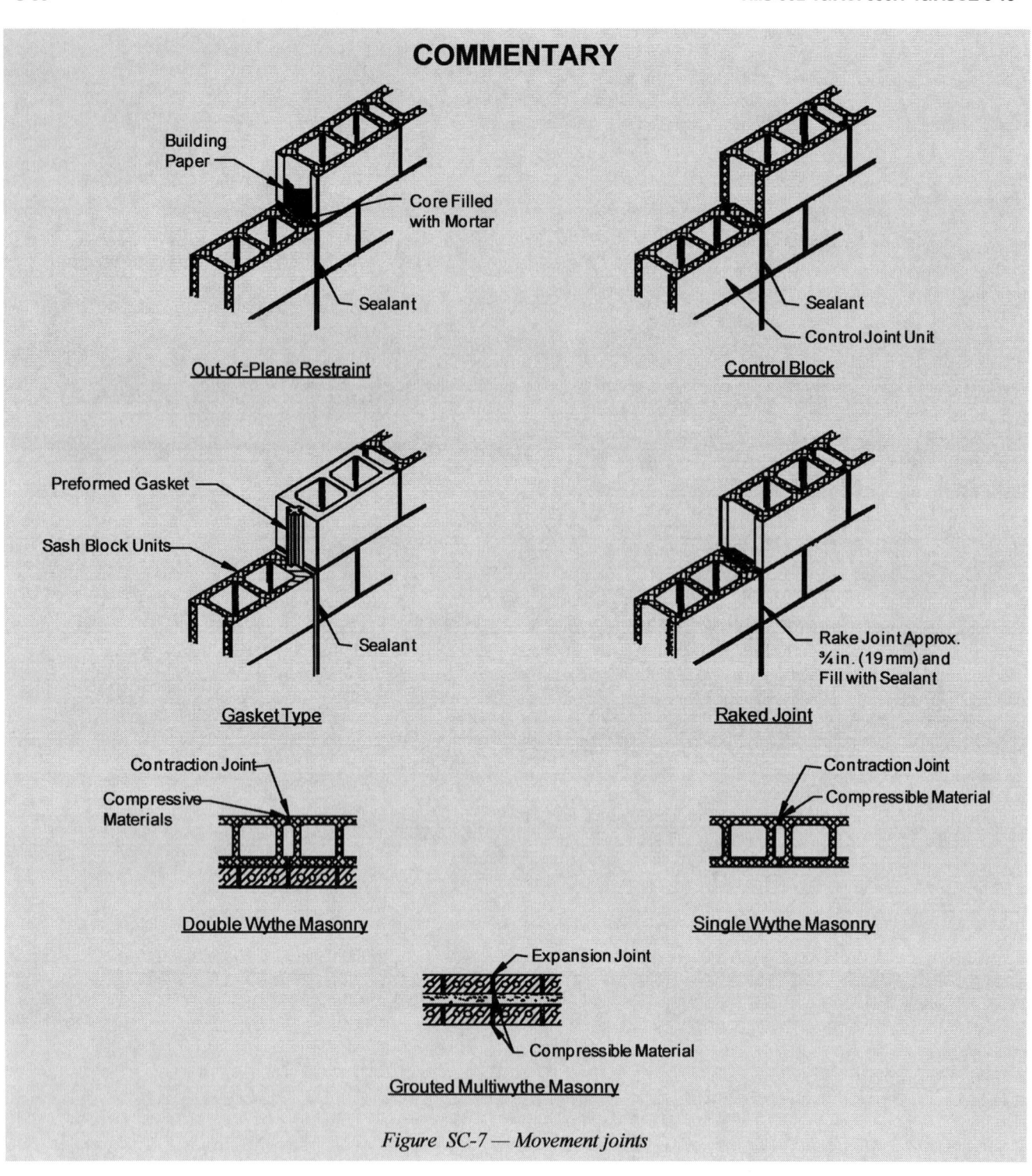

Figure SC-7 — Movement joints

SPECIFICATION

2.6 — Mixing

2.6 A. *Mortar*

1. Mix cementitious materials and aggregates between 3 and 5 minutes in a mechanical batch mixer with a sufficient amount of water to produce a workable consistency. Unless acceptable, do not hand mix mortar. Maintain workability of mortar by remixing or retempering. Discard mortar which has begun to stiffen or is not used within $2\frac{1}{2}$ hr after initial mixing.

2. Limit the weight of mineral oxide or carbon black pigments added to project-site prepared mortar to the following maximum percentages by weight of cement:

   a. Pigmented portland cement-lime mortar

       1) Mineral oxide pigment   10 percent

       2) Carbon black pigment   2 percent

   b. Pigmented mortar cement mortar

       1) Mineral oxide pigment   5 percent

       2) Carbon black pigment   1 percent

   c. Pigmented masonry cement mortar

       1) Mineral oxide pigment   5 percent

       2) Carbon black pigment   1 percent

   Do not add mineral oxide or carbon black pigment to preblended colored mortar or colored cement without the approval of the Architect/Engineer.

3. Do not use admixtures containing more than 0.2 percent chloride ions.

4. *Glass unit masonry* — Reduce the amount of water to account for the lack of absorption. Do not retemper mortar after initial set. Discard unused mortar within $1\frac{1}{2}$ hr after initial mixing.

COMMENTARY

2.6 — Mixing

2.6 A. *Mortar* — Caution must be exercised when adding color pigment in field-prepared mortar so that the proportions comply with the Specification requirements.

Preblended products are typically certified to the applicable ASTM Standard and the addition of color at the project site may impact mortar performance.

SPECIFICATION

2.6 B. *Grout*

1. Except for self-consolidating grout, mix grout in accordance with the requirements of ASTM C476.

2. Unless otherwise required, mix grout other than self-consolidating grout to a consistency that has a slump between 8 and 11 in. (203 and 279 mm).

3. Proportioning of self-consolidating grout at the project site is not permitted. Do not add water at the project site except in accordance with the self-consolidating grout manufacturer's recommendations.

COMMENTARY

2.6 B. *Grout* — The two types of grout are fine grout and coarse grout, which are defined by aggregate size. ASTM C476 requires the grout type to be specified by proportion or strength requirements, but not by both methods. ASTM proportion requirements are given in Table SC-7.

The permitted ranges in the required proportions of fine and coarse aggregates are intended to accommodate variations in aggregate type and gradation. As noted in Specification Table 7, the selection of the grout type depends on the size of the space to be grouted. Fine grout is selected for grout spaces with restricted openings. Coarse grout specified under ASTM C476 has a maximum aggregate size that will pass through a ½ in. (12.7 mm) and at least 85% that will pass through a 3/8 in. (9.5 mm) opening.

Grout meeting the proportion specifications of ASTM C476 typically has compressive strength ranges shown in Table SC-8 when measured by ASTM C1019. Grout compressive strength is influenced by the water cement ratio, aggregate content, and the type of units used.

Because grout is placed in an absorptive form made of masonry units, a high water content is required. A slump of at least 8 in. (203 mm) provides a mix fluid enough to be properly placed and supplies sufficient water to satisfy the water demand of the masonry units.

Small cavities or cells require grout with a higher slump than larger cavities or cells. As the surface area and unit shell thickness in contact with the grout decrease in relation to the volume of the grout, the slump of the grout should be reduced. Segregation of materials should not occur.

The grout in place will have a lower water-cement ratio than when mixed. This concept of high slump and absorptive forms is different from that of concrete.

Proportioning of self-consolidating grout at the project site is not permitted because the mixes can be sensitive to variations in proportions, and tighter quality control on the mix is required than can be achieved in the field. Typically, self-consolidating grout comes ready mixed from the manufacturer. Self-consolidating grout may also be available as a preblended dry mix requiring the addition of water at the project site. Manufacturers provide instructions on proper mixing techniques and amount of water to be added. Slump values for self-consolidating grout are expressed as a slump flow because they exceed the 8 in. to 11 in. (203 to 279 mm) slump range for non-self-consolidating grouts.

COMMENTARY

Table SC-7 — Grout proportions by volume

Grout type	Cement	Lime	Aggregate damp, loose[1] Fine	Aggregate damp, loose[1] Coarse
Fine	1	0 to 1/10	2¼ to 3	—
Coarse	1	0 to 1/10	2¼ to 3	1 to 2

[1] Times the sum of the volumes of the cementitious materials

Table SC-8 — Grout strengths

Grout type	Location	Compressive strength, psi (MPa) Low	Compressive strength, psi (MPa) Mean	Compressive strength, psi (MPa) High	Reference
Coarse	Lab	1,965 (13.55)	3,106 (21.41)	4,000 (27.58)	ACI-SEASC, 1982
Coarse	Lab	3,611 (24.90)	4,145 (28.58)	4,510 (31.10)	Li and Neis, 1986
Coarse	Lab	5,060 (34.89)	5,455 (37.61)	5,940 (40.96)	ATL, 1982

SPECIFICATION

2.6 C. *Thin-bed mortar for AAC* – Mix thin-bed mortar for AAC masonry as specified by the thin-bed mortar manufacturer.

2.7 — Fabrication

2.7 A. *Reinforcement*

1. Fabricate reinforcing bars in accordance with the fabricating tolerances of ACI 117.
2. Unless otherwise required, bend bars cold and do not heat bars.
3. The minimum inside diameter of bend for stirrups shall be five bar diameters.
4. Do not bend Grade 40 bars in excess of 180 degrees. The minimum inside diameter of bend is five bar diameters.
5. The minimum inside bend diameter for other bars is as follows:

   a. No. 3 through No. 8 (M#10 through 25) .... 6 bar diameters

   b. No. 9 through No. 11 (M#29 through 36) .... 8 bar diameters

6. Provide standard hooks that conform to the following:

   a. A standard 180-degree hook: 180-degree bend plus a minimum extension of 4 bar diameters or 2½ in. (64 mm), whichever is greater.

   b. A standard 90-degree hook: 90-degree bend plus a minimum extension of 12 bar diameters.

   c. For stirrups and tie hooks for a No. 5 (M#16) bar and smaller: a 90- or 135-degree bend plus a minimum of 6 bar diameters or 2½ in. (64 mm), whichever is greater.

COMMENTARY

2.7 — Fabrication

2.7 A. *Reinforcement* — ACI 117 Specifications for Tolerances for Concrete Construction and Materials and Commentary contains fabrication tolerances for steel reinforcement. Recommended methods and standards for preparing design drawings, typical details, and drawings for the fabrications and placing of reinforcing steel in reinforced concrete structures are given in ACI 315 (1999) and may be used as a reference in masonry design and construction.

SPECIFICATION

2.7 B. *Prefabricated masonry*

1. Unless otherwise required, provide prefabricated masonry that conforms to the provisions of ASTM C901.

2. Unless otherwise required, provide prefabricated masonry lintels that have an appearance similar to the masonry units used in the wall surrounding each lintel.

3. Mark prefabricated masonry for proper location and orientation.

COMMENTARY

2.7 B. *Prefabricated masonry* — ASTM C901 addresses the requirements for prefabricated masonry panels, including materials, structural design, dimensions and variations, workmanship, quality control, identification, shop drawings, and handling.

PART 3 — EXECUTION

SPECIFICATION

3.1 — Inspection

3.1 A. Prior to the start of masonry construction, the Contractor shall verify:

1. That foundations are constructed within a level alignment tolerance of $\pm^1/_2$ in. (12.7 mm).

2. That reinforcing dowels are positioned in accordance with the Project Drawings.

3.1 B. If stated conditions are not met, notify the Architect/Engineer.

COMMENTARY

3.1 — Inspection

3.1 A. The tolerances in this Article are taken from ACI 117 (1990). The dimensional tolerances of the supporting concrete are important because they control such aspects as mortar joint thickness and bearing area dimensions, which influence the performance of the masonry. Tolerances for variation in grade or elevation are shown in Figure SC-8. The specified width of the foundation is obviously more critical than its specified length. A foundation wider than specified will not normally cause structural problems.

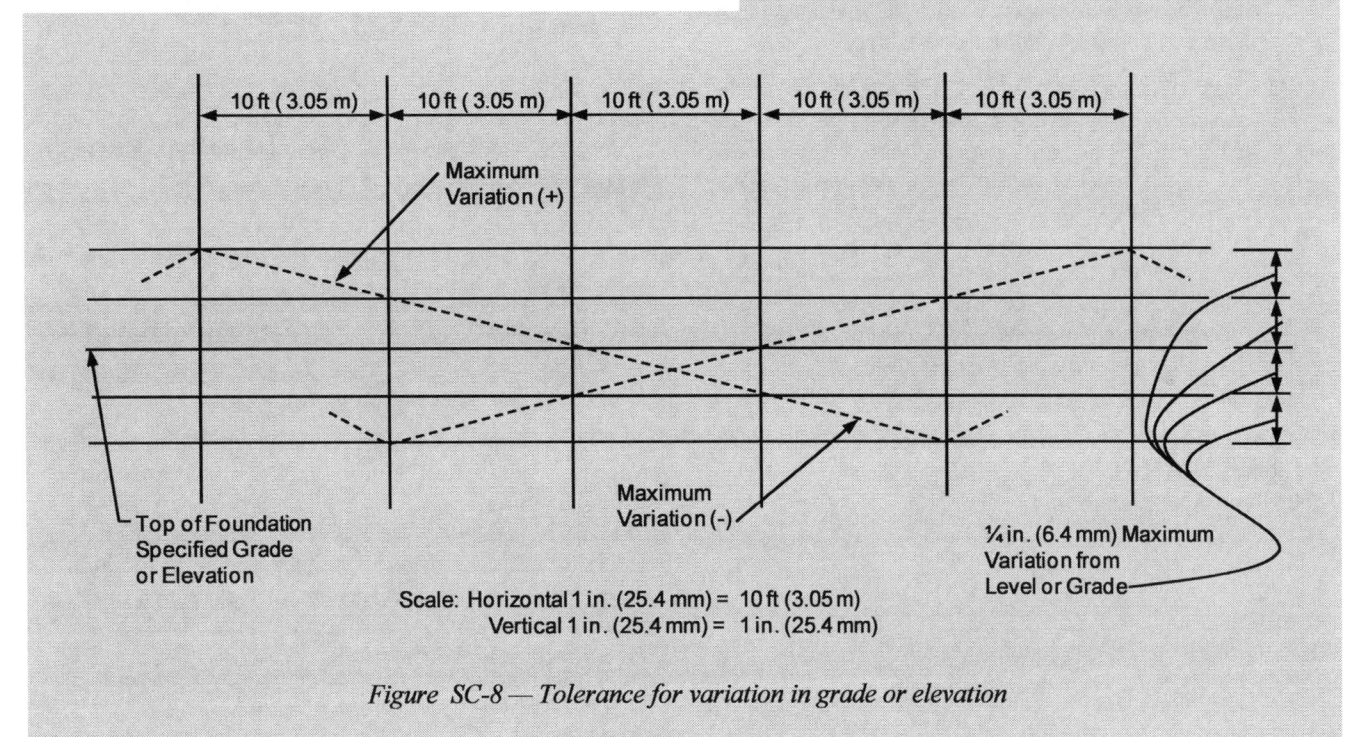

Figure SC-8 — Tolerance for variation in grade or elevation

SPECIFICATION

3.2 — Preparation

3.2 A. Clean reinforcement and shanks of anchor bolts by removing mud, oil, or other materials that will adversely affect or reduce bond at the time mortar or grout is placed. Reinforcement with rust, mill scale, or both are acceptable without cleaning or brushing provided that the dimensions and weights, including heights of deformations, of a cleaned sample are not less than required by the ASTM specification that governs this reinforcement.

3.2 B. Prior to placing masonry, remove laitance, loose aggregate, and anything else that would prevent mortar from bonding to the foundation.

3.2 C. *Wetting masonry units*

1. *Concrete masonry* — Unless otherwise required, do not wet concrete masonry or AAC masonry units before laying. Wet cutting is permitted.

2. *Clay or shale masonry* — Wet clay or shale masonry units having initial absorption rates in excess of 1 g per min. per in.^{2} (0.0016 g per min. per mm^2), when measured in accordance with ASTM C67, so the initial rate of absorption will not exceed 1 g per min. per in.^{2} (0.0016 g per min. per mm^2) when the units are used. Lay wetted units when surface dry. Do not wet clay or shale masonry units having an initial absorption rate less than 0.2 g per min. per in.^{2} (0.00031 g per min. per mm^2).

3.2 D. *Debris* — Construct grout spaces free of mortar dropping, debris, loose aggregates, and any material deleterious to masonry grout.

3.2 E. *Reinforcement* — Place reinforcement and ties in grout spaces prior to grouting.

3.2 F. *Cleanouts* — Provide cleanouts in the bottom course of masonry for each grout pour when the grout pour height exceeds 5 ft 4 in. (1.63 m).

1. Construct cleanouts so that the space to be grouted can be cleaned and inspected. In solid grouted masonry, space cleanouts horizontally a maximum of 32 in. (813 mm) on center.

2. Construct cleanouts with an opening of sufficient size to permit removal of debris. The minimum opening dimension shall be 3 in. (76.2 mm).

3. After cleaning, close cleanouts with closures braced to resist grout pressure.

COMMENTARY

3.2 C. *Wetting masonry units* — Concrete masonry units increase in volume when wetted and shrink upon subsequent drying. Water introduced during wet cutting is localized and does not significantly affect the shrinkage potential of concrete masonry. Clay masonry units with high absorption rates dry the mortar/unit interface. This may result in a lower extent of bond between the units and mortar, which may create paths for moisture intrusion. Selection of compatible units and mortar can mitigate this effect.

3.2 D. *Debris* — Continuity in the grout is critical for uniform stress distribution. A reasonably clean space to receive the grout is necessary for this continuity. Cells need not be vacuumed to achieve substantial cleanliness. Inspection of the bottom of the space prior to grouting is critical to ensure that it is substantially clean and does not have accumulations of deleterious materials that would prevent continuity of the grout.

3.2 E. *Reinforcement* — Loss of bond and misalignment of the reinforcement can occur if it is not placed prior to grouting.

3.2 F. *Cleanouts* — Cleanouts can be constructed by removing the exposed face shell of units in hollow unit grouted masonry or individual units when grouting between wythes. The purpose of cleanouts is to allow the grout space to be adequately cleaned prior to grouting. They can also be used to verify reinforcement placement and tying.

SPECIFICATION

3.3 — Masonry erection

3.3 A. *Bond pattern* — Unless otherwise required, lay masonry in running bond.

3.3 B. *Placing mortar and units*

1. *Bed joints at foundations* — In the starting course on foundations and other supporting members, construct bed joints so that the bed joint thickness is at least ¼ in. (6.4 mm) and not more than:

    a. ¾ in. (19.1 mm) when the masonry is ungrouted or partially grouted.

    b. 1¼ in. (31.8 mm) when the first course of masonry is solid grouted and supported by a concrete foundation.

COMMENTARY

3.3 B. *Placing mortar and units* — Article 3.3 B applies to masonry construction in which the units support their own weight.

1. *Bed joints at foundations* — The range of permitted mortar bed joint thickness at foundations for solid grouted masonry walls is compatible with the foundation tolerances of Article 3.1 A.1. Figure SC-9 shows the allowable foundation tolerance of ± ½ in. and the relationship of the mortar bed joint. The contractor should coordinate the mortar bed joint at foundations with the coursing requirements so that the intended masonry module is met at critical points, such as story height and top of wall, window and door openings. Either fine or coarse grout for the first course of masonry may be placed when normal masonry grouting is performed for fully grouted masonry, or may be placed after the first course is laid and prior to placement of additional courses when the masonry is not fully grouted.

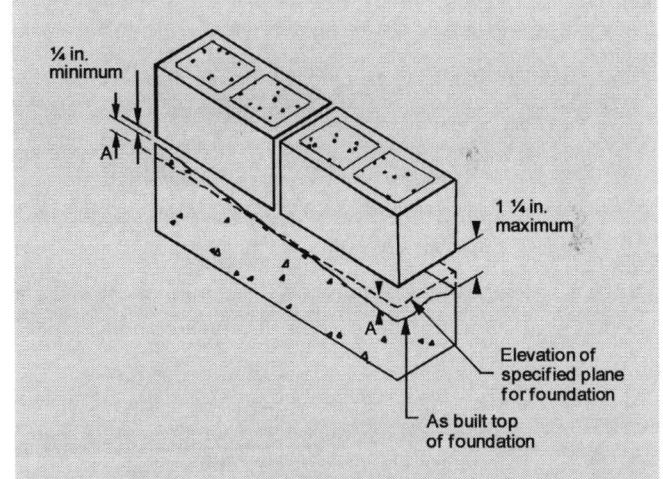

Figure SC-9 Mortar bed joint thickness for solid grouted walls on a foundation

SPECIFICATION

3.3 B. *Placing mortar and units* (Continued)

2. *Bed and head joints* — Unless otherwise required, construct $^3/_8$-in. (9.5-mm) thick bed and head joints, except at foundation or with glass unit masonry. Provide glass unit masonry bed and head joint thicknesses in accordance with Article 3.3 B.7.c. Provide AAC masonry bed and head joint thicknesses in accordance with Article 3.3 B.9.b. Construct joints that also conform to the following:

   a. Fill holes not specified in exposed and below grade masonry with mortar.

   b. Unless otherwise required, tool joint with a round jointer when the mortar is thumbprint hard.

   c. Remove masonry protrusions extending $^1/_2$ in. (12.7 mm) or more into cells or cavities to be grouted.

3. *Collar joints* — Unless otherwise required, solidly fill collar joints less than $^3/_4$ in. (19.1 mm) wide with mortar as the project progresses.

4. *Hollow units* — Place hollow units so:

   a. Face shells of bed joints are fully mortared.

   b. Webs are fully mortared in:

      1) all courses of piers, columns and pilasters;

      2) when necessary to confine grout or insulation.

   c. Head joints are mortared, a minimum distance from each face equal to the face shell thickness of the unit.

   d. Vertical cells to be grouted are aligned and unobstructed openings for grout are provided in accordance with the Project Drawings.

COMMENTARY

4. *Hollow units* — Face shell mortar bedding of hollow units is standard, except in locations detailed in Article 3.3 B.4.b. Figure SC-10 shows the typical placement of mortar for hollow-unit masonry walls. In partially grouted walls, however, cross webs next to cells that are to be grouted are usually mortared. If full mortar beds throughout are required for structural capacity, for example, the specifier must so stipulate in the Project Specifications or Project Drawings.

SPECIFICATION

3.3 B. *Placing mortar and units* (Continued)

5. *Solid units* — Unless otherwise required, place mortar so that bed and head joints are fully mortared and:

   a. Do not fill head joints by slushing with mortar.

   b. Construct head joints by shoving mortar tight against the adjoining unit.

   c. Do not deeply furrow bed joints.

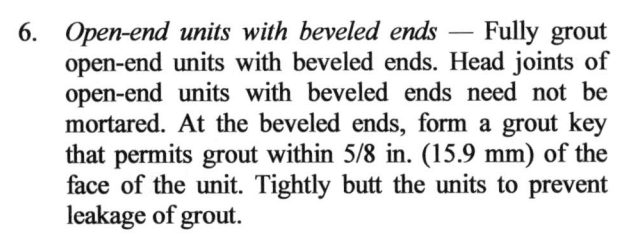

6. *Open-end units with beveled ends* — Fully grout open-end units with beveled ends. Head joints of open-end units with beveled ends need not be mortared. At the beveled ends, form a grout key that permits grout within 5/8 in. (15.9 mm) of the face of the unit. Tightly butt the units to prevent leakage of grout.

7. *Glass units*

   a. Apply a complete coat of asphalt emulsion, not exceeding $1/8$ in. (3.2 mm) in thickness, to panel bases.

   b. Lay units so head and bed joints are filled solidly. Do not furrow mortar.

   c. Unless otherwise required, construct head and bed joints of glass unit masonry $1/4$ in. (6.4 mm) thick, except that vertical joint thickness of radial panels shall not be less than $1/8$ in. (3.2 mm). The bed-joint thickness tolerance shall be minus $1/16$ in. (1.6 mm) and plus $1/8$ in. (3.2 mm). The head-joint thickness tolerance shall be plus or minus $1/8$ in. (3.2 mm).

   d. Do not cut glass units.

COMMENTARY

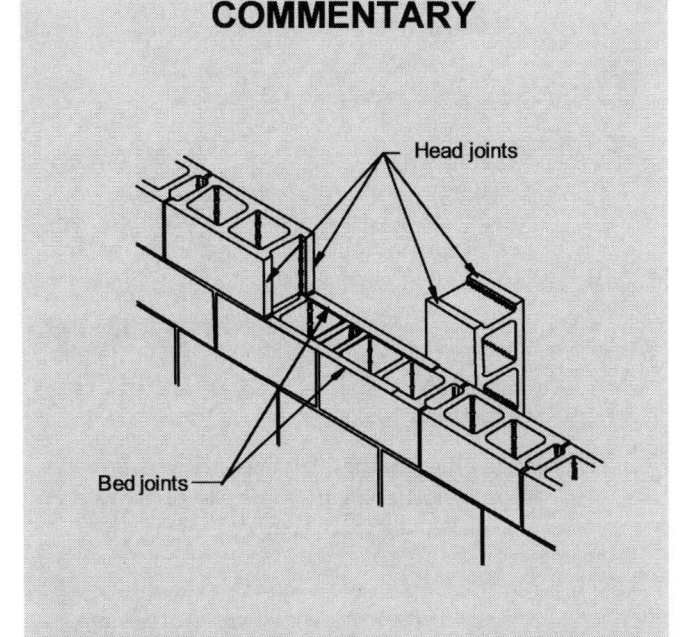

Figure SC-10 — Mortar placement of hollow units in walls

SPECIFICATION

3.3 B. *Placing mortar and units* (Continued)

8. *All units*

   a. Place clean units while the mortar is soft and plastic. Remove and re-lay in fresh mortar any unit disturbed to the extent that initial bond is broken after initial positioning.

   b. Except for glass units, cut exposed edges or faces of masonry units smooth, or position so that exposed faces or edges are unaltered manufactured surfaces.

   c. When the bearing of a masonry wythe on its support is less than two-thirds of the wythe thickness, notify the Architect/Engineer.

9. *AAC masonry*

   a. Place mortar for leveling bed joint in accordance with the requirements of Article 3.3 B.1.

   b. Lay subsequent courses using thin-bed mortar. Use special notched trowels manufactured for use with thin-bed mortar to spread thin-bed mortar so that it completely fills the bed joints. Unless otherwise specified in the Contract Documents, similarly fill the head joints. Spread mortar and place the next unit before the mortar dries. Place each AAC unit as close to head joint as possible before lowering the block onto the bed joint. Avoid excessive movement along bed joint. Make adjustments while thin-bed mortar is still soft and plastic by tapping to plumb and bring units into alignment. Set units into final position, in mortar joints at least 1/16-in. (1.5-mm) thick, by striking on the end and top with a rubber mallet.

   c. Lay units in alignment with the plane of the wall. Align vertically and plumb using the first course for reference. Make minor adjustments by sanding the exposed faces of the units and the bed joint surface with a sanding board manufactured for use with AAC masonry.

COMMENTARY

3.3 B.9 *AAC masonry* — AAC masonry can be cut, shaped and drilled with tools that are capable of cutting wood; however, saws, sanding boards, and rasps manufactured for use with AAC are recommended for field use. Because thin-bed mortar joints do not readily allow for plumbing of a wall, the ability of AAC masonry to be easily cut and shaped allows for field adjustment to attain required tolerances.

SPECIFICATION

3.3 C. *Placing adhered veneer*

1. Brush a paste of neat portland cement on the backing and on the back of the veneer unit.

2. Apply Type S mortar to the backing and to the veneer unit.

3. Tap the veneer unit into place, completely filling the space between the veneer unit and the backing. Sufficient mortar shall be used to create a slight excess to be forced out between the edges of the veneer units. The resulting thickness of the mortar in back of the veneer unit shall not be less than $^3/_8$ in. (9.5 mm) nor more than 1¼ in. (31.8 mm).

4. Tool the mortar joint with a round jointer when the mortar is thumbprint hard.

3.3 D. *Embedded items and accessories* — Install embedded items and accessories as follows:

1. Construct chases as masonry units are laid.

2. Install pipes and conduits passing horizontally through masonry partitions.

3. Place pipes and conduits passing horizontally through piers, pilasters, or columns.

4. Place horizontal pipes and conduits in and parallel to plane of walls.

5. Install and secure connectors, flashing, weep holes, weep vents, nailing blocks, and other accessories.

6. Install movement joints.

7. Aluminum — Do not embed aluminum conduits, pipes, and accessories in masonry, grout, or mortar, unless they are effectively coated or isolated to prevent chemical reaction between aluminum and cement or electrolytic action between aluminum and steel.

COMMENTARY

3.3 C *Placing adhered veneer* — Article 3.3 C applies to adhered veneer in which the backing supports the weight of the units. This basic method has served satisfactorily since the early 1950s. Properly filled and tooled joints (3.3 C.4) are essential for proper performance of adhered veneer.

SPECIFICATION

3.3 E. *Bracing of masonry* — Design, provide, and install bracing that will assure stability of masonry during construction.

3.3 F. *Site tolerances* — Erect masonry within the following tolerances from the specified dimensions.

1. Dimension of elements

    a. In cross section or elevation
    ....-¼ in. (6.4 mm), +½ in. (12.7 mm)

    b. Mortar joint thickness

    bed joints between masonry courses
    ....±⅛ in. (3.2 mm)

    bed joint between flashing and masonry
    .... - ½ in. (12.7 mm), +⅛ in. (3.2 mm)

    head ....- ¼ in. (6.4 mm), + ⅜ in. (9.5 mm)

    collar............-¼ in. (6.4 mm), + ⅜ in. (9.5 mm)

    glass unit masonry .... see Article 3.3 B.7.c

    AAC thin-bed mortar joint thickness.................
    .... -0, + 1/8 in. (3.2 mm)

    c. Grout space or cavity width, except for masonry walls passing framed construction
    ....-¼ in. (6.4 mm), + ⅜ in. (9.5 mm)

COMMENTARY

3.3 E. *Bracing of masonry* — For guidance on bracing of masonry walls for wind, consult Standard Practice for Bracing Masonry Walls Under Construction (MCAA, 2012).

3.3 F. *Site tolerances* — Tolerances are established to limit eccentricity of applied load. Because masonry is usually used as an exposed material, it is subjected to tighter dimensional tolerances than those for structural frames. The tolerances given are based on structural performance, not aesthetics.

The provisions for cavity width shown are for the space between wythes of non-composite masonry. The provisions do not apply to situations where masonry extends past floor slabs, spandrel beams, or other structural elements.

The remaining provisions set the standard for quality of workmanship and ensure that the structure is not overloaded during construction.

Mortar is required to bond masonry courses, but it is not required when masonry is laid on top of flashing.

SPECIFICATION

3.3 F. *Site tolerances* (Continued)

2. Elements

    a. Variation from level:

    bed joints
    ....±$^1/_4$ in. (6.4 mm) in 10 ft (3.05 m)
    ....±$^1/_2$ in. (12.7 mm) maximum

    top surface of load-bearing walls
    ....±$^1/_4$ in. (6.4 mm) in 10 ft (3.05 m)
    ....±$^1/_2$ in. (12.7 mm) maximum

    b. Variation from plumb
    ....±$^1/_4$ in. (6.4 mm) in 10 ft (3.05 m)
    ....±$^3/_8$ in. (9.5 mm) in 20 ft (6.10 m)
    ....±$^1/_2$ in. (12.7 mm) maximum

    c. True to a line
    ....±$^1/_4$ in. (6.4 mm) in 10 ft (3.05 m)
    ....±$^3/_8$ in. (9.5 mm) in 20 ft (6.10 m)
    ....±$^1/_2$ in. (12.7 mm) maximum

    d. Alignment of columns and walls
    (bottom versus top)
    .... ±$^1/_2$ in. (12.7 mm) for
    load-bearing walls and columns
    .±$^3/_4$ in. (19.1 mm) for non-load-bearing walls

3. Location of elements

    a. Indicated in plan
    ....±$^1/_2$ in. (12.7 mm) in 20 ft (6.10 m)
    ....±$^3/_4$ in. (19.1 mm) maximum

    b. Indicated in elevation
    ....±$^1/_4$ in. (6.4 mm) in story height
    ....±$^3/_4$ in. (19.1 mm) maximum

4. If the above conditions cannot be met due to previous construction, notify the Architect/ Engineer.

COMMENTARY

SPECIFICATION

3.4 — Reinforcement, tie, and anchor installation

3.4 A. *Basic requirements* — Place reinforcement, wall ties, and anchors in accordance with the sizes, types, and locations indicated on the Project Drawings and as specified. Do not place dissimilar metals in contact with each other.

3.4 B. *Reinforcement*

1. Support reinforcement to prevent displacement caused by construction loads or by placement of grout or mortar, beyond the allowable tolerances.

2. Completely embed reinforcing bars in grout in accordance with Article 3.5.

3. Maintain clear distance between reinforcing bars and the interior of masonry unit or formed surface of at least 1/4 in. (6.4 mm) for fine grout and 1/2 in. (12.7 mm) for coarse grout, except where cross webs of hollow units are used as supports for horizontal reinforcement.

4. Place reinforcing bars maintaining the following minimum cover:

    a. Masonry face exposed to earth or weather: 2 in. (50.8 mm) for bars larger than No. 5 (M #16); 1½ in. (38.1 mm) for No. 5 (M #16) bars or smaller.

    b. Masonry not exposed to earth or weather: 1½ in. (38.1 mm).

5. Maintain minimum clear distance between parallel bars of the nominal bar size or 1 in. (25.4 mm), whichever is greater.

6. In columns and pilasters, maintain minimum clear distance between vertical bars of one and one-half times the nominal bar size or 1½ in. (38.1 mm), whichever is greater.

7. Splice only where indicated on the Project Drawings, unless otherwise acceptable. When splicing by welding, provide welds in conformance with the provisions of AWS D 1.4.

8. Unless accepted by the Architect/Engineer, do not bend reinforcement after it is embedded in grout or mortar.

9. *Noncontact lap splices* — Position bars spliced by noncontact lap splice no farther apart transversely than one-fifth the specified length of lap nor more than 8 in. (203 mm)

COMMENTARY

3.4 — Reinforcement, tie, and anchor installation

The requirements given ensure that:

a. galvanic action is inhibited,

b. location is as assumed in the design,

c. there is sufficient clearance for grout and mortar to surround reinforcement, ties, and anchors so stresses are properly transferred,

d. corrosion is delayed, and

e. compatible lateral deflection of wythes is achieved.

Tolerances for placement of reinforcement in masonry first appeared in the 1985 Uniform Building Code (UBC, 1985). Reinforcement location obviously influences structural performance of the member. Figure SC-11 illustrates several devices used to secure reinforcement.

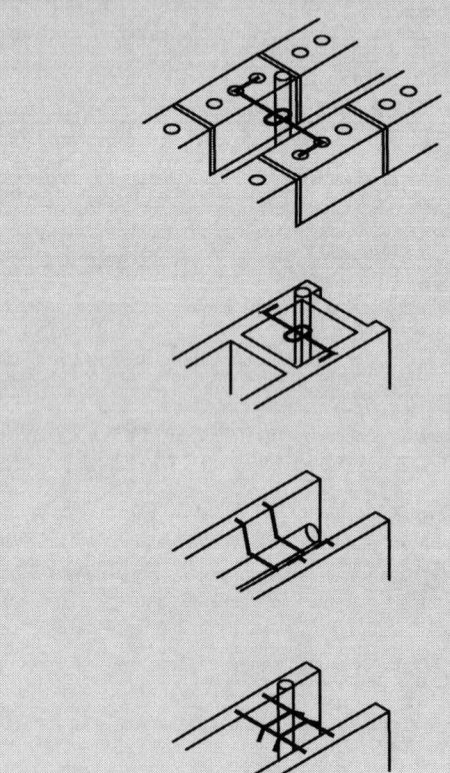

Figure SC-11 — Typical reinforcing bar positioners

9. *Noncontact lap splices* — Lap splices may be constructed with the bars in adjacent grouted cells if the requirements of this section are met.

SPECIFICATION

3.4 B. *Reinforcement* (Continued)

10. *Joint reinforcement*

   a. Place joint reinforcement so that longitudinal wires are embedded in mortar with a minimum cover of $1/2$ in. (12.7 mm) when not exposed to weather or earth; or $5/8$ in. (15.9 mm) when exposed to weather or earth.

   b. Provide minimum 6-in. (152-mm) lap splices for joint reinforcement.

   c. Ensure that all ends of longitudinal wires of joint reinforcement at laps are embedded in mortar or grout.

COMMENTARY

10. *Joint reinforcement* — There must be a minimum protective cover for the joint reinforcement as shown in Figure SC-12. Deeply tooled mortar joints, which provide inadequate protective cover for joint reinforcement, should be avoided.

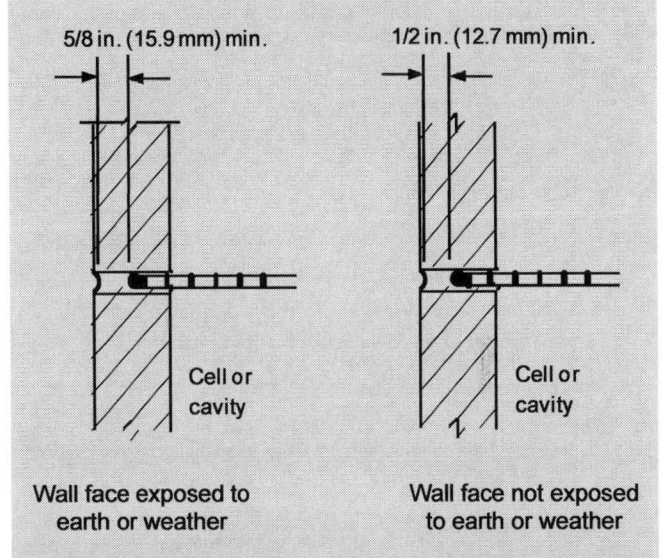

Figure SC-12 Joint Reinforcement Cover Requirements

c. Where laps occur in longitudinal wires of joint reinforcement the minimum embedment provisions of Article 3.4 B.10.a apply. Figure SC-13 shows typical joint reinforcement lap splices in mortar or grout.

SPECIFICATION

COMMENTARY

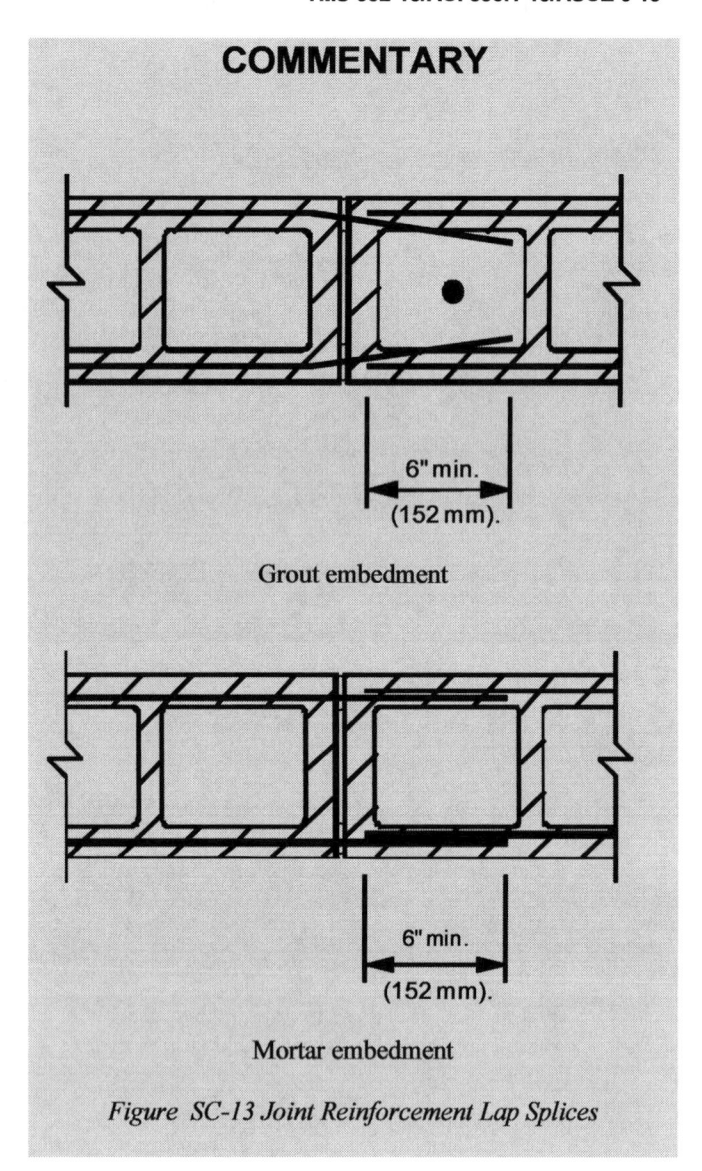

Grout embedment

Mortar embedment

Figure SC-13 Joint Reinforcement Lap Splices

SPECIFICATION

3.4 B. *Reinforcement* (Continued)

11. *Placement tolerances*

   a. Place reinforcing bars in walls and flexural elements within a tolerance of $\pm \frac{1}{2}$ in. (12.7 mm) when the distance from the centerline of reinforcing bars to the opposite face of masonry, d, is equal to 8 in. (203 mm) or less, ± 1 in. (25.4 mm) for d equal to 24 in. (610 mm) or less but greater than 8 in. (203 mm), and $\pm 1\frac{1}{4}$ in. (31.8 mm) for d greater than 24 in. (610 mm).

   b. Place vertical bars within:

   1) 2 in. (50.8 mm) of the required location along the length of the wall when the wall segment length exceeds 24 in. (610 mm).

   2) 1 in. (25.4 mm) of the required location along the length of the wall when the wall segment length does not exceed 24 in. (610 mm)

   c. If it is necessary to move bars more than one bar diameter or a distance exceeding the tolerance stated above to avoid interference with other reinforcing steel, conduits, or embedded items, notify the Architect/Engineer for acceptance of the resulting arrangement of bars.

COMMENTARY

11. *Placement tolerances*

   a. Ways to measure d distance in various common masonry elements are shown in Figures SC-14 through SC-16 (Chrysler, 2010). The maximum permissible tolerance for placement of reinforcement in a wall, beam, and column is based on the d dimension of that element.

   In masonry, the d dimension is measured perpendicular to the length of the element and is defined in the Specification as the distance from the center of the reinforcing bar to the compression face of masonry.

   In a wall subject to out-of-plane loading, the distance, d, to the compression face is normally the larger distance when reinforcing bars are offset from the center of the wall, as shown in Figure SC-14.

   The d dimension in masonry columns will establish the maximum allowable tolerance for placement of the vertical reinforcement. As shown in Figure SC-15, two dimensions for each vertical bar must be considered to establish the allowable tolerance for placement of the vertical reinforcement in each primary direction.

   The d dimension in a masonry beam will establish the maximum allowable tolerance for placement of the horizontal reinforcement within the depth of the beam. As shown in Figure SC-16, the distance to the top of beam is used to establish the allowable tolerance for placement of the reinforcement.

   b. The tolerance for placement of vertical reinforcing bars along the length of the wall is shown in Figure SC-14. As shown, the allowable tolerance is +/- 2 in. (50.8 mm), except for wall segments not exceeding 24 in. (610 mm) where the allowable tolerance is decreased to +/- 1 in. (25.4 mm). This tolerance applies to each reinforcing bar relative to the specified location in the wall. An accumulation of tolerances could result in bar placement that interferes with cross webs in hollow masonry units.

COMMENTARY

when $d \leq 8$ in. (203 mm), tolerance = ± ½ in. (12.7 mm)
when 8 in. (203 mm) < $d \leq 24$ in. (610 mm), tolerance = ± 1 in. (25.4 mm)
when $d > 24$ in. (610 mm), tolerance = ± 1 ¼ in. (31.8 mm)

Reinforcement on one or both faces

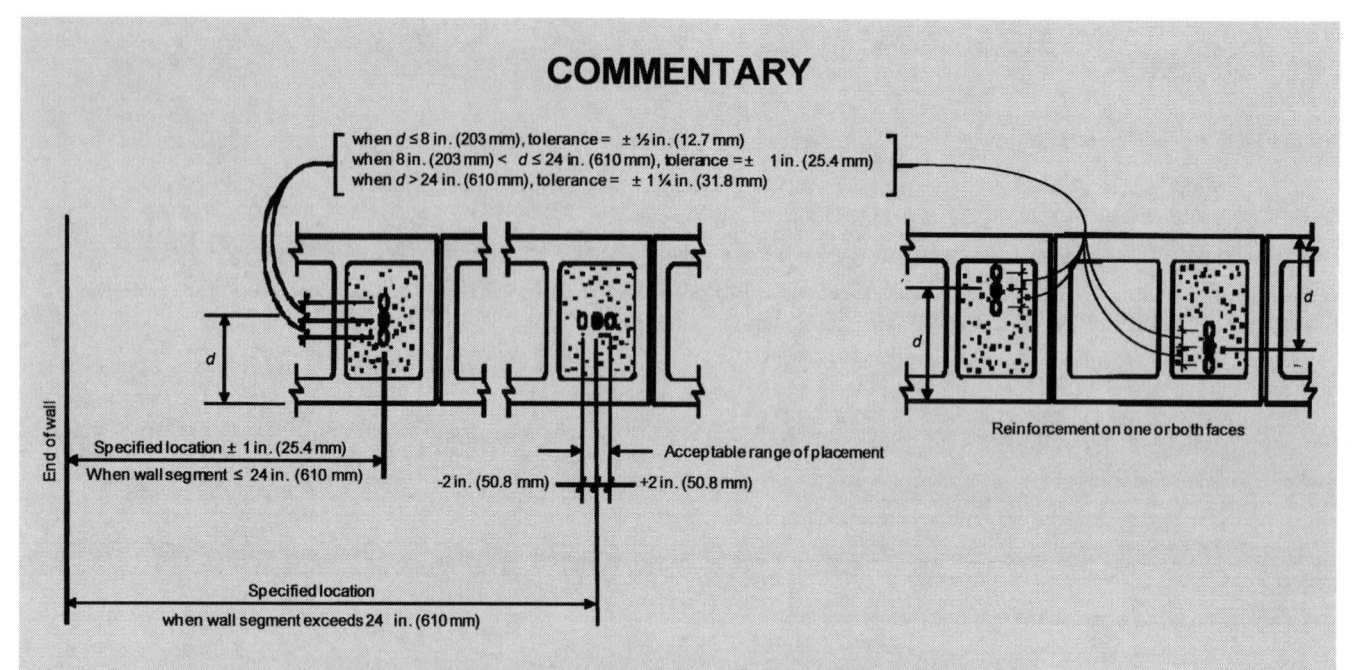

End of wall
Specified location ± 1 in. (25.4 mm)
When wall segment ≤ 24 in. (610 mm)
Acceptable range of placement
-2 in. (50.8 mm)   +2 in. (50.8 mm)
Specified location
when wall segment exceeds 24 in. (610 mm)

Figure SC-14 — Typical 'd' distance in a wall

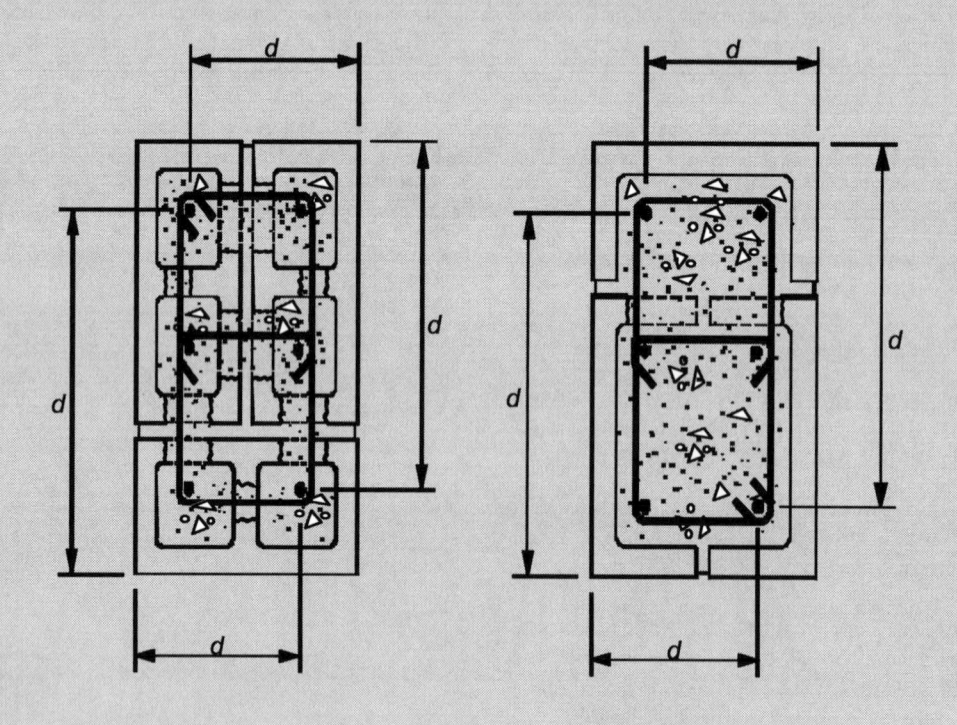

Figure SC-15 — Typical 'd' distance in a column

SPECIFICATION

3.4 B.11. *Reinforcement, Placement tolerances*
(Continued)

d. Foundation dowels that interfere with unit webs are permitted to be bent to a maximum of 1 in. (25.4 mm) horizontally for every 6 in. (152 mm) of vertical height.

COMMENTARY

d. Misaligned foundation dowels may interfere with placement of the masonry units. Interfering dowels may be bent in accordance with this provision (see Figure SC-17) (Stecich et al, 1984; NCMA TEK 3-2A, 2005). Removing a portion of the web to better accommodate the dowel may also be acceptable as long as the dowel is fully encapsulated in grout and masonry cover is maintained.

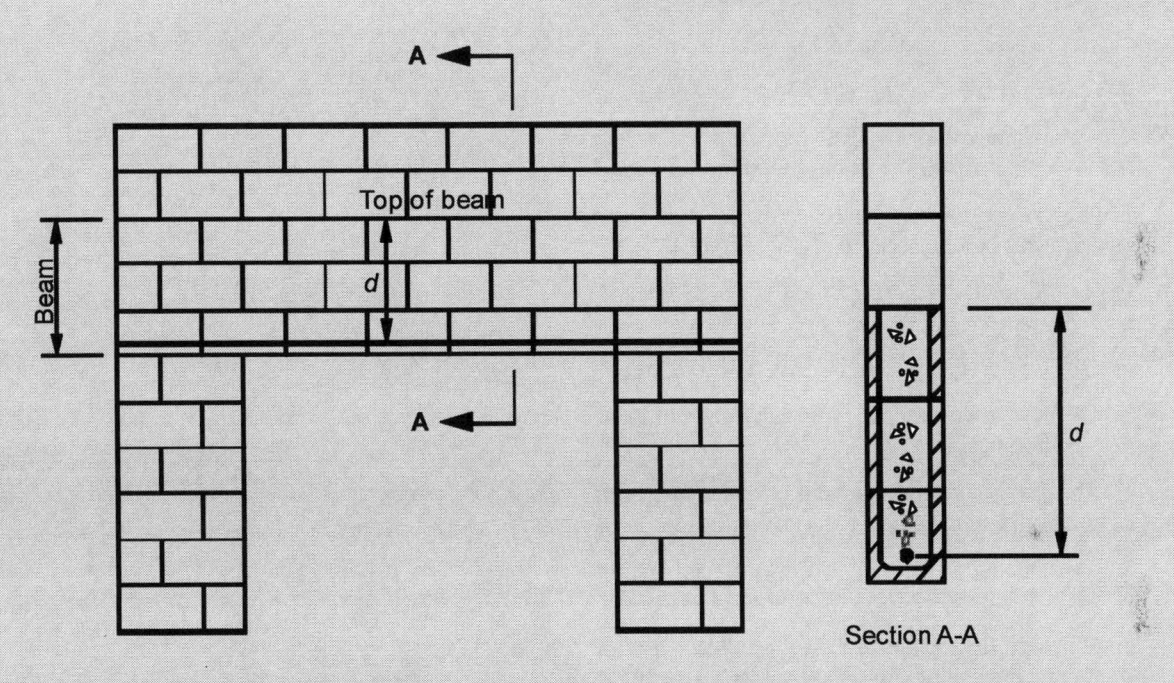

Figure SC-16 — Typical 'd' distance in a beam

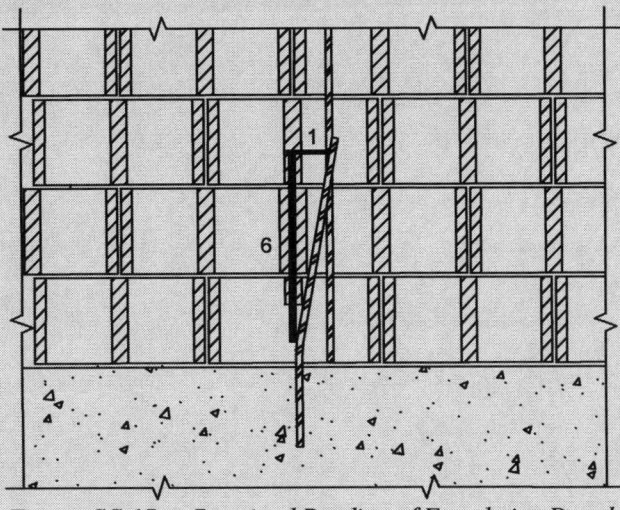

Figure SC-17 — Permitted Bending of Foundation Dowels

SPECIFICATION

3.4 C. Wall ties

1. Embed the ends of wall ties in mortar joints. Embed wall tie ends at least 1/2 in. (12.7 mm) into the outer face shell of hollow units. Embed wire wall ties at least 1 1/2 in. (38.1 mm) into the mortar bed of solid masonry units or solid grouted hollow units.

2. Unless otherwise required, bond wythes not bonded by headers with wall ties as follows:

Wire size	Minimum number of wall ties required
W1.7 (MW11)	One per 2.67 ft² (0.25 m²)
W2.8 (MW18)	One per 4.50 ft² (0.42 m²)

   The maximum spacing between ties is 36 in. (914 mm) horizontally and 24 in. (610 mm) vertically.

3. Unless accepted by the Architect/Engineer, do not bend wall ties after being embedded in grout or mortar.

4. Unless otherwise required, install adjustable ties in accordance with the following requirements:

   a. One tie for each 1.77 ft² (0.16 m²) of wall area.

   b. Do not exceed 16 in. (406 mm) horizontal or vertical spacing.

   c. The maximum misalignment of bed joints from one wythe to the other is 1 1/4 in. (31.8 mm).

   d. The maximum clearance between connecting parts of the ties is 1/16 in. (1.6 mm)

   e. When pintle anchors are used, provide ties with one or more pintle leg made of wire size W2.8 (MW18).

COMMENTARY

3.4 C. Wall ties — The Code does not permit the use of cavity wall ties with drips, nor the use of Z-ties in ungrouted, hollow unit masonry. The requirements for adjustable ties are shown in Figure SC-18.

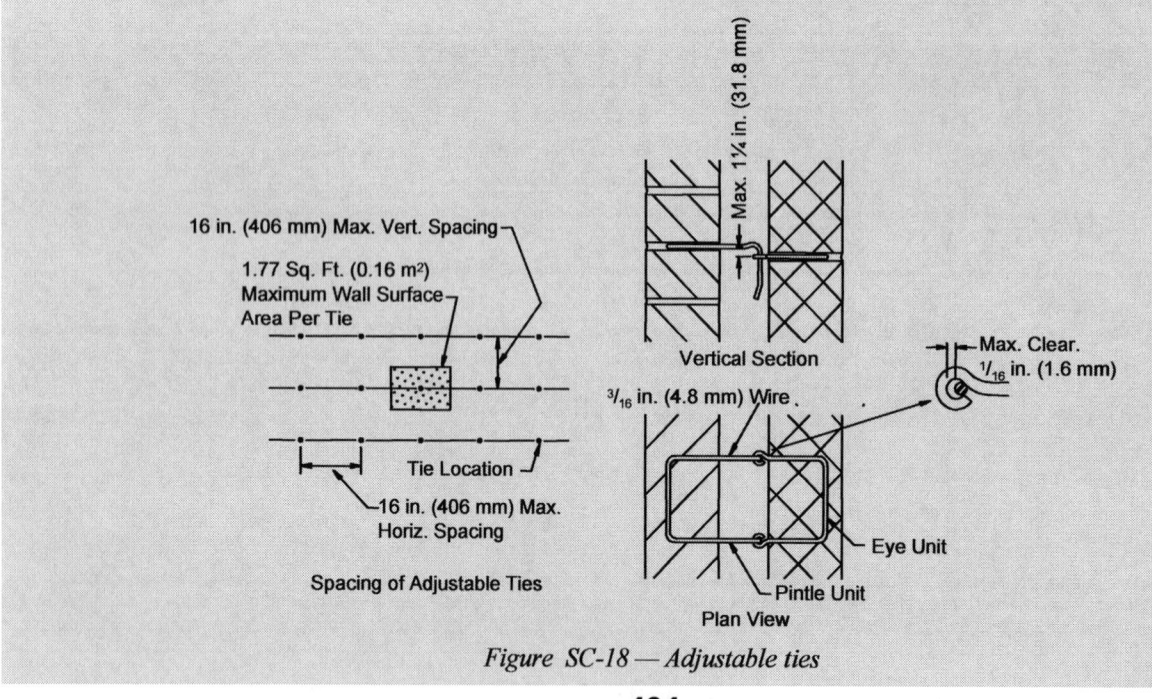

Figure SC-18 — Adjustable ties

SPECIFICATION

3.4 C. *Wall ties* (Continued)

5. Install wire ties perpendicular to a vertical line on the face of the wythe from which they protrude. Where one-piece ties or joint reinforcement are used, the bed joints of adjacent wythes shall align.

6. Unless otherwise required, provide additional unit ties around openings larger than 16 in. (406 mm) in either dimension. Space ties around perimeter of opening at a maximum of 3 ft (0.91 m) on center. Place ties within 12 in. (305 mm) of opening.

7. Unless otherwise required, provide unit ties within 12 in. (305 mm) of unsupported edges at horizontal or vertical spacing given in Article 3.4 C.2.

3.4 D. *Anchor bolts*

1. Embed headed and bent-bar anchor bolts larger than ¼ in. (6.4 mm) diameter in grout that is placed in accordance with Article 3.5 A and Article 3.5 B. Anchor bolts of ¼ in. (6.4 mm) diameter or less are permitted to be placed in grout or mortar bed joints that have a specified thickness of at least ½ in. (12.7 mm) thickness.

2. For anchor bolts placed in the top of grouted cells and bond beams, maintain a clear distance between the bolt and the face of masonry unit of at least ¼ in. (6.4 mm) when using fine grout and at least ½ in. (12.7 mm) when using coarse grout.

3. For anchor bolts placed through the face shell of a hollow masonry unit, drill a hole that is tight-fitting to the bolt or provide minimum clear distance that conforms to Article 3.4 D.2 around the bolt and through the face shell. For the portion of the bolt that is within the grouted cell, maintain a clear distance between the bolt and the face of masonry unit and between the head or bent leg of the bolt and the formed surface of grout of at least ¼ in. (6.4 mm) when using fine grout and at least ½ in. (12.7 mm) when using coarse grout.

4. Place anchor bolts with a clear distance between parallel anchor bolts not less than the nominal diameter of the anchor bolt, nor less than 1 in. (25.4 mm).

COMMENTARY

3. Quality assurance/control (QA/QC) procedures should assure that there is sufficient clearance around the bolts prior to grout placement. These procedures should also include observation during grout placement to assure that grout completely surrounds the bolts, as required by the QA Tables in Article 1.6.A

The clear distance requirement for grout to surround an anchor bolt does not apply where the bolt fits tightly in the hole of the face shell, but is required where the bolt is placed in an oversized hole in the face shell and where grout surrounds the anchor bolt in a grouted cell or cavity. See Figure SC-19.

COMMENTARY

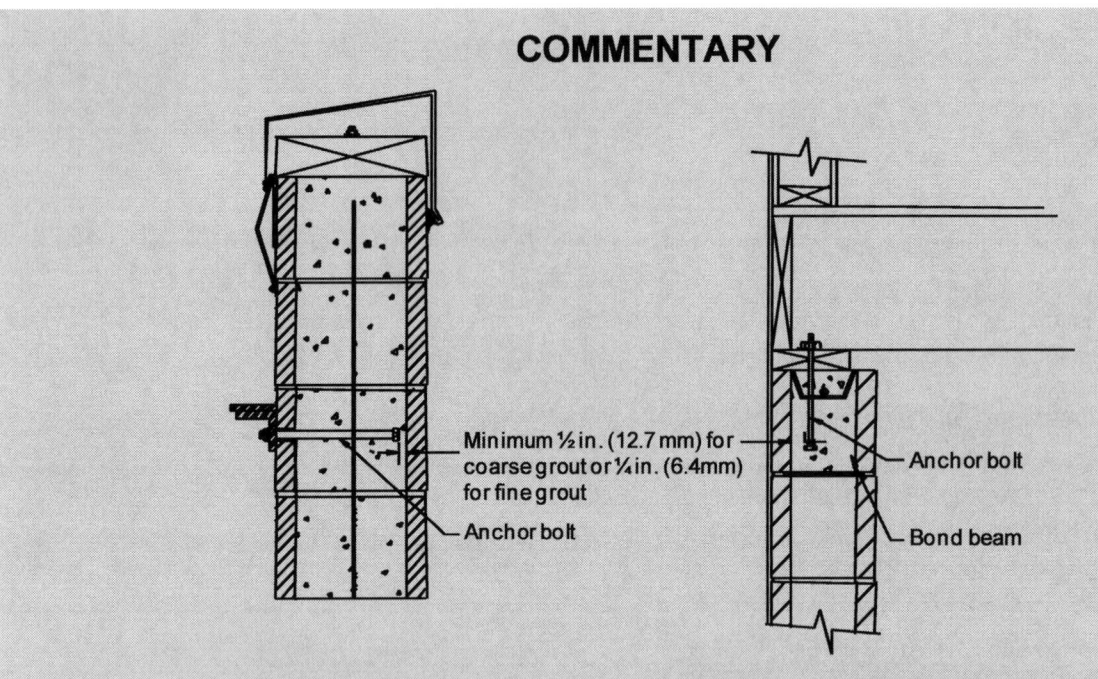

Figure SC-19 — Anchor bolt clearance requirements for headed anchor bolts – bent-bars are similar

SPECIFICATION

3.4 E. *Veneer anchors* — Place corrugated sheet-metal anchors, sheet-metal anchors, and wire anchors as follows:

1. With solid units, embed anchors in mortar joint and extend into the veneer a minimum of 1½ in. (38.1 mm), with at least $^5/_8$ in. (15.9 mm) mortar cover to the outside face.

2. With hollow units, embed anchors in mortar or grout and extend into the veneer a minimum of 1 ½ in. (38.1 mm), with at least $^5/_8$ in. (15.9 mm) mortar or grout cover to outside face.

3. Install adjustable anchors in accordance with the requirements of Articles 3.4 C.4.c, d, and e.
4. Provide at least one adjustable two-piece anchor, anchor of wire size W 1.7 (MW11), or 22 gage (0.8 mm) corrugated sheet-metal anchor for each 2.67 ft² (0.25 m²) of wall area.
5. Provide at least one anchor of other types for each 3.5 ft² (0.33 m²) of wall area.
6. Space anchors at a maximum of 32 in. (813 mm) horizontally and 25 in. (635 mm) vertically, but not to exceed the applicable requirement of Article 3.4 E.4 or 3.4 E.5.
7. Provide additional anchors around openings larger than 16 in. (406 mm) in either dimension. Space anchors around the perimeter of opening at a maximum of 3 ft (0.9 m) on center. Place anchors within 12 in. (305 mm) of opening.

COMMENTARY

3.4 E. *Veneer anchors* — Minimum embedment requirements have been established for each of the anchor types to ensure load resistance against push-through or pullout of the mortar joint.

2. Proper anchorage of veneer anchors into veneers using hollow masonry units can be satisfied by mortaring anchors in bed joints or on the cross-webs of the units; by grouting the cells or cores adjacent to the anchor; or by following the anchor manufacturer's requirements for installing the anchor into the cell or core above or below the bed joint and filling the cell or core containing the anchor with mortar or grout.

SPECIFICATION

3.4 F. *Glass unit masonry panel anchors* — When used instead of channel-type restraints, install panel anchors as follows:

1. Unless otherwise required, space panel anchors at 16 in. (406 mm) in both the jambs and across the head.

2. Embed panel anchors a minimum of 12 in. (305 mm), except for panels less than 2 ft (0.61 m) in the direction of embedment. When a panel dimension is less than 2 ft (0.61 m), embed panel anchors in the short direction a minimum of 6 in. (152 mm), unless otherwise required.

3. Provide two fasteners, capable of resisting the required loads, per panel anchor.

3.5 — Grout placement

3.5 A. *Placing time* — Place grout within $1\frac{1}{2}$ hr from introducing water in the mixture and prior to initial set.

1. Discard site-mixed grout that does not meet the specified slump without adding water after initial mixing.

2. For ready-mixed grout:

   a. Addition of water is permitted at the time of discharge to adjust slump.

   b. Discard ready-mixed grout that does not meet the specified slump without adding water, other than the water that was added at the time of discharge.

   The time limitation is waived as long as the ready-mixed grout meets the specified slump.

3.5 B. *Confinement* — Confine grout to the areas indicated on the Project Drawings. Use material to confine grout that permits bond between masonry units and mortar.

COMMENTARY

3.5 — Grout placement

Grout may be placed by pumping or pouring from large or small buckets. The amount of grout to be placed and contractor experience influence the choice of placement method.

The requirements of this Article do not apply to prestressing grout.

3.5 A. *Placing time* — Grout placement is often limited to 1½ hours after initial mixing, but this time period may be too long in hot weather (initial set may occur) and may be unduly restrictive in cooler weather. One indicator that the grout has not reached initial set is a stable and reasonable grout temperature. However, sophisticated equipment and experienced personnel are required to determine initial set with absolute certainty.

Article 3.5 A.2 permits water to be added to ready-mixed grout to compensate for evaporation that has occurred prior to discharge. Replacement of evaporated water is not detrimental to ready-mixed grout. However, water may not be added to ready-mixed grout after discharge.

3.5 B. *Confinement* — Certain locations in the wall may not be grouted in order to reduce dead loads or allow placement of other materials such as insulation or wiring. Cross webs adjacent to cells to be grouted can be bedded with mortar to confine the grout. Metal lath, plastic screening, or other items can be used to plug cells below bond beams.

SPECIFICATION

3.5 C. *Grout pour height* — Do not exceed the maximum grout pour height given in Table 7.

COMMENTARY

3.5 C. *Grout pour height* — Table 7 in the Specification has been developed as a guide for grouting procedures. The designer can impose more stringent requirements if so desired. The recommended maximum height of grout pour (see Figure SC-20) corresponds with the least clear dimension of the grout space. The minimum width of grout space is used when the grout is placed between wythes. The minimum cell dimensions are used when grouting cells of hollow masonry units. As the height of the pour increases, the minimum grout space increases. The grout space dimensions are clear dimensions. See the Commentary for Section 3.2.1 of the Code for additional information.

Grout pour heights and minimum dimensions that meet the requirements of Table 7 do not automatically mean that the grout space will be filled.

Grout spaces smaller than specified in Table 7 have been used successfully in some areas. When the contractor asks for acceptance of a grouting procedure that does not meet the limits in Table 7, construction of a grout demonstration panel is required. Destructive or non-destructive evaluation can confirm that filling and adequate consolidation have been achieved. The Architect/Engineer should establish criteria for the grout demonstration panel to assure that critical masonry elements included in the construction will be represented in the demonstration panel. Because a single grout demonstration panel erected prior to masonry construction cannot account for all conditions that may be encountered during construction, the Architect/Engineer should establish inspection procedures to verify grout placement during construction. These inspection procedures should include destructive or non-destructive evaluation to confirm that filling and adequate consolidation have been achieved.

Table 7 — Grout space requirements

Grout type[1]	Maximum grout pour height, ft (m)	Minimum clear width of grout space,[2,3] in. (mm)	Minimum clear grout space dimensions for grouting cells of hollow units,[3,4,5] in. x in. (mm x mm)
Fine	1 (0.30)	3/4 (19.1)	1 1/2 x 2 (38.1 x 50.8)
Fine	5.33 (1.63)	2 (50.8)	2 x 3 (50.8 x 76.2)
Fine	12.67 (3.86)	2 1/2 (63.5)	2 1/2 x 3 (63.5 x 76.2)
Fine	24 (7.32)	3 (76.2)	3 x 3 (76.2 x 76.2)
Coarse	1 (0.30)	1 1/2 (38.1)	1 1/2 x 3 (38.1 x 76.2)
Coarse	5.33 (1.63)	2 (50.8)	2 1/2 x 3 (63.5 x 76.2)
Coarse	12.67 (3.86)	2 1/2 (63.5)	3 x 3 (76.2 x 76.2)
Coarse	24 (7.32)	3 (76.2)	3 x 4 (76.2 x 102)

[1] Fine and coarse grouts are defined in ASTM C476.

[2] For grouting between masonry wythes.

[3] Minimum clear width of grout space and minimum clear grout space dimension are the net dimension of the space determined by subtracting masonry protrusions and the diameters of horizontal bars from the as-built cross-section of the grout space. Select the grout type and maximum grout pour height based on the minimum clear space.

[4] Area of vertical reinforcement shall not exceed 6 percent of the area of the grout space.

[5] Minimum grout space dimension for AAC masonry units shall be 3 in. (76.2 mm) x 3 in. (76.2 mm) or a 3 in. (76.2 mm) diameter cell.

SPECIFICATION

3.5 D. *Grout lift height*

1. For grout conforming to Article 2.2 A:

   a. Where the following conditions are met, place grout in lifts not exceeding 12 ft 8 in. (3.86 m).

      i. The masonry has cured for at least 4 hours.

      ii. The grout slump is maintained between 10 and 11 in. (254 and 279 mm).

      iii. No intermediate reinforced bond beams are placed between the top and the bottom of the pour height.

   b. When the conditions of Articles 3.5 D.1.a.i and 3.5 D.1.a.ii are met but there are intermediate bond beams within the grout pour, limit the grout lift height to the bottom of the lowest bond beam that is more than 5 ft 4 in. (1.63 m) above the bottom of the lift, but do not exceed a grout lift height of 12 ft 8 in. (3.86 m).

   c. When the conditions of Article 3.5 D.1.a.i or Article 3.5 D.1.a.ii are not met, place grout in lifts not exceeding 5 ft 4 in. (1.63 m).

2. For self-consolidating grout conforming to Article 2.2:

   a. When placed in masonry that has cured for at least 4 hours, place in lifts not exceeding the grout pour height.

   b. When placed in masonry that has not cured for at least 4 hours, place in lifts not exceeding 5 ft 4 in. (1.63 m) or the grout pour height, whichever is less.

3.5 E. *Consolidation*

1. Consolidate grout at the time of placement.

   a. Consolidate grout pours 12 in. (305 mm) or less in height by mechanical vibration or by puddling.

   b. Consolidate pours exceeding 12 in. (305 mm) in height by mechanical vibration, and reconsolidate by mechanical vibration after initial water loss and settlement has occurred.

2. Consolidation or reconsolidation is not required for self-consolidating grout.

COMMENTARY

3.5 D. *Grout lift height* — A lift is the height to which grout is placed into masonry in one continuous operation (see Figure SC-20). After placement of a grout lift, water is absorbed by the masonry units. Following this water loss, a subsequent lift may be placed on top of the still plastic grout.

Grouted construction develops fluid pressure in the grout space. Grout pours composed of several lifts may develop this fluid pressure for the full pour height. The faces of hollow units with unbraced ends can break out. Wythes may separate. The wire ties between wythes may not be sufficient to prevent this from occurring. Higher lifts may be used with self-consolidating grout because its fluidity and its lower initial water-cement ratio result in reduced potential for fluid pressure problems.

The 4-hour time period is stipulated for grout lifts over 5 ft 4 in. (1.63 m) to provide sufficient curing time to minimize potential displacement of units during the consolidation and reconsolidation process. The 4 hours is based on typical curing conditions and may be increased based on local climatic conditions at the time of construction. For example, during cold weather construction, consider increasing the 4-hour curing period. When a wall is to be grouted with self-consolidating grout, the grout lift height is not restricted by intermediate, reinforced bond beam locations because self-consolidating grout easily flows around reinforcing bars (NCMA MR29, 2006; NCMA MR31, 2007)

3.5 E. *Consolidation* — Except for self-consolidating grout, consolidation is necessary to achieve complete filling of the grout space. Reconsolidation returns the grout to a plastic state and eliminates the voids resulting from the water loss from the grout by the masonry units. It is possible to have a height loss of 8 in. (203 mm) in 8 ft (2.44 m).

Consolidation and reconsolidation are normally achieved with a mechanical vibrator. A low velocity vibrator with a ¾ in. (19.1 mm) head is used. The vibrator is activated for one to two seconds in each grouted cell of hollow unit masonry. When double open-end units are used, one cell is considered to be formed by the two open ends placed together. When grouting between wythes, the vibrator is placed in the grout at points spaced 12 to 16 in. (305 to 406 mm) apart. Excess vibration does not improve consolidation and may blow out the face shells of hollow units or separate the wythes when grouting between wythes.

COMMENTARY

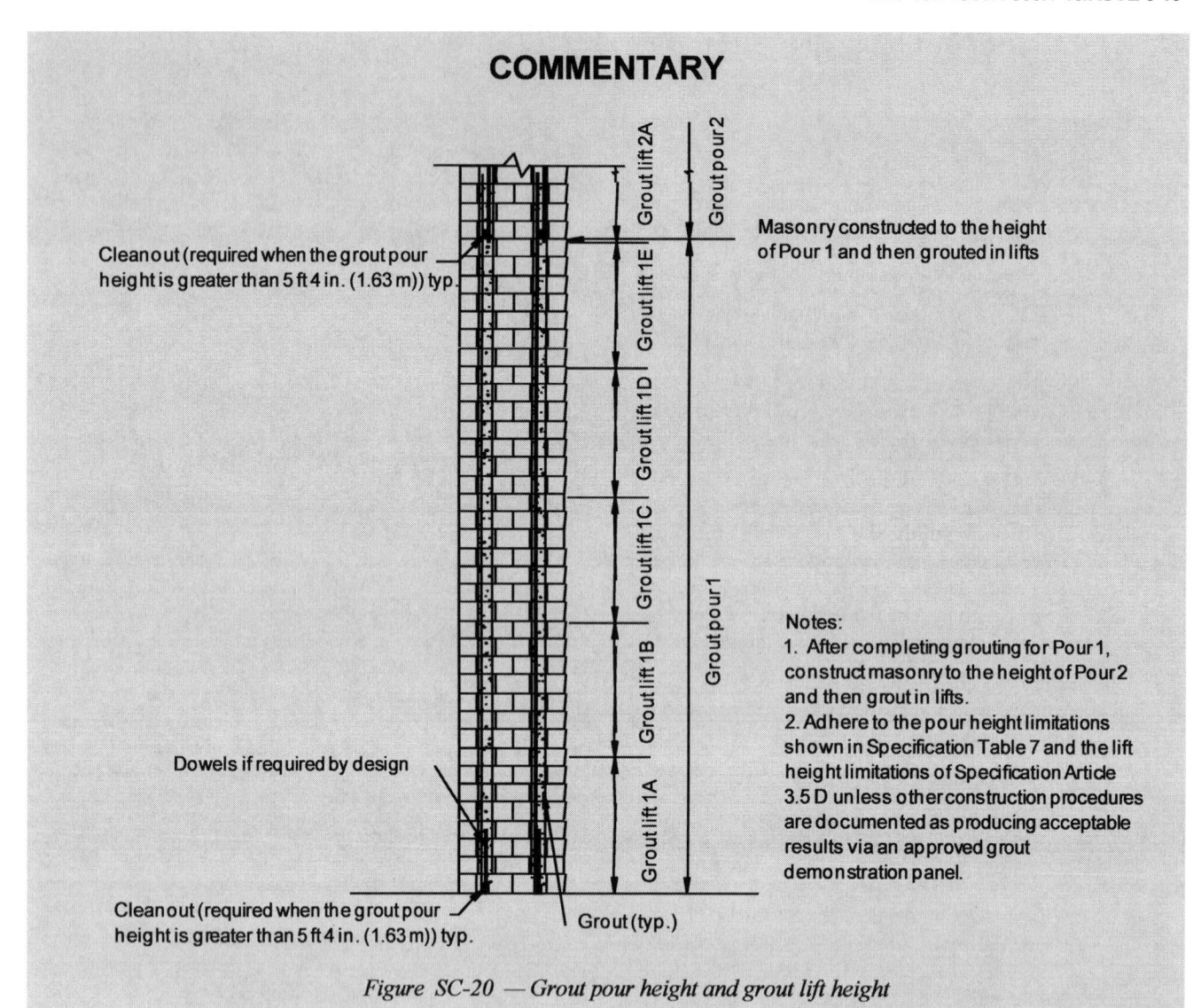

Figure SC-20 — Grout pour height and grout lift height

SPECIFICATION

3.5 F. *Grout key* — When grouting, form grout keys between grout pours. Form grout keys between grout lifts when the first lift is permitted to set prior to placement of the subsequent lift

1. Form a grout key by terminating the grout a minimum of 1½ in. (38.1 mm) below a mortar joint.

2. Do not form grout keys within beams.

3. At beams or lintels laid with closed bottom units, terminate the grout pour at the bottom of the beam or lintel without forming a grout key.

3.5 G. *Alternate grout placement* — Place masonry units and grout using construction procedures employed in the accepted grout demonstration panel.

3.5 H. *Grouting AAC masonry* — Wet AAC masonry thoroughly before grouting to ensure that the grout flows to completely fill the space to be grouted.

COMMENTARY

3.5 F. *Grout key* — The top of a grout pour should not be located at the top of a unit, but at a minimum of 1½ in. (38 mm) below the bed joint.

If a lift of grout is permitted to set prior to placing the subsequent lift, a grout key is required within the grout pour. This setting normally occurs if the grouting is stopped for more than one hour.

SPECIFICATION

3.6 — Prestressing tendon installation and stressing procedure

3.6 A. *Site tolerances*

1. Tolerance for prestressing tendon placement in the out-of-plane direction in walls shall be ± $^1/_4$ in. (6.4 mm) for masonry cross-sectional dimensions less than nominal 8 in. (203 mm) and ± $^3/_8$ in. (9.5 mm) for masonry cross-sectional dimensions equal to or greater than nominal 8 in. (203 mm).

2. Tolerance for prestressing tendon placement in the in-plane direction of walls shall be ± 1 in. (25.4 mm).

3. If prestressing tendons are moved more than one tendon diameter or a distance exceeding the tolerances stated in Articles 3.6 A.1 and 3.6 A.2 to avoid interference with other tendons, reinforcement, conduits, or embedded items, notify the Architect/Engineer for acceptance of the resulting arrangement of prestressing tendons.

3.6 B. *Application and measurement of prestressing force*

1. Determine the prestressing force by both of the following methods:

   a. Measure the prestressing tendon elongation and compare it with the required elongation based on average load-elongation curves for the prestressing tendons.

   b. Observe the jacking force on a calibrated gage or load cell or by use of a calibrated dynamometer. For prestressing tendons using bars of less than 150 ksi (1034 MPa) tensile strength, Direct Tension Indicator (DTI) washers complying with ASTM F959 or ASTM F959M are acceptable.

2. Ascertain the cause of the difference in force determined by the two methods described in Article 3.6 B.1 when the difference exceeds 5 percent for pretensioned elements or 7 percent for post-tensioned elements, and correct the cause of the difference.

3. When the total loss of prestress due to unreplaced broken prestressing tendons exceeds 2 percent of total prestress, notify the Architect/Engineer.

COMMENTARY

3.6 — Prestressing tendon installation and stressing procedure

Installation of tendons with the specified tolerances is common practice. The methods of application and measurement of prestressing force are common techniques for prestressed concrete and masonry members. Designer, contractor, and inspector should be experienced with prestressing and should consult the Post-Tensioning Institute's Field Procedures Manual for Unbonded Single Strand Tendons (PTI, 1994) or similar literature before conducting the Work. Critical aspects of the prestressing operation that require inspection include handling and storage of the prestressing tendons and anchorages, installation of the anchorage hardware into the foundation and capping members, integrity and continuity of the corrosion-protection system for the prestressing tendons and anchorages, and the prestressing tendon stressing and grouting procedures.

The design method in Code Chapter 10 is based on an accurate assessment of the level of prestress. Tendon elongation and tendon force measurements with a calibrated gauge or load cell or by use of a calibrated dynamometer have proven to provide the required accuracy. For tendons using steels of less than 150 ksi (1034 MPa) strength, Direct Tension Indicator (DTI) washers also provide adequate accuracy. These washers have dimples that are intended to compress once a predetermined force is applied on them by the prestressing force. These washers were first developed by the steel industry for use with high-strength bolts and have been modified for use with prestressed masonry. The designer should verify the actual accuracy of DTI washers and document it in the design.

Burning and welding operations in the vicinity of prestressing tendons must be carefully performed because the heat may lower the tendon strength and cause failure of the stressed tendon.

SPECIFICATION

3.6 C. *Grouting bonded tendons*

1. Mix prestressing grout in equipment capable of continuous mechanical mixing and agitation so as to produce uniform distribution of materials, pass through screens, and pump in a manner that will completely fill tendon ducts.

2. Maintain temperature of masonry above 35°F (1.7°C) at time of grouting and until field-cured 2 in. (50.8 mm) cubes of prestressing grout reach a minimum compressive strength of 800 psi (5.52 MPa).

3. Keep prestressing grout temperatures below 90°F (32.2°C) during mixing and pumping.

3.6 D. *Burning and welding operations* — Carefully perform burning and welding operations in the vicinity of prestressing tendons so that tendons and sheathings, if used, are not subjected to excessive temperatures, welding sparks, or grounding currents.

3.7 — Field quality control

3.7 A. Verify f'_m and f'_{AAC} in accordance with Article 1.6.

3.7 B. Sample and test grout as required by Articles 1.4 B and 1.6.

3.8 — Cleaning

Clean exposed masonry surfaces of stains, efflorescence, mortar and grout droppings, and debris using methods that do not damage the masonry.

COMMENTARY

3.7 — Field quality control

3.7 A. The specified frequency of testing must equal or exceed the minimum requirements of the quality assurance tables.

3.7 B. ASTM C1019 requires a mold for the grout specimens made from the masonry units that will be in contact with the grout. Thus, the water absorption from the grout by the masonry units is simulated. Sampling and testing frequency may be based on the volume of grout to be placed rather than the wall area. Alternative forming methods can also be used provided a conversion factor based on comparative testing of 10 sets of specimens has been established as required by ASTM C1019, Section 6.2

3.8 — Cleaning

Use of undiluted cleaning products, especially acids, and failing to pre-wet the masonry or to adequately rinse the masonry after cleaning can cause damage. In some situations, cleaning without chemicals may be appropriate.

FOREWORD TO SPECIFICATION CHECKLISTS

SPECIFICATION

F1. This Foreword is included for explanatory purposes only; it does not form a part of Specification TMS 602–13/ACI 530.1–13/ASCE 6–13.

F2. Specification TMS 602–13/ACI 530.1–13/ASCE 6–13 may be referenced by the Architect/Engineer in the Project Specification for any building project, together with supplementary requirements for the specific project. Responsibilities for project participants must be defined in the Project Specification.

F3. Checklists do not form a part of Specification TMS 602–13/ACI 530.1–13/ASCE 6–13. Checklists are provided to assist the Architect/Engineer in selecting and specifying project requirements in the Project Specification. The checklists identify the Sections, Parts, and Articles of the reference Specification and the action required or available to the Architect/Engineer.

F4. The Architect/Engineer must make adjustments to the Specification based on the needs of a particular project by reviewing each of the items in the checklists and including the items the Architect/Engineer selects as mandatory requirements in the Project Specification.

F5. The Mandatory Requirements Checklist indicates work requirements regarding specific qualities, procedures, materials, and performance criteria that are not defined in Specification TMS 602–13/ACI 530.1–13/ASCE 6–13 or requirements for which the Architect/Engineer must define which of the choices apply to the project.

F6. The Optional Requirements Checklist identifies Architect/Engineer choices and alternatives.

COMMENTARY

F1. No Commentary

F2. Building codes (of which this standard is a part by reference) set minimum requirements necessary to protect the public. Project specifications may stipulate requirements more restrictive than the minimum. Adjustments to the needs of a particular project are intended to be made by the Architect/Engineer by reviewing each of the items in the Checklists and then including the Architect/Engineer's decision on each item as a mandatory requirement in the project specifications.

F3. The Checklists are addressed to each item of this Specification where the Architect/Engineer must or may make a choice of alternatives; may add provisions if not indicated; or may take exceptions. The Checklists consist of two columns; the first identifies the sections, parts, and articles of the Specification, and the second column contains notes to the Architect/Engineer to indicate the type of action that may be required by the Architect/Engineer. Checklist items that are not applicable to a project should not be included in the Project Specifications.

MANDATORY REQUIREMENTS CHECKLIST

Section/Part/Article	Notes to the Architect/Engineer
PART 1 — GENERAL	
1.4 A Compressive strength requirements	Specify f'_m and f'_{AAC}, except for veneer, glass unit masonry, prescriptively designed partition walls, and empirically designed masonry. Specify f'_{mi} for prestressed masonry.
1.4 B.2 Unit strength method	Specify when strength of grout is to be determined by test.
1.5 Submittals	Define the submittal reporting and review procedure.
1.6 A.1 Testing Agency's services and duties	Specify which of Tables 3, 4, or 5 applies to the project. Specify which portions of the masonry were designed in accordance with the prescriptive partition wall, empirical, veneer, or glass unit masonry provisions of this Code and are, therefore, exempt from verification of f'_m.
1.6 B.1 Inspection Agency's services and duties	Specify which of Tables 3, 4, or 5 applies to the project. Specify which portions of the masonry were designed in accordance with the prescriptive partition wall, empirical, veneer, or glass unit masonry provisions of this Code and are, therefore, exempt from verification of f'_m.
1.6 D Sample panels	Specify requirements for sample panels.
PART 2 — PRODUCTS	
2.1 Mortar materials	Specify type, color, and cementitious materials to be used in mortar and mortar to be used for the various parts of the project and the type of mortar to be used with each type of masonry unit.
2.3 Masonry unit materials	Specify the masonry units to be used for the various parts of the projects.
2.4 Reinforcement, prestressing tendons, and metal accessories	Specify type and grade of reinforcement, tendons, connectors, and accessories.
2.4 A Reinforcing Steel	When deformed reinforcing bars conforming to ASTM A615/A615M or ASTM A996/A996M are required by strength design in accordance with Code Chapter 9 or Chapter 11, specify that the actual yield strength must not exceed the specified yield strength multiplied by 1.3.
2.4 C.1 Joint reinforcement	Specify joint reinforcement wire size and number of longitudinal wires when joint reinforcement is to be used as shear reinforcement.
2.4 C.3 Welded wire reinforcement	Specify when welded wire reinforcement is to be plain.
2.4 E Stainless steel	Specify when stainless steel joint reinforcement, anchors, ties, and/or accessories are required.
2.4 F Coating for corrosion protection	Specify the types of corrosion protection that are required for each portion of the masonry construction.
2.4 G Corrosion protection for tendons	Specify the corrosion protection method.

MANDATORY REQUIREMENTS CHECKLIST (Continued)

Section/Part/Article		Notes to the Architect/Engineer
2.4 H	Prestressing anchorages, couplers, and end blocks	Specify the anchorages and couplers and their corrosion protection.
2.5 E	Joint fillers	Specify size and shape of joint fillers.
2.7 B	Prefabricated masonry	Specify prefabricated masonry and requirements in supplement of those of ASTM C901.

PART 3 — EXECUTION

3.3 D.2-4	Pipes and conduits	Specify sleeve sizes and spacing.
3.3 D.5	Accessories	Specify accessories not indicated on the project drawings.
3.3 D.6	Movement joints	Indicate type and location of movement joints on the project drawings.
3.4 B.11	Placement tolerances	Indicate d distance for beams on drawings or as a schedule in the project specifications.
3.4 E	Veneer anchors	Specify type of anchor required.

OPTIONAL REQUIREMENTS CHECKLIST

Section/Part/Article		Notes to the Architect/Engineer
PART 1 — GENERAL		
1.5 B		Specify additional required submittals.
1.6	Quality assurance	Define who will retain the Testing Agency and Inspection Agency, if other than the Owner.
PART 2 — PRODUCTS		
2.2		Specify grout requirements at variance with TMS 602/ACI 530.1/ASCE 6. Specify admixtures.
2.5 A and 2.5 B	Movement joint	Specify requirements at variance with TMS 602/ACI 530.1/ASCE 6.
2.5 D	Masonry cleaner	Specify where acid or caustic solutions are allowed and how to neutralize them.
2.6 A	Mortar	Specify if hand mixing is allowed and the method of measurement of material.
2.6 B.2	Grout consistency	Specify requirements at variance with TMS 602/ACI 530.1/ASCE 6
PART 3 — EXECUTION		
3.2 C	Wetting masonry units	Specify when units are to be wetted.
3.3 A	Bond pattern	Specify bond pattern if not running bond.
3.3 B.2	Bed and head joints	Specify thickness and tooling differing from TMS 602/ACI 530.1/ASCE 6.
3.3 B.3	Collar joints	Specify the filling of collar joints less than $3/4$ in. (19.1 mm) thick differing from TMS 602/ACI 530.1/ASCE 6.
3.3 B.4	Hollow units	Specify when cross webs are to be mortar bedded.
3.3 B.5	Solid units	Specify mortar bedding at variance with TMS 602/ACI 530.1/ASCE 6.
3.3 B.7	Glass units	Specify mortar bedding at variance with TMS 602/ACI 530.1/ASCE 6.
3.3 B.9.b	AAC Masonry	Specify when mortar may be omitted from AAC running bond masonry head joints that are less than 8 in. (200 mm) (nominal) tall.
3.3 D.2	Embedded items and accessories	Specify locations where sleeves are required for pipes or conduits.
3.4 B.10	Joint reinforcement	When joint reinforcement is used as shear reinforcement, specify a lap length of $48d_b$ instead of 6 inches.
3.4 C.2, 3, and 4		Specify requirements at variance with TMS 602/ACI 530.1/ASCE 6.

TEXT FROM

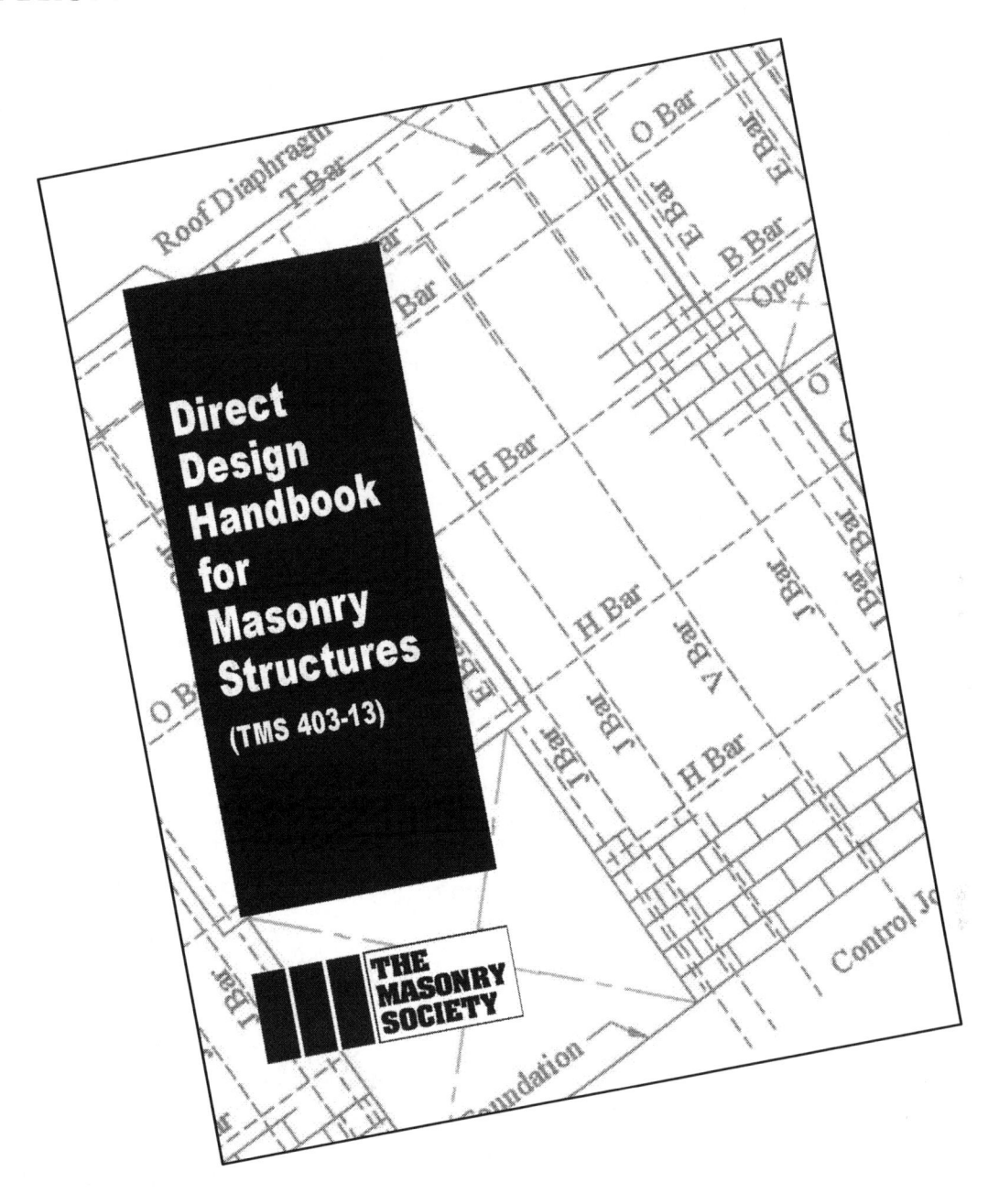

"*Direct Design Handbook for Masonry Structures (TMS 403-13)*, copyright © 2013, is reproduced with the permission of the publisher, The Masonry Society."

The Masonry Society
105 South Sunset Street, Suite Q
Longmont, CO 80501
www.masonrysociety.org

Forward to the Reference Guide

The reference guide that follows does not form a part of the *Handbook*. The purpose of the reference guide is to assist the user in properly following each step in the direct design procedure or verifying that each step was completed. This reference guide is not a substitute for reading and understanding the requirements of the direct design procedure.

Direct Design Procedure Reference Guide

1. From ASCE 7, obtain Risk Category, p_g, V, Exposure Category, S_S, and S_1.
2. Verify site conditions meet Section 2.1 and create preliminary architectural plans in accordance with Chapter 2.
3. Obtain SDC from Table 3.2-1.
4. Obtain LFRS options from Table 3.2-2.
5. In each principal plan direction, divide the roof plan into rectangular diaphragms.
   - 5A Calculate A
   - 5B Obtain C_W from Table 3.2-3.
   - 5C $V_{LFRS\text{-}wind} = C_W A$
   - 5D Calculate W
   - 5E Obtain C_S from Table 3.2-4.
   - 5F $V_{LFRS\text{-}seismic} = C_S W$
   - 5G V_{LFRS} = greater of $V_{LFRS\text{-}wind}$ and $V_{LFRS\text{-}seismic}$.
   - 5H For each wall line parallel to the direction under consideration:
      - 5H.1 Identify maximum L_{joist} and maximum h_{max} from plans.
      - 5H.2 Obtain s_{V1}, from Table 3.2-5.
      - 5H.3 Obtain s_{V2}, from Table 3.2-6.
   - 5I To determine L_{req} for each line of resistance parallel to the direction under consideration, for each diaphragm:
      - 5I.1 Using the largest L_{joist}, identify applicable Table 3.2-8 using Table 3.2-7.
      - 5I.2 Select L_{seg} for design based on plans.
      - 5I.3 Obtain k_1 and k_2 from Table 3.2-8, based on $s_V \leq s_{V1}$; $s_V \leq s_{V2}$; and verify s_V is permitted by Tables 3.2-5 and 3.2-6.
      - 5I.4 L_{req} = greatest of L_1 and L_2.
         $L_1 = V_{LFRS} / k_1$
         $L_2 = V_{LFRS}\, h_{max} / k_2$
      - 5I.5 For each line of resistance, adjust openings and control joints if necessary so that $\Sigma L_{seg} \geq L_{req}$. Alternatively, change assumptions in Step 5I.3 and/or Step 5E, and redo from those Steps.
   - 5J For non-designated shear walls, verify that s_V and s_H is the same as that for designated shear walls.
   - 5K For non-participating walls, verify that $s_V \leq s_{V1}$; $s_V \leq s_{V2}$, and details required by Chapter 6 are met.
   - 5L Determine the number of C Bars required by Table 3.2-9.
6. For diaphragms sharing a common shear wall line, verify that $L_{req} = (L_{req1} + L_{req2})$.
7. Detail J Bars. Provide symmetrical layout of vertical bars in each wall segment.
8. At header panels with parapets, detail V Bars.
9. At headers and sill panels:
   - 9A Detail control joints per Chapter 6.
   - 9B Obtain B bars from Table 3.2-10. Detail grouting and reinforcement at bottom and mirror up to joist bearing.
   - 9C Detail O Bars per Table 3.2-11.
10. Provide required design information on construction documents.
11. Provide required specifications on construction documents in accordance with Chapter 5.
12. Provide required details on construction documents in accordance with Chapter 6.

DIRECT DESIGN HANDBOOK FOR MASONRY STRUCTURES

Direct Design Handbook for Masonry Structures (TMS 0403-13)

**Prepared by the
TMS Design Practices Committee**

THE MASONRY SOCIETY
Longmont, Colorado

Direct Design Handbook for Masonry Structures (TMS 403-13)

Prepared by   TMS Design Practices Committee
Published by   The Masonry Society
                  105 South Sunset Street, Suite Q
                  Longmont, CO 80501-6172
                  Phone: 303-939-9700
                  Fax: 303-541-9215
                  Website: www.masonrysociety.org

ABSTRACT

TMS 403-13 *Direct Design Handbook for Masonry Structures* (hereinafter referred to as the *Handbook*) was developed by The Masonry Society's Design Practices Committee. This *Handbook* provides a direct procedure for the structural design of single-story, reinforced and unreinforced concrete masonry structures. The procedure is based on the strength design provisions of TMS 402-11/ACI 530-11/ASCE 5-11 *Building Code Requirements for Masonry Structures* and ASCE 7-10 *Minimum Design Loads for Buildings and Other Structures*. The document is applicable to both residential and commercial structures. So that users are required to do only minimal calculation, parameters are limited and design options are dictated. This *Handbook* applies to common structures over the vast majority of the United States including mapped ground snow loads up to 60 lb/ft^2 (2.9 kPa), mapped basic wind speeds up to 200 mph (89.4 m/s), mapped seismic 0.2 second spectral response accelerations up to 3.0g, and mapped seismic 1.0 second spectral response accelerations up to 1.25g. This *Handbook* was developed as a consensus standard and written in mandatory language so that it may form a part of a legally adopted building code as an alternative to standards that address a much broader range of masonry construction. This *Handbook* was written so that architects, engineers, contractors, building officials, researchers, educators, suppliers, manufacturers and others may use this *Handbook* in their practice for various purposes. Among the topics covered are reference standards, definitions and notations, site limitations, architectural limitations, loading limitations, material and construction requirements, direct design procedure, specifications, and details. The Commentary to this *Handbook* presents background analysis, details and committee considerations used to develop this *Handbook*. While not part of the legal requirements of this standardized *Handbook*, an Appendix providing an example of how to use the direct design procedure for a typical masonry building is provided following the Commentary.

Copyright © 2013, The Masonry Society.

All rights reserved including rights of reproduction and use in any form or by any means, including the making of copies by any photo process, or by any electronic or mechanical device, printed, written or oral, or recording for sound or visual reproduction or for any use in any knowledge retrieval system or device, unless permission in writing is obtained from The Masonry Society.

ISBN 978-1-929081-44-8
TMS Order No. TMS 403-13

Direct Design Handbook for Masonry Structures (TMS 403-13)

Developed by The Masonry Society's Design Practices Committee

Benchmark H. Harris, Chairman
Jason J. Thompson, Secretary

Voting Members[1]

Peter M. Babaian	Dennis W. Graber	J. Eric Peterson
Craig V. Baltimore	Charles A. Haynes	Kurtis K. Siggard
David T. Biggs	Richard E. Klingner	Christine A. Subasic
James A. Farny	W. Mark McGinley	John E. Swink
Andrew E. Geister	Raymond T. Miller	

Corresponding Members[2]

Subhash C. Anand	Edwin T. Huston	Malcolm E. Phipps
Jefferson W. Asher	Rochelle C. Jaffe	Sarah L. Rogers
Thomas B. Brady	Brian E. Johnson	Joseph E. Saliba
John M. Bufford	Eric N. Johnson	David G. Sommer
Charles B. Clark	Gregory R. Kingsley	Narendra Taly
Richard Filloramo	Michael D. Lewis	John G. Tawresey
Fernando Fonseca	John H. Matthys	Itzhak Tepper
Thomas A. Gangel	Michael C. Mota	Tyler W. Witthuhn
Thomas A. Hagood	Vilas Mujumdar	Terrence A. Weigel
Matthew Hamann	Javeed A. Munshi	David B. Woodham
Frederick A. Herget	Craig Parrino	Daniel Zechmeister

1. Voting members fully participate in Committee activities including responding to correspondence and voting on revisions to this document.
2. Corresponding members monitor Committee activities, but do not have voting privileges.

Technical Activities Committee

The Technical Activities Committee (TAC) of The Masonry Society (TMS) is responsible for reviewing, and approving the work of all TMS Technical Committees. As such, they reviewed the drafts of this *Direct Design Handbook for Masonry Structures*, and have approved revisions based on comments submitted by both TAC and the public. Members of the Technical Activities Committee who assisted with review, development, and approval of this *Handbook* are:

David I. McLean, TAC Chairman

Peter M. Babaian	Jams P. Mwangi	Jason J. Thompson
Sunup S. Mathew	John J. Myers	A. Rhett Whitlock
Darrell W. McMillian	Sarah L. Rogers	
Raymond T. Miller	Phillip J. Samblanet, Staff	

DIRECT DESIGN HANDBOOK FOR MASONRY STRUCTURES

Synopsis

This *Handbook* provides a direct procedure for the structural design of single-story concrete masonry structures. The procedure is based on the strength design provisions of TMS 402-11/ACI 530-11/ASCE 5-11 and the corresponding loading requirements of ASCE 7-10. It is written in such a form that it may be adopted by reference in a general building code.

Among the topics covered are reference standards, definitions and notations, site limitations, architectural limitations, loading limitations, material and construction requirements, direct design procedure, optional modifications to the direct design procedure, specifications, and details.

The Commentary to this Standard presents background analysis, details, and Committee considerations used to develop the Standard. While not part of the legal requirements of this Standard, an Appendix providing an example of how to use the direct design procedure for a typical masonry building is provided following the Commentary.

Keywords

Beams; building codes; cements; clay brick; compressive strength; concrete block; concrete masonry; construction; control joints; design handbook; detailing; direct design; grout; grouting; handbook; high wind; joints; loads (forces); masonry; masonry cement; masonry load-bearing walls; masonry mortars; masonry units; masonry walls; mortars; portland cement; portland cement-lime mortar; reinforced masonry; rebar; reinforcing steel; seismic requirements; shear strength; simplified design; specifications; splicing; stresses; strength design; structural analysis; structural design; unreinforced masonry; veneers; walls

Future Updates and Possible Errata

To access information on possible future updates or errata on this handbook, see http://www.masonrysociety.org/html/resources/directdesign/.

DIRECT DESIGN HANDBOOK FOR MASONRY STRUCTURES

Table of Contents

Direct Design Procedure Reference Guide .... Inside, Front Cover

Chapter 1 General 1
1.1 – Scope 1
1.2 – Standards cited 1
1.3 – Definitions 1
1.4 – Notations 2

Chapter 2 Limitations 5
2.1 – Site Conditions 5
2.2 – Architectural Conditions 7
2.3 – Loading Conditions 8
2.4 – Material and Construction Requirements 9

Chapter 3 Procedure 11
3.1 – General 11
3.2 – Direct Design Procedure 11

Chapter 4 Clay Masonry (Future) 59

Chapter 5 Specification 61

Chapter 6 Details 63

DIRECT DESIGN HANDBOOK FOR MASONRY STRUCTURES

Commentary Chapter 1　General **C-1**
C1.1 – Scope C-1
C1.2 – Standards cited C-2
C1.3 – Definitions C-2
C1.4 – Notations C-5
References C-5

Commentary Chapter 2　Limitations **C-7**
C2.1 – Site Conditions C-7
C2.2 – Architectural Conditions C-8
C2.3 – Loading Conditions C-13
C2.4 – Materials and Construction Requirements C-14
References C-17

Commentary Chapter 3　Procedure **C-19**
C3.1 – General C-19
C3.2 – Direct Design Procedure C-19
References C-35

Commentary Chapter 4　Clay Masonry (Future) **C-37**

Commentary Chapter 5　Specification **C-39**

Commentary Chapter 6　Details **C-41**

Appendix　　Direct Design Procedure Design Example **A-1**

DIRECT DESIGN HANDBOOK FOR MASONRY STRUCTURES

Chapter 1
General

1.1 – Scope

This *Direct Design Handbook for Masonry Structures*, herein referred to as the *Handbook*, provides a direct procedure for the structural design of single-story, concrete masonry structures. The procedure shall be permitted to be used to design concrete masonry subjected to factored combinations of dead, roof live, wind, seismic, snow and rain loads. The procedure outlined in this *Handbook* is based on the strength design provisions of the 2011 *Building Code Requirements for Masonry Structures* (TMS 402-11/ACI 530-11/ASCE 5-11) and the 2010 ASCE 7 *Minimum Design Loads for Buildings and Other Structures* (ASCE 7-10).

1.2 – Standards cited

Standards of the American Concrete Institute, the Structural Engineering Institute of the American Society of Civil Engineers, ASTM International, and The Masonry Society cited in this document are listed below with their serial designations, including year of adoption or revision, and are declared to be part of this *Handbook* as if fully set forth in this document.

TMS 402-11/ACI 530-11/ASCE 5-11 – *Building Code Requirements for Masonry Structures*, referenced herein as the MSJC Code

TMS 602-11/ACI 530.1-11/ASCE 6-11 – *Specification for Masonry Structures*, referenced herein as the MSJC Specification

ASCE 7-10 – *Minimum Design Loads for Buildings and Other Structures*, referenced herein as ASCE 7

ASTM A615/A615M-09 – *Standard Specification for Deformed and Plain Carbon-Steel Bars for Concrete Reinforcement*

ASTM A706/A706M-08a – *Standard Specification for Low-Alloy Steel Deformed and Plain Bars for Concrete Reinforcement*

ASTM A996/A996M-09 – *Standard Specification for Rail-Steel and Axle-Steel Deformed Bars for Concrete Reinforcement*

ASTM C90-08 – *Standard Specification for Loadbearing Concrete Masonry Units*

ASTM C270-08 – *Standard Specification for Mortar for Unit Masonry*

ASTM C476-09 – *Standard Specification for Grout for Masonry*

ASTM C1314-07 – *Standard Test Method for Compressive Strength of Masonry Prisms*

1.3 – Definitions

Terms defined in the MSJC Code and MSJC Specification shall apply to the design of masonry designed in accordance with this *Handbook*. The following terms, as used in this *Handbook*, shall have the following meaning:

B Bars – horizontal reinforcing bars located at the bottom of masonry beams over openings, mirrored at the diaphragm level over the openings.

C Bars – reinforcing bars located at the level at which the diaphragm is connected and used to resist diaphragm chord tension.

Designated shear wall segment – a portion of a wall that is continuous in plan, uninterrupted from the foundation to the diaphragm elevation, and is used to resist in-plane shear loads.

E Bars – vertical reinforcing bars located in the first cell adjacent to the edge of openings.

Exposure Category – Wind exposure category as defined in Section 26.7 of ASCE 7.

H Bars – horizontal reinforcing bars.

Header panel – the portion of masonry above an opening, but is not relied upon in the direct design method to resist in-plane shear loads.

J Bars – vertical reinforcing bars located at the jamb of an opening, required when H Bars are interrupted by an opening, and mirrored at the opposite end of the wall segment.

LFRS – Lateral force-resisting system.

Non-designated shear wall segment – a portion of a wall that is continuous in plan, uninterrupted from the foundation to the diaphragm elevation, but is not relied upon in the direct design procedure to resist in-plane shear loads.

Non-participating walls – walls that are isolated so that in-plane lateral forces are not imparted to these elements.

O Bars – horizontal reinforcing bars located in header panels above an opening or sill panels below an opening.

OPMSW – Ordinary plain (unreinforced) masonry shear wall.

ORMSW – Ordinary reinforced masonry shear wall.

Risk Category – Risk category of the structure determined in accordance with Table 1.5-1 of ASCE 7.

SDC – Seismic design category.

Sill panel – the portion of masonry below an opening, but is not relied upon in the direct design method to resist in-plane shear loads.

SRMSW – Special reinforced masonry shear wall.

T Bars – horizontal reinforcing bars located at the top of a parapet.

V Bars – vertical reinforcing bars.

1.4 – Notations

A = area of a building elevation perpendicular to the principal plan direction under consideration, ft² (m²).

C_s = seismic response coefficient as determined by Table 3.2-4.

C_w = wind response coefficient as determined by Table 3.2-3, lb/ft² (kPa).

d_s = depth of water on the undeflected roof up to the inlet of the secondary drainage system when the primary drainage system is blocked; also referred to as the static head, in. (mm).

d_h = additional depth of water on the undeflected roof above the inlet of the secondary drainage system at its design flow; also referred to as the hydraulic head, in. (mm).

g = acceleration due to gravity, ft/sec² (m/s²).

DIRECT DESIGN HANDBOOK FOR MASONRY STRUCTURES

h_{max} = maximum permitted height from the lateral bracing point at the foundation to the diaphragm attachment, ft (m).

k_1 = coefficient for calculating the length of shear-dominated shear walls, lb/ft (kN/m).

k_2 = coefficient for calculating the length of flexure-dominated shear walls, lb (kN).

L_1 = length of shear wall required for shear-dominated performance, ft (m).

L_2 = length of shear wall required for flexure-dominated performance, ft (m).

L_{joist} = maximum design span of roof joists bearing on a wall, ft (m).

L_{req} = minimum required total plan length of designated shear wall segments along each line of lateral resistance, ft (m).

L_{seg} = minimum plan length of designated shear wall segments on a particular line of resistance, ft (m).

p_g = ground snow load, determined in accordance with Figure 7-1 or Table 7-1 of ASCE 7, lb/ft^2 (kPa).

S_H = spacing of H Bars, in. (mm).

S_S = mapped risk-targeted maximum considered earthquake ground motion response acceleration, 5 percent damped, spectral response acceleration parameter at short periods, determined in accordance with Section 11.4.1 of ASCE 7.

S_V = spacing of V Bars, in. (mm).

S_{V1} = spacing of vertical reinforcement as governed by required resistance to out-of-plane wind loads, in. (mm).

S_{V2} = spacing of vertical reinforcement as governed by required resistance to out-of-plane seismic loads, in. (mm).

S_1 = mapped risk-targeted maximum considered earthquake ground motion response acceleration, 5 percent damped, spectral response acceleration parameter at a period of 1 second, determined in accordance with Section 11.4.1 of ASCE 7.

V = 3-second gust basic wind speed, determined in accordance with Figure 26.5-1 of ASCE 7, miles per hour (kilometers per hour).

$V_{LFRS\text{-}seismic}$ = factored lateral seismic force acting on the lateral force-resisting system, lb (kN).

$V_{LFRS\text{-}wind}$ = factored lateral wind force acting on the lateral force-resisting system, lb (kN).

V_{LFRS} = the larger of $V_{LFRS\text{-}seismic}$ and $V_{LFRS\text{-}wind}$, lb (kN).

W = effective seismic weight resisted by each line of resistance, lb (kN).

W_{tot} = total effective seismic weight of the structure, lb (kN).

DIRECT DESIGN HANDBOOK FOR MASONRY STRUCTURES

This Page Intentionally Left Blank

Chapter 2
Limitations

If the limitations specified in Chapter 2 are satisfied, use of this *Handbook* to perform the structural design of concrete masonry buildings shall be permitted.

Any segment, member, or portion of a masonry structure that does not meet the limitations of Chapter 2 shall be designed in accordance with the legally adopted building code provided each of the following conditions are met:

a) The strength and stiffness compatibility between the elements designed in accordance with the legally adopted building code and the masonry designed in accordance with this *Handbook* is verified;

b) The load path through the masonry designed in accordance with this *Handbook* is not interrupted; and

c) Loads are not transferred from the elements designed in accordance with the legally adopted building code into the masonry designed in accordance with this *Handbook*.

2.1– Site Conditions

2.1.1 *Ground Snow Load* – The ground snow load, p_g, as given in ASCE 7 Figure 7-1 or ASCE 7 Table 7-1, shall not exceed 60 lb/ft^2 (2.9 kPa).

2.1.2 *Basic Wind Speed* – The basic wind speed (3-second gust), V, as given in Figure 26.5-1 of ASCE 7, shall not exceed 200 mph (89.4 m/s).

2.1.3 *Exposure Category* – The exposure category, as defined in Section 26.7 of ASCE 7, shall be Exposure Category B or C.

2.1.4 *Topography* – The location of the structure shall comply with Section 2.1.4.1 or the basic wind speed shall be modified in accordance with Section 2.1.4.2.

2.1.4.1 The structure shall not be located in the upper one-half of a hill or ridge or near the crest of an escarpment whose height is greater than 60 ft (18.3 m) for Exposure Category B conditions nor greater than 15 ft (4.6 m) for Exposure Category C conditions.

2.1.4.2 The basic wind speed (3-second gust), V, shall be determined in accordance with Table 2.1-1.

2.1.5 *Mapped Spectral Acceleration for Short Periods* – The mapped spectral acceleration for short period structures, S_S, determined in accordance with Section 11.4.1 of ASCE 7, shall not exceed 3.0g.

2.1.6 *Mapped Spectral Acceleration for 1-Second Periods* – The mapped spectral acceleration for 1-second period structures, S_1, determined in accordance with Section 11.4.1 of ASCE 7, shall not exceed 1.25g.

2.1.7 *Site Class* – The Site Class, determined in accordance with Chapter 20 of ASCE 7, shall be A, B, C, or D.

Table 2.1-1 – Basic Wind Speed Modification for Topographic Wind Speed Up Effect [a, b]

Topographic Feature	Basic Wind Speed from ASCE 7 Figure 26.5-1	Average Slope of the Top Half of the Topographic Feature:						
		0.10	0.125	0.15	0.175	0.20	0.23	0.25 or greater
		Required Basic Wind Speed, Modified for Topographic Wind Speed Up:						
Ridge	110	142	150	158	167	174	182	190
	115	149	157	165	174	182	190	198
	120	155	164	172	182	190	198	207
	130	168	177	186	197	206	215	NP
	140	181	191	201	212	NP	NP	NP
	150	194	204	NP	NP	NP	NP	NP
	160	207	NP	NP	NP	NP	NP	NP
Escarpment	110	129	134	139	143	148	152	158
	115	135	140	145	150	155	159	165
	120	141	146	152	156	161	166	172
	130	153	158	164	169	175	180	186
	140	164	170	177	182	188	194	201
	150	176	182	189	195	201	207	NP
	160	188	194	202	208	NP	NP	NP
	180	211	218	NP	NP	NP	NP	NP
Hill	110	134	139	146	151	157	162	169
	115	140	145	152	158	164	170	176
	120	146	152	159	165	171	177	184
	130	158	164	172	179	185	192	199
	140	170	177	185	192	199	206	215
	150	182	189	198	206	213	NP	NP
	160	194	202	212	NP	NP	NP	NP
	180	218	NP	NP	NP	NP	NP	NP

[a] Table values shall be permitted to be interpolated.
[b] The provisions of this *Handbook* shall not be used where the modified basic wind speed exceeds 200 mph (89.4 m/s). Values of the modified basic wind speed greater than 200 mph (89.4 m/s) are shown only to aid in interpolation.

2.2 – Architectural Conditions

2.2.1 *Number of Stories* – The building shall be a one-story structure.

2.2.2 *Walls* – All masonry walls shall be single-wythe, constructed of concrete masonry units having a nominal thickness of 8 in. (203 mm). Walls shall be provided in at least two lines in each of the two principal, perpendicular plan directions.

2.2.3 *Height* – Tops of walls, parapets and roof peaks shall not be higher than 30 ft (9.1 m) above the finished grade elevation adjacent to the structure. The vertical span of walls, measured between points of lateral bracing, shall not be less than 4 ft (1.2 m) and shall not exceed 30 ft (9.1 m).

2.2.4 *Parapets* – The cantilevered height of parapets, measured from diaphragm attachment to the top of the parapet, shall not exceed 4 ft (1.2 m). For walls with a vertical span less than 12 ft (3.7 m), the height of parapets shall not exceed the vertical span of the wall divided by 3. Parapets shall not be supported by unreinforced masonry walls. A single T Bar shall be provided in the top course of parapets.

2.2.5 *Openings* – Openings in masonry walls designed in accordance with this *Handbook* shall meet the requirements of Section 2.2.5.1 or 2.2.5.2.

2.2.5.1 Openings in walls shall be rectangular. The sides of openings shall be oriented vertically. Vertically spanning wall segments between openings and horizontally spanning wall segments above and below openings shall be designed in accordance with Chapter 3. The location of openings in plan shall not overlap. Control joints shall be provided on both sides of each opening in a masonry element designed in accordance with this *Handbook*.

2.2.5.2 Openings shall not exceed 6 in. (152 mm) in any dimension at the face of the wall and shall not interrupt reinforcement required by Chapter 3. The cumulative area of openings shall not exceed 144 in.2 (0.093 m2) in any 10 ft^2 (0.93 m^2) of wall surface area.

2.2.6 *Roof Diaphragms* – The roof system shall consist of one or more flexible diaphragms and shall meet the following requirements:

- The maximum plan dimension of a single diaphragm shall not exceed 200 ft (61.0 m).
- Each roof diaphragm shall be rectangular in plan dimensions.
- The larger plan dimension of a diaphragm shall not exceed four times the shorter plan dimension of the diaphragm.
- A roof area shall not be designated as a diaphragm unless the area is surrounded by, and connected to, masonry walls along all four sides of the diaphragm. Anchorage design shall be in accordance with the MSJC Code to transfer the required forces from the diaphragm into the walls and from the walls into the diaphragm.
- Roof diaphragms shall be anchored to masonry walls at a location that coincides with a reinforced masonry bond beam.
- A designated shear wall segment shall be provided at each corner of a diaphragm in both principal plan directions.
- The maximum area of each opening in the roof diaphragm shall not exceed ten percent of the gross roof diaphragm area.
- The maximum dimension of each opening in the roof diaphragm shall not exceed ten percent of the smaller plan dimension of the roof diaphragm.
- The sum of the opening areas in a roof diaphragm shall not exceed 20 percent of the gross roof diaphragm area.

2.2.7 *Roof Slope* – The slope of the finished roof surface shall not be less than $^1/_4$ in./ft (20.8 mm/m) and shall not exceed 12 in./ft (1,000 mm/m). Where the ground snow load is greater than 25 lb/ft^2 (1.2 kPa), the roof shall be designed to shed snow and shall not have a curved, multiple folded plate, sawtooth, barrel vault, or dome configuration.

2.2.8 *Changes in Diaphragm Elevation* – It shall not be permitted to step a roof diaphragm within the perimeter of the diaphragm. The difference in elevation between adjacent diaphragms plus the height of the upper parapet shall not create a projection above the lower roof that exceeds the maximum permitted parapet height of 4 ft (1.2 m). The projection shall be measured vertically from the top-of-roof on the lower diaphragm to the top-of-masonry on the adjacent diaphragm. Where the ground snow load does not exceed 25 lb/ft^2 (1.2 kPa), non-slippery upper roofs with a slope greater than 2:12 that could shed snow on a lower roof and slippery roofs with a slope greater than 1/4:12 that could shed snow on a lower roof shall have a maximum horizontal distance from the eave to the ridge of 42 ft. Where the ground snow load exceeds 25 lb/ft^2 (1.2 kPa), the upper roof shall not be permitted to shed snow on a lower roof.

2.2.9 *Joists* – The spacing of roof joists shall not exceed 6 ft (1.8 m). Joist span lengths, from the centerline of support to the opposing centerline of support, shall not exceed 60 ft (18.3 m). Joists shall not support a tributary area greater than the span length multiplied by the joist spacing. Masonry shall not be used to support reactions from tributary areas greater than that supported by a joist.

2.2.10 *Roof Drainage* – The elevation of all secondary drainage inlets shall be established so that the sum of *ds* and *dh*, as defined in Section 8.1 of ASCE 7, does not exceed 5.5 in. (139 mm).

2.2.11 *Isolation of Features* – Features such as canopies, signs, and overhangs shall be structurally isolated from the masonry. Wall finishes shall be permitted to be attached, anchored, or adhered to the masonry wall system provided that:

- the finish does not extend farther than 6 in. (152 mm) from the face of the masonry; and
- the average installed weight of all finishes, excluding masonry veneers complying with Section 2.2.13, does not exceed 3 lb/ft^2 (0.14 kPa) over the area to which it is attached.

The weight of all installed finishes shall be included in the effective seismic weight, W.

2.2.12 *Simplified Wind Load Procedure Limitations* – The limitations of Section 28.6.2 of ASCE 7 shall be met.

2.2.13 *Veneers* – Masonry veneers shall comply with the requirements of Chapter 6 of the MSJC Code and shall have an installed weight that does not exceed 35 lb/ft^2 (1.68 kPa).

2.3 – Loading Conditions

2.3.1 *Load Types* – The provisions outlined in this *Handbook* shall not be applied to the design of structures that include design loads other than roof dead loads, roof live loads, snow loads, wind loads, seismic loads, and rain loads, except as permitted by Section 2.3.1.1.

2.3.1.1 *Soil Loads* – Portions of masonry that extend below grade and are subjected to balanced lateral loading conditions shall be permitted.

2.3.2 *Roof Dead Load* – The roof dead load shall not be less than 2 lb/ft^2 (0.1 kPa) and shall not exceed 30 lb/ft^2 (1.4 kPa).

2.3.3 *Roof Live Load* – The roof live load, as determined in accordance with Section 4.8 of ASCE 7, shall not exceed 20 lb/ft^2 (1.0 kPa).

2.3.4 *Eccentricity of Roof Loads* – The maximum eccentricity of applied roof loads shall not exceed 1.25 in. (31.8 mm). Eccentrically applied roof loads from opposite sides of interior load-bearing walls shall not be applied to the same half of the wall cross-section.

2.4 –Material and Construction Requirements

2.4.1 *Units* – Concrete masonry units having a nominal thickness of 8 in. (203 mm) and complying with ASTM C90 shall be used. The units shall be laid in running bond construction. The density of the concrete masonry units shall not exceed 135 lb/ft^2 (2,162 kg/m^3).

2.4.2 *Mortar* – Type S Mortar complying with ASTM C 270 and Table 2.4-1 shall be used. The specified mortar joint thickness shall be $^3/_8$ in. (9.5 mm).

Table 2.4-1 – Permitted Mortar for Seismic Design Category

SDC A, B, and C	SDC D, E, and F
Masonry cement mortar	Mortar cement mortar
Mortar cement mortar	Non-air-entrained portland cement-lime mortar
Portland cement-lime mortar	

2.4.3 *Reinforcement* – Unless designing unreinforced masonry, the size of reinforcement, including dowels but excluding bed joint reinforcement and lintel stirrups, shall be No. 5 (M#16), Grade 60 (420 MPa) reinforcement. The spacing of No. 5 (M#16) bars shall not be less than 16 in. (406 mm) and shall not exceed 10 ft (3.0 m). Vertical reinforcement shall extend the full height of the masonry element in which it is required, less cover distances. Horizontal reinforcement shall extend the full length of the masonry element in which it is required, less cover distances. Shear stirrups used in the construction of lintels shall be No. 3 (M#10), Grade 60 (420 MPa) reinforcement. Stirrups shall terminate with either a 90 degree or 135 degree hook having a minimum extension of 2.5 in. (64 mm). The spacing of lintel stirrups shall not exceed 8 in. (203 mm).

2.4.4 *Grout* – Grout shall comply with ASTM C476. Walls shall be fully grouted if the spacing of either the vertical reinforcement or the horizontal bond beam reinforcement is 16 in. (406 mm). Walls shall be partially grouted if the spacing of the vertical reinforcement and the horizontal bond beam reinforcement are both greater than 16 in. (406 mm).

2.4.5 *Specifications* – The construction documents shall contain specifications that meet or exceed the requirements of Chapter 5.

2.4.6 *Details* – The detailing requirements of Chapter 6 appropriate to the project shall be provided in the contract documents.

Chapter 3
Procedure

3.1 – General

Design of masonry in accordance with this *Handbook* shall comply with the procedure given in Chapter 3. In each step, the design variables in boxes shall be determined by the user. Table references shown in shaded highlight denote tables containing design information. Tabulated values shall not be interpolated unless explicitly permitted by a footnote to the table.

Walls designed in accordance with this *Handbook* shall be configured and detailed in accordance with the tables of Chapter 3 and, when reinforcement is required by those tables, reinforcement shall be detailed as illustrated in Figure 3.1-1.

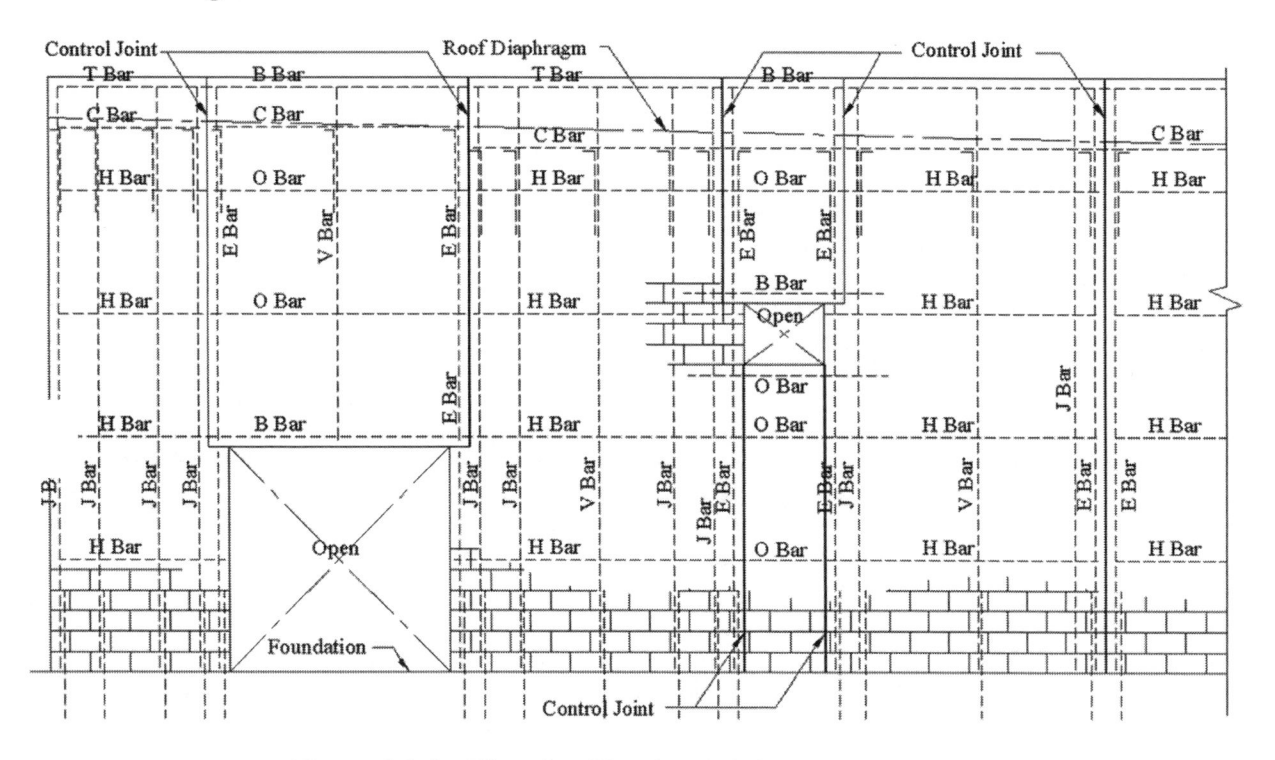

Figure 3.1-1 – Elevation Showing Reinforcement Designation

3.2 – Direct Design Procedure

1. Using ASCE 7, the following design requirements shall be determined:

    1A The Risk Category, based on ASCE 7 Table 1.5-1.

    1B The Ground Snow Load, p_g, using ASCE 7 Figure 7-1 or ASCE 7 Table 7-1.

    1C The 3-Second Gust Basic Wind Speed, V, using ASCE 7 Figure 26.5-1.

    1D The Exposure Category, using Section 26.7 of ASCE 7.

1E The mapped spectral acceleration for short periods, S_S, in units of %g, by dividing the values on ASCE 7 Figure 22-1, 22-3, 22-5, or 22-6 by 100.

1F The mapped spectral acceleration for a 1-second period, S_1, in units of %g, by dividing the values on ASCE 7 Figure 22-2, 22-4, 22-5, or 22-6 by 100.

2. If the site-specific condition limitations of Section 2.1 of this *Handbook* are not met, the procedure in this *Handbook* shall not be used. If Section 2.1 is satisfied, a building configuration that satisfies the architectural, loading, and construction conditions of Chapter 2 shall be developed.

3. Based on S_S and S_1, the Seismic Design Category, SDC, shall be determined using:
   - Table 3.2-1(1) for Site Class A;
   - Table 3.2-1(2) for Site Class B;
   - Table 3.2-1(3) for Site Class C; or
   - Table 3.2-1(4) for site Class D.

4. Using Table 3.2-2 of this *Handbook*, and based on the SDC, determine which Lateral Force Resisting Systems, or LFRS options, are permitted. In each principal plan direction of each rectangular roof diaphragm of the building, a single LFRS option must be selected.

5. In each principal plan direction of the building, divide the roof plan into rectangular roof diaphragms with masonry wall lines on each side. The designation of diaphragms shall be permitted to be different in each principal plan direction. Walls inside the perimeter of a diaphragm are permitted to be connected to that diaphragm only if they are perpendicular to the plan direction under consideration or if they are detailed as non-participating walls. In each principal plan direction, and for each diaphragm designated for that direction, complete the following steps:

   5A Calculate the area of the projected building elevation perpendicular to the principal plan direction under consideration, A, in units of square feet (square meters). This area is the product of the plan dimension of the diaphragm perpendicular to the direction under consideration, and the average vertical dimension to the top of the masonry walls defining that diaphragm and oriented perpendicular to the direction under consideration or the top of the roof diaphragm, whichever is greater.

   5B Using Table 3.2-3 of this *Handbook*, and based on V from Step 1, obtain the Wind Response Coefficient, C_W.

   5C Multiply C_W by A to calculate the lateral force on each line of lateral resistance in the Lateral Force Resisting System due to wind, $V_{LFRS-wind}$.

   5D Calculate the effective seismic weight, W, resisted by each line of lateral resistance.

   5E With S_S and S_1 from Step 1, the Seismic Design Category from Step 3, and an assumed LFRS from the LFRS Options, use Table 3.2-4 of this *Handbook* to obtain the corresponding Seismic Response Coefficient, C_S. Table 3.2-4(1) shall be used for Risk Category I or II structures. Table 3.2-4(2) shall be used for Risk Category III structures. Tables 3.2-4(3) shall be used for Risk Category IV structures.

This direct design procedure permits the following LFRS Options, listed in order of increasing prescriptive reinforcement along with the value of the seismic response modification coefficient, R, corresponding to each option.

Ordinary Plain (Unreinforced) Masonry Shear Walls	($R = 1.5$)
Ordinary Reinforced Masonry Shear Walls	($R = 2.0$)
Special Reinforced Masonry Shear Walls	($R = 5.0$)

5F   Multiply C_S by W to calculate the lateral force on each of the two lines of resistance in the Lateral Force Resisting System due to seismic loads, $V_{LFRS\text{-}seismic}$.

5G   Determine the governing lateral load on the Lateral Force Resisting System, V_{LFRS}, as the greatest of $V_{LFRS\text{-}wind}$ and $V_{LFRS\text{-}seismic}$.

5H   In the following sub-steps, determine the maximum permitted spacing of the V Bars for each wall, as governed by out-of-plane wind and seismic loads. Unreinforced masonry is permitted only if a special mechanical anchorage system is provided to resist all net axial uplift. The mechanical anchorage system is not permitted to transfer force to the masonry wall. Unreinforced masonry walls are not permitted to have parapets.

       5H.1   For each wall line that is parallel to the direction under consideration, the maximum span of the joists bearing on that wall line, L_{joist}, and the maximum wall height, h_{max} shall be determined. For walls with no joist bearing, if the first parallel joist is located such that the roof deck cantilevers past the joist to resist wind uplift, L_{joist} shall be permitted to be considered equal to 0 ft (0 m).

       5H.2   Using the appropriate Table 3.2-5 for the project conditions, the maximum permitted spacing of vertical reinforcement as governed by out-of-plane wind loads, S_{V1}, shall be determined. The largest spacing of the vertical reinforcement that permits a vertical span greater than or equal to the actual vertical span of the wall under consideration shall be the maximum permitted spacing of the V Bars as governed by out-of-plane wind load, S_{V1}. Horizontal H Bars shall be spaced at a maximum of 120 in. (3,048 mm) on center where vertical reinforcement is used.

       5H.3   Using the appropriate Table 3.2-6, for the project conditions, the maximum permitted spacing of vertical reinforcement as governed by out-of-plane seismic load, S_{V2}, shall be determined. The largest spacing of the vertical reinforcement that permits a vertical span greater than or equal to the actual vertical span of the wall under consideration shall be the maximum permitted spacing of the V Bars as governed by out-of-plane seismic load, S_{V2}. Horizontal H Bars shall be spaced at a maximum of 120 in. (3,048 mm) on center where vertical reinforcement is used.

5I   In the following sub-steps, select the spacing of the V Bars and of the H Bars for each line of resistance against in-plane seismic and wind loads.

       5I.1   Using Table 3.2-7, determine the applicable Lateral Force Coefficients from Table 3.2-8. This determination is based on V_{wind} and the largest value of L_{joist} on the line of resistance. Assume interior conditions, unless the entire plan length of the designated shear walls along the line of resistance is exterior.

5I.2   Pick an appropriate value for L_{seg}. The minimum length of shear wall segments shall not be less than 2 ft (0.61 m).

5I.3   The resistance coefficients k_1 and k_2 shall be selected from Table 3.2-8 for the selected shear wall segment length, L_{seg}, comprising the line of resistance. For shear wall segment lengths not shown in Table 3.2-8, the value for the resistance coefficient k_2 shall be determined using a tabulated shear wall segment length less than the designated shear wall segment length.

Select a value of S_V less than or equal to the smaller of S_{V1} and S_{V2} that is permitted in Tables 3.2-5 and 3.2-6.

For the selected LFRS Option and shear wall segment length, L_{seg}, select a pair of vertical and horizontal spacing values, S_V and S_H, that appear in Table 3.2-8. If a particular combination of spacing values does not appear in Table 3.2-8, that combination is not permitted.

5I.4   Based on the above information, determine the required length of designated wall segments on each of the two lines of resistance for each diaphragm, L_{req}, as the greatest of L_1 and L_2.

$$L_1 = V_{LFRS} / k_1$$

$$L_2 = V_{LFRS} \, h_{max} / k_2$$

5I.5   Each line of resistance shall have a total length of designated shear wall segments at least equal to L_{req}. Each line of resistance shall be permitted to include non-designated shear walls and non-participating walls.

5J   For non-designated shear wall segments the spacing of the vertical and horizontal reinforcement and the grouting schedule shall be the same as for the designated shear wall segments on that line of resistance.

5K   For non-participating walls, the maximum permitted spacing of vertical reinforcement shall not exceed the smaller of S_{V1} and S_{V2}.

5L   Using Table 3.2-9, based upon the value of V_{LFRS} determined in Step 5G, the required number of horizontal bond beam courses shall be determined. Each bond beam course required by Table 3.2-9 shall contain two No. 5 (M#16) C Bars. The first bond beam course shall be located at the level where the diaphragm is connected. Where additional bond beam courses are required by Table 3.2-9, they shall be placed in subsequent courses below the first bond beam. The same number of bond-beam courses shall be provided on all four sides of the diaphragm. C Bars shall be provided in all masonry walls that are part of the lateral force-resisting system.

When more than one diaphragm is connected to a common shear wall, the required number of C Bars shall be the largest of the values determined considering each diaphragm independently.

6. Along a wall between two diaphragms, the sum of the lengths of all designated shear wall segments shall equal or exceed ($L_{req1} + L_{req2}$), where L_{req1} and L_{req2} are the required segment lengths associated with each diaphragm. If the two diaphragms have different heights, L_{req1} and L_{req2} shall be provided along the plan lengths of wall defining diaphragm 1 and diaphragm 2 respectively.

7. Panels of masonry above and below openings are termed header panels and sill panels respectively. On both sides of those panels, provide additional J Bars from the foundation to the C Bars. The number of J Bars on each side shall be at least equal to the one-half the plan length of the opening (units of in.) divided by the lesser of the required S_V for the two wall segments adjacent to the opening. A single E Bar shall be located in the cell closest to the opening. E Bars shall terminate with a standard 90-degree hook embedded in the bond-beam course containing the C Bars or shall extend vertically past the bond beam a minimum of 26 in. (660 mm) when permitted by the presence of a parapet of sufficient size. E Bars need not be continuous through the beam above the opening. In the wall segments on each side of the header panel, place J Bars in the cell adjacent to the E Bar, then in the next eligible cell, and so on, until all required J Bars are placed. There shall not be more than one vertical bar in each cell. The spacing of V Bars and J Bars shall not be closer than 16 in. (406 mm) on center. It shall be permitted to place V Bars and J Bars within 8 in. (203 mm) of E Bars. Ensure that the wall segment on each side of the opening is long enough in plan to accommodate the required bars. Not including E Bars, the vertical reinforcement pattern shall be symmetrical about the centerline of the plan length of the wall segment, which will require additional bars on the opposite end of the wall unless there are openings of equal plan length on each end of the wall.

To determine the permissible h_{max} for jamb strips of unreinforced masonry wall segments with openings, start with the maximum h_{max} permitted by this Procedure for an otherwise identical wall segment without openings, and then divide that maximum permitted h_{max} by the square root of the quotient (W_T/W_S) for the jamb strip. The values of W_T and W_S shall be as defined in Section 5.5.1 of the MSJC Code.

8. E Bars shall be provided in header panels in the vertical cells adjacent to the control joints defining the panel. In cases where the lintel bearing length adjacent to openings is greater than 8 in. (203 mm), only a single E Bar shall be required in the cell closest to the opening. For each line of resistance, header panels with parapets shall be vertically reinforced at the V Bar spacing determined in accordance with Step 5H. Parapets shall not be permitted to be supported by unreinforced masonry designed in accordance with this *Handbook*. Reinforcement in header panels shall also satisfy the prescriptive seismic requirements of Chapter 6.

9. At header panels and sill panels, provide the following additional reinforcement:

    9A At each end of header panels and sill panels, provide horizontal reinforcement crossing the control joint, de-bonded on one side of the control joint as required in Chapter 6.

    9B Determine which of Tables 3.2-10(1) to Table 3.2-10(8) is the appropriate Table 3.2-10, based on comparing p_g from Step 1 to the design p_g in the table headings. For header panels, use Table 3.2-10 to determine the minimum number of grouted courses required at the bottom of the panel, and provide the bottom course with two No. 5 (M#16) B Bars. Provide the same number of grouted courses, identically reinforced, just below the elevation of joist bearing or diaphragm attachment. These courses need not be in addition to courses of C Bars. Alternatively, these upper courses of B Bars may occur at the top of a parapet. If the depth of the header panel does not permit the required number of courses below and above, it is permitted to provide a total number of courses equal to the minimum required from Table 3.2-10, provided that all required C Bars are provided.

For unreinforced masonry walls, the same requirements apply for grouted courses and B Bars.

9C. Table 3.2-11(1) and Table 3.2-11(2) shall be used to determine the maximum permitted spacing of O Bars for header panels and sill panels as the lesser of the required reinforcement spacing as governed by out-of-plane wind and seismic loading.

O Bars shall be required for header panels, but sill panels shall be permitted to be unreinforced.

10. Provide the following information on the plans with values that do not exceed the values below. This information and loading criteria form the basis for this direct design procedure for masonry structures. If other systems are designed with lower values for any of the following criteria, where permitted by ASCE 7 and the legally adopted building code, then those values shall be provided on the plans instead of the following values.

Roof Live Load = 20 lb/ft^2 (1.0 kPa)

Roof Snow Load Data:

Snow Exposure Factor = 1.2
Thermal Factor = 1.2

For p_g = 0 lb/ft^2 (0 kPa):
Roof Snow Load = 0 lb/ft^2 (0 kPa)
For $0 < p_g \leq 20$ lb/ft^2 ($0 < p_g \leq 1.0$ kPa):
Roof Snow Load = 52 lb/ft^2 (2.5 kPa)
For $20 < p_g \leq 40$ lb/ft^2 ($1.0 < p_g \leq 1.9$ kPa):
Roof Snow Load = 65 lb/ft^2 (3.1 kPa)
For $40 < p_g \leq 60$ lb/ft^2 ($1.9 < p_g \leq 2.9$ kPa):
Roof Snow Load = 76 lb/ft^2 (3.6 kPa)

For Risk Categories I and II:
Snow Load Importance Factor = 1.0
For Risk Categories III and IV:
Snow Load Importance Factor = 1.2

Wind Design Data:
Basic Wind Speed = V from Step 1
Wind Importance Factor =
1.0 for Risk Categories I and II
1.15 for Risk Categories III and IV
Wind Exposure Category: Exposure Category from Step 1.
Applicable Internal Pressure Coefficient: +/- 0.18

Earthquake Design Data:
Seismic Importance Factor =
1.0 for Risk Category I and II Structures
1.25 for Risk Category III Structures
1.5 for Risk Category IV Structures
Mapped 0.2 Second Spectral Response Acceleration = S_S

Mapped 1.0 Second Spectral Response Acceleration = S_1
Site Class:

Seismic Design Category: SDC
Basic Seismic-Force-Resisting-System: LFRS

Design Base Seismic Shear:
   The total base shear shall be calculated as:
   (2) $(W_{tot})(C_s)$
Seismic Response Coefficient = $(2)C_S$
Response Modification Factor = R
Analysis Procedure Used: Equivalent Lateral Force Method

Flood Design Data:
   This building has not been designed for flood loads.

Special Loads:
   This building has not been designed for any special loads.

11. For Specifications, reference the MSJC *Specification*, and use Chapter 5 to make the choices in the Mandatory Requirements Checklist of that document.

12. Provide the standard details from Chapter 6 on the construction documents.

Table 3.2-1(1): Seismic Design Category (SDC) for Site Class A

S_S	S_1	Seismic Design Category	
		Risk Categories I, II, III	Risk Category IV
$S_S < 0.314$	$S_1 < 0.126$	A	A
	$0.126 \leq S_1 < 0.250$	B	C
	$0.250 \leq S_1 < 0.375$	C	D
	$0.375 \leq S_1 < 0.750$	D	D
	$0.75 \leq S_1 \leq 1.250$	E	F
$0.314 \leq S_S < 0.619$	$S_1 < 0.250$	B	C
	$0.250 \leq S_1 < 0.375$	C	D
	$0.375 \leq S_1 < 0.750$	D	D
	$0.75 \leq S_1 \leq 1.250$	E	F
$0.619 \leq S_S < 0.938$	$S_1 < 0.375$	C	D
	$0.375 \leq S_1 < 0.750$	D	D
	$0.75 \leq S_1 \leq 1.250$	E	F
$0.938 \leq S_S \leq 3.00$	$S_1 < 0.750$	D	D
	$0.75 \leq S_1 \leq 1.250$	E	F

Table 3.2-1(2): Seismic Design Category (SDC) for Site Class B

S_S	S_1	Seismic Design Category Risk Categories I, II, III	Seismic Design Category Risk Category IV
$S_S < 0.250$	$S_1 < 0.101$	A	A
	$0.101 \leq S_1 < 0.200$	B	C
	$0.200 \leq S_1 < 0.300$	C	D
	$0.300 \leq S_1 < 0.750$	D	D
	$0.75 \leq S_1 \leq 1.250$	E	F
$0.250 \leq S_S < 0.495$	$S_1 < 0.200$	B	C
	$0.200 \leq S_1 < 0.300$	C	D
	$0.300 \leq S_1 < 0.750$	D	D
	$0.75 \leq S_1 \leq 1.250$	E	F
$0.495 \leq S_S < 0.7508$	$S_1 < 0.300$	C	D
	$0.300 \leq S_1 < 0.750$	D	D
	$0.75 \leq S_1 \leq 1.250$	E	F
$0.750 \leq S_S \leq 3.00$	$S_1 < 0.750$	D	D
	$0.75 \leq S_1 \leq 1.250$	E	F

Table 3.2-1(3): Seismic Design Category (SDC) for Site Class C

S_S	S_1	Seismic Design Category Risk Categories I, II, III	Seismic Design Category Risk Category IV
$S_S < 0.209$	$S_1 < 0.060$	A	A
	$0.060 \leq S_1 < 0.119$	B	C
	$0.119 \leq S_1 < 0.186$	C	D
	$0.186 \leq S_1 < 0.750$	D	D
	$0.75 \leq S_1 \leq 1.250$	E	F
$0.209 \leq S_S < 0.413$	$S_1 < 0.119$	B	C
	$0.119 \leq S_1 < 0.186$	C	D
	$0.186 \leq S_1 < 0.750$	D	D
	$0.75 \leq S_1 \leq 1.250$	E	F
$0.413 \leq S_S < 0.661$	$S_1 < 0.186$	C	D
	$0.186 \leq S_1 < 0.750$	D	D
	$0.75 \leq S_1 \leq 1.250$	E	F
$0.661 \leq S_S \leq 3.00$	$S_1 < 0.750$	D	D
	$0.75 \leq S_1 \leq 1.250$	E	F

DIRECT DESIGN HANDBOOK FOR MASONRY STRUCTURES

Table 3.2-1(4): Seismic Design Category (SDC) for Site Class D

S_S	S_1	Seismic Design Category Risk Categories I, II, III	Seismic Design Category Risk Category IV
$S_S < 0.156$	$S_1 < 0.041$	A	A
	$0.041 \leq S_1 < 0.083$	B	C
	$0.083 \leq S_1 < 0.132$	C	D
	$0.132 \leq S_1 < 0.750$	D	D
	$0.75 \leq S_1 \leq 1.250$	E	F
$0.156 \leq S_S < 0.321$	$S_1 < 0.083$	B	C
	$0.083 \leq S_1 < 0.132$	C	D
	$0.132 \leq S_1 < 0.750$	D	D
	$0.75 \leq S_1 \leq 1.250$	E	F
$0.321 \leq S_S < 0.553$	$S_1 < 0.132$	C	D
	$0.132 \leq S_1 < 0.750$	D	D
	$0.75 \leq S_1 \leq 1.250$	E	F
$0.553 \leq S_S \leq 3.00$	$S_1 < 0.750$	D	D
	$0.75 \leq S_1 \leq 1.250$	E	F

Table 3.2-2: LFRS Options

Seismic Design Category (SDC)	Lateral Force Resisting System Options
A or B	Ordinary Plain (Unreinforced) Masonry Shear Walls (OPMSW) Ordinary Reinforced Masonry Shear Walls (ORMSW) Special Reinforced Masonry Shear Walls (SRMSW)
C	Ordinary Reinforced Masonry Shear Walls (ORMSW) Special Reinforced Masonry Shear Walls (SRMSW)
D, E, or F	Special Reinforced Masonry Shear Walls (SRMSW)

Table 3.2-3: C_w Values

Basic Wind Speed, V	Exposure Category	C_w
$V \leq 110$ mph	B	7.1 lb/ft² (0.34 kPa)
($V \leq 49.2$ m/s)	C	9.9 lb/ft² (0.48 kPa)
$110 < V \leq 115$ mph	B	7.8 lb/ft² (0.37 kPa)
($177 < V \leq 51.4$ m/s)	C	10.9 lb/ft² (0.52 kPa)
$115 < V \leq 100$ mph	B	8.4 lb/ft² (0.40 kPa)
($185 < V \leq 53.6$ m/s)	C	11.8 lb/ft² (0.57 kPa)
$120 < V \leq 130$ mph	B	9.9 lb/ft² (0.47 kPa)
($193 < V \leq 58.1$ m/s)	C	13.8 lb/ft² (0.66 kPa)
$130 < V \leq 140$ mph	B	11.5 lb/ft² (0.55 kPa)
($209 < V \leq 62.5$ m/s)	C	16.1 lb/ft² (0.77 kPa)
$140 < V \leq 150$ mph	B	13.2 lb/ft² (0.63 kPa)
($225 < V \leq 66.9$ m/s)	C	18.4 lb/ft² (0.88 kPa)
$150 < V \leq 160$ mph	B	15.0 lb/ft² (0.72 kPa)
($241 < V \leq 71.7$ m/s)	C	21.0 lb/ft² (1.01 kPa)
$160 < V \leq 180$ mph	B	19.0 lb/ft² (0.91 kPa)
($258 < V \leq 80.6$ m/s)	C	26.6 lb/ft² (1.27 kPa)
$180 < V \leq 200$ mph	B	23.4 lb/ft² (1.12 kPa)
($290 < V \leq 89.4$ m/s)	C	32.8 lb/ft² (1.57 kPa)

DIRECT DESIGN HANDBOOK FOR MASONRY STRUCTURES

Table 3.2-4(1): C_s Values for Risk Category I and II Structures

Site Class				C_s		
A	B	C	D	OPMSW	ORMSW	SRMSW
$S_S < 0.314$	$S_S < 0.251$	$S_S < 0.209$	$S_S < 0.157$	0.112	0.084	0.034
$0.314 \le S_S < 0.619$	$0.251 \le S_S < 0.495$	$0.209 \le S_S < 0.413$	$0.157 \le S_S < 0.321$	0.220	0.165	0.067
$0.619 \le S_S < 0.938$	$0.495 \le S_S < 0.750$	$0.413 \le S_S < 0.661$	$0.321 \le S_S < 0.553$	NP	0.250	0.100
$0.938 \le S_S < 1.125$	$0.750 \le S_S < 0.900$	$0.661 \le S_S < 0.849$	$0.553 \le S_S < 0.750$	NP	NP	0.120
$1.125 \le S_S < 1.407$	$0.900 \le S_S < 1.125$	$0.849 \le S_S < 1.125$	$0.750 \le S_S < 1.023$	NP	NP	0.150
$1.407 \le S_S < 1.875$	$1.125 \le S_S < 1.500$	$1.125 \le S_S < 1.500$	$1.023 \le S_S < 1.500$	NP	NP	0.200
$1.875 \le S_S < 2.344$	$1.500 \le S_S < 1.875$	$1.500 \le S_S < 1.875$	$1.500 \le S_S < 1.875$	NP	NP	0.250
$2.344 \le S_S < 2.813$	$1.875 \le S_S < 2.250$	$1.875 \le S_S < 2.250$	$1.875 \le S_S < 2.250$	NP	NP	0.300
$2.813 \le S_S < 3.000$	$2.250 \le S_S < 2.400$	$2.250 \le S_S < 2.400$	$2.250 \le S_S < 2.400$	NP	NP	0.320
	$2.400 \le S_S < 2.625$	$2.400 \le S_S < 2.625$	$2.400 \le S_S < 2.625$	NP	NP	0.350
	$2.625 \le S_S < 3.000$	$2.625 \le S_S < 3.000$	$2.625 \le S_S < 3.000$	NP	NP	0.400

NP = Not Permitted

Table 3.2-4(2): C_s Values for Risk Category III Structures

Site Class				C_s		
A	B	C	D	OPMSW	ORMSW	SRMSW
$S_S < 0.314$	$S_S < 0.251$	$S_S < 0.209$	$S_S < 0.157$	0.140	0.105	0.043
$0.314 \le S_S < 0.619$	$0.251 \le S_S < 0.495$	$0.209 \le S_S < 0.413$	$0.157 \le S_S < 0.321$	0.275	0.206	0.084
$0.619 \le S_S < 0.938$	$0.495 \le S_S < 0.750$	$0.413 \le S_S < 0.661$	$0.321 \le S_S < 0.553$	NP	0.313	0.125
$0.938 \le S_S < 1.125$	$0.750 \le S_S < 0.900$	$0.661 \le S_S < 0.849$	$0.553 \le S_S < 0.750$	NP	NP	0.150
$1.125 \le S_S < 1.407$	$0.900 \le S_S < 1.125$	$0.849 \le S_S < 1.125$	$0.750 \le S_S < 1.023$	NP	NP	0.188
$1.407 \le S_S < 1.875$	$1.125 \le S_S < 1.500$	$1.125 \le S_S < 1.500$	$1.023 \le S_S < 1.500$	NP	NP	0.250
$1.875 \le S_S < 2.344$	$1.500 \le S_S < 1.875$	$1.500 \le S_S < 1.875$	$1.500 \le S_S < 1.875$	NP	NP	0.313
$2.344 \le S_S < 2.813$	$1.875 \le S_S < 2.250$	$1.875 \le S_S < 2.250$	$1.875 \le S_S < 2.250$	NP	NP	0.375
$2.813 \le S_S < 3.000$	$2.250 \le S_S < 2.400$	$2.250 \le S_S < 2.400$	$2.250 \le S_S < 2.400$	NP	NP	0.400
	$2.400 \le S_S < 2.625$	$2.400 \le S_S < 2.625$	$2.400 \le S_S < 2.625$	NP	NP	0.438
	$2.625 \le S_S < 3.000$	$2.625 \le S_S < 3.000$	$2.625 \le S_S < 3.000$	NP	NP	0.500

NP = Not Permitted

Table 3.2-4(3): C_s Values for Risk Category IV Structures

Site Class				C_s		
A	B	C	D	OPMSW	ORMSW	SRMSW
$S_S < 0.314$	$S_S < 0.251$	$S_S < 0.209$	$S_S < 0.157$	**0.168**	**0.126**	**0.051**
$0.314 \leq S_S < 0.619$	$0.251 \leq S_S < 0.495$	$0.209 \leq S_S < 0.413$	$0.157 \leq S_S < 0.321$	NP	0.248	0.101
$0.619 \leq S_S < 0.938$	$0.495 \leq S_S < 0.750$	$0.413 \leq S_S < 0.661$	$0.321 \leq S_S < 0.553$	NP	NP	0.150
$0.938 \leq S_S < 1.125$	$0.750 \leq S_S < 0.900$	$0.661 \leq S_S < 0.849$	$0.553 \leq S_S < 0.750$	NP	NP	0.180
$1.125 \leq S_S < 1.407$	$0.900 \leq S_S < 1.125$	$0.849 \leq S_S < 1.125$	$0.750 \leq S_S < 1.023$	NP	NP	0.225
$1.407 \leq S_S < 1.875$	$1.125 \leq S_S < 1.500$	$1.125 \leq S_S < 1.500$	$1.023 \leq S_S < 1.500$	NP	NP	0.300
$1.875 \leq S_S < 2.344$	$1.500 \leq S_S < 1.875$	$1.500 \leq S_S < 1.875$	$1.500 \leq S_S < 1.875$	NP	NP	0.325
$2.344 \leq S_S < 2.813$	$1.875 \leq S_S < 2.250$	$1.875 \leq S_S < 2.250$	$1.875 \leq S_S < 2.250$	NP	NP	0.450
$2.813 \leq S_S < 3.000$	$2.250 \leq S_S < 2.400$	$2.250 \leq S_S < 2.400$	$2.250 \leq S_S < 2.400$	NP	NP	0.480
	$2.400 \leq S_S < 2.625$	$2.400 \leq S_S < 2.625$	$2.400 \leq S_S < 2.625$	NP	NP	0.525
	$2.625 \leq S_S < 3.000$	$2.625 \leq S_S < 3.000$	$2.625 \leq S_S < 3.000$	NP	NP	0.600

NP = Not Permitted

Table 3.2-5(1a): Maximum Vertical Spans for Wall Segments (ft) if 40 psf < p_s ≤ 60 psf

Wind V, mph (kph)	Risk Category	Maximum L_{joist} (ft)	Unreinforced PCL Mortar Ungrouted Exterior	Unreinforced PCL Mortar Ungrouted Interior	Unreinforced PCL Mortar Fully Grouted Exterior	Unreinforced PCL Mortar Fully Grouted Interior	Unreinforced Masonry Ungrouted Exterior	Unreinforced Masonry Ungrouted Interior	Unreinforced Masonry Fully Grouted Exterior	Unreinforced Masonry Fully Grouted Interior	Vertical No. 5 at 120" oc Exterior	Vertical No. 5 at 120" oc Interior	Vertical No. 5 at 96" oc Exterior	Vertical No. 5 at 96" oc Interior	Vertical No. 5 at 72" oc Exterior	Vertical No. 5 at 72" oc Interior	Vertical No. 5 at 48" oc Exterior	Vertical No. 5 at 48" oc Interior	Vertical No. 5 at 32" oc Exterior	Vertical No. 5 at 32" oc Interior	Vertical No. 5 at 24" oc Exterior	Vertical No. 5 at 24" oc Interior	Vertical No. 5 at 16" oc Exterior	Vertical No. 5 at 16" oc Interior
V ≤ 110 (V ≤ 177)	B	NJ	8'-8"	18'-8"	16'-8"	30'-0"	6'-8"	14'-8"	16'-8"	30'-0"	10'-0"	22'-0"	15'-4"	24'-0"	17'-4"	26'-0"	21'-4"	30'-0"	25'-4"	30'-0"	28'-0"	30'-0"	30'-0"	30'-0"
		30	NP	NP	NP	NP	NP	NP	NP	NP	10'-0"	14'-8"	16'-0"	16'-0"	17'-4"	17'-4"	20'-0"	20'-0"	23'-4"	20'-0"	26'-0"	20'-0"	28'-8"	20'-0"
		60	NP	NP	NP	NP	NP	NP	NP	NP	NP	NP	14'-8"	12'-0"	16'-8"	13'-4"	20'-0"	16'-0"	20'-0"	16'-8"	20'-0"	18'-0"	NP	NP
	C	NJ	8'-0"	16'-8"	14'-8"	30'-0"	6'-0"	13'-4"	14'-8"	30'-0"	10'-0"	22'-0"	14'-0"	24'-0"	16'-0"	26'-0"	18'-8"	30'-0"	22'-0"	30'-0"	24'-8"	30'-0"	28'-0"	30'-0"
		30	NP	NP	NP	NP	NP	NP	NP	NP	10'-0"	14'-8"	14'-8"	16'-0"	17'-4"	17'-4"	18'-8"	20'-0"	21'-4"	20'-0"	23'-4"	20'-0"	26'-0"	20'-0"
		60	NP	NP	NP	NP	NP	NP	NP	NP	NP	NP	13'-4"	12'-0"	16'-0"	13'-4"	18'-0"	16'-0"	20'-0"	16'-8"	20'-0"	18'-0"	NP	NP
110 < V ≤ 115 (177 < V ≤ 185)	B	NJ	8'-0"	18'-0"	16'-0"	30'-0"	6'-0"	14'-0"	16'-0"	30'-0"	10'-0"	21'-4"	14'-8"	22'-8"	16'-8"	25'-4"	20'-8"	29'-4"	24'-0"	30'-0"	27'-4"	30'-0"	30'-0"	30'-0"
		30	NP	NP	NP	NP	NP	NP	NP	NP	10'-0"	14'-8"	15'-4"	16'-0"	16'-8"	17'-4"	19'-4"	20'-0"	22'-8"	20'-0"	25'-4"	20'-0"	27'-4"	20'-0"
		60	NP	NP	NP	NP	NP	NP	NP	NP	NP	NP	14'-8"	12'-0"	16'-0"	13'-4"	18'-0"	16'-0"	20'-0"	16'-8"	20'-0"	18'-0"	NP	NP
	C	NJ	7'-4"	16'-0"	14'-8"	29'-4"	6'-0"	12'-8"	14'-0"	28'-8"	10'-0"	21'-4"	14'-0"	22'-8"	16'-0"	25'-4"	18'-0"	29'-4"	21'-4"	30'-0"	24'-0"	30'-0"	27'-4"	30'-0"
		30	NP	NP	NP	NP	NP	NP	NP	NP	10'-0"	14'-8"	14'-8"	16'-0"	15'-4"	17'-4"	18'-0"	20'-0"	20'-8"	20'-0"	22'-8"	20'-0"	25'-4"	20'-0"
		60	NP	NP	NP	NP	NP	NP	NP	NP	NP	NP	13'-4"	12'-0"	15'-4"	13'-4"	18'-0"	16'-0"	20'-0"	16'-8"	20'-0"	18'-0"	NP	NP
115 < V ≤ 120 (185 < V ≤ 193)	B	NJ	8'-0"	17'-4"	15'-4"	30'-0"	6'-0"	13'-4"	14'-8"	30'-0"	10'-0"	20'-8"	14'-0"	22'-0"	16'-0"	24'-8"	19'-4"	28'-8"	23'-4"	30'-0"	26'-0"	30'-0"	30'-0"	30'-0"
		30	NP	NP	NP	NP	NP	NP	NP	NP	10'-0"	14'-8"	14'-8"	16'-0"	16'-8"	17'-4"	19'-4"	20'-0"	22'-0"	20'-0"	24'-8"	20'-0"	26'-8"	20'-0"
		60	NP	NP	NP	NP	NP	NP	NP	NP	NP	NP	14'-0"	12'-0"	16'-0"	13'-4"	18'-8"	16'-0"	20'-0"	16'-8"	20'-0"	18'-0"	NP	NP
	C	NJ	7'-4"	15'-4"	14'-0"	28'-0"	5'-4"	12'-0"	13'-4"	27'-4"	10'-0"	20'-8"	14'-0"	22'-0"	14'-8"	24'-8"	17'-4"	28'-8"	20'-8"	30'-0"	22'-8"	30'-0"	24'-8"	30'-0"
		30	NP	NP	NP	NP	NP	NP	NP	NP	10'-0"	14'-8"	14'-0"	16'-0"	14'-8"	17'-4"	17'-4"	20'-0"	20'-0"	20'-0"	22'-0"	20'-0"	24'-8"	20'-0"
		60	NP	NP	NP	NP	NP	NP	NP	NP	NP	NP	13'-4"	12'-0"	14'-8"	13'-4"	17'-4"	16'-0"	19'-4"	16'-8"	20'-0"	18'-0"	NP	NP
120 < V ≤ 130 (193 < V ≤ 209)	B	NJ	7'-4"	16'-0"	14'-0"	30'-0"	5'-4"	12'-0"	14'-0"	30'-0"	10'-0"	19'-4"	13'-4"	21'-4"	15'-4"	23'-4"	18'-0"	27'-4"	22'-0"	30'-0"	24'-8"	30'-0"	28'-0"	30'-0"
		30	NP	NP	NP	NP	NP	NP	NP	NP	10'-0"	14'-8"	14'-0"	16'-0"	15'-4"	17'-4"	18'-0"	20'-0"	20'-8"	20'-0"	23'-4"	20'-0"	26'-0"	20'-0"
		60	NP	NP	NP	NP	NP	NP	NP	NP	NP	NP	13'-4"	12'-0"	15'-4"	13'-4"	18'-0"	16'-0"	19'-4"	16'-8"	20'-0"	18'-0"	NP	NP
	C	NJ	6'-8"	14'-8"	12'-8"	26'-0"	5'-4"	11'-4"	12'-8"	25'-4"	10'-0"	19'-4"	13'-4"	21'-4"	14'-0"	23'-4"	16'-0"	27'-4"	19'-4"	30'-0"	21'-4"	30'-0"	24'-8"	30'-0"
		30	NP	NP	NP	NP	NP	NP	NP	NP	10'-0"	14'-0"	13'-4"	16'-0"	14'-0"	17'-4"	16'-8"	20'-0"	18'-8"	20'-0"	20'-8"	20'-0"	23'-4"	20'-0"
		60	NP	NP	NP	NP	NP	NP	NP	NP	NP	NP	12'-8"	12'-0"	14'-0"	13'-4"	16'-8"	16'-0"	18'-8"	16'-8"	20'-0"	18'-0"	NP	NP

Table values shall not be permitted to be interpolated.
NJ = No Joist
NP = Not Permitted

ft = 0.3048 m
in. = 25.4 mm
lb/ft² = 0.0479 kPa

Table 3.2-5(1b): Maximum Vertical Spans for Wall Segments (ft) if 40 psf < p_s ≤ 60 psf

Wind V, mph (kph)	Risk Category	Maximum L_joist (ft)	Unreinforced PCL Mortar Ungrouted Exterior	Ungrouted Interior	Fully Grouted Exterior	Fully Grouted Interior	Unreinforced Masonry Ungrouted Exterior	Ungrouted Interior	Fully Grouted Exterior	Fully Grouted Interior	Vertical No. 5 at 120" oc Exterior	Interior	Vertical No. 5 at 96" oc Exterior	Interior	Vertical No. 5 at 72" oc Exterior	Interior	Vertical No. 5 at 48" oc Exterior	Interior	Vertical No. 5 at 32" oc Exterior	Interior	Vertical No. 5 at 24" oc Exterior	Interior	Vertical No. 5 at 16" oc Exterior	Interior
130 < V ≤ 140 (209 < V ≤ 225)	B	NJ	6'-8"	14'-8"	13'-4"	28'-8"	5'-4"	11'-4"	12'-8"	28'-0"	10'-0"	10'-0"	10'-0"	20'-0"	14'-0"	22'-0"	16'-8"	26'-0"	20'-0"	28'-8"	22'-8"	30'-0"	26'-8"	30'-0"
		30	NP	NP	NP	NP	NP	NP	NP	NP	10'-0"	10'-0"	10'-0"	16'-0"	14'-8"	17'-4"	18'-0"	20'-0"	18'-0"	20'-0"	22'-0"	20'-0"	24'-8"	20'-0"
		60	NP	NP	NP	NP	NP	NP	NP	NP	NP	NP	10'-0"	NP	13'-4"	13'-4"	16'-8"	16'-0"	19'-4"	16'-8"	20'-0"	18'-0"	NP	NP
	C	NJ	6'-0"	13'-4"	12'-0"	24'-8"	4'-8"	10'-0"	11'-4"	24'-0"	9'-4"	10'-0"	10'-0"	20'-0"	12'-8"	22'-0"	15'-4"	26'-0"	18'-0"	28'-0"	20'-0"	30'-0"	23'-4"	30'-0"
		30	NP	NP	NP	NP	NP	NP	NP	NP	8'-8"	10'-0"	10'-0"	16'-0"	13'-4"	17'-4"	15'-4"	20'-0"	18'-0"	20'-0"	19'-4"	20'-0"	22'-0"	20'-0"
		60	NP	NP	NP	NP	NP	NP	NP	NP	NP	NP	8'-0"	NP	11'-4"	13'-4"	15'-4"	16'-0"	17'-4"	16'-8"	19'-4"	18'-0"	NP	NP
140 < V ≤ 150 (225 < V ≤ 241)	B	NJ	6'-0"	14'-0"	12'-0"	26'-8"	4'-8"	10'-8"	12'-0"	26'-0"	9'-4"	10'-0"	10'-0"	18'-8"	13'-4"	21'-4"	16'-0"	24'-8"	18'-8"	26'-8"	21'-4"	28'-8"	25'-4"	30'-0"
		30	NP	NP	NP	NP	NP	NP	NP	NP	9'-4"	10'-0"	10'-0"	15'-4"	13'-4"	17'-4"	16'-0"	20'-0"	18'-8"	20'-0"	20'-0"	20'-0"	24'-0"	20'-0"
		60	NP	NP	NP	NP	NP	NP	NP	NP	NP	NP	8'-8"	NP	12'-0"	NP	16'-0"	16'-0"	18'-0"	16'-8"	20'-0"	18'-0"	NP	NP
	C	NJ	5'-4"	12'-8"	11'-4"	22'-8"	4'-0"	9'-4"	10'-8"	22'-0"	8'-8"	10'-0"	10'-0"	18'-8"	10'-8"	21'-4"	14'-8"	24'-8"	16'-8"	26'-8"	19'-4"	28'-8"	22'-0"	30'-0"
		30	NP	NP	NP	NP	NP	NP	NP	NP	7'-4"	NP	9'-4"	12'-8"	13'-4"	17'-4"	14'-8"	20'-0"	16'-8"	20'-0"	18'-8"	20'-0"	21'-4"	20'-0"
		60	NP	NP	NP	NP	NP	NP	NP	NP	NP	NP	6'-8"	NP	10'-0"	NP	14'-0"	16'-0"	16'-8"	16'-8"	18'-0"	17'-4"	NP	NP
150 < V ≤ 160 (241 < V ≤ 258)	B	NJ	6'-0"	12'-8"	11'-4"	25'-4"	4'-8"	10'-0"	11'-4"	24'-0"	8'-8"	10'-0"	10'-0"	18'-0"	12'-0"	20'-0"	14'-8"	23'-4"	18'-0"	26'-0"	20'-0"	27'-4"	24'-0"	30'-0"
		30	NP	NP	NP	NP	NP	NP	NP	NP	8'-0"	8'-0"	10'-0"	12'-8"	12'-8"	17'-4"	15'-4"	20'-0"	18'-0"	20'-0"	19'-4"	20'-0"	22'-8"	20'-0"
		60	NP	NP	NP	NP	NP	NP	NP	NP	NP	NP	7'-4"	NP	10'-8"	NP	15'-4"	16'-0"	17'-4"	16'-8"	18'-0"	17'-4"	NP	NP
	C	NJ	5'-4"	11'-4"	10'-8"	21'-4"	4'-0"	8'-8"	10'-0"	20'-8"	8'-0"	10'-0"	9'-4"	18'-0"	10'-0"	20'-0"	13'-4"	23'-4"	16'-0"	26'-0"	18'-0"	27'-4"	20'-8"	30'-0"
		30	NP	NP	NP	NP	NP	NP	NP	NP	NP	NP	9'-4"	9'-4"	10'-0"	16'-0"	14'-0"	20'-0"	16'-0"	20'-0"	17'-4"	20'-0"	20'-0"	20'-0"
		60	NP	NP	NP	NP	NP	NP	NP	NP	NP	NP	NP	NP	8'-0"	NP	12'-8"	15'-4"	16'-0"	16'-8"	17'-4"	17'-4"	NP	NP
160 < V ≤ 180 (258 < V ≤ 290)	B	NJ	5'-4"	11'-4"	10'-0"	22'-0"	4'-0"	8'-8"	10'-0"	21'-4"	8'-0"	10'-0"	9'-4"	18'-8"	10'-0"	18'-8"	13'-4"	21'-4"	16'-0"	23'-4"	18'-0"	25'-4"	21'-4"	28'-8"
		30	NP	NP	NP	NP	NP	NP	NP	NP	6'-0"	NP	8'-8"	7'-4"	10'-0"	13'-4"	14'-0"	20'-0"	14'-0"	20'-0"	18'-8"	20'-0"	20'-8"	20'-0"
		60	NP	NP	NP	NP	NP	NP	NP	NP	NP	NP	NP	NP	8'-0"	NP	12'-0"	14'-0"	14'-8"	16'-8"	17'-4"	17'-4"	NP	NP
	C	NJ	4'-8"	10'-0"	9'-4"	19'-4"	NP	8'-0"	8'-8"	18'-8"	7'-4"	10'-0"	8'-8"	18'-8"	8'-8"	18'-8"	12'-0"	21'-4"	14'-8"	23'-4"	16'-0"	25'-4"	18'-8"	28'-0"
		30	NP	NP	NP	NP	NP	NP	NP	NP	4'-0"	NP	6'-0"	NP	8'-8"	10'-8"	12'-0"	19'-4"	14'-8"	20'-0"	16'-0"	20'-0"	18'-8"	20'-0"
		60	NP	NP	NP	NP	NP	NP	NP	NP	NP	NP	NP	NP	5'-4"	NP	10'-0"	NP	14'-0"	16'-8"	14'-0"	17'-4"	NP	NP
190 < V ≤ 200 (306 < V ≤ 322)	B	NJ	4'-8"	10'-0"	9'-4"	20'-0"	NP	7'-4"	8'-0"	19'-4"	7'-4"	9'-4"	8'-0"	10'-0"	9'-4"	17'-4"	12'-0"	20'-0"	14'-8"	22'-0"	16'-0"	23'-4"	19'-4"	26'-0"
		30	NP	NP	NP	NP	NP	NP	NP	NP	4'-0"	NP	6'-0"	NP	8'-8"	9'-4"	10'-0"	17'-4"	14'-0"	20'-0"	16'-0"	20'-0"	18'-8"	20'-0"
		60	NP	NP	NP	NP	NP	NP	NP	NP	NP	NP	NP	NP	5'-4"	NP	10'-0"	NP	14'-0"	16'-8"	16'-0"	17'-4"	18'-8"	NP
	C	NJ	4'-0"	9'-4"	8'-0"	17'-4"	NP	7'-4"	8'-0"	16'-8"	6'-8"	9'-4"	7'-4"	10'-0"	8'-8"	17'-4"	10'-8"	20'-0"	13'-4"	22'-0"	14'-8"	23'-4"	17'-4"	26'-0"
		30	NP	NP	NP	NP	NP	NP	NP	NP	4'-0"	NP	4'-0"	NP	6'-8"	NP	10'-0"	14'-8"	13'-4"	20'-0"	14'-8"	20'-0"	16'-8"	20'-0"
		60	NP	NP	NP	NP	NP	NP	NP	NP	NP	NP	NP	NP	NP	NP	8'-0"	NP	12'-0"	NP	NP	17'-4"	NP	NP

Table values shall not be permitted to be interpolated.

NJ = No Joist
NP = Not Permitted

ft = 0.3048 m
in. = 25.4 mm
lb/ft² = 0.0479 kPa

DIRECT DESIGN HANDBOOK FOR MASONRY STRUCTURES

Table 3.2-5(2a): Maximum Vertical Spans for Wall Segments (ft) if 20 psf < p_s ≤ 40 psf

Wind V, mph (kph)	Risk Category	Maximum L_{joist} (ft)	Unreinforced PCL Mortar Ungrouted Exterior	Ungrouted Interior	Fully Grouted Exterior	Fully Grouted Interior	Unreinforced Masonry Ungrouted Exterior	Ungrouted Interior	Fully Grouted Exterior	Fully Grouted Interior	Vertical No. 5 at 120" oc Exterior	Interior	Vertical No. 5 at 96" oc Exterior	Interior	Vertical No. 5 at 72" oc Exterior	Interior	Vertical No. 5 at 48" oc Exterior	Interior	Vertical No. 5 at 32" oc Exterior	Interior	Vertical No. 5 at 24" oc Exterior	Interior	Vertical No. 5 at 16" oc Exterior	Interior
V ≤ 110 (V ≤ 177)	B	NJ	8'-8"	18'-8"	16'-8"	30'-0"	6'-8"	14'-8"	16'-8"	30'-0"	10'-0"	22'-0"	15'-4	24'-0"	17'-4	26'-0"	21'-4	30'-0"	25'-4	30'-0"	28'-0"	30'-0"	30'-0"	30'-0"
		30	NP	NP	NP	NP	NP	NP	NP	NP	10'-0"	15'-4	16'-0"	16'-8"	17'-4	18'-8"	20'-8"	20'-0"	23'-4	20'-0"	26'-0"	20'-0"	29'-4	20'-0"
		60	NP	NP	NP	NP	NP	NP	NP	NP	10'-0"	NP	15'-4	12'-8"	16'-8"	14'-0"	20'-0"	16'-0"	20'-0"	18'-0"	20'-0"	18'-0"	NP	NP
	C	NJ	8'-0"	16'-8"	14'-8"	30'-0"	6'-0"	13'-4	14'-8"	30'-0"	10'-0"	22'-0"	14'-0"	24'-0"	15'-4	26'-0"	18'-8"	30'-0"	22'-0"	30'-0"	24'-8"	30'-0"	28'-0"	30'-0"
		30	NP	NP	NP	NP	NP	NP	NP	NP	10'-0"	15'-4	14'-8"	20'-0"	16'-0"	20'-0"	18'-8"	30'-0"	21'-4	30'-0"	23'-4	30'-0"	27'-4	30'-0"
		60	NP	NP	NP	NP	NP	NP	NP	NP	10'-0"	NP	13'-4	19'-4	16'-0"	14'-0"	18'-8"	20'-0"	21'-4	20'-0"	20'-0"	20'-0"	26'-8"	20'-0"
110 < V ≤ 115 (177 < V ≤ 185)	B	NJ	8'-0"	18'-0"	16'-0"	30'-0"	6'-0"	14'-0"	16'-0"	30'-0"	10'-0"	21'-4	14'-8"	22'-8"	16'-8"	25'-4	20'-8"	30'-0"	24'-0"	30'-0"	27'-4	30'-0"	30'-0"	30'-0"
		30	NP	NP	NP	NP	NP	NP	NP	NP	10'-0"	15'-4	15'-4	16'-8"	16'-8"	18'-0"	18'-0"	20'-0"	22'-8"	20'-0"	25'-4	20'-0"	28'-8"	20'-0"
		60	NP	NP	NP	NP	NP	NP	NP	NP	10'-0"	NP	14'-8"	12'-8"	16'-8"	14'-0"	19'-4	16'-8"	20'-0"	18'-8"	20'-0"	20'-0"	NP	NP
	C	NJ	7'-4	16'-0"	14'-0"	29'-4	6'-0"	12'-8"	14'-0"	28'-8"	10'-0"	21'-4	10'-0"	22'-8"	14'-8"	25'-4	18'-0"	30'-0"	21'-4	30'-0"	24'-0"	30'-0"	27'-4	30'-0"
		30	NP	NP	NP	NP	NP	NP	NP	NP	10'-0"	19'-4	10'-0"	20'-0"	15'-4	18'-8"	18'-0"	29'-4	20'-8"	30'-0"	22'-8"	30'-0"	26'-8"	30'-0"
		60	NP	NP	NP	NP	NP	NP	NP	NP	10'-0"	NP	10'-0"	16'-8"	15'-4	14'-0"	18'-0"	20'-0"	20'-8"	20'-0"	22'-0"	20'-0"	26'-0"	20'-0"
115 < V ≤ 120 (185 < V ≤ 193)	B	NJ	8'-0"	17'-4	15'-4	30'-0"	6'-0"	13'-4	14'-8"	30'-0"	10'-0"	20'-8"	14'-8"	22'-0"	16'-0"	24'-8"	19'-4	30'-0"	23'-4	30'-0"	26'-0"	30'-0"	30'-0"	30'-0"
		30	NP	NP	NP	NP	NP	NP	NP	NP	10'-0"	15'-4	14'-8"	16'-8"	16'-0"	18'-0"	19'-4	20'-0"	22'-0"	20'-0"	24'-8"	20'-0"	27'-4	20'-0"
		60	NP	NP	NP	NP	NP	NP	NP	NP	10'-0"	NP	14'-0"	12'-8"	16'-0"	14'-0"	18'-8"	16'-8"	20'-0"	18'-0"	20'-0"	20'-0"	NP	NP
	C	NJ	7'-4	15'-4	14'-0"	28'-0"	5'-4	12'-0"	13'-4	27'-4	10'-0"	20'-8"	10'-0"	22'-0"	14'-8"	24'-8"	17'-4	28'-8"	20'-8"	30'-0"	22'-8"	30'-0"	25'-4	30'-0"
		30	NP	NP	NP	NP	NP	NP	NP	NP	10'-0"	17'-4	10'-0"	20'-0"	14'-8"	20'-0"	17'-4	28'-8"	20'-8"	30'-0"	22'-0"	30'-0"	25'-4	30'-0"
		60	NP	NP	NP	NP	NP	NP	NP	NP	9'-4	NP	10'-0"	15'-4	14'-8"	14'-0"	17'-4	20'-0"	20'-0"	20'-0"	21'-4	20'-0"	24'-8"	20'-8"
120 < V ≤ 130 (193 < V ≤ 209)	B	NJ	7'-4	16'-0"	14'-0"	30'-0"	5'-4	12'-0"	14'-0"	30'-0"	10'-0"	19'-4	10'-0"	21'-4	15'-4	23'-4	18'-0"	27'-4	22'-0"	30'-0"	24'-8"	30'-0"	28'-0"	30'-0"
		30	NP	NP	NP	NP	NP	NP	NP	NP	10'-0"	15'-4	14'-0"	16'-8"	15'-4	18'-0"	18'-0"	20'-0"	20'-8"	20'-0"	23'-4	20'-0"	26'-0"	20'-0"
		60	NP	NP	NP	NP	NP	NP	NP	NP	9'-4	NP	10'-0"	12'-8"	15'-4	14'-0"	16'-8"	16'-8"	20'-0"	20'-0"	20'-0"	20'-0"	NP	NP
	C	NJ	6'-8"	14'-8"	12'-8"	26'-0"	5'-4	11'-4	12'-8"	25'-4	10'-0"	19'-4	10'-0"	21'-4	14'-0"	23'-4	16'-0"	27'-4	19'-4	30'-0"	21'-4	30'-0"	24'-8"	30'-0"
		30	NP	NP	NP	NP	NP	NP	NP	NP	10'-0"	14'-0"	10'-0"	18'-8"	14'-0"	20'-0"	16'-0"	27'-4	19'-4	30'-0"	20'-8"	30'-0"	24'-8"	30'-0"
		60	NP	NP	NP	NP	NP	NP	NP	NP	7'-4	NP	10'-0"	NP	12'-8"	17'-4	16'-0"	20'-0"	18'-8"	20'-0"	20'-0"	20'-0"	24'-0"	20'-0"

Table values shall not be permitted to be interpolated.
NJ = No Joist
NP = Not Permitted

ft = 0.3048 m
in. = 25.4 mm
lb/ft² = 0.0479 kPa

TMS 403-13

Table 3.2-5(2b): Maximum Vertical Spans for Wall Segments (ft) if 20 psf < p_x ≤ 40 psf

Wind V, mph (kph)	Risk Category	Maximum L_{joist} (ft)	Unreinforced PCL Mortar Ungrouted Exterior	Ungrouted Interior	Fully Grouted Exterior	Fully Grouted Interior	Unreinforced Masonry Ungrouted Exterior	Ungrouted Interior	Fully Grouted Exterior	Fully Grouted Interior	Vertical No. 5 at 120" oc Exterior	Interior	Vertical No. 5 at 96" oc Exterior	Interior	Vertical No. 5 at 72" oc Exterior	Interior	Vertical No. 5 at 48" oc Exterior	Interior	Vertical No. 5 at 32" oc Exterior	Interior	Vertical No. 5 at 24" oc Exterior	Interior	Vertical No. 5 at 16" oc Exterior	Interior
130 < V ≤ 140 (209 < V ≤ 225)	B	NJ	6'-8"	14'-8"	13'-4"	28'-8"	5'-4"	11'-4"	12'-8"	28'-0"	10'-0"	10'-0"	10'-0"	20'-0"	14'-0"	22'-0"	16'-8"	26'-0"	20'-0"	28'-0"	22'-8"	30'-0"	26'-8"	30'-0"
		30	NP	NP	NP	NP	NP	NP	NP	NP	10'-0"	10'-0"	10'-0"	16'-8"	14'-8"	18'-0"	17'-4"	20'-0"	20'-0"	20'-0"	22'-0"	20'-0"	25'-4"	20'-0"
		60	NP	NP	NP	NP	NP	NP	NP	NP	NP	NP	10'-0"	NP	13'-4"	14'-0"	17'-4"	16'-8"	19'-4"	17'-4"	20'-0"	18'-8"	NP	NP
	C	NJ	6'-0"	13'-4"	12'-0"	24'-8"	4'-8"	10'-0"	11'-4"	24'-0"	9'-4"	10'-0"	10'-0"	20'-0"	12'-8"	22'-0"	15'-4"	26'-0"	18'-0"	28'-0"	20'-0"	30'-0"	23'-4"	30'-0"
		30	NP	NP	NP	NP	NP	NP	NP	NP	8'-8"	10'-0"	10'-0"	16'-0"	13'-4"	20'-0"	15'-4"	26'-8"	18'-0"	30'-0"	20'-0"	30'-0"	22'-8"	30'-0"
		60	NP	NP	NP	NP	NP	NP	NP	NP	5'-4"	NP	8'-0"	NP	11'-4"	NP	15'-4"	20'-0"	18'-0"	20'-0"	20'-0"	20'-0"	22'-8"	20'-0"
140 < V ≤ 150 (225 < V ≤ 241)	B	NJ	6'-0"	14'-0"	12'-0"	26'-8"	4'-8"	10'-8"	12'-0"	26'-0"	9'-4"	10'-0"	10'-0"	18'-8"	13'-4"	21'-4"	16'-0"	24'-8"	18'-8"	26'-8"	21'-4"	28'-8"	25'-4"	30'-0"
		30	NP	NP	NP	NP	NP	NP	NP	NP	9'-4"	10'-0"	10'-0"	15'-4"	13'-4"	18'-0"	16'-0"	20'-0"	18'-8"	20'-0"	20'-8"	20'-0"	24'-0"	20'-0"
		60	NP	NP	NP	NP	NP	NP	NP	NP	NP	NP	8'-8"	NP	11'-4"	NP	16'-0"	16'-8"	18'-0"	17'-4"	20'-0"	18'-8"	NP	NP
	C	NJ	5'-4"	12'-8"	11'-4"	22'-8"	4'-0"	9'-4"	10'-8"	22'-0"	8'-8"	10'-0"	10'-0"	18'-8"	12'-0"	21'-4"	14'-8"	24'-8"	16'-8"	26'-8"	19'-4"	28'-8"	21'-0"	30'-0"
		30	NP	NP	NP	NP	NP	NP	NP	NP	7'-4"	NP	9'-4"	NP	12'-0"	18'-8"	14'-8"	25'-4"	16'-8"	28'-8"	18'-8"	30'-0"	21'-4"	30'-0"
		60	NP	NP	NP	NP	NP	NP	NP	NP	NP	NP	6'-8"	NP	NP	NP	14'-0"	20'-0"	16'-8"	20'-0"	18'-8"	20'-0"	NP	20'-0"
150 < V ≤ 160 (241 < V ≤ 258)	B	NJ	6'-0"	12'-8"	11'-4"	25'-4"	4'-8"	10'-0"	11'-4"	24'-0"	8'-8"	10'-0"	10'-0"	18'-0"	12'-0"	20'-0"	14'-8"	23'-4"	18'-0"	26'-0"	20'-0"	27'-4"	24'-0"	30'-0"
		30	NP	NP	NP	NP	NP	NP	NP	NP	8'-0"	NP	10'-0"	18'-0"	12'-8"	20'-0"	15'-4"	20'-0"	18'-0"	20'-0"	19'-4"	20'-0"	22'-8"	20'-0"
		60	NP	NP	NP	NP	NP	NP	NP	NP	NP	NP	7'-4"	NP	10'-8"	NP	15'-4"	16'-8"	17'-4"	17'-4"	19'-4"	18'-8"	NP	NP
	C	NJ	5'-4"	11'-4"	10'-8"	21'-4"	4'-0"	8'-8"	10'-0"	20'-8"	8'-0"	9'-4"	10'-0"	18'-0"	10'-0"	20'-0"	13'-4"	24'-8"	16'-0"	26'-8"	18'-0"	30'-0"	20'-8"	30'-0"
		30	NP	NP	NP	NP	NP	NP	NP	NP	6'-0"	NP	8'-8"	NP	10'-0"	16'-0"	14'-0"	24'-8"	16'-0"	30'-0"	18'-0"	30'-0"	20'-0"	30'-0"
		60	NP	NP	NP	NP	NP	NP	NP	NP	NP	NP	NP	NP	NP	NP	12'-8"	16'-8"	16'-0"	20'-0"	18'-0"	20'-0"	20'-0"	20'-0"
160 < V ≤ 180 (258 < V ≤ 290)	B	NJ	4'-8"	10'-0"	10'-0"	22'-0"	4'-0"	8'-8"	8'-8"	21'-4"	6'-8"	10'-0"	9'-4"	18'-8"	10'-0"	18'-8"	13'-4"	21'-4"	16'-0"	23'-4"	18'-0"	25'-4"	21'-4"	28'-0"
		30	NP	NP	NP	NP	NP	NP	NP	NP	6'-0"	NP	8'-0"	NP	10'-0"	13'-4"	14'-0"	20'-0"	14'-8"	20'-0"	17'-4"	20'-0"	20'-8"	20'-0"
		60	NP	NP	NP	NP	NP	NP	NP	NP	NP	NP	NP	NP	8'-0"	NP	12'-0"	14'-0"	14'-8"	17'-4"	16'-0"	18'-8"	NP	NP
	C	NJ	4'-0"	10'-0"	9'-4"	19'-4"	NP	8'-0"	8'-8"	18'-8"	7'-4"	10'-0"	8'-8"	18'-8"	10'-0"	18'-8"	12'-0"	19'-4"	14'-8"	23'-4"	16'-0"	25'-4"	18'-8"	28'-0"
		30	NP	NP	NP	NP	NP	NP	NP	NP	4'-0"	NP	6'-0"	NP	10'-0"	10'-8"	12'-0"	19'-4"	14'-0"	20'-0"	16'-0"	20'-0"	18'-8"	20'-0"
		60	NP	NP	NP	NP	NP	NP	NP	NP	NP	NP	NP	NP	5'-4"	NP	10'-8"	NP	14'-8"	16'-8"	16'-0"	18'-0"	NP	NP
190 < V ≤ 200 (306 < V ≤ 322)	B	NJ	4'-8"	10'-0"	9'-4"	20'-0"	NP	8'-0"	8'-8"	19'-4"	7'-4"	10'-0"	8'-0"	20'-0"	9'-4"	17'-4"	12'-0"	20'-0"	14'-8"	22'-0"	16'-0"	23'-4"	19'-4"	26'-0"
		30	NP	NP	NP	NP	NP	NP	NP	NP	4'-0"	NP	6'-0"	NP	8'-8"	9'-4"	12'-0"	17'-4"	14'-8"	20'-0"	16'-0"	20'-0"	18'-8"	20'-0"
		60	NP	NP	NP	NP	NP	NP	NP	NP	NP	NP	NP	NP	5'-4"	NP	10'-8"	NP	14'-0"	16'-8"	16'-0"	18'-0"	NP	NP
	C	NJ	4'-0"	9'-4"	8'-0"	17'-4"	NP	7'-4"	8'-0"	16'-8"	6'-8"	10'-0"	7'-4"	20'-0"	8'-8"	17'-4"	10'-8"	20'-0"	13'-4"	22'-0"	14'-8"	23'-4"	17'-4"	26'-0"
		30	NP	NP	NP	NP	NP	NP	NP	NP	NP	NP	4'-0"	NP	6'-8"	NP	10'-0"	14'-8"	13'-4"	20'-0"	14'-8"	20'-0"	16'-8"	20'-0"
		60	NP	NP	NP	NP	NP	NP	NP	NP	NP	NP	NP	NP	NP	NP	8'-0"	NP	12'-0"	NP	14'-8"	18'-0"	NP	NP

Table values shall not be permitted to be interpolated.

NJ = No Joist
NP = Not Permitted

ft = 0.3048 m
in. = 25.4 mm
lb/ft² = 0.0479 kPa

DIRECT DESIGN HANDBOOK FOR MASONRY STRUCTURES

Table 3.2-5(3a): Maximum Vertical Spans for Wall Segments (ft) if 0 psf < p_s ≤ 20 psf

| Wind V, mph (kph) | Risk Category | Maximum L_{joist} (ft) | Unreinforced PCL Mortar |||| Unreinforced Masonry Fully Grouted |||| Vertical No. 5 at 120" oc || Vertical No. 5 at 96" oc || Vertical No. 5 at 72" oc || Vertical No. 5 at 48" oc || Vertical No. 5 at 32" oc || Vertical No. 5 at 24" oc || Vertical No. 5 at 16" oc ||
| | | | Ungrouted || Fully Grouted || Ungrouted || Fully Grouted || Ext | Int | Ext | Int | Ext | Int | Ext | Int | Ext | Int | Ext | Int | Ext | Int |
			Ext	Int	Ext	Int	Ext	Int	Ext	Int														
V ≤ 110 (V ≤ 177)	B	NJ	8'-8"	18'-8"	16'-8"	30'-0"	6'-8"	14'-8"	16'-8"	30'-0"	10'-0"	22'-0"	15'-4"	24'-0"	17'-4"	26'-0"	21'-4"	30'-0"	25'-4"	30'-0"	28'-0"	30'-0"	30'-0"	30'-0"
		30	NP	NP	NP	NP	NP	NP	NP	NP	10'-0"	16'-0"	16'-0"	17'-4"	17'-4"	19'-4"	20'-8"	20'-0"	24'-0"	20'-0"	26'-0"	20'-0"	29'-4"	20'-0"
		60	NP	NP	NP	NP	NP	NP	NP	NP	10'-0"	12'-0"	15'-4"	13'-4"	17'-4"	15'-4"	20'-0"	18'-0"	20'-0"	20'-0"	20'-0"	20'-0"	NP	NP
	C	NJ	8'-0"	16'-8"	14'-8"	30'-0"	6'-0"	13'-4"	14'-8"	30'-0"	10'-0"	22'-0"	14'-0"	24'-0"	15'-4"	26'-0"	18'-8"	30'-0"	22'-0"	30'-0"	24'-8"	28'-0"	28'-0"	30'-0"
		30	NP	NP	NP	NP	NP	NP	NP	NP	10'-0"	16'-0"	14'-8"	17'-4"	16'-0"	19'-4"	18'-8"	20'-0"	21'-4"	20'-0"	23'-4"	20'-0"	26'-8"	20'-0"
		60	NP	NP	NP	NP	NP	NP	NP	NP	10'-0"	NP	13'-4"	15'-4"	16'-0"	15'-4"	18'-8"	17'-4"	20'-0"	20'-0"	20'-0"	20'-0"	NP	NP
110 < V ≤ 115 (177 < V ≤ 185)	B	NJ	8'-0"	18'-0"	16'-0"	30'-0"	6'-0"	14'-0"	16'-0"	30'-0"	10'-0"	21'-4"	14'-8"	22'-8"	16'-8"	25'-4"	20'-0"	29'-4"	24'-0"	30'-0"	27'-4"	30'-0"	30'-0"	30'-0"
		30	NP	NP	NP	NP	NP	NP	NP	NP	10'-0"	16'-0"	15'-4"	17'-4"	17'-4"	19'-4"	20'-8"	20'-0"	23'-4"	20'-0"	25'-4"	20'-0"	28'-8"	20'-0"
		60	NP	NP	NP	NP	NP	NP	NP	NP	10'-0"	12'-0"	14'-8"	13'-4"	16'-8"	15'-4"	20'-0"	18'-0"	20'-0"	20'-0"	20'-0"	20'-0"	NP	NP
	C	NJ	7'-4"	16'-0"	14'-8"	29'-4"	6'-0"	12'-8"	14'-0"	28'-8"	10'-0"	21'-4"	14'-8"	22'-8"	14'-8"	25'-4"	19'-4"	29'-4"	21'-4"	30'-0"	24'-0"	30'-0"	27'-4"	30'-0"
		30	NP	NP	NP	NP	NP	NP	NP	NP	10'-0"	16'-0"	14'-8"	17'-4"	15'-4"	19'-4"	18'-0"	20'-0"	20'-8"	20'-0"	22'-8"	20'-0"	26'-0"	20'-0"
		60	NP	NP	NP	NP	NP	NP	NP	NP	10'-0"	NP	10'-0"	13'-4"	15'-4"	15'-4"	18'-0"	17'-4"	20'-0"	20'-0"	20'-0"	20'-0"	NP	NP
115 < V ≤ 120 (185 < V ≤ 193)	B	NJ	8'-0"	17'-4"	15'-4"	30'-0"	6'-0"	13'-4"	14'-8"	30'-0"	10'-0"	20'-8"	14'-0"	22'-0"	16'-0"	24'-8"	19'-4"	28'-8"	23'-4"	30'-0"	26'-0"	30'-0"	30'-0"	30'-0"
		30	NP	NP	NP	NP	NP	NP	NP	NP	9'-4"	16'-0"	14'-8"	17'-4"	16'-8"	19'-4"	19'-4"	20'-0"	22'-8"	20'-0"	24'-8"	20'-0"	28'-8"	20'-0"
		60	NP	NP	NP	NP	NP	NP	NP	NP	10'-0"	NP	10'-0"	13'-4"	14'-8"	15'-4"	19'-4"	18'-0"	20'-0"	20'-0"	20'-0"	20'-0"	NP	NP
	C	NJ	7'-4"	15'-4"	14'-0"	28'-0"	5'-4"	12'-0"	13'-4"	27'-4"	10'-0"	20'-8"	14'-0"	21'-4"	14'-8"	24'-8"	17'-4"	28'-8"	20'-8"	30'-0"	23'-4"	30'-0"	26'-8"	30'-0"
		30	NP	NP	NP	NP	NP	NP	NP	NP	10'-0"	16'-0"	14'-8"	17'-4"	14'-8"	19'-4"	17'-4"	20'-0"	20'-0"	20'-0"	22'-0"	20'-0"	24'-8"	20'-0"
		60	NP	NP	NP	NP	NP	NP	NP	NP	9'-4"	NP	10'-0"	NP	14'-8"	15'-4"	17'-4"	17'-4"	20'-0"	18'-8"	20'-0"	20'-0"	NP	NP
120 < V ≤ 130 (193 < V ≤ 209)	B	NJ	7'-4"	16'-0"	14'-0"	30'-0"	5'-4"	12'-0"	14'-0"	30'-0"	10'-0"	19'-4"	10'-0"	21'-4"	15'-4"	23'-4"	18'-0"	27'-4"	22'-0"	30'-0"	24'-8"	30'-0"	30'-0"	30'-0"
		30	NP	NP	NP	NP	NP	NP	NP	NP	10'-0"	16'-0"	10'-0"	17'-4"	15'-4"	19'-4"	18'-0"	20'-0"	20'-0"	20'-0"	23'-4"	20'-0"	26'-8"	20'-0"
		60	NP	NP	NP	NP	NP	NP	NP	NP	9'-4"	NP	10'-0"	NP	14'-0"	15'-4"	16'-8"	17'-4"	20'-0"	20'-0"	20'-0"	20'-0"	NP	NP
	C	NJ	6'-8"	14'-8"	12'-8"	26'-0"	5'-4"	11'-4"	12'-8"	25'-4"	10'-0"	19'-4"	10'-0"	21'-4"	14'-0"	23'-4"	16'-8"	27'-4"	19'-4"	30'-0"	21'-4"	30'-0"	24'-8"	30'-0"
		30	NP	NP	NP	NP	NP	NP	NP	NP	10'-0"	14'-0"	10'-0"	17'-4"	14'-0"	19'-4"	16'-8"	20'-0"	18'-8"	20'-0"	20'-8"	20'-0"	24'-0"	20'-0"
		60	NP	NP	NP	NP	NP	NP	NP	NP	7'-4"	NP	10'-0"	NP	12'-8"	14'-8"	16'-8"	17'-4"	18'-8"	18'-8"	20'-0"	20'-0"	NP	NP

Table values shall not be permitted to be interpolated.
NJ = No Joist
NP = Not Permitted

ft = 0.3048 m
in. = 25.4 mm
lb/ft² = 0.0479 kPa

TMS 403-13

DIRECT DESIGN HANDBOOK FOR MASONRY STRUCTURES

DIRECT DESIGN HANDBOOK FOR MASONRY STRUCTURES

Table 3.2-5(3b): Maximum Vertical Spans for Wall Segments (ft) if 0 psf < p_k ≤ 20 psf

Wind V, mph (kph)	Risk Category	Maximum L_{joist} (ft)	Unreinforced PCL Mortar Ungrouted Exterior	Unreinforced PCL Mortar Ungrouted Interior	Unreinforced PCL Mortar Fully Grouted Exterior	Unreinforced PCL Mortar Fully Grouted Interior	Unreinforced Masonry Ungrouted Exterior	Unreinforced Masonry Ungrouted Interior	Unreinforced Masonry Fully Grouted Exterior	Unreinforced Masonry Fully Grouted Interior	Vertical No. 5 at 120" oc Exterior	Vertical No. 5 at 120" oc Interior	Vertical No. 5 at 96" oc Exterior	Vertical No. 5 at 96" oc Interior	Vertical No. 5 at 72" oc Exterior	Vertical No. 5 at 72" oc Interior	Vertical No. 5 at 48" oc Exterior	Vertical No. 5 at 48" oc Interior	Vertical No. 5 at 32" oc Exterior	Vertical No. 5 at 32" oc Interior	Vertical No. 5 at 24" oc Exterior	Vertical No. 5 at 24" oc Interior	Vertical No. 5 at 16" oc Exterior	Vertical No. 5 at 16" oc Interior
130 < V ≤ 140 (209 < V ≤ 225)	B	NJ	6'-8"	14'-8"	13'-4"	28'-8"	5'-4	11'-4	12'-8"	28'-0"	10'-0"	10'-0"	10'-0"	20'-0"	14'-0"	22'-0"	16'-8"	26'-0"	20'-0"	28'-0"	22'-8"	30'-0"	26'-8"	30'-0"
		30	NP	NP	NP	NP	NP	NP	NP	NP	10'-0"	10'-0"	10'-0"	17'-4	14'-8"	19'-4	17'-4	20'-0"	20'-0"	20'-0"	22'-8"	20'-0"	25'-4	20'-0"
		60	NP	NP	NP	NP	NP	NP	NP	NP	8'-0"	NP	10'-0"	NP	13'-4	14'-8"	17'-4	17'-4	19'-4	18'-8"	20'-0"	20'-0"	NP	NP
	C	NJ	6'-0"	13'-4	12'-0"	24'-8"	4'-8"	10'-0"	11'-4	24'-0"	9'-4	10'-0"	10'-0"	20'-0"	12'-8"	22'-0"	15'-4	26'-0"	18'-0"	28'-0"	20'-0"	30'-0"	23'-4	30'-0"
		30	NP	NP	NP	NP	NP	NP	NP	NP	8'-8"	10'-0"	10'-0"	16'-0"	13'-4	19'-4	15'-4	20'-0"	18'-0"	20'-0"	20'-0"	20'-0"	22'-8"	20'-0"
		60	NP	NP	NP	NP	NP	NP	NP	NP	5'-4	NP	8'-0"	NP	11'-4	NP	15'-4	17'-4	17'-4	18'-8"	19'-4	20'-0"	NP	NP
140 < V ≤ 150 (225 < V ≤ 241)	B	NJ	6'-0"	14'-0"	12'-0"	26'-8"	4'-8"	10'-8"	12'-0"	26'-0"	9'-4	10'-0"	10'-0"	18'-8"	13'-4	21'-4	16'-0"	24'-8"	18'-8"	26'-8"	21'-4	28'-8"	25'-4	30'-0"
		30	NP	NP	NP	NP	NP	NP	NP	NP	9'-4	10'-0"	10'-0"	15'-4	13'-4	19'-4	16'-0"	20'-0"	18'-8"	20'-0"	20'-8"	20'-0"	24'-0"	20'-0"
		60	NP	NP	NP	NP	NP	NP	NP	NP	6'-0"	NP	8'-8"	NP	12'-0"	NP	14'-8"	17'-4	18'-8"	18'-8"	20'-0"	19'-4	NP	NP
	C	NJ	5'-4	12'-8"	11'-4	22'-8"	4'-0"	9'-4	10'-8"	22'-0"	8'-8"	10'-0"	10'-0"	18'-8"	12'-0"	21'-4	14'-8"	24'-8"	16'-8"	26'-8"	19'-4	28'-8"	22'-0"	30'-0"
		30	NP	NP	NP	NP	NP	NP	NP	NP	7'-4	NP	9'-4	12'-8"	12'-0"	18'-0"	14'-8"	20'-0"	16'-8"	20'-0"	18'-8"	20'-0"	21'-4	20'-0"
		60	NP	NP	NP	NP	NP	NP	NP	NP	NP	NP	6'-8"	NP	10'-0"	NP	14'-0"	17'-4	16'-8"	18'-3"	18'-8"	19'-4	NP	NP
150 < V ≤ 160 (241 < V ≤ 258)	B	NJ	6'-0"	12'-8"	11'-4	25'-4	4'-8"	10'-0"	11'-4	24'-0"	8'-8"	10'-0"	10'-0"	18'-8"	12'-8"	20'-0"	14'-8"	23'-4	18'-0"	26'-0"	20'-0"	27'-4	24'-0"	30'-0"
		30	NP	NP	NP	NP	NP	NP	NP	NP	8'-0"	8'-0"	10'-0"	12'-8"	12'-8"	18'-0"	15'-4	20'-0"	18'-0"	20'-0"	20'-0"	20'-0"	22'-8"	20'-0"
		60	NP	NP	NP	NP	NP	NP	NP	NP	NP	NP	7'-4	NP	10'-8"	NP	15'-4	17'-4	17'-4	18'-8"	19'-4	19'-4	NP	NP
	C	NJ	5'-4	11'-4	10'-8"	21'-4	4'-0"	8'-8"	10'-0"	20'-8"	8'-0"	NP	10'-0"	18'-0"	10'-0"	20'-0"	13'-4	23'-4	16'-0"	26'-0"	18'-0"	27'-4	20'-8"	30'-0"
		30	NP	NP	NP	NP	NP	NP	NP	NP	6'-0"	NP	8'-0"	NP	10'-0"	16'-0"	14'-0"	20'-0"	16'-0"	20'-0"	18'-0"	20'-0"	20'-0"	20'-0"
		60	NP	NP	NP	NP	NP	NP	NP	NP	NP	NP	NP	NP	8'-0"	NP	12'-8"	16'-8"	16'-0"	18'-0"	17'-4	19'-4	NP	NP
160 < V ≤ 180 (258 < V ≤ 290)	B	NJ	5'-4	11'-4	10'-0"	22'-0"	4'-0"	8'-8"	10'-0"	21'-4	8'-0"	NP	9'-4	10'-0"	10'-0"	18'-8"	13'-4	21'-4	16'-0"	23'-4	18'-0"	25'-4	21'-4	28'-8"
		30	NP	NP	NP	NP	NP	NP	NP	NP	6'-0"	NP	8'-0"	7'-4	10'-0"	13'-4	14'-0"	20'-0"	14'-0"	20'-0"	18'-0"	20'-0"	20'-8"	20'-0"
		60	NP	NP	NP	NP	NP	NP	NP	NP	NP	NP	NP	NP	8'-0"	NP	12'-0"	14'-0"	16'-0"	18'-0"	17'-4	19'-4	18'-8"	NP
	C	NJ	4'-8"	10'-0"	9'-4	19'-4	NP	8'-0"	8'-8"	18'-8"	7'-4	NP	8'-8"	10'-0"	10'-0"	18'-8"	12'-0"	21'-4	14'-8"	23'-4	16'-0"	25'-4	18'-8"	28'-0"
		30	NP	NP	NP	NP	NP	NP	NP	NP	4'-0"	NP	6'-0"	NP	8'-0"	10'-8"	12'-0"	19'-4	14'-8"	20'-0"	16'-0"	20'-0"	18'-8"	20'-0"
		60	NP	NP	NP	NP	NP	NP	NP	NP	NP	NP	NP	NP	5'-4	NP	10'-0"	NP	14'-0"	18'-0"	16'-0"	19'-4	NP	NP
190 < V ≤ 200 (306 < V ≤ 322)	B	NJ	4'-8"	10'-0"	9'-4	20'-0"	NP	8'-0"	8'-8"	19'-4	7'-4	NP	8'-0"	10'-0"	9'-4	17'-4	12'-0"	20'-0"	14'-8"	22'-0"	16'-0"	23'-4	19'-4	26'-0"
		30	NP	NP	NP	NP	NP	NP	NP	NP	4'-0"	NP	6'-0"	NP	8'-8"	9'-4	12'-0"	17'-4	14'-8"	20'-0"	16'-0"	20'-0"	18'-8"	20'-0"
		60	NP	NP	NP	NP	NP	NP	NP	NP	NP	NP	NP	NP	5'-4	NP	10'-0"	NP	14'-0"	18'-0"	16'-0"	19'-4	NP	NP
	C	NJ	4'-0"	9'-4	8'-0"	17'-4	NP	7'-4	8'-0"	16'-8"	6'-8"	NP	7'-4	10'-0"	8'-8"	17'-4	10'-8"	20'-0"	13'-4	22'-0"	14'-8"	23'-4	17'-4	25'-0"
		30	NP	NP	NP	NP	NP	NP	NP	NP	NP	NP	4'-0"	NP	6'-8"	NP	10'-0"	14'-8"	13'-4	20'-0"	14'-8"	20'-0"	16'-8"	20'-0"
		60	NP	NP	NP	NP	NP	NP	NP	NP	NP	NP	NP	NP	NP	NP	8'-0"	NP	12'-0"	NP	14'-8"	19'-4	NP	NP

Table values shall not be permitted to be interpolated.
NJ = No Joist
NP = Not Permitted

ft = 0.3048 m
in. = 25.4 mm
lb/ft² = 0.0479 kPa

TMS 403-13

DIRECT DESIGN HANDBOOK FOR MASONRY STRUCTURES

DIRECT DESIGN HANDBOOK FOR MASONRY STRUCTURES

Table 3.2-5(4a): Maximum Vertical Spans for Wall Segments (ft) if $p_4 = 0$ psf

Wind V, mph (kph)	Risk Category	Maximum L_{joist} (ft)	Unreinforced PCL Mortar — Ungrouted Exterior	Ungrouted Interior	Fully Grouted Exterior	Fully Grouted Interior	Unreinforced Masonry — Ungrouted Exterior	Ungrouted Interior	Fully Grouted Exterior	Fully Grouted Interior	Vertical No. 5 at 120" oc Exterior	Interior	Vertical No. 5 at 96" oc Exterior	Interior	Vertical No. 5 at 72" oc Exterior	Interior	Vertical No. 5 at 48" oc Exterior	Interior	Vertical No. 5 at 32" oc Exterior	Interior	Vertical No. 5 at 24" oc Exterior	Interior	Vertical No. 5 at 16" oc Exterior	Interior
V ≤ 110 (V ≤ 177)	B	NJ	8'-8"	18'-8"	16'-8"	30'-0"	6'-8"	14'-8"	16'-8"	30'-0"	10'-0"	22'-0"	15'-4"	24'-0"	17'-4	26'-0"	21'-4	30'-0"	25'-4	30'-0"	28'-0"	30'-0"	30'-0"	30'-0"
		30	NP	NP	NP	NP	NP	NP	NP	NP	10'-0"	20'-0"	15'-4	20'-0"	18'-0"	20'-0"	21'-4	20'-0"	24'-8"	30'-0"	26'-8"	30'-0"	30'-0"	30'-0"
		60	NP	NP	NP	NP	NP	NP	NP	NP	10'-0"	16'-8"	16'-0"	20'-0"	17'-4	20'-0"	20'-8"	20'-0"	24'-0"	20'-0"	26'-0"	22'-0"	29'-4	30'-0"
	C	NJ	8'-0"	16'-8"	14'-8"	30'-0"	6'-0"	13'-4	14'-8"	30'-0"	10'-0"	22'-0"	14'-0"	24'-0"	15'-4	26'-0"	18'-8"	30'-0"	22'-0"	30'-0"	24'-8"	30'-0"	28'-0"	30'-0"
		30	NP	NP	NP	NP	NP	NP	NP	NP	10'-0"	20'-0"	14'-8"	20'-0"	16'-0"	20'-0"	18'-8"	30'-0"	21'-4	20'-0"	24'-0"	30'-0"	27'-4	30'-0"
		60	NP	NP	NP	NP	NP	NP	NP	NP	10'-0"	20'-0"	13'-4	19'-4	16'-0"	20'-0"	18'-0"	20'-0"	21'-4	20'-0"	23'-0"	20'-0"	26'-8"	30'-0"
110 < V ≤ 115 (177 < V ≤ 185)	B	NJ	8'-0"	18'-0"	16'-0"	30'-0"	6'-0"	14'-0"	16'-0"	30'-0"	10'-0"	21'-4	14'-8"	22'-8"	16'-8"	25'-4	20'-8"	29'-4	24'-0"	30'-0"	27'-4	30'-0"	30'-0"	30'-0"
		30	NP	NP	NP	NP	NP	NP	NP	NP	10'-0"	20'-0"	15'-4	20'-0"	17'-4	20'-0"	20'-0"	20'-0"	23'-4	30'-0"	26'-0"	30'-0"	28'-8"	30'-0"
		60	NP	NP	NP	NP	NP	NP	NP	NP	10'-0"	14'-8"	14'-8"	20'-0"	17'-4	20'-0"	20'-0"	20'-0"	22'-8"	20'-0"	25'-4	20'-0"	28'-8"	30'-0"
	C	NJ	7'-4	16'-0"	14'-8"	29'-4	6'-0"	12'-8"	14'-8"	28'-8"	10'-0"	21'-4	10'-0"	22'-8"	14'-8"	25'-4	18'-0"	29'-4	21'-4	30'-0"	24'-0"	30'-0"	27'-4	30'-0"
		30	NP	NP	NP	NP	NP	NP	NP	NP	10'-0"	19'-4	10'-0"	20'-0"	15'-4	29'-1	18'-0"	29'-1	20'-8"	30'-0"	23'-4	30'-0"	26'-8"	30'-0"
		60	NP	NP	NP	NP	NP	NP	NP	NP	10'-0"	14'-8"	10'-0"	20'-0"	15'-4	20'-0"	18'-0"	20'-0"	20'-8"	20'-0"	22'-8"	20'-0"	26'-0"	30'-0"
115 < V ≤ 120 (185 < V ≤ 193)	B	NJ	8'-0"	17'-4	15'-4	30'-0"	6'-0"	13'-4	14'-8"	30'-0"	10'-0"	20'-8"	10'-0"	22'-0"	16'-0"	24'-8"	19'-4	28'-8"	23'-4	30'-0"	26'-0"	30'-0"	30'-0"	30'-0"
		30	NP	NP	NP	NP	NP	NP	NP	NP	10'-0"	19'-4	14'-8"	20'-0"	16'-8"	24'-8"	19'-4	28'-8"	22'-8"	30'-0"	25'-4	30'-0"	28'-8"	30'-0"
		60	NP	NP	NP	NP	NP	NP	NP	NP	10'-0"	NP	14'-8"	18'-0"	16'-8"	20'-0"	19'-4	20'-0"	22'-0"	20'-0"	24'-8"	20'-0"	28'-0"	20'-8"
	C	NJ	7'-4	15'-4	14'-0"	28'-0"	5'-4	12'-0"	13'-4	27'-4	9'-4	20'-8"	10'-0"	22'-0"	14'-8"	24'-8"	17'-4	28'-8"	20'-8"	30'-0"	22'-8"	30'-0"	25'-4	30'-0"
		30	NP	NP	NP	NP	NP	NP	NP	NP	10'-0"	17'-4	10'-0"	22'-0"	14'-8"	20'-0"	17'-4	28'-8"	20'-8"	30'-0"	22'-8"	30'-0"	24'-8"	30'-0"
		60	NP	NP	NP	NP	NP	NP	NP	NP	9'-4	NP	10'-0"	NP	14'-8"	20'-0"	17'-4	20'-0"	22'-0"	20'-0"	22'-0"	20'-0"	24'-8"	20'-8"
120 < V ≤ 130 (193 < V ≤ 209)	B	NJ	7'-4	16'-0"	14'-0"	30'-0"	5'-4	12'-0"	14'-0"	30'-0"	10'-0"	19'-4	10'-0"	21'-4	15'-4	23'-4	18'-0"	27'-4	22'-0"	30'-0"	24'-8"	30'-0"	28'-0"	30'-0"
		30	NP	NP	NP	NP	NP	NP	NP	NP	10'-0"	16'-8"	10'-0"	20'-0"	15'-4	20'-0"	18'-0"	27'-4	21'-4	30'-0"	24'-0"	30'-0"	27'-4	30'-0"
		60	NP	NP	NP	NP	NP	NP	NP	NP	10'-0"	19'-4	10'-0"	21'-4	14'-0"	23'-4	16'-0"	27'-4	19'-4	30'-0"	23'-4	30'-0"	26'-8"	30'-0"
	C	NJ	6'-8"	14'-8"	12'-8"	26'-0"	5'-4	11'-4	12'-8"	25'-4	10'-0"	19'-4	10'-0"	21'-4	14'-0"	23'-4	16'-0"	27'-4	19'-4	30'-0"	21'-4	30'-0"	24'-8"	30'-0"
		30	NP	NP	NP	NP	NP	NP	NP	NP	10'-0"	14'-0"	10'-0"	18'-8"	14'-0"	20'-0"	16'-0"	27'-4	19'-4	30'-0"	20'-8"	30'-0"	24'-8"	30'-0"
		60	NP	NP	NP	NP	NP	NP	NP	NP	7'-4	NP	10'-0"	NP	12'-8"	17'-4	16'-0"	20'-0"	18'-8"	20'-0"	20'-8"	20'-0"	24'-0"	20'-0"

Table values shall not be permitted to be interpolated.
NJ = No Joist
NP = Not Permitted

ft = 0.3048 m
in. = 25.4 mm
lb/ft² = 0.0479 kPa

TMS 403-13

DIRECT DESIGN HANDBOOK FOR MASONRY STRUCTURES

453

Table 3.2-5(4b): Maximum Vertical Spans for Wall Segments (ft) if $p_{fl} = 0$ psf

Wind V, mph (kph)	Risk Category	Maximum L_{joist} (ft)	Unreinforced PCL Mortar Ungrouted Exterior	Ungrouted Interior	Fully Grouted Exterior	Fully Grouted Interior	Unreinforced Masonry Ungrouted Exterior	Ungrouted Interior	Fully Grouted Exterior	Fully Grouted Interior	Vertical No. 5 at 120" oc Exterior	Interior	Vertical No. 5 at 96" oc Exterior	Interior	Vertical No. 5 at 72" oc Exterior	Interior	Vertical No. 5 at 48" oc Exterior	Interior	Vertical No. 5 at 32" oc Exterior	Interior	Vertical No. 5 at 24" oc Exterior	Interior	Vertical No. 5 at 16" oc Exterior	Interior
130 < V ≤ 140 (209 < V ≤ 225)	B	NJ	6'-8"	14'-8"	13'-4"	28'-8"	5'-4"	11'-4"	12'-8"	28'-0"	10'-0"	10'-0"	10'-0"	20'-0"	14'-0"	22'-0"	16'-8"	26'-0"	20'-0"	28'-0"	22'-8"	30'-0"	26'-8"	30'-0"
		30	NP	NP	NP	NP	NP	NP	NP	NP	10'-0"	10'-0"	10'-0"	18'-0"	14'-0"	20'-0"	17'-4"	26'-8"	20'-0"	30'-0"	22'-8"	30'-0"	26'-0"	30'-0"
		60	NP	NP	NP	NP	NP	NP	NP	NP	8'-0"	NP	10'-0"	16'-8"	13'-4"	16'-8"	17'-4"	20'-0"	20'-0"	20'-0"	22'-0"	20'-0"	25'-4"	22'-8"
	C	NJ	6'-0"	13'-4"	12'-0"	24'-8"	4'-8"	10'-0"	11'-4"	24'-0"	9'-4"	10'-0"	10'-0"	20'-0"	12'-8"	22'-0"	15'-4"	26'-0"	18'-0"	28'-0"	20'-0"	30'-0"	23'-4"	30'-0"
		30	NP	NP	NP	NP	NP	NP	NP	NP	8'-8"	10'-0"	10'-0"	16'-0"	13'-4"	20'-0"	15'-4"	26'-8"	18'-0"	30'-0"	20'-0"	30'-0"	22'-8"	30'-0"
		60	NP	NP	NP	NP	NP	NP	NP	NP	5'-4"	NP	8'-0"	NP	11'-4"	NP	15'-4"	20'-0"	18'-0"	20'-0"	20'-0"	20'-0"	22'-8"	20'-0"
140 < V ≤ 150 (225 < V ≤ 241)	B	NJ	6'-0"	14'-0"	12'-0"	26'-8"	4'-8"	10'-8"	12'-0"	26'-0"	9'-4"	10'-0"	10'-0"	18'-8"	13'-4"	21'-4"	16'-0"	24'-8"	18'-8"	26'-8"	21'-4"	28'-8"	25'-4"	30'-0"
		30	NP	NP	NP	NP	NP	NP	NP	NP	9'-4"	10'-0"	10'-0"	15'-4"	13'-4"	20'-0"	16'-0"	25'-4"	18'-8"	28'-8"	21'-4"	30'-0"	24'-8"	30'-0"
		60	NP	NP	NP	NP	NP	NP	NP	NP	6'-0"	NP	8'-8"	NP	12'-0"	NP	16'-0"	20'-0"	20'-8"	20'-0"	20'-8"	20'-0"	24'-0"	20'-0"
	C	NJ	5'-4"	12'-8"	11'-4"	22'-8"	4'-0"	9'-4"	10'-8"	22'-0"	8'-8"	10'-0"	10'-0"	18'-8"	10'-0"	21'-4"	14'-8"	24'-8"	16'-8"	26'-8"	19'-4"	28'-8"	22'-0"	30'-0"
		30	NP	NP	NP	NP	NP	NP	NP	NP	7'-4"	NP	9'-4"	12'-8"	10'-0"	18'-8"	14'-8"	25'-4"	16'-8"	28'-8"	18'-8"	30'-0"	21'-4"	30'-0"
		60	NP	NP	NP	NP	NP	NP	NP	NP	NP	NP	6'-8"	NP	NP	NP	14'-0"	20'-0"	16'-8"	20'-0"	18'-8"	20'-0"	21'-4"	20'-0"
150 < V ≤ 160 (241 < V ≤ 258)	B	NJ	6'-0"	12'-8"	11'-4"	25'-4"	4'-0"	10'-0"	11'-4"	24'-0"	8'-8"	10'-0"	10'-0"	18'-0"	12'-8"	20'-0"	14'-8"	23'-4"	18'-0"	26'-0"	20'-0"	27'-4"	24'-0"	30'-0"
		30	NP	NP	NP	NP	NP	NP	NP	NP	8'-0"	8'-0"	10'-0"	12'-8"	12'-8"	18'-0"	15'-4"	24'-8"	18'-0"	28'-0"	20'-0"	30'-0"	23'-4"	30'-0"
		60	NP	NP	NP	NP	NP	NP	NP	NP	NP	NP	7'-4"	NP	10'-8"	NP	15'-4"	16'-8"	18'-0"	20'-0"	20'-0"	20'-0"	22'-8"	20'-0"
	C	NJ	5'-4"	11'-4"	10'-8"	21'-4"	4'-0"	8'-8"	10'-0"	20'-8"	8'-0"	10'-0"	10'-0"	18'-0"	10'-0"	20'-0"	13'-4"	23'-4"	16'-0"	26'-0"	18'-0"	27'-4"	20'-8"	30'-0"
		30	NP	NP	NP	NP	NP	NP	NP	NP	6'-0"	NP	8'-0"	9'-4"	10'-0"	16'-0"	14'-0"	24'-8"	16'-0"	28'-0"	18'-0"	30'-0"	20'-8"	30'-0"
		60	NP	NP	NP	NP	NP	NP	NP	NP	NP	NP	8'-0"	NP	8'-0"	NP	12'-8"	16'-8"	16'-0"	20'-0"	18'-0"	20'-0"	20'-0"	20'-0"
160 < V ≤ 180 (258 < V ≤ 290)	B	NJ	5'-4"	11'-4"	10'-0"	22'-0"	NP	8'-8"	10'-0"	21'-4"	8'-0"	10'-0"	10'-0"	18'-0"	10'-0"	18'-8"	13'-4"	21'-4"	16'-0"	23'-4"	18'-0"	25'-4"	21'-4"	28'-0"
		30	NP	NP	NP	NP	NP	NP	NP	NP	7'-4"	NP	8'-8"	NP	8'-8"	13'-4"	13'-4"	21'-4"	16'-0"	24'-8"	18'-0"	28'-8"	20'-8"	30'-0"
		60	NP	NP	NP	NP	NP	NP	NP	NP	4'-0"	NP	6'-0"	NP	10'-8"	NP	12'-0"	14'-0"	14'-8"	20'-0"	16'-0"	20'-0"	18'-8"	20'-0"
	C	NJ	4'-8"	10'-0"	9'-4"	19'-4"	NP	8'-0"	8'-8"	18'-8"	7'-4"	9'-4"	10'-0"	10'-0"	9'-4"	18'-8"	12'-0"	21'-4"	14'-8"	23'-4"	16'-0"	25'-4"	18'-8"	28'-8"
		30	NP	NP	NP	NP	NP	NP	NP	NP	4'-0"	NP	6'-0"	NP	10'-8"	NP	12'-0"	19'-4"	14'-8"	24'-8"	16'-0"	28'-8"	18'-8"	30'-0"
		60	NP	NP	NP	NP	NP	NP	NP	NP	NP	NP	NP	NP	5'-4"	NP	10'-0"	NP	14'-0"	18'-0"	16'-0"	20'-0"	18'-8"	20'-0"
190 < V ≤ 200 (306 < V ≤ 322)	B	NJ	4'-8"	10'-0"	9'-4"	20'-0"	NP	8'-0"	8'-8"	19'-4"	7'-4"	9'-4"	8'-8"	10'-0"	9'-4"	17'-4"	12'-0"	20'-0"	14'-8"	22'-0"	16'-0"	23'-4"	19'-4"	26'-0"
		30	NP	NP	NP	NP	NP	NP	NP	NP	4'-0"	NP	6'-0"	NP	9'-4"	NP	12'-0"	17'-4"	14'-8"	23'-4"	16'-0"	26'-8"	19'-4"	30'-0"
		60	NP	NP	NP	NP	NP	NP	NP	NP	NP	NP	NP	NP	5'-4"	NP	10'-0"	NP	14'-0"	18'-0"	16'-0"	20'-0"	18'-8"	20'-0"
	C	NJ	4'-0"	9'-4"	8'-0"	17'-4"	NP	7'-4"	8'-0"	16'-8"	6'-8"	9'-4"	7'-4"	10'-0"	8'-8"	17'-4"	10'-8"	20'-0"	13'-4"	22'-0"	14'-8"	23'-4"	17'-4"	26'-0"
		30	NP	NP	NP	NP	NP	NP	NP	NP	NP	NP	4'-0"	NP	6'-8"	NP	10'-0"	14'-8"	13'-4"	20'-0"	14'-8"	21'-4"	16'-8"	30'-0"
		60	NP	NP	NP	NP	NP	NP	NP	NP	NP	NP	NP	NP	NP	NP	8'-0"	NP	12'-0"	NP	14'-8"	20'-0"	16'-8"	20'-0"

Table values shall not be permitted to be interpolated.

NJ = No Joist
NP = Not Permitted

ft = 0.3048 m
in. = 25.4 mm
lb/ft² = 0.0479 kPa

DIRECT DESIGN HANDBOOK FOR MASONRY STRUCTURES

Table 3.2-6(1a): Maximum Vertical Spans (ft-in.) for Walls without Openings Constructed using Lightweight Concrete Masonry Units for Seismic Conditions and for Ground Snow Loads, p_g, up to 60 psf

Seismic S_S	Risk Category	Max. L_{joist} (ft)	Unreinforced Ungrouted PCL Mortar Exterior	Unreinforced Ungrouted PCL Mortar Interior	Unreinforced Fully Grouted PCL Mortar Exterior	Unreinforced Fully Grouted PCL Mortar Interior	Unreinforced Ungrouted MC Mortar Exterior	Unreinforced Ungrouted MC Mortar Interior	Unreinforced Fully Grouted MC Mortar Exterior	Unreinforced Fully Grouted MC Mortar Interior	Vertical No.5 at 120" oc Exterior	Vertical No.5 at 120" oc Interior	Vertical No.5 at 96" oc Exterior	Vertical No.5 at 96" oc Interior	Vertical No.5 at 72" oc Exterior	Vertical No.5 at 72" oc Interior	Vertical No.5 at 48" oc Exterior	Vertical No.5 at 48" oc Interior	Vertical No.5 at 32" oc Exterior	Vertical No.5 at 32" oc Interior	Vertical No.5 at 24" oc Exterior	Vertical No.5 at 24" oc Interior	Vertical No.5 at 16" oc Exterior	Vertical No.5 at 16" oc Interior
0.156	I or II	NJ	18'-8"	18'-8"	30'-0"	30'-0"	14'-8"	14'-8"	30'-0"	30'-0"	24'-0"	24'-0"	26'-0"	26'-0"	28'-8"	28'-8"	30'-0"	30'-0"	30'-0"	30'-0"	30'-0"	30'-0"	30'-0"	30'-0"
		30	NP	NP	NP	NP	NP	NP	NP	NP	21'-4"	18'-8"	23'-4"	20'-8"	26'-0"	24'-0"	30'-0"	26'-8"	30'-0"	30'-0"	30'-0"	30'-0"	30'-0"	30'-0"
		60	NP	NP	NP	NP	NP	NP	NP	NP	19'-4"	14'-8"	20'-8"	16'-0"	24'-0"	18'-8"	27'-4"	23'-4"	30'-0"	26'-0"	30'-0"	28'-0"	30'-0"	30'-0"
	III or IV	NJ	18'-8"	18'-8"	30'-0"	30'-0"	14'-8"	14'-8"	30'-0"	30'-0"	24'-0"	24'-0"	26'-0"	26'-0"	28'-8"	28'-8"	30'-0"	30'-0"	30'-0"	30'-0"	30'-0"	30'-0"	30'-0"	30'-0"
		30	NP	NP	NP	NP	NP	NP	NP	NP	21'-4"	18'-8"	23'-4"	20'-8"	26'-0"	24'-0"	30'-0"	26'-8"	30'-0"	28'-8"	30'-0"	30'-0"	30'-0"	30'-0"
		60	NP	NP	NP	NP	NP	NP	NP	NP	19'-4"	14'-8"	20'-8"	16'-0"	24'-0"	18'-8"	27'-4"	23'-4"	30'-0"	26'-0"	30'-0"	28'-0"	30'-0"	30'-0"
0.32	I or II	NJ	16'-0"	16'-0"	30'-0"	30'-0"	12'-8"	12'-8"	30'-0"	30'-0"	21'-4"	21'-4"	23'-4"	23'-4"	26'-0"	26'-0"	30'-0"	30'-0"	30'-0"	30'-0"	30'-0"	30'-0"	30'-0"	30'-0"
		30	NP	NP	NP	NP	NP	NP	NP	NP	19'-4"	17'-4"	21'-4"	19'-4"	24'-0"	22'-0"	28'-0"	26'-8"	30'-0"	28'-8"	30'-0"	30'-0"	30'-0"	30'-0"
		60	NP	NP	NP	NP	NP	NP	NP	NP	17'-4"	14'-0"	19'-4"	16'-0"	22'-0"	18'-0"	26'-8"	21'-4"	28'-8"	24'-8"	30'-0"	27'-4"	30'-0"	30'-0"
	III or IV	NJ	13'-4"	13'-4"	26'-0"	26'-0"	10'-0"	10'-0"	24'-8"	24'-8"	10'-0"	10'-0"	20'-0"	20'-0"	22'-8"	22'-8"	26'-0"	26'-0"	28'-0"	28'-0"	29'-4"	29'-4"	30'-0"	30'-0"
		30	NP	NP	NP	NP	NP	NP	NP	NP	10'-0"	10'-0"	18'-8"	17'-4"	21'-4"	20'-0"	24'-8"	23'-4"	26'-8"	25'-4"	28'-0"	26'-8"	30'-0"	28'-8"
		60	NP	NP	NP	NP	NP	NP	NP	NP	10'-0"	10'-0"	17'-4"	15'-4"	20'-0"	17'-4"	23'-4"	21'-4"	25'-4"	23'-4"	26'-8"	24'-8"	29'-4"	27'-4"
0.553	I or II	NJ	NP	NP	NP	NP	NP	NP	NP	NP	10'-0"	10'-0"	20'-0"	20'-0"	22'-8"	22'-8"	26'-0"	26'-0"	27'-4"	27'-4"	28'-8"	28'-8"	30'-0"	30'-0"
		30	NP	NP	NP	NP	NP	NP	NP	NP	10'-0"	10'-0"	18'-8"	17'-4"	20'-8"	19'-4"	24'-8"	23'-4"	26'-8"	25'-4"	28'-0"	26'-8"	30'-0"	28'-8"
		60	NP	NP	NP	NP	NP	NP	NP	NP	10'-0"	10'-0"	17'-4"	14'-8"	20'-0"	17'-8"	23'-4"	20'-8"	25'-4"	23'-4"	26'-8"	24'-8"	27'-4"	27'-4"
	III or IV	NJ	NP	NP	NP	NP	NP	NP	NP	NP	10'-0"	10'-0"	17'-4"	17'-4"	19'-4"	19'-4"	22'-8"	22'-8"	24'-0"	24'-0"	25'-4"	25'-4"	26'-8"	26'-8"
		30	NP	NP	NP	NP	NP	NP	NP	NP	10'-0"	10'-0"	16'-0"	15'-4"	18'-0"	17'-4"	21'-4"	20'-8"	22'-8"	22'-0"	24'-8"	23'-4"	26'-0"	25'-4"
		60	NP	NP	NP	NP	NP	NP	NP	NP	10'-0"	10'-0"	15'-4"	13'-4"	17'-4"	15'-4"	20'-8"	18'-8"	22'-0"	20'-8"	23'-4"	22'-0"	25'-4"	24'-0"
1.000	I or II	NJ	NP	NP	NP	NP	NP	NP	NP	NP	10'-0"	10'-0"	16'-8"	16'-8"	19'-4"	19'-4"	22'-8"	22'-8"	24'-0"	24'-0"	25'-4"	25'-4"	26'-8"	26'-8"
		30	NP	NP	NP	NP	NP	NP	NP	NP	10'-0"	10'-0"	16'-0"	15'-4"	18'-8"	17'-4"	22'-0"	20'-8"	23'-4"	22'-0"	24'-8"	23'-4"	26'-0"	25'-4"
		60	NP	NP	NP	NP	NP	NP	NP	NP	10'-0"	10'-0"	15'-4"	13'-4"	17'-4"	16'-8"	20'-8"	18'-8"	22'-8"	20'-8"	23'-4"	22'-0"	25'-4"	24'-0"
	III or IV	NJ	NP	NP	NP	NP	NP	NP	NP	NP	10'-0"	10'-0"	10'-0"	10'-0"	16'-8"	16'-8"	19'-4"	19'-4"	20'-8"	20'-8"	22'-0"	22'-0"	23'-4"	23'-4"
		30	NP	NP	NP	NP	NP	NP	NP	NP	10'-0"	10'-0"	10'-0"	10'-0"	16'-0"	15'-4"	18'-8"	18'-0"	20'-8"	19'-4"	21'-4"	20'-8"	22'-8"	22'-0"
		60	NP	NP	NP	NP	NP	NP	NP	NP	10'-0"	10'-0"	10'-0"	10'-0"	15'-4"	14'-0"	18'-0"	16'-8"	19'-4"	18'-3"	20'-8"	19'-4"	22'-0"	21'-4"

Table values shall not be permitted to be interpolated.
NP = Not Permitted
NJ = No Joist
ft = 0.3048 m
in. = 25.4 mm
lb/ft² = 0.0479 kPa
Unreinforced Ungrouted PCL or Mortar Cement Mortar
Unreinforced Fully Grouted PCL or Mortar Cement Mortar
The density of concrete masonry units shall be less than 105 lb/ft³ (1680 kg/m³).

Table 3.2-6(1b): Maximum Vertical Spans (ft-in.) for Walls without Openings Constructed using Lightweight Concrete Masonry Units for Seismic Conditions and for Ground Snow Loads, p_g, up to 60 psf

Seismic S_S	Risk Category	Max. L_{joist} (ft)	Unreinforced Ungrouted PCL Mortar Exterior	Unreinforced Ungrouted PCL Mortar Interior	Unreinforced Fully Grouted PCL Mortar Exterior	Unreinforced Fully Grouted PCL Mortar Interior	Unreinforced Ungrouted MC Mortar Exterior	Unreinforced Ungrouted MC Mortar Interior	Unreinforced Fully Grouted MC Mortar Exterior	Unreinforced Fully Grouted MC Mortar Interior	Vertical No.5 at 120 in. oc Exterior	Vertical No.5 at 120 in. oc Interior	Vertical No.5 at 96 in. oc Exterior	Vertical No.5 at 96 in. oc Interior	Vertical No.5 at 72 in. oc Exterior	Vertical No.5 at 72 in. oc Interior	Vertical No.5 at 48 in. oc Exterior	Vertical No.5 at 48 in. oc Interior	Vertical No.5 at 32 in. oc Exterior	Vertical No.5 at 32 in. oc Interior	Vertical No.5 at 24 in. oc Exterior	Vertical No.5 at 24 in. oc Interior	Vertical No.5 at 16 in. oc Exterior	Vertical No.5 at 16 in. oc Interior
1.500	I or II	NJ	NP	NP	NP	NP	NP	NP	NP	NP	10'-0"	10'-0"	10'-0"	10'-0"	17'-4"	17'-4"	20'-0"	20'-0"	21'-4"	21'-4"	22'-8"	22'-8"	24'-0"	24'-0"
1.500	I or II	30	NP	NP	NP	NP	NP	NP	NP	NP	10'-0"	10'-0"	10'-0"	10'-0"	16'-8"	16'-0"	19'-4"	18'-8"	20'-8"	20'-0"	22'-0"	21'-4"	24'-4"	22'-8"
1.500	I or II	60	NP	NP	NP	NP	NP	NP	NP	NP	10'-0"	10'-0"	10'-0"	10'-0"	16'-0"	14'-0"	18'-8"	17'-4"	20'-0"	18'-8"	21'-4"	20'-0"	22'-8"	22'-0"
1.500	III or IV	0	NP	NP	NP	NP	NP	NP	NP	NP	8'-8"	8'-8"	10'-0"	10'-0"	10'-0"	10'-0"	17'-4"	17'-4"	18'-8"	18'-8"	19'-4"	19'-4"	20'-8"	20'-8"
1.500	III or IV	30	NP	NP	NP	NP	NP	NP	NP	NP	8'-8"	8'-8"	10'-0"	10'-0"	10'-0"	10'-0"	16'-8"	16'-0"	18'-0"	17'-4"	19'-4"	18'-8"	20'-8"	20'-0"
1.500	III or IV	60	NP	NP	NP	NP	NP	NP	NP	NP	8'-8"	8'-8"	10'-0"	10'-0"	10'-0"	10'-0"	16'-8"	15'-4"	17'-4"	16'-8"	18'-8"	18'-0"	20'-0"	20'-0"
2.000	I or II	NJ	NP	NP	NP	NP	NP	NP	NP	NP	8'-8"	8'-8"	10'-0"	10'-0"	15'-4"	15'-4"	18'-0"	18'-0"	19'-4"	19'-4"	20'-0"	20'-0"	21'-4"	21'-4"
2.000	I or II	30	NP	NP	NP	NP	NP	NP	NP	NP	8'-8"	8'-8"	10'-0"	10'-0"	14'-8"	14'-0"	17'-4"	16'-8"	18'-8"	18'-0"	20'-0"	19'-4"	21'-4"	20'-8"
2.000	III or IV	60	NP	NP	NP	NP	NP	NP	NP	NP	8'-8"	8'-8"	10'-0"	10'-0"	14'-8"	13'-4"	16'-8"	16'-0"	18'-0"	17'-4"	19'-4"	18'-8"	20'-8"	20'-0"
2.500	I or II	NJ	NP	NP	NP	NP	NP	NP	NP	NP	7'-4"	7'-4"	9'-4"	9'-4"	10'-0"	10'-0"	15'-4"	15'-4"	16'-8"	16'-8"	17'-4"	17'-4"	18'-8"	18'-8"
2.500	I or II	30	NP	NP	NP	NP	NP	NP	NP	NP	7'-4"	7'-4"	9'-4"	9'-4"	10'-0"	10'-0"	15'-4"	14'-8"	16'-0"	16'-0"	17'-4"	16'-8"	18'-8"	18'-0"
2.500	III or IV	60	NP	NP	NP	NP	NP	NP	NP	NP	7'-4"	7'-4"	9'-4"	9'-4"	10'-0"	10'-0"	14'-8"	14'-0"	16'-0"	15'-4"	16'-8"	16'-0"	18'-0"	17'-4"
2.500	I or II	NJ	NP	NP	NP	NP	NP	NP	NP	NP	8'-0"	8'-0"	10'-0"	10'-0"	10'-0"	10'-0"	16'-8"	16'-8"	18'-0"	18'-0"	18'-8"	18'-8"	20'-0"	20'-0"
2.500	I or II	30	NP	NP	NP	NP	NP	NP	NP	NP	8'-0"	8'-0"	10'-0"	10'-0"	10'-0"	10'-0"	16'-0"	15'-4"	17'-4"	16'-8"	18'-0"	18'-0"	19'-4"	19'-4"
2.500	I or II	60	NP	NP	NP	NP	NP	NP	NP	NP	8'-0"	8'-0"	10'-0"	10'-0"	10'-0"	10'-0"	16'-0"	14'-8"	16'-8"	16'-0"	18'-0"	17'-4"	19'-4"	18'-8"
2.500	III or IV	NJ	NP	NP	NP	NP	NP	NP	NP	NP	6'-8"	6'-8"	8'-0"	8'-0"	10'-0"	10'-0"	14'-8"	14'-8"	15'-4"	15'-4"	16'-0"	16'-0"	17'-4"	17'-4"
2.500	III or IV	30	NP	NP	NP	NP	NP	NP	NP	NP	6'-8"	6'-8"	8'-0"	8'-0"	10'-0"	10'-0"	14'-0"	13'-4"	15'-4"	14'-8"	16'-0"	15'-4"	16'-8"	16'-8"
2.500	III or IV	60	NP	NP	NP	NP	NP	NP	NP	NP	6'-8"	6'-8"	8'-0"	8'-0"	10'-0"	10'-0"	14'-0"	12'-8"	14'-0"	14'-0"	15'-4"	14'-8"	16'-8"	16'-0"
3.000	I or II	NJ	NP	NP	NP	NP	NP	NP	NP	NP	7'-4"	7'-4"	9'-4"	9'-4"	10'-0"	10'-0"	15'-4"	15'-4"	16'-8"	16'-8"	17'-4"	17'-4"	18'-8"	18'-8"
3.000	I or II	30	NP	NP	NP	NP	NP	NP	NP	NP	7'-4"	7'-4"	9'-4"	9'-4"	10'-0"	10'-0"	15'-4"	14'-8"	16'-0"	16'-0"	17'-4"	16'-8"	18'-0"	18'-0"
3.000	I or II	60	NP	NP	NP	NP	NP	NP	NP	NP	7'-4"	7'-4"	9'-4"	9'-4"	10'-0"	10'-0"	14'-8"	14'-0"	16'-0"	15'-4"	16'-8"	16'-0"	18'-0"	17'-4"
3.000	III or IV	NJ	NP	NP	NP	NP	NP	NP	NP	NP	6'-0"	6'-0"	7'-4"	7'-4"	10'-0"	10'-0"	13'-4"	13'-4"	14'-8"	14'-8"	15'-4"	15'-4"	16'-0"	16'-0"
3.000	III or IV	30	NP	NP	NP	NP	NP	NP	NP	NP	6'-0"	6'-0"	7'-4"	7'-4"	10'-0"	10'-0"	13'-4"	12'-8"	14'-0"	14'-0"	14'-8"	14'-8"	16'-0"	15'-4"
3.000	III or IV	60	NP	NP	NP	NP	NP	NP	NP	NP	6'-0"	6'-0"	7'-4"	7'-4"	10'-0"	10'-0"	12'-8"	12'-0"	14'-0"	13'-4"	14'-8"	14'-0"	15'-4"	15'-4"

Table values shall not be permitted to be interpolated.
NP = Not Permitted
NJ = No Joist
ft = 0.3048 m
in. = 25.4 mm
lb/ft² = 0.0479 kPa
Unreinforced Ungrouted PCL or Mortar Cement Mortar
Unreinforced Fully Grouted PCL or Mortar Cement Mortar
The density of concrete masonry units shall be less than 105 lb/ft³ (1680 kg/m³).

DIRECT DESIGN HANDBOOK FOR MASONRY STRUCTURES

Table 3.2-6(2a): Maximum Vertical Spans (ft-in.) for Walls without Openings Constructed using Medium and Normal Weight Concrete Masonry Units for Seismic Conditions and for Ground Snow Loads, p_g, up to 60 psf

Seismic S_s	Risk Category	Max. L_{joist} (ft)	Unreinforced Ungrouted PCL Mortar Exterior	Unreinforced Ungrouted PCL Mortar Interior	Unreinforced Fully Grouted PCL Mortar Exterior	Unreinforced Fully Grouted PCL Mortar Interior	Unreinforced Ungrouted MC Mortar Exterior	Unreinforced Ungrouted MC Mortar Interior	Unreinforced Fully Grouted MC Mortar Exterior	Unreinforced Fully Grouted MC Mortar Interior	Vertical No. 5 at 120 in. oc Exterior	Vertical No. 5 at 120 in. oc Interior	Vertical No. 5 at 96 in. oc Exterior	Vertical No. 5 at 96 in. oc Interior	Vertical No. 5 at 72 in. oc Exterior	Vertical No. 5 at 72 in. oc Interior	Vertical No. 5 at 48 in. oc Exterior	Vertical No. 5 at 48 in. oc Interior	Vertical No. 5 at 32 in. oc Exterior	Vertical No. 5 at 32 in. oc Interior	Vertical No. 5 at 24 in. oc Exterior	Vertical No. 5 at 24 in. oc Interior	Vertical No. 5 at 16 in. oc Exterior	Vertical No. 5 at 16 in. oc Interior
0.156	I or II	NJ	18'-8"	18'-8"	24'-8"	24'-8"	14'-8"	14'-8"	24'-0"	24'-0"	22'-0"	22'-0"	24'-0"	24'-0"	26'-8"	26'-8"	30'-0"	30'-0"	30'-0"	30'-0"	30'-0"	30'-0"	30'-0"	30'-0"
		30	NP	NP	NP	NP	NP	NP	NP	NP	20'-0"	18'-0"	22'-0"	20'-0"	24'-8"	22'-8"	29'-8"	26'-8"	30'-0"	30'-0"	30'-0"	30'-0"	30'-0"	30'-0"
		60	NP	NP	NP	NP	NP	NP	NP	NP	18'-0"	14'-0"	20'-0"	16'-0"	22'-8"	18'-0"	27'-4"	22'-0"	29'-4"	25'-4"	30'-0"	27'-4"	30'-0"	30'-0"
	III or IV	NJ	18'-8"	18'-8"	24'-8"	24'-8"	14'-8"	14'-8"	24'-0"	24'-0"	22'-0"	22'-0"	24'-0"	24'-0"	26'-8"	26'-8"	30'-0"	30'-0"	30'-0"	30'-0"	30'-0"	30'-0"	30'-0"	30'-0"
		30	NP	NP	NP	NP	NP	NP	NP	NP	20'-0"	18'-0"	22'-0"	20'-0"	24'-8"	22'-8"	29'-4"	26'-8"	30'-0"	29'-4"	30'-0"	30'-0"	30'-0"	30'-0"
		60	NP	NP	NP	NP	NP	NP	NP	NP	18'-0"	14'-0"	20'-0"	16'-0"	22'-8"	18'-0"	27'-4"	22'-0"	29'-4"	25'-4"	30'-0"	27'-4"	30'-0"	30'-0"
0.32	I or II	NJ	16'-0"	16'-0"	21'-4"	21'-4"	12'-8"	12'-8"	20'-8"	20'-8"	20'-0"	20'-0"	22'-0"	22'-0"	24'-8"	24'-8"	28'-0"	28'-0"	30'-0"	30'-0"	30'-0"	30'-0"	30'-0"	30'-0"
		30	NP	NP	NP	NP	NP	NP	NP	NP	18'-8"	16'-8"	20'-0"	18'-8"	22'-8"	21'-4"	26'-8"	25'-4"	30'-0"	30'-0"	30'-0"	29'-4"	30'-0"	29'-4"
		60	NP	NP	NP	NP	NP	NP	NP	NP	16'-8"	14'-0"	18'-8"	15'-4"	21'-4"	18'-0"	25'-4"	21'-4"	27'-4"	24'-8"	29'-4"	26'-8"	30'-0"	29'-4"
	III or IV	NJ	12'-8"	12'-8"	17'-4"	17'-4"	10'-0"	10'-0"	16'-8"	16'-8"	20'-0"	20'-0"	18'-8"	18'-8"	21'-4"	21'-4"	24'-8"	24'-8"	26'-0"	26'-0"	27'-4"	27'-4"	29'-4"	29'-4"
		30	NP	NP	NP	NP	NP	NP	NP	NP	10'-0"	10'-0"	18'-0"	16'-8"	20'-0"	18'-8"	23'-4"	21'-4"	25'-4"	24'-0"	26'-8"	25'-4"	28'-8"	28'-0"
		60	NP	NP	NP	NP	NP	NP	NP	NP	10'-0"	10'-0"	16'-8"	14'-8"	18'-8"	16'-8"	22'-0"	20'-0"	24'-0"	22'-0"	25'-4"	24'-0"	28'-0"	26'-0"
0.553	I or II	NJ	NP	NP	NP	NP	NP	NP	NP	NP	10'-0"	10'-0"	18'-8"	18'-8"	20'-8"	20'-8"	24'-8"	24'-8"	26'-0"	26'-0"	27'-4"	27'-4"	29'-4"	29'-4"
		30	NP	NP	NP	NP	NP	NP	NP	NP	10'-0"	10'-0"	17'-4"	16'-8"	20'-0"	18'-8"	23'-4"	21'-4"	25'-4"	24'-0"	26'-8"	25'-4"	28'-8"	27'-4"
		60	NP	NP	NP	NP	NP	NP	NP	NP	10'-0"	10'-0"	16'-8"	14'-8"	18'-8"	16'-8"	22'-0"	20'-0"	24'-0"	22'-0"	25'-4"	24'-0"	27'-4"	26'-0"
	III or IV	NJ	NP	NP	NP	NP	NP	NP	NP	NP	10'-0"	10'-0"	18'-0"	16'-8"	18'-8"	18'-0"	21'-4"	21'-4"	23'-4"	22'-8"	24'-0"	23'-4"	25'-4"	24'-8"
		30	NP	NP	NP	NP	NP	NP	NP	NP	10'-0"	10'-0"	16'-8"	14'-8"	17'-4"	16'-0"	20'-8"	19'-4"	22'-0"	21'-4"	23'-4"	22'-8"	24'-8"	23'-4"
		60	NP	NP	NP	NP	NP	NP	NP	NP	10'-0"	10'-0"	16'-8"	14'-8"	16'-8"	14'-8"	19'-4"	18'-0"	21'-4"	20'-0"	22'-8"	21'-4"	24'-8"	23'-4"
1.000	I or II	NJ	NP	NP	NP	NP	NP	NP	NP	NP	10'-0"	10'-0"	10'-0"	10'-0"	18'-0"	18'-0"	21'-4"	21'-4"	22'-8"	22'-8"	24'-0"	24'-0"	25'-4"	25'-4"
		30	NP	NP	NP	NP	NP	NP	NP	NP	10'-0"	10'-0"	10'-0"	10'-0"	17'-4"	16'-0"	20'-8"	19'-4"	22'-0"	21'-4"	23'-4"	22'-8"	24'-8"	24'-8"
		60	NP	NP	NP	NP	NP	NP	NP	NP	8'-8"	8'-8"	10'-0"	10'-0"	16'-8"	15'-4"	19'-4"	18'-0"	21'-4"	20'-0"	22'-8"	21'-4"	24'-8"	23'-4"
	III or IV	NJ	NP	NP	NP	NP	NP	NP	NP	NP	8'-8"	8'-8"	10'-0"	10'-0"	15'-4"	15'-4"	18'-0"	17'-4"	19'-4"	19'-4"	20'-8"	20'-8"	22'-0"	22'-0"
		30	NP	NP	NP	NP	NP	NP	NP	NP	8'-8"	8'-8"	10'-0"	10'-0"	15'-4"	14'-8"	17'-4"	17'-4"	19'-4"	18'-8"	20'-0"	19'-4"	22'-0"	21'-4"
		60	NP	NP	NP	NP	NP	NP	NP	NP	8'-8"	8'-8"	10'-0"	10'-0"	14'-8"	13'-4"	17'-4"	16'-0"	18'-8"	17'-4"	19'-4"	18'-8"	21'-4"	20'-8"

Table values shall not be permitted to be interpolated.
NP = Not Permitted
NJ = No Joist
ft = 0.3048 m
in. = 25.4 mm
lb/ft² = 0.0479 kPa
Unreinforced Ungrouted PCL or Mortar Cement Mortar
Unreinforced Fully Grouted PCL or Mortar Cement Mortar
The density of concrete masonry units shall be less than 135 lb/ft³ (2162 kg/m³).

TMS 403-13

Table 3.2-6(2b): Maximum Vertical Spans (ft-in.) for Walls without Openings Constructed using Medium and Normal Weight Concrete Masonry Units for Seismic Conditions and for Ground Snow Loads, p_g, up to 60 psf

Seismic S_s	Risk Category	Max. L_{joist} (ft)	Unreinforced Ungrouted PCL Mortar Exterior	Interior	Unreinforced Fully Grouted PCL Mortar Exterior	Interior	Unreinforced Ungrouted MC Mortar Exterior	Interior	Unreinforced Fully Grouted MC Mortar Exterior	Interior	Vertical No. 5 at 120 in. oc Exterior	Interior	Vertical No. 5 at 96 in. oc Exterior	Interior	Vertical No. 5 at 72 in. oc Exterior	Interior	Vertical No. 5 at 48 in. oc Exterior	Interior	Vertical No. 5 at 32 in. oc Exterior	Interior	Vertical No. 5 at 24 in. oc Exterior	Interior	Vertical No. 5 at 16 in. oc Exterior	Interior
1.500	I or II	NJ	NP	NP	NP	NP	NP	NP	NP	NP	9'-4"	9'-4"	10'-0"	10'-0"	16'-0"	16'-0"	18'-8"	18'-8"	20'-0"	20'-0"	21'-4"	21'-4"	22'-8"	22'-8"
		30	NP	NP	NP	NP	NP	NP	NP	NP	9'-4"	9'-4"	10'-0"	10'-0"	15'-4"	14'-8"	18'-0"	17'-4"	20'-0"	19'-4"	21'-8"	20'-0"	22'-8"	22'-0"
		60	NP	NP	NP	NP	NP	NP	NP	NP	9'-4"	9'-4"	10'-0"	10'-0"	14'-8"	13'-4"	17'-4"	16'-8"	19'-4"	18'-0"	20'-0"	19'-4"	22'-0"	20'-8"
	III or IV	NJ	NP	NP	NP	NP	NP	NP	NP	NP	8'-0"	8'-0"	10'-0"	10'-0"	16'-0"	16'-0"	16'-0"	16'-0"	17'-4"	17'-4"	18'-8"	18'-8"	20'-0"	20'-0"
		30	NP	NP	NP	NP	NP	NP	NP	NP	8'-0"	8'-0"	10'-0"	10'-0"	10'-0"	10'-0"	16'-0"	15'-4"	17'-4"	16'-8"	18'-0"	17'-4"	19'-4"	19'-4"
		60	NP	NP	NP	NP	NP	NP	NP	NP	8'-0"	8'-0"	10'-0"	10'-0"	10'-0"	10'-0"	15'-4"	14'-8"	16'-8"	16'-0"	17'-4"	16'-8"	19'-4"	18'-8"
2.000	I or II	NJ	NP	NP	NP	NP	NP	NP	NP	NP	8'-0"	8'-0"	10'-0"	10'-0"	10'-0"	10'-0"	16'-8"	16'-8"	18'-0"	18'-0"	19'-4"	19'-4"	20'-8"	20'-8"
		30	NP	NP	NP	NP	NP	NP	NP	NP	8'-0"	8'-0"	10'-0"	10'-0"	10'-0"	10'-0"	16'-8"	16'-0"	18'-0"	17'-4"	18'-8"	18'-0"	20'-0"	20'-0"
		60	NP	NP	NP	NP	NP	NP	NP	NP	8'-0"	8'-0"	10'-0"	10'-0"	10'-0"	10'-0"	15'-4"	14'-8"	17'-4"	16'-8"	18'-8"	17'-4"	20'-0"	19'-4"
	III or IV	NJ	NP	NP	NP	NP	NP	NP	NP	NP	6'-8"	6'-8"	8'-8"	8'-8"	10'-0"	10'-0"	14'-8"	14'-0"	16'-0"	16'-0"	16'-8"	16'-8"	18'-0"	18'-0"
		30	NP	NP	NP	NP	NP	NP	NP	NP	6'-8"	6'-8"	8'-8"	8'-8"	10'-0"	10'-0"	14'-0"	14'-0"	15'-4"	15'-4"	16'-0"	16'-0"	17'-4"	17'-4"
		60	NP	NP	NP	NP	NP	NP	NP	NP	6'-8"	6'-8"	8'-8"	8'-8"	10'-0"	10'-0"	14'-0"	13'-4"	15'-4"	14'-8"	16'-0"	15'-4"	17'-4"	16'-8"
2.500	I or II	NJ	NP	NP	NP	NP	NP	NP	NP	NP	7'-4"	7'-4"	9'-4"	9'-4"	10'-0"	10'-0"	15'-4"	15'-4"	16'-8"	16'-8"	18'-0"	18'-0"	18'-8"	18'-8"
		30	NP	NP	NP	NP	NP	NP	NP	NP	7'-4"	7'-4"	9'-4"	9'-4"	10'-0"	10'-0"	15'-4"	14'-8"	16'-8"	16'-0"	17'-4"	16'-8"	18'-8"	18'-0"
		60	NP	NP	NP	NP	NP	NP	NP	NP	7'-4"	7'-4"	9'-4"	9'-4"	10'-0"	10'-0"	14'-8"	14'-0"	16'-0"	15'-4"	16'-8"	16'-0"	18'-0"	18'-0"
	III or IV	NJ	NP	NP	NP	NP	NP	NP	NP	NP	6'-0"	6'-0"	7'-4"	7'-4"	10'-0"	10'-0"	13'-4"	13'-4"	14'-8"	14'-8"	15'-4"	15'-4"	16'-8"	16'-8"
		30	NP	NP	NP	NP	NP	NP	NP	NP	6'-0"	6'-0"	7'-4"	7'-4"	10'-0"	10'-0"	13'-4"	12'-8"	14'-0"	14'-0"	14'-8"	14'-8"	16'-0"	16'-0"
		60	NP	NP	NP	NP	NP	NP	NP	NP	6'-0"	6'-0"	7'-4"	7'-4"	10'-0"	10'-0"	12'-8"	12'-0"	14'-0"	13'-4"	14'-8"	14'-0"	16'-0"	15'-4"
3.000	I or II	NJ	NP	NP	NP	NP	NP	NP	NP	NP	6'-8"	6'-8"	8'-8"	8'-8"	10'-0"	10'-0"	14'-8"	14'-8"	16'-0"	16'-0"	16'-8"	16'-8"	18'-0"	18'-0"
		30	NP	NP	NP	NP	NP	NP	NP	NP	6'-8"	6'-8"	8'-8"	8'-8"	10'-0"	10'-0"	14'-0"	14'-0"	15'-4"	15'-4"	16'-0"	16'-0"	17'-4"	17'-4"
		60	NP	NP	NP	NP	NP	NP	NP	NP	6'-8"	6'-8"	8'-8"	8'-8"	10'-0"	10'-0"	14'-0"	13'-4"	15'-4"	14'-0"	16'-0"	15'-4"	17'-4"	16'-8"
	III or IV	NJ	NP	NP	NP	NP	NP	NP	NP	NP	5'-4"	5'-4"	6'-8"	6'-8"	9'-4"	9'-4"	10'-0"	10'-0"	13'-4"	13'-4"	14'-0"	14'-0"	15'-4"	15'-4"
		30	NP	NP	NP	NP	NP	NP	NP	NP	5'-4"	5'-4"	6'-8"	6'-8"	9'-4"	9'-4"	10'-0"	10'-0"	13'-4"	13'-4"	14'-0"	14'-0"	14'-8"	14'-8"
		60	NP	NP	NP	NP	NP	NP	NP	NP	5'-4"	5'-4"	6'-8"	6'-8"	9'-4"	9'-4"	10'-0"	10'-0"	13'-4"	12'-8"	14'-0"	14'-0"	14'-8"	14'-8"

Table values shall not be permitted to be interpolated.
NP = Not Permitted
NJ = No Joist
ft = 0.3048 m
in. = 25.4 mm
lb/ft² = 0.0479 kPa
Unreinforced Ungrouted PCL or Mortar Cement Mortar
Unreinforced Fully Grouted PCL or Mortar Cement Mortar
The density of concrete masonry units shall be less than 135 lb/ft³ (2162 kg/m³).

Table 3.2-6(3a): Maximum Vertical Spans (ft-in.) without Openings for Walls Constructed of Lightweight Concrete Masonry Units with Veneer Cladding for Seismic Conditions for Ground Snow Loads, p_g, up to 60 psf

Seismic S_s	Risk Category	Max. L_{joist} (ft)	Unreinforced Ungrouted PCL Mortar Exterior	Unreinforced Ungrouted PCL Mortar Interior	Unreinforced Fully Grouted PCL Mortar Exterior	Unreinforced Fully Grouted PCL Mortar Interior	Unreinforced Ungrouted MC Mortar Exterior	Unreinforced Ungrouted MC Mortar Interior	Unreinforced Fully Grouted MC Mortar Exterior	Unreinforced Fully Grouted MC Mortar Interior	Vertical No. 5 at 120 in. oc Exterior	Vertical No. 5 at 120 in. oc Interior	Vertical No. 5 at 96 in. oc Exterior	Vertical No. 5 at 96 in. oc Interior	Vertical No. 5 at 72 in. oc Exterior	Vertical No. 5 at 72 in. oc Interior	Vertical No. 5 at 48 in. oc Exterior	Vertical No. 5 at 48 in. oc Interior	Vertical No. 5 at 32 in. oc Exterior	Vertical No. 5 at 32 in. oc Interior	Vertical No. 5 at 24 in. oc Exterior	Vertical No. 5 at 24 in. oc Interior	Vertical No. 5 at 16 in. oc Exterior	Vertical No. 5 at 16 in. oc Interior
0.156	I or II	NJ	14'-0"	14'-0"	21'-4"	21'-4"	11'-4"	11'-4"	20'-8"	20'-8"	18'-8"	18'-8"	20'-8"	20'-8"	22'-8"	22'-8"	26'-8"	26'-8"	28'-8"	28'-8"	30'-0"	30'-0"	30'-0"	30'-0"
		30	NP	NP	NP	NP	NP	NP	NP	NP	17'-4"	16'-0"	19'-4"	18'-0"	21'-4"	20'-0"	25'-4"	24'-0"	27'-4"	26'-0"	29'-4"	28'-0"	30'-0"	30'-0"
		60	NP	NP	NP	NP	NP	NP	NP	NP	16'-0"	13'-4"	18'-0"	15'-4"	20'-0"	18'-0"	24'-0"	21'-4"	26'-0"	24'-0"	28'-0"	26'-0"	30'-0"	28'-8"
	III or IV	NJ	14'-0"	14'-0"	21'-4"	21'-4"	11'-4"	11'-4"	20'-8"	20'-8"	18'-8"	18'-8"	20'-8"	20'-8"	22'-8"	22'-8"	26'-8"	26'-8"	28'-8"	28'-8"	30'-0"	30'-0"	30'-0"	30'-0"
		30	NP	NP	NP	NP	NP	NP	NP	NP	17'-4"	16'-0"	19'-4"	18'-0"	21'-4"	20'-0"	25'-4"	24'-0"	27'-4"	26'-0"	29'-4"	28'-0"	30'-0"	30'-0"
		60	NP	NP	NP	NP	NP	NP	NP	NP	16'-0"	13'-4"	18'-0"	15'-4"	20'-0"	18'-0"	24'-0"	21'-4"	26'-0"	24'-0"	28'-0"	26'-0"	30'-0"	28'-8"
0.32	I or II	NJ	12'-0"	12'-0"	18'-0"	18'-0"	9'-4"	9'-4"	17'-4"	17'-4"	16'-0"	16'-0"	18'-8"	18'-8"	20'-8"	20'-8"	24'-0"	24'-0"	26'-0"	26'-0"	28'-0"	28'-0"	30'-0"	30'-0"
		30	NP	NP	NP	NP	NP	NP	NP	NP	10'-0"	10'-0"	17'-4"	16'-0"	19'-4"	18'-8"	23'-4"	22'-0"	25'-4"	24'-0"	26'-8"	26'-0"	29'-4"	28'-0"
		60	NP	NP	NP	NP	NP	NP	NP	NP	10'-0"	10'-0"	16'-0"	14'-0"	18'-8"	16'-8"	22'-0"	20'-0"	24'-0"	22'-0"	26'-0"	24'-0"	28'-0"	26'-8"
	III or IV	NJ	9'-4"	9'-4"	14'-8"	14'-8"	7'-4"	7'-4"	14'-0"	14'-0"	10'-0"	10'-0"	10'-0"	10'-0"	18'-0"	18'-0"	20'-8"	20'-8"	22'-8"	22'-8"	24'-0"	24'-0"	26'-0"	26'-0"
		30	NP	NP	NP	NP	NP	NP	NP	NP	10'-0"	10'-0"	10'-0"	10'-0"	17'-4"	16'-0"	20'-0"	19'-4"	22'-0"	21'-4"	23'-4"	22'-8"	25'-4"	24'-8"
		60	NP	NP	NP	NP	NP	NP	NP	NP	10'-0"	10'-0"	10'-0"	10'-0"	16'-0"	14'-8"	19'-4"	18'-0"	21'-4"	20'-0"	22'-8"	21'-4"	24'-8"	23'-4"
0.553	I or II	NJ	NP	NP	NP	NP	NP	NP	NP	NP	10'-0"	10'-0"	10'-0"	10'-0"	18'-0"	18'-0"	20'-8"	20'-8"	22'-8"	22'-8"	24'-0"	24'-0"	26'-0"	26'-0"
		30	NP	NP	NP	NP	NP	NP	NP	NP	8'-8"	8'-8"	10'-0"	10'-0"	16'-8"	16'-8"	20'-0"	19'-4"	22'-0"	21'-4"	23'-4"	22'-8"	25'-4"	24'-8"
		60	NP	NP	NP	NP	NP	NP	NP	NP	8'-8"	8'-8"	10'-0"	10'-0"	16'-0"	14'-8"	19'-4"	18'-0"	21'-4"	20'-0"	22'-8"	21'-4"	24'-8"	23'-4"
	III or IV	NJ	NP	NP	NP	NP	NP	NP	NP	NP	10'-0"	10'-0"	10'-0"	10'-0"	15'-4"	15'-4"	18'-0"	18'-0"	19'-4"	19'-4"	20'-8"	20'-8"	22'-8"	22'-8"
		30	NP	NP	NP	NP	NP	NP	NP	NP	8'-8"	8'-8"	10'-0"	10'-0"	14'-8"	14'-0"	17'-4"	16'-8"	18'-8"	18'-8"	20'-0"	19'-4"	22'-0"	21'-4"
		60	NP	NP	NP	NP	NP	NP	NP	NP	8'-8"	8'-8"	10'-0"	10'-0"	14'-0"	13'-4"	16'-8"	16'-0"	18'-8"	17'-4"	19'-4"	18'-8"	21'-4"	20'-8"
1.000	I or II	NJ	NP	NP	NP	NP	NP	NP	NP	NP	8'-8"	8'-8"	10'-0"	10'-0"	15'-4"	15'-4"	18'-0"	18'-0"	19'-4"	19'-4"	20'-8"	20'-8"	22'-8"	22'-8"
		30	NP	NP	NP	NP	NP	NP	NP	NP	8'-8"	8'-8"	10'-0"	10'-0"	14'-8"	14'-0"	17'-4"	16'-8"	18'-8"	18'-8"	20'-0"	20'-0"	22'-0"	21'-4"
		60	NP	NP	NP	NP	NP	NP	NP	NP	8'-8"	8'-8"	10'-0"	10'-0"	14'-0"	13'-4"	16'-8"	16'-0"	18'-8"	17'-4"	20'-0"	18'-8"	21'-4"	20'-8"
	III or IV	NJ	NP	NP	NP	NP	NP	NP	NP	NP	7'-4"	7'-4"	9'-4"	9'-4"	10'-0"	10'-0"	15'-4"	15'-4"	16'-8"	16'-8"	18'-0"	18'-8"	19'-4"	19'-4"
		30	NP	NP	NP	NP	NP	NP	NP	NP	7'-4"	7'-4"	9'-4"	9'-4"	10'-0"	10'-0"	15'-4"	14'-8"	16'-8"	16'-0"	17'-4"	17'-4"	19'-4"	18'-8"
		60	NP	NP	NP	NP	NP	NP	NP	NP	7'-4"	7'-4"	9'-4"	9'-4"	10'-0"	10'-0"	14'-8"	14'-0"	16'-0"	15'-4"	17'-4"	16'-8"	18'-8"	18'-8"

Table values shall not be permitted to be interpolated.
NP = Not Permitted
NJ = No Joist
ft = 0.3048 m
in. = 25.4 mm
lb/ft² = 0.0479 kPa
Unreinforced Ungrouted PCL or Mortar Cement Mortar
Unreinforced Fully Grouted PCL or Mortar Cement Mortar
The density of concrete masonry units shall be less than 105 lb/ft³ (1680 kg/m³).

Table 3.2-6(3h): Maximum Vertical Spans (ft-in.) without Openings for Walls Constructed of Lightweight Concrete Masonry Units with Veneer Cladding for Seismic Conditions for Ground Snow Loads, p_g, up to 60 psf

Seismic S_s	Risk Category	Max. L_{joist} (ft)	Unreinforced Ungrouted PCL Mortar Exterior	Unreinforced Ungrouted PCL Mortar Interior	Unreinforced Fully Grouted PCL Mortar Exterior	Unreinforced Fully Grouted PCL Mortar Interior	Unreinforced Ungrouted MC Mortar Exterior	Unreinforced Ungrouted MC Mortar Interior	Unreinforced Fully Grouted MC Mortar Exterior	Unreinforced Fully Grouted MC Mortar Interior	Vertical No. 5 at 120 in. oc Exterior	Vertical No. 5 at 120 in. oc Interior	Vertical No. 5 at 96 in. oc Exterior	Vertical No. 5 at 96 in. oc Interior	Vertical No. 5 at 72 in. oc Exterior	Vertical No. 5 at 72 in. oc Interior	Vertical No. 5 at 48 in. oc Exterior	Vertical No. 5 at 48 in. oc Interior	Vertical No. 5 at 32 in. oc Exterior	Vertical No. 5 at 32 in. oc Interior	Vertical No. 5 at 24 in. oc Exterior	Vertical No. 5 at 24 in. oc Interior	Vertical No. 5 at 16 in. oc Exterior	Vertical No. 5 at 16 in. oc Interior
1.500	I or II	NJ	NP	NP	NP	NP	NP	NP	NP	NP	7'-4"	7'-4"	9'-4"	9'-4"	10'-0"	10'-0"	16'-0"	16'-0"	17'-4"	17'-4"	18'-8"	18'-8"	20'-0"	20'-0"
		30	NP	NP	NP	NP	NP	NP	NP	NP	7'-4"	7'-4"	9'-4"	9'-4"	10'-0"	10'-0"	15'-4"	15'-4"	17'-4"	16'-8"	18'-0"	18'-0"	20'-0"	19'-4"
		60	NP	NP	NP	NP	NP	NP	NP	NP	7'-4"	7'-4"	9'-4"	9'-4"	10'-0"	10'-0"	15'-4"	14'-0"	16'-8"	16'-0"	18'-0"	16'-8"	19'-4"	18'-8"
	III or IV	NJ	NP	NP	NP	NP	NP	NP	NP	NP	6'-0"	6'-0"	8'-0"	8'-0"	10'-0"	10'-0"	14'-0"	14'-0"	14'-8"	14'-8"	16'-0"	16'-0"	17'-4"	17'-4"
		30	NP	NP	NP	NP	NP	NP	NP	NP	6'-0"	6'-0"	8'-0"	8'-0"	10'-0"	10'-0"	13'-4"	13'-4"	14'-8"	14'-8"	16'-0"	15'-4"	17'-4"	16'-8"
		60	NP	NP	NP	NP	NP	NP	NP	NP	6'-0"	6'-0"	8'-0"	8'-0"	10'-0"	10'-0"	13'-4"	12'-8"	14'-8"	14'-0"	15'-4"	14'-8"	16'-8"	16'-0"
2.000	I or II	NJ	NP	NP	NP	NP	NP	NP	NP	NP	6'-8"	6'-8"	8'-0"	8'-0"	10'-0"	10'-0"	14'-8"	14'-8"	15'-4"	15'-4"	16'-8"	16'-8"	18'-0"	18'-0"
		30	NP	NP	NP	NP	NP	NP	NP	NP	6'-8"	6'-8"	8'-0"	8'-0"	10'-0"	10'-0"	14'-0"	13'-4"	15'-4"	14'-8"	16'-8"	16'-0"	18'-0"	17'-4"
		60	NP	NP	NP	NP	NP	NP	NP	NP	6'-8"	6'-8"	8'-0"	8'-0"	10'-0"	10'-0"	14'-0"	12'-8"	15'-4"	14'-0"	16'-0"	15'-4"	17'-4"	16'-8"
	III or IV	NJ	NP	NP	NP	NP	NP	NP	NP	NP	5'-4"	5'-4"	6'-8"	6'-8"	9'-4"	9'-4"	10'-0"	10'-0"	13'-4"	13'-4"	14'-0"	14'-0"	15'-4"	15'-4"
		30	NP	NP	NP	NP	NP	NP	NP	NP	5'-4"	5'-4"	6'-8"	6'-8"	9'-4"	9'-4"	10'-0"	10'-0"	13'-4"	13'-4"	14'-0"	14'-0"	15'-4"	15'-4"
		60	NP	NP	NP	NP	NP	NP	NP	NP	5'-4"	5'-4"	6'-8"	6'-8"	9'-4"	9'-4"	10'-0"	10'-0"	13'-4"	12'-8"	14'-0"	14'-0"	14'-8"	14'-8"
2.500	I or II	NJ	NP	NP	NP	NP	NP	NP	NP	NP	6'-0"	6'-0"	7'-4"	7'-4"	10'-0"	10'-0"	13'-4"	13'-4"	14'-8"	14'-8"	15'-4"	15'-4"	16'-8"	16'-8"
		30	NP	NP	NP	NP	NP	NP	NP	NP	6'-0"	6'-0"	7'-4"	7'-4"	10'-0"	10'-0"	12'-8"	12'-8"	14'-0"	14'-0"	14'-8"	14'-8"	16'-8"	16'-0"
		60	NP	NP	NP	NP	NP	NP	NP	NP	6'-0"	6'-0"	7'-4"	7'-4"	10'-0"	10'-0"	12'-8"	12'-0"	14'-0"	13'-4"	14'-8"	14'-0"	16'-0"	16'-0"
	III or IV	NJ	NP	NP	NP	NP	NP	NP	NP	NP	4'-8"	4'-8"	6'-0"	6'-0"	8'-0"	8'-0"	10'-0"	10'-0"	12'-8"	12'-8"	13'-4"	13'-4"	14'-8"	14'-8"
		30	NP	NP	NP	NP	NP	NP	NP	NP	4'-8"	4'-8"	6'-0"	6'-0"	8'-0"	8'-0"	10'-0"	10'-0"	12'-0"	12'-0"	13'-4"	12'-8"	14'-0"	14'-0"
		60	NP	NP	NP	NP	NP	NP	NP	NP	4'-8"	4'-8"	6'-0"	6'-0"	8'-0"	8'-0"	10'-0"	10'-0"	12'-0"	11'-4"	12'-8"	12'-8"	14'-0"	14'-0"
3.000	I or II	NJ	NP	NP	NP	NP	NP	NP	NP	NP	5'-4"	5'-4"	6'-8"	6'-8"	9'-4"	9'-4"	10'-0"	10'-0"	13'-4"	13'-4"	14'-8"	14'-8"	15'-4"	15'-4"
		30	NP	NP	NP	NP	NP	NP	NP	NP	5'-4"	5'-4"	6'-8"	6'-8"	9'-4"	9'-4"	10'-0"	10'-0"	13'-4"	12'-8"	14'-0"	14'-0"	15'-4"	15'-4"
		60	NP	NP	NP	NP	NP	NP	NP	NP	5'-4"	5'-4"	6'-8"	6'-8"	9'-4"	9'-4"	10'-0"	10'-0"	12'-8"	12'-8"	14'-0"	13'-4"	15'-4"	14'-8"
	III or IV	NJ	NP	NP	NP	NP	NP	NP	NP	NP	4'-8"	4'-8"	5'-4"	5'-4"	7'-4"	7'-4"	10'-0"	10'-0"	11'-4"	11'-4"	12'-8"	12'-8"	13'-4"	13'-4"
		30	NP	NP	NP	NP	NP	NP	NP	NP	4'-8"	4'-8"	5'-4"	5'-4"	7'-4"	7'-4"	10'-0"	10'-0"	11'-4"	11'-4"	12'-0"	12'-0"	13'-4"	13'-4"
		60	NP	NP	NP	NP	NP	NP	NP	NP	4'-8"	4'-8"	5'-4"	5'-4"	7'-4"	7'-4"	10'-0"	10'-0"	11'-4"	10'-8"	12'-0"	12'-0"	13'-4"	12'-8"

Table values shall not be permitted to be interpolated.
NP = Not Permitted
NJ = No Joist
ft = 0.3048 m
in. = 25.4 mm
lb/ft² = 0.0479 kPa
Unreinforced Ungrouted PCL or Mortar Cement Mortar
Unreinforced Fully Grouted PCL or Mortar Cement Mortar
The density of concrete masonry units shall be less than 105 lb/ft³ (1680 kg/m³).

DIRECT DESIGN HANDBOOK FOR MASONRY STRUCTURES

Table 3.2-6(4a): Maximum Vertical Spans (ft-in.) without Openings for Walls Constructed Using Medium and Normal Weight Concrete Masonry Units with Veneer Cladding for Seismic Conditions for Ground Snow Loads, p_g, up to 60 psf

Seismic S_s	Risk Category	Max. L_{joist} (ft)	Unreinforced Ungrouted PCL Mortar Exterior	Unreinforced Ungrouted PCL Mortar Interior	Unreinforced Fully Grouted PCL Mortar Exterior	Unreinforced Fully Grouted PCL Mortar Interior	Unreinforced Ungrouted MC Mortar Exterior	Unreinforced Ungrouted MC Mortar Interior	Unreinforced Fully Grouted MC Mortar Exterior	Unreinforced Fully Grouted MC Mortar Interior	Vertical No. 5 at 120 in. oc Exterior	Vertical No. 5 at 120 in. oc Interior	Vertical No. 5 at 96 in. oc Exterior	Vertical No. 5 at 96 in. oc Interior	Vertical No. 5 at 72 in. oc Exterior	Vertical No. 5 at 72 in. oc Interior	Vertical No. 5 at 48 in. oc Exterior	Vertical No. 5 at 48 in. oc Interior	Vertical No. 5 at 32 in. oc Exterior	Vertical No. 5 at 32 in. oc Interior	Vertical No. 5 at 24 in. oc Exterior	Vertical No. 5 at 24 in. oc Interior	Vertical No. 5 at 16 in. oc Exterior	Vertical No. 5 at 16 in. oc Interior
0.156	I or II	NJ	13'-4"	13'-4"	20'-8"	20'-8"	10'-8"	10'-8"	20'-0"	20'-0"	10'-0"	10'-0"	20'-0"	20'-0"	22'-0"	22'-0"	26'-0"	26'-0"	28'-0"	28'-0"	29'-4"	29'-4"	30'-0"	30'-0"
0.156	I or II	30	NP	NP	NP	NP	NP	NP	NP	NP	10'-0"	10'-0"	18'-8"	17'-4"	20'-8"	19'-4"	24'-8"	23'-4"	26'-8"	25'-4"	28'-8"	27'-4"	30'-0"	30'-0"
0.156	I or II	60	NP	NP	NP	NP	NP	NP	NP	NP	10'-0"	10'-0"	17'-4"	14'-8"	19'-4"	17'-4"	23'-4"	20'-8"	25'-4"	23'-4"	27'-4"	25'-4"	30'-0"	28'-0"
0.156	III or IV	NJ	13'-4"	13'-4"	20'-8"	20'-8"	10'-8"	10'-8"	20'-0"	20'-0"	10'-0"	10'-0"	20'-0"	20'-0"	22'-0"	22'-0"	26'-0"	26'-0"	28'-0"	28'-0"	29'-4"	29'-4"	30'-0"	30'-0"
0.156	III or IV	30	NP	NP	NP	NP	NP	NP	NP	NP	10'-0"	10'-0"	18'-8"	17'-4"	20'-8"	19'-4"	24'-8"	23'-4"	26'-8"	25'-4"	28'-8"	27'-4"	30'-0"	30'-0"
0.156	III or IV	60	NP	NP	NP	NP	NP	NP	NP	NP	10'-0"	10'-0"	17'-4"	14'-8"	19'-4"	17'-4"	23'-4"	20'-8"	25'-4"	23'-4"	27'-4"	25'-4"	30'-0"	28'-0"
0.32	I or II	NJ	11'-4"	11'-4"	17'-4"	17'-4"	8'-8"	8'-8"	17'-4"	17'-4"	10'-0"	10'-0"	18'-0"	18'-0"	20'-0"	20'-0"	23'-4"	23'-4"	25'-4"	25'-4"	26'-8"	26'-8"	29'-4"	29'-4"
0.32	I or II	30	NP	NP	NP	NP	NP	NP	NP	NP	10'-0"	10'-0"	16'-8"	16'-0"	18'-8"	18'-0"	22'-0"	21'-4"	24'-0"	23'-4"	26'-0"	25'-4"	28'-8"	27'-4"
0.32	I or II	60	NP	NP	NP	NP	NP	NP	NP	NP	10'-0"	10'-0"	16'-0"	14'-0"	18'-0"	16'-0"	21'-4"	19'-4"	23'-4"	21'-4"	25'-4"	23'-4"	27'-4"	26'-0"
0.32	III or IV	NJ	9'-4"	9'-4"	14'-0"	14'-0"	7'-4"	7'-4"	14'-0"	14'-0"	10'-0"	10'-0"	16'-0"	16'-0"	17'-4"	17'-4"	20'-0"	20'-0"	22'-0"	22'-0"	23'-4"	23'-4"	25'-4"	25'-4"
0.32	III or IV	30	NP	NP	NP	NP	NP	NP	NP	NP	10'-0"	10'-0"	14'-8"	14'-0"	16'-8"	16'-0"	19'-4"	18'-8"	21'-4"	20'-8"	22'-8"	22'-0"	24'-8"	24'-0"
0.32	III or IV	60	NP	NP	NP	NP	NP	NP	NP	NP	10'-0"	10'-0"	14'-0"	12'-8"	16'-0"	14'-8"	18'-8"	17'-4"	20'-8"	19'-4"	22'-0"	20'-8"	24'-0"	22'-8"
0.553	I or II	NJ	NP	NP	NP	NP	NP	NP	NP	NP	10'-0"	10'-0"	10'-0"	10'-0"	17'-4"	17'-4"	20'-0"	20'-0"	22'-0"	22'-0"	23'-4"	23'-4"	25'-4"	25'-4"
0.553	I or II	30	NP	NP	NP	NP	NP	NP	NP	NP	10'-0"	10'-0"	10'-0"	10'-0"	16'-8"	16'-0"	19'-4"	18'-8"	21'-4"	20'-8"	22'-8"	22'-0"	24'-8"	24'-0"
0.553	I or II	60	NP	NP	NP	NP	NP	NP	NP	NP	10'-0"	10'-0"	10'-0"	10'-0"	16'-0"	14'-8"	18'-8"	17'-4"	20'-8"	19'-4"	22'-0"	20'-8"	24'-0"	22'-8"
0.553	III or IV	NJ	NP	NP	NP	NP	NP	NP	NP	NP	10'-0"	10'-0"	10'-0"	10'-0"	10'-0"	10'-0"	17'-4"	17'-4"	18'-8"	18'-8"	20'-0"	20'-0"	22'-0"	22'-0"
0.553	III or IV	30	NP	NP	NP	NP	NP	NP	NP	NP	8'-0"	8'-0"	10'-0"	10'-0"	10'-0"	10'-0"	16'-8"	16'-0"	18'-0"	18'-0"	19'-4"	19'-4"	21'-4"	20'-8"
0.553	III or IV	60	NP	NP	NP	NP	NP	NP	NP	NP	8'-0"	8'-0"	10'-0"	10'-0"	10'-0"	10'-0"	16'-0"	15'-4"	18'-0"	16'-8"	19'-4"	18'-0"	20'-8"	20'-0"
1.000	I or II	NJ	NP	NP	NP	NP	NP	NP	NP	NP	8'-8"	8'-8"	10'-0"	10'-0"	10'-0"	10'-0"	14'-8"	14'-8"	18'-8"	18'-8"	20'-0"	20'-0"	22'-0"	22'-0"
1.000	I or II	30	NP	NP	NP	NP	NP	NP	NP	NP	8'-8"	8'-8"	10'-0"	10'-0"	10'-0"	10'-0"	14'-8"	14'-0"	18'-8"	18'-0"	19'-4"	19'-4"	21'-4"	20'-8"
1.000	I or II	60	NP	NP	NP	NP	NP	NP	NP	NP	8'-8"	8'-8"	10'-0"	10'-0"	10'-0"	10'-0"	14'-8"	13'-4"	18'-0"	16'-8"	19'-4"	18'-0"	20'-8"	20'-0"
1.000	III or IV	NJ	NP	NP	NP	NP	NP	NP	NP	NP	6'-8"	6'-8"	8'-8"	8'-8"	10'-0"	10'-0"	14'-8"	14'-0"	16'-0"	16'-0"	17'-4"	17'-4"	18'-8"	18'-8"
1.000	III or IV	30	NP	NP	NP	NP	NP	NP	NP	NP	6'-8"	6'-8"	8'-8"	8'-8"	10'-0"	10'-0"	14'-8"	14'-0"	16'-0"	15'-4"	16'-8"	16'-8"	18'-8"	18'-8"
1.000	III or IV	60	NP	NP	NP	NP	NP	NP	NP	NP	6'-8"	6'-8"	8'-8"	8'-8"	10'-0"	10'-0"	14'-0"	13'-4"	15'-4"	14'-8"	16'-8"	16'-0"	18'-0"	17'-4"

Table values shall not be permitted to be interpolated.
NP = Not Permitted
NJ = No Joist
ft = 0.3048 m
in. = 25.4 mm
lb/ft² = 0.0479 kPa
Unreinforced Ungrouted PCL or Mortar Cement Mortar
Unreinforced Fully Grouted PCL or Mortar Cement Mortar
The density of concrete masonry units shall be less than 135 lb/ft³ (2,162 kg/m³).

Table 3.2-6(4b): Maximum Vertical Spans (ft-in.) without Openings for Walls Constructed Using Medium and Normal Weight Concrete Masonry Units with Veneer Cladding for Seismic Conditions for Ground Snow Loads, p_g, up to 60 psf

Seismic S_S	Risk Category	Max. L_{joist} (ft)	Unreinforced Ungrouted PCL Mortar Exterior	Unreinforced Ungrouted PCL Mortar Interior	Unreinforced Fully Grouted PCL Mortar Exterior	Unreinforced Fully Grouted PCL Mortar Interior	Unreinforced Ungrouted MC Mortar Exterior	Unreinforced Ungrouted MC Mortar Interior	Unreinforced Fully Grouted MC Mortar Exterior	Unreinforced Fully Grouted MC Mortar Interior	Vertical No. 5 at 120 in. oc Exterior	Vertical No. 5 at 120 in. oc Interior	Vertical No. 5 at 96 in. oc Exterior	Vertical No. 5 at 96 in. oc Interior	Vertical No. 5 at 72 in. oc Exterior	Vertical No. 5 at 72 in. oc Interior	Vertical No. 5 at 48 in. oc Exterior	Vertical No. 5 at 48 in. oc Interior	Vertical No. 5 at 32 in. oc Exterior	Vertical No. 5 at 32 in. oc Interior	Vertical No. 5 at 24 in. oc Exterior	Vertical No. 5 at 24 in. oc Interior	Vertical No. 5 at 16 in. oc Exterior	Vertical No. 5 at 16 in. oc Interior
1.500	I or II	NJ	NP	NP	NP	NP	NP	NP	NP	NP	7'-4"	7'-4"	9'-4"	9'-4"	10'-0"	10'-0"	15'-4"	15'-4"	16'-8"	16'-8"	18'-0"	18'-0"	19'-4"	19'-4"
		30	NP	NP	NP	NP	NP	NP	NP	NP	7'-4"	7'-4"	9'-4"	9'-4"	10'-0"	10'-0"	15'-4"	14'-8"	16'-8"	16'-0"	17'-4"	17'-4"	19'-4"	18'-8"
		60	NP	NP	NP	NP	NP	NP	NP	NP	7'-4"	7'-4"	9'-4"	9'-4"	10'-0"	10'-0"	14'-8"	14'-0"	16'-0"	15'-4"	17'-4"	16'-8"	18'-0"	18'-0"
	III or IV	NJ	NP	NP	NP	NP	NP	NP	NP	NP	6'-0"	6'-0"	7'-4"	7'-4"	10'-0"	10'-0"	13'-4"	13'-4"	14'-8"	14'-8"	15'-4"	15'-4"	16'-8"	16'-8"
		30	NP	NP	NP	NP	NP	NP	NP	NP	6'-0"	6'-0"	7'-4"	7'-4"	10'-0"	10'-0"	12'-8"	12'-8"	14'-0"	14'-0"	14'-8"	14'-8"	16'-8"	16'-0"
		60	NP	NP	NP	NP	NP	NP	NP	NP	6'-0"	6'-0"	7'-4"	7'-4"	10'-0"	10'-0"	12'-8"	12'-0"	14'-0"	13'-4"	14'-8"	14'-8"	16'-0"	16'-0"
2.000	I or II	NJ	NP	NP	NP	NP	NP	NP	NP	NP	6'-0"	6'-0"	8'-0"	8'-0"	10'-0"	10'-0"	14'-0"	14'-0"	15'-4"	15'-4"	16'-0"	16'-0"	17'-4"	17'-4"
		30	NP	NP	NP	NP	NP	NP	NP	NP	6'-0"	6'-0"	8'-0"	8'-0"	10'-0"	10'-0"	13'-4"	13'-4"	14'-8"	14'-8"	15'-4"	15'-4"	17'-4"	17'-4"
		60	NP	NP	NP	NP	NP	NP	NP	NP	6'-0"	6'-0"	8'-0"	8'-0"	10'-0"	10'-0"	13'-4"	12'-8"	14'-8"	14'-0"	15'-4"	14'-8"	16'-8"	16'-8"
	III or IV	NJ	NP	NP	NP	NP	NP	NP	NP	NP	5'-4"	5'-4"	6'-8"	6'-8"	8'-8"	8'-8"	10'-0"	10'-0"	12'-8"	12'-8"	14'-0"	14'-0"	15'-4"	15'-4"
		30	NP	NP	NP	NP	NP	NP	NP	NP	5'-4"	5'-4"	6'-8"	6'-8"	8'-8"	8'-8"	10'-0"	10'-0"	12'-8"	12'-8"	13'-4"	13'-4"	14'-8"	14'-8"
		60	NP	NP	NP	NP	NP	NP	NP	NP	5'-4"	5'-4"	6'-8"	6'-8"	8'-8"	8'-8"	10'-0"	10'-0"	12'-0"	12'-0"	13'-4"	13'-4"	14'-8"	14'-8"
2.500	I or II	NJ	NP	NP	NP	NP	NP	NP	NP	NP	6'-0"	6'-0"	7'-4"	7'-4"	9'-4"	9'-4"	10'-0"	10'-0"	14'-0"	14'-0"	14'-8"	14'-8"	16'-0"	16'-0"
		30	NP	NP	NP	NP	NP	NP	NP	NP	6'-0"	6'-0"	7'-4"	7'-4"	9'-4"	9'-4"	10'-0"	10'-0"	13'-4"	13'-4"	14'-8"	14'-8"	16'-0"	15'-4"
		60	NP	NP	NP	NP	NP	NP	NP	NP	6'-0"	6'-0"	7'-4"	7'-4"	9'-4"	9'-4"	10'-0"	10'-0"	13'-4"	12'-8"	14'-0"	14'-0"	15'-4"	15'-4"
	III or IV	NJ	NP	NP	NP	NP	NP	NP	NP	NP	4'-8"	4'-8"	6'-0"	6'-0"	8'-0"	8'-0"	10'-0"	10'-0"	12'-0"	12'-0"	12'-8"	12'-8"	14'-0"	14'-0"
		30	NP	NP	NP	NP	NP	NP	NP	NP	4'-8"	4'-8"	6'-0"	6'-0"	8'-0"	8'-0"	10'-0"	10'-0"	12'-0"	11'-4"	12'-8"	12'-8"	14'-0"	13'-4"
		60	NP	NP	NP	NP	NP	NP	NP	NP	4'-8"	4'-8"	6'-0"	6'-0"	8'-0"	8'-0"	10'-0"	10'-0"	11'-4"	11'-4"	12'-8"	12'-0"	13'-4"	13'-4"
3.000	I or II	NJ	NP	NP	NP	NP	NP	NP	NP	NP	5'-4"	5'-4"	6'-8"	6'-8"	8'-8"	8'-8"	10'-0"	10'-0"	12'-8"	12'-8"	14'-0"	14'-0"	15'-4"	15'-4"
		30	NP	NP	NP	NP	NP	NP	NP	NP	5'-4"	5'-4"	6'-8"	6'-8"	8'-8"	8'-8"	10'-0"	10'-0"	12'-8"	12'-8"	13'-4"	13'-4"	14'-8"	14'-8"
		60	NP	NP	NP	NP	NP	NP	NP	NP	5'-4"	5'-4"	6'-8"	6'-8"	8'-8"	8'-8"	10'-0"	10'-0"	12'-8"	12'-0"	13'-4"	13'-4"	14'-0"	14'-0"
	III or IV	NJ	NP	NP	NP	NP	NP	NP	NP	NP	4'-0"	4'-0"	5'-4"	5'-4"	7'-4"	7'-4"	10'-0"	10'-0"	11'-4"	11'-4"	12'-0"	12'-0"	12'-8"	12'-8"
		30	NP	NP	NP	NP	NP	NP	NP	NP	4'-0"	4'-0"	5'-4"	5'-4"	7'-4"	7'-4"	10'-0"	10'-0"	11'-4"	10'-8"	12'-0"	11'-4"	12'-8"	12'-8"
		60	NP	NP	NP	NP	NP	NP	NP	NP	4'-0"	4'-0"	5'-4"	5'-4"	7'-4"	7'-4"	10'-0"	9'-4"	10'-8"	10'-8"	11'-4"	11'-4"	12'-8"	12'-8"

Table values shall not be permitted to be interpolated.
NP = Not Permitted
NJ = No Joist
ft = 0.3048 m
in. = 25.4 mm
lb/ft² = 0.0479 kPa
Unreinforced Ungrouted PCL or Mortar Cement Mortar
Unreinforced Fully Grouted PCL or Mortar Cement Mortar
The density of concrete masonry units shall be less than 135 lb/ft³ (2,162 kg/m³).

Table 3.2-7: List of Lateral Force Coefficients Tables

Basic Wind Speed, V	Maximum L_{joist}, ft (m)	Wall Location	LFRS Options OPMSW	ORMSW	SRMSW
V ≤ 110 mph (V ≤ 177 kph)	NJ	exterior	Table 3.2-8(1)	Table 3.2-8(2)	Table 3.2-8(15)
		interior	Table 3.2-8(1)	Table 3.2-8(2)	Table 3.2-8(15)
	30	exterior	NP	Table 3.2-8(3)	Table 3.2-8(16)
		interior	NP	Table 3.2-8(4)	Table 3.2-8(18)
	60	exterior	NP	Table 3.2-8(5)	Table 3.2-8(18)
		interior	NP	Table 3.2-8(6)	Table 3.2-8(19)
110 < V ≤ 115 mph (177 < V ≤ 185 kph)	NJ	exterior	Table 3.2-8(1)	Table 3.2-8(2)	Table 3.2-8(15)
		interior	Table 3.2-8(1)	Table 3.2-8(2)	Table 3.2-8(15)
	30	exterior	NP	Table 3.2-8(3)	Table 3.2-8(16)
		interior	NP	Table 3.2-8(4)	Table 3.2-8(18)
	60	exterior	NP	Table 3.2-8(5)	Table 3.2-8(18)
		interior	NP	Table 3.2-8(6)	Table 3.2-8(19)
115 < V ≤ 120 mph (185 < V ≤ 193 kph)	NJ	exterior	Table 3.2-8(1)	Table 3.2-8(2)	Table 3.2-8(15)
		interior	Table 3.2-8(1)	Table 3.2-8(2)	Table 3.2-8(15)
	30	exterior	NP	Table 3.2-8(3)	Table 3.2-8(16)
		interior	NP	Table 3.2-8(4)	Table 3.2-8(18)
	60	exterior	NP	Table 3.2-8(5)	Table 3.2-8(18)
		interior	NP	Table 3.2-8(6)	Table 3.2-8(19)
120 < V ≤ 130 mph (193 < V ≤ 209 kph)	NJ	exterior	Table 3.2-8(1)	Table 3.2-8(2)	Table 3.2-8(15)
		interior	Table 3.2-8(1)	Table 3.2-8(2)	Table 3.2-8(15)
	30	exterior	NP	Table 3.2-8(3)	Table 3.2-8(16)
		interior	NP	Table 3.2-8(4)	Table 3.2-8(20)
	60	exterior	NP	Table 3.2-8(5)	Table 3.2-8(20)
		interior	NP	Table 3.2-8(6)	Table 3.2-8(21)
130 < V ≤ 140 mph (209 < V ≤ 225 kph)	NJ	exterior	Table 3.2-8(1)	Table 3.2-8(2)	Table 3.2-8(15)
		interior	Table 3.2-8(1)	Table 3.2-8(2)	Table 3.2-8(15)
	30	exterior	NP	Table 3.2-8(3)	Table 3.2-8(16)
		interior	NP	Table 3.2-8(7)	Table 3.2-8(20)
	60	exterior	NP	Table 3.2-8(7)	Table 3.2-8(20)
		interior	NP	Table 3.2-8(8)	Table 3.2-8(22)
140 < V ≤ 150 mph (225 < V ≤ 241 kph)	NJ	exterior	Table 3.2-8(1)	Table 3.2-8(2)	Table 3.2-8(15)
		interior	Table 3.2-8(1)	Table 3.2-8(2)	Table 3.2-8(15)
	30	exterior	NP	Table 3.2-8(3)	Table 3.2-8(17)
		interior	NP	Table 3.2-8(7)	Table 3.2-8(23)
	60	exterior	NP	Table 3.2-8(9)	Table 3.2-8(23)
		interior	NP	Table 3.2-8(8)	Table 3.2-8(22)
150 < V ≤ 160 mph (241 < V ≤ 258 kph)	NJ	exterior	Table 3.2-8(1)	Table 3.2-8(2)	Table 3.2-8(15)
		interior	Table 3.2-8(1)	Table 3.2-8(2)	Table 3.2-8(15)
	30	exterior	NP	Table 3.2-8(3)	Table 3.2-8(17)
		interior	NP	Table 3.2-8(7)	Table 3.2-8(23)
	60	exterior	NP	Table 3.2-8(9)	Table 3.2-8(23)
		interior	NP	Table 3.2-8(10)	Table 3.2-8(24)
160 < V ≤ 180 mph (258 < V ≤ 290 kph)	NJ	exterior	Table 3.2-8(1)	Table 3.2-8(2)	Table 3.2-8(15)
		interior	Table 3.2-8(1)	Table 3.2-8(2)	Table 3.2-8(15)
	30	exterior	NP	Table 3.2-8(3)	Table 3.2-8(17)
		interior	NP	Table 3.2-8(11)	Table 3.2-8(25)
	60	exterior	NP	Table 3.2-8(11)	Table 3.2-8(25)
		interior	NP	Table 3.2-8(13)	Table 3.2-8(26)
180 < V ≤ 200 mph (290 < V ≤ 322 kph)	NJ	exterior	Table 3.2-8(1)	Table 3.2-8(2)	Table 3.2-8(15)
		interior	Table 3.2-8(1)	Table 3.2-8(2)	Table 3.2-8(15)
	30	exterior	NP	Table 3.2-8(3)	Table 3.2-8(17)
		interior	NP	Table 3.2-8(12)	Table 3.2-8(27)
	60	exterior	NP	Table 3.2-8(12)	Table 3.2-8(27)
		interior	NP	Table 3.2-8(14)	Table 3.2-8(28)

NJ = No Joist
NP = Not Permitted

Table 3.2-8(1): Lateral Force Coefficients for Ordinary Plain (Unreinforced) Masonry Shear Walls Under any of the Following Conditions:
Condition 1: $V \leq 200$ mph (322 kph); $L_{joist} = 0$ ft (0 m); Exterior Location
Condition 2: $V \leq 200$ mph (322 kph); $L_{joist} = 0$ ft (0 m); Interior Location

Vertical Reinforcement Schedule	Horizontal Reinforcement Schedule	Grouting Schedule	k_1	k_2 $L_{seg}=2'-0"$	$L_{seg}=2'-8"$	$L_{seg}=3'-4"$	$L_{seg}=4'-0"$	$L_{seg}=8'-0"$	$L_{seg}=12'-0"$	$L_{seg}=16'-0"$	$L_{seg}=20'-0"$	$L_{seg}=60'-0"$	$L_{seg}=100'-0"$	$L_{seg}=200'-0"$
None	None	None	2,688	1,435	2,552	3,987	5,741	22,964	51,670	91,858	143,528	1,291,751	3,588,197	14,352,787
None or No. 5 at 48, 32, 24, or 16 in.	None or No. 5 at 48, 32, 24, or 16 in.	Fully	13,176	17,626	31,334	48,960	70,502	282,010	634,522	1,128,038	1,762,560	15,863,040	44,064,000	176,256,000
No. 5 at 48 in.	None or No. 5 at 48, 32, or 24, in.	Partially	3,647	6,141	10,255	14,954	20,174	61,569	119,691	194,384	285,617	2,106,675	5,579,630	21,488,800
No. 5 at 32 in.	None or No. 5 at 48, 32, or 24, in.	Partially	4,122	7,077	11,818	17,233	23,249	70,954	137,937	224,016	329,156	2,427,815	6,430,183	24,764,532
No. 5 at 24 in.	None or No. 5 at 48, 32, or 24, in.	Partially	4,596	8,013	13,382	19,513	26,324	80,340	156,183	253,648	372,695	2,748,954	7,280,737	28,040,264

Table values shall not be permitted to be interpolated.
ft = 0.3048 m
in. = 25.4 mm

Table 3.2-8(2): Lateral Force Coefficients for Ordinary Reinforced Masonry Shear Walls Under any of the Following Conditions:
Condition 1: $V \leq 200$ mph (322 kph); $L_{joist} = 0$ ft (0 m); Exterior Location
Condition 2: $V \leq 200$ mph (322 kph); $L_{joist} = 0$ ft (0 m); Interior Location

Vertical Reinforcement Schedule	Horizontal Reinforcement Schedule	Grouting Schedule	k_1	$L_{seg}=2'-0"$	$L_{seg}=2'-8"$	$L_{seg}=3'-4"$	$L_{seg}=4'-0"$	$L_{seg}=8'-0"$	$L_{seg}=12'-0"$	$L_{seg}=16'-0"$	$L_{seg}=20'-0"$	$L_{seg}=60'-0"$	$L_{seg}=100'-0"$	$L_{seg}=200'-0"$
No. 5 at 120 in.	BJR at 16 in.	Partially	4,782	20,902	24,047	25,933	27,191	30,336	31,384	31,908	32,222	98,899	165,575	332,266
No. 5 at 96 in.	BJR at 16 in.	Partially	4,922	20,902	24,047	25,933	27,191	30,336	31,384	31,908	40,233	123,490	206,747	414,890
No. 5 at 72 in.	BJR at 16 in.	Partially	5,173	20,902	24,047	25,933	27,191	30,336	31,384	42,465	53,546	164,358	275,171	552,201
No. 5 at 48 in.	BJR at 16 in.	Partially	5,675	20,902	24,047	25,933	27,191	30,336	46,898	63,461	80,024	245,651	411,279	825,348
No. 5 at 48 in.	No. 5 at 48, 32, or 24 in.	Partially	10,088	20,902	24,047	25,933	27,191	30,336	46,898	63,461	80,024	245,651	411,279	825,348
No. 5 at 48 in.	No. 5 at 16 in.	Fully	22,680	20,902	24,047	25,933	27,191	30,336	46,898	63,461	80,024	245,651	411,279	825,348
No. 5 at 32 in.	BJR at 16 in.	Partially	6,414	20,902	24,047	25,933	27,191	45,237	69,949	94,660	119,371	366,483	613,596	1,231,376
No. 5 at 32 in.	No. 5 at 48, 32, or 24 in.	Partially	11,402	20,902	24,047	25,933	27,191	45,237	69,949	94,660	119,371	366,483	613,596	1,231,376
No. 5 at 32 in.	No. 5 at 16 in.	Fully	22,680	20,902	24,047	25,933	27,191	45,237	69,949	94,660	119,371	366,483	613,596	1,231,376
No. 5 at 24 in.	BJR at 16 in.	Partially	7,153	20,902	24,047	25,933	27,191	59,962	92,733	125,504	158,275	485,986	813,697	1,632,973
No. 5 at 24 in.	No. 5 at 48, 32, or 24 in.	Partially	12,716	20,902	24,047	25,933	27,191	59,962	92,733	125,504	158,275	485,986	813,697	1,632,973
No. 5 at 24 in.	No. 5 at 16 in.	Fully	22,680	20,902	24,047	25,933	27,191	59,962	92,733	125,504	158,275	485,986	813,697	1,632,973
No. 5 at 16 in.	BJR at 16 in.	Fully	12,758	20,902	24,047	32,151	40,255	88,880	137,505	186,130	234,754	721,003	1,207,252	2,422,875
No. 5 at 16 in.	No. 5 at 48, 32, 24, or 16 in.	Fully	22,058	20,902	24,047	32,151	40,255	88,880	137,505	186,130	234,754	721,003	1,207,252	2,422,875

Table values shall not be permitted to be interpolated.
ft = 0.3048 m
in. = 25.4 mm

Table 3.2-8(3): Lateral Force Coefficients for Ordinary Reinforced Masonry Shear Walls Under any of the Following Conditions:
Condition 1: $V \leq 200$ mph (322 kph); 0 ft (0 m) $< L_{joist} \leq 30$ ft (9.1 m); Exterior Location

Vertical Reinforcement Schedule	Horizontal Reinforcement Schedule	Grouting Schedule	k_1	$L_{seg}=2'-0"$	$L_{seg}=2'-8"$	$L_{seg}=3'-4"$	$L_{seg}=4'-0"$	$L_{seg}=8'-0"$	$L_{seg}=12'-0"$	$L_{seg}=16'-0"$	$L_{seg}=20'-0"$	$L_{seg}=60'-0"$	$L_{seg}=100'-0"$	$L_{seg}=200'-0"$
No. 5 at 120 in.	BJR at 16 in.	Partially	5,950	21,879	25,584	28,031	29,849	36,357	40,767	44,654	48,331	148,349	248,367	498,413
No. 5 at 96 in.	BJR at 16 in.	Partially	6,089	21,879	25,584	28,031	29,849	36,357	40,767	44,654	56,307	172,834	289,361	580,679
No. 5 at 72 in.	BJR at 16 in.	Partially	6,340	21,879	25,584	28,031	29,849	36,357	40,767	55,164	69,560	213,524	357,487	717,397
No. 5 at 48 in.	BJR at 16 in.	Partially	6,842	21,879	25,584	28,031	29,849	36,357	56,211	76,065	95,919	294,460	493,001	989,354
No. 5 at 48 in.	No. 5 at 48, 32, or 24 in.	Partially	10,088	21,879	25,584	28,031	29,849	36,357	56,211	76,065	95,919	294,460	493,001	989,354
No. 5 at 48 in.	No. 5 at 16 in.	Fully	22,680	21,879	25,584	28,031	29,849	36,357	56,211	76,065	95,919	294,460	493,001	989,354
No. 5 at 32 in.	BJR at 16 in.	Partially	7,581	21,879	25,584	28,031	29,849	51,187	79,154	107,121	135,088	414,757	694,426	1,393,598
No. 5 at 32 in.	No. 5 at 48, 32, or 24 in.	Partially	11,402	21,879	25,584	28,031	29,849	51,187	79,154	107,121	135,088	414,757	694,426	1,393,598
No. 5 at 32 in.	No. 5 at 16 in.	Fully	22,680	21,879	25,584	28,031	29,849	51,187	79,154	107,121	135,088	414,757	694,426	1,393,598
No. 5 at 24 in.	BJR at 16 in.	Partially	8,320	21,879	25,584	28,031	29,849	65,840	101,832	137,823	173,814	533,724	893,635	1,793,411
No. 5 at 24 in.	No. 5 at 48, 32, or 24 in.	Partially	12,716	21,879	25,584	28,031	29,849	65,840	101,832	137,823	173,814	533,724	893,635	1,793,411
No. 5 at 24 in.	No. 5 at 16 in.	Fully	22,680	21,879	25,584	28,031	29,849	65,840	101,832	137,823	173,814	533,724	893,635	1,793,411
No. 5 at 16 in.	BJR at 16 in.	Fully	13,925	21,879	25,584	34,213	42,842	94,615	146,389	198,162	249,936	767,671	1,285,407	2,579,745
No. 5 at 16 in.	No. 5 at 48, 32, 24, or 16 in.	Fully	22,680	21,879	25,584	34,213	42,842	94,615	146,389	198,162	249,936	767,671	1,285,407	2,579,745

Table values shall not be permitted to be interpolated.
ft = 0.3048 m
in. = 25.4 mm

DIRECT DESIGN HANDBOOK FOR MASONRY STRUCTURES

Table 3.2-8(4): Lateral Force Coefficients for Ordinary Reinforced Masonry Shear Walls Under any of the Following Conditions:
Condition 1: V ≤ 130 mph (209 kph); 0 ft (0 m) < L_{joist} ≤ 30 ft (9.1 m); Interior Location

Vertical Reinforcement Schedule	Horizontal Reinforcement Schedule	Grouting Schedule	k_1	k_2 L_{seg}=2'-0"	L_{seg}=2'-8"	L_{seg}=3'-4"	L_{seg}=4'-0"	L_{seg}=8'-0"	L_{seg}=12'-0"	L_{seg}=16'-0"	L_{seg}=20'-0"	L_{seg}=60'-0"	L_{seg}=100'-0"	L_{seg}=200'-0"
No. 5 at 120 in.	BJR at 16 in.	Partially	7,117	22,530	26,609	29,431	31,623	40,377	47,034	53,167	59,090	181,379	303,668	609,390
No. 5 at 96 in.	BJR at 16 in.	Partially	7,257	22,530	26,609	29,431	31,623	40,377	47,034	53,167	67,042	205,792	344,542	691,418
No. 5 at 72 in.	BJR at 16 in.	Partially	7,508	22,530	26,609	29,431	31,623	40,377	47,034	63,645	80,255	246,362	412,469	827,736
No. 5 at 48 in.	BJR at 16 in.	Partially	8,010	22,530	26,609	29,431	31,623	40,377	62,429	84,482	106,534	327,059	547,584	1,098,897
No. 5 at 48 in.	No. 5 at 48, 32, or 24 in.	Partially	10,088	22,530	26,609	29,431	31,623	40,377	62,429	84,482	106,534	327,059	547,584	1,098,897
No. 5 at 48 in.	No. 5 at 16 in.	Fully	22,680	22,530	26,609	29,431	31,623	40,377	62,429	84,482	106,534	327,059	547,584	1,098,897
No. 5 at 32 in.	BJR at 16 in.	Partially	8,749	22,530	26,609	29,431	31,623	55,160	85,301	115,442	145,584	446,997	748,411	1,501,945
No. 5 at 32 in.	No. 5 at 48, 32, or 24 in.	Partially	11,402	22,530	26,609	29,431	31,623	55,160	85,301	115,442	145,584	446,997	748,411	1,501,945
No. 5 at 32 in.	No. 5 at 16 in.	Fully	22,680	22,530	26,609	29,431	31,623	55,160	85,301	115,442	145,584	446,997	748,411	1,501,945
No. 5 at 24 in.	BJR at 16 in.	Partially	9,488	22,530	26,609	29,431	31,623	69,765	107,907	146,048	184,190	565,606	947,022	1,900,562
No. 5 at 24 in.	No. 5 at 48, 32, or 24 in.	Partially	12,716	22,530	26,609	29,431	31,623	69,765	107,907	146,048	184,190	565,606	947,022	1,900,562
No. 5 at 24 in.	No. 5 at 16 in.	Fully	22,680	22,530	26,609	29,431	31,623	69,765	107,907	146,048	184,190	565,606	947,022	1,900,562
No. 5 at 16 in.	BJR at 16 in.	Fully	15,093	22,530	26,609	35,589	44,568	98,444	152,321	206,197	260,073	798,835	1,337,598	2,684,504
No. 5 at 16 in.	No. 5 at 48, 32, 24, or 16 in.	Fully	22,680	22,530	26,609	35,589	44,568	98,444	152,321	206,197	260,073	798,835	1,337,598	2,684,504

Table values shall not be permitted to be interpolated.
ft = 0.3048 m
in. = 25.4 mm

Table 3.2-8(5): Lateral Force Coefficients for Ordinary Reinforced Masonry Shear Walls Under any of the Following Conditions:
Condition 1: V ≤ 130 mph (209 kph); 30 ft (9.1 m) < L_{joist} ≤ 60 ft (18.3 m); Exterior Location

Vertical Reinforcement Schedule	Horizontal Reinforcement Schedule	Grouting Schedule	k_1	L_{seg}=2'-0"	L_{seg}=2'-8"	L_{seg}=3'-4"	L_{seg}=4'-0"	L_{seg}=8'-0"	L_{seg}=12'-0"	L_{seg}=16'-0"	L_{seg}=20'-0"	L_{seg}=60'-0"	L_{seg}=100'-0"	L_{seg}=200'-0"
No. 5 at 120 in.	BJR at 16 in.	Partially	7,117	22,660	26,814	29,710	31,977	41,179	48,284	54,865	61,237	187,970	314,703	631,536
No. 5 at 96 in.	BJR at 16 in.	Partially	7,257	22,660	26,814	29,710	31,977	41,179	48,284	54,865	69,184	212,369	355,553	713,515
No. 5 at 72 in.	BJR at 16 in.	Partially	7,508	22,660	26,814	29,710	31,977	41,179	48,284	65,337	82,389	252,915	423,440	849,754
No. 5 at 48 in.	BJR at 16 in.	Partially	8,010	22,660	26,814	29,710	31,977	41,179	63,670	86,161	108,652	333,564	558,476	1,120,755
No. 5 at 48 in.	No. 5 at 48, 32, or 24 in.	Partially	10,088	22,660	26,814	29,710	31,977	41,179	63,670	86,161	108,652	333,564	558,476	1,120,755
No. 5 at 48 in.	No. 5 at 16 in.	Fully	22,680	22,660	26,814	29,710	31,977	41,179	63,670	86,161	108,652	333,564	558,476	1,120,755
No. 5 at 32 in.	BJR at 16 in.	Partially	8,749	22,660	26,814	29,710	31,977	55,952	86,527	117,103	147,678	453,430	759,183	1,523,564
No. 5 at 32 in.	No. 5 at 48, 32, or 24 in.	Partially	11,402	22,660	26,814	29,710	31,977	55,952	86,527	117,103	147,678	453,430	759,183	1,523,564
No. 5 at 32 in.	No. 5 at 16 in.	Fully	22,680	22,660	26,814	29,710	31,977	55,952	86,527	117,103	147,678	453,430	759,183	1,523,564
No. 5 at 24 in.	BJR at 16 in.	Partially	9,488	22,660	26,814	29,710	31,977	70,548	109,119	147,689	186,260	571,967	957,674	1,921,942
No. 5 at 24 in.	No. 5 at 48, 32, or 24 in.	Partially	12,716	22,660	26,814	29,710	31,977	70,548	109,119	147,689	186,260	571,967	957,674	1,921,942
No. 5 at 24 in.	No. 5 at 16 in.	Fully	22,680	22,660	26,814	29,710	31,977	70,548	109,119	147,689	186,260	571,967	957,674	1,921,942
No. 5 at 16 in.	BJR at 16 in.	Fully	15,093	22,660	26,814	35,863	44,912	99,208	153,504	207,800	262,096	805,053	1,348,011	2,705,406
No. 5 at 16 in.	No. 5 at 48, 32, 24, or 16 in.	Fully	22,680	22,660	26,814	35,863	44,912	99,208	153,504	207,800	262,096	805,053	1,348,011	2,705,406

Table values shall not be permitted to be interpolated.
ft = 0.3048 m
in. = 25.4 mm

Table 3.2-8(6): Lateral Force Coefficients for Ordinary Reinforced Masonry Shear Walls Under any of the Following Conditions:
Condition 1: V ≤ 130 mph (209 kph); 30 ft (9.1 m) < L_{joist} ≤ 60 ft (18.3 m); Interior Location

Vertical Reinforcement Schedule	Horizontal Reinforcement Schedule	Grouting Schedule	k_1	L_{seg}=2'-0"	L_{seg}=2'-8"	L_{seg}=3'-4"	L_{seg}=4'-0"	L_{seg}=8'-0"	L_{seg}=12'-0"	L_{seg}=16'-0"	L_{seg}=20'-0"	L_{seg}=60'-0"	L_{seg}=100'-0"	L_{seg}=200'-0"
No. 5 at 120 in.	BJR at 16 in.	Partially	8,502	24,138	29,145	32,895	36,016	50,338	62,564	74,266	85,758	263,258	440,759	884,510
No. 5 at 96 in.	BJR at 16 in.	Partially	8,750	24,138	29,145	32,895	36,016	50,338	62,564	74,266	93,650	287,493	481,335	965,942
No. 5 at 72 in.	BJR at 16 in.	Partially	9,196	24,138	29,145	32,895	36,016	50,338	62,564	84,664	106,764	327,765	548,765	1,101,267
No. 5 at 48 in.	BJR at 16 in.	Partially	10,088	24,138	29,145	32,895	36,016	50,338	77,840	105,342	132,844	407,866	682,887	1,370,441
No. 5 at 48 in.	No. 5 at 48, 32, or 24 in.	Partially	10,088	24,138	29,145	32,895	36,016	50,338	77,840	105,342	132,844	407,866	682,887	1,370,441
No. 5 at 48 in.	No. 5 at 16 in.	Fully	22,680	24,138	29,145	32,895	36,016	50,338	77,840	105,342	132,844	407,866	682,887	1,370,441
No. 5 at 32 in.	BJR at 16 in.	Partially	11,084	24,138	29,145	32,895	36,016	65,001	100,533	136,064	171,596	526,910	882,224	1,770,509
No. 5 at 32 in.	No. 5 at 48, 32, or 24 in.	Partially	11,402	24,138	29,145	32,895	36,016	65,001	100,533	136,064	171,596	526,910	882,224	1,770,509
No. 5 at 32 in.	No. 5 at 16 in.	Fully	22,680	24,138	29,145	32,895	36,016	65,001	100,533	136,064	171,596	526,910	882,224	1,770,509
No. 5 at 24 in.	BJR at 16 in.	Partially	11,823	24,138	29,145	32,895	36,016	79,488	122,960	166,432	209,904	644,625	1,079,345	2,166,147
No. 5 at 24 in.	No. 5 at 48, 32, or 24 in.	Partially	12,716	24,138	29,145	32,895	36,016	79,488	122,960	166,432	209,904	644,625	1,079,345	2,166,147
No. 5 at 24 in.	No. 5 at 16 in.	Fully	22,680	24,138	29,145	32,895	36,016	79,488	122,960	166,432	209,904	644,625	1,079,345	2,166,147
No. 5 at 16 in.	BJR at 16 in.	Fully	17,428	24,138	29,145	38,993	48,841	107,929	167,016	226,104	285,191	876,066	1,466,941	2,944,129
No. 5 at 16 in.	No. 5 at 48, 32, 24, or 16 in.	Fully	22,680	24,138	29,145	38,993	48,841	107,929	167,016	226,104	285,191	876,066	1,466,941	2,944,129

Table values shall not be permitted to be interpolated.
ft = 0.3048 m
in. = 25.4 mm

TMS 403-13

DIRECT DESIGN HANDBOOK FOR MASONRY STRUCTURES

Table 3.2-8(7): Lateral Force Coefficients for Ordinary Reinforced Masonry Shear Walls Under any of the Following Conditions:
Condition 1: 130 (209 kph) < V ≤ 160 mph (258 kph); 0 ft (0 m) < L_{joist} ≤ 30 ft (9.1 m); Interior Location
Condition 2: 130 (209 kph) < V ≤ 140 mph (225 kph); 30 ft (9.1 m) < L_{joist} ≤ 60 ft (18.3 m); Exterior Location

Vertical Reinforcement Schedule	Horizontal Reinforcement Schedule	Grouting Schedule	k_1	k_2 L_{seg}=2'-0"	L_{seg}=2'-8"	L_{seg}=3'-4"	L_{seg}=4'-0"	L_{seg}=8'-0"	L_{seg}=12'-0"	L_{seg}=16'-0"	L_{seg}=20'-0"	L_{seg}=60'-0"	L_{seg}=100'-0"	L_{seg}=200'-0"
No. 5 at 120 in.	BJR at 16 in.	Partially	7,212	23,603	28,302	31,742	34,554	47,021	57,393	67,240	76,877	235,990	395,103	792,886
No. 5 at 96 in.	BJR at 16 in.	Partially	7,351	23,603	28,302	31,742	34,554	47,021	57,393	67,240	84,789	260,284	435,779	874,516
No. 5 at 72 in.	BJR at 16 in.	Partially	7,602	23,603	28,302	31,742	34,554	47,021	57,393	77,665	97,936	300,656	503,375	1,010,173
No. 5 at 48 in.	BJR at 16 in.	Partially	8,104	23,603	28,302	31,742	34,554	47,021	72,709	98,396	124,083	380,956	637,828	1,280,010
No. 5 at 48 in.	No. 5 at 48, 32, or 24 in.	Partially	10,088	23,603	28,302	31,742	34,554	47,021	72,709	98,396	124,083	380,956	637,828	1,280,010
No. 5 at 48 in.	No. 5 at 16 in.	Fully	22,680	23,603	28,302	31,742	34,554	47,021	72,709	98,396	124,083	380,956	637,828	1,280,010
No. 5 at 32 in.	BJR at 16 in.	Partially	8,843	23,603	28,302	31,742	34,554	61,725	95,461	129,198	162,934	500,298	837,662	1,681,073
No. 5 at 32 in.	No. 5 at 48, 32, or 24 in.	Partially	11,402	23,603	28,302	31,742	34,554	61,725	95,461	129,198	162,934	500,298	837,662	1,681,073
No. 5 at 32 in.	No. 5 at 16 in.	Fully	22,680	23,603	28,302	31,742	34,554	61,725	95,461	129,198	162,934	500,298	837,662	1,681,073
No. 5 at 24 in.	BJR at 16 in.	Partially	9,582	23,603	28,302	31,742	34,554	76,251	117,948	159,645	201,342	618,311	1,035,281	2,077,705
No. 5 at 24 in.	No. 5 at 48, 32, or 24 in.	Partially	12,716	23,603	28,302	31,742	34,554	76,251	117,948	159,645	201,342	618,311	1,035,281	2,077,705
No. 5 at 24 in.	No. 5 at 16 in.	Fully	22,680	23,603	28,302	31,742	34,554	76,251	117,948	159,645	201,342	618,311	1,035,281	2,077,705
No. 5 at 16 in.	BJR at 16 in.	Fully	15,187	23,603	28,302	37,860	47,419	104,771	162,123	219,476	276,828	850,350	1,423,872	2,857,677
No. 5 at 16 in.	No. 5 at 48, 32, 24, or 16 in.	Fully	22,680	23,603	28,302	37,860	47,419	104,771	162,123	219,476	276,828	850,350	1,423,872	2,857,677

Table values shall not be permitted to be interpolated.
ft = 0.3048 m
in. = 25.4 mm

Table 3.2-8(8): Lateral Force Coefficients for Ordinary Reinforced Masonry Shear Walls Under any of the Following Conditions:
Condition 1: 130 (209 kph) < V ≤ 150 mph (241 kph); 30 ft (9.1 m) < L_{joist} ≤ 60 ft (18.3 m); Interior Location

Vertical Reinforcement Schedule	Horizontal Reinforcement Schedule	Grouting Schedule	k_1	k_2 L_{seg}=2'-0"	L_{seg}=2'-8"	L_{seg}=3'-4"	L_{seg}=4'-0"	L_{seg}=8'-0"	L_{seg}=12'-0"	L_{seg}=16'-0"	L_{seg}=20'-0"	L_{seg}=60'-0"	L_{seg}=100'-0"	L_{seg}=200'-0"
No. 5 at 120 in.	BJR at 16 in.	Partially	8,502	26,249	32,483	37,458	41,805	63,485	83,068	102,127	120,976	371,413	621,850	1,247,942
No. 5 at 96 in.	BJR at 16 in.	Partially	8,750	26,249	32,483	37,458	41,805	63,485	83,068	102,127	128,789	395,409	662,029	1,328,579
No. 5 at 72 in.	BJR at 16 in.	Partially	9,196	26,249	32,483	37,458	41,805	63,485	83,068	112,419	141,771	435,284	728,798	1,462,581
No. 5 at 48 in.	BJR at 16 in. or No. 5 at 48, 32, or 24 in.	Partially	10,088	26,249	32,483	37,458	41,805	63,485	98,185	132,886	167,586	514,591	861,596	1,729,108
No. 5 at 48 in.	No. 5 at 16 in.	Fully	22,680	26,249	32,483	37,458	41,805	63,485	98,185	132,886	167,586	514,591	861,596	1,729,108
No. 5 at 32 in.	BJR at 16 in.	Partially	11,272	26,249	32,483	37,458	41,805	77,989	120,640	163,290	205,941	632,444	1,058,948	2,125,207
No. 5 at 32 in.	No. 5 at 48, 32, or 24 in.	Partially	11,402	26,249	32,483	37,458	41,805	77,989	120,640	163,290	205,941	632,444	1,058,948	2,125,207
No. 5 at 32 in.	No. 5 at 16 in.	Fully	22,680	26,249	32,483	37,458	41,805	77,989	120,640	163,290	205,941	632,444	1,058,948	2,125,207
No. 5 at 24 in.	BJR at 16 in.	Partially	12,011	26,249	32,483	37,458	41,805	92,317	142,829	193,340	243,852	748,968	1,254,084	2,516,874
No. 5 at 24 in.	No. 5 at 48, 32, or 24 in.	Partially	12,716	26,249	32,483	37,458	41,805	92,317	142,829	193,340	243,852	748,968	1,254,084	2,516,874
No. 5 at 24 in.	No. 5 at 16 in.	Fully	22,680	26,249	32,483	37,458	41,805	92,317	142,829	193,340	243,852	748,968	1,254,084	2,516,874
No. 5 at 16 in.	BJR at 16 in.	Fully	17,616	26,249	32,483	43,477	54,472	120,440	186,409	252,377	318,345	978,028	1,637,710	3,286,916
No. 5 at 16 in.	No. 5 at 48, 32, 24, or 16 in.	Fully	22,680	26,249	32,483	43,477	54,472	120,440	186,409	252,377	318,345	978,028	1,637,710	3,286,916

Table values shall not be permitted to be interpolated.
ft = 0.3048 m
in. = 25.4 mm

Table 3.2-8(9): Lateral Force Coefficients for Ordinary Reinforced Masonry Shear Walls Under any of the Following Conditions:
Condition 1: 140 (225 kph) < V ≤ 160 mph (258 kph); 30 ft (9.1 m) < L_{joist} ≤ 60 ft (18.3 m); Exterior Location

Vertical Reinforcement Schedule	Horizontal Reinforcement Schedule	Grouting Schedule	k_1	k_2 L_{seg}=2'-0"	L_{seg}=2'-8"	L_{seg}=3'-4"	L_{seg}=4'-0"	L_{seg}=8'-0"	L_{seg}=12'-0"	L_{seg}=16'-0"	L_{seg}=20'-0"	L_{seg}=60'-0"	L_{seg}=100'-0"	L_{seg}=200'-0"
No. 5 at 120 in.	BJR at 16 in.	Partially	7,248	24,244	29,312	33,123	36,305	50,995	63,588	75,657	87,516	268,658	449,800	902,654
No. 5 at 96 in.	BJR at 16 in.	Partially	7,387	24,244	29,312	33,123	36,305	50,995	63,588	75,657	95,405	292,880	490,356	984,046
No. 5 at 72 in.	BJR at 16 in.	Partially	7,638	24,244	29,312	33,123	36,305	50,995	63,588	86,050	108,512	333,133	557,754	1,119,306
No. 5 at 48 in.	BJR at 16 in.	Partially	8,140	24,244	29,312	33,123	36,305	50,995	78,856	106,718	134,579	413,194	691,810	1,388,348
No. 5 at 48 in.	No. 5 at 48, 32, or 24 in.	Partially	10,088	24,244	29,312	33,123	36,305	50,995	78,856	106,718	134,579	413,194	691,810	1,388,348
No. 5 at 48 in.	No. 5 at 16 in.	Fully	22,680	24,244	29,312	33,123	36,305	50,995	78,856	106,718	134,579	413,194	691,810	1,388,348
No. 5 at 32 in.	BJR at 16 in.	Partially	8,879	24,244	29,312	33,123	36,305	65,650	101,537	137,424	173,311	532,179	891,048	1,788,219
No. 5 at 32 in.	No. 5 at 48, 32, or 24 in.	Partially	11,402	24,244	29,312	33,123	36,305	65,650	101,537	137,424	173,311	532,179	891,048	1,788,219
No. 5 at 32 in.	No. 5 at 16 in.	Fully	22,680	24,244	29,312	33,123	36,305	65,650	101,537	137,424	173,311	532,179	891,048	1,788,219
No. 5 at 24 in.	BJR at 16 in.	Partially	9,618	24,244	29,312	33,123	36,305	80,129	123,952	167,776	211,599	649,835	1,088,070	2,183,659
No. 5 at 24 in.	No. 5 at 48, 32, or 24 in.	Partially	12,716	24,244	29,312	33,123	36,305	80,129	123,952	167,776	211,599	649,835	1,088,070	2,183,659
No. 5 at 24 in.	No. 5 at 16 in.	Fully	22,680	24,244	29,312	33,123	36,305	80,129	123,952	167,776	211,599	649,835	1,088,070	2,183,659
No. 5 at 16 in.	BJR at 16 in.	Fully	15,223	24,244	29,312	39,217	49,123	108,554	167,985	227,416	286,847	881,158	1,475,469	2,961,247
No. 5 at 16 in.	No. 5 at 48, 32, 24, or 16 in.	Fully	22,680	24,244	29,312	39,217	49,123	108,554	167,985	227,416	286,847	881,158	1,475,469	2,961,247

Table values shall not be permitted to be interpolated.
ft = 0.3048 m
in. = 25.4 mm

DIRECT DESIGN HANDBOOK FOR MASONRY STRUCTURES

Table 3.2-8(10): Lateral Force Coefficients for Ordinary Reinforced Masonry Shear Walls Under any of the Following Conditions:
Condition 1: 150 (241 kph) < V ≤ 160 mph (258 kph); 30 ft (9.1 m) < L_{joist} ≤ 60 ft (18.3 m); Interior Location

Vertical Reinforcement Schedule	Horizontal Reinforcement Schedule	Grouting Schedule	k_1	k_2 L_{seg}=2'-0"	L_{seg}=2'-8"	L_{seg}=3'-4"	L_{seg}=4'-0"	L_{seg}=8'-0"	L_{seg}=12'-0"	L_{seg}=16'-0"	L_{seg}=20'-0"	L_{seg}=60'-0"	L_{seg}=100'-0"	L_{seg}=200'-0"
No. 5 at 120 in.	BJR at 16 in.	Partially	8,502	27,892	35,087	41,023	46,331	73,774	99,122	123,945	148,558	456,135	763,712	1,532,654
No. 5 at 96 in.	BJR at 16 in.	Partially	8,750	27,892	35,087	41,023	46,331	73,774	99,122	123,945	156,308	479,943	803,577	1,612,664
No. 5 at 72 in.	BJR at 16 in.	Partially	9,196	27,892	35,087	41,023	46,331	73,774	99,122	134,153	169,185	519,504	869,823	1,745,620
No. 5 at 48 in.	BJR at 16 in. or No. 5 at 48, 32, or 24 in.	Partially	10,088	27,892	35,087	41,023	46,331	73,774	114,113	154,453	194,792	598,184	1,001,575	2,010,055
No. 5 at 48 in.	No. 5 at 16 in.	Fully	22,680	27,892	35,087	41,023	46,331	73,774	114,113	154,453	194,792	598,184	1,001,575	2,010,055
No. 5 at 32 in.	BJR at 16 in. or No. 5 at 48, 32, or 24 in.	Partially	11,402	27,892	35,087	41,023	46,331	88,154	136,380	184,606	232,832	715,095	1,197,358	2,403,016
No. 5 at 32 in.	No. 5 at 16 in.	Fully	22,680	27,892	35,087	41,023	46,331	88,154	136,380	184,606	232,832	715,095	1,197,358	2,403,016
No. 5 at 24 in.	BJR at 16 in. or No. 5 at 48, 32, or 24 in.	Partially	12,160	27,892	35,087	41,023	46,331	102,355	158,380	214,405	270,430	830,678	1,390,926	2,791,545
No. 5 at 24 in.	No. 5 at 16 in.	Fully	22,680	27,892	35,087	41,023	46,331	102,355	158,380	214,405	270,430	830,678	1,390,926	2,791,545
No. 5 at 16 in.	BJR at 16 in.	Fully	17,765	27,892	35,087	46,979	58,872	130,228	201,584	272,940	344,296	1,057,855	1,771,414	3,555,312
No. 5 at 16 in.	No. 5 at 48, 32, 24, or 16 in.	Fully	22,680	27,892	35,087	46,979	58,872	130,228	201,584	272,940	344,296	1,057,855	1,771,414	3,555,312

Table values shall not be permitted to be interpolated.
ft = 0.3048 m
in. = 25.4 mm

Table 3.2-8(11): Lateral Force Coefficients for Ordinary Reinforced Masonry Shear Walls Under any of the Following Conditions:
Condition 1: 160 (258 kph) < V ≤ 180 mph (290 kph); 0 ft (0 m) < L_{joist} ≤ 30 ft (9.1 m); Interior Location
Condition 2: 160 (258 kph) < V ≤ 180 mph (290 kph); 30 ft (9.1 m) < L_{joist} ≤ 60 ft (18.3 m); Exterior Location

Vertical Reinforcement Schedule	Horizontal Reinforcement Schedule	Grouting Schedule	k_1	L_{seg}=2'-0"	L_{seg}=2'-8"	L_{seg}=3'-4"	L_{seg}=4'-0"	L_{seg}=8'-0"	L_{seg}=12'-0"	L_{seg}=16'-0"	L_{seg}=20'-0"	L_{seg}=60'-0"	L_{seg}=100'-0"	L_{seg}=200'-0"
No. 5 at 120 in.	BJR at 16 in.	Partially	7,372	25,418	31,168	35,660	39,523	58,299	74,979	91,135	107,082	328,740	550,398	1,104,544
No. 5 at 96 in.	BJR at 16 in.	Partially	7,511	25,418	31,168	35,660	39,523	58,299	74,979	91,135	114,926	352,830	590,734	1,185,495
No. 5 at 72 in.	BJR at 16 in.	Partially	7,762	25,418	31,168	35,660	39,523	58,299	74,979	101,469	127,960	392,862	657,765	1,320,021
No. 5 at 48 in.	BJR at 16 in.	Partially	8,264	25,418	31,168	35,660	39,523	58,299	90,159	122,020	153,880	472,484	791,087	1,587,596
No. 5 at 48 in.	No. 5 at 48, 32, or 24 in.	Partially	10,088	25,418	31,168	35,660	39,523	58,299	90,159	122,020	153,880	472,484	791,087	1,587,596
No. 5 at 48 in.	No. 5 at 16 in.	Fully	22,680	25,418	31,168	35,660	39,523	58,299	90,159	122,020	153,880	472,484	791,087	1,587,596
No. 5 at 32 in.	BJR at 16 in.	Partially	9,003	25,418	31,168	35,660	39,523	72,867	112,708	152,550	192,392	590,808	989,224	1,985,265
No. 5 at 32 in.	No. 5 at 48, 32, or 24 in.	Partially	11,402	25,418	31,168	35,660	39,523	72,867	112,708	152,550	192,392	590,808	989,224	1,985,265
No. 5 at 32 in.	No. 5 at 16 in.	Fully	22,680	25,418	31,168	35,660	39,523	72,867	112,708	152,550	192,392	590,808	989,224	1,985,265
No. 5 at 24 in.	BJR at 16 in.	Partially	9,742	25,418	31,168	35,660	39,523	87,257	134,991	182,726	230,460	707,803	1,185,146	2,378,503
No. 5 at 24 in.	No. 5 at 48, 32, or 24 in.	Partially	12,716	25,418	31,168	35,660	39,523	87,257	134,991	182,726	230,460	707,803	1,185,146	2,378,503
No. 5 at 24 in.	No. 5 at 16 in.	Fully	22,680	25,418	31,168	35,660	39,523	87,257	134,991	182,726	230,460	707,803	1,185,146	2,378,503
No. 5 at 16 in.	BJR at 16 in.	Fully	15,347	25,418	31,168	41,710	52,252	115,506	178,760	242,014	305,267	937,805	1,570,343	3,151,688
No. 5 at 16 in.	No. 5 at 48, 32, 24, 16 in.	Fully	22,680	25,418	31,168	41,710	52,252	115,506	178,760	242,014	305,267	937,805	1,570,343	3,151,688

Table values shall not be permitted to be interpolated.
ft = 0.3048 m
in. = 25.4 mm

Table 3.2-8(12): Lateral Force Coefficients for Ordinary Reinforced Masonry Shear Walls Under any of the Following Conditions:
Condition 1: 180 (290 kph) < V ≤ 200 mph (322 kph); 0 ft (0 m) < L_{joist} ≤ 30 ft (9.1 m); Interior Location
Condition 2: 180 (290 kph) < V ≤ 200 mph (322 kph); 30 ft (9.1 m) < L_{joist} ≤ 60 ft (18.3 m); Exterior Location

Vertical Reinforcement Schedule	Horizontal Reinforcement Schedule	Grouting Schedule	k_1	L_{seg}=2'-0"	L_{seg}=2'-8"	L_{seg}=3'-4"	L_{seg}=4'-0"	L_{seg}=8'-0"	L_{seg}=12'-0"	L_{seg}=16'-0"	L_{seg}=20'-0"	L_{seg}=60'-0"	L_{seg}=100'-0"	L_{seg}=200'-0"
No. 5 at 120 in.	BJR at 16 in.	Partially	7,468	26,467	32,828	37,931	42,405	64,847	85,193	105,015	124,628	382,628	640,628	1,285,629
No. 5 at 96 in.	BJR at 16 in.	Partially	7,607	26,467	32,828	37,931	42,405	64,847	85,193	105,015	132,432	406,599	680,766	1,366,184
No. 5 at 72 in.	BJR at 16 in.	Partially	7,858	26,467	32,828	37,931	42,405	64,847	85,193	115,297	145,400	446,433	747,466	1,500,048
No. 5 at 48 in.	BJR at 16 in.	Partially	8,360	26,467	32,828	37,931	42,405	64,847	100,294	135,741	171,188	525,657	880,126	1,766,299
No. 5 at 48 in.	No. 5 at 48, 32, or 24 in.	Partially	10,088	26,467	32,828	37,931	42,405	64,847	100,294	135,741	171,188	525,657	880,126	1,766,299
No. 5 at 48 in.	No. 5 at 16 in.	Fully	22,680	26,467	32,828	37,931	42,405	64,847	100,294	135,741	171,188	525,657	880,126	1,766,299
No. 5 at 32 in.	BJR at 16 in.	Partially	9,099	26,467	32,828	37,931	42,405	79,335	122,724	166,112	209,501	643,386	1,077,271	2,161,984
No. 5 at 32 in.	No. 5 at 48, 32, or 24 in.	Partially	11,402	26,467	32,828	37,931	42,405	79,335	122,724	166,112	209,501	643,386	1,077,271	2,161,984
No. 5 at 32 in.	No. 5 at 16 in.	Fully	22,680	26,467	32,828	37,931	42,405	79,335	122,724	166,112	209,501	643,386	1,077,271	2,161,984
No. 5 at 24 in.	BJR at 16 in.	Partially	9,838	26,467	32,828	37,931	42,405	93,646	144,888	196,129	247,371	759,786	1,272,200	2,553,237
No. 5 at 24 in.	No. 5 at 48, 32, or 24 in.	Partially	12,716	26,467	32,828	37,931	42,405	93,646	144,888	196,129	247,371	759,786	1,272,200	2,553,237
No. 5 at 24 in.	No. 5 at 16 in.	Fully	22,680	26,467	32,828	37,931	42,405	93,646	144,888	196,129	247,371	759,786	1,272,200	2,553,237
No. 5 at 16 in.	BJR at 16 in.	Fully	15,443	26,467	32,828	43,941	55,055	121,737	188,418	255,100	321,781	988,597	1,655,412	3,322,452
No. 5 at 16 in.	No. 5 at 48, 32, or 24 in.	Fully	22,680	26,467	32,828	43,941	55,055	121,737	188,418	255,100	321,781	988,597	1,655,412	3,322,452
No. 5 at 16 in.	No. 5 at 16 in.	Fully	22,680	26,467	32,828	43,941	55,055	121,737	188,418	255,100	321,781	988,597	1,655,412	3,322,452

Table values shall not be permitted to be interpolated.
ft = 0.3048 m
in. = 25.4 mm

Table 3.2-8(13): Lateral Force Coefficients for Ordinary Reinforced Masonry Shear Walls Under any of the Following Conditions:
Condition 1: 160 (258 kph) < V ≤ 180 mph (290 kph); 30 ft (9.1 m) < L_{joist} ≤ 60 ft (18.3 m); Interior Location

Vertical Reinforcement Schedule	Horizontal Reinforcement Schedule	Grouting Schedule	k_1	k_2 L_{seg}=2'-0"	L_{seg}=2'-8"	L_{seg}=3'-4"	L_{seg}=4'-0"	L_{seg}=8'-0"	L_{seg}=12'-0"	L_{seg}=16'-0"	L_{seg}=20'-0"	L_{seg}=60'-0"	L_{seg}=100'-0"	L_{seg}=200'-0"
No. 5 at 120 in.	BJR at 16 in.	Partially	8,502	29,776	38,078	45,123	51,539	85,632	117,629	149,101	180,364	553,851	927,337	1,861,053
No. 5 at 96 in.	BJR at 16 in.	Partially	8,750	29,776	38,078	45,123	51,539	85,632	117,629	149,101	188,041	577,439	966,837	1,940,333
No. 5 at 72 in.	BJR at 16 in.	Partially	9,196	29,776	38,078	45,123	51,539	85,632	117,629	159,212	200,796	616,636	1,032,475	2,072,073
No. 5 at 48 in.	BJR at 16 in. or No. 5 at 48, 32, or 24 in.	Partially	10,088	29,776	38,078	45,123	51,539	85,632	132,474	179,317	226,160	694,585	1,163,011	2,334,075
No. 5 at 48 in.	No. 5 at 16 in.	Fully	22,680	29,776	38,078	45,123	51,539	85,632	132,474	179,317	226,160	694,585	1,163,011	2,334,075
No. 5 at 32 in.	BJR at 16 in. or No. 5 at 48, 32, or 24 in.	Partially	11,402	29,776	38,078	45,123	51,539	99,865	154,522	209,179	263,835	810,402	1,356,969	2,723,386
No. 5 at 32 in.	No. 5 at 16 in.	Fully	22,680	29,776	38,078	45,123	51,539	99,865	154,522	209,179	263,835	810,402	1,356,969	2,723,386
No. 5 at 24 in.	BJR at 16 in.	Partially	12,332	29,776	38,078	45,123	51,539	113,921	176,303	238,685	301,068	924,889	1,548,711	3,108,265
No. 5 at 24 in.	No. 5 at 48, 32, or 24 in.	Partially	12,716	29,776	38,078	45,123	51,539	113,921	176,303	238,685	301,068	924,889	1,548,711	3,108,265
No. 5 at 24 in.	No. 5 at 16 in.	Fully	22,680	29,776	38,078	45,123	51,539	113,921	176,303	238,685	301,068	924,889	1,548,711	3,108,265
No. 5 at 16 in.	BJR at 16 in.	Fully	17,937	29,776	38,078	51,006	63,934	141,501	219,069	296,636	374,203	1,149,877	1,925,550	3,864,733
No. 5 at 16 in.	No. 5 at 48, 32, 24, or 16 in.	Fully	22,680	29,776	38,078	51,006	63,934	141,501	219,069	296,636	374,203	1,149,877	1,925,550	3,864,733

Table values shall not be permitted to be interpolated.
ft = 0.3048 m
in. = 25.4 mm

Table 3.2-8(14): Lateral Force Coefficients for Ordinary Reinforced Masonry Shear Walls Under any of the Following Conditions:
Condition 1: 180 (290 kph) < V ≤ 200 mph (322 kph); 30 ft (9.1 m) < L_{joist} ≤ 60 ft (18.3 m); Interior Location

Vertical Reinforcement Schedule	Horizontal Reinforcement Schedule	Grouting Schedule	k_1	k_2 L_{seg}=2'-0"	L_{seg}=2'-8"	L_{seg}=3'-4"	L_{seg}=4'-0"	L_{seg}=8'-0"	L_{seg}=12'-0"	L_{seg}=16'-0"	L_{seg}=20'-0"	L_{seg}=60'-0"	L_{seg}=100'-0"	L_{seg}=200'-0"
No. 5 at 120 in.	BJR at 16 in.	Partially	8,502	31,790	41,287	49,526	57,136	98,393	137,554	176,191	214,618	659,113	1,103,608	2,214,846
No. 5 at 96 in.	BJR at 16 in.	Partially	8,750	31,790	41,287	49,526	57,136	98,393	137,554	176,191	222,216	682,464	1,142,712	2,293,332
No. 5 at 72 in.	BJR at 16 in.	Partially	9,196	31,790	41,287	49,526	57,136	98,393	137,554	186,197	234,839	721,263	1,207,688	2,423,748
No. 5 at 48 in.	BJR at 16 in. or No. 5 at 48, 32, or 24 in.	Partially	10,088	31,790	41,287	49,526	57,136	98,393	152,241	206,089	259,938	798,419	1,336,900	2,683,104
No. 5 at 48 in.	No. 5 at 16 in.	Fully	22,680	31,790	41,287	49,526	57,136	98,393	152,241	206,089	259,938	798,419	1,336,900	2,683,104
No. 5 at 32 in.	BJR at 16 in. or No. 5 at 48, 32, or 24 in.	Partially	11,402	31,790	41,287	49,526	57,136	112,468	174,050	235,633	297,216	913,045	1,528,873	3,068,444
No. 5 at 32 in.	No. 5 at 16 in.	Fully	22,680	31,790	41,287	49,526	57,136	112,468	174,050	235,633	297,216	913,045	1,528,873	3,068,444
No. 5 at 24 in.	BJR at 16 in.	Partially	12,523	31,790	41,287	49,526	57,136	126,365	195,594	264,823	334,052	1,026,341	1,718,631	3,449,354
No. 5 at 24 in.	No. 5 at 48, 32, or 24 in.	Partially	12,716	31,790	41,287	49,526	57,136	126,365	195,594	264,823	334,052	1,026,341	1,718,631	3,449,354
No. 5 at 24 in.	No. 5 at 16 in.	Fully	22,680	31,790	41,287	49,526	57,136	126,365	195,594	264,823	334,052	1,026,341	1,718,631	3,449,354
No. 5 at 16 in.	BJR at 16 in.	Fully	18,128	31,790	41,287	55,330	69,372	153,627	237,883	322,138	406,393	1,248,946	2,091,499	4,197,882
No. 5 at 16 in.	No. 5 at 48, 32, 24, or 16 in.	Fully	22,680	31,790	41,287	55,330	69,372	153,627	237,883	322,138	406,393	1,248,946	2,091,499	4,197,882

Table values shall not be permitted to be interpolated.
ft = 0.3048 m
in. = 25.4 mm

DIRECT DESIGN HANDBOOK FOR MASONRY STRUCTURES

Table 3.2-8(15): Lateral Force Coefficients for Special Reinforced Masonry Shear Walls Under any of the Following Conditions:
Condition 1: V ≤ 200 mph (322 kph); L_{joist} = 0 ft (0 m); Exterior Location
Condition 2: V ≤ 200 mph (322 kph); L_{joist} = 0 ft (0 m); Interior Location

Vertical Reinforcement Schedule	Horizontal Reinforcement Schedule	Grouting Schedule	k_1	k_2										
				L_{seg}=2'-0"	L_{seg}=2'-8"	L_{seg}=3'-4"	L_{seg}=4'-0"	L_{seg}=8'-0"	L_{seg}=12'-0"	L_{seg}=16'-0"	L_{seg}=20'-0"	L_{seg}=60'-0"	L_{seg}=100'-0"	L_{seg}=200'-0"
No. 5 at 48 in.	No. 5 at 48, 32, or 24 in.	Partially	4,035	20,902	24,047	25,933	27,191	30,336	46,898	63,461	80,024	245,651	411,279	825,348
No. 5 at 48 in.	No. 5 at 16 in.	Fully	9,072	20,902	24,047	25,933	27,191	30,336	46,898	63,461	80,024	245,651	411,279	825,348
No. 5 at 32 in.	No. 5 at 48, 32, or 24 in.	Partially	4,561	20,902	24,047	25,933	27,191	45,237	69,949	94,660	119,371	366,483	613,596	1,231,376
No. 5 at 32 in.	No. 5 at 16 in.	Fully	9,072	20,902	24,047	25,933	27,191	45,237	69,949	94,660	119,371	366,483	613,596	1,231,376
No. 5 at 24 in.	No. 5 at 48, 32, or 24 in.	Partially	5,086	20,902	24,047	25,933	27,191	59,962	92,733	125,504	158,275	485,986	813,697	1,632,973
No. 5 at 24 in.	No. 5 at 16 in.	Fully	9,072	20,902	24,047	25,933	27,191	59,962	92,733	125,504	158,275	485,986	813,697	1,632,973
No. 5 at 16 in.	No. 5 at 48, 32, 24, or 16 in.	Fully	8,823	20,902	24,047	32,151	40,255	88,880	137,505	186,130	234,754	721,003	1,207,252	2,422,875

Table values shall not be permitted to be interpolated.
ft = 0.3048 m
in. = 25.4 mm

Table 3.2-8(16): Lateral Force Coefficients for Special Reinforced Masonry Shear Walls Under any of the Following Conditions:
Condition 1: V ≤ 140 mph (225 kph); 0 ft (0 m) < L_{joist} ≤ 30 ft (9.1 m); Exterior Location

Vertical Reinforcement Schedule	Horizontal Reinforcement Schedule	Grouting Schedule	k_1	k_2										
				L_{seg}=2'-0"	L_{seg}=2'-8"	L_{seg}=3'-4"	L_{seg}=4'-0"	L_{seg}=8'-0"	L_{seg}=12'-0"	L_{seg}=16'-0"	L_{seg}=20'-0"	L_{seg}=60'-0"	L_{seg}=100'-0"	L_{seg}=200'-0"
No. 5 at 48 in.	No. 5 at 48, 32, or 24 in.	Partially	4,035	21,879	25,584	28,031	29,849	36,357	56,211	76,065	95,919	294,460	493,001	989,354
No. 5 at 48 in.	No. 5 at 16 in.	Fully	9,072	21,879	25,584	28,031	29,849	36,357	56,211	76,065	95,919	294,460	493,001	989,354
No. 5 at 32 in.	No. 5 at 48, 32, or 24 in.	Partially	4,561	21,879	25,584	28,031	29,849	51,187	79,154	107,121	135,088	414,757	694,426	1,393,598
No. 5 at 32 in.	No. 5 at 16 in.	Fully	9,072	21,879	25,584	28,031	29,849	51,187	79,154	107,121	135,088	414,757	694,426	1,393,598
No. 5 at 24 in.	No. 5 at 48, 32, or 24 in.	Partially	5,086	21,879	25,584	28,031	29,849	65,840	101,832	137,823	173,814	533,724	893,635	1,793,411
No. 5 at 24 in.	No. 5 at 16 in.	Fully	9,072	21,879	25,584	28,031	29,849	65,840	101,832	137,823	173,814	533,724	893,635	1,793,411
No. 5 at 16 in.	No. 5 at 48, 32, 24, or 16 in.	Fully	9,072	21,879	25,584	34,213	42,842	94,615	146,389	198,162	249,936	767,671	1,285,407	2,579,745

Table values shall not be permitted to be interpolated.
ft = 0.3048 m
in. = 25.4 mm

Table 3.2-8(17): Lateral Force Coefficients for Special Reinforced Masonry Shear Walls Under any of the Following Conditions:
Condition 1: 140 mph (225 kph) < V ≤ 200 mph (322 kph); 0 ft (0 m) < L_{joist} ≤ 30 ft (9.1 m); Exterior Location

Vertical Reinforcement Schedule	Horizontal Reinforcement Schedule	Grouting Schedule	k_1	k_2										
				L_{seg}=2'-0"	L_{seg}=2'-8"	L_{seg}=3'-4"	L_{seg}=4'-0"	L_{seg}=8'-0"	L_{seg}=12'-0"	L_{seg}=16'-0"	L_{seg}=20'-0"	L_{seg}=60'-0"	L_{seg}=100'-0"	L_{seg}=200'-0"
No. 5 at 48 in.	No. 5 at 48, 32, or 24 in.	Partially	4,035	22,758	26,969	29,922	32,246	41,789	64,614	87,438	110,263	338,511	566,758	1,137,377
No. 5 at 48 in.	No. 5 at 16 in.	Fully	9,072	22,758	26,969	29,922	32,246	41,789	64,614	87,438	110,263	338,511	566,758	1,137,377
No. 5 at 32 in.	No. 5 at 48, 32, or 24 in.	Partially	4,561	22,758	26,969	29,922	32,246	56,555	87,460	118,365	149,270	458,322	767,374	1,540,005
No. 5 at 32 in.	No. 5 at 16 in.	Fully	9,072	22,758	26,969	29,922	32,246	56,555	87,460	118,365	149,270	458,322	767,374	1,540,005
No. 5 at 24 in.	No. 5 at 48, 32, or 24 in.	Partially	5,086	22,758	26,969	29,922	32,246	71,143	110,040	148,937	187,834	576,805	965,775	1,938,201
No. 5 at 24 in.	No. 5 at 16 in.	Fully	9,072	22,758	26,969	29,922	32,246	71,143	110,040	148,937	187,834	576,805	965,775	1,938,201
No. 5 at 16 in.	No. 5 at 48, 32, 24, or 16 in.	Fully	9,072	22,758	26,969	36,072	45,174	99,789	154,404	209,019	263,633	809,782	1,355,930	2,721,301

Table values shall not be permitted to be interpolated.
ft = 0.3048 m
in. = 25.4 mm

Table 3.2-8(18): Lateral Force Coefficients for Special Reinforced Masonry Shear Walls Under any of the Following Conditions:
Condition 1: V ≤ 120 mph (193 kph); 0 ft (0 m) < L_{joist} ≤ 30 ft (9.1 m); Interior Location
Condition 1: V ≤ 120 mph (193 kph); 30 ft (9.1 m) < L_{joist} ≤ 60 ft (18.3 m); Exterior Location

Vertical Reinforcement Schedule	Horizontal Reinforcement Schedule	Grouting Schedule	k_1	k_2										
				L_{seg}=2'-0"	L_{seg}=2'-8"	L_{seg}=3'-4"	L_{seg}=4'-0"	L_{seg}=8'-0"	L_{seg}=12'-0"	L_{seg}=16'-0"	L_{seg}=20'-0"	L_{seg}=60'-0"	L_{seg}=100'-0"	L_{seg}=200'-0"
No. 5 at 48 in.	No. 5 at 48, 32, or 24 in.	Partially	4,035	22,530	26,609	29,431	31,623	40,377	62,429	84,482	106,534	327,059	547,584	1,098,897
No. 5 at 48 in.	No. 5 at 16 in.	Fully	9,072	22,530	26,609	29,431	31,623	40,377	62,429	84,482	106,534	327,059	547,584	1,098,897
No. 5 at 32 in.	No. 5 at 48, 32, or 24 in.	Partially	4,561	22,530	26,609	29,431	31,623	55,160	85,301	115,442	145,584	446,997	748,411	1,501,945
No. 5 at 32 in.	No. 5 at 16 in.	Fully	9,072	22,530	26,609	29,431	31,623	55,160	85,301	115,442	145,584	446,997	748,411	1,501,945
No. 5 at 24 in.	No. 5 at 48, 32, or 24 in.	Partially	5,086	22,530	26,609	29,431	31,623	69,765	107,907	146,048	184,190	565,606	947,022	1,900,562
No. 5 at 24 in.	No. 5 at 16 in.	Fully	9,072	22,530	26,609	29,431	31,623	69,765	107,907	146,048	184,190	565,606	947,022	1,900,562
No. 5 at 16 in.	No. 5 at 48, 32, 24, or 16 in.	Fully	9,072	22,530	26,609	35,589	44,568	98,444	152,321	206,197	260,073	798,835	1,337,598	2,684,504

Table values shall not be permitted to be interpolated.
ft = 0.3048 m
in. = 25.4 mm

Table 3.2-8(19): Lateral Force Coefficients for Special Reinforced Masonry Shear Walls Under any of the Following Conditions:
Condition 1: V < 120 mph (193 kph); 30 ft (9.1 m) < L_{joist} ≤ 60 ft (18.3 m); Interior Location

Vertical Reinforcement Schedule	Horizontal Reinforcement Schedule	Grouting Schedule	k_1	k_2										
				L_{seg}=2'-0"	L_{seg}=2'-8"	L_{seg}=3'-4"	L_{seg}=4'-0"	L_{seg}=8'-0"	L_{seg}=12'-0"	L_{seg}=16'-0"	L_{seg}=20'-0"	L_{seg}=60'-0"	L_{seg}=100'-0"	L_{seg}=200'-0"
No. 5 at 48 in.	No. 5 at 48, 32, or 24 in.	Partially	4,035	NP	29,145	32,895	36,016	50,338	77,840	105,342	132,844	407,866	682,887	1,370,441
No. 5 at 48 in.	No. 5 at 16 in.	Fully	9,072	NP	29,145	32,895	36,016	50,338	77,840	105,342	132,844	407,866	682,887	1,370,441
No. 5 at 32 in.	No. 5 at 48, 32, or 24 in.	Partially	4,561	NP	29,145	32,895	36,016	65,001	100,533	136,064	171,596	526,910	882,224	1,770,509
No. 5 at 32 in.	No. 5 at 16 in.	Fully	9,072	NP	29,145	32,895	36,016	65,001	100,533	136,064	171,596	526,910	882,224	1,770,509
No. 5 at 24 in.	No. 5 at 48, 32, or 24 in.	Partially	5,086	NP	29,145	32,895	36,016	79,488	122,960	166,432	209,904	644,625	1,079,345	2,166,147
No. 5 at 24 in.	No. 5 at 16 in.	Fully	9,072	NP	29,145	32,895	36,016	79,488	122,960	166,432	209,904	644,625	1,079,345	2,166,147
No. 5 at 16 in.	No. 5 at 48, 32, 24, or 16 in.	Fully	9,072	NP	29,145	38,993	48,841	107,929	167,016	226,104	285,191	876,066	1,466,941	2,944,129

Table values shall not be permitted to be interpolated.
ft = 0.3048 m
in. = 25.4 mm

Table 3.2-8(20): Lateral Force Coefficients for Special Reinforced Masonry Shear Walls Under any of the Following Conditions:
Condition 1: 120 mph (193 kph) < V ≤ 140 mph (225 kph); 0 ft (0 m) < L_{joist} ≤ 30 ft (9.1 m); Interior Location
Condition 1: 120 mph (193 kph) < V ≤ 140 mph (225 kph); 30 ft (9.1 m) < L_{joist} ≤ 60 ft (18.3 m); Exterior Location

Vertical Reinforcement Schedule	Horizontal Reinforcement Schedule	Grouting Schedule	k_1	k_2										
				L_{seg}=2'-0"	L_{seg}=2'-8"	L_{seg}=3'-4"	L_{seg}=4'-0"	L_{seg}=8'-0"	L_{seg}=12'-0"	L_{seg}=16'-0"	L_{seg}=20'-0"	L_{seg}=60'-0"	L_{seg}=100'-0"	L_{seg}=200'-0"
No. 5 at 48 in.	No. 5 at 48, 32, or 24 in.	Partially	4,035	23,215	27,689	30,905	33,492	44,614	68,984	93,355	117,725	361,427	605,130	1,214,386
No. 5 at 48 in.	No. 5 at 16 in.	Fully	9,072	23,215	27,689	30,905	33,492	44,614	68,984	93,355	117,725	361,427	605,130	1,214,386
No. 5 at 32 in.	No. 5 at 48, 32, or 24 in.	Partially	4,561	23,215	27,689	30,905	33,492	59,346	91,780	124,214	156,648	480,986	805,324	1,616,170
No. 5 at 32 in.	No. 5 at 16 in.	Fully	9,072	23,215	27,689	30,905	33,492	59,346	91,780	124,214	156,648	480,986	805,324	1,616,170
No. 5 at 24 in.	No. 5 at 48, 32, or 24 in.	Partially	5,086	23,215	27,689	30,905	33,492	73,901	114,310	154,719	195,128	599,215	1,003,303	2,013,522
No. 5 at 24 in.	No. 5 at 16 in.	Fully	9,072	23,215	27,689	30,905	33,492	73,901	114,310	154,719	195,128	599,215	1,003,303	2,013,522
No. 5 at 16 in.	No. 5 at 48, 32, 24, or 16 in.	Fully	9,072	23,215	27,689	37,038	46,387	102,479	158,572	214,665	270,758	831,686	1,392,614	2,794,935

Table values shall not be permitted to be interpolated.
ft = 0.3048 m
in. = 25.4 mm

Table 3.2-8(21): Lateral Force Coefficients for Special Reinforced Masonry Shear Walls Under any of the Following Conditions:
Condition 1: 120 mph (193 kph) < V ≤ 130 mph (209 kph); 30 ft (9.1 m) < L_{joist} ≤ 60 ft (18.3 m); Interior Location

Vertical Reinforcement Schedule	Horizontal Reinforcement Schedule	Grouting Schedule	k_1	k_2										
				L_{seg}=2'-0"	L_{seg}=2'-8"	L_{seg}=3'-4"	L_{seg}=4'-0"	L_{seg}=8'-0"	L_{seg}=12'-0"	L_{seg}=16'-0"	L_{seg}=20'-0"	L_{seg}=60'-0"	L_{seg}=100'-0"	L_{seg}=200'-0"
No. 5 at 48 in.	No. 5 at 48, 32, or 24 in.	Partially	4035	NP	31277	35809	39713	58730	90827	122923	155019	475984	796948	1599358
No. 5 at 48 in.	No. 5 at 16 in.	Fully	9072	NP	31277	35809	39713	58730	90827	122923	155019	475984	796948	1599358
No. 5 at 32 in.	No. 5 at 48, 32, or 24 in.	Partially	4561	NP	31277	35809	39713	73293	113368	153443	193518	594269	995020	1996897
No. 5 at 32 in.	No. 5 at 16 in.	Fully	9072	NP	31277	35809	39713	73293	113368	153443	193518	594269	995020	1996897
No. 5 at 24 in.	No. 5 at 48, 32, or 24 in.	Partially	5086	NP	31277	35809	39713	87678	135643	183608	231573	711225	1190876	2390005
No. 5 at 24 in.	No. 5 at 16 in.	Fully	9072	NP	31277	35809	39713	87678	135643	183608	231573	711225	1190876	2390005
No. 5 at 16 in.	No. 5 at 48, 32, 24, or 16 in.	Fully	9072	NP	31277	41857	52437	115916	179396	242875	306355	941149	1575943	3162929

Table values shall not be permitted to be interpolated.
ft = 0.3048 m
in. = 25.4 mm

Table 3.2-8(22): Lateral Force Coefficients for Special Reinforced Masonry Shear Walls Under any of the Following Conditions:
Condition 1: 130 mph (209 kph) < V ≤ 150 mph (241 kph); 30 ft (9.1 m) < L_{joist} ≤ 60 ft (18.3 m); Interior Location

Vertical Reinforcement Schedule	Horizontal Reinforcement Schedule	Grouting Schedule	k_1	k_2										
				L_{seg}=2'-0"	L_{seg}=2'-8"	L_{seg}=3'-4"	L_{seg}=4'-0"	L_{seg}=8'-0"	L_{seg}=12'-0"	L_{seg}=16'-0"	L_{seg}=20'-0"	L_{seg}=60'-0"	L_{seg}=100'-0"	L_{seg}=200'-0"
No. 5 at 48 in.	No. 5 at 48, 32, or 24 in.	Partially	4,035	NP	32,483	37,458	41,805	63,485	98,185	132,886	167,586	514,591	861,596	1,729,108
No. 5 at 48 in.	No. 5 at 16 in.	Fully	9,072	NP	32,483	37,458	41,805	63,485	98,185	132,886	167,586	514,591	861,596	1,729,108
No. 5 at 32 in.	No. 5 at 48, 32, or 24 in.	Partially	4,561	NP	32,483	37,458	41,805	77,989	120,640	163,290	205,941	632,444	1,058,948	2,125,207
No. 5 at 32 in.	No. 5 at 16 in.	Fully	9,072	NP	32,483	37,458	41,805	77,989	120,640	163,290	205,941	632,444	1,058,948	2,125,207
No. 5 at 24 in.	No. 5 at 48, 32, or 24 in.	Partially	5,086	NP	32,483	37,458	41,805	92,317	142,829	193,340	243,852	748,968	1,254,084	2,516,874
No. 5 at 24 in.	No. 5 at 16 in.	Fully	9,072	NP	32,483	37,458	41,805	92,317	142,829	193,340	243,852	748,968	1,254,084	2,516,874
No. 5 at 16 in.	No. 5 at 48, 32, 24, or 16 in.	Fully	9,072	NP	32,483	43,477	54,472	120,440	186,409	252,377	318,345	978,028	1,637,710	3,286,916

Table values shall not be permitted to be interpolated.
ft = 0.3048 m
in. = 25.4 mm

DIRECT DESIGN HANDBOOK FOR MASONRY STRUCTURES

Table 3.2-8(23): Lateral Force Coefficients for Special Reinforced Masonry Shear Walls Under any of the Following Conditions:
Condition 1: 140 mph (225 kph) < V ≤ 160 mph (258 kph); 0 ft (0 m) < L_{joist} ≤ 30 ft (9.1 m); Interior Location
Condition 1: 140 mph (225 kph) < V ≤ 160 mph (258 kph); 30 ft (9.1 m) < L_{joist} ≤ 60 ft (18.3 m); Exterior Location

Vertical Reinforcement Schedule	Horizontal Reinforcement Schedule	Grouting Schedule	k_1	k_2 L_{seg}=2'-0"	L_{seg}=2'-8"	L_{seg}=3'-4"	L_{seg}=4'-0"	L_{seg}=8'-0"	L_{seg}=12'-0"	L_{seg}=16'-0"	L_{seg}=20'-0"	L_{seg}=60'-0"	L_{seg}=100'-0"	L_{seg}=200'-0"
No. 5 at 48 in.	No. 5 at 48, 32, or 24 in.	Partially	4,035	24,008	28,940	32,614	35,660	49,531	76,591	103,651	130,712	401,315	671,918	1,348,425
No. 5 at 48 in.	No. 5 at 16 in.	Fully	9,072	24,008	28,940	32,614	35,660	49,531	76,591	103,651	130,712	401,315	671,918	1,348,425
No. 5 at 32 in.	No. 5 at 48, 32, or 24 in.	Partially	4,561	24,008	28,940	32,614	35,660	64,204	99,298	134,393	169,487	520,431	871,376	1,748,736
No. 5 at 32 in.	No. 5 at 16 in.	Fully	9,072	24,008	28,940	32,614	35,660	64,204	99,298	134,393	169,487	520,431	871,376	1,748,736
No. 5 at 24 in.	No. 5 at 48, 32, or 24 in.	Partially	5,086	24,008	28,940	32,614	35,660	78,700	121,740	164,780	207,820	638,219	1,068,618	2,144,616
No. 5 at 24 in.	No. 5 at 16 in.	Fully	9,072	24,008	28,940	32,614	35,660	78,700	121,740	164,780	207,820	638,219	1,068,618	2,144,616
No. 5 at 16 in.	No. 5 at 48, 32, or 24 in.	Fully	9,072	24,008	28,940	38,718	48,495	107,160	165,825	224,490	283,155	869,806	1,456,457	2,923,083

Table values shall not be permitted to be interpolated.
ft = 0.3048 m
in. = 25.4 mm

Table 3.2-8(24): Lateral Force Coefficients for Special Reinforced Masonry Shear Walls Under any of the Following Conditions:
Condition 1: 150 mph (241 kph) < V ≤ 160 mph (258 kph); 30 ft (9.1 m) < L_{joist} ≤ 60 ft (18.3 m); Interior Location

Vertical Reinforcement Schedule	Horizontal Reinforcement Schedule	Grouting Schedule	k_1	k_2 L_{seg}=2'-0"	L_{seg}=2'-8"	L_{seg}=3'-4"	L_{seg}=4'-0"	L_{seg}=8'-0"	L_{seg}=12'-0"	L_{seg}=16'-0"	L_{seg}=20'-0"	L_{seg}=60'-0"	L_{seg}=100'-0"	L_{seg}=200'-0"
No. 5 at 48 in.	No. 5 at 48, 32, or 24 in.	Partially	4,035	NP	35,087	41,023	46,331	73,774	114,113	154,453	194,792	598,184	1,001,575	2,010,055
No. 5 at 48 in.	No. 5 at 16 in.	Fully	9,072	NP	35,087	41,023	46,331	73,774	114,113	154,453	194,792	598,184	1,001,575	2,010,055
No. 5 at 32 in.	No. 5 at 48, 32, or 24 in.	Partially	4,561	NP	35,087	41,023	46,331	88,154	136,380	184,606	232,832	715,095	1,197,358	2,403,016
No. 5 at 32 in.	No. 5 at 16 in.	Fully	9,072	NP	35,087	41,023	46,331	88,154	136,380	184,606	232,832	715,095	1,197,358	2,403,016
No. 5 at 24 in.	No. 5 at 48, 32, or 24 in.	Partially	5,086	NP	35,087	41,023	46,331	102,355	158,380	214,405	270,430	830,678	1,390,926	2,791,545
No. 5 at 24 in.	No. 5 at 16 in.	Fully	9,072	NP	35,087	41,023	46,331	102,355	158,380	214,405	270,430	830,678	1,390,926	2,791,545
No. 5 at 16 in.	No. 5 at 48, 32, 24, or 16 in.	Fully	9,072	NP	35,087	46,979	58,872	130,228	201,584	272,940	344,296	1,057,855	1,771,414	3,555,312

Table values shall not be permitted to be interpolated.
ft = 0.3048 m
in. = 25.4 mm

Table 3.2-8(25): Lateral Force Coefficients for Special Reinforced Masonry Shear Walls Under any of the Following Conditions:
Condition 1: 160 mph (258 kph) < V ≤ 180 mph (290 kph); 0 ft (0 m) < L_{joist} ≤ 30 ft (9.1 m); Interior Location
Condition 2: 160 mph (258 kph) < V ≤ 180 mph (290 kph); 30 ft (9.1 m) < L_{joist} ≤ 60 ft (18.3 m); Exterior Location

Vertical Reinforcement Schedule	Horizontal Reinforcement Schedule	Grouting Schedule	k_1	k_2 L_{seg}=2'-0"	L_{seg}=2'-8"	L_{seg}=3'-4"	L_{seg}=4'-0"	L_{seg}=8'-0"	L_{seg}=12'-0"	L_{seg}=16'-0"	L_{seg}=20'-0"	L_{seg}=60'-0"	L_{seg}=100'-0"	L_{seg}=200'-0"
No. 5 at 48 in.	No. 5 at 48, 32, or 24 in.	Partially	4,035	25,418	31,168	35,660	39,523	58,299	90,159	122,020	153,880	472,484	791,087	1,587,596
No. 5 at 48 in.	No. 5 at 16 in.	Fully	9,072	25,418	31,168	35,660	39,523	58,299	90,159	122,020	153,880	472,484	791,087	1,587,596
No. 5 at 32 in.	No. 5 at 48, 32, or 24 in.	Partially	4,561	25,418	31,168	35,660	39,523	72,867	112,708	152,550	192,392	590,808	989,224	1,985,265
No. 5 at 32 in.	No. 5 at 16 in.	Fully	9,072	25,418	31,168	35,660	39,523	72,867	112,708	152,550	192,392	590,808	989,224	1,985,265
No. 5 at 24 in.	No. 5 at 48, 32, or 24 in.	Partially	5,086	25,418	31,168	35,660	39,523	87,257	134,991	182,726	230,460	707,803	1,185,146	2,378,503
No. 5 at 24 in.	No. 5 at 16 in.	Fully	9,072	25,418	31,168	35,660	39,523	87,257	134,991	182,726	230,460	707,803	1,185,146	2,378,503
No. 5 at 16 in.	No. 5 at 48, 32, 24, or 16 in.	Fully	9,072	25,418	31,168	41,710	52,252	115,506	178,760	242,014	305,267	937,805	1,570,343	3,151,688

Table values shall not be permitted to be interpolated.
ft = 0.3048 m
in. = 25.4 mm

Table 3.2-8(26): Lateral Force Coefficients for Special Reinforced Masonry Shear Walls Under any of the Following Conditions:
Condition 1: 160 mph (258 kph) < V ≤ 180 mph (290 kph); 30 ft (9.1 m) < L_{joist} ≤ 60 ft (18.3 m); Interior Location

Vertical Reinforcement Schedule	Horizontal Reinforcement Schedule	Grouting Schedule	k_1	k_2 L_{seg}=2'-0"	L_{seg}=2'-8"	L_{seg}=3'-4"	L_{seg}=4'-0"	L_{seg}=8'-0"	L_{seg}=12'-0"	L_{seg}=16'-0"	L_{seg}=20'-0"	L_{seg}=60'-0"	L_{seg}=100'-0"	L_{seg}=200'-0"
No. 5 at 48 in.	No. 5 at 48, 32, or 24 in.	Partially	4,035	NP	NP	45,123	51,539	85,632	132,474	179,317	226,160	694,585	1,163,011	2,334,075
No. 5 at 48 in.	No. 5 at 16 in.	Fully	9,072	NP	NP	45,123	51,539	85,632	132,474	179,317	226,160	694,585	1,163,011	2,334,075
No. 5 at 32 in.	No. 5 at 48, 32, or 24 in.	Partially	4,561	NP	NP	45,123	51,539	99,865	154,522	209,179	263,835	810,402	1,356,969	2,723,386
No. 5 at 32 in.	No. 5 at 16 in.	Fully	9,072	NP	NP	45,123	51,539	99,865	154,522	209,179	263,835	810,402	1,356,969	2,723,386
No. 5 at 24 in.	No. 5 at 48, 32, or 24 in.	Partially	5,086	NP	NP	45,123	51,539	113,921	176,303	238,685	301,068	924,889	1,548,711	3,108,265
No. 5 at 24 in.	No. 5 at 16 in.	Fully	9,072	NP	NP	45,123	51,539	113,921	176,303	238,685	301,068	924,889	1,548,711	3,108,265
No. 5 at 16 in.	No. 5 at 48, 32, 24, or 16 in.	Fully	9,072	NP	NP	NP	NP	NP	NP	NP	NP	NP	NP	NP

Table values shall not be permitted to be interpolated.
ft = 0.3048 m
in. = 25.4 mm

Table 3.2-8(27): Lateral Force Coefficients for Special Reinforced Masonry Shear Walls Under any of the Following Conditions:
Condition 1: 180 mph (290 kph) < V ≤ 200 mph (322 kph); 0 ft (0 m) < L_{joist} ≤ 30 ft (9.1 m); Interior Location
Condition 2: 180 mph (290 kph) < V ≤ 200 mph (322 kph); 30 ft (9.1 m) < L_{joist} ≤ 60 ft (18.3 m); Exterior Location

Vertical Reinforcement Schedule	Horizontal Reinforcement Schedule	Grouting Schedule	k_1	k_2										
				L_{seg}=2'-0"	L_{seg}=2'-8"	L_{seg}=3'-4"	L_{seg}=4'-0"	L_{seg}=8'-0"	L_{seg}=12'-0"	L_{seg}=16'-0"	L_{seg}=20'-0"	L_{seg}=60'-0"	L_{seg}=100'-0"	L_{seg}=200'-0"
No. 5 at 48 in.	No. 5 at 48, 32, or 24 in.	Partially	4,035	26,467	32,828	37,931	42,405	64,847	100,294	135,741	171,188	525,657	880,126	1,766,299
No. 5 at 48 in.	No. 5 at 16 in.	Fully	9,072	26,467	32,828	37,931	42,405	64,847	100,294	135,741	171,188	525,657	880,126	1,766,299
No. 5 at 32 in.	No. 5 at 48, 32, or 24 in.	Partially	4,561	26,467	32,828	37,931	42,405	79,335	122,724	166,112	209,501	643,386	1,077,271	2,161,984
No. 5 at 32 in.	No. 5 at 16 in.	Fully	9,072	26,467	32,828	37,931	42,405	79,335	122,724	166,112	209,501	643,386	1,077,271	2,161,984
No. 5 at 24 in.	No. 5 at 48, 32, or 24 in.	Partially	5,086	26,467	32,828	37,931	42,405	93,646	144,888	196,129	247,371	759,786	1,272,200	2,553,237
No. 5 at 24 in.	No. 5 at 16 in.	Fully	9,072	26,467	32,828	37,931	42,405	93,646	144,888	196,129	247,371	759,786	1,272,200	2,553,237
No. 5 at 16 in.	No. 5 at 48, 32, 24, or 16 in.	Fully	9,072	26,467	32,828	43,941	55,055	121,737	188,418	255,100	321,781	988,597	1,655,412	3,322,452

Table values shall not be permitted to be interpolated.
ft = 0.3048 m
in. = 25.4 mm

Table 3.2-8(28): Lateral Force Coefficients for Special Reinforced Masonry Shear Walls Under any of the Following Conditions:
Condition 1: 180 mph (290 kph) < V ≤ 200 mph (322 kph); 30 ft (9.1 m) < L_{joist} ≤ 60 ft (18.3 m); Interior Location

Vertical Reinforcement Schedule	Horizontal Reinforcement Schedule	Grouting Schedule	k_1	k_2										
				L_{seg}=2'-0"	L_{seg}=2'-8"	L_{seg}=3'-4"	L_{seg}=4'-0"	L_{seg}=8'-0"	L_{seg}=12'-0"	L_{seg}=16'-0"	L_{seg}=20'-0"	L_{seg}=60'-0"	L_{seg}=100'-0"	L_{seg}=200'-0"
No. 5 at 48 in.	No. 5 at 48, 32, or 24 in.	Partially	4,035	NP	NP	49,526	57,136	98,393	152,241	206,089	259,938	798,419	1,336,900	2,683,104
No. 5 at 48 in.	No. 5 at 16 in.	Fully	9,072	NP	NP	49,526	57,136	98,393	152,241	206,089	259,938	798,419	1,336,900	2,683,104
No. 5 at 32 in.	No. 5 at 48, 32, or 24 in.	Partially	4,561	NP	NP	49,526	57,136	112,468	174,050	235,633	297,216	913,045	1,528,873	3,068,444
No. 5 at 32 in.	No. 5 at 16 in.	Fully	9,072	NP	NP	49,526	57,136	112,468	174,050	235,633	297,216	913,045	1,528,873	3,068,444
No. 5 at 24 in.	No. 5 at 48, 32, or 24 in.	Partially	5,086	NP	NP	49,526	57,136	126,365	195,594	264,823	334,052	1,026,341	1,718,631	3,449,354
No. 5 at 24 in.	No. 5 at 16 in.	Fully	9,072	NP	NP	49,526	57,136	126,365	195,594	264,823	334,052	1,026,341	1,718,631	3,449,354
No. 5 at 16 in.	No. 5 at 48, 32, or 24 in.	Fully	9,072	NP	NP	NP	NP	NP	NP	NP	NP	NP	NP	NP

Table values shall not be permitted to be interpolated.
ft = 0.3048 m
in. = 25.4 mm

Table 3.2-9: Diaphragm Chord Reinforcement (C Bars)

Ratio of Larger to Shorter Roof Diaphragm Plan Dimensions	Maximum Permitted V_{LFRS}, lb (N)			
	2 No. 5 C Bars (1 course)	4 No. 5 C Bars (2 courses)	6 No. 5 C Bars (3 courses)	8 No. 5 C Bars (4 courses)
4:1	33,480 lb (148,926 N)	66,960 lb (297,853 N)	100,440 lb (446,770 N)	133,920 lb (595,706 N)
3:1	44,640 lb (198,569 N)	89,280 lb (397,137 N)	133,920 lb (595,706 N)	178,560 lb (794,274 N)
2:1	66,960 lb (297,873 N)	133,920 lb (595,706 N)	200,880 lb (893,559 N)	267,840 lb (1,191,412 N)
1:1	133,920 lb (595,706 N)	267,840 lb (1,191,412 N)	401,760 lb (1,787,118 N)	535,680 lb (2,382,823 N)

Note: Linear interpolation of values in Table 3.2-9 shall be permitted.

Table 3.2-10(1) - Lintel Design Tables for $p_g \leq 10$ psf (No Shear Reinforcement Provided)*

Basic Wind Speed (mph)	Wall Location	L_{joist} (ft)	\multicolumn{6}{c}{Minimum Grouted Depth (in.)/Minimum Bearing Length (in.) for the Maximum Clear Opening Span (ft)}
			4	8	12	16	20	24
Any	Exterior or Interior	NJ	8 / 8	8 / 8	8 / 8	8 / 8	8 / 8	8 / 8
≤ 115	Exterior	30	8 / 8	16 / 8	16 / 8	24 / 8	32 / 8	32 / 8[A]
		60	16 / 8	24 / 8	32 / 8	40 / 8	56 / 8	64 / 8[A]
	Interior	30	16 / 8	24 / 8	32 / 8	40 / 8	56 / 8	64 / 8[A]
		60	24 / 8	40 / 8	56 / 8	80 / 16	104 / 16	128 / 16[A]
$> 115 \leq 140$	Exterior	30	8 / 8	16 / 8	24 / 8	24 / 8	32 / 8	40 / 8[A]
		60	16 / 8	24 / 8	40 / 8	48 / 8	64 / 8	80 / 16[A]
	Interior	30	16 / 8	24 / 8	40 / 8	48 / 8	64 / 8	80 / 16[A]
		60	24 / 8	48 / 8	80 / 16	104 / 16	128 / 16	160 / 24[A]
$> 140 \leq 165$	Exterior	30	8 / 8	16 / 8	24 / 8	32 / 8	40 / 8	48 / 8[A]
		60	16 / 8	32 / 8	48 / 8	64 / 8	80 / 16	96 / 16[A]
	Interior	30	16 / 8	32 / 8	48 / 8	64 / 8	80 / 16	96 / 16[A]
		60	32 / 8	56 / 8	96 / 16	120 / 16	160 / 24[A]	208 / 32[B]
$> 165 \leq 190$	Exterior	30	16 / 8	24 / 8	32 / 8	40 / 8	48 / 8	56 / 8[A]
		60	24 / 8	40 / 8	56 / 8	80 / 16	96 / 16	120 / 16[A]
	Interior	30	24 / 8	40 / 8	56 / 8	80 / 16	96 / 16	120 / 16[A]
		60	40 / 8	80 / 16	112 / 16	160 / 24[A]	200 / 24[A]	248 / 32[B]

*The following conditions shall apply:
    1) Two No. 5 bars (B Bars) in the bottom course and top course shall be provided, unless otherwise noted.
    2) Effective depth to reinforcement = beam depth - 2 in.
[A] Requires 3 No. 5 Bars (B Bars)
[B] Requires 4 No. 5 Bars (B Bars)
Table values shall not be permitted to be interpolated.
NJ = No Joist
mph = 0.447 m/s
ft = 0.3048 m
in. = 25.4 mm
lb/ft² = 0.0479 kPa

Table 3.2-10(2) - Lintel Design Tables 10 < p_g ≤ 20 psf (No Shear Reinforcement Provided)*

Basic Wind Speed (mph)	Wall Location	L_{joist} (ft)	\multicolumn{6}{c}{Minimum Grouted Depth (in.)/Minimum Bearing Length (in.) for the Maximum Clear Opening Span (ft)}
			4	8	12	16	20	24
Any	Exterior or Interior	NJ	8 / 8	8 / 8	8 / 8	8 / 8	8 / 8	8 / 8
≤ 115	Exterior	30	16 / 8	24 / 8	32 / 8	48 / 8	56 / 8	72 / 16[A]
		60	24 / 8	48 / 8	64 / 8	96 / 16	120 / 16	152 / 24[A]
	Interior	30	24 / 8	48 / 8	64 / 8	96 / 16	120 / 16	152 / 24[A]
		60	48 / 8	96 / 16	144 / 24[A]	192 / 24[A]	248 / 32[B]	312 / 40[C]
> 115 ≤ 140	Exterior	30	16 / 8	24 / 8	40 / 8	48 / 8	64 / 8	80 / 16[A]
		60	24 / 8	48 / 8	72 / 16	96 / 16	128 / 16	160 / 24[A]
	Interior	30	24 / 8	48 / 8	72 / 16	96 / 16	128 / 16	160 / 24[A]
		60	48 / 8	96 / 16	152 / 24[A]	208 / 32[B]	272 / 40[B]	344 / 48[D]
> 140 ≤ 165	Exterior	30	16 / 8	24 / 8	40 / 8	48 / 8	64 / 8	80 / 16[A]
		60	32 / 8	48 / 8	80 / 16	104 / 16	128 / 16	168 / 24[A]
	Interior	30	32 / 8	48 / 8	80 / 16	104 / 16	128 / 16	168 / 24[A]
		60	56 / 8	104 / 16	160 / 24[A]	224 / 32[B]	288 / 40[C]	NP
> 165 ≤ 190	Exterior	30	16 / 8	32 / 8	40 / 8	56 / 8	72 / 16	88 / 16[A]
		60	32 / 8	56 / 8	80 / 16	112 / 16	144 / 24[A]	176 / 24[A]
	Interior	30	32 / 8	56 / 8	80 / 16	112 / 16	144 / 24[A]	176 / 24[A]
		60	56 / 8	112 / 16	168 / 24[A]	232 / 32[B]	304 / 40[C]	NP

*The following conditions shall apply:
    1) Two No. 5 bars (B Bars) in the bottom course and top course shall be provided, unless otherwise noted.
    2) Effective depth to reinforcement = beam depth - 2 in.

[A]Requires 3 No. 5 Bars (B Bars)
[B]Requires 4 No. 5 Bars (B Bars)
[C]Requires 5 No. 5 Bars (B Bars)
[D]Requires 6 No. 5 Bars (B Bars)
Table values shall not be permitted to be interpolated.
NJ = No Joist
NP = Not permitted. Required beam depth exceeds maximum wall height permitted by the direct design procedure.
mph = 0.447 m/s
ft = 0.3048 m
in. = 25.4 mm
lb/ft² = 0.0479 kPa

DIRECT DESIGN HANDBOOK FOR MASONRY STRUCTURES

Table 3.2-10(3) - Lintel Design Tables $20 < p_g \leq 40$ psf (No Shear Reinforcement Provided)*

Basic Wind Speed (mph)	Wall Location	L_{joist} (ft)	\multicolumn{6}{c}{Minimum Grouted Depth (in.)/Minimum Bearing Length (in.) for the Maximum Clear Opening Span (ft)}
			4	8	12	16	20	24
Any	Exterior or Interior	NJ	8/8	8/8	8/8	8/8	8/8	8/8
≤ 115	Exterior	30	16/8	32/8	40/8	56/8	72/16	88/16[A]
		60	32/8	56/8	80/16	112/16	144/24[A]	176/24[A]
	Interior	30	32/8	56/8	80/16	112/16	144/24[A]	176/24[A]
		60	56/8	112/16	168/24[A]	232/32[B]	304/40[C]	NP
$>115 \leq 140$	Exterior	30	16/8	32/8	40/8	56/8	72/16	88/16[A]
		60	32/8	56/8	88/16	112/16	152/24[A]	184/24[A]
	Interior	30	32/8	56/8	88/16	112/16	152/24[A]	184/24[A]
		60	56/8	112/16	176/24[A]	240/32[B]	312/40[C]	NP
$>140 \leq 165$	Exterior	30	16/8	32/8	48/8	56/8	80/16	96/16[A]
		60	32/8	56/8	88/16	120/16	160/24[A]	192/24[A]
	Interior	30	32/8	56/8	88/16	120/16	160/24[A]	192/24[A]
		60	56/8	120/16	184/24[A]	256/32[B]	344/48[D]	NP
$>165 \leq 190$	Exterior	30	16/8	32/8	48/8	64/8	80/16	96/16[A]
		60	32/8	64/8	96/16	128/16	168/24[A]	208/32[B]
	Interior	30	32/8	64/8	96/16	128/16	168/24[A]	208/32[B]
		60	64/8	128/16	192/24[A]	280/40[C]	NP	NP

*The following conditions shall apply:
    1) Two No. 5 bars (B Bars) in the bottom course and top course shall be provided, unless otherwise noted.
    2) Effective depth to reinforcement = beam depth - 2 in.
[A]Requires 3 No. 5 Bars (B Bars)
[B]Requires 4 No. 5 Bars (B Bars)
[C]Requires 5 No. 5 Bars (B Bars)
[D]Requires 6 No. 5 Bars (B Bars)
Table values shall not be permitted to be interpolated.
NJ = No Joist
NP = Not permitted. Required beam depth exceeds maximum wall height permitted by the direct design procedure.
mph = 0.447 m/s
ft = 0.3048 m
in. = 25.4 mm
lb/ft^2 = 0.0479 kPa

DIRECT DESIGN HANDBOOK FOR MASONRY STRUCTURES

Table 3.2-10(4) - Lintel Design Tables $40 < p_g \leq 60$ psf (No Shear Reinforcement Provided)*

Basic Wind Speed (mph)	Wall Location	L_{joist} (ft)	\multicolumn{6}{c}{Minimum Grouted Depth (in.)/Minimum Bearing Length (in.) for the Maximum Clear Opening Span (ft)}
			4	8	12	16	20	24
Any	Exterior or Interior	NJ	8 / 8	8 / 8	8 / 8	8 / 8	8 / 8	8 / 8
≤ 115	Exterior	30	16 / 8	32 / 8	48 / 8	64 / 8	80 / 16	96 / 16[A]
		60	32 / 8	56 / 8	88 / 16	120 / 16	160 / 24[A]	192 / 24[A]
	Interior	30	32 / 8	56 / 8	88 / 16	120 / 16	160 / 24[A]	192 / 24[A]
		60	64 / 8	120 / 16	184 / 24[A]	256 / 32[B]	352 / 48[D]	NP
$> 115 \leq 140$	Exterior	30	16 / 8	32 / 8	48 / 8	64 / 8	80 / 16	104 / 16[A]
		60	32 / 8	64 / 8	96 / 16	128 / 16	168 / 24[A]	208 / 32[B]
	Interior	30	32 / 8	64 / 8	96 / 16	128 / 16	168 / 24[A]	208 / 32[B]
		60	64 / 8	128 / 16	192 / 24[A]	280 / 40[C]	NP	NP
$> 140 \leq 165$	Exterior	30	16 / 8	32 / 8	48 / 8	64 / 8	88 / 16	104 / 16[A]
		60	32 / 8	64 / 8	96 / 16	128 / 16	176 / 24[A]	216 / 32[B]
	Interior	30	32 / 8	64 / 8	96 / 16	128 / 16	176 / 24[A]	216 / 32[B]
		60	64 / 8	128 / 16	216 / 32[B]	296 / 40[C]	NP	NP
$> 165 \leq 190$	Exterior	30	24 / 8	32 / 8	48 / 8	64 / 8	88 / 16	112 / 16[A]
		60	40 / 8	64 / 8	104 / 16	144 / 24[A]	184 / 24[A]	232 / 32[B]
	Interior	30	40 / 8	64 / 8	104 / 16	144 / 24[A]	184 / 24[A]	232 / 32[B]
		60	72 / 8	152 / 24[A]	224 / 32[B]	304 / 40[C]	NP	NP

*The following conditions shall apply:
    1) Two No. 5 bars (B Bars) in the bottom course and top course shall be provided, unless otherwise noted.
    2) Effective depth to reinforcement = beam depth - 2 in.
[A] Requires 3 No. 5 Bars (B Bars)
[B] Requires 4 No. 5 Bars (B Bars)
[C] Requires 5 No. 5 Bars (B Bars)
[D] Requires 6 No. 5 Bars (B Bars)
Table values shall not be permitted to be interpolated.
NJ = No Joist
NP = Not permitted. Required beam depth exceeds maximum wall height permitted by the direct design procedure.
mph = 0.447 m/s
ft = 0.3048 m
in. = 25.4 mm
lb/ft^2 = 0.0479 kPa

DIRECT DESIGN HANDBOOK FOR MASONRY STRUCTURES

DIRECT DESIGN HANDBOOK FOR MASONRY STRUCTURES

Table 3.2-10(5) - Lintel Design Tables for $p_g \leq 10$ psf (Shear Reinforcement Provided)*

Basic Wind Speed (mph)	Wall Location	L_{joist} (ft)	\multicolumn{6}{c}{Minimum Grouted Depth (in.)/Minimum Bearing Length (in.) for the Maximum Clear Opening Span (ft)}
			4	8	12	16	20	24
Any	Exterior or Interior	NJ	NP	NP	NP	NP	NP	NP
≤ 115	Exterior	30	NP	NP	NP	NP	NP	NP
		60	NP	NP	NP	24 / 8	24 / 8	32 / 8[A]
	Interior	30	NP	NP	NP	24 / 8	24 / 8	32 / 8[A]
		60	NP	24 / 8	32 / 8	32 / 16	48 / 16	56 / 16[A]
$> 115 \leq 140$	Exterior	30	NP	NP	NP	NP	NP	24 / 8[A]
		60	NP	NP	24 / 8	24 / 8	32 / 8	32 / 16[A]
	Interior	30	NP	NP	24 / 8	24 / 8	32 / 8	32 / 16[A]
		60	NP	24 / 8	32 / 16	48 / 16	56 / 16	64 / 24[A]
$> 140 \leq 165$	Exterior	30	NP	NP	NP	NP	24 / 8	24 / 8[A]
		60	NP	24 / 8	24 / 8	32 / 8	32 / 16	48 / 16[A]
	Interior	30	NP	24 / 8	24 / 8	32 / 8	32 / 16	48 / 16[A]
		60	NP	32 / 8	40 / 16	56 / 16	64 / 24[A]	80 / 32[B]
$> 165 \leq 190$	Exterior	30	NP	NP	NP	24 / 8	24 / 8	24 / 8[A]
		60	NP	24 / 8	24 / 8	32 / 16	48 / 16	48 / 16[A]
	Interior	30	NP	24 / 8	24 / 8	32 / 16	48 / 16	48 / 16[A]
		60	24 / 8	32 / 16	48 / 16	64 / 24[A]	80 / 24[A]	NP

*The following conditions shall apply:
    1) Two No. 5 bars (B Bars) in the bottom course and top course shall be provided, unless otherwise noted.
    2) Effective depth to reinforcement = beam depth - 2 in.

[A] Requires 3 No. 5 Bars (B Bars)
[B] Requires 4 No. 5 Bars (B Bars)
NP – Not Permitted. See Table 3.2-10(1) for design alternatives.
Table values shall not be permitted to be interpolated.
NJ = No Joist
NP = Not permitted. Required beam depth exceeds maximum wall height permitted by the direct design procedure.
mph = 0.447 m/s
ft = 0.3048 m
in. = 25.4 mm
lb/ft^2 = 0.0479 kPa

DIRECT DESIGN HANDBOOK FOR MASONRY STRUCTURES

Table 3.2-10(6) - Lintel Design Tables 10 < p_g ≤ 20 psf (Shear Reinforcement Provided)*

Basic Wind Speed (mph)	Wall Location	L_{joist} (ft)	\multicolumn{6}{c}{Minimum Grouted Depth (in.)/Minimum Bearing Length (in.) for the Maximum Clear Opening Span (ft)}					
			4	8	12	16	20	24
Any	Exterior or Interior	NJ	NP	NP	NP	NP	NP	NP
≤ 115	Exterior	30	NP	NP	NP	NP	NP	NP
		60	NP	NP	NP	24 / 8	24 / 8	32 / 8[A]
	Interior	30	NP	NP	NP	24 / 8	24 / 8	32 / 8[A]
		60	NP	24 / 8	32 / 8	32 / 16	48 / 16	56 / 16[A]
> 115 ≤ 140	Exterior	30	NP	NP	NP	NP	NP	24 / 8[A]
		60	NP	NP	24 / 8	24 / 8	32 / 8	32 / 16[A]
	Interior	30	NP	NP	24 / 8	24 / 8	32 / 8	32 / 16[A]
		60	NP	24 / 8	32 / 16	48 / 16	56 / 16	64 / 24[A]
> 140 ≤ 165	Exterior	30	NP	NP	NP	NP	24 / 8	24 / 8[A]
		60	NP	24 / 8	24 / 8	32 / 8	32 / 16	48 / 16[A]
	Interior	30	NP	24 / 8	24 / 8	32 / 8	32 / 16	48 / 16[A]
		60	NP	32 / 8	40 / 16	56 / 16	64 / 24[A]	80 / 32[B]
> 165 ≤ 190	Exterior	30	NP	NP	NP	24 / 8	24 / 8	24 / 8[A]
		60	NP	24 / 8	24 / 8	32 / 16	48 / 16	48 / 16[A]
	Interior	30	NP	24 / 8	24 / 8	32 / 16	48 / 16	48 / 16[A]
		60	24 / 8	32 / 16	48 / 16	64 / 24[A]	80 / 24[A]	NP

*The following conditions shall apply:
    1) Two No. 5 bars (B Bars) in the bottom course and top course shall be provided, unless otherwise noted.
    2) Effective depth to reinforcement = beam depth - 2 in.
[A]Requires 3 No. 5 Bars (B Bars)
[B]Requires 4 No. 5 Bars (B Bars)
NP – Not Permitted. See Table 3.2-10(2) for design alternatives.
Table values shall not be permitted to be interpolated.
NJ = No Joist
NP = Not permitted. Required beam depth exceeds maximum wall height permitted by the direct design procedure.
mph = 0.447 m/s
ft = 0.3048 m
in. = 25.4 mm
lb/ft² = 0.0479 kPa

DIRECT DESIGN HANDBOOK FOR MASONRY STRUCTURES

Table 3.2-10(7) - Lintel Design Tables $20 < p_g \leq 40$ psf (Shear Reinforcement Provided)*

Basic Wind Speed (mph)	Wall Location	L_{joist} (ft)	\multicolumn{6}{c	}{Minimum Grouted Depth (in.)/Minimum Bearing Length (in.) for the Maximum Clear Opening Span (ft)}
			4	8	12	16	20	24
Any	Exterior or Interior	NJ	NP	NP	NP	NP	NP	NP
≤ 115	Exterior	30	NP	NP	24 / 8	24 / 8	32 / 16	40 / 16[A]
		60	NP	24 / 8	40 / 16	48 / 16	56 / 24[A]	72 / 24[A]
	Interior	30	NP	24 / 8	40 / 16	48 / 16	56 / 24[A]	72 / 24[A]
		60	24 / 8	48 / 16	72 / 24[A]	NP	NP	NP
$> 115 \leq 140$	Exterior	30	NP	NP	24 / 8	24 / 8	32 / 16	40 / 16[A]
		60	NP	24 / 8	40 / 16	48 / 16	64 / 24[A]	72 / 24[A]
	Interior	30	NP	24 / 8	40 / 16	48 / 16	64 / 24[A]	72 / 24[A]
		60	24 / 8	48 / 16	72 / 24[A]	NP	NP	NP
$> 140 \leq 165$	Exterior	30	NP	NP	24 / 8	32 / 8	32 / 16	40 / 16[A]
		60	NP	32 / 8	40 / 16	56 / 16	64 / 24[A]	80 / 24[A]
	Interior	30	NP	32 / 8	40 / 16	56 / 16	64 / 24[A]	80 / 24[A]
		60	32 / 8	56 / 16	80 / 24[A]	NP	NP	NP
$> 165 \leq 190$	Exterior	30	NP	NP	24 / 8	32 / 8	40 / 16	40 / 16[A]
		60	NP	32 / 8	40 / 16	56 / 16	72 / 24[A]	80 / 32[B]
	Interior	30	NP	32 / 8	40 / 16	56 / 16	72 / 24[A]	80 / 32[B]
		60	32 / 8	56 / 16	80 / 24[A]	NP	NP	NP

*The following conditions shall apply:
    1) Two No. 5 bars (B Bars) in the bottom course and top course shall be provided, unless otherwise noted.
    2) Effective depth to reinforcement = beam depth - 2 in.
[A]Requires 3 No. 5 Bars (B Bars)
[B]Requires 4 No. 5 Bars (B Bars)
NP – Not Permitted. See Table 3.2-10(3) for design alternatives.
Table values shall not be permitted to be interpolated.
NJ = No Joist
NP = Not permitted. Required beam depth exceeds maximum wall height permitted by the direct design procedure.
mph = 0.447 m/s
ft = 0.3048 m
in. = 25.4 mm
lb/ft² = 0.0479 kPa

Table 3.2-10(8) - Lintel Design Tables $40 < p_g \leq 60$ psf (Shear Reinforcement Provided)*

Basic Wind Speed (mph)	Wall Location	L_{joist} (ft)	\multicolumn{6}{c}{Minimum Grouted Depth (in.)/Minimum Bearing Length (in.) for the Maximum Clear Opening Span (ft)}
			4	8	12	16	20	24
Any	Exterior or Interior	NJ	NP	NP	NP	NP	NP	NP
≤ 115	Exterior	30	NP	NP	24 / 8	32 / 8	32 / 16	40 / 16[A]
		60	NP	32 / 8	40 / 16	56 / 16	64 / 24[A]	80 / 24[A]
	Interior	30	NP	32 / 8	40 / 16	56 / 16	64 / 24[A]	80 / 24[A]
		60	32 / 8	56 / 16	80 / 24[A]	NP	NP	NP
$> 115 \leq 140$	Exterior	30	NP	NP	24 / 8	32 / 8	40 / 16	40 / 16[A]
		60	NP	32 / 8	40 / 16	56 / 16	72 / 24[A]	80 / 32[B]
	Interior	30	NP	32 / 8	40 / 16	56 / 16	72 / 24[A]	80 / 32[B]
		60	32 / 8	56 / 16	80 / 24[A]	NP	NP	NP
$> 140 \leq 165$	Exterior	30	NP	NP	24 / 8	32 / 8	40 / 16	48 / 16[A]
		60	NP	32 / 8	48 / 16	56 / 16	72 / 24[A]	NP
	Interior	30	NP	32 / 8	48 / 16	56 / 16	72 / 24[A]	NP
		60	32 / 8	56 / 16	NP	NP	NP	NP
$> 165 \leq 190$	Exterior	30	NP	NP	24 / 8	32 / 8	40 / 16	48 / 16[A]
		60	24 / 8	32 / 8	48 / 16	64 / 24[A]	80 / 24[A]	NP
	Interior	30	24 / 8	32 / 8	48 / 16	64 / 24[A]	80 / 24[A]	NP
		60	32 / 8	64 / 24[A]	NP	NP	NP	NP

*The following conditions shall apply:
    1) Two No. 5 bars (B Bars) in the bottom course and top course shall be provided, unless otherwise noted.
    2) Effective depth to reinforcement = beam depth - 2 in.
[A] Requires 3 No. 5 Bars (B Bars)
[B] Requires 4 No. 5 Bars (B Bars)
NP – Not Permitted. See Table 3.2-10(4) for design alternatives.
Table values shall not be permitted to be interpolated.
NJ = No Joist
NP = Not permitted. Required beam depth exceeds maximum wall height permitted by the direct design procedure.
mph = 0.447 m/s
ft = 0.3048 m
in. = 25.4 mm
lb/ft^2 = 0.0479 kPa

DIRECT DESIGN HANDBOOK FOR MASONRY STRUCTURES

Table 3.2-11(1): Maximum Horizontal Spans for Walls above & below Openings (ft) for Wind Conditions

Wind V (mph)	Risk Category	Exposure Category	Unreinforced* Ungrouted exterior	Unreinforced* Ungrouted interior	Unreinforced* Fully Grouted exterior	Unreinforced* Fully Grouted interior	Horizontal No. 5 Bars at 48" oc exterior	Horizontal No. 5 Bars at 48" oc interior	Horizontal No. 5 Bars at 32" oc exterior	Horizontal No. 5 Bars at 32" oc interior	Horizontal No. 5 Bars at 24" oc exterior	Horizontal No. 5 Bars at 24" oc interior	Horizontal No. 5 Bars at 16" oc exterior	Horizontal No. 5 Bars at 16" oc interior
up to 90	I or II	B	12'-8"	22'-0"	20'-0"	30'-0"	20'-8"	30'-0"	24'-0"	30'-0"	26'-0"	30'-0"	28'-8"	30'-0"
up to 90	I or II	C	10'-8"	22'-0"	16'-8"	30'-0"	18'-8"	30'-0"	21'-4"	30'-0"	22'-8"	30'-0"	25'-4"	30'-0"
up to 90	III or IV	B	12'-0"	22'-0"	18'-8"	30'-0"	19'-4"	30'-0"	22'-8"	30'-0"	24'-8"	30'-0"	26'-8"	30'-0"
up to 90	III or IV	C	10'-0"	22'-0"	15'-4"	30'-0"	17'-4"	30'-0"	20'-8"	30'-0"	22'-8"	30'-0"	24'-0"	30'-0"
91 to 110	I or II	B	10'-8"	20'-0"	16'-0"	30'-0"	17'-4"	29'-4"	20'-8"	30'-0"	22'-0"	30'-0"	24'-8"	30'-0"
91 to 110	I or II	C	8'-8"	20'-0"	13'-4"	30'-0"	15'-4"	29'-4"	18'-0"	30'-0"	20'-0"	30'-0"	22'-0"	30'-0"
91 to 110	III or IV	B	10'-0"	20'-0"	15'-4"	30'-0"	16'-0"	29'-4"	19'-4"	30'-0"	20'-8"	30'-0"	23'-4"	30'-0"
91 to 110	III or IV	C	8'-0"	20'-0"	12'-8"	30'-0"	14'-8"	29'-4"	16'-8"	30'-0"	19'-4"	30'-0"	20'-8"	30'-0"
111 to 130	I or II	B	8'-8"	18'-8"	13'-4"	28'-8"	14'-8"	28'-0"	17'-4"	30'-0"	19'-4"	30'-0"	21'-4"	30'-0"
111 to 130	I or II	C	7'-4"	18'-8"	11'-4"	28'-8"	13'-4"	28'-0"	16'-0"	30'-0"	17'-4"	30'-0"	19'-4"	30'-0"
111 to 130	III or IV	B	8'-0"	18'-8"	12'-8"	28'-8"	13'-4"	28'-0"	16'-0"	30'-0"	18'-8"	30'-0"	20'-8"	30'-0"
111 to 130	III or IV	C	6'-8"	18'-8"	10'-8"	28'-8"	12'-8"	28'-0"	14'-8"	30'-0"	16'-0"	30'-0"	18'-8"	30'-0"
131 to 150	I or II	B	7'-4"	17'-4"	12'-0"	26'-8"	12'-8"	26'-8"	15'-4"	30'-0"	17'-4"	30'-0"	19'-4"	30'-0"
131 to 150	I or II	C	6'-8"	17'-4"	10'-0"	26'-8"	11'-4"	26'-8"	14'-0"	30'-0"	15'-4"	30'-0"	17'-4"	30'-0"
131 to 150	III or IV	B	7'-4"	17'-4"	10'-8"	26'-8"	12'-0"	26'-8"	14'-0"	30'-0"	16'-0"	30'-0"	18'-0"	30'-0"
131 to 150	III or IV	C	6'-0"	17'-4"	9'-4"	26'-8"	10'-8"	26'-8"	12'-8"	30'-0"	14'-8"	30'-0"	16'-8"	30'-0"

Table values shall not be permitted to be interpolated.
NP = Not Permitted
*See Commentary

mph = 1.61 kph
ft = 0.3048 m
in. = 25.4 mm

Table 3.2-11(2): Maximum Horizontal Spans for Walls above and below Openings (ft) for Seismic Conditions

Seismic S_S	Unreinforced* Ungrouted exterior	Unreinforced* Ungrouted interior	Unreinforced* Ungrouted exterior	Unreinforced* Ungrouted interior	Unreinforced* Fully Grouted exterior	Unreinforced* Fully Grouted interior	Horizontal No. 5 Bars at 48" oc exterior	Horizontal No. 5 Bars at 48" oc interior	Horizontal No. 5 Bars at 32" oc exterior	Horizontal No. 5 Bars at 32" oc interior	Horizontal No. 5 Bars at 24" oc exterior	Horizontal No. 5 Bars at 24" oc interior	Horizontal No. 5 Bars at 16" oc exterior	Horizontal No. 5 Bars at 16" oc interior
0.156	26'-8"	26'-8"	30'-0"	30'-0"	30'-0"	30'-0"	30'-0"	30'-0"	30'-0"	30'-0"	30'-0"	30'-0"	30'-0"	30'-0"
0.32	18'-8"	18'-0"	30'-0"	30'-0"	30'-0"	30'-0"	30'-0"	30'-0"	30'-0"	30'-0"	30'-0"	30'-0"	30'-0"	30'-0"
0.553	-	-	-	-	28'-0"	28'-0"	30'-0"	30'-0"	30'-0"	30'-0"	30'-0"	30'-0"	29'-4"	29'-4"
1.000	-	-	-	-	23'-4"	23'-4"	26'-8"	26'-8"	28'-8"	28'-8"	30'-0"	30'-0"	26'-0"	26'-0"
1.500	-	-	-	-	20'-0"	20'-0"	22'-8"	22'-8"	24'-8"	24'-8"	26'-8"	26'-8"	23'-4"	23'-4"
2.000	-	-	-	-	17'-4"	17'-4"	20'-0"	20'-0"	21'-4"	21'-4"	23'-4"	23'-4"	21'-4"	21'-4"
2.500	-	-	-	-	15'-4"	15'-4"	18'-0"	18'-0"	19'-4"	19'-4"	21'-4"	21'-4"	20'-0"	20'-0"
3.000	-	-	-	-	14'-0"	14'-0"	16'-0"	16'-0"	17'-4"	17'-4"	19'-4"	19'-4"	18'-8"	18'-8"

Table values shall not be permitted to be interpolated.
NP = Not Permitted
*See Commentary

mph = 1.61 kph
ft = 0.3048 m
in. = 25.4 mm

This Page Intentionally Left Blank

Chapter 4
Clay Masonry (Future)

This Page Intentionally Left Blank

Chapter 5
Specification

Project specifications for masonry designed in accordance with this *Handbook* shall meet or exceed the requirements of the MSJC Specification. Design decisions required by the Mandatory Requirements Checklist of the MSJC Specification that are applicable to masonry designed in accordance with this *Handbook* include the following:

Specification Article	Description	Design Requirement
1.4 A	Compressive strength requirements	$f'_m = 1500$ lb/in.^{2} (10.3 MPa)
1.4	Masonry Compressive Strength	Verify compliance with f'_m
1.6	Quality Assurance	Define the submittal reporting and review procedure and specify the required level of quality assurance as defined in Tables 4 or 5 of the MSJC Specification based on the building risk category
2.1	Mortar materials	Mortar conforming to ASTM C270 Type S, cement-lime, mortar cement, or masonry cement as permitted by Section 2.4.2
2.2	Grout Materials	Verify grout meets the requirements of ASTM C476
2.3	Masonry unit materials	Concrete masonry units complying with ASTM C90 and having a nominal thickness of 8 in. (203 mm)
2.4	Reinforcement	Reinforcement conforming to ASTM A615 (Grade 60 (420 MPa)), ASTM A706 (Grade 60 (420 MPa)), or ASTM A996 (Grade 60 (420 MPa))

This Page Intentionally Left Blank

Chapter 6
Details

Where applicable to the project under consideration, the following details shall be required.

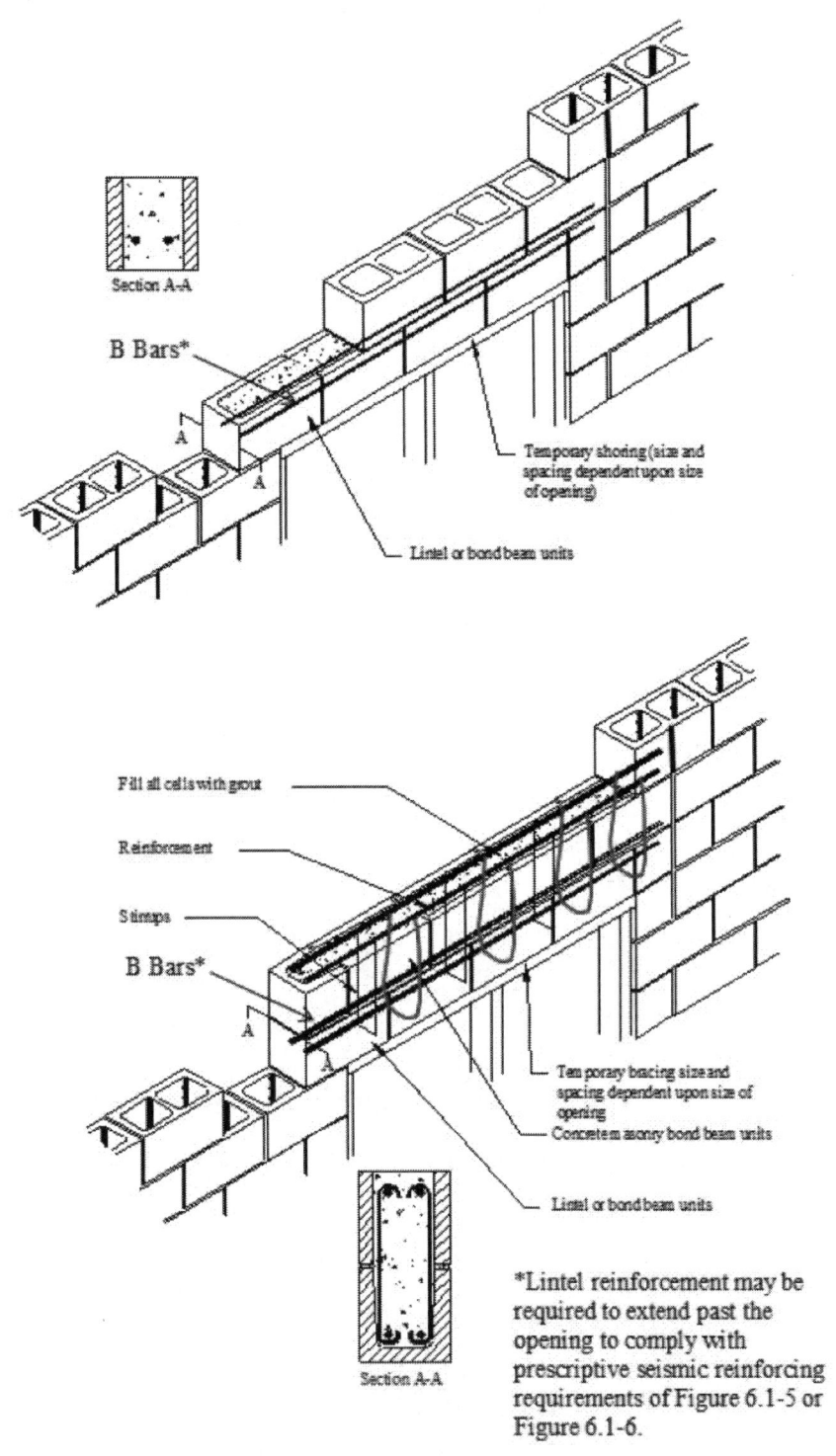

*Lintel reinforcement may be required to extend past the opening to comply with prescriptive seismic reinforcing requirements of Figure 6.1-5 or Figure 6.1-6.

Figure 6.1-1 – Masonry Lintels

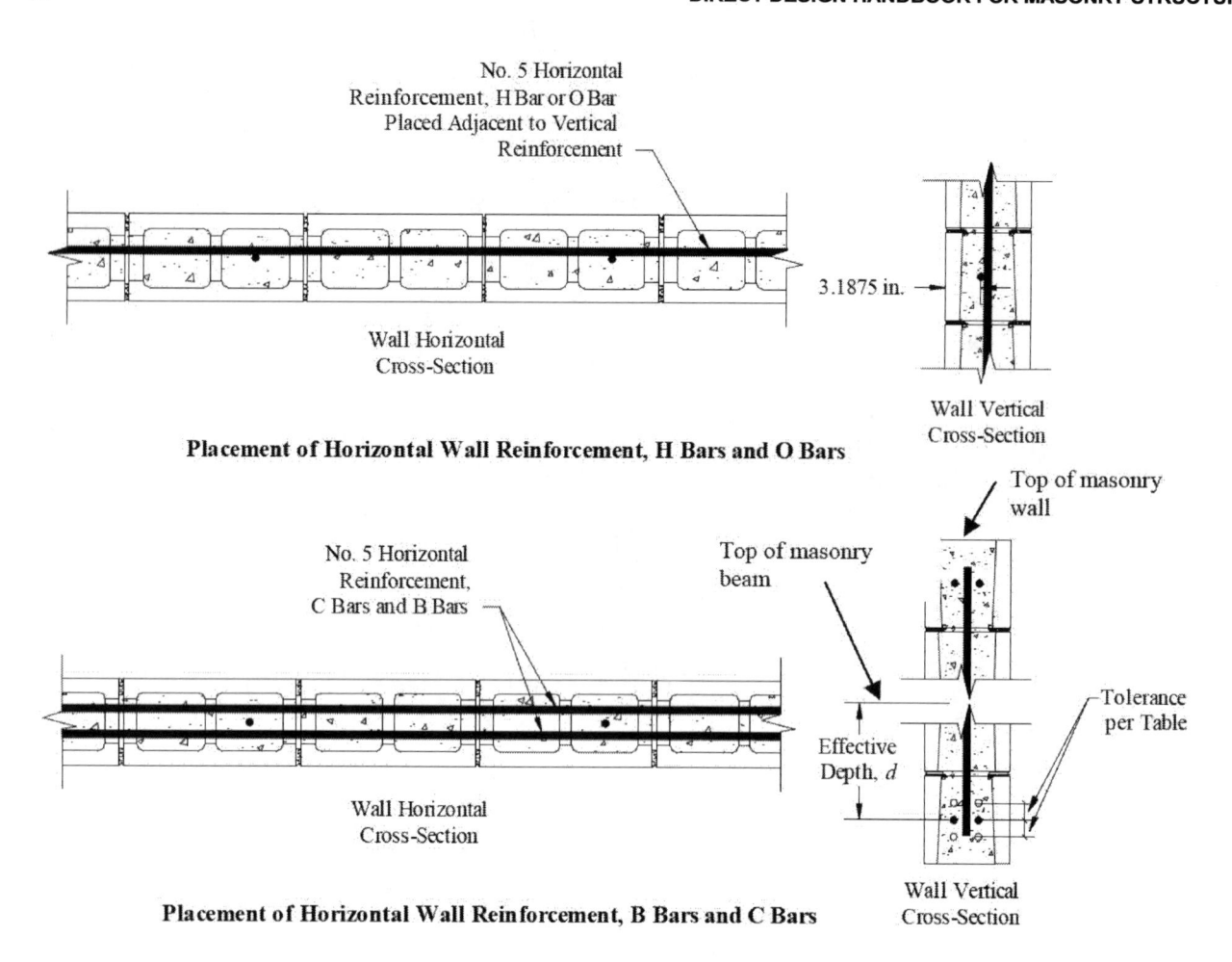

Specified Distance d From Top of Lintel to Center of Reinforcement	Allowable Tolerance
$d \leq 8$ in. (203 mm)	± ½ in. (13 mm)
8 in. (203 mm) < $d \leq 24$ in. (607 mm)	± 1 in. (25 mm)
$d > 24$ in. (607 mm)	± 1 ¼ in. (32 mm)

1 in. = 25.4 mm

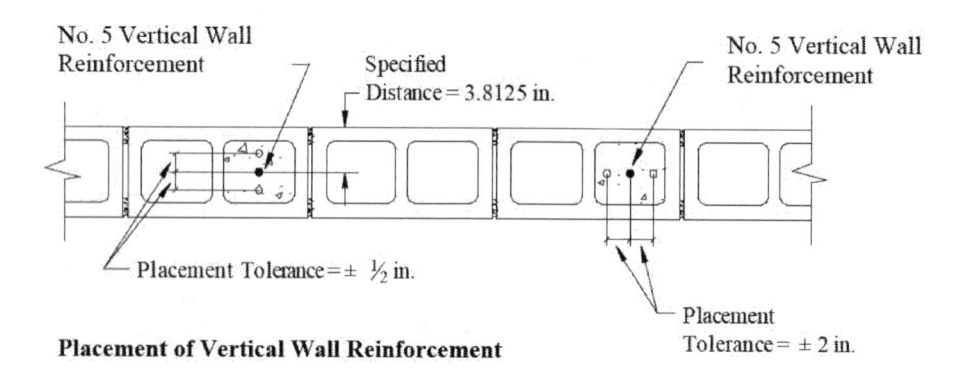

Figure 6.1-2 – Placement Tolerances for Reinforcement

DIRECT DESIGN HANDBOOK FOR MASONRY STRUCTURES

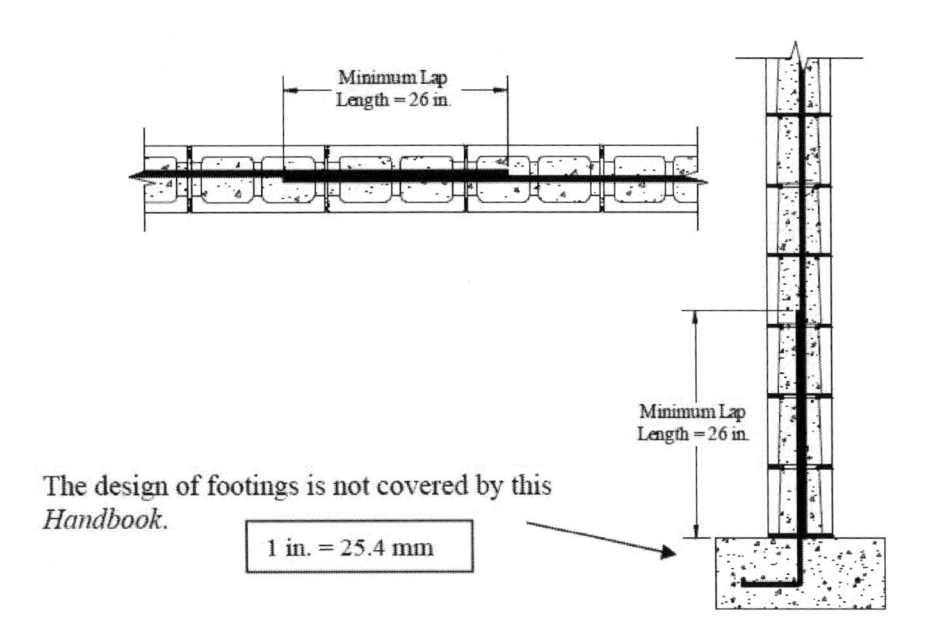

Figure 6.1-3 – Lap Splices of Reinforcement

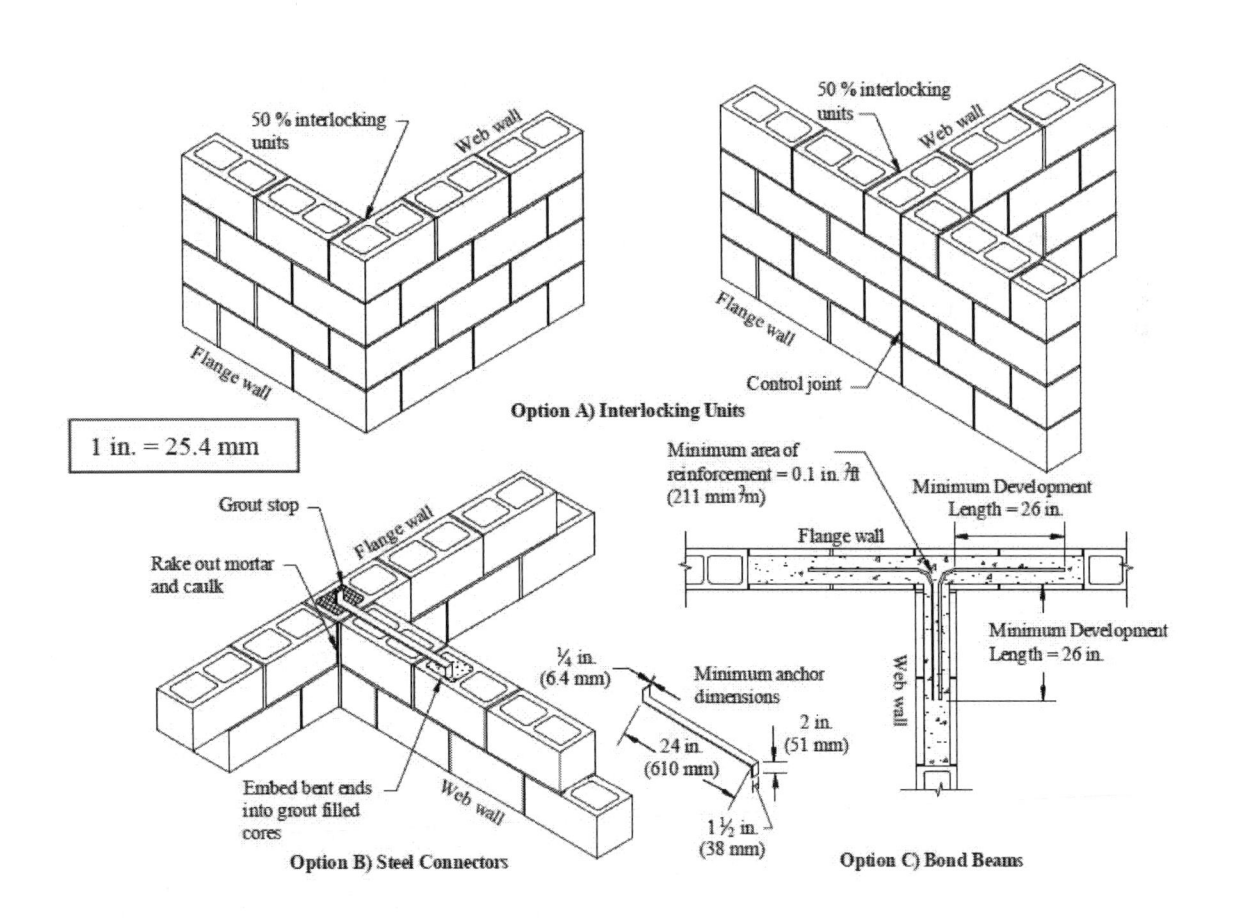

Figure 6.1-4 – Details of Designated and Non-Designated Shear Flanged Wall Intersections

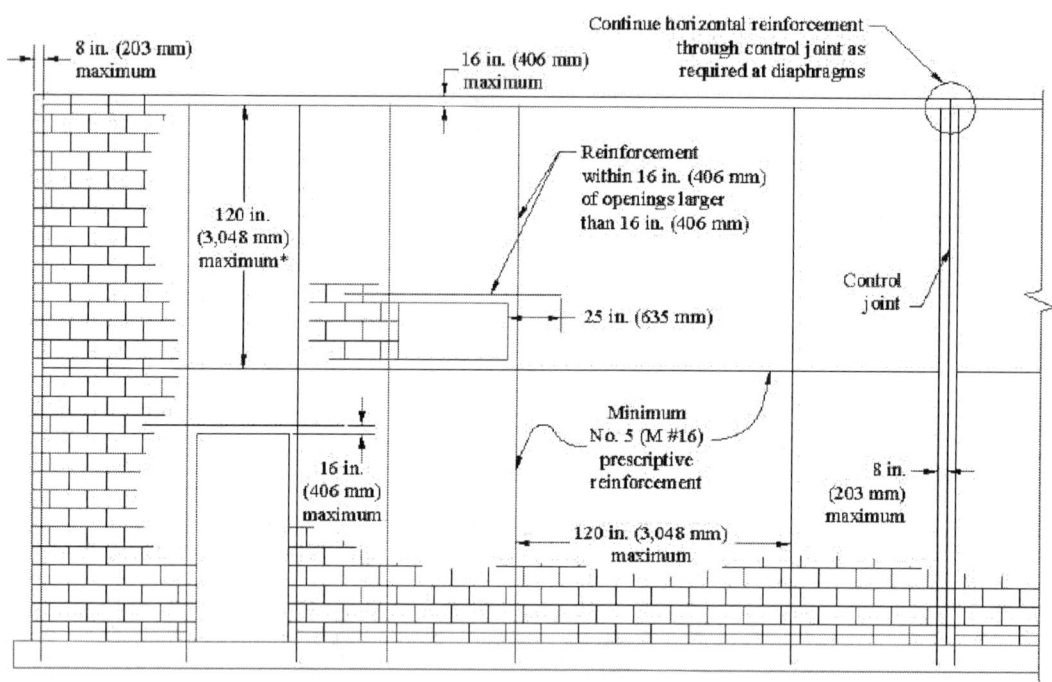

*In lieu of bond beams with No. 5 bars (M#16) at 120 in. (3,048 mm) on center, provide two wires of wire size W1.7 (MW 11) joint reinforcement at 16 in. (406 mm) on center

Figure 6.1-5 – Minimum Prescriptive Seismic Reinforcement for Ordinary Reinforced Masonry Shear Walls as Required by the MSJC Code

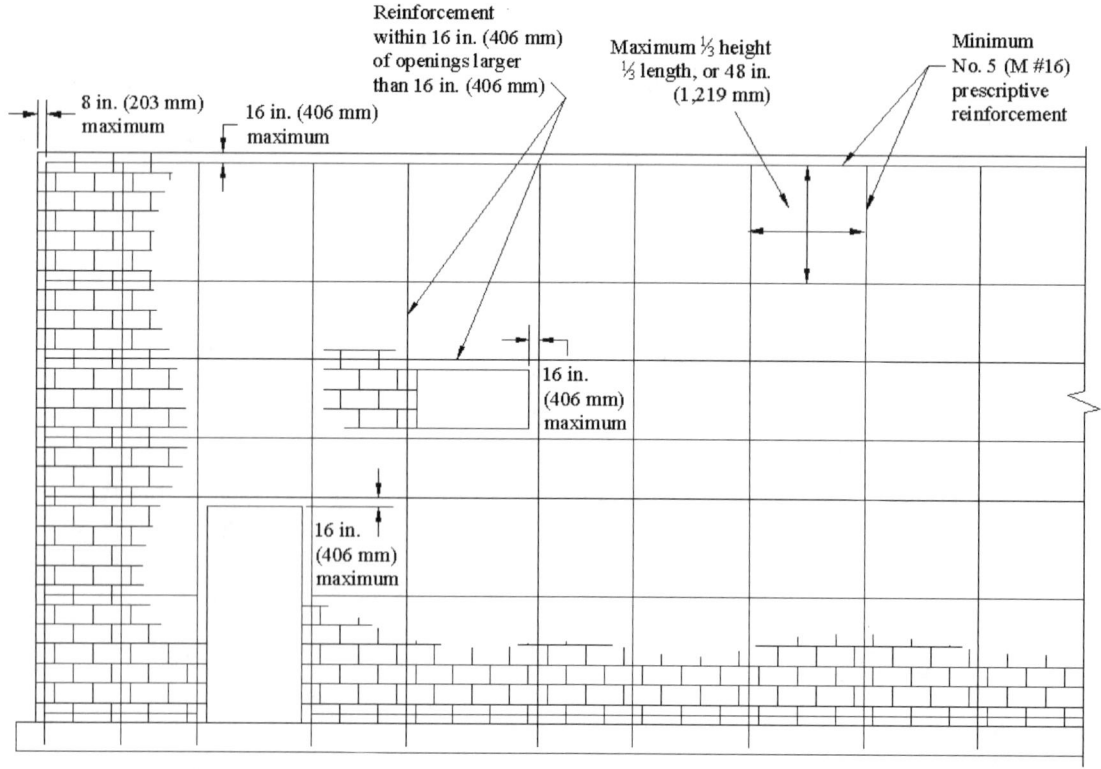

Figure 6.1-6 – Minimum Prescriptive Seismic Reinforcement for Special Reinforced Masonry Shear Walls as Required by the MSJC Code

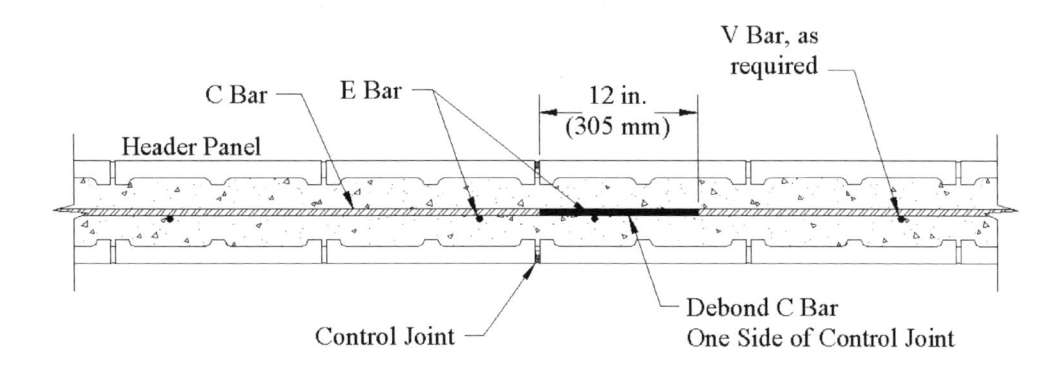

Figure 6.1-7 – Control Joint Detail at C Bar Locations

This Page Intentionally Left Blank

Commentary Chapter 1
General

C1.1 – Scope

This *Handbook* is written in mandatory language to permit adoption by reference in standards and codes. The procedures in this *Handbook* were developed based on loading conditions as defined by ASCE 7 and procedures defined by the strength design method of the MSJC Code.

Although the current edition of this *Handbook* is limited to single-story concrete masonry structures, changes to future editions are contemplated to include a broader array of design variables and site conditions. Some, but not all of the alternative design and construction variables currently under consideration for future editions of this *Handbook* include:

- Introduce the use of loadbearing clay masonry construction as an alternative to concrete masonry construction.
- Design of masonry elements having a nominal thickness other than 8 in. (203 mm).
- Include reinforcement requirements using bars of different diameter.
- Permit the use of Type N masonry mortar where permitted by the MSJC Code.
- Introduce changes in roof elevations that are larger than those permitted by parapet heights.
- Add options for multi-story construction.
- Include design options that would permit roof overhangs.
- Increase the maximum plan dimensions of diaphragms.
- Incorporate options for using rigid diaphragms.
- Introduce design options for using larger values of the specified compressive strength of masonry, f'_m.
- Permit non-shear wall elements (columns, piers, etc.) to be used to resist lateral forces.
- Permit alternative loads, such as those resulting from mechanical systems, soil backfill, or canopies.
- Permit parapets heights greater than 4 ft (1.2 m).
- Add design options for partially enclosed structures.

The direct design procedure is a table-based structural design method that permits the user, following a specific series of steps, to design and specify relatively simple, single-story concrete masonry bearing-wall structures complying with the MSJC Code and Specification and ASCE 7. See Commentary Section C2.3 for additional discussion on permitted load types and limitations.

The direct design procedure outlined herein embodies three principal phases:

- In the first phase, the designer compiles information to be used later in the calculation of design loads, including identification of loading requirements and critical loading combinations, and identification of permissible shear wall types. This phase is covered by Steps 1 through 4 of the direct design procedure.

- In the second phase, gravity and lateral loads are computed based on the information gathered in the first phase and the spacing of each scheduled type of reinforcement is determined. Computation of loads is based on the concept of flexible diaphragms, consistent with the distribution of loads by tributary area. Computation of lateral loads is based on the concepts of projected frontal area for wind loads, and projected frontal area plus diaphragm area for seismic loads. Lateral loads are distributed along lines of lateral resistance, which must be selected by the designer. Each designated line of resistance is composed of wall segments whose lengths are also identified by the designer. These global steps, in effect, require the designer to designate horizontal diaphragms, to

- specify how gravity loads are transferred from the diaphragms to supporting walls, and to specify how lateral loads are transferred from the walls that are perpendicular to the line of action to the in-plane lines of lateral load resistance. As necessary, reinforcement is provided to resist in-plane and out-of-plane bending moments in vertically spanning masonry strips due to gravity, wind, and earthquake loads; to resist gravity loads over openings; to resist wind uplift; to resist out-of-plane loads on masonry above or under openings; and to resist diaphragm chord forces. The remaining steps are identical to those that would be followed in the conventional strength-based design of each scheduled type of reinforcement. This phase is covered by Steps 5 through 9 of the direct design procedure.

- In the third phase, the designer details the required quantities and placement of reinforcement on the project drawings, lists the information required by the MSJC Specification on the design drawings; and prepares a project specification. Chapter 6 provides the minimum detailing requirements for structures designed using the provisions of this *Handbook*. This phase is covered by Steps 10 through 12 of the direct design procedure.

The direct design procedure is primarily oriented toward reinforced masonry, but the procedure permits unreinforced masonry if tension tie-downs (for example, threaded rods unbonded from the surrounding masonry) are used to resist wind uplift. This requirement is necessary because the MSJC Code does not permit unreinforced masonry to resist axial tensile stresses, which would likely be present under common design scenarios covered by this *Handbook*.

The direct design procedure originated with the members of the Veneer, Glass Block and Empirical Subcommittee of the Masonry Standards Joint Committee (MSJC), who were in search of a design approach that would be as simple to use and implement as the existing empirical design approach, but without corresponding code-imposed limits. The resulting direct design procedure in this *Handbook* can be used in virtually any seismic design category and design wind speed area, whereas the empirical design method cannot.

C1.2 – Standards cited

Because the procedures of this *Handbook* are based upon the combined requirements of ASCE 7 and the MSJC Code, users should verify that locally-adopted and enforced building code requirements are consistent with the methodologies incorporated into this *Handbook*. As changes in the referenced documents evolve, future editions of this *Handbook* are contemplated to correspond to each new edition of ASCE 7 and the MSJC.

C1.3 – Definitions

For consistent application in this *Handbook*, terms that have particular meanings in the context of this *Handbook* are defined. The definitions given are for the unique application in this *Handbook* and may not correspond to ordinary usage. Glossaries of masonry terminology are available from several sources, including:

Glossary of Concrete Masonry Terms, NCMA TEK 1-4, National Concrete Masonry Association, Herndon, VA, 2004.

B Bars are located in the masonry lintel above openings. As illustrated in Figure C1.3-1, the header panel above the opening is symmetrically reinforced with B Bars to accommodate uplift due to wind. For clarity, no other reinforcing bars are shown in Figure C1.3-1. When B Bars and C Bars would be located in the same masonry course at the top of the header, the *Handbook* permits the use of C Bars alone to resist uplift and diaphragm chord tension provided that the number of C Bars, required by Table 3.2-9, is greater than or equal to the number of B Bars required by Table 3.2-10. The number of B Bars required for a given opening, as determined in accordance with Table 3.2-10, depends upon the opening size, basic wind speed, snow loads, joist span, and whether the opening is located in an exterior or interior wall.

Because the upper B Bars located at the diaphragm level are required only for uplift resistance due to negative bending of the header panel, the portion of the masonry between the upper and lower B Bar locations may not need to be grouted unless required by Table 3.2-10. For example, for the condition shown in Figure C1.3-1, if Table 3.2-10 requires only the first two courses of masonry to be grouted then the *Handbook* would require the top two courses to be similarly grouted and reinforced. In this example, the course of masonry between the upper and lower two courses of masonry in the header panel would not need to be grouted, although for ease of construction the direct design procedure permits solid grouting when the appropriate design tables are used.

Vertical reinforcement in masonry structures designed in accordance with this *Handbook* consists of V Bars, J Bars, and E Bars. The spacing of the V Bars (vertical bars) is determined in accordance with Table 3.2-5, 3.2-6, or 3.2-8, as appropriate, and can vary from 16 in. (406 mm) to 120 in. (3,048 mm). E Bars (edge bars) are required at the edges of each masonry panel. Common examples of E Bars include vertical bars located in cells adjacent to control joints or adjacent to openings. In accordance with the *Handbook* provisions, where the spacing of the V Bars is such that the end vertical cell of a masonry panel contains a V Bar, a single reinforcing bar satisfies the requirement for both the V Bar and E Bar, precluding the need to provide two vertical reinforcing bars in a single cell. J Bars (jamb bars) are only required where the location or size of an opening interrupts the placement of the V Bars. In such cases, J Bars are placed on both sides of the opening in accordance with the provisions of this *Handbook*. The number of J Bars required depends upon the number of V Bars displaced by the opening.

Header panels and sill panels designed in accordance with the direct design method are structural elements in that they transfer in-plane loads to the lateral force-resisting system and out-of-plane loads to their respective supports, if applicable. Header panels and sill panels are not considered to contribute to the in-plane strength and stiffness of a line of resistance, however, they must meet the minimum prescriptive reinforcement required for the wall in which they are located.

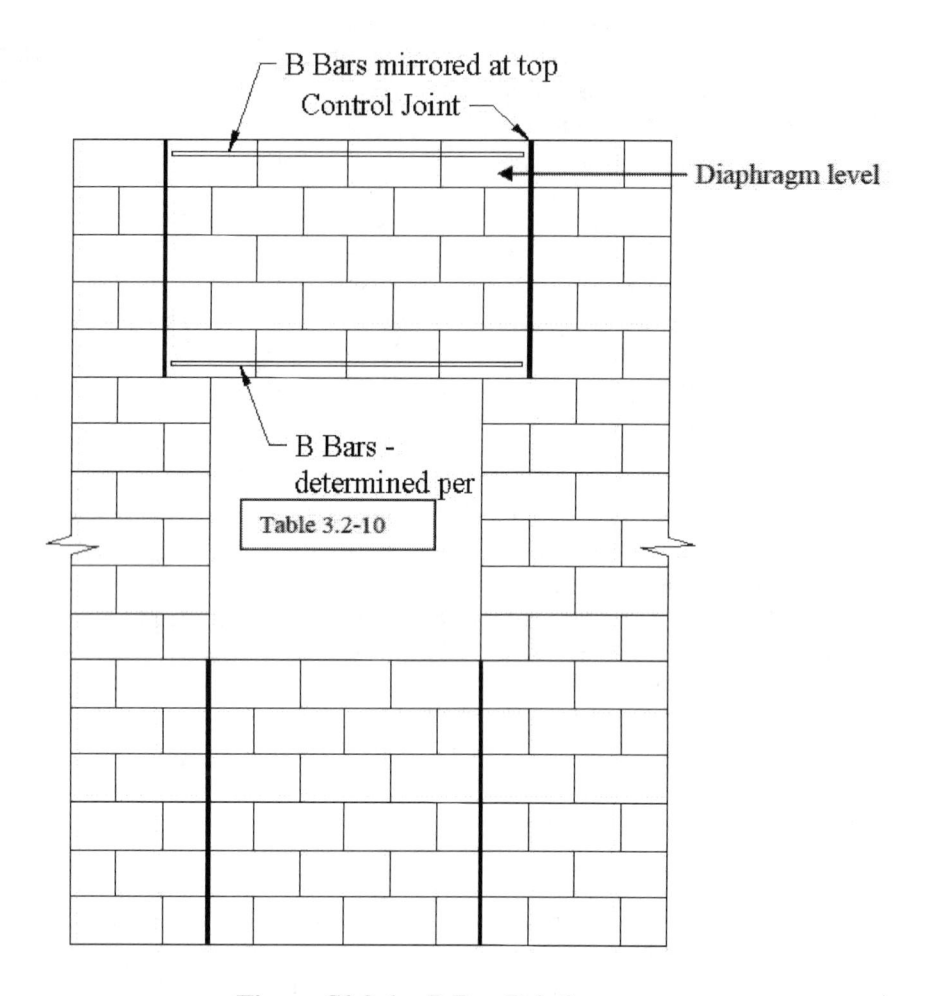

Figure C1.3-1 – B Bar Reinforcement

In the horizontal direction, H Bars are analogous to V Bars and O Bars are analogous to J Bars. B Bars are analogous to E Bars but also serve to provide primary flexural reinforcement in lintels above openings and at the top of a wall where wind uplift can result in the development of flexural tension in the masonry.

Per Figure 6.1-5, C Bars are required to be continuous through control joints to maintain structural continuity at diaphragm connection locations. Other reinforcing bars should be discontinued at control joint locations in accordance with industry practices for crack control[C1.1, C1.2].

C1.4 – Notations

Notations used in this *Handbook* are summarized in this section. Each symbol is unique, and where possible, matches the notation used in other standards or reference documents.

The Commentary to this *Handbook* also introduces notations that are unique in context and undefined in Section 1.4 of the *Handbook*. Definitions for these notations are as follows:

a = the width of the pressure coefficient zone used in determining wind pressures in accordance with ASCE 7, ft (m).
M_n = nominal moment strength of a section, in.-lb (N-mm).

References

C1.1 NCMA TEK 10-2B, *Control Joints for Concrete Masonry Walls – Empirical Method*, National Concrete Masonry Association, Herndon, VA, 2008.

C1.2 NCMA TEK 10-3, *Control Joints for Concrete Masonry Walls – Alternative Engineered Method*, National Concrete Masonry Association, Herndon, VA, 2003.

This Page Intentionally Left Blank

Commentary Chapter 2
Limitations

Use of this *Handbook* is permitted if the limitations of Chapter 2 are satisfied.

Many common masonry design and construction conditions would not comply with the limitations established by this *Handbook*. To maximize the use and flexibility of this design approach, portions of a masonry structure may be designed using the engineering provisions presented in the legally adopted building code, of which the MSJC Code forms a part, or for non-masonry materials, designed in accordance with the legally adopted building code. This modification is permitted because the conditions require strength and stiffness compatibility, require a continuous load path, and prohibit the transfer of loads into masonry designed in accordance with this *Handbook*.

An example of a common design condition where engineering analysis could be considered is the bearing of a concentrated load from a transfer girder on a masonry wall. Under these conditions, the portion of masonry subjected to this concentrated load would be designed using the provisions of the MSJC Code while the remainder of the masonry structure could be designed using the procedures of this *Handbook* provided the conditions of this section are met.

C2.1– Site Conditions

C2.1.1 *Ground Snow Load* – Permitting ground snow loads up to 60 lb/ft^2 (2.9 kPa) is intended to allow the use of this *Handbook* throughout most of the United States. Many jurisdictions throughout the U.S. require the use of local snow load design values based either on past experience or site-specific case studies. Such local criterion can be used with this *Handbook* provided the upper limit of 60 lb/ft^2 (2.9 kPa) is not exceeded. The calculations performed in developing the *Handbook* tables are conservatively based on an exposure factor of 1.2, a thermal factor of 1.2, an importance factor of 1.2, and a slope factor of 1.0.

C2.1.2 *Basic Wind Speed* – The upper limit of 200 mph (89.4 m/s) on the basic wind speed effectively covers all U.S. States and Territories, except the Florida Keys, Guam, and areas that are designated special wind regions where the basic wind speed exceeds 200 mph (89.4 m/s).

C2.1.3 *Exposure Category* – The wind design loads in this *Handbook* were derived using wind Exposure Category B and C site characteristics. The exposure category is a measure of the surface roughness and topographical irregularities adjacent to a project site. This *Handbook* cannot be used for projects located within Exposure Category D. ASCE 7 Commentary Section C26.7 provides additional information and background on exposure categories and surface roughness.

C2.1.4 *Topography* – The procedure in this *Handbook* is based on a Topographic Factor, K_{zt}, as defined in ASCE 7, of 1.0. If the limitations of Section 2.1.4.1 are satisfied, the Topographic Factor will be 1.0. Section 26.8 of ASCE 7 provides more exceptions that may be considered in evaluating K_{zt} if the limitations of Section 2.1.4.1 are not satisfied. Section 2.1.4.2 provides alternative methods of determining basic wind speeds for values of K_{zt} greater than 1.0 when the conditions of Section 2.1.4.1 are not met.

Section 2.1.4.2 allows the provisions of this *Handbook* to be used to design buildings where the regional topography adjacent to a project location results in wind speed-up effects that must be accounted for in design. Table 2.1-1 provides basic wind speeds modified by the Topographic Factor, K_{zt}, for use in the design tables presented in this *Handbook*. The basic wind speed as modified by K_{zt} and presented in Table 2.1-1 cannot exceed 200 mph (89.4 m/s) as limited by Section 2.1.2 for use in this *Handbook*. The modified values of the basic wind speed shown in Table 2.1-1 that exceed 200 mph (89.4 m/s) are provided solely to aid the user in interpolating values of the modified basic wind speed.

C2.1.5 In daily practice, mapped seismic acceleration parameters are often determined from databases that are searchable by zip code or latitude and longitude particularly in regions where the mapped seismic acceleration parameters vary considerably over relatively short distances. The use of such resources may provide a more accurate determination of local site conditions for seismic design.

C2.1.6 See commentary to Section C2.1.5.

C2.1.7 *Site Class* – Section 20.1 of ASCE 7 states that "Where the soil properties are not known in sufficient detail to determine the site class, Site Class D shall be used unless the authority having jurisdiction or geotechnical data determine Site Class E or F soils are present at the site." This *Handbook* does not permit structures to be located on Site Class E or F soils.

C2.2 – Architectural Conditions

C2.2.1 *Number of Stories* – This procedure is limited to the design of one-story, above-grade structures. Floor loads, such as those created by a mezzanine, are not permitted. Because Section 2.3.1 excludes soil loads, except as permitted by Section 2.3.1.1, below-grade masonry subjected to unbalanced lateral loads, including basements, is not permitted to be designed using this procedure. See Commentary Section C2.3.1.1 for additional discussion on below grade masonry.

C2.2.2 *Walls* – The analysis and design used to develop the tables in this *Handbook* are based on wall cross-sectional properties that are achieved when concrete masonry units having a nominal thickness of 8 in. (203 mm) are used. Although the MSJC Code permits the use of multi-wythe, composite construction that could result in the same nominal wall thickness as the single-wythe construction, the section properties of composite masonry may differ from a single-wythe assembly and are, therefore, not permitted.

C2.2.3 *Height* – The upper limit of 30 ft (9.1 m) on the total building height is driven by the wind analysis procedures used in developing this *Handbook*. When the wall height is less than 4 ft (1.2 m), the shear-dominated performance of the wall may change, which in turn could affect the values of k_1 in Table 3.2-8. Therefore, a lower bound of 4 ft (1.2 m) was placed on the wall height, which was considered a practical lower limit for building configurations covered by this *Handbook*. The 4 ft (1.2 m) minimum does not apply to segments of a wall, such as header or sill panels, which are permitted to be smaller than 4 ft (1.2 m).

Wall height and vertical wall span are not necessarily the same value for a given wall. Wall height is measured relative to finished grade and includes the parapet height, if present. Vertical wall span is measured from the point of lateral bracing at the base of the wall to the point of lateral bracing near or at the top of the wall, excluding the parapet height, if present.

C2.2.4 *Parapets-* The maximum permitted parapet height of 4 ft (1.2 m) is generally tall enough to accommodate the roof slopes associated with buildings that would meet the limitations of Chapter 2. A maximum height was necessary to limit the design snow load in the calculations on which the procedure is based, because snow can drift and accumulate against a parapet. In addition, as discussed in the Commentary in Section C27.6.2 of ASCE 7, there can be higher out-of-plane wind loads on parapets than on the wall below the diaphragm. Limiting the ratio of the parapet to the vertical span of the wall below the roof diaphragm to 1:3 effectively permits more than twice the design out-of-plane wall wind load and seismic load to occur at the parapet support without exceeding the moment of the wall below. The procedure in this *Handbook* assumes that the masonry supporting the parapet acts as a simple span from foundation to the roof diaphragm. For walls that support parapets meeting the limitations of this section, including the requirement that the supporting walls be designed as reinforced masonry, the resulting loads and stresses induced in the support wall are comparatively small relative to other controlling design loads. This finding is based on analyses of various buildings incorporating parapets constrained by the limitations of the direct design procedure.

C2.2.5 *Openings* – Openings in masonry walls can range in size depending upon their intended purpose. Section 2.2.5.1 addresses requirements for relatively large openings, such as windows and doors. Conversely, Section 2.2.5.2 provides options for smaller openings, or a grouping of smaller openings, to be located in masonry without the limits of Section 2.2.5.1. Openings that do not comply with the requirements of Section 2.2.5.1 or Section 2.2.5.2 would need to be analyzed using the provisions of the MSJC Code for their impact on the strength of the assembly in which they are located.

C2.2.5.1 Except as permitted by Section 2.2.5.2, the procedure in this *Handbook* does not account for multiple openings that are stacked vertically. Such openings may interrupt the load path of masonry segments, thereby potentially reducing the resistance to out-of-plane wind and seismic loads. Therefore, this procedure does not permit vertically aligned or partially aligned openings. To simplify the design assumptions, distribution of loads, and detailing requirements around openings, this *Handbook* requires that control joints be provided on both sides of an opening in a masonry wall. Providing control joints on both sides of an opening isolates the panels above and below the opening so as not to inadvertently transfer in-plane loads from designated shear wall segments to these portions of the masonry system.

In some instances it may be desirable to remove one or both of the control joints adjacent to an opening. In such cases, users of this *Handbook* have the option of removing or relocating the control joints required by Section 2.2.5 provided that the constraints stipulated in the introduction to Chapter 2 are met. This includes ensuring, at a minimum, that the structural performance of the system is not compromised in such a way as to no longer meet the minimum requirements of the legally adopted building code. By removing one or both control joints adjacent to an opening, additional reinforcement not required by this *Handbook* may be necessary.

C2.2.5.2 Relatively small openings, such as through-wall penetrations to accommodate electrical conduit or other utilities, are common in virtually all buildings. Some openings are relatively small and have negligible impact on the strength or performance of the element in which they are located. Other, larger, openings can impact the strength of an assembly. The provisions of Section 2.2.5.2 permit relatively minor openings to be placed in masonry walls designed in accordance with this *Handbook* provided that the openings, regardless of their size, do not interrupt the reinforcement within the assembly as this could have significant impact on the strength of the masonry.

C2.2.6 *Roof Diaphragms* – Each rectangular roof diaphragm must be surrounded and supported along each edge by masonry walls as illustrated in Figure C2.2-1 because the procedure in this *Handbook* assigns chord forces to a bond beam in the masonry. Openings in the supporting masonry wall are permitted provided the conditions of this *Handbook* are met. Outside of stated limitations, this *Handbook* does not address diaphragm design. A designated shear wall segment is required in both principal directions at each corner of each diaphragm because the procedure in this *Handbook* assumes that the perpendicular wall segment provides additional out-of-plane bracing and tension reinforcement. The additional stiffening is sufficient such that the critical wall strip will be located beyond a distance that is $2a$ from the corner, where a is defined by Figure 28.6-1 and Figure 30.5-1 of ASCE 7. The difference between end zone loads and interior zone loads is compared to the difference in capacities between a corner wall system and a simple vertically spanning wall strip. The value of $2a$ can be as large as 24 ft (7.3 m) depending on the height and the plan dimensions of the building. Further, a designated shear wall segment can have a much smaller plan length than $2a$. The procedure in this *Handbook*, however, requires that segments with smaller plan lengths have more reinforcement for taller buildings and for buildings with larger plan dimensions, which increases the required length. Theoretically, a wall could have narrow openings with many small designated shear wall segments, which may create a condition where the stiffening effect of the corner may not be enough and the end load zones may govern. Considering the impracticality of these conditions, however, this design assumption is deemed reasonable and appropriate because of the conservatism of the procedure (e.g. pinned-pinned assumption) and the inherent redundancy of masonry.

This *Handbook* does not address the design of connections. Transferring forces between horizontal diaphragms and vertical walls is critical to the performance of the structure. Anchors used to achieve this force transfer are required to be designed in accordance with the MSJC Code.

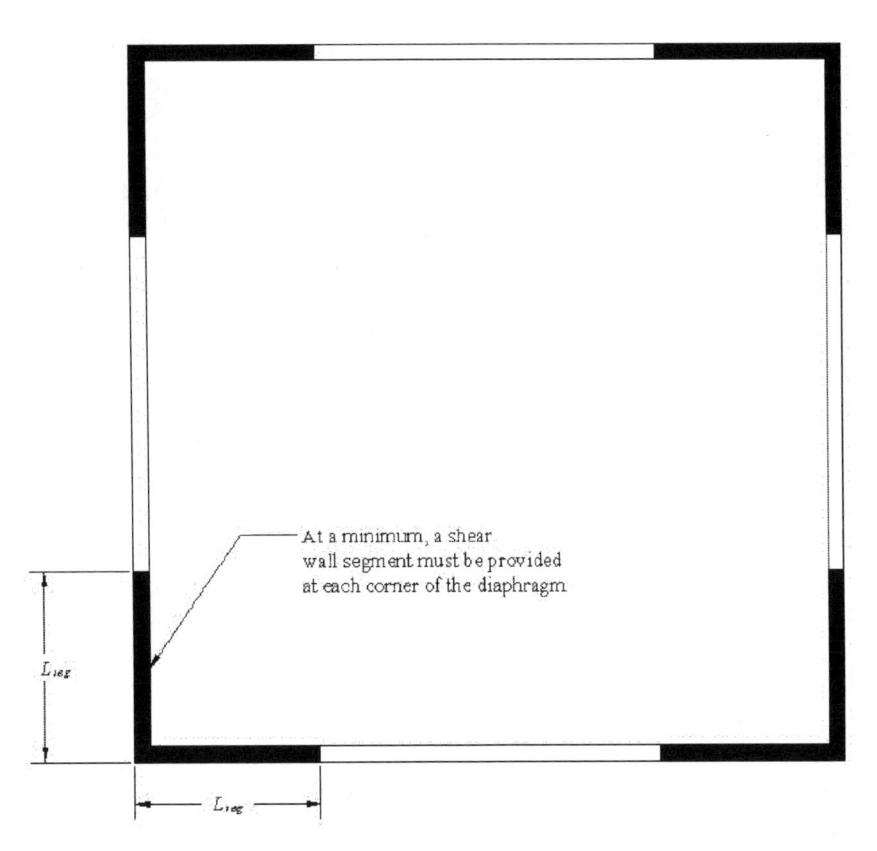

Figure C2.2-1 – Shear Wall Layout at Corners

C2.2.7 *Roof Slope* – The required minimum roof slope is to assure the roof will drain water. Chapter 8 of ASCE 7 outlines minimum design criteria for roof drainage. Because different roofing systems shed snow in different ways, there is no single trigger for a minimum roof slope that would shed snow off of a structure. When the ground snow load exceeds 25 lb/ft^2 (1.2 kPa), this *Handbook* requires that the roof be designed to shed snow. In developing the design tables of this *Handbook*, this limit was felt necessary in order to keep the design loads to a reasonable level for the broad range of ground snow loads considered in this document.

ASCE 7 Section 7.6 requires that certain unbalanced snow load conditions be accounted for when designing hip, gable, curved, multiple folded plate, sawtooth or barrel vault roofs. The calculations performed when developing the direct design tables were based on conservative uniform snow loads that accounted for snow drift with a C_s coefficient equal to 1 on top of the balanced snow loads. These conservative uniform snow loads were associated with a blanket of snow that is as high as the tallest parapet permitted by the *Handbook*. Analysis determined that the peak snow load for unbalanced snow conditions on curved, multiple folded plate, sawtooth, and barrel vault roofs is less than or equal to the blanket load used in the calculations when the ground snow load is less than 25 lb/ft^2 (1.2 kPa). For hip and gable roofs, ASCE 7 Figure 7-5 requires that a relatively narrow portion of the roof at the ridge be designed for a peak load that could exceed the blanket load used in the calculations. However, calculations that considered this condition were based on joists that were either 30 ft (9.1 m) or 60 ft (18.3 m) long and the analysis determined that the overall load effect caused by snow on masonry elements will be less than the load used in the calculations.

C2.2.8 *Changes in Diaphragm Elevation* – The maximum projection between adjacent roof diaphragms is limited to the maximum permitted parapet height to account for snow drift, which cannot exceed the height of the parapet. The procedure in this *Handbook* is based on the maximum snow drift that can accumulate due to the full height of the maximum permitted parapet height. The snow drift is assumed to act over the entire area of the roof diaphragm assuming the minimum permitted slope. These conditions represent the greatest snow load that would be imposed on the masonry system.

ASCE 7 Section 7.9 requires that sliding snow be accounted for in addition to the balanced snow load if the roof is not slippery with a slope that is greater than 2:12, and if the roof is slippery with a slope that is greater than 1/4:12. An example of a slippery roof is a standing seam metal roof. Analysis determined that the provisions of this *Handbook* are applicable when the horizontal distance between the eave of the upper roof and the ridge of the upper roof is not greater than 42 ft (12.8 m) and the ground snow load does not exceed 25 lb/ft^2 (1.2 kPa).

C2.2.9 *Joists – Joists* – The tables provided in this *Handbook* cover three different joist span lengths: No Joist (NJ), 30 ft (9.1 m), and 60 ft (18.3 m). The No Joist option corresponds to the case of an interior or exterior wall wherein the joist does not impose axial load on the wall. For reasons detailed in Commentary Section C3.1, linear interpolation (or extrapolation) is not permitted for joist spans other than those explicitly listed. Therefore, designs must be based on the larger tabulated joist span length, when actual joist spans are between those listed. Limiting the tributary area supported by each joist ensures that the resulting load applied to the masonry is relatively uniform in magnitude and distribution.

The requirement that masonry elements are not permitted to support reactions from tributary areas greater than that supported by a single joist is to preclude the application of loads from collector elements, such as joist girders, that would result in the application of a load on the masonry that is not considered in this *Handbook*.

The analyses and modeling assumptions used in developing the provisions of this *Handbook* are based on an assumption that the joists transmit a uniform load along the plan length of the bearing walls. Considering the limitations of the design solutions permitted by the *Handbook*, this design criterion can safely be applied to joists spaced up to 6 feet (1.8 m) on center. In cases where this joist spacing is exceeded, the *Handbook* can still be used to perform preliminary designs; however, the final solution will need to be checked using conventional structural analysis techniques.

C2.2.10 *Roof Drainage* – The requirements for roof drainage are only intended for roofs with low slopes, which have overflow drains or scuppers. The value d_s refers to the difference in elevation from the top of the roofing material to the inlet elevation for an overflow drain or the flow line of a scupper. The value d_h refers to the hydraulic head above the inlet elevation for an overflow drain or the flow line of a scupper. The hydraulic head must be determined by a hydraulic analysis that accounts for local rainfall intensity-duration-frequency curves, the tributary area that flows into the overflow drain or scupper, and the geometry of the drainage system. The hydraulic head can be reduced by increasing the number and/or area of the drains. The procedure in this *Handbook* is based on a roof live load of 20 lb/ft^2 (1.0 kPa).

In the load combinations required by ASCE 7, the greatest of the roof live load, snow load, and rain load will govern in combinations that include those loads. Therefore, the rain load will not govern design if it is less than either the roof live load or the snow load. Section

2.2.10 effectively limits the rain load to 20 lb/ft^2 (1.0 kPa) when the roof slope is at its minimum allowable and the joist span is at least 30 ft. The hydraulic pressure is greater than 20 lb/ft^2 (1.0 kPa) at the drain; but reduces radially from the drain due to the slope in the roof and constant top elevation of the accumulated water. The design assumption of a 30 ft (9.1 m) joist span is conservative for the three span options permitted by this *Handbook*.

C2.2.11 *Isolation of Features* - The *Handbook* procedures for designing masonry walls do not have a mechanism for considering the additional loads imposed by canopies, signs, or overhangs that are connected to them. Consequently, such features are prohibited from being connected to the masonry unless they meet the requirements of Section C2.2.11.

The exceptions noted in *Handbook* Section 2.2.11 are intended to accommodate common finishes such as vinyl siding, plank siding, signs, and small lights and similar fixtures that are regularly attached to masonry walls. The wall finish projection is limited to 6 in. (152 mm) to minimize the magnitude of increase in wind load acting on the structure as a result of increased projected area. The 3 lb/ft^2 (0.14 kPa) wall finish weight limit is established to keep the resulting out-of-plane seismic forces acting on the wall as a result of increased dead load to a safe level.

C2.2.12 *Simplified Wind Load Procedure Limitations* – The procedure in this *Handbook* is based on the Main Wind Force Resisting System Envelope Procedure, Part 1 of ASCE 7. That procedure is based on assumptions stated in Chapter 28 of ASCE 7. A masonry structure that satisfies all the other limitations of Chapter 2 will generally satisfy the ASCE 7 limitations as long as the building is classified as an enclosed building by Section 26.2 of ASCE 7. That section states that a building is enclosed if it is not open and not partially enclosed. Generally, a building is open if each exterior wall is more than 80% open and a building is partially enclosed if one exterior wall has more open area than all the other exterior walls combined. Openings are generally defined by ASCE 7 as holes that are designed to be open during design winds.

C2.2.13 *Veneers* – A masonry veneer, whether anchored or adhered, would impart additional out-of-plane loads onto the structural wythe of masonry during a seismic event. Adhered masonry veneer would additionally impart a vertical load onto the structural wythe of masonry because the dead load of the adhered veneer would be supported by the masonry backup. The 35 lb/ft^2 (1.68 kPa) upper limit on the installed weight of the masonry veneer is a practical limit intended to cover the majority of commonly used masonry veneer systems.

C2.3 – Loading Conditions

Within the limitations established by Chapter 2, design of masonry in accordance with this *Handbook* accounts for gravity loads and for lateral loads in each principal plan direction.

C2.3.1 *Load Types* – The procedure in this *Handbook* is based on the assumption that the structure is not ice-sensitive, as defined in Section 10.2 of ASCE 7. A masonry structure meeting the limitations in Chapter 2 would not be ice-sensitive by this definition. No other load types are permitted on the masonry, including but not limited to floor dead loads, floor live loads, hydrostatic pressure, unbalanced soil loads, flood loads, or crane loads.

C2.3.1.1 *Soil Loads* – Exterior masonry walls and interior loadbearing masonry walls often extend below grade to provide adequate and stable soil bearing, to mitigate potential frost heave, or both. This *Handbook* permits such construction provided that the out-of-plane lateral loads, including the effects of soil and surcharge loads, are balanced. Unbalanced lateral loading conditions would result in flexural tension stresses that are not accounted for in this *Handbook*.

C2.3.2 *Roof Dead Load* – The Commentary for Chapter 3, Step 5D, provides assistance in calculating the roof dead load. A minimum dead load value is provided because the masonry must be designed to resist uplift due to wind. The procedure in this *Handbook* is based on a 2 lb/ft^2 (0.1 kPa) roof dead load when using load combinations in which dead load counteracts uplift forces. This minimum roof dead load is further reduced by the load factor in the considered load combination. The maximum dead load value was set at 30 lb/ft^2 (1.4 kPa) because that is an upper bound on common roof dead load values used by designers for typical flexible diaphragm structures. The combined limitations of the maximum roof dead load and flexible diaphragm criterion precludes the use of many concrete roof systems.

Users can verify that the actual roof dead loads do not exceed the permitted roof dead loads using several different methods, even accounting for the presence of mechanical rooftop equipment. Consider a simple span diaphragm measuring 20 ft (6.1 m) by 20 ft (6.1 m) in plan. This *Handbook* designs the bearing walls of this structure for a superimposed axial dead load of 40 lb/ft (584 N/m) and 600 lb/ft (8,756 N/m), corresponding to the lower 2 lb/ft^2 (0.1 kPa) and upper 30 lb/ft^2 (1.4 kPa) roof dead load, respectively. If the actual dead load of the roof was only 12 lb/ft^2 (0.6 kPa), resulting in an axial dead load on the bearing walls of 240 lb/ft (3,502 N/m), the weight and location of the rooftop equipment would need to be limited in such a manner so that the axial dead load contribution from the rooftop equipment does not exceed 360 lb/ft (5,254 N/m) at any location. Note, however, that this analysis does not consider the vertical profile of the rooftop equipment and the corresponding wind and snow drifting loads that may result due to the presence of the rooftop equipment, which requires additional analysis.

C2.3.3 *Roof Live Load* – The upper limit of 20 lb/ft^2 (1.0 kPa) for the roof live load is a practical value for the majority of the types of buildings intended to be covered by this *Handbook*. Because this *Handbook* only considers single story buildings, floor live loads are not considered.

C2.3.4 *Eccentricity of Roof Loads* – Refer to Commentary Section 3.2, Step 5H for additional discussion on permitted eccentricity of applied roof loads. Interior load-bearing walls are designed to carry roof loads from both sides of the wall. The eccentricity of these applied loads is permitted to vary up to 1.25 in. (31.8 mm) from the centroid of the wall cross-section, but the resultant of each eccentrically applied load is not permitted to be applied to the same half of the wall cross-section. Doing so would result in bending moments (P-delta effects) that are cumulative, which was not accounted for in this *Handbook*. Concentrically applied roof loads are permitted.

C2.4 – Material and Construction Requirements

C2.4.1 *Units* – The procedure in this *Handbook* requires that all masonry components designed using the direct design procedure be constructed of concrete masonry units having a nominal thickness of 8 in. (203 mm). Clay masonry may not be designed using the procedure in this *Handbook* because it has a lower design modulus of elasticity for a given specified compressive strength compared to concrete masonry construction used in developing this procedure. The lower modulus of elasticity would yield larger deflections, which may create a condition that is not permitted by ASCE 7 and/or MSJC Code. In addition, hollow clay masonry units commonly have a thinner face shell than the minimum required for concrete units. Because a T-beam analysis based on the minimum face shell thickness of concrete masonry units was performed to determine the bending capacities of partially grouted walls, the values in this *Handbook* should not be used for thinner face shells. For these reasons, masonry constructed of clay masonry units cannot be designed using this *Handbook*. Although clay masonry typically has a higher compressive strength than concrete masonry, several of the underlying modeling assumptions that form the basis of the *Handbook* would be violated if clay masonry were designed using the provisions of the *Handbook*. In the future, modifications are planned for Chapter 4 of this *Handbook* to permit clay masonry construction to be designed by the direct design process.

The procedures of Chapter 3 permit the use of lightweight, medium weight, and normal weight concrete masonry units.

C2.4.2 *Mortar* – The procedure in this *Handbook* is based on the strength design method according to Chapter 3 of the MSJC Code. Section 3.1.8.1.1 of the MSJC Code requires that the specified compressive strength of masonry equal or exceed 1,500 lb/in.^{2} (10.3 MPa). Concrete masonry units are commonly specified to comply with ASTM C90, including the minimum unit compressive strength of 1,900 lb/in.^{2} (13.1 MPa). Chapter 5 of this *Handbook* is philosophically consistent with this practice because it does not require the compressive strength of the concrete masonry units to be higher than the minimum required by ASTM C90. Article 1.4 B of the MSJC Specification permits verification of the specified compressive strength of masonry (f'_m) by either the unit strength method, which is based on a table that requires knowledge of the unit strength and mortar type, or the prism strength method, which requires testing of masonry assemblages consisting of units, mortar, and grout (if applicable) to verify conformance with the specified compressive strength of masonry used in design assumptions. For relatively simple structures that do not need high strength materials, such as those addressed through this procedure, the prism strength method is not as commonly used as the unit strength method because the prism strength method requires compressive strength testing, which increases the cost and complexity of this inspection task. The combination of minimum material properties required by this *Handbook* ensure that the minimum specified compressive strength of the masonry will be at least 1,500 lb/in.^{2} (10.3 MPa), which can be easily verified through the use of the unit strength method.

To achieve the required 1,500 lb/in.^{2} (10.3 MPa) specified compressive strength of masonry using the unit strength method with the minimum 1,900 lb/in.^{2} (13.1 MPa) concrete unit strength permitted by ASTM C90, Type M or S mortar is required. While a Type M mortar could be used without any impact on the structural design provisions of this *Handbook*, Type S mortar is more commonly used and generally more appropriate for masonry structures covered by this *Handbook*. Type N mortar, conversely, cannot be used to construct structures designed by the procedures of this *Handbook* because both the compressive strength of the masonry and the mortar-to-unit bond strength (modulus of rupture values in accordance with Table 3.1.8.2 of the MSJC Code) would be lower. For structures assigned to Seismic Design Categories A, B, or C, any Type S masonry mortar can be used (i.e., masonry cement mortar, cement-lime mortar, or mortar cement mortar). In accordance with the MSJC Code, however, neither Type N nor masonry cement mortars are permitted to be used in Seismic Design

Categories D, E, or F. Because air-entrained cement-lime mortars exhibit a reduced bond strength, which was not taken into account in establishing the tables of this *Handbook*, cement-lime mortars are not permitted to contain entrained air when used in the construction of structures assigned to Seismic Design Categories D, E, or F.

C2.4.3 *Reinforcement* – No. 5 (M#16), Grade 60 (420 MPa) reinforcing bars are commonly used in modern reinforced masonry construction because they are readily available in most markets; the lap splice length requirement is modest compared to larger-diameter reinforcement, which facilitates and economizes the use of low-lift grouting procedures; and because they are small enough that multiple layers can be conveniently placed in the cells of an 8 x 8 x 16 inch (203 x 203 x 406 mm) hollow concrete masonry unit with sufficient tolerance and coverage to permit horizontal reinforcement and vertical reinforcement to pass each other. The procedure in this *Handbook* requires all bars to be the same size for simplicity. Bars other than No. 5 (M# 16) are not permitted, except for shear reinforcement used in the construction of masonry lintels, because there are many provisions in the MSJC Code that are satisfied by the tables in the procedure without the user's explicit notification. An example is the various prescriptive minimum reinforcement requirements in the MSJC Code. Another example is the maximum reinforcement limits imposed by the MSJC Code that a user might exceed if an alternative reinforcing bar size with an equivalent or larger area were to be used.

C2.4.4 *Grout* – Either fine or coarse grout complying with the requirements of ASTM C476 is permitted to be used. For practical and economical reasons, it is common practice to solidly grout a masonry wall when the reinforcement is closely spaced, even if all of the grouted cells do not contain reinforcement. If the spacing of either the vertical or horizontal reinforcement is 16 in. (406 mm) on center, the procedures of this *Handbook* assume a solid grouted cross-section, except where the horizontal reinforcement consists of bed joint reinforcement only, in which case solid grouting is not required. Because the cross-section is assumed to be solid grouted in these cases when determining the design strengths of the masonry elements, it would not be appropriate or conservative to construct the masonry as partially grouted even though reinforcement is not present in each cell or bond beam. As required by the MSJC Code, cells containing reinforcement must be grouted.

C2.4.5 *Specifications* – Chapter 5 outlines the minimum material, testing, inspection, and construction requirements required by this *Handbook*.

C2.4.6 *Details* – Details presented in Chapter 6 address minimum structural stability and life safety provisions as required by the MSJC and ASCE 7. Not all of these details are applicable to every building designed by this *Handbook*. The user should select only those details that are applicable to the structure being designed. Serviceability and aesthetic criteria, including crack control and water penetration mitigation, are not addressed by this *Handbook*. Recommended industry practices should be followed for detailing the structure to prevent or minimize water infiltration and potential cracking.

Section 2.2.5 of the *Handbook* requires control joints to be placed on both sides of openings located in walls designed in accordance with this *Handbook*. This requirement is necessary to be consistent with the modeling and load distribution assumptions inherent in the direct design procedure. The *Handbook* does not require locating and spacing control joints for crack control, which are covered by industry guidelines[C2.1, C2.2]. These industry guidelines also provide for options to increase the amount of horizontal reinforcement to minimize or eliminate the need for control joints for crack control.

Because control joints are required on either side of openings in accordance with this *Handbook* are tied to structural performance, as opposed to crack control, the control joints required by this *Handbook* are required regardless of the amount of horizontal reinforcement provided. This *Handbook* requirement does not apply to control joints that may or may not be provided at locations that are not adjacent to openings.

References

C2.1 NCMA TEK 10-2B, *Control Joints for Concrete Masonry Walls – Empirical Method*, National Concrete Masonry Association, Herndon, VA, 2008.

C2.2 NCMA TEK 10-3, *Control Joints for Concrete Masonry Walls – Alternative Engineered Method*, National Concrete Masonry Association, Herndon, VA, 2003.

This Page Intentionally Left Blank

Commentary Chapter 3
Procedure

C3.1 – General

The procedure in this *Handbook* is based on a comparison of factored design loads and design strengths in accordance with Chapter 3 "Strength Design of Masonry" of the MSJC Code for the limiting envelope of conditions associated with each value in each table. Linear interpolation is not permitted unless specifically stated as being permitted because there are many factors that cause a non-linear relationship between values in the table; in addition, so as to have a practical number of options in each table, the limiting conditions are typically so different that a linear approximation is not valid.

This *Handbook* covers only the design of masonry walls meeting the limitations of Chapter 2. The design of diaphragms, foundations, and other components critical to the performance of buildings is not covered by this *Handbook*.

The reinforcement bar designations in Figure 3.1-1 represent the reinforcing bar designations defined in Chapter 1.

A structure designed in accordance with this *Handbook* may not require every type of reinforcing bar shown in Figure 3.1-1. For example, a structure where the diaphragm is connected to the top of the masonry wall (no parapet is present) is not required to have T Bars, because the C Bars located in the top course provide the necessary structural continuity that would otherwise be provided by T Bars at the top of a parapet. Similarly, unless the tables of Chapter 3 require more than one reinforcing bar per cell or course, only a single reinforcing bar is required. For example, if two B Bars are required in a lintel above an opening, two B Bars must be provided in the top course of the same header panel. For the remaining perimeter of the top masonry course exclusive of this header panel, however, only a single T Bar is required (for cases where a parapet is present and no other openings in the masonry occur). Likewise, an O Bar provided in a panel above or below an opening is not provided in addition to H Bars in the panel, but in lieu of these horizontal reinforcing bars over the length of the panel.

As required by this *Handbook* and the MSJC Code, structural continuity must be maintained between the application of load and the final point of resistance. For reinforcement, this is most commonly achieved through lap splicing of the reinforcement. Chapter 6 provides standardized details for lap splicing that is applicable to the design of structures in accordance with this *Handbook*.

Additional commentary discussion is provided for each type of reinforcing bar throughout Chapter 3 in the corresponding section where it is addressed.

C3.2 – Direct Design Procedure

Commentary on Step 1

The information obtained in this step is required for design. Both wind and seismic design criteria are necessary because ASCE 7 requires that all buildings be designed for both wind and seismic loading conditions, and meet all prescriptive seismic requirements even if seismic loads do not govern the design of any elements.

The procedure in this *Handbook* is valid for any of the four Risk Categories listed in Table 1.5-1 of ASCE 7. Where the Risk Category is not required for input in a design table in this *Handbook*, the content of the design table is based on Risk Category IV, which requires the highest importance factors.

Commentary on Step 2

Creation of preliminary plans will be more efficient when based on experience with the procedure in this *Handbook*. The direct design procedure requires selection of options that affect architectural limitations. For example, a preliminary floor plan containing many small openings or a few large openings could result in a series of shear wall segments that are not long enough to accommodate the minimum spacing of vertical reinforcement required. The procedure in this *Handbook* is designed so that users may quickly learn the limitations relevant to a particular geographic region.

Commentary on Step 3

The Seismic Design Categories in Table 3.2-1 were determined in accordance with Sections 11.4 to 11.6 of ASCE 7 based upon the Site Class soil profile. The site soil is classified in accordance with Chapter 20 of ASCE 7 as Site Class A, B, C, or D, corresponding to hard rock, rock, very dense soil and soft rock, and stiff soil, respectively. The direct design procedure is not permitted for Site Class E (soft clay) or Site Class F (soils requiring site response analysis) soil profiles. Where applicable, accounting for the actual soil conditions at the project location may reduce the Seismic Design Category and therefore increase the economy of the structure. See Commentary Section C2.1.7 for additional discussion on seismic site class options.

Commentary on Step 4

The LFRS Options in Table 3.2-2 satisfy the provisions of Section 1.18 of the MSJC Code. The lateral force-resisting system is used in Step 5E to determine C_s. The MSJC Code permits other types of Lateral Force Resisting Systems, such as Intermediate Reinforced Masonry Shear Walls. The three shear wall types included in this *Handbook*, however, are only a subset of the shear wall types permitted by the MSJC Code. The limitation on the number of shear wall types addressed by this *Handbook* is to minimize the number of tables required to cover all the necessary lateral force coefficients unique to each combination of shear wall type, shear wall length, and reinforcement schedule.

Commentary on Step 5

The procedure in this *Handbook* is based on a flexible diaphragm analysis, which is appropriate for the limitations of Chapter 2. Designation of rectangular roof diaphragms within the entire roof plan of a building is a critical part of the process, and there is often more than one way to designate these diaphragms. Refer to Figure C3.2-1 for an example roof plan. For this example, the walls are shown and there is no change in diaphragm elevation at the interior wall. When considering a North-South wind and seismic loading direction, one may wish to divide this example roof plan into two designated diaphragms. As shown in Figure C3.2-2 a, however, the dimensions of the northern rectangle may not meet the limitations of Section 2.1 of this *Handbook*, so it may be necessary to combine areas that are defined by wall lines and designate their combined area as one diaphragm. In Figure C3.2-2 b, the North-South walls (wall lines A and B) on each side of Diaphragm 1NS are lines of resistance that must have sufficient designated shear wall segments as required by the procedure in this *Handbook*. When considering an East-West wind and seismic loading direction, one may wish to divide this example roof plan into two designated diaphragms, as shown in Figure C3.2-3 a. In this case, the interior wall (wall line G) is a line of resistance for both diaphragms and must have sufficient designated shear wall segments for both diaphragms combined as required by the procedure in this *Handbook*. The plan dimensions of Diaphragm 1EW are approximately 1:5, as defined by Section 2.1 of this *Handbook*, which is permitted because this is much less than 4:1. Alternatively, one may wish to designate the entire roof as one diaphragm, as shown in Figure C3.2-3 b. This is permitted; however, note that the interior wall (wall line F) must then be detailed as a non-participating wall. It is permitted for the diaphragm designations to be different for the two principal plan directions because the walls parallel to the plan direction under consideration are designed in Steps 5A through 5L.

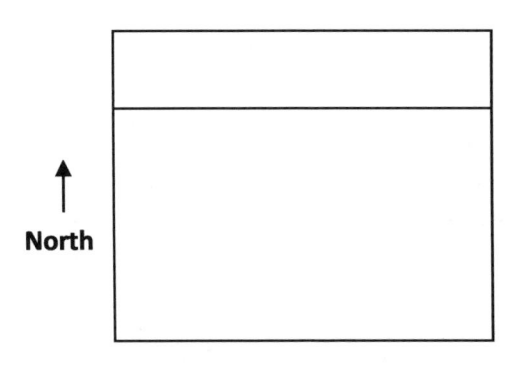

COMMENTARY TO DIRECT DESIGN HANDBOOK FOR MASONRY STRUCTURES C-21

Figure C3.2-1 – Example Roof Plan

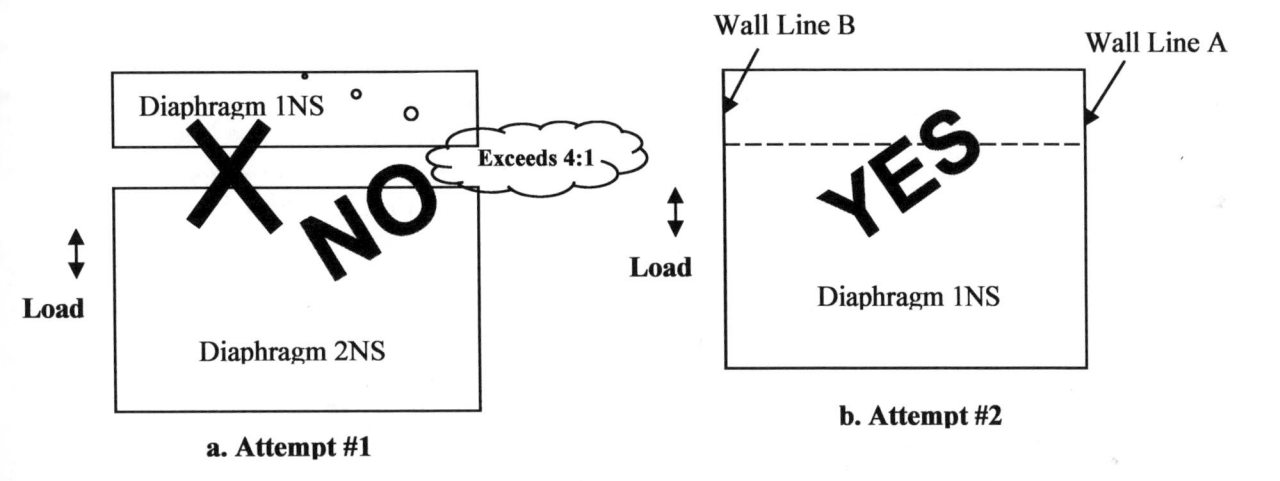

Figure C3.2-2 – Examples of Designations for North-South Wind and Seismic Loading

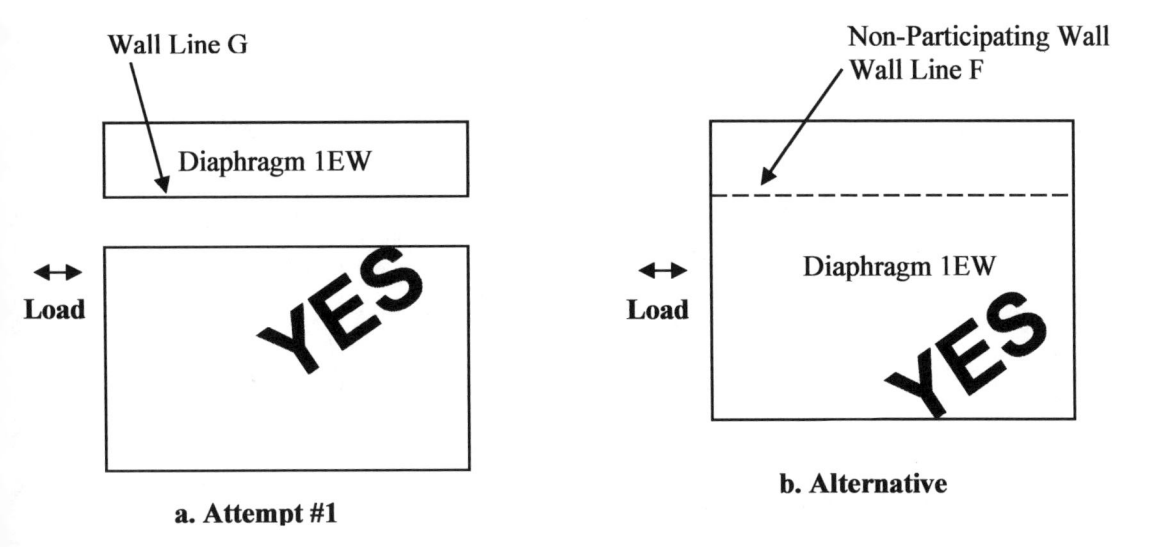

Figure C3.2-3 – Examples of Designations for East-West Wind and Seismic Loading

DIRECT DESIGN HANDBOOK FOR MASONRY STRUCTURES

Commentary on Steps 5A through 5C

The area of the building elevation is directly proportional to the cumulative length of shear wall segments that will be required in later steps. (For additional guidance on calculating the projected surface area of a building for wind loads, refer to Figure 28.6-1 of ASCE 7.) This area is to be calculated separately for each designated rectangular roof diaphragm, for each principal direction of loading. The total area is determined in Step 5A, even though a portion of the load applied to the building elevations will be transferred directly into the foundation, because Table 3.2-3 incorporates a 0.667 area reduction factor for this purpose. The 0.667 coefficient accounts for the presence of a parapet when determining design wind pressures. For a simply supported wall with no parapet, this value would be 0.50. However, the maximum value of 0.667 occurs when a wall spans 12 ft (3.7 m) or less and has the maximum parapet height permitted by Chapter 2. Table 3.2-3 is also based on the net Main Wind Force Resisting System pressures from Section 27.4 of ASCE 7. The calculations assume that the end zone pressures, which are greater than the interior zone pressures, would be applied over 40% of the area. In ASCE 7, the length of the end pressure coefficient zone (a) ranges up to 40% of the total length. Table 3.2-3 incorporates the maximum adjustment factor for building height and exposure (λ) as defined in ASCE 7 that could occur, based on Exposure Category and the maximum permitted height of 30 ft. Table 3.2-3 incorporates the wind importance factors associated with the Risk Categories. In addition, the greatest load factor required for wind cases in all of the required load combinations is 1.6. The values of C_w were therefore calculated by multiplying A times the 0.667 area reduction factor times the weighted average MWFRS pressure from ASCE 7 times the importance factor times the maximum λ that could occur times a 1.6 load factor and dividing the product by 2 since there are two lines of resistance required for each roof diaphragm.

The calculation of the projected area, A, for determining the lateral wind force in each principle plan direction is based on the full height of the building under consideration. Considering the building elevations shown in Figure C3.2-4, the projected area, A, for each direction is calculated as follows:

For north-south elevations: $A = (56 \text{ ft})(16.667 \text{ ft}) = 933.3 \text{ ft}^2$

For east-west elevations: $A = (10 \text{ ft})(40 \text{ ft}) + (20 \text{ ft})(6.667 \text{ ft}) = 533.3 \text{ ft}^2$

The values for A calculated above would in turn be multiplied by the appropriate C_w to determine the factored lateral wind force, $V_{LFRS\text{-}wind}$.

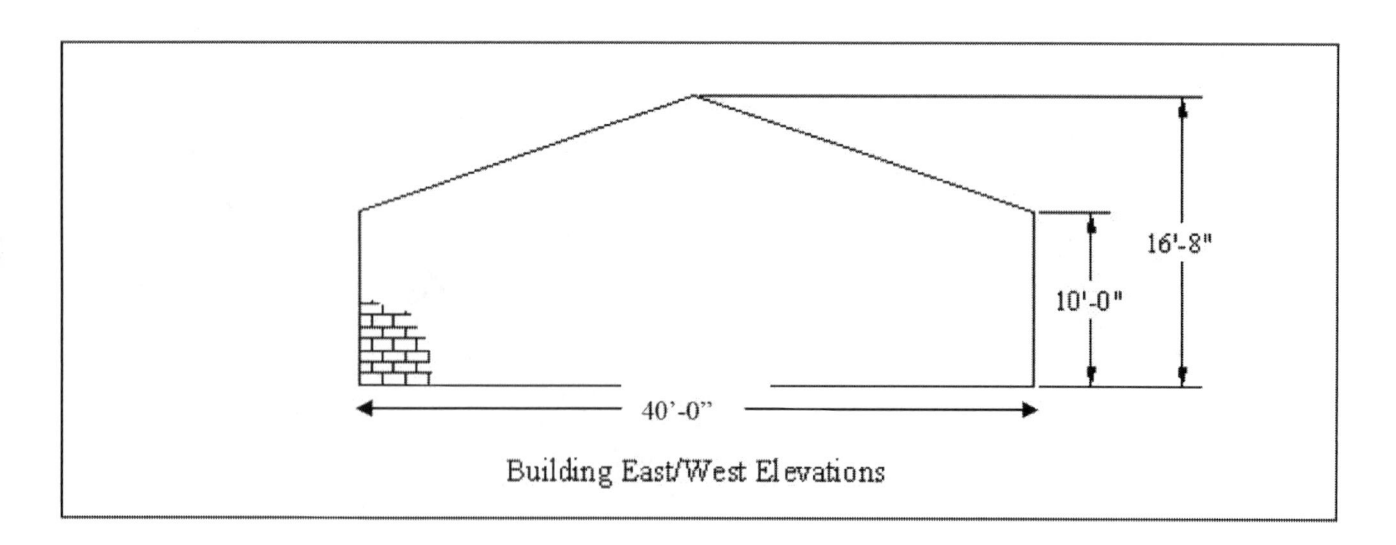

Building East/West Elevations

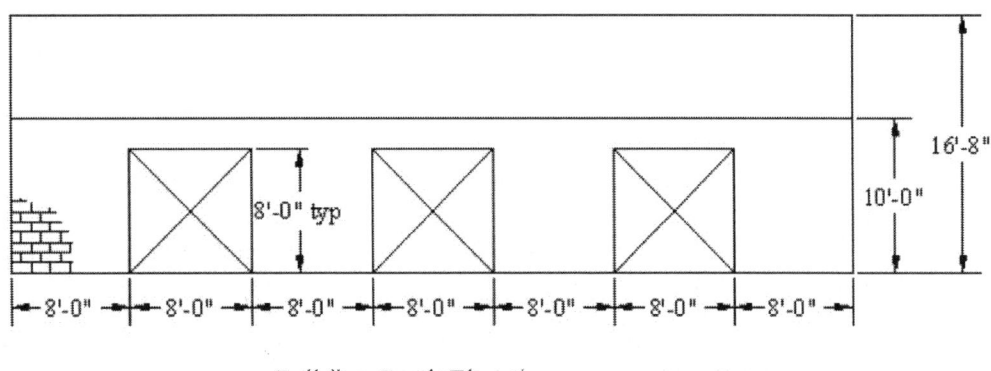

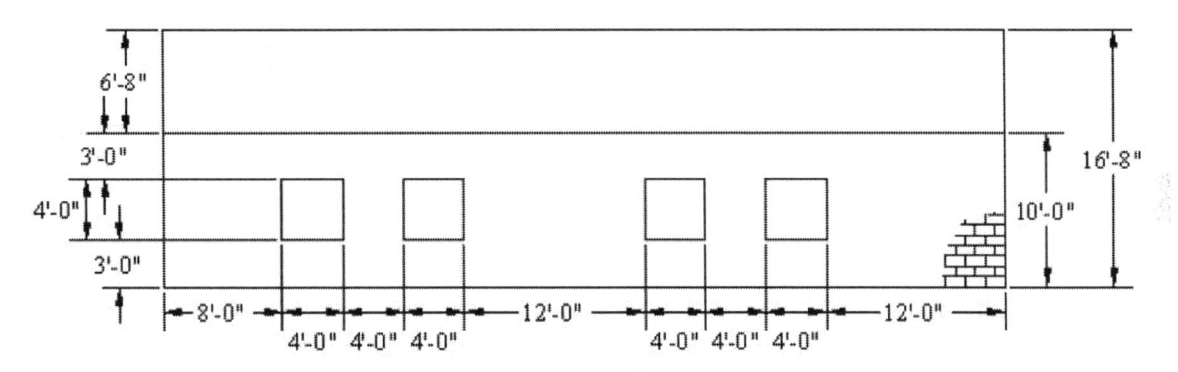

Figure C3.2-4: Example Building Elevations

Example calculation for C_w values provided in Table 3.2-3:

From Figure 28.6-1 of ASCE 7, the critical horizontal wind pressures applied to the walls for a 115 mph (51.4 m/s) basic wind speed are 29.0 lb/ft² (1.39 kPa) for the end zone and 19.4 lb/ft² (0.93 kPa) for the interior zone. For these loading conditions and an Exposure Category B structure, the value for C_w is calculated as follows:

C_w = [(0.6)(interior zone pressure) + (0.4)(end zone pressure)](area reduction factor)(λ)/2

C_w = [(0.6)(19.4 lb/ft²) + (0.4)(29.0 lb/ft²)](0.667)(1)/2 = 7.8 lb/ft²

To convert from lb/ft² to kPa, multiply by 0.0479.

Commentary on Step 5D

According to the requirements of Section 12.7.2 of ASCE 7, the effective seismic weight must include the roof dead load, the portion of the wall dead loads that transfer to the roof diaphragm, the dead load of any permanent equipment on the roof, and 20 percent of the uniform design snow load where the flat roof snow load is greater than 30 lb/ft² (1.4 kPa).

Table C3-1 and Table C3-2 in Chapter C3 of ASCE 7 provides the weights of many commonly used building materials. One could use these tables to sum the weights of the dead loads to determine the roof dead load in units of lb/ft^2 (kPa). Generally speaking, a bar joist roof with metal deck and a built-up minimum slope roof or a wood-framed roof with shingles on a pitch weighs approximately 10 to 15 lb/ft^2 (0.48 to 0.72 kPa); however, it is common for a higher dead load to be used to account for future roofing layers that may be added and for miscellaneous dead loads such as ceiling materials, ductwork, and lights. It is not uncommon for a dead load of 20 or 25 lb/ft^2 (0.96 to 1.20 kPa) to be used for such systems. However, be aware that pre-engineered metal building roof systems can be extremely light. It is not uncommon for a dead load of 2 lb/ft^2 (0.10 kPa) to be used in the design of pre-engineered metal buildings. It is possible for a masonry structure to have a Z-purlin and standing seam roof system with no added dead loads for some applications such as storage buildings or warehouses. Note that such a roof system would have to be through-fastened for it to be considered a simple diaphragm that meets the limitations of Chapter 2.

It may be reasonable to assign 0.667 of the wall heights to the effective seismic weight, to account for the effect of parapets as discussed in the Commentary above for calculating C_w. This should include all walls that are attached to the diaphragm without a detail permitting slip in the direction under consideration. Resources for the estimation of concrete masonry wall weights are available in NCMA TEK 14-13B[C.3.1].

The initial estimate of the seismic weight of the structure is based on an assumed spacing of the vertical and horizontal reinforcement, which may change from this initial assumption in subsequent steps of the direct design procedure. As one becomes more familiar with the direct design procedure, a more accurate initial estimate of reinforcement spacing (and therefore seismic weight) can be obtained for a given set of design variables. If there is uncertainty as to the quantity of required grout and reinforcement that may be necessary for a particular design, an initial estimate can be obtained by determining the maximum permitted spacing of the vertical reinforcement in accordance with Tables 3.2-5 and 3.2-6.

If the initially assumed seismic weight is less than that required by the resulting design, the design must be rechecked to verify that the resulting seismic design forces can be safely resisted.

Generally speaking, the flat roof snow load can be more than 30 lb/ft^2 (1.4 kPa) if the ground snow load is greater than 25 lb/ft^2 (1.2 kPa). The flat roof snow load can be calculated by Equation 7.3-1 of ASCE 7. The procedure in this *Handbook* is based on the assumption that all factors in this formula are taken equal to the maximum permitted values, which reduces this formula so that it is known that the flat roof snow load cannot exceed 1.21 times the ground snow load. If one did not want to calculate Equation 7.3-1 of ASCE 7, it would be conservative to assume the flat roof snow load is 1.21 times the ground snow load.

Commentary on Steps 5E through 5F

The values in Table 3.2-4 were determined by Section 12.8 of ASCE 7, according to the Equivalent Lateral Force Method, using the importance factor corresponding to each Risk Category. The legally adopted building code often requires that the seismic response modification coefficients, R, be shown on the construction documents. The response modification coefficients indicated in Section 3.2, Step 5E are taken from ASCE 7 Table 12.2-1 for the corresponding lateral force-resisting system listed.

Commentary on Step 5G

In this step, the greater lateral load will govern design for that designated roof diaphragm in that direction of consideration.

Commentary on Step 5H

This step provides the maximum permitted spacing of vertical reinforcement for wall strips with out-of-plane loading conditions. The selection of a permitted reinforcement schedule in accordance with this *Handbook* is based on the design assumptions, models, and inherent conservatism outlined in these provisions. Engineering analyses conducted in accordance with the provisions of the MSJC may result in slightly different solutions as a result of differing assumptions or less conservatism than employed by this *Handbook*.

Table 3.2-5 is based on a comprehensive load and resistance analysis incorporating the maximum loads that could occur within the limitations of Chapter 2. Each wall is designed individually, running all the load combinations in Section 2.3 of ASCE 7 and using the Strength Design method in Chapter 3 of the MSJC Code.

In the analyses, both a dead load of 2 lb/ft^2 (0.10 kPa) and a dead load of 30 lb/ft^2 (1.4 kPa) were checked for all load combinations.

In the analyses, the maximum possible snow load obtainable by Chapter 7 of ASCE 7 for each ground snow load case was determined to be the following, including the effects of snow drift accumulation against parapets. For a 0 lb/ft^2 (0 kPa) ground snow load, 0 lb/ft^2 (0 kPa) was used. For a 20 lb/ft^2 (1.0 kPa) ground snow load, 52 lb/ft^2 (2.5 kPa) was used. For a 40 lb/ft^2 (1.9 kPa) ground snow load, 65 lb/ft^2 (3.1 kPa) was used. For a 60 lb/ft^2 (2.9 kPa) ground snow load, 76 lb/ft^2 (3.6 kPa) was used. This is conservative and should account for drift from roof top units as well.

In the analysis, as discussed in the Commentary of this *Handbook* for Section 2.2.12, the wind pressures were determined using Chapters 28 and 30 of ASCE 7. The following wind load scenarios were analyzed for all load combinations:

>Positive Components & Cladding Wind Load on the Roof
>Negative Components & Cladding Wind Load on the Roof
>Positive Components & Cladding Wind Load on the Wall
>Negative Components & Cladding Wind Load on the Wall
>Positive Main Wind Force Resisting System Wind Load on the Roof and Wall
>Negative Main Wind Force Resisting System Wind Load on the Roof and Wall

For the Main Wind Force Resisting System wind pressures on the walls, 85% of the net pressures were assumed to be positive pressures and 70% of the net pressures were assumed to be negative pressures. This was based on the coefficients permitted in the Analytical Method of ASCE 7 because the simplified method tables in ASCE 7 provide net pressures, not MWFRS pressures on individual walls.

For wind design, interior pressures of 5.6 lb/ft^2 (0.27 kPa) for up to 110 mph (49.2 m/s) wind speed, 6.1 lb/ft^2 (0.29 kPa) for up to 115 mph (51.4 m/s) wind speed, 6.6 lb/ft^2 (0.32 kPa) for up to 120 mph (53.6 m/s) wind speed, 7.8 lb/ft^2 (0.37 kPa) for up to 130 mph (58.1 m/s) wind speed, 9.1 lb/ft^2 (0.44 kPa) for up to 140 mph (62.5 m/s) wind speed, 10.4 lb/ft^2 (0.50 kPa) for up to 150 mph (66.9 m/s) wind speed, 11.8 lb/ft^2 (0.56 kPa) for up to 160 mph (71.7 m/s) wind speed, 14.9 lb/ft^2 (0.71 kPa) for up to 180 mph (80.6 m/s) wind speed, and 18.4 lb/ft^2 (0.88 kPa) for up to 200 mph (89.4 m/s) wind speed were used for out-of-plane loads on interior walls. These pressures were compared to the out-of-plane seismic forces generated as a function of the weight of the wall, with the controlling load selected for determining the spacing of the vertical reinforcement.

For exterior walls, it was assumed there is a joist on one side of the wall only. For all analyses, it was assumed that the axial loads of exterior walls were applied at an eccentricity of 1.25 in. (31.8 mm) toward the interior from the centerline of the wall.

For interior walls, it was assumed there is a joist bearing on both sides of the wall. Although this may be conservative in some cases, it is intended to capture the majority of typical applications. To accomplish this, for all analyses, it was assumed that the full axial load associated with joists on both sides of the wall was applied but at an eccentricity of 0.625 in. (15.9 mm) to one side, which is half the exterior wall eccentricity. This models the moment associated with half the axial load applied at the full eccentricity, which could occur.

In reinforced masonry, the masonry must span horizontally to transfer loads to the vertical bars. The direct design procedure is based on an arch model, commonly referred to as arching action. The model is supported by research[C3.2]. This model complies with the Strength Design provisions of the MSJC and applies universally to vertical bars at any spacing, including bars greater than and less than six times the nominal wall thickness, and is described below.

In walls with vertical reinforcement resisting out-of-plane loads, for a free body diagram of a horizontal section of masonry that is as wide as the spacing of the vertical bars, with a vertical bar at the center, internal stresses are required to resist any unbalanced loading condition for there to be static equilibrium. An example of an unbalanced loading condition can occur at the jamb of a cased opening with no door, where there is masonry on only one side of the vertical bar. The masonry must be able to safely resist these internal stresses or the masonry would fail, causing the masonry to blow in between the vertical bars. This is not a commonly observed phenomenon. However, many designers limit the spacing of vertical reinforcement to six times the nominal wall thickness. For normal structures, it appears that it would be difficult for there to be sufficient horizontal pressure to cause a blow in failure with bars that close. Nonetheless, it is theoretically possible for a wall with sufficiently high out-of-plane pressures to fail in horizontal load transfer with vertical bars at six times the nominal wall thickness, before the vertical strips of masonry would fail. Therefore, masonry design should be based on a reliable mechanism of transferring loads horizontally to the vertical bars in the masonry.

The above mechanism must occur regardless of the spacing of the vertical bars. In addition, there is no limit on bar spacing for walls in Chapter 3 of the MSJC Code. (The limit of Section 1.9.6 pertains to the compressive width per bar due to shear lag.) Therefore, design for this mechanism can and should be universal for any vertical bar spacing imaginable, including more and less than six times the nominal wall thickness.

This mechanism cannot be modeled using reinforced masonry that incorporates bed joint reinforcement because the minimum yield strength of the wires required by the mandatory ASTM's referenced by the MSJC Specification exceeds the permitted specified yield strength for reinforcement according to MSJC Section 3.1.8.3. While the MSJC Code, and in turn this *Handbook*, do not prohibit the use of bed joint reinforcement, the contribution of bed joint reinforcement to the flexural and shear strength of a masonry assembly is neglected.

In masonry with vertical bars and no bed joint reinforcement, it has been observed that the masonry will be able to resist some out-of-plane pressure horizontally spanning as an unreinforced element until a vertical crack forms between the vertical bars and an arch is formed. The arch creates an in-plane thrust. The thrust is either resisted by an opposing thrust from a continuous masonry wall, or at discontinuous ends it is resisted by the section of masonry between the vertical crack and the end of the wall acting in-plane as a beam which transfers this load to horizontal tie beams which resist this horizontal tension.

The MSJC Code Section 3.2.1.3 states that "Unreinforced (plain) masonry members shall be designed to remain uncracked." A strict interpretation of this provision prohibits designing masonry as reinforced masonry to resist loads in one direction and then as unreinforced masonry to resist loads in another direction. Because vertical bars are required to resist out-of-plane bending or axial tension, the direct design procedure is based upon a reinforced masonry design model. In cases where reinforcement is required in only one direction (either horizontal or vertical) the masonry spanning in the perpendicular direction is designed as an arch in accordance with the design requirements of Section 3.3 (strength design of reinforced masonry) in the MSJC Code.

The MSJC Code Section 3.3.1 states that Section 3.3 "Reinforced Masonry" applies to "masonry design in which reinforcement is used to resist tensile forces". Because masonry is not relied upon to resist any tension in the arch model and all tension is resisted by reinforcement through the vertical bars resisting in-plane thrust and the horizontal bars acting as tie beams, there is no reason that the arched model cannot be classified as reinforced masonry and the provisions of MSJC Section 3.3 applied to the model. Further, there is no prohibition on such a model in the MSJC.

The strict interpretation referenced above also can be applied to shear walls. Therefore, for direct design, shear walls with any vertical reinforcement for structural purposes should be designed as reinforced masonry when resisting in-plane forces. The direct design procedure provides three LFRS Options (Ordinary Unreinforced Masonry Shear Walls, Ordinary Reinforced Masonry Shear Walls, and Specially Reinforced Shear Walls). Therefore, for walls with V bars, it can be assumed that there is at least one No. 5 (M#16) horizontal bar at 10 ft (3.0 m) on center, which is the minimum prescriptive horizontal reinforcement for Ordinary Reinforced Shear Walls. These H bars can be relied on to act as tension tie bars in the arch model of the mechanism, without adding any cost to the structure and permitting bars spaced further than 48 in. (1,219 mm).

Due to the numerous discontinuous conditions that will occur in real buildings, it is conservative and is a simplification to assume that all panels are discontinuous with respect to arching action loads, which creates the most tension in both V and H bars for that case, but continuous for out-of-plane and axial loading. Therefore the following interaction equation is appropriate for direct design:

(Moment due to out-of-plane and axial) / (ϕM_n out-of-plane)

+ (Moment due to in-plane-thrust) / (ϕM_n in-plane-thrust) ≤ 1.0.

In the analysis, the tallest height was determined for the reinforced wall systems that satisfied the following 16 conditions.

Condition No. 1: The ratio of factored load to gross area, not the net area, is required by Section 3.3.5.3 of the MSJC Code to be less than 20% of the specified compressive strength of masonry, which as discussed in the Commentary for Section 2.4.2 was assumed to be 1500 lb/in.^{2} (10.3 MPa).

Condition No. 2: The ratio of factored load to net area is required by Section 3.3.5.3 of the MSJC Code to be less than 5% of the specified compressive strength of masonry.

Condition No. 3: The factored axial load is required by MSJC Section 3.3.4.1.1 to be less than the nominal strength times a phi factor of 0.9.

Condition No. 4: The factored axial load and moment at the top of the wall are required to be within the permitted curve on a nominal moment-axial load capacity interaction diagram. This includes a tension check as well as compression check.

Condition No. 5: The factored axial load and moment at the middle of the wall, ignoring P-delta effects, are required to be within the permitted curve on a nominal moment-axial load capacity interaction diagram, accounting for phi.

DIRECT DESIGN HANDBOOK FOR MASONRY STRUCTURES

Condition No. 6: A P-delta analysis was performed with 10 iterations because the factored moment at the middle of the wall is required by Section 3.3.5 of the MSJC Code to be less than the nominal moment capacity times phi.

Condition No. 7: The out-of-plane deflection under service loads was determined based on service loads instead of factored loads by another *P*-delta analysis with 10 iterations because the service deflection is required by Section 3.3.5.5 of the MSJC Code to be less than 0.007 times the vertical span of the wall.

Condition No. 8: The out-of-plane deflection under service load from the previous condition was also checked for cases with components and cladding wind loads to be less than 0.010 times the vertical span of the wall because Section 1604.3.1 of the IBC[C3.3] requires that the out-of-plane deflection of such walls be limited to the vertical span divided by 240. Footnote f to Table 1604.3, referenced by Section 1604.3.1 of the IBC permits determining the deflection by the components and cladding wind load times a 0.42 reduction factor. 1/240 further divided by 0.42 equals 0.010.

Condition No. 9: The factored shear at the top of the wall is required by MSJC Code Section 3.3.4.1.2.1 to be less than the nominal shear strength times a phi factor of 0.8. Note that the shear was checked at the top of the wall because there is less axial load, and less axial load will be associated with less nominal shear capacity. Note also that the value for P_u in Equation 3-23 was included regardless of whether this load was downward or upward (due to wind uplift). Also note that $M_u/(V_u d_v)$ in that same equation was taken as 1.0 as permitted by that section of the MSJC.

Where:
d_v = actual depth of a member in direction of shear considered, in. (mm)
M_u = factored moment, in.-lb (N-mm)
P_u = factored axial load, lb (N)
V_u = factored shear force, lb (N)

Condition No. 10: The area of reinforcement was compared to the maximum area of reinforcement permitted by Section 3.3.3.5.1 of the MSJC Code, which is also required for out-of-plane wall design by 3.3.3.5.2, 3.3.3.5.3, and 3.3.3.5.4. The lambda values for wind pressures in Exposure C Categories vary depending on building height according to Figure 28.6-1 and Figure 30.5-1 of ASCE 7. The above analyses were performed for each wall assuming they span 30 ft (9.1 m). Then, if the maximum span permitted was less than 30 ft (9.1 m), the analyses were re-performed with the λ value associated with a building height equal to the maximum span permitted by the analysis. This process was based on the assumption that the bottom of the wall is at or very near the ground level. Structures where this is not the case, such as structures that are significantly elevated, should be evaluated independently.

Condition No. 11: For arching action between vertical bars, the tension was checked in the tie bars, assuming there is just one No. 5 (M#16) taking 10 ft (3.0 m) of tributary in-place thrust reaction, and assuming there are two discontinuous ends on each side of a panel that is as wide as the bar spacing, which represents a worst case scenario that could occur under the limitations and assumptions upon which this method is based.

Condition No. 12: For arching action, arch crushing of the masonry was checked.

Condition No. 13: With arching action taken into account, the tension in the vertical bars under biaxial bending and eccentric axial loading was checked using the conservative and simplifying interaction equation above.

Condition No. 14: The stability of the arch was checked to prevent snap-through buckling.

DIRECT DESIGN HANDBOOK FOR MASONRY STRUCTURES

Condition No. 15: The out-of-plane deflection of the horizontal span (the flat arch) was compared to the 0.007 times the horizontal span. This is similar to the vertical span deflection check above.

Condition No. 16: The out-of-plane deflection of the horizontal span (the flat arch) resulting from a components and cladding wind load was compared the serviceability check of 0.007 times the horizontal span. This is similar to the vertical span deflection check above.

The values in Table 3.2-6 were determined by a similar procedure. Table 3.2-6(1) applies to walls constructed of lightweight concrete masonry units. Table 3.2-6(2) applies to walls constructed of medium weight or normal weight concrete masonry units. In accordance with ASTM C90, lightweight concrete masonry units are required to have a density less than 105 lb/ft^3 (1680 kg/m^3). In order to establish an upper limit on unit density for use with Table 3.2-6(2), units are not permitted to have a density in excess of 135 lb/ft^3 (2162 kg/m^3) when used to construct masonry designed in accordance with this *Handbook*.

The increased weight of medium weight and normal weight concrete masonry units compared to lightweight concrete masonry units increases the corresponding out-of-plane loading due to seismic response. These larger seismic loads are accounted for in the permitted wall heights and reinforcement schedules provided in Table 3.2-6. Although the increased dead load of walls constructed of medium weight and normal weight concrete masonry units could be used to offset wind uplift and overturning effects, such modeling is conservatively neglected in the direct design procedure. The permitted wall heights of Table 3.2-6(b) were determined by increasing the installed weight of the masonry wall by 8 lb/ft^2 (0.38 kPa) over the corresponding lightweight concrete masonry unit wall weight. This increase in wall weight corresponds to a unit density of 135 lb/ft^3 (2162 kg/m^3), which captures the majority of normal weight concrete masonry units and can be conservatively applied to any medium weight concrete masonry unit.

Commentary on Step 5I

In this *Handbook*, a designated shear wall segment is defined as a portion of wall that is continuous in plan and uninterrupted from the foundation to the diaphragm elevation. A designated shear wall segment cannot contain an opening, a control joint, an expansion joint, or a construction joint. Because designated shear wall segments with shorter plan lengths typically require more reinforcement, it may be more economical to treat short wall segments as non-designated shear walls or as non-participating walls, rather than as designated shear wall segments.

The values of k_1 and k_2 depend on L_{seg}, the selected LFRS Option, and the spacing of the horizontal and vertical reinforcement. The wall reinforcement schedule requires that the designer specify the spacing of the V Bars, S_V.

The wall reinforcement schedule requires that the designer specify the spacing of the H Bars, S_H, and the grouting schedule. Walls with either vertical or horizontal bars spaced at 16 in. (406 mm) or less are assumed to be fully grouted. Walls with reinforcement at a greater spacing are assumed to be partially grouted. In some cases, a lower value of S_H may increase the values of k_1 and k_2.

L_{req} decreases as the resistance coefficients increase. If the sum of the plan lengths of designated shear wall segments on each line of resistance is less than L_{req}, either select a smaller S_V than the one selected in Step 5I.3, or select an LFRS Option with more prescriptive reinforcement than the one chosen in Step 5E, or revise the preliminary floor plan created in Step 2, and redo the procedure from the earliest revised Step.

The in-plane design of masonry shear walls is provided for in Table 3.2-8. Three different groups are provided, each inherently complying with the minimum prescriptive reinforcement, design, and detailing requirements for Ordinary Plain (Unreinforced) Masonry Shear Walls, Ordinary Reinforced Masonry Shear

Walls, and Special Reinforced Masonry Shear Walls per the provisions of the MSJC Code. As such, detailing options that may be permitted, for example for Ordinary Plain Reinforced Masonry Shear Walls, may not be permitted for Special Reinforced Masonry Shear Walls and so on.

The design coefficients k_1 and k_2 are based on the limiting shear and flexural strength, respectively, per the strength design requirements of the MSJC for the wall configuration and reinforcement detailing listed. Based upon the critical design shear load for each wall configuration, the inelastic story drift (elastic deformation multiplied by the appropriate C_d value) is also checked for each wall configuration. In all cases, the permitted deflection is limited to 1 percent of the wall height. Further, in the case of the reinforced shear walls, the maximum tensile reinforcement limits are also checked. Only those walls configurations that provide a k_1 and k_2 value meet all of the strength design requirements per the MSJC Code. Shear wall design coefficients listed as not permitted (NP) in Table 3.2-8 either exceed the permitted strength of the wall segment, exceed the in-plane deflection permitted by the MSJC Code or ASCE 7, or exceed the permitted maximum flexural tension reinforcement values allowed by the MSJC Code for the specific design conditions considered by each table. The units for k_1 are force per unit length because the value for k_1 must be multiplied by the wall height to obtain a shear wall segment length design strength. The units for k_2, which are not multiplied by wall height, are force.

For simplicity, the factored shear force, V_u, is limited to 40 percent of the nominal shear strength, V_n, for shear critical special reinforced masonry shear walls to meet the requirements of Section 1.18.3.2.6.1.1 of the MSJC Code. Likewise, for flexurally dominated special reinforced masonry shear walls, the limiting design shear strength, ϕV_n, is checked against the shear corresponding to the development of 125 percent of the nominal flexural strength. Limiting the design shear strengths based on these criterion is conservative. More economical designs can be achieved through a more rigorous analysis taking into account specific design requirements.

Note that Section 3.2.1.3 of the MSJC Code requires that unreinforced masonry members be designed to remain uncracked. The lateral force coefficients provided in Table 3.2-8 for Ordinary Plain (Unreinforced) Masonry Shear Walls were therefore determined for the uncracked sections. The tables show reinforced options for these systems only because the presence of reinforcement is permitted, even though it has no explicit structural minimum building code purpose. Reinforcement may be desired for redundancy against progressive collapse, for security of valuables, for detention of inmates, for future uses, and for emergency hurricane and/or tornado shelters.

The selection of the minimum shear wall segment length can have a significant impact on the required horizontal and vertical reinforcement requirements and resulting design economy. In general, the longer the length of L_{seg} selected for a given lateral force, the less reinforcement required to resist the in-plane load. Consider the north elevation shown in Figure C3.2-5, which consists of one 8'-0" (2.4 m) wall segment; two 4'-0" (1.2 m) wall segments; and two 12'-0" (3.7 m) wall segments. Because Section 2.2.6 requires that a designated shear wall segment be provided at each corner of the roof diaphragm, the selection of which wall segments are permitted to be designated shear wall segments is limited. In the case of the elevation shown in Figure C3.2-5, the designated shear wall segment length for this line of resistance cannot exceed 8'-0" (2.4 m) as this is the controlling length set by the left corner of the building. Permitted shear wall segment length options include:

- L_{seg} = 8'-0" (2.4 m): In this scenario, the two 4'-0" (1.2 m) wall segments would be classified as non-designated shear wall segments and their in-plane strength subsequently neglected. The line of resistance would then consist of three designated shear wall segments each having a design length of 8'-0" (2.4 m). Under this option, the strength assigned to each of the 12'-0" (3.7 m) walls is conservatively assumed to be taken equal to an 8'-0" (2.4 m) shear wall length. Therefore, the resulting total effective designated shear wall length of the three designated shear wall segments considered under this option would be taken equal to 24'-0" (7.3 m) when verifying the required length of designated shear wall under Step 5I.4.

DIRECT DESIGN HANDBOOK FOR MASONRY STRUCTURES

- L_{seg} = 4'-0" (1.2 m): In this scenario, all five wall segments are classified as designated shear wall segments providing a total effective shear wall length of 20'-0" (6.1 m) when checking the required length of designated shear wall under Step 5I.4. Compared to the first scenario, classifying all of the wall segments as designated shear walls would likely result in a more conservative design as more reinforcement would be required under this option to provide a strength equivalent to the first option. If the elevation contained several more 4'-0" (1.2 m) wall segments, however, it is possible that this scenario could prove to be more economical that the first scenario.

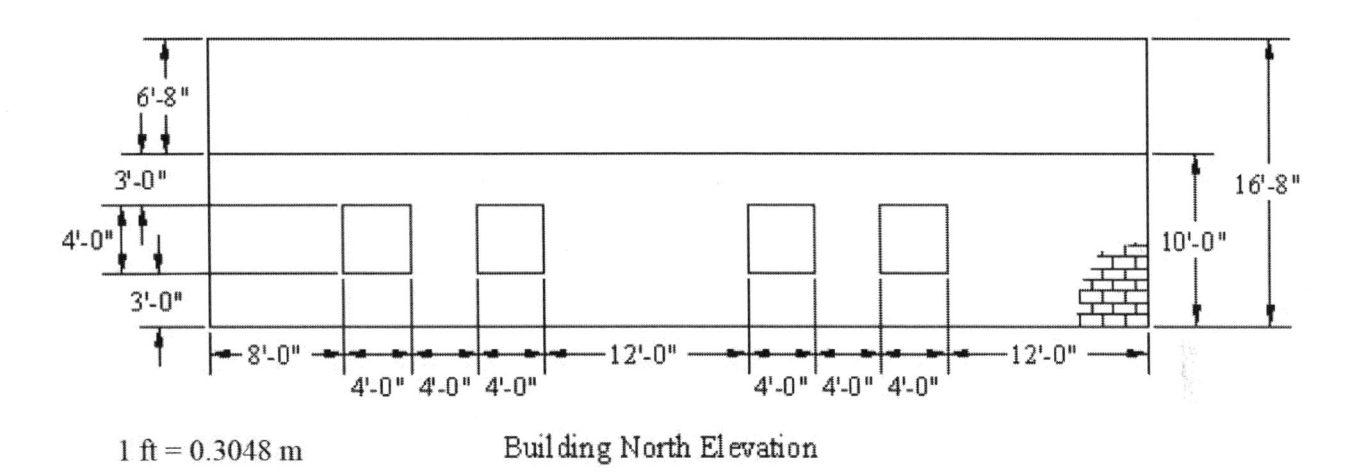

Figure C3.2-5: Example Building Elevation

Commentary on Step 5I.3

Table 3.2-8 only provides discrete values for L_{seg} that may not coincide with the actual length of the designated shear wall segment selected. For example, Table 3.2-8 provides resistance coefficients for shear wall segment lengths of 4 ft (1.2 m) and 8 ft (2.4 m), but no resistance coefficients for 6 ft (1.8 m) shear wall segment lengths. Because linear interpolation is not permitted with Table 3.2-8, this *Handbook* requires that the resistance coefficient k_2 be selected using a tabulated shear wall segment length smaller than the actual length of the designated shear wall segment length.

Commentary on Step 5J

Non-designated shear walls are attached to the diaphragm so that lateral load parallel to the wall will be transferred onto the wall, however, they are not designated as shear walls so that a longer L_{seg} may be used in design, ignoring the smaller wall, so that a more economical wall may be achieved. For example, if there is a large building with many 20 ft (6.1 m) long shear walls but one 2-foot (610 mm) long shear wall, it would be very conservative to use the tables with L_{seg} = 2 ft (610 mm), which would be assuming all shear wall segments have the same stiffness and capacity as the shorter wall segment. By analysis of model structures with different shear wall layouts, it was determined that reinforcing the non-designated shear walls the same as the designated shear wall segments was still conservative.

Commentary on Step 5K

Reinforcement requirements for non-participating walls are governed by out-of-plane wind and seismic loading. Mechanically attaching a non-participating wall segment to the lateral force-resisting system would likely violate the requirement that non-participating walls be isolated from the lateral force-resisting system. One option is to provide a detail that supports the non-participating wall while permitting differential displacement between the top of the wall and its support so as to prevent the transfer of in-plane loads to the non-participating wall. Depending upon the project-specific variables, however, such details may be more costly compared to providing a conventional mechanical attachment, reclassifying the non-participating wall as a participating wall, and providing the necessary reinforcing details required for a participating wall.

Figure C3.2-6 illustrates one method of connecting a non-participating wall to the roof construction. Support is provided at the top and bottom of the wall to resist out-of-plane loads while preventing in-plane loads from being transferred to the non-participating wall. The isolation joints between the nonparticipating walls and the structure must be designed to accommodate the design story drift and vertical deflection from the roof above.

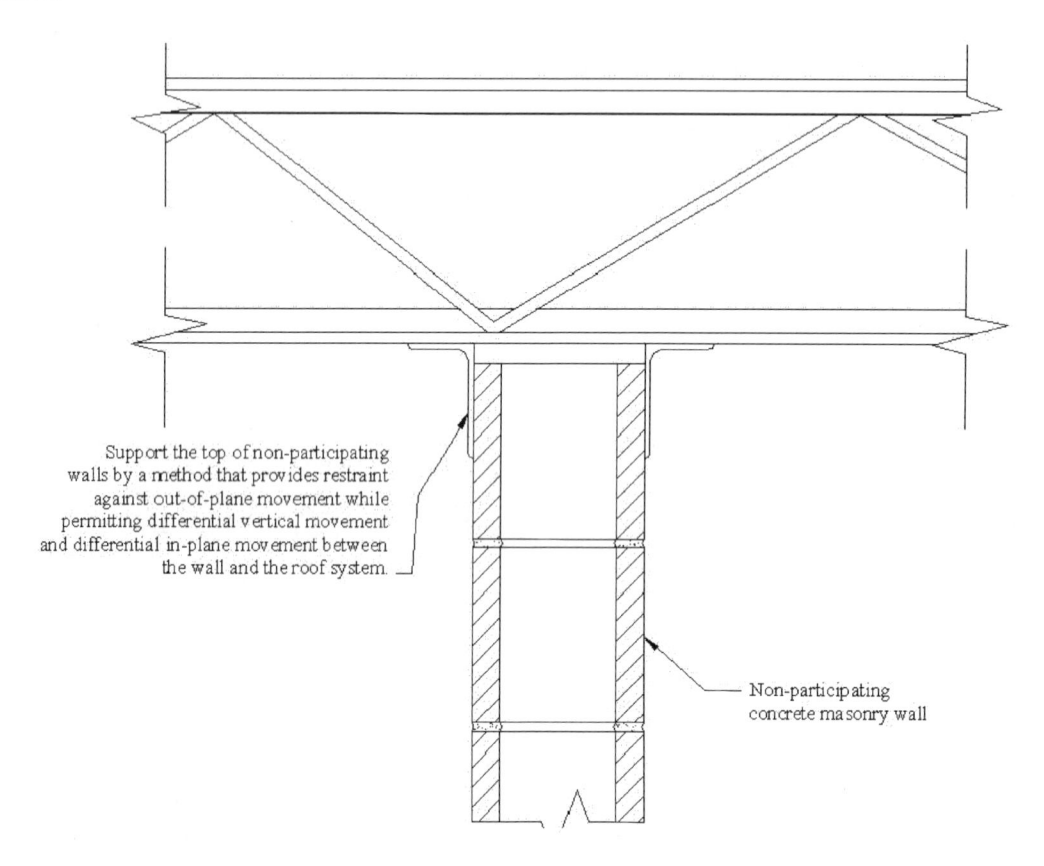

Figure C3.2-6 – Non-participating Wall Detailing Option

Commentary on Step 5L

Horizontal bond beams are assumed to act as tension and compression chord elements for the roof diaphragm. Table 3.2-9 accounts for both the critical seismic conditions and wind conditions by using V_{LFRS}, which is the greatest of $V_{LFRS\text{-}seismic}$ and $V_{LFRS\text{-}wind}$. The values in Table 3.2-9 are based on an analysis that assumes there is a uniformly distributed load applied along the length of the simple span of the flexible roof diaphragm, oriented in the plane of the roof diaphragm. The magnitude of the uniformly distributed load is assumed to be equal to V_{LFRS} times two divided by the length of the roof diaphragm. The maximum tension in the C Bars is equal to the maximum moment divided by the distance separating the tension and compression chords. In the resulting equation, length terms cancel each other out so that it is only essential to know the aspect ratio of the roof diaphragm plan dimensions and V_{LFRS} to determine the minimum number of C Bars required. Users are required to conservatively apply the critical conditions for both directions and use the same number of C Bars on all four sides of the diaphragm.

Consider the plan view layout of three Diaphragms A, B, and C shown in Figure C3.2-7 that share a common shear wall. Using Table 3.2-9 it is determined that Diaphragm A requires two courses (four No. 5 (M#16)) C Bars and Diaphragms B and C each require one course (two No. 5 (M#16)) C Bars. For the interior shear walls connected to more than one diaphragm, only the critical number of C Bars is required to be provided; not the cumulative number of C Bars required for all common shear walls. In this example, two courses (four No. 5 (M#16)) C Bars would be required in the shear wall common to all three diaphragms and one course (two No. 5 (M#16)) C Bars would be required in the shear wall common to Diaphragms B and C.

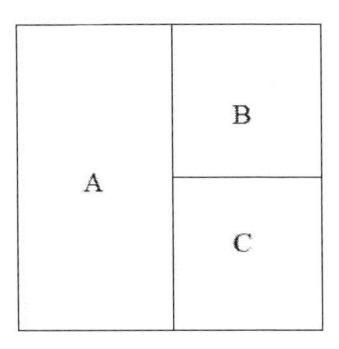

Figure C3.2-7 – Internal Shear Wall Layout

Commentary on Step 6

For walls with different diaphragm elevations on each side, it is necessary to make sure that the required shear walls are provided individually for each diaphragm system because a transfer mechanism would be required if a diaphragm was not attached to the shear walls designed to resist the lateral loads for that diaphragm.

Commentary on Step 7

Section 3.3.3.5 of the MSJC Code establishes maximum reinforcement limits (ρ_{max}) for elements subjected to in-plane and out-of-plane loads. These limits are included in the analyses used to generate the tables of this *Handbook*.

In checking the maximum reinforcement limits for walls subjected to in-plane loads, the MSJC Code permits the vertical reinforcement in the compression zone of the wall to be taken into consideration even if that vertical reinforcement is not laterally tied. The procedures of this *Handbook* take into consideration the vertical reinforcement in the compression zone in verifying compliance with Section 3.3.3.5 of the MSJC Code. This *Handbook* further requires that the vertical reinforcement be symmetrically arranged about the mid-length of the wall to help balance the vertical reinforcement in the compression zone with the vertical reinforcement in the tension zone. This prescriptive reinforcement detailing requirement may only be partially effective at balancing the reinforcement about the neutral axis, depending upon the location of the neutral axis along the length of the wall.

In checking the maximum reinforcement limits for walls subjected to out-of-plane loads, however, there is no vertical reinforcement located in the compression zone of the wall's cross-section. As such, the vertical reinforcement cannot be balanced about the neutral axis in the out-of-plane direction. Using the procedure in this *Handbook*, reinforcement is not permitted to be spaced closer than 16 in. (406 mm) as this would violate the maximum reinforcement limits of the MSJC Code. This limitation does not apply to E bars, which may be located within 8 in. of J Bars and V Bars, because the E bars are the primary load path reinforcement for the header panels and are not relied upon to resist the out-of-plane loads applied to the wall.

To reduce the number of iterations required to determine a design solution that satisfies all of the requirements of this *Handbook*, it may be helpful to note that the design of lintels over openings in accordance with Step 9B may require a minimum bearing length greater than 8 in. (203 mm) for large openings or roof loads. As the lintel bearing length on either side of an opening increases, the first J Bar must be shifted further away from the opening accordingly.

Commentary on Step 8

Vertical bars are required in header panels with parapets to accommodate the out-of-plane moment created by the parapet.

Commentary on Step 9

There will be uplift forces as well as downward forces acting on the header panel. The grouting and reinforcement at the bottom of the header panel per Table 3.2-10 was designed for the downward forces. It was determined that the critical design uplift forces were always less than the critical design downward forces. Therefore, providing the same reinforcement and grouting for reverse curvature bending should be sufficient. It is permitted for the grouting in compression in the positive moment case to overlap the grouting in compression in the negative moment case because the two cases do not occur simultaneously.

O Bars are required at header panels because the panels are assumed to be reinforced cracked sections when resisting the downward and upward forces on the header panel. Section 3.2.1.3 of the MSJC Code requires that unreinforced masonry members be designed to remain uncracked.

Commentary on Step 10

The legally adopted building code often requires that the design base shear be shown on the project plans. The design tables of this *Handbook*, however, are not based on the total seismic base shear, but only the lateral seismic load applied at the top of a wall, which is based on the applicable roof loads and the dead load of the upper half of the walls. For the total base shear required to be shown on the plans, the total effective seismic weight of the structure (W_{tot}) must be calculated. This quantity is in turn multiplied by $2C_s$ to determine the total base shear for the structure. Because the design tables of this *Handbook* quantify in-plane seismic design loads for each line of resistance, instead of total base shear, the value of C_s from Table 3.2-4 must be doubled.

Commentary on Steps 11 and 12

Refer to the Chapter 5 and Chapter 6 commentary discussion for additional information on specification requirements and standardized detailing, respectively.

References

C3.1 NCMA TEK 14-13B, *Concrete Masonry Wall Weights*, National Concrete Masonry Association, Herndon, VA, 2008.

C3.2 McGinley, W. M., *Spacing of Reinforcing Bars in Partially Grouted Masonry*, National Concrete Masonry Association Education and Research Foundation, Herndon, VA, 2007.

C3.3 International Building Code (IBC), International Code Council, Falls Church, VA, 2012.

This Page Intentionally Left Blank

Commentary Chapter 4
Clay Masonry (Future)

The direct design procedure currently applies only to concrete masonry construction for reasons discussed in Commentary Section C2.4.1. Chapter 4 of the *Handbook* is intended to serve as a placeholder until such time as provisions applicable to clay masonry are developed.

This Page Intentionally Left Blank

Commentary Chapter 5
Specification

Chapter 5 defines the required materials in order to use the design provisions of this *Handbook*, and lists the issues that require a decision by the designer in order to comply with the MSJC Code and Specification. The designer may choose to add additional or more stringent requirements to the project specification as permitted by the MSJC Specification. Specifying the use of an integral water repellent or surface-applied coating is one example of an additional requirement that the designer might consider.

The MSJC *Specification* is a reference specification. The Mandatory Requirements Checklist of the MSJC Specification lists the choices that must be made by the designer, and the Optional Requirements Checklist lists the choices that are permitted to be made (but do not have to be made) if the designer wishes to invoke a requirement other than the default requirement where such a choice is permitted.

The combination of requiring Type S mortar and concrete masonry units complying with ASTM C90 stipulated by this *Handbook* is intended to provide a minimum compressive strength of the masonry that will equal or exceed 1,500 lb/in.^{2} (10.3 MPa). For varying reasons, some projects or jurisdictions may prefer the use of the prism test method over the unit strength method for verifying conformance with the specified compressive strength of masonry. Because the method of verifying masonry compressive strength does not affect the resulting design strength, provided that the compressive strength of the masonry equals or exceeds 1,500 lb/in.^{2} (10.3 MPa), either option is permitted with no modification required to the analysis procedure.

This Page Intentionally Left Blank

Commentary Chapter 6
Details

The detailing requirements covered by this *Handbook* address minimum requirements for life safety. Many construction details that are common to many masonry structures are not explicitly addressed by this *Handbook* including modular layout, weeps and flashing, and control joints for crack control. Users are directed to other publications that offer guidance in this area, including:
- *Masonry Designers' Guide, 6^h Edition*, The Masonry Society, Longmont, CO, 2010. www.masonrysociety.org
- *Annotated Design and Construction Details for Concrete Masonry*, National Concrete Masonry Association, Herndon, VA, 2003. www.ncma.org

Commentary on Figure 6.1-1 – Masonry Lintels

Figure 6.1-1 illustrates a typical masonry lintel constructed of concrete masonry units. Temporary shoring contains the grout, if solid bottomed units are not used for lintel construction. Weep holes should be spaced at a maximum of 32 in. (813 mm) on top of lintel.

Concrete masonry lintels are sometimes constructed as a portion of a continuous bond beam. This construction provides several benefits: it is considered to be more beneficial in high seismic areas or in areas where high winds may be expected; control of wall movement due to shrinkage or temperature differentials is more easily accomplished; and lintel deflection is often substantially reduced. Concrete masonry lintels have the advantages of easily maintaining the bond pattern, color and surface texture of the surrounding masonry. They can also be placed in the wall without the need for special lifting equipment, which is common for precast items. Shear reinforcement, usually in the form of stirrups, is used to control diagonal cracking in the lintel. As required by the MSJC Code, hooks for stirrups incorporate either 90 or 135 degree bends, with minimum extensions of either six bar diameters or 2.5 in. (64 mm), the latter of which controls for the No. 3 (M#10) stirrups required by this *Handbook*.

Commentary on Figure 6.1-2 – Placement Tolerances for Reinforcement

In accordance with the requirements of Chapter 1 for compliance with the provisions of the MSJC Code and Specification, reinforcement for masonry designed in accordance with this *Handbook* must conform to the placement tolerances stipulated in the MSJC Specification. In accordance with the MSJC Specification, the tolerance for the placement of reinforcement in walls and other flexural elements is ± ½ inch (13 mm) when the specified distance (*d*), measured from the centerline of the reinforcement to the opposite compression face of the masonry, is 8 in. (203 mm) or less. The tolerance increases to ± 1 inch (25 mm) for *d* equal to 24 in. (610 mm) or less but greater than 8 in. (203 mm). For *d* greater than 24 in. (610 mm), the tolerance for the placement of reinforcement is ±1 ¼ in. (32 mm).

Vertical bars must be placed within 2 in. (51 mm) of their specified location measured parallel to the length of the wall for all applications. The placement tolerances for such reinforcement are larger because slight deviations from specified locations have a negligible impact on the structural performance of an assemblage.

To facilitate the placement of reinforcement and achieve the required placement tolerances, reinforcing bar positioners may be used for both horizontal and vertical reinforcement, although bar positioners may hinder high lift grouting procedures. Reinforcing bar positioners are not required by the MSJC Code or Specification.

Commentary on Figure 6.1-3 – Lap Splices of Reinforcement

Using the material properties and placement conditions required by this *Handbook*, the minimum length of lap for spliced No. 5 (M#16) reinforcement is 26 in. (660 mm) as shown in Figure 6.1-3.

Commentary on Figure 6.1-4 – Details of Flanged Wall Intersections

The design of wall intersections generally falls into one of two categories; those in which shear is designed to be transferred between two intersecting walls and those in which shear is prevented from being transferred from one wall to another. Detailing for shear transfer between intersecting walls can substantially increase the flexural and axial load capacity of two intersecting walls, and is required by this *Handbook* for shear walls located at the corners of diaphragms.

When the design relies upon two intersecting walls to act compositely to resist applied loads, the MSJC Code stipulates three options to transfer stresses from one wall to the other, each requiring the masonry to be laid in running bond. When any of these conditions are not met, the transfer of shear forces between walls is required to be prevented.

Option A: Walls are constructed such that 50 percent of the units interlock at the interface. This option, while easily accomplished at corners, results in bond interruption at 'T' intersections. As such, it is generally good practice to install a control joint in the flange wall to minimize cracking at this location, unless horizontal reinforcement or other detailing is provided to eliminate the need for a control joint. If a control joint is constructed, the portion of the flange wall separated from the intersection by the control joint generally should not be considered effective in resisting applied loads from the web wall.

Option B: Walls are anchored together by steel connectors spaced at vertical intervals not exceeding 48 in. (1,219 mm) on center. While not required by Code, it is generally good practice to construct a control joint (to minimize cracking potential) at the intersection of two walls anchored in such a manner.

Option C: Bond beams are incorporated into the intersecting walls. The bond beams are required to contain at least 0.1 in.^{2} of reinforcement per foot (211 mm^2/m) of wall height, and be spaced no further than 48 in. (1,219 mm) on center vertically.

Commentary on Figure 6.1-5 – Ordinary Reinforced Masonry Shear Walls Prescriptive Reinforcement

Ordinary reinforced masonry shear walls, which are designed in accordance with reinforced masonry procedures, rely upon the reinforcement to carry and distribute anticipated tensile stresses, while the masonry carries the compressive stresses. Although such walls contain some reinforcement, to ensure a minimum level of performance during a design level earthquake, a minimum amount of prescriptive reinforcement is also mandated by the MSJC Code. The reinforcement required by design may also serve as the minimum prescriptive reinforcement, which entails:

Vertical Reinforcement: The prescriptive vertical reinforcement is required to consist of at least one No. 5 bar (M #16) at each corner, within 16 in. (406 mm) of each side of openings, within 8 in. (203 mm) of each side of control joints, within 8 in. (203 mm) of the ends of walls, and at a maximum spacing of 120 in. (3,048 mm). Although the MSJC Code only requires No. 4 (M#13) bars to comply with the minimum prescriptive seismic reinforcement requirements, this *Handbook* opts to require the use of No. 5 (M#16) reinforcing bars for prescriptive seismic reinforcement for consistency with the provisions of Chapter 1 and ease of detailing.

Horizontal Reinforcement: The minimum prescriptive horizontal reinforcement consists of at least two wires of wire size W1.7 (MW 11) joint reinforcement spaced not more than 16 in. (406 mm) on center or bond beams containing no less than one No. 5 (M #16) bar spaced not more than 120 in. (3,048 mm) apart. Horizontal reinforcement is also required at the bottom and top of wall openings. Such reinforcement shall extend at least 25 in. (635 mm) past the opening. Structural reinforcement located at roof and floor levels is required to be continuous. The horizontal reinforcing bar located closest to the top of the wall shall be placed within 16 in. (406 mm) of the top of the wall.

Neither horizontal nor vertical prescriptive reinforcement is required for openings smaller than 16 in. (406 mm) in either the horizontal or vertical direction, unless the required prescriptive reinforcement is interrupted by such openings.

Commentary on Figure 6.1-6 – Special Reinforced Masonry Shear Walls Prescriptive Reinforcement

The prescriptive reinforcement for special reinforced masonry shear walls is required to comply with the requirements for ordinary reinforced masonry shear walls and the following:

The sum of the cross-sectional area of horizontal and vertical reinforcement shall be at least 0.002 times the gross cross-sectional area of the wall, and the minimum cross-sectional area in each direction shall be not less than 0.0007 times the gross cross-sectional area of the wall. The maximum spacing of vertical and horizontal reinforcement shall be the smaller of one-third the length of the shear wall, one-third the height of the shear wall or 48 in. (1,219 mm) and shall be uniformly distributed. The minimum cross-sectional area of vertical reinforcement shall be one-third of the required horizontal reinforcement. Horizontal reinforcement required to resist shear forces or part of the minimum prescriptive seismic reinforcement is required by the MSJC Code to be anchored around the vertical reinforcement with a standard hook.

For clarity, Figures 6.1-5 and 6.1-6 illustrate only the prescriptive seismic detailing provisions as required for Ordinary and Special Reinforced Masonry Shear Walls by the MSJC Code. Designs complying with this *Handbook* will likely require additional detailing or reinforcement not shown in Figures 6.1-5 and 6.1-6. This may include O Bars above or below openings, J Bars when vertical reinforcement is interrupted by openings, additional control joints on both sides of openings, or T Bars or B Bars when a parapet is present.

No Commentary on Figure 6.1-7

This Page Intentionally Left Blank

DIRECT DESIGN HANDBOOK FOR MASONRY STRUCTURES

Forward to the Appendix

The design example that follows does not form a part of the *Handbook*. The purpose of this non-mandatory appendix is to familiarize the user with the proper procedure for designing a structuring using the direct design procedure. This design example is not a substitute for reading and understanding the requirements of the direct design procedure. For clarity, SI equivalents are not shown in this design example.

Appendix
Direct Design Procedure Design Example
Retail Center in St. Louis, MO

1. From ASCE 7:

    Risk Category = II

    p_g = 20 psf

    V = 115 mph

    Exposure Category = B

    S_s = 0.50

    S_1 = 0.15

2. The site-specific limitations of Section 2.1 are met. Assume Site Class D for this example. If a geotechnical investigation will not be performed at the site, the default site classification should be supported by local geologic maps and/or other geotechnical investigations.

3. From Table 3.2-1(4), SDC = C

4. From Table 3.2-2, LFRS options = Ordinary Reinforced Masonry Shear Walls (ORMSW)

5. Only one rectangular diaphragm. Analyze each principal plan direction.

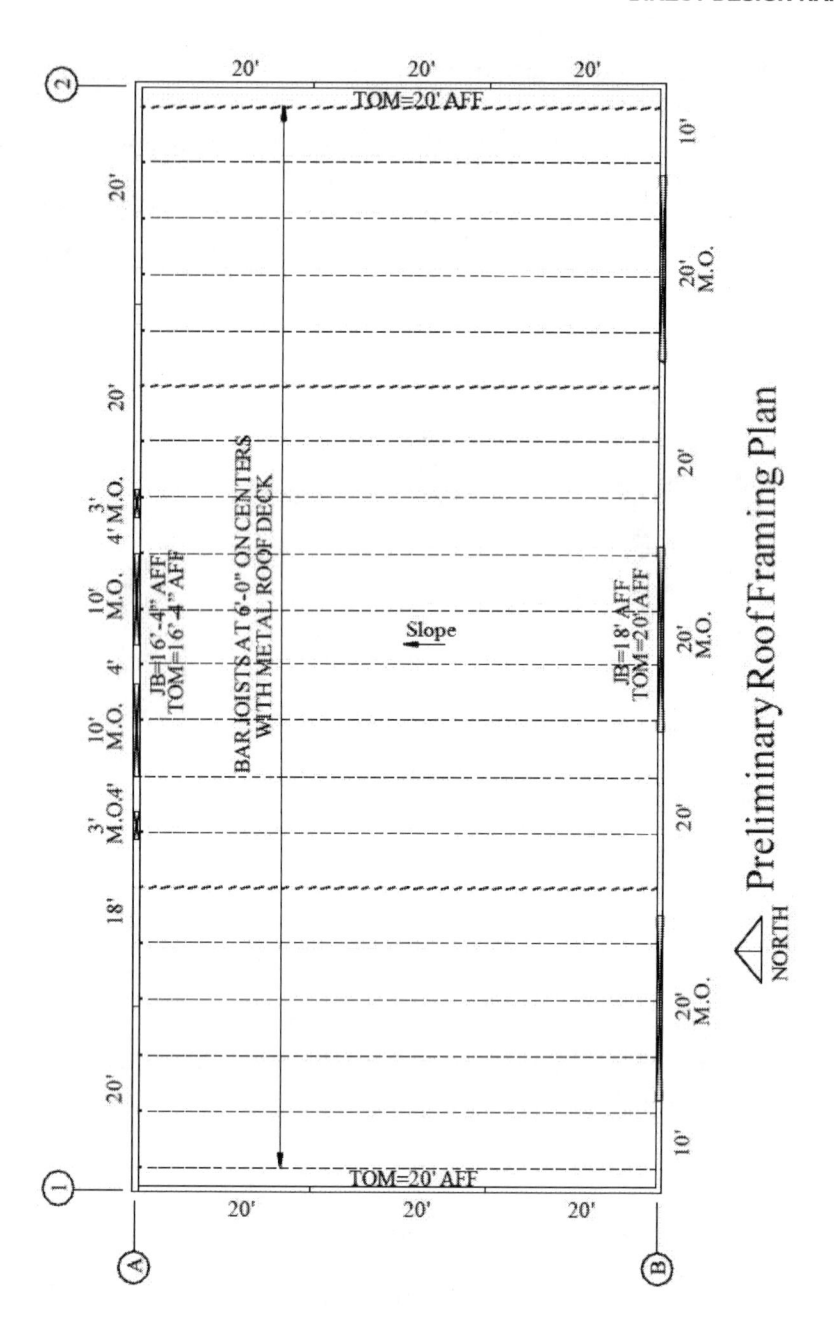

NORTH-SOUTH WIND AND SEISMIC LOADING DIRECTION:

5A A = 120 ft x 20 ft = 2,400 ft²
5B   From Table 3.2-3, C_W = 7.8 lb/ft²
5C $V_{LFRS\text{-}wind}$ = $(C_W)(A)$ = (7.8 lb/ft²)(2,400 ft²) = 18,720 lb
5D   Calculate W

W is the effective seismic weight that imparts load onto each line of resistance on each side of the diaphragm.

DIRECT DESIGN HANDBOOK FOR MASONRY STRUCTURES

For the roof itself, using a roof dead load of 20 lb/ft² multiplied by the area of the roof assigned to each line of resistance, the effective weight is (20 lb/ft²) (120 ft / 2) (60 ft) = 72,000 lb.

For the walls, using an estimated value of 50 lb/ft² for the wall weight, it is conservative to assign 12 ft of wall height to the diaphragm all around the building. Therefore the effective weight assigned to each line of resistance is (50 lb/ft²) (120 ft + 60 ft) x 12 ft = 108,000 lb.

Assuming no interior non-participating elements will impart effective load, W = 72,000 lb + 108,000 lb = 180,000 lb. This assumption would be valid in an open store with no interior non-participating elements; however, other types of stores may require that their weight be included in W.

5E   From Table 3.2-4(1), C_S = 0.25

5F $V_{LFRS\text{-}seismic}$ = $(C_S)(W)$ = (0.25) (180,000 lb) = 45,000 lb

5G V_{LFRS} = 45,000 lb, greater of $V_{LFRS\text{-}wind}$ and $V_{LFRS\text{-}seismic}$.

5H   Analyze each wall line parallel to the direction under consideration assuming lightweight concrete masonry units will be specified.

       For wall line 1:

5H.1 L_{joist} = 0 ft ; h_{max} = 18 ft ; exterior
5H.2   From Table 3.2-5(3a), S_{V1} = 48 in.
5H.3   From Table 3.2-6(1a), S_{V2} = 96 in.

The spacing of the vertical reinforcement for wall line 1 is controlled by out-of-plane wind; S_V = 48 in. For wall line 2, all parameters are identical.

5I   Determine L_{req} for wall lines 1 and 2, with all wall segments being "designated shear wall segments".

5I.1   From Table 3.2-7, Table 3.2-8(2) applies.

5I.2   Based on the preliminary plan, L_{seg} = 20 ft

5I.3   From Table 3.2-8(2), based on S_V = 48 in. and S_H = BJR (Bed Joint Reinforcement) 16 in., then:

k_1 = 5,675 plf and k_2 = 80,024 lb

5I.4   Determine the required length of designated shear wall for each line of resistance.

$L_1 = V_{LFRS} / k_1$ = 45,000 lb / 5,675 plf = 7.93 ft

$L_2 = V_{LFRS}\, h_{max} / k_2$ = (45,000 lb) (18 ft) / 80,024 lb = 10.13 ft

L_{req} = 10.13 ft, which is the greatest of L_1 and L_2.

5I.5   For wall line 1: $\Sigma L_{seg} = 60$ ft $\geq L_{req} = 10.13$ ft. …design satisfied.
For wall line 2: $\Sigma L_{seg} = 60$ ft $\geq L_{req} = 10.13$ ft. …design satisfied.

5J   There are no "participating non-designated shear walls".

5K   There are no "non-participating walls"

5L   The ratio of the roof diaphragm plan dimensions is 120:60, or 2:1. From Table 3.2-9, 2 No. 5 C Bars are required for North-South loading because $V_{LFRS} = 45,000$ lb $< 66,960$ lb. (2 No. 5 bars in one course at the diaphragm perimeter.) Must verify that a greater number of C Bars are not required for East-West loading.

EAST-WEST WIND AND SEISMIC LOADING DIRECTION:

5A $A = (60$ ft$)(20$ ft$) = 1,200$ ft^2

5B   From Table 3.2-3, $C_W = 7.8$ lb/ft^2

5C $V_{LFRS\text{-}wind} = (C_W)(A) = (7.8$ lb/ft$^2)(1,200$ ft$^2) = 9,360$ lb

5D W is the same as for the North-South direction. $W = 180,000$ lb

5E   From Table 3.2-4(1), $C_S = 0.25$

5F $V_{LFRS\text{-}seismic} = (C_S)(W) = 0.25 \times 180,000$ lb $= 45,000$ lb

5G $V_{LFRS} = 28,800$ lb; greater of $V_{LFRS\text{-}wind}$ and $V_{LFRS\text{-}seismic}$.

5H   Analyze each wall line parallel to the direction under consideration.

For wall line A:

5H.1 $L_{joist} = 60$ ft ; $h_{max} = 16.3$ ft. ; exterior

5H.2   From Table 3.2-5(3a), $S_{V1} = 72$ in.

5H.3   From Table 3.2-6(1a), $S_{V2} = 96$ in.

For wall line B,

5H.1 $L_{joist} = 60$ ft ; $h_{max} = 18$ ft. ; exterior

5H.2   From Table 3.2-5(3a), $S_{V1} = 48$ in.

5H.3   From Table 3.2-6, $S_{V2} = 72$ in.

5I   Determine L_{req} for wall lines A and B, with all wall segments being "designated shear wall segments".

5I.1   From Table 3.2-7, Table 3.2-8(5) applies.

DIRECT DESIGN HANDBOOK FOR MASONRY STRUCTURES

5I.2 Based on the preliminary plan, $\boxed{L_{seg}}$ = 4 ft.

5I.3 From Table 3.2-8(5), based on S_V = 48 in. and S_H = BJR (Bed Joint Reinforcement) 16 in.:

$\boxed{k_1}$ = 8,010 lb/ft and $\boxed{k_2}$ = 31,977 lb

5I.4 Determine the required length of designated shear wall for each line of resistance.

$L_1 = V_{LFRS} / k_1$ = 45,000 lb / 8,010 plf = 5.62 ft

$L_2 = V_{LFRS} h_{max} / k_2$ = (45,000 lb) (18 ft) / 31,977 lb = 25.33 ft

$\boxed{L_{req}}$ = 25.33 ft, which is the greatest of L_1 and L_2.

5I.5 For wall line A: ΣL_{seg} = 90 ft ≥ L_{req} = 25.33 ft. ...design satisfied.
For wall line B: ΣL_{seg} = 60 ft ≥ L_{req} = 25.33 ft. ...design satisfied.

5J There are no "participating non-designated shear walls".

5K There are no "non-participating walls"

5L From Table 3.2-9, 2 No. 5 "C" Bars are required. (2 No. 5 bars in one course at the diaphragm perimeter.) This is not less than the number of "C" Bars required for North-South loading; therefore, provide 2 No. 5 "C" Bars on all four sides of the diaphragm.

6. Because there is only one rectangular diaphragm, there are no shear wall lines between two diaphragms to consider.

7. Detail "J" Bars. Verify jambs are sufficient. Provide symmetrical layout of vertical bars in each wall segment.

There are no openings to consider on wall lines 1 and 2.

On wall line A, there is a 4 ft segment between two 10 ft masonry openings to consider. This has 14 ft of tributary width (5 ft + 4 ft + 5 ft = 14 ft). The structural jamb of the 4 ft shear wall segment, however, is only 2.67 ft because there is assumed to be 8 in. of bearing for the headers on each side of the jamb panel. The quantity of bars assigned to the tributary width is (14 ft) (12 in./ft) / 48 in per bar = 3.5 bars. Therefore, 4 bars will need to be able to fit in the jamb panel, between the headers. There are, however, only 4 cells available to place the vertical reinforcing bars, which would require the spacing to be less than 16 in., which is not permitted. Therefore the length of the wall segment between the two openings will need to be increased. If there is a wall segment that is 8 ft long between the openings, the tributary width is 18 ft and the structural jamb is 6.6 ft. The quantity of bars assigned to the tributary width is 18 ft / 48 in = 4.5 bars. Therefore 5 bars will need to be able to fit in the jamb panel, between the headers. Because there are 10 cells and 5 bars, the bars will be spaced at 16 in on center, which is the minimum spacing permitted, and the revision is acceptable. In this example, this revision does not require any reworking of the above steps.

On wall line B, there is a 20 ft segment between two 20 ft masonry openings to consider. This has 40 ft of tributary width (10 ft + 20 ft + 10 ft = 40 ft). The structural jamb is 18.6 ft. The quantity of bars assigned to the tributary width is 40 ft (12 in./ft) / 48 in. per bar = 10 bars. Because there are 28

DIRECT DESIGN HANDBOOK FOR MASONRY STRUCTURES

cells and 10 bars there is sufficient wall length to accommodate the required number of bars without spacing the bars closer together than 6 in. on center.

8. At the header panels with parapets, install V Bars at the same spacing required for S_V for each wall line.

9. At headers and sill panels:

   9A   Refer to Chapter 6 details at control joints.

   9B   From Table 3.2-10(2), the B bars required are as follows.

   20 ft opening: 128 in. deep beam reinforced with 2 No. 5 bars top and bottom with 16 in. of bearing each side. If the door height would not accommodate this size of lintel above the opening, consider the use of shear reinforcement to reduce the depth of the beam.

   10 ft opening: 80 in. deep beam reinforced with 2 No. 5 bars top and bottom with 16 in. of bearing each side. Note that Table 3.2-10(2) does not have a 10 ft opening, thus the 12 ft opening is used.

   3 ft opening: 32 in. deep beam reinforced with 2 No. 5 bars top and bottom with 8 in. of bearing each side. Note that Table 3.2-10(2) does not have a 3 ft opening, thus the 4 ft opening is used.

   Because the header over the 20 ft and 10 ft openings will require 16 in. of bearing, not the assumed 8 in, at each end, the spacing of the bars in the jamb panel must be rechecked.

   9C   Based on Table 3.2-11(1) and 3.2-11(2) and the discussion in the commentary regarding cracked section behavior of reinforced masonry, in this example No. 5 bars at 48 in. on center are used at all headers for the horizontal O Bars.

10. Put the following design data on the plans.

    Roof Live Load = 20 psf

    Roof Snow Load Data:
        Flat Roof Snow Load = 72.5 psf
        Snow Exposure Factor = 1.2
        Snow Load Importance Factor = 1.2
        Thermal Factor = 1.2

    Wind Design Data:
        Basic Wind Speed = 115 mph
        Wind Importance Factor = 1.0
        Wind Exposure Category: B
        Applicable Internal Pressure Coefficient: +/- 0.18

    Earthquake Design Data:
        Seismic Importance Factor = 1.5
        Mapped 0.2 Second Spectral Response Acceleration = 0.50
        Mapped 1.0 Second Spectral Response Acceleration = 0.15
        Seismic Site Class: D
        Seismic Design Category: C
        Basic Seismic-Force-Resisting-System: Ordinary Reinforced Masonry Shear Wall

Design Seismic Shear: (2) (45,000 lb) = 90,000 lb.
Seismic Response Coefficient = 0.25
Response Modification Factor = 2.5
Analysis Procedure Used: Equivalent Lateral Force Method

Flood Design Data:
This building has not been designed for flood loads.

Special Loads:
This building has not been designed for any special loads.

11. Specifications per Chapter 5.

12. Put the required details from Chapter 6 on the construction documents. Verify that the minimum prescriptive reinforcement requirements per Figure 6.1-5 for ordinary reinforced masonry shear walls are met using the reinforcement schedule determined above.

EXCERPTS FROM

DETERMINING FIRE RESISTANCE OF CONCRETE AND MASONRY CONSTRUCTION ASSEMBLIES

An ACI/TMS Standard

Code Requirements for Determining Fire Resistance of Concrete and Masonry Construction Assemblies

Reported by ACI/TMS Committee 216

ACI/TMS 216.1-14

"Portions of this work are reproduced from the *Code Requirements for Determining Fire Resistance of Concrete and Masonry Construction Assemblies*, copyright © 2014, with the permission of the publishers, the American Concrete Institute and The Masonry Society."

American Concrete Institute
38800 Country Club Drive
Farmington Hills, MI 48331
www.concrete.org

The Masonry Society
105 South Sunset Street, Suite Q
Longmont, CO 80501
www.masonrysociety.org

CODE REQUIREMENTS FOR DETERMINING FIRE RESISTANCE OF CONCRETE AND MASONRY CONSTRUCTION ASSEMBLIES

ACI/TMS STANDARD

CHAPTER 1—GENERAL

1.1—Scope

This standard describes acceptable methods for determining the fire resistance of concrete and masonry building assemblies and structural elements, including walls, floor and roof slabs, beams, columns, lintels, and masonry fire protection for structural steel columns. These methods shall be used for design and analysis purposes and shall be based on the fire exposure and applicable end-point criteria of ASTM E119. This standard does not apply to composite metal deck floor or roof assemblies.

The primary intended use of this standard is for determining the design requirements for concrete and masonry elements to resist fire and provide fire protection. Tolerance compliance to the provisions for concrete shall be based on information provided in ACI 117. Masonry construction shall comply with TMS 402/ACI 530.1/ASCE 6.

The provisions of this standard establish fire resistance based on calculations. The fire resistance associated with an element or assembly shall be deemed acceptable when established by the calculation procedures in this standard or when established in accordance with 1.2.

1.2—Alternative methods

Methods other than those presented in this standard shall be permitted for use in assessing the fire resistance of concrete and masonry building assemblies and structural elements if the methods are based on the fire exposure and applicable end-point criteria specified in ASTM E119. Computer models, when used, shall be validated and supported by published literature to substantiate their accuracy. Alternative methods include:

Qualification by testing—Materials and assemblies of materials of construction tested in accordance with the requirements set forth in ASTM E119 shall be classified for fire resistance in accordance with the results and conditions of such tests.

Approval through past performance—The application of fire resistance ratings to elements and assemblies that have been applied in the past and have been proven through performance shall be permitted.

Other methods—The provisions of this standard are not intended to prevent the application of new and emerging technology for predicting the life safety and property protection implications of buildings and structures.

CHAPTER 2—NOTATION AND DEFINITIONS

2.1—Notation

A_1, A_2, and A_n = air factor for each continuous air space having a distance of 1/2 to 3-1/2 in. between wythes
A_{ps} = cross-sectional area of prestressing tendons, in.^{2}
A_s = cross-sectional area of nonprestressed longitudinal tension reinforcement, in.^{2}
A_{st} = cross-sectional area of the steel column, in.^{2}
a = depth of equivalent rectangular concrete compressive stress block at nominal flexural strength, in.
a_θ = depth of equivalent concrete rectangular stress block at elevated temperature, in.
B = least dimension of rectangular concrete column, in.
b = width of concrete slab or beam, in.
b_f = width of flange, in.
C = compressive force due to unfactored dead load and live load, kip
c_c = ambient temperature specific heat of concrete, Btu/(lb-°F)
d = effective depth, distance from centroid of tension reinforcement to extreme compressive fiber or depth of steel column, in.
D = for hollow structural steel columns, outside diameter for circular columns, in.; outside dimension for square columns, in.; and least outside dimension for rectangular columns, in.
D_c = oven-dried density of concrete, lb/ft^3
d_{ef} = distance from centroid of tension reinforcement to most extreme concrete compressive fiber at which point temperature does not exceed 1400°F, in.
d_l = thickness of fire-exposed concrete layer, in.
d_{st} = column width, in.
f_c = measured compressive strength of concrete test cylinders at ambient temperature, psi
f_c' = specified compressive strength of concrete, psi
$f_{c\theta}'$ = reduced compressive strength of concrete at elevated temperature, psi
f_{ps} = stress in prestressing steel at nominal flexural strength, psi
$f_{ps\theta}$ = reduced stress of prestressing steel at elevated temperature, psi
f_{pu} = specified tensile strength of prestressing tendons, psi
f_y = specified yield strength of nonprestressed reinforcing steel, psi
$f_{y\theta}$ = reduced yield strength of nonprestressed reinforcing steel at elevated temperature, psi
H = specified height of masonry unit, in.
H_s = ambient temperature thermal capacity of steel column, Btu/(ft-°F)
h = average thickness of concrete cover, in.
KL = column effective length, ft
k_c = thermal conductivity of concrete at room temperature, Btu/(h-ft-°F)
k_{cm} = thermal conductivity of concrete masonry at room temperature, Btu/(h-ft-°F)
L = specified length of masonry unit or interior dimension of rectangular concrete box protection for steel column, in.
ℓ = clear span between supports, ft
M = moment due to full service load on member, lb-ft

Note: Only portions of this Standard are shown which are particularly applicable to masonry construction.

CODE REQUIREMENTS FOR DETERMINING FIRE RESISTANCE OF CONCRETE AND MASONRY CONSTRUCTION ASSEMBLIES

DETERMINING FIRE RESISTANCE OF CONCRETE AND MASONRY CONSTRUCTION ASSEMBLIES

M_n = nominal moment capacity at section, lb-ft

$M_{n\theta}$ = nominal moment capacity of section at elevated temperature, lb-ft

$M_{n\theta}^+$ = nominal positive moment capacity of section at elevated temperature, lb-ft

$M_{n\theta}^-$ = nominal negative moment capacity of section at elevated temperature, lb-ft

M_{x1} = maximum value of redistributed positive moment at some distance x_1, lb-ft

m = equivalent moisture content of the concrete by volume (percent)

p = inner perimeter of concrete masonry protection, in.

p_s = heated perimeter of steel column, in.

R = fire resistance of assembly, hours

R_0 = fire resistance at zero moisture content, minutes

R_1, R_2, R_n = fire resistance of layer 1, 2,...n, respectively, hours

s = center-to-center spacing of items such as ribs or undulations, in.

t = time, minutes

t_e = equivalent thickness of a ribbed or undulating concrete section, in.

t_{e2} = equivalent thickness t_e calculated by dividing the net cross-sectional area by the panel width

t_{min} = minimum thickness, in.

t_{tot} = total slab thickness, in.

t_w = thickness of web, in.

T = specified thickness of concrete masonry and clay masonry unit, in.

T_e = equivalent thickness of concrete, concrete masonry and clay masonry unit, in.

T_{ea} = equivalent thickness of concrete masonry assembly, in.

T_{ef} = equivalent thickness of finishes, in.

u = average thickness of concrete between the center of main reinforcing steel and fire-exposed surface, in.

u_{ef} = an adjusted value of u to accommodate beam geometry where fire exposure to concrete surfaces is from three sides, in.

V_n = net volume of masonry unit, in.³

W = average weight of the steel column, lb/ft

w = sum of unfactored dead and live service loads, lb/ft

w_c = density of concrete, lb/ft³

w_{cm} = density of masonry protection, lb/ft³

x_0 = distance from inflection point to location of first interior support, measured after moment redistribution has occurred, ft

x_1 = distance at which maximum value of redistributed positive moment occurs measured from: (a) outer support for continuity over one support; and (b) either support where continuity extends over two supports, ft

x_2 = in continuous span, distance between adjacent inflection points, ft

θ = subscript denoting changes of parameter due to elevated temperature

ρ = reinforcement ratio (A_s/bd)

ρ_g = ratio of total reinforcement area to cross-sectional area of column

ω_p = reinforcement index for concrete beam reinforced with prestressing steel

$\omega_{p\theta}$ = reinforcement index for concrete beam reinforced with prestressing steel at elevated temperature

ω_r = reinforcement index for concrete beam reinforced with nonprestressed steel

ω_θ = reinforcement index for concrete beam at elevated temperature

ψ = modification factor reflecting type of column infill

2.2—Definitions

The following terms are defined for general use in this code.

bar, high-strength alloy steel—reinforcement conforming to the requirements of ASTM A722.

barrier element—a building member that performs as a barricade to the spread of fire (for example, walls, floors, and roofs).

beam—a structural member subjected to axial load and flexure but primarily to flexure.

building code—a legal document that establishes the minimum requirements necessary for building design and construction to provide for public health and safety.

building official—(1) the official charged with administration and enforcement of the applicable building code; (2) the duly authorized representative of the official.

carbonate aggregate concrete—concrete made with coarse aggregate consisting mainly of calcium carbonate or a combination of calcium and magnesium carbonate (for example, limestone or dolomite).

ceramic fiber blanket—mineral wool insulating material made of alumina-silica fibers and having a density of 4 to 8 lb/ft³.

cellular concrete—a low-density product consisting of portland cement, cement silica, cement pozzolan, lime pozzolan, lime silica pastes, or pastes containing a blend of these ingredients and having a homogeneous void or cell structure attained with gas-forming chemicals or foaming agents (for cellular concretes containing binder ingredients other than, or in addition to, portland cement, autoclave curing is usually used).

cementitious materials—pozzolans and hydraulic cements.

clay masonry unit—solid or hollow unit (brick or tile) composed of clay, shale, or similar naturally occurring earthen substance shaped into prismatic units and subjected to heat treatment at elevated temperature (firing), meeting requirements of ASTM C34, C56, C62, C126, C212, C216, C652, or C1088.

CODE REQUIREMENTS FOR DETERMINING FIRE RESISTANCE OF CONCRETE AND MASONRY CONSTRUCTION ASSEMBLIES

cold-drawn wire reinforcement—steel wire made from rods that have been hot rolled from billets and cold-drawn through a die.

concrete masonry unit—either a hollow or solid unit (block) composed of portland-cement concrete; often referred to by indicating the type of mineral aggregate incorporated (for example, lightweight or sand-gravel block), meeting the requirements of ASTM C55, C73, C90, C129, C744, or C1634.

continuous slab or beam—a slab or beam that extends as a unit over three or more supports in a given direction.

critical temperature—temperature of reinforcing steel in unrestrained flexural members during fire exposure at which the nominal flexural strength of a member is reduced to the moment produced by application of service loads to that member.

end-point criteria—conditions of acceptance for an ASTM E119 fire test.

fire endurance—a measure of the elapsed time during which a material or assembly continues to exhibit fire resistance; as applied to elements of buildings with respect to this standard, it shall be measured by the methods and criteria contained in ASTM E119.

fire resistance—the property of a material or assembly to withstand fire or provide protection from it. As applied to elements of buildings, it is characterized by the ability to confine a fire or, when exposed to fire, to continue to perform a given structural function, or both.

fire resistance rating—a legal term defined in building codes, usually based on fire endurance; fire-resistance ratings are assigned by building codes for various types of construction and occupancies, and are usually given in half-hour or hourly increments.

glass fiberboard—fibrous glass insulation board complying with ASTM C612.

gypsum type X wallboard—mill-fabricated product, complying with ASTM C1396, Type X, made of a gypsum core containing special minerals and encased in a smooth, finished paper on the face side and liner paper on the back.

heat transmission end-point—An acceptance criterion of ASTM E119 limiting the temperature rise of the unexposed surface to an average of 250°F for all measuring points or a maximum of 325°F at any one point.

hollow brick or tile of clay or shale—clay or shale masonry units in which the net cross-sectional area in any plane parallel to the surface, containing the cores, cells, or frogs, is less than 75 percent of its gross cross-sectional area measured in the same plane.

hot-rolled steel—steel used for reinforcing bars or structural steel members.

integrity end-point—an acceptance criterion of ASTM E119 prohibiting the passage of flame or gases hot enough to ignite cotton waste before the end of the desired fire-endurance period. The term also applies to the hose-stream test of a fire-exposed wall.

intumescent mastic—spray-applied coating that reacts to heat at approximately 300°F by foaming to a multicellular structure having 10 to 15 times its initial thickness.

joist—a comparatively narrow beam, used in closely spaced arrangements to support floor or roof slabs (that require no reinforcement except that required for temperature and shrinkage stresses); also a horizontal structural member such as that which supports deck form sheathing.

lightweight-aggregate concrete—concrete made with aggregates conforming to ASTM C330 or C331.

mineral board—board made of mineral fiber insulation complying with ASTM C726.

normalweight concrete—concrete containing only aggregate that conforms to ASTM C33.

perlite concrete—nonstructural lightweight insulating concrete having a density of approximately 30 lb/ft^3, made by mixing perlite aggregate complying with ASTM C332 with portland cement slurry.

plain concrete—structural concrete with no reinforcement or less reinforcement than the minimum amount specified in ACI 318 for reinforced concrete.

reinforced concrete—structural concrete reinforced with no less than the minimum amount of prestressing steel or nonprestressed reinforcement as specified in ACI 318.

reinforced masonry—masonry in which reinforcement is used to resist tensile forces.

semi-lightweight concrete—concrete made with a combination of lightweight aggregates (expanded clay, shale, slag, slate, or sintered fly ash) and normalweight aggregates, having an equilibrium density of 105 to 120 lb/ft^3 in accordance with ASTM C567.

siliceous-aggregate concrete—concrete made with normal-density aggregates having constituents composed mainly of silica or silicates (such as quartz or granite).

solid brick of clay or shale—clay or shale units in which the net cross-sectional area in any plane parallel to the surface, containing the cores or frogs, is at least 75 percent of the gross cross-sectional area measured in the same plane

sprayed mineral fiber—a blend of refined mineral fibers and inorganic binders to which water is added during the spraying operation.

standard fire exposure—the time-temperature relationship defined by ASTM E119.

standard fire test—the test prescribed by ASTM E119.

steel temperature end-point—an acceptance criterion of ASTM E119 defining the limiting steel temperatures for unrestrained assembly classifications.

strand—an assembly of wires twisted about a center wire or core.

structural concrete—plain or reinforced concrete in a member that is part of a structural system required to transfer gravity, lateral loads, or both, along a load path to the ground.

CODE REQUIREMENTS FOR DETERMINING FIRE RESISTANCE OF CONCRETE AND MASONRY CONSTRUCTION ASSEMBLIES

structural end-point—the acceptance criterion of ASTM E119 that states that the specimen shall sustain with applied load without collapse.

tendon—an assembly consisting of a tensioned element (such as a wire, bar, rod, strand, or a bundle of these elements) used to impart compressive stress in concrete, along with any associated components used to enclose and anchor the tensioned element.

unreinforced (plain) masonry—masonry in which the tensile resistance of masonry is taken into consideration and the resistance of the reinforcing steel, if present, is neglected.

vermiculite cementitious material—material containing mill-mixed vermiculite to which water is added to form a mixture suitable for spraying.

vermiculite concrete—concrete in which the aggregate consists of exfoliated vermiculite.

CHAPTER 3—REFERENCED STANDARDS

American Concrete Institute
ACI 117-10—Specification for Tolerances for Concrete Construction and Materials and Commentary
ACI 318-11—Building Code Requirements for Structural Concrete and Commentary

American Institute of Steel Construction
ANSI/AISC 360-10—Specification for Structural Steel Buildings

ASTM International
A722/A722M-12—Standard Specification for Uncoated High-Strength Steel Bars for Prestressing Concrete
ASTM C33/C33M-13—Standard Specification for Concrete Aggregates
ASTM C34-13—Standard Specification for Structural Clay Load-Bearing Wall Tile
ASTM C55-14—Standard Specification for Concrete Building Brick
ASTM C56-13—Standard Specification for Structural Clay Nonloadbearing Tile
ASTM C62-13a—Standard Specification for Building Brick (Solid Masonry Units Made from Clay or Shale)
ASTM C73-10—Standard Specification for Calcium Silicate Brick (Sand-Lime Brick)
ASTM C90-14—Standard Specification for Load-Bearing Concrete Masonry Units
ASTM C126-14—Standard Specification for Ceramic Glazed Structural Clay Facing Tile, Facing Brick, and Solid Masonry Units
ASTM C129-14—Standard Specification for Nonload-bearing Concrete Masonry Units
ASTM C140/C140M-14a—Standard Test Methods for Sampling and Testing Concrete Masonry Units and Related Units
ASTM C212-14—Standard Specification for Structural Clay Facing Tile
ASTM C216-14—Standard Specification for Facing Brick (Solid Masonry Units Made from Clay or Shale)
ASTM C330/C330M-14—Standard Specification for Lightweight Aggregates for Structural Concrete
ASTM C331/C331M-14—Standard Specification for Lightweight Aggregates for Concrete Masonry Units
ASTM C332-09—Standard Specification for Lightweight Aggregates for Insulating Concrete
ASTM C516-08(2013)[ε1]—Standard Specification for Vermiculite Loose Fill Thermal Insulation
ASTM C549-06 (2012)—Standard Specification for Perlite Loose Fill Insulation
ASTM C567/C567M-14—Standard Test Method for Determining Density of Structural Lightweight Concrete
ASTM C612-14—Standard Specification for Mineral Fiber Block and Board Thermal Insulation
ASTM C652-14—Standard Specification for Hollow Brick (Hollow Masonry Units Made from Clay or Shale)
ASTM C726-12—Standard Specification for Mineral Wool Roof Insulation Board
ASTM C744-14—Standard Specification for Prefaced Concrete and Calcium Silicate Masonry Units
ASTM C1088-14—Standard Specification for Thin Veneer Brick Units Made from Clay or Shale
ASTM C1396/C1396M-14—Standard Specification for Gypsum Board
ASTM C1405-14—Standard Specification for Glazed Brick (Single Fired, Brick Units)
ASTM C1634-11—Standard Specification for Concrete Building Brick
ASTM E119-12a—Standard Test Methods for Fire Tests of Building Construction and Materials

The Masonry Society
TMS 402-13/ACI 530-13/ASCE 5-13—Building Code Requirements for Masonry Structures
TMS 602-13/ACI 530.1-13/ASCE 6-13—Specification for Masonry Structures

CHAPTER 5—CONCRETE MASONRY

5.1—General

The fire resistance of concrete masonry assemblies shall be determined in accordance with the provisions of this chapter. The minimum equivalent thicknesses of concrete masonry assemblies required to provide fire resistance of 1 to 4 hours shall conform to values given in Tables 5.1a, 5.1b, or 5.1c, as is appropriate for the assembly being considered. Except where the provisions of this chapter are more stringent, the design, construction, and material requirements of concrete masonry including units, mortar, grout, control joint materials, and reinforcement shall comply with TMS 402/ACI 530/ASCE 5 and TMS 602/ACI 530.1/ASCE 6. Concrete masonry units shall comply with ASTM C55, C73, C90, C129, C744, or C1634.

Table 5.1a—Fire-resistance rating of concrete masonry assemblies

Aggregate type	Minimum equivalent thickness T_{ea} for fire-resistance rating, in.[*][†]						
	1/2 hour	3/4 hour	1 hour	1-1/2 hours	2 hours	3 hours	4 hours
Calcareous or siliceous gravel (other than limestone)	2.0	2.4	2.8	3.6	4.2	5.3	6.2
Limestone, cinders, or air-cooled slag	1.9	2.3	2.7	3.4	4.0	5.0	5.9
Expanded clay, expanded shale, or expanded slate	1.8	2.2	2.6	3.3	3.6	4.4	5.1
Expanded slag or pumice	1.5	1.9	2.1	2.7	3.2	4.0	4.7

[*]Fire-resistance ratings between the hourly fire-resistance rating periods listed shall be determined by linear interpolation based on the equivalent thickness value of the concrete masonry assembly.

[†]Minimum required equivalent thickness corresponding to the fire-resistance rating for units made with a combination of aggregates shall be determined by linear interpolation based on the percent by dry-rodded volume of each aggregate used in manufacturing the units.

Table 5.1b—Reinforced masonry columns

Fire resistance, h	1	2	3	4
Minimum nominal column dimensions, in.	8	10	12	14

Table 5.1c—Reinforced masonry lintels

Nominal lintel width, in.	Minimum longitudinal reinforcement cover for fire-resistance rating, in.			
	1 hour	2 hours	3 hours	4 hours
6	1-1/2	2	NP*	NP*
8	1-1/2	1-1/2	1-3/4	3
10 or more	1-1/2	1-1/2	1-1/2	1-3/4

*Not permitted without a more detailed analysis

5.2—Equivalent thickness

The equivalent thickness of concrete masonry construction shall be determined in accordance with the provisions of this section. The equivalent thickness of concrete masonry assemblies, T_{ea}, shall be calculated as the sum of the equivalent thickness of the concrete masonry unit, T_e, as determined by 5.2.1, 5.2.2, or 5.2.3, plus the equivalent thickness of finishes, T_{ef}, determined in accordance with Chapter 7.

$$T_{ea} = T_e + T_{ef} \quad (5\text{-}2a)$$

$$T_e = V_n/LH \quad (5\text{-}2b)$$

where T_{ea} is the equivalent thickness of concrete masonry assembly, in.; T_e is the equivalent thickness of concrete masonry unit, in.; T_{ef} is the equivalent thickness of finishes, in.; V_n is the net volume of masonry unit, in.³; L is the specified length of masonry unit, in.; and H is the specified height of masonry unit, in.

V_n, L, and H shall be determined in accordance with the procedures contained in ASTM C140.

R_1 = fire resistance rating of wythe 1
R_2 = fire resistance rating of wythe 2
A_1 = air space factor = 0.3

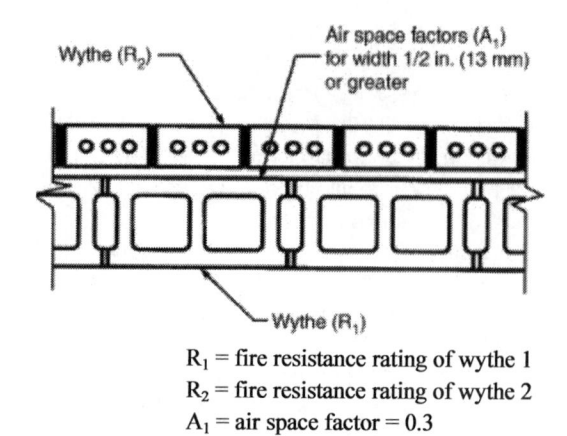

Fig. 5.3.2—Multi-wythe walls.

5.2.1 *Ungrouted or partially grouted construction*—The equivalent thickness T_e of an ungrouted or partially grouted concrete masonry assemblage shall be taken equal to the value determined by Eq. (5.2b).

5.2.2 *Solid grouted construction*—The equivalent thickness T_e of solid grouted concrete masonry units shall be taken equal to the thickness of the unit determined in accordance with ASTM C140.

5.2.3 *Air spaces and cells filled with loose fill material*—The equivalent thickness T_e of hollow concrete masonry units completely filled with loose material shall be taken as the thickness of the unit determined in accordance with ASTM C140 when loose fill materials are sand, pea gravel, crushed stone, or slag that meet ASTM C33 requirements; pumice, scoria, expanded shale, expanded clay, expanded slate, expanded slag, expanded fly ash, or cinders that comply with ASTM C331; perlite meeting the requirements of ASTM C549; or vermiculite meeting the requirements of ASTM C516.

5.3—Concrete masonry wall assemblies

The minimum equivalent thickness of various types of plain or reinforced concrete masonry bearing or nonbearing walls required to provide fire-resistance ratings of 1 to 4 hours shall conform to Table 5.1a.

5.3.1 *Single-wythe wall assemblies*—The fire-resistance rating of single-wythe concrete masonry walls shall be determined in accordance with Table 5.1a.

5.3.2 *Multi-wythe wall assemblies*—The fire resistance of multi-wythe walls (Fig. 5.3.2) shall be calculated using the fire resistance of each wythe and any air space between each wythe in accordance with Eq. (4.2.5.3).

5.3.3 *Expansion or contraction joints*—Expansion or contraction joints in fire-rated masonry wall assemblies in which openings are not permitted, or in wall assemblies where openings are required to be protected, shall comply

---

CODE REQUIREMENTS FOR DETERMINING FIRE RESISTANCE OF CONCRETE AND MASONRY CONSTRUCTION ASSEMBLIES

DETERMINING FIRE RESISTANCE OF CONCRETE AND MASONRY CONSTRUCTION ASSEMBLIES

with Fig. 5.3.3. The amount of ceramic fiber felt (alumina silica fibers) where required in Fig. 5.3.3 shall be in accordance with Fig. 4.2.7.2 using the overall thickness of the assembly as follows.

a) Calcareous or siliceous gravel, limestone, cinders, or air-cooled slag aggregate concrete masonry assemblies shall meet the requirements denoted by the solid-line curves of Fig. 4.2.7.2.

b) Expanded clay, expanded shale, expanded slate, expanded slag, or pumice aggregate concrete masonry assemblies shall meet the requirements denoted by the dashed-line curves of Fig. 4.2.7.2.

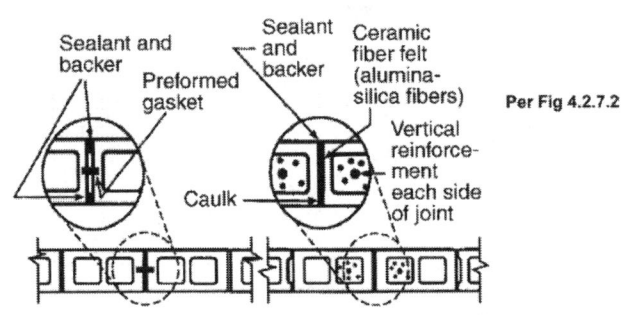

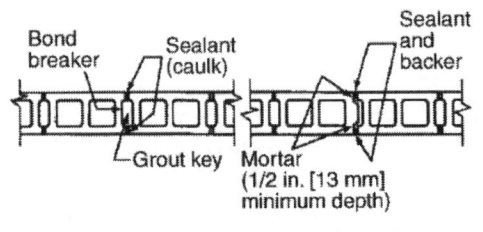

Fig. 5.3.3—Expansion or construction joints in masonry walls with 1/2 in. maximum width having 1- to 4-hour fire resistance. Exception: maximum joint width of ceramic fiber felt option in accordance with Fig. 4.2.7.2 is 1 in.

5.3.4 *Effects of finish materials on fire resistance*—The use of finish materials to increase the fire-resistance rating shall be permitted. The effects of the finish materials, whether on the fire-exposed side or the non-fire-exposed side, shall be evaluated in accordance with the provisions of Chapter 7.

5.4—Reinforced concrete masonry columns

The fire resistance of reinforced concrete masonry columns shall be determined using the least plan dimension of the column in accordance with the requirements of Table 5.1b. The minimum cover for longitudinal reinforcement shall be 2 in.

5.5—Reinforced concrete masonry lintels

The fire resistance of concrete masonry lintels shall be established based on the nominal width of the lintel and the minimum cover of longitudinal reinforcement in accordance with Table 5.1c.

5.6—Structural steel columns protected by concrete masonry

The fire resistance of structural steel columns protected by concrete masonry shall be determined using the following equation

$$R = 0.401(A_{st}/p_s)^{0.7} + [0.285(T_{ea}^{1.6}/k_{cm}^{0.2})] \times [1.0 + 42.7\{(A_{st}/w_{cm}T_{ea})/(0.25p + T_{ea})\}^{0.8}] \quad (5.6a)$$

where p_s is calculated according to Eq. (5.6b) through (5.6d); k_{cm} is calculated according to Eq. (5.6e); A_{st} is the cross-sectional area of the steel column, in.^{2}; k_{cm} is thermal conductivity of concrete masonry at room temperature, Btu/(h-ft-°F); and w_{cm} is the density of masonry protection, lb/ft^3.

$$p_s = 2(b_f + d_{st}) + 2(b_f - t_w) \text{ [W-section]} \quad (5.6b)$$

$$p_s = \pi d_{st} \text{ [pipe section]} \quad (5.6c)$$

$$p_s = 4d_{st} \text{ [square structural tube section]} \quad (5.6d)$$

where d_{st} is illustrated in Fig. 5.6, and t_w is illustrated in Fig. 5.6(a) (w-shape).

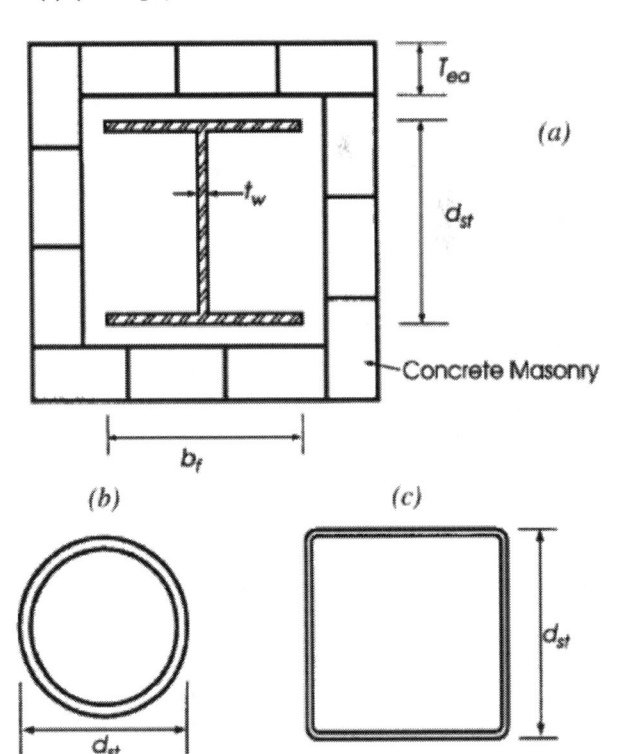

Fig. 5.6—Structural steel shapes protected by concrete masonry.

CODE REQUIREMENTS FOR DETERMINING FIRE RESISTANCE OF CONCRETE AND MASONRY CONSTRUCTION ASSEMBLIES

ACI/TMS STANDARD

It shall be permitted to calculate the thermal conductivity of concrete masonry for use in Eq. (5.6a) as

$$k_{cm} = 0.0417 e^{0.02 W_{cm}} \quad (5.6e)$$

The minimum required equivalent thickness of concrete masonry units for specified fire-resistance ratings of several commonly used column shapes and sizes is shown in Table 5.6.

Table 5.6—Fire resistance of concrete-masonry-protected steel columns

Square structural tubing

Nominal tube size, in.	Concrete masonry density, lb/ft³	Minimum equivalent thickness for fire-resistance rating of concrete masonry protection assembly T_e, in.			
		1 hour	2 hours	3 hours	4 hours
4 x 4-1/2 wall thickness	80	0.93	1.90	2.71	3.43
	100	1.08	2.13	2.99	3.76
	110	1.16	2.24	3.13	3.91
	120	1.22	2.34	3.26	4.06
4 x 4-3/8 wall thickness	80	1.05	2.03	2.84	3.57
	100	1.20	2.25	3.11	3.88
	110	1.27	2.35	3.24	4.02
	120	1.34	2.45	3.37	4.17
4 x 4-1/4 wall thickness	80	1.21	2.20	3.01	3.73
	100	1.35	2.40	3.26	4.02
	110	1.41	2.50	3.38	4.16
	120	1.48	2.59	3.50	4.30
6 x 6-1/2 wall thickness	80	0.82	1.75	2.54	3.25
	100	0.98	1.99	2.84	3.59
	110	1.05	2.10	2.98	3.75
	120	1.12	2.21	3.11	3.91
6 x 6-3/8 wall thickness	80	0.96	1.91	2.71	3.42
	100	1.12	2.14	3.00	3.75
	110	1.19	2.25	3.13	3.90
	120	1.26	2.35	3.26	4.05
6 x 6-1/4 wall thickness	80	1.14	2.11	2.92	3.63
	100	1.29	2.32	3.18	3.93
	110	1.36	2.43	3.30	4.08
	120	1.42	2.52	3.43	4.22
8 x 8-1/2 wall thickness	80	0.77	1.66	2.44	3.13
	100	0.92	1.61	2.75	3.49
	110	1.00	2.02	2.89	3.66
	120	1.07	2.14	3.03	3.82
8 x 8-3/8 wall thickness	80	0.91	1.84	2.63	3.33
	100	1.07	2.08	2.92	3.67
	110	1.14	2.19	3.06	3.83
	120	1.21	2.29	3.13	3.87
8 x 8-1/4 wall thickness	80	1.10	2.06	2.86	3.57
	100	1.25	2.28	3.19	3.98
	110	1.32	2.38	3.25	4.02
	120	1.39	2.48	3.38	4.17

Steel pipe

Column size	Concrete masonry density, lb/ft³	Minimum equivalent thickness for fire-resistance rating of concrete masonry protection assembly T_e, in.			
		1 hour	2 hours	3 hours	4 hours
Four double extra-strong 0.674 wall thickness	80	0.80	1.75	2.56	3.28
	100	0.95	1.99	2.85	3.62
	110	1.02	2.10	2.99	3.78
	120	1.09	2.20	3.12	3.93
Four extra-strong 0.337 wall thickness	80	1.12	2.11	2.93	3.65
	100	1.26	2.32	3.19	3.95
	110	1.33	2.42	3.31	4.09
	120	1.40	2.52	3.43	4.23
Four standard 0.237 wall thickness	80	1.26	2.25	3.07	3.79
	100	1.40	2.45	3.31	4.07
	110	1.46	2.55	3.43	4.21
	120	1.53	2.64	3.54	4.34
Five double extra-Strong 0.750 wall thickness	80	0.70	1.61	2.40	3.12
	100	1.85	1.86	2.71	3.47
	110	0.91	1.97	2.85	3.63
	120	0.98	2.02	2.99	3.79
Five extra-strong 0.375 wall thickness	80	1.04	2.01	2.83	3.54
	100	1.19	2.23	3.09	3.85
	110	1.26	2.34	3.22	4.00
	120	1.32	2.44	3.34	4.14
Five standard 0.258 wall thickness	80	1.20	2.19	3.00	3.72
	100	1.34	2.39	3.25	4.00
	110	1.41	2.49	3.37	4.14
	120	1.47	2.58	3.49	4.28
Six double extra-strong 0.864 wall thickness	80	0.59	1.46	2.23	3.29
	100	0.73	1.71	2.54	3.29
	110	0.80	1.82	2.69	3.47
	120	0.86	1.93	2.83	3.63
Six extra-strong 0.432 wall thickness	80	0.94	1.90	2.70	3.42
	100	1.10	2.13	2.98	3.74
	110	1.17	2.23	3.11	3.89
	120	1.24	2.34	3.24	4.04
Six standard 0.280 wall thickness	80	1.14	2.12	2.93	3.64
	100	1.29	2.33	3.19	3.94
	110	1.36	2.43	3.31	4.08
	120	1.42	2.53	3.43	4.22

Note: Tabulated values assume 1 in. air gap between masonry and steel section.

CODE REQUIREMENTS FOR DETERMINING FIRE RESISTANCE OF CONCRETE AND MASONRY CONSTRUCTION ASSEMBLIES

DETERMINING FIRE RESISTANCE OF CONCRETE AND MASONRY CONSTRUCTION ASSEMBLIES

Table 5.6 (cont.)—Fire resistance of concrete-masonry-protected steel columns

W shapes

Column size	Concrete masonry density, lb/ft³	1 hour	2 hours	3 hours	4 hours	Column size	Concrete masonry density, lb/ft³	1 hour	2 hours	3 hours	4 hours
W14 x 82	80	0.73	1.59	2.31	2.98	W10 x 68	80	0.72	1.58	2.33	3.01
	100	0.89	1.82	2.63	3.35		100	0.87	1.83	2.65	3.38
	110	0.96	1.94	2.78	3.53		110	0.94	1.95	2.79	3.55
	120	1.03	2.06	2.93	3.70		120	1.01	2.06	2.94	3.72
W14 x 68	80	0.83	1.69	2.45	3.13	W10 x 54	80	0.88	1.76	2.53	3.21
	100	0.98	1.94	2.76	3.49		100	1.04	2.01	2.83	3.57
	110	1.06	2.06	2.91	3.66		110	1.11	2.12	2.98	3.73
	120	1.13	2.17	3.05	3.82		120	1.19	2.24	3.12	3.90
W14 x 53	80	0.91	1.81	2.58	3.27	W10 x 45	80	0.92	1.83	2.60	3.30
	100	1.07	2.05	2.88	3.62		100	1.08	2.07	2.90	3.64
	110	1.15	2.17	3.02	3.78		110	1.16	2.18	3.04	3.80
	120	1.22	2.28	3.16	3.94		120	1.23	2.29	2.18	3.96
W14 x 43	80	1.01	1.93	2.71	3.41	W10 x 33	80	1.06	2.00	2.79	3.49
	100	1.17	2.17	3.00	3.74		100	1.22	2.23	3.07	3.81
	110	1.25	2.28	3.14	3.90		110	1.30	2.34	3.20	3.96
	120	1.32	2.38	3.16	3.94		120	1.37	2.44	3.33	4.12
W12 x 72	80	0.81	1.66	3.27	4.05	W8 x 40	80	0.94	1.85	2.63	3.33
	100	0.91	1.88	2.70	3.43		100	1.10	2.10	2.93	3.67
	110	0.99	1.99	2.84	3.60		110	1.18	2.21	3.07	3.83
	120	1.06	2.10	2.98	3.76		120	1.25	2.32	3.20	3.99
W12 x 58	80	0.88	1.76	2.52	3.21	W8 x 31	80	1.06	2.00	2.78	3.49
	100	1.04	2.01	2.83	3.56		100	1.22	2.23	3.07	3.81
	110	1.11	2.12	2.97	3.73		110	1.29	2.33	3.20	3.97
	120	1.19	2.23	3.11	3.89		120	1.36	2.44	3.33	4.12
W12 x 50	80	0.91	1.81	2.58	3.27	W8 x 24	80	1.14	2.09	2.89	3.59
	100	1.07	2.05	2.88	3.62		100	1.29	2.31	3.16	3.90
	110	1.15	2.17	3.02	3.78		110	1.36	2.42	3.28	4.05
	120	1.22	2.28	3.16	3.94		120	1.43	2.52	3.41	4.20
W12 x 40	80	1.01	1.94	2.72	3.41	W8 x 18	80	1.22	2.20	3.01	3.72
	100	1.17	2.17	3.01	3.75		100	1.36	2.40	3.25	4.01
	110	1.25	2.28	3.14	3.90		110	1.42	2.50	3.37	4.14
	120	1.32	2.39	3.27	4.06		120	1.48	2.59	3.49	4.28

Note: Tabulated values assume 1 in. air gap between masonry and steel section.

CHAPTER 6—CLAY BRICK AND TILE MASONRY

6.1—General

The calculated fire resistance of clay masonry assemblies shall be determined based on the provisions of this chapter. Except where the provisions of this chapter are more stringent, the design, construction, and material requirements of clay masonry including units, mortar, grout, control joint materials, and reinforcement shall comply with TMS 402/ ACI 530/ASCE 5 and TMS 602/ACI 530.1/ASCE 6. Clay masonry units shall comply with ASTM C34, C56, C62, C73, C126, C212, C216, C652, or C1405.

6.2—Equivalent thickness

The equivalent thickness of clay masonry assemblies shall be determined in accordance with the provisions of this section.

6.2.1 The equivalent thickness of masonry construction of solid brick of clay or shale shall be the actual thickness of the unit.

6.2.2 The equivalent thickness of masonry construction of hollow brick or tile of clay or shale shall be based on the equivalent thickness of the clay masonry unit as determined by 6.2.3, 6.2.4, 6.2.5, and Eq. (6.2.2).

$$T_e = V_n/LH \quad (6.2.2)$$

where T_e is equivalent thickness of clay masonry unit, in.; V_n is the net volume of masonry unit, in.3; L is the specified length of masonry unit, in.; and H is the specified height of masonry unit, in.

6.2.3 *Ungrouted or partially grouted construction*—The equivalent thickness T_e of an ungrouted or partially grouted unit of hollow brick or tile of clay or shale shall be taken as the value determined by Eq. (6.2.2).

6.2.4 *Solid grouted construction*—The equivalent thickness T_e of a solidly grouted unit of hollow brick or tile of clay or shale shall be taken as the actual thickness of the unit.

6.2.5 *Air spaces and cells filled with loose fill material*—The equivalent thickness T_e of a hollow brick or tile of clay or shale completely filled shall be taken as the actual thickness of the unit when loose fill materials are sand, pea gravel, crushed stone, or slag that meet ASTM C33 requirements; pumice, scoria, expanded shale, expanded clay, expanded slate, expanded slag, expanded fly ash, or cinders in compliance with ASTM C331; perlite meeting the requirements of ASTM C549; or vermiculite meeting the requirements of ASTM C516.

6.3—Clay brick and tile masonry wall assemblies

The fire resistance of clay brick and tile masonry wall assemblies shall be determined in accordance with the provisions of this section.

6.3.1 *Filled and unfilled clay brick and tile masonry*—The fire resistance of clay brick and tile walls shall be determined from Table 6.3.1 using the equivalent thickness calculation procedure prescribed in 6.2.

6.3.2 *Single-wythe walls*—The fire resistance of clay brick and tile masonry walls shall be determined from Table 6.3.1.

6.3.3 *Multi-wythe walls*—The fire resistance of multi-wythe walls shall be determined in accordance with the provisions of this section and Table 6.3.1.

6.3.3.1 *Multi-wythe clay masonry walls with dimensionally dissimilar wythes*—The fire resistance of multi-wythe clay masonry walls consisting of two or more dimensionally dissimilar wythes shall be based on the fire resistance of each wythe. Equation (4.2.5.3) shall be used to determine fire resistance of the wall assembly.

6.3.3.2 *Multi-wythe walls with dissimilar materials*—For multi-wythe walls consisting of two or more wythes of dissimilar materials (concrete or concrete masonry units), the fire resistance of the dissimilar wythes, R_n, shall be determined in accordance with 4.2, Fig. 4.2.5.1 for concrete, and 5.3 and Table 5.1a for concrete masonry units. Equation (4.2.5.3) shall be used to determine fire resistance of the wall assembly.

6.3.3.3 *Continuous air spaces*—The fire resistance of multi-wythe clay brick and tile masonry walls separated by continuous air spaces between each wythe shall be determined using Eq. (4.2.5.3).

6.3.4 *Effects of finish materials on fire resistance*—The use of finish materials to increase the fire-resistance rating shall be permitted. The effects of the finish materials, whether on the fire-exposed side or the non-fire-exposed side, shall be evaluated in accordance with the provisions of Chapter 7.

Table 6.3.1—Fire resistance of clay masonry walls

Material type	Minimum equivalent thickness for fire resistance, in.[*][†][‡]			
	1 hour	2 hours	3 hours	4 hours
Solid brick of clay or shale[§]	2.7	3.8	4.9	6.0
Hollow brick or tile of clay or shale, unfilled	2.3	3.4	4.3	5.0
Hollow brick or tile of clay or shale, grouted or filled with materials specified in 6.2.3	3.0	4.4	5.5	6.6

[*]Equivalent thickness as determined from 6.2.
[†]Calculated fire resistance between the hourly increments listed shall be determined by linear interpolation.
[‡]Where combustible members are framed into the wall, the thickness of solid material between the end of each member and the opposite face of the wall, or between members set in from opposite sides, shall not be less than 93 percent of the thickness shown.
[§]Units in which the net cross-sectional area of cored or frogged brick in any plane parallel to the surface containing the cores or frog is at least 75 percent of the gross cross-sectional area measured in the same plane.

6.4—Reinforced clay masonry columns

The fire resistance of reinforced clay masonry columns shall be based on the least plan dimension of the column in accordance with the requirements of Table 5.1b. The minimum cover for longitudinal reinforcement shall be 2 in.

6.5—Reinforced clay masonry lintels

The fire resistance of clay masonry lintels shall be determined based on the nominal width of the lintel and the minimum cover for the longitudinal reinforcement in accordance with Table 5.1c.

6.6—Expansion or contraction joints

Expansion or contraction joints in fire-rated clay masonry wall assemblies shall be in accordance with 5.3.3.

DETERMINING FIRE RESISTANCE OF CONCRETE AND MASONRY CONSTRUCTION ASSEMBLIES

Table 6.7.1—Fire resistance of clay-masonry-protected steel columns

Square structural tubing

Nominal tube size, in.	Clay masonry density, lb/ft³	Minimum equivalent thickness for fire-resistance rating of clay masonry protection assembly T_e, in.			
		1 hour	2 hours	3 hours	4 hours
4 x 4-1/2 wall thickness	120	1.44	2.72	3.76	4.68
	130	1.62	3.00	4.12	5.11
4 x 4-3/8 wall thickness	120	1.56	2.84	3.88	4.78
	130	1.74	3.12	4.23	5.21
4 x 4-1/4 wall thickness	120	1.72	2.99	4.02	4.92
	130	1.89	3.26	4.37	5.34
6 x 6-1/2 wall thickness	120	1.33	2.58	3.62	4.52
	130	1.50	2.86	3.98	4.96
6 x 6-3/8 wall thickness	120	1.48	2.74	3.76	4.67
	130	1.65	3.01	4.13	5.10
6 x 6-1/4 wall thickness	120	1.66	2.91	3.94	4.84
	130	1.83	3.19	4.30	5.27
8 x 8-1/2 wall thickness	120	1.27	2.50	3.52	4.42
	130	1.44	2.78	3.89	4.86
8 x 8-3/8 wall thickness	120	1.43	2.67	3.69	4.59
	130	1.60	2.95	4.05	5.02
8 x 8-1/4 wall thickness	120	1.62	2.87	3.89	4.78
	130	1.79	3.14	4.24	5.21

Steel pipe

Column size	Concrete masonry density, lb/ft³	Minimum equivalent thickness for fire-resistance rating of clay masonry protection assembly T_e, in.			
		1 hour	2 hours	3 hours	4 hours
Four double extra-strong 0.674 wall thickness	120	1.26	2.55	3.60	4.52
	130	1.42	2.82	3.96	4.95
Four extra-strong 0.337 wall thickness	120	1.60	2.89	3.92	4.83
	130	1.77	3.16	4.28	5.25
Four standard 0.237 wall thickness	120	1.74	3.02	4.05	4.95
	130	1.92	3.29	4.40	5.37
Four double extra-strong 0.750 wall thickness	120	1.17	2.44	3.48	4.40
	130	1.33	2.72	3.84	4.83
Four extra-strong 0.375 wall thickness	120	1.55	2.82	3.85	4.76
	130	1.72	3.09	4.21	5.18
Four standard 0.258 wall thickness	120	1.71	2.97	4.00	4.90
	130	1.88	3.24	4.35	5.32
Six double extra-strong 0.864 wall thickness	120	1.04	2.28	3.32	4.23
	130	1.19	2.60	3.68	4.67
Six extra-strong 0.432 wall thickness	120	1.45	2.71	3.75	4.67
	130	1.62	2.99	4.10	5.08
Six standard 0.280 wall thickness	120	1.65	2.91	3.94	4.84
	130	1.82	3.19	4.30	5.27

W shapes

Column size	Clay masonry density, lb/ft³	Minimum equivalent thickness for fire-resistance rating of clay masonry protection assembly T_e, in.			
		1 hour	2 hours	3 hours	4 hours
W14 x 82	120	1.23	2.42	3.41	4.29
	130	1.40	2.70	3.78	4.74
W14 x 68	120	1.34	2.54	3.54	4.43
	130	1.51	2.82	3.91	4.87
W14 x 53	120	1.43	2.65	3.65	4.54
	130	1.61	2.93	4.02	4.98
W14 x 43	120	1.54	2.76	3.77	4.66
	130	1.72	3.04	4.13	5.09
W12 x 72	120	1.32	2.52	3.51	4.40
	130	1.50	2.80	3.88	4.84
W12 x 58	120	1.40	2.61	3.61	4.50
	130	1.57	2.89	3.98	4.94
W12 x 50	120	1.43	2.65	3.66	4.55
	130	1.61	2.93	4.02	4.99
W12 x 40	120	1.54	2.77	3.78	4.67
	130	1.72	3.05	4.14	5.10

Column size	Clay masonry density, lb/ft³	Minimum equivalent thickness for fire-resistance rating of clay masonry protection assembly T_e, in.			
		1 hour	2 hours	3 hours	4 hours
W10 x 68	120	1.27	2.46	3.46	4.35
	130	1.44	2.75	3.83	4.80
W10 x 54	120	1.40	2.61	3.62	4.51
	130	1.58	2.89	3.98	4.95
W10 x 45	120	1.44	2.66	3.67	4.57
	130	1.62	2.95	4.04	5.01
W10 x 33	120	1.59	2.82	3.84	4.73
	130	1.77	3.10	4.20	5.13
W8 x 40	120	1.47	2.70	3.71	4.61
	130	1.65	2.98	4.08	5.04
W8 x 31	120	1.59	2.82	3.84	4.73
	130	1.77	3.10	4.20	5.17
W8 x 24	120	1.66	2.90	3.92	4.82
	130	1.84	3.18	4.82	5.25
W8 x 18	120	1.75	3.00	4.01	4.91
	130	1.93	3.27	4.37	5.34

Note: Tabulated values assume 1 in. air gap between masonry and steel section.

CODE REQUIREMENTS FOR DETERMINING FIRE RESISTANCE OF CONCRETE AND MASONRY CONSTRUCTION ASSEMBLIES

6.7—Structural steel columns protected by clay masonry

6.7.1 *Calculation of fire resistance*—It shall be permitted to calculate fire resistance of a structural steel column protected of clay masonry necessary for meeting a fire-resistance requirement, following the methods of 5.6. For this calculation, the thermal conductivity of the clay masonry shall be taken as follows.

$$\text{density} = 120 \text{ lb/ft}^3; \quad k_{cm} = 1.25 \text{ Btu/(h-ft-°F)}$$

$$\text{density} = 130 \text{ lb/ft}^3; \quad k_{cm} = 2.25 \text{ Btu/(h-ft-°F)}$$

The minimum required equivalent thicknesses of clay masonry for specified fire resistance of several commonly used column shapes and sizes are shown in Table 6.7.1.

CHAPTER 7—EFFECTS OF FINISH MATERIALS ON FIRE RESISTANCE

7.1—General

The contribution of additional fire resistance provided by finish materials installed on concrete or masonry assemblies shall be determined in accordance with the provisions of this chapter. The increase in fire resistance of the assembly shall be based strictly on the influence of the finish material's ability to extend the heat transmission end-point in an ASTM E119 test fire.

7.2—Calculation procedure

The fire-resistance rating of walls or slabs of cast-in-place or precast concrete, or walls of concrete or clay masonry with finishes of gypsum wallboard or plaster applied to one or both sides of the wall or slab shall be determined in accordance with this section.

7.2.1 *Fire exposure*—For a wall or slab having no finish on one side or having different types, thicknesses, or both, of finish on each side, the calculation procedures in 7.2.2 and 7.2.3 shall be performed, assuming that each side of the wall or slab is the fire-exposed side. The resulting fire resistance of the wall or slab, including finishes, shall not exceed the smaller of the two values calculated, except in the case of the building code requiring that walls or slabs only be rated for fire exposure from one side of the wall or slab.

7.2.2 *Calculation for non-fire-exposed side*—Where the finish is applied to the non-fire-exposed side of the slab or wall, the fire resistance of the entire assembly shall be determined as follows. The thickness of the finish shall be adjusted by multiplying the actual thickness of the finish by the applicable factor from Table 7.2.2 based on the type of aggregate in the concrete or concrete masonry units, or the type of clay masonry. The adjusted finish thickness shall be added to the actual thickness or equivalent thickness of the wall or slab, then determine the fire resistance of the concrete or masonry, including the effect of finish, from Table 4.2, Fig. 4.2.3, or Fig. 4.2.5.1 for concrete; from Table 5.1a for concrete masonry; or from Table 6.3.1 for clay masonry.

7.2.3 *Calculation for fire-exposed side*—Where the finish is applied to the fire-exposed side of the slab or wall, the fire resistance of the entire assembly shall be determined as follows. The time assigned to the finish in Table 7.2.3 shall be added to the fire resistance determined from Table 4.2, Fig. 4.2.3, or Fig. 4.2.5.1 for the concrete alone; from Table 5.1a for concrete masonry; from Table 6.3.1 for clay masonry; or to the fire resistance as determined in accordance with 7.2.2 for the concrete or masonry and finish on the non-fire-exposed side.

Table 7.2.2—Multiplying factor for finishes on non-fire-exposed side of concrete slabs and concrete and masonry walls

Type of material used in slab or wall	Portland cement-sand plaster* or terrazzo	Gypsum-sand plaster	Gypsum-vermiculite or perlite plaster	Gypsum washboard
Concrete slab or wall				
Concrete—Siliceous, carbonate, air-cooled blast-furnace slag	1.00	1.25	1.75	3.00
Concrete—semi-lightweight	0.75	1.00	1.50	2.25
Concrete—lightweight, insulating concrete	0.75	1.00	1.25	2.25
Concrete masonry wall				
Concrete masonry—siliceous, calcareous, limestone, cinders, air-cooled blast-furnace slag	1.00	1.25	1.75	3.00
Concrete masonry— made with 80 percent or more by volume of expanded shale, slate or clay, expanded slag, or pumice	0.75	1.00	1.25	2.25
Clay masonry wall				
Clay masonry—solid brick of clay or shale	1.00	1.25	1.75	3.00
Clay masonry—hollow brick or tile of clay or shale	0.75	1.00	1.50	2.25

*For portland cement-sand plaster 5/8 in. or less in thickness and applied directly to concrete or masonry on the non-fire-exposed side of the wall, multiplying factor shall be 1.0.

Table 7.2.3—Time assigned to finish materials on fire-exposed side of concrete and masonry walls

Finish description	Time, min.
Gypsum wallboard	
3/8 in.	10
1/2 in.	15
5/8 in.	20
Two layers of 3/8 in.	25
One layer of 3/8 in. and one layer of 1/2 in.	35
Two layers of 1/2 in.	40
Type "X" gypsum wallboard	
1/2 in.	25
5/8 in.	40
Direct-applied portland cement-sand plaster*	
Portland cement-sand plaster on metal lath	
3/4 in.	20
7/8 in.	25
1 in.	30
Gypsum-sand plaster on 3/8 in. gypsum lath	
1/2 in.	35
5/8 in.	40
3/4 in.	50
Gypsum-sand plaster on metal lath	
3/4 in.	50
7/8 in.	60
1 in.	80

*For purposes of determining the contribution of portland cement-sand plaster to the equivalent thickness of concrete or masonry for use in Tables 4.2, 5.1a, or 6.3.1, it shall be permitted to use the actual thickness of the plaster or 5/8 in., whichever is smaller.

7.2.4 *Minimum fire resistance provided by concrete or masonry*—Where the finish applied to a concrete slab or a concrete or masonry wall contributes to the fire resistance, the concrete or masonry alone shall provide not less than half of the total required fire resistance. In addition, the contribution to fire resistance of the finish on the non-fire-exposed side of the wall shall not exceed half the contribution of the concrete or masonry alone.

7.3—Installation of finishes

Finishes on concrete slabs and concrete and masonry walls that are assumed to contribute to the total fire resistance shall comply with the installation requirements of 7.3.1, 7.3.2, and other applicable provisions of the building code. Plaster and terrazzo shall be applied directly to the slab or wall. Gypsum wallboard shall be permitted to be attached to wood or steel furring members or attached directly to walls by adhesives.

7.3.1 *Gypsum wallboard*—Gypsum wallboard and gypsum lath shall be attached to concrete slabs and concrete and masonry walls in accordance with the requirements of this section or as otherwise permitted by the building code.

7.3.1.1 *Furring*—Gypsum wallboard and gypsum lath shall be attached to wood or steel furring members spaced not more than 24 in. on center. Gypsum wallboard and gypsum lath shall be attached in accordance with one of the methods in 7.3.1.1.1 or 7.3.1.1.2.

7.3.1.1.1 Self-tapping drywall screws shall be spaced at a maximum of 12 in. on center and shall penetrate 3/8 in. into resilient steel furring channels running horizontally and spaced at a maximum of 24 in. on center.

7.3.1.1.2 Lath nails shall be spaced at a maximum of 12 in. on center and shall penetrate 3/4 in. into nominal 1 x 2 in. wood furring strips that are secured to the masonry by 2 in. concrete nails, and spaced at a maximum of 16 in. on center.

7.3.1.2 *Adhesive attachment to concrete and clay masonry*—Place a 3/8 in. bead of panel adhesive around the perimeter of the wallboard and across the diagonals. After the wall board is laminated to the masonry surface, secure it with one masonry nail for each 2 ft^2 of panel.

7.3.1.3 *Gypsum wallboard orientation*— Gypsum wallboard shall be installed with the long dimension parallel to furring members and with all horizontal and vertical joints supported and finished. An exception is 5/8 in. thick Type "X" gypsum wallboard is permitted to be installed horizontally on walls with the horizontal joints unsupported.

7.3.2 *Plaster and stucco*—Plaster and stucco attached to a concrete or masonry surface for the purpose of increasing fire resistance shall be applied in accordance with provisions of the building code.

ACI/TMS STANDARD

CODE REQUIREMENTS FOR DETERMINING FIRE RESISTANCE OF CONCRETE AND MASONRY CONSTRUCTION ASSEMBLIES

TEXT FROM

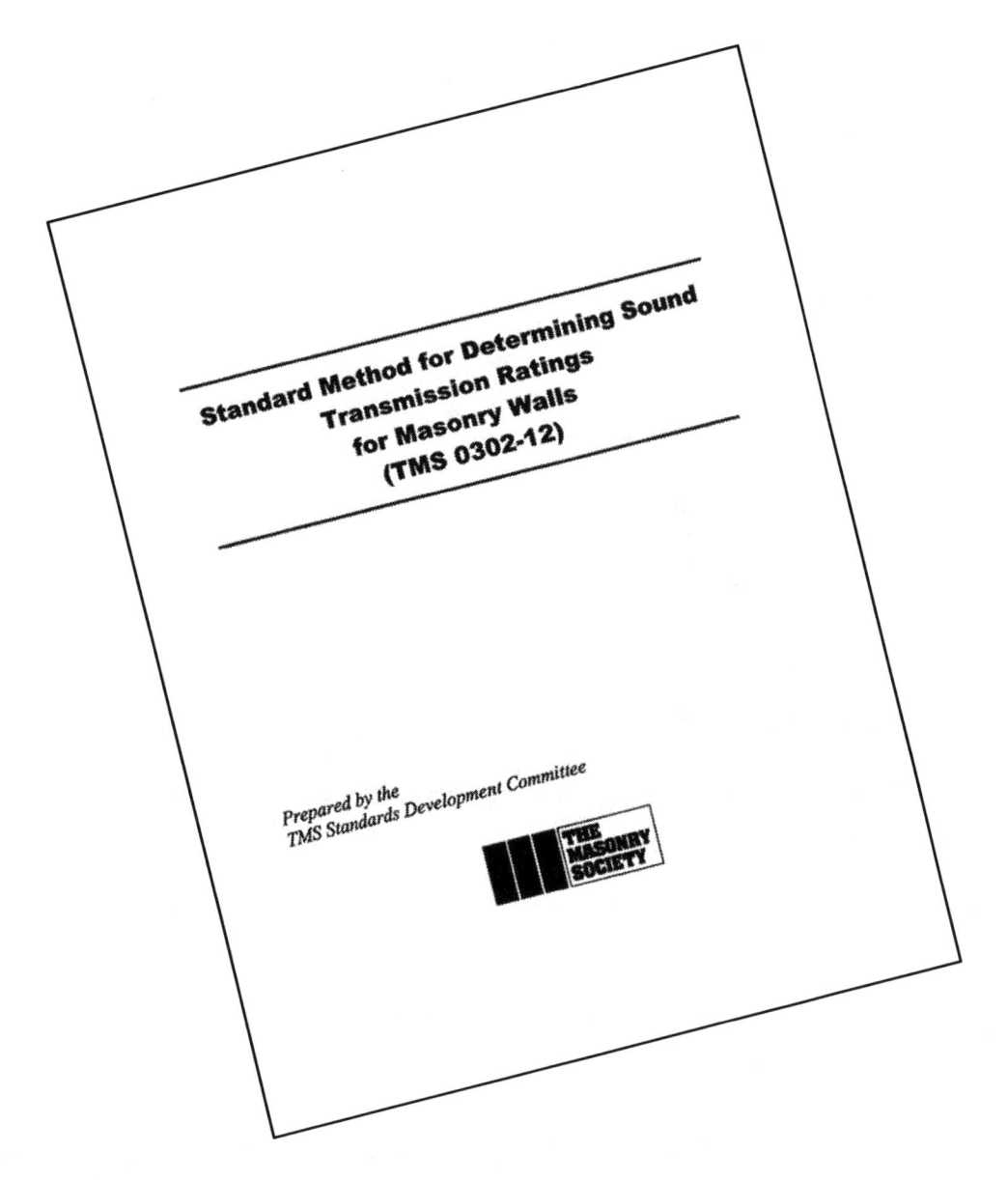

"*Standard Method for Determining Sound Transmission Ratings for Masonry Walls (TMS 0302-12)*, copyright © 2012, is reproduced with the permission of the publisher, The Masonry Society."

The Masonry Society
105 South Sunset Street, Suite Q
Longmont, CO 80501
www.masonrysociety.org

STANDARD METHOD FOR DETERMINING SOUND TRANSMISSION RATINGS FOR MASONRY WALLS (TMS 0302-12)

Standard Method for Determining Sound Transmission Ratings for Masonry Walls (TMS 0302-12)

Prepared by TMS Standards Development Committee
Published by The Masonry Society
　　　　　　　105 South Sunset Street, Suite Q
　　　　　　　Longmont, CO USA 80501-6172
　　　　　　　Phone: 303-939-9700
　　　　　　　Fax:　　303-541-9215
　　　　　　　E-Mail: info@masonrysociety.org
　　　　　　　Website: www.masonrysociety.org

ABSTRACT

Standard Method for Determining Sound Transmission Ratings for Masonry Walls (TMS 0302-12) (hereinafter referred to as the Standard) was developed by The Masonry Society's Standards Development Committee. The Standard provides minimum requirements for rating masonry walls for a sound transmission class, *STC*, and outdoor-indoor transmission class, *OITC*, based on field or laboratory testing or based on a calculation procedure. Some of the topics covered include reference standards, definitions and notations, materials, construction, quality assurance and sound transmission class rating. The Standard is written as a legal document in mandatory language so that it may be incorporated into a legally-adopted building code. The Commentary presents background details, committee considerations, and research data used to develop the Standard.

Copyright © 2012, The Masonry Society.

All rights reserved including rights of reproduction and use in any form or by any means, including the making of copies by any photo process, or by any electronic or mechanical device, printed, written or oral, or recording for sound or visual reproduction or for any use in any knowledge retrieval system or device, unless permission in writing is obtained from The Masonry Society.

ISBN 1-929081-40-5
TMS Order No. TMS 0302-12

STANDARD METHOD FOR DETERMINING SOUND TRANSMISSION RATINGS FOR MASONRY WALLS (TMS 0302-12)

Standard Method for Determining Sound Transmission Ratings for Masonry Walls
(TMS 0302-12)

Developed by The Masonry Society's Standards Development Committee

Craig V. Baltimore, Chairman
Dennis W. Graber, Secretary

Voting Members[1]

Thomas A. Hagood Jr.	David I. McLean	Craig Parrino
Lawrence F. Kahn	Raymond T. Miller	J. Eric Peterson
W. Mark McGinley	Jerry M. Painter	Max L. Porter

Corresponding Members[2]

Bechara E. Abboud	Thomas F. Herrell	Jennifer E. Tanner
Christine Beall	Lee S. Johnson	Michael Tate
Olene L. Bigelow	Sunup S. Mathew	Bruce Weems
Charles B. Clark Jr.	John H. Matthys	Terence Allan Weigel
Thomas A. Gangel	Vilas Mujumdar	Daniel Zechmeister

Technical Activities Committee

The Technical Activities Committee (TAC) of The Masonry Society (TMS) is responsible for reviewing, and approving the work of TMS Technical Committees. As such, TAC monitored Committee balloting, reviewed the draft of this *Standard*, reviewed and approved responses to TAC and public comments, and gave final approval of the *Standard*. Members of the Technical Activities Committee who specifically assisted with review of this *Standard* are shown below.

Peter M. Babaian	Darrell W. McMillian	Phillip J. Samblanet
David I. McLean, Chairman	Sarah Lowe Rogers	Jason J. Thompson

Synopsis

This Standard provides minimum requirements for rating masonry walls for a sound transmission class, STC, and outdoor-indoor transmission class, OITC, based on field or laboratory testing or based on a calculation procedure. It is written in such a form that it may be adopted by reference in a legally-adopted building code.

Among the topics covered are reference standards, definitions and notations, materials, construction, quality assurance and sound transmission class rating. The Standard is written as a legal document in mandatory language so that it may be incorporated into a legally-adopted building code. The Commentary to this Standard presents background details, committee considerations, and research data used to develop the Standard.

Keywords

clay brick; clay tile; coatings; concrete masonry; construction; control joints; expansion joints; field sound transmission class; grout; joint sealants; joints; masonry; masonry units; mortar; and outdoor-indoor transmission class; sealants; sound; sound transmission; sound transmission class; sound transmission loss; wall weight

---

1. Voting members fully participate in Committee activities including responding to correspondence and voting.
2. Corresponding members may participate in Committee Activities, but do not have voting privileges.

STANDARD METHOD FOR DETERMINING SOUND TRANSMISSION RATINGS FOR MASONRY WALLS (TMS 0302-12)

Standard Method for Determining Sound Transmission Ratings for Masonry Walls
TMS Standard TMS 0302-12

STANDARD METHOD FOR DETERMINING SOUND TRANSMISSION RATINGS FOR MASONRY WALLS (TMS 0302-12)

Standard Method for Determining Sound Transmission Ratings for Masonry Walls
TMS Standard TMS 0302-12

1 — Scope

This Standard provides minimum requirements for rating masonry walls for a sound transmission class, *STC*, and outdoor-indoor transmission class *OITC*. These ratings are for masonry walls in structures erected under the requirements of the legally-adopted building code of which this Standard forms a part. In areas without a legally-adopted building code, this Standard defines minimum acceptable methods to determine the *STC* and *OITC* ratings of masonry wall assemblies. These ratings of masonry walls are based on field or laboratory testing in accordance with standard test methods or are based on calculation procedures. All masonry dimensions referred to in this standard are nominal unless indicated otherwise.

2 — Reference Standards

ASTM C 34-10 Specification for Structural Clay Loadbearing Wall Tile

ASTM C 55-11a Specification for Concrete Brick

ASTM C 56-10 Specification for Structural Clay Nonloadbearing Tile

ASTM C 62-10 Specification for Building Brick (Solid Masonry Units Made from Clay or Shale)

ASTM C 73-10 Specification for Calcium Silicate Face Brick (Sand-Lime Brick)

ASTM C 90-11a Specification for Loadbearing Concrete Masonry Units

ASTM C 126-11 Specification for Ceramic Glazed Structural Clay Facing Tile, Facing Brick and Solid Masonry Units

ASTM C 129-11 Specification for Nonloadbearing Concrete Masonry Units

ASTM C 212-11 Specification for Structural Clay Facing Tile

ASTM C 216-11 Specification for Facing Brick (Solid Masonry Units Made from Clay or Shale)

ASTM C 270-10 Specification for Mortar for Unit Masonry

ASTM C 476-10 Standard Specification for Grout for Masonry

ASTM C 652-11 Specification for Hollow Brick (Hollow Masonry Units Made from Clay or Shale)

ASTM C 744-11 Specification for Prefaced Concrete and Calcium Silicate Masonry Units

ASTM C 920-11 Specification for Elastomeric Joint Sealants

ASTM C 1405-10 Specification for Glazed Brick (Single Fired, Brick Units)

ASTM C 1634-11 Specification for Concrete Facing Brick

ASTM C1714 / C1714M - 10 Specification for Preblended Dry Mortar Mix for Unit Masonry

ASTM E 90-09 Test Method for Laboratory Measurement of Airborne Sound Transmission Loss of Building Partitions and Elements

ASTM E 336-11 Test Method for Measurement of Airborne Sound Attenuation between Rooms in Buildings

ASTM E 413-10 Classification for Rating Sound Insulation

ASTM E 996-04 Guide for Field Measurements of Airborne Sound Insulation of Building Facades and Facade Elements

ASTM E 1332-10a Standard Classification for Rating Outdoor-Indoor Sound Attenuation

3 — Notations

DSTC = the change in *STC* rating from a bare concrete masonry wall
d = the thickness of the furring space
FSTC = Field Sound Transmission Class
OITC = Outdoor-Indoor Transmission Class
STC = Sound Transmission Class
STL = Sound Transmission Loss
W = Wall Weight, psf. (kg/m^2)

Standard Method for Determining Sound Transmission Ratings for Masonry Walls
TMS Standard TMS 0302-12

4 — Definitions

Coarse textured — a relative term referring to the porosity of the matrix through the thickness of the masonry which is related to the airflow through a masonry unit.

Field Sound Transmission Class, FSTC — Sound transmission class calculated using values of field transmission loss.

Outdoor-Indoor Transmission Class, OITC — A single-number rating calculated in accordance with ASTM E 1332 using values of sound transmission loss.

Sound absorbing material — Fibrous materials, such as cellulose fiber, glass fiber, or rock wool insulation.

Sound Transmission Class, STC — A single-number rating calculated in accordance with ASTM E 413 using values of sound transmission loss.

Sound Transmission Loss, STL — A measure equal to ten times the common logarithm of the ratio of the airborne sound power, in a specified frequency band, incident on the wall to the sound power transmitted by the wall and radiated on the opposite side of the wall.

Wall weight, W — The average wall weight based on the weight of the masonry units; the weight of mortar, grout and loose fill material in voids within the wall and the weight of plaster, stucco, and paint. The weight of drywall shall not be included.

5 — Materials

5.1 Masonry Units
Masonry units shall comply with the requirements of one of the following standards: ASTM C 34, ASTM C 55, ASTM C 56, ASTM C 62, ASTM C 73, ASTM C 90, ASTM C 126, ASTM C 129, ASTM C 212, ASTM C 216, ASTM C 652, ASTM C 744, ASTM C 1405 or ASTM C 1634.

5.2 Mortar
Mortar shall comply with the requirements of ASTM C 270.

5.3 Grout
Grout shall comply with the requirements of ASTM C 476.

5.4 Joint Sealants
Joint sealants shall comply with the requirements of ASTM C 920.

6 — Construction
Construction shall conform to the requirements of TMS 602 for concrete masonry and clay masonry and shall conform to the requirements of this Standard.

6.1 Sealing openings and joints

6.1.1 Seal through-wall openings with joint sealant, mortar, or grout. Prior to sealing around through-wall openings, fill gaps with foam, cellulose fiber, glass fiber, ceramic fiber, or mineral wool.

6.1.2 Seal membrane penetration openings and inserts with joint sealant, mortar, or grout.

6.1.3 Seal control joints, expansion joints, and joints between the top of walls and roof or floor assemblies with joint sealant. Fill the joint space behind the sealant backing with mortar, grout, foam, cellulose fiber, glass fiber, or mineral wool. Where roof or floor construction is metal deck, use special shape foam filler strips to seal the top of the wall. For fire rated walls and smoke containment walls, safing insulation shall be used.

6.1.4 Noncompressible materials that is, mortar and grout shall not be used for fillers for expansion joints, but they shall be permitted in control joints.

6.2 Surface coatings
Seal coarse-textured concrete masonry walls on one or both faces with at least one coat of acrylic latex, alkyd or cement based paint, plaster, or other suitable coating.

Standard Method for Determining Sound Transmission Ratings for Masonry Walls
TMS Standard TMS 0302-12

7 — Sound Transmission Class Ratings

The sound transmission class, *STC*, ratings of masonry walls shall be determined in accordance with Section 7.1, 7.2, or 7.3.

7.1 Laboratory testing

The *STC* ratings of masonry walls based on laboratory testing of an assembly that is representative of the actual wall construction shall be tested and determined in accordance with the requirements of ASTM E 90.

7.2 Field testing

The *STC* ratings of masonry walls first by field testing the completed construction shall be determined in accordance with the requirements of ASTM E336. The measured values shall then be used to determine the field *STC* for the masonry in accordance with ASTM E413.

7.3 Calculation

Except for walls having a thickness less than 3 inches (75 mm), the *STC* ratings for masonry walls shall be determined based on calculation using Eq. 1 for clay masonry or Eq. 2 for concrete masonry. For multiwythe walls consisting of concrete masonry and clay masonry wythes, the sound transmission class shall be determined by Eq. 1 and Eq. 2 utilizing the total of the multiwythes for wall weight, *W* and then interpolate between the two values according to the relative weights of the two materials in the wall.

$$STC = 19.6W^{0.230} \quad \text{Eq. 1.}$$
$$SI \quad STC = 13.6W^{0.230}$$

$$STC = 20.5W^{0.234} \quad \text{Eq. 2.}$$
$$SI \quad STC = 14.1W^{0.234}$$

7.3.1 Effect of Drywall on *STC* Ratings of Concrete Masonry Walls

7.3.1.1 When *STC* ratings are determined by the calculation method, drywall attached directly to the concrete masonry shall be assumed to not change the, *STC*, rating.

7.3.1.2 The change in the sound transmission class, *STC*, ratings for 1/2-inch (13 mm) or 5/8-inch (16 mm) thick drywall attached to concrete masonry walls with resilient metal furring or a combination of resilient metal furring and wood furring shall be determined using Eq. 3, 4, 5, or 6 as appropriate. Where sound absorbing material is used, it shall fill the entire furring space. When drywall is applied to both sides of the wall, the cavity depth shall be identical on each side.

For drywall on one side of the wall with no sound absorbing material in the furring space:
$$DSTC = 2.8d - 1.22 \quad \text{Eq. 3}$$
$$SI \quad DSTC = 0.11d - 1.22$$

For drywall on both sides of the wall and no sound absorbing material in the furring spaces:
$$DSTC = 3.6d - 2.78 \quad \text{Eq. 4}$$
$$SI \quad DSTC = 0.14d - 2.78$$

For drywall on one side of the wall with sound absorbing material in the furring space:
$$DSTC = 3.0d + 1.87 \quad \text{Eq. 5}$$
$$SI \quad DSTC = 0.12d + 1.87$$

For drywall on both sides of the wall and sound absorbing material in the furring spaces:
$$DSTC = 11.2d - 7.37 \quad \text{Eq. 6}$$
$$SI \quad DSTC = 0.44d - 7.37$$

8 — Outdoor-Indoor Transmission Class Ratings

The Outdoor-Indoor Transmission Class, *OITC*, ratings of masonry walls shall be determined in accordance with Section 8.1, 8.2, or 8.3.

8.1 Laboratory testing

The *OITC* ratings of masonry walls shall be determined based on laboratory testing of an assembly that is representative of the actual wall construction tested in accordance with the requirements of ASTM E 90 and calculated in accordance with ASTM E 1332.

8.2 Field testing

The *OITC* ratings of masonry walls shall be determined by first field testing the completed construction in accordance with the requirements of ASTM E 966. The measured values shall then be used to determine the field *OITC* for the masonry in accordance with ASTM E1332.

8.3 Calculation

Except for walls having a thickness less than 3 inches (75 mm), the *OITC* ratings for masonry walls shall be determined based on calculation using Eq. 7 for clay masonry or Eq. 8 for concrete masonry. For multiwythe walls consisting of concrete masonry and clay masonry wythes, the outdoor-indoor transmission class shall be determined by Eq. 7 and Eq. 8 utilizing the total of all the wythes for wall weight, *W* and then interpolate between the two values according to the relative weights of the two materials in the wall.

$$OITC = 17.4 W^{0.224}$$
$$SI \quad OITC = 12.2 W^{0.224}$$
Eq. 7.

$$OITC = 14.7 W^{0.290}$$
$$SI \quad OITC = 9.28 W^{0.290}$$
Eq. 8.

STANDARD METHOD FOR DETERMINING SOUND TRANSMISSION RATINGS FOR MASONRY WALLS (TMS 0302-12)

Commentary
Standard Method for Determining Sound Transmission Ratings for Masonry Walls
(TMS 0302-12)

This commentary accompanies the Standard and provides an explanation of and justification for the requirements of the Standard. This commentary is not intended to be part of the Standard. The Standard is a concise statement of requirements and is intended to be adopted by reference in construction documents, building codes, and other standards. The commentary, on the other hand, provides background information including illustrations, example applications, and clarifications of the requirements of the Standard and is not intended to be adopted by reference in other documents. The commentary is intended to assist the designer and other users of the Standard in applying the Standard and in understanding the basis for specific requirements of the Standard.

STANDARD METHOD FOR DETERMINING SOUND TRANSMISSION RATINGS FOR MASONRY WALLS (TMS 0302-12)

Commentary to TMS Standard TMS 0302-12

1 — Scope

Ratings of masonry walls are based on field or laboratory testing in accordance with standard test methods or by calculation. Performance of walls in resisting sound transmission depends on:

(1) sound in the noisier room,
(2) sound transmission loss of the wall, and
(3) background noise in the quieter room.

Sound transmission loss, *STL*, is the decrease or attenuation in sound energy expressed in decibels (dB) of air borne sound as it passes through a wall. In general, *STL* increases as the frequency of the sound increases.

Sound transmission class, *STC*, is determined by ASTM E 90 and ASTM E 413. It provides an estimate of the performance of a wall in certain common sound insulation applications. The *STC* of masonry walls is approximately 4 dB greater than the *STL* at a frequency of 500 Hz (cycles per second). Although *STC* is a convenient index to relative sound transmission, the *STL* spectra should be studied in order to meet particular sound transmission requirements.

Outdoor-indoor transmission class, *OITC*, is determined in accordance with ASTM E1332. ASTM E 1332 presents a standard procedure to determine *OITC* based on measured sound transmission loss, *STL*, across a wall or wall element at frequencies from 80 to 4,000 Hz. *OITC* is calculated using tested *STL* values and the sound spectrum of a reference sound source. This reference sound spectrum is an average of typical spectra from three transportation noise sources: aircraft takeoff, freeway, and railroad passby. The reference sound spectrum is A-weighted to better correlate with human hearing (A-weighting is a frequency response adjustment that accounts for the changes in human hearing sensitivity as a function of frequency).

2 — Reference Standards

No commentary.

3 — Notations

No commentary.

4 — Definitions

The weight of the drywall is not to be included in the calculated sound rating of the wall assembly. The effect of drywall, a cavity and sound absorbing insulation is considered as an adjustment to the STC rating of the bare masonry wall. The air space between the drywall can resonate somewhat like the skin of a drum and actually reduce the STC rating of the wall as reflected by Equations 3 and 4 and Commentary Table 7.3-5. When drywall is attached directly to the surface of coarse-textured concrete masonry, it provides the same benefit for sound transmission loss as for sealing the surface but provides no additional benefit due to its mass (ref. 7).

5 — Materials

No commentary.

6 — Construction

Review the provisions of TMS 602 and identify any provisions that may conflict with the provisions of this Standard. Resolve conflicts between TMS 602 and this Standard prior to completing the Contract Documents.

6.1 Sealing openings and joints

The type of hole, crack, void and wall penetration may greatly affect the transmission loss of a wall (see Figure 6.2.1). This variation is hard to quantify and it is only prudent to seal all holes, cracks, voids, and wall penetrations (see Figure 6.2.2).

To act as an effective sound barrier, partitions should be carried to the underside of the floor or roof. The joint between the underside of a slab and top of a partition should provide for slab deflection and be sealed against sound transmission. Fire-rated assemblies are also required to meet fire resistive construction requirements including fire stopping of through-wall penetrations and fire-resistive sealing materials in accordance with the legally adopted building code. If roof or floor construction is metal deck rather than concrete, it is not feasible to use joint sealants alone to seal top of wall because of the shape of the deck flutes. For fire and smoke containment walls, safing insulation is used instead of foam filler strips.

STANDARD METHOD FOR DETERMINING SOUND TRANSMISSION RATINGS FOR MASONRY WALLS (TMS 0302-12)

Commentary to
TMS Standard TMS 0302-12

Transmission of Airborne Noise

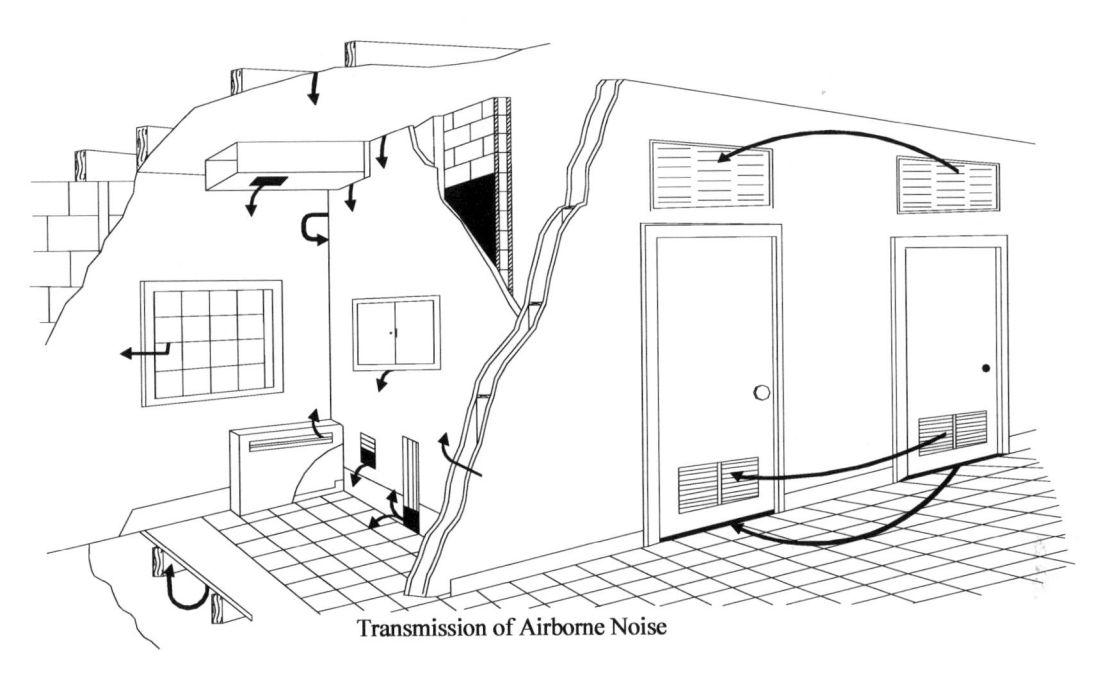

Figure 6.2.1 — Acoustical Leaks (Ref. 1)

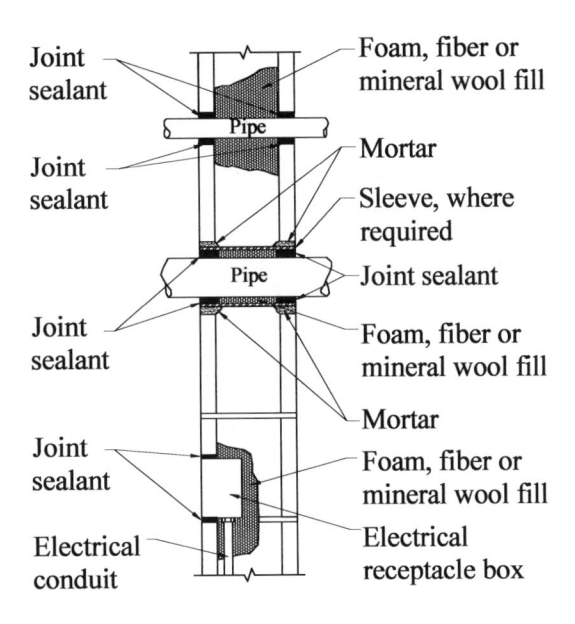

Figure 6.2.2 — Sealing Around Penetrations and Fixtures

STANDARD METHOD FOR DETERMINING SOUND TRANSMISSION RATINGS FOR MASONRY WALLS (TMS 0302-12)

6.2 Surface coatings

Fine and medium textured concrete masonry unit walls and fired clay masonry unit walls do not require additional surface treatments, however coarse textured concrete masonry unit walls which may allow airborne sound to enter the wall require a surface treatment to seal at least one surface of the wall. Coatings of acrylic latex, alkyd or cement-based paint; or of plaster are acceptable. Other coatings are also acceptable provided they effectively seal the surface of coarse-textured concrete masonry units.

There is substantial discussion about the effect of porosity of concrete masonry units in Reference 7. This reference included both lightweight aggregate and what was termed very porous (wood aggregate) blocks in their study for the purposes of comparison. The report indicated that leakage of sound was somewhat related to the airflow resistivity of the units and also that sealing of coarse-textured units on only one surface to be effective. Normal weight blocks showed little or no improvement in sound transmission resistivity after sealing. Texture as used in this context does not refer to the surface roughness of the block but the matrix of the mix used in manufacturing the block.

The committee reasoned that, in most cases, sound rated walls with coarse-textured units would have a surface treatment on at least one surface. Therefore the data for unsealed coarse-textured masonry units was not included in developing the concrete masonry equations for calculated sound transmission ratings as it would have had a negative effect on the equations.

7 — Sound Transmission Class Ratings

7.1 Laboratory testing

Representative masonry materials need not be from the same manufacturer.

7.2 Field testing

No commentary.

7.3 Calculation

Sound transmission class, STC, data of clay masonry walls (Ref. 2) are plotted against wall weight, W, in Figure 7.3-1. The equation for the curve best fitting the data is $STC = 19.6W^{0.230}$ ($STC = 13.6W^{0.230}$) with a correlation coefficient of 0.885. Figure 7.3-1 also shows that a power curve fit is better than a linear fit of the data.

STC data of concrete masonry walls (Ref. 3, 4, 5, and 6) are plotted against wall weight, W, in Figure 7.3-2. The equation for the curve best fitting the data is $STC = 20.5W^{0.234}$ ($STC = 14.1W^{0.234}$) with a correlation coefficient of 0.863. Figure 7.3-2 also shows that a power curve fit is better than a linear fit of the data.

The density of fired-clay products ranges from 103 to 142 pcf (1659 to 2275 kg/m^3) and averages 123 pcf (1970 kg/m^3). Table 7.3-1 lists calculated STC values for clay masonry having a density of 120 pcf (1922 kg/m^3). Similar tables could be developed for clay masonry of other densities.

The density of concrete in concrete masonry units typically ranges from 85 to 140 pcf (1362 to 2243 kg/m^3). Table 7.3-2 lists calculated STC values for concrete masonry of various densities.

The amount of acoustical testing on multi-wythe walls containing wythes of concrete masonry and clay masonry is not sufficient to develop a separate equation for this type of wall system. Additionally, the increased number of variables involved also makes it more difficult to establish an equation. Much higher STC values can be achieved by using materials other than wire ties to connect the two wythes, varying the cavity depth and the type of insulation used in the cavity (Ref. 6).

Commentary to TMS Standard TMS 0302-12

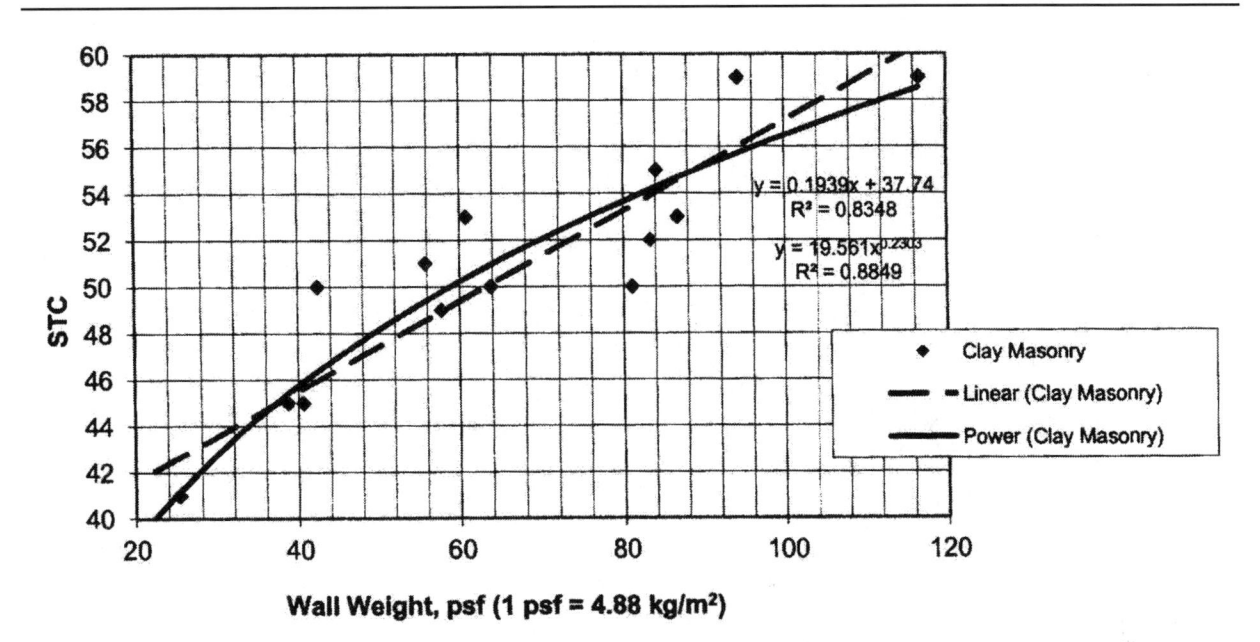

[Figure: STC vs Wall Weight plot with linear fit $y = 0.1939x + 37.74$, $R^2 = 0.8348$ and power fit $y = 19.561x^{0.2303}$, $R^2 = 0.8849$]

Figure 7.3-1 — Curve Fit for Clay Masonry

Table 7.3-1 — Data for Clay Masonry (Ref. 2)

Weight psf, (kg/m²)	Reported *STC*
22.3 (109)	39
25.3 (124)	41
38.7 (189)	45
40.6 (198)	45
42.4 (207)	50
55.8 (272)	51
57.7 (282)	49
60.8 (297)	53
63.8 (311)	50
81 (395)	50
83.3 (407)	52
84.1 (411)	55
86.7 (423)	53
94.2 (460)	59
116.7 (570)	59

STANDARD METHOD FOR DETERMINING SOUND TRANSMISSION RATINGS FOR MASONRY WALLS (TMS 0302-12)

**Commentary to
TMS Standard TMS 0302-12**

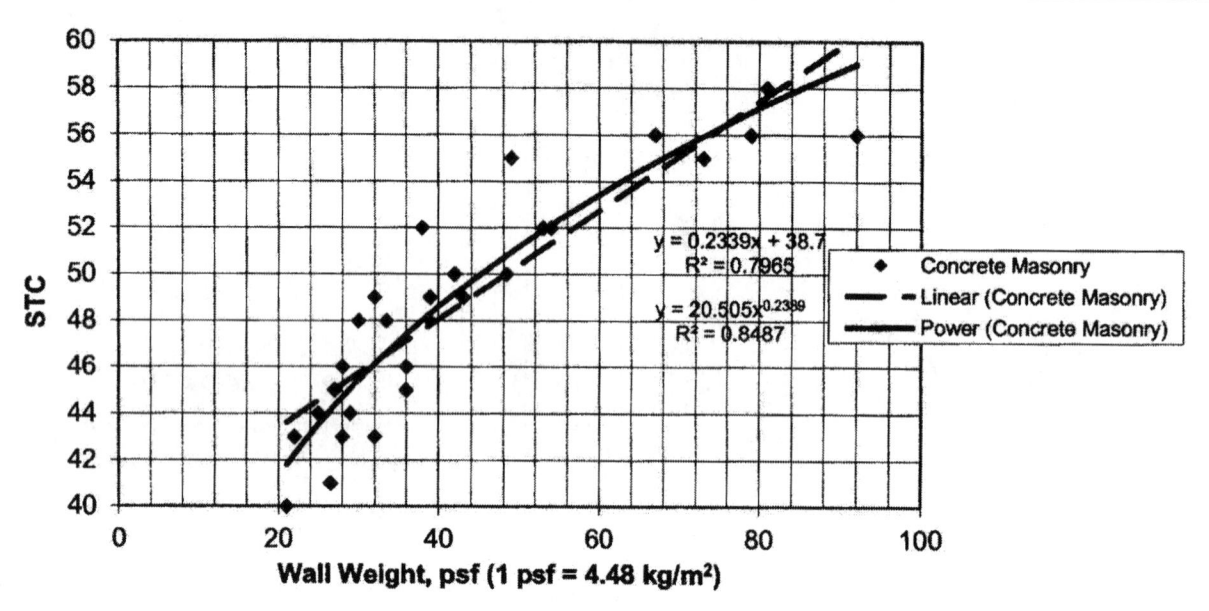

Figure 7.3-2 — Curve Fit for Concrete Masonry

**STANDARD METHOD FOR DETERMINING SOUND TRANSMISSION
RATINGS FOR MASONRY WALLS (TMS 0302-12)**

Commentary to TMS Standard TMS 0302-12

Table 7.3-2— Data for Concrete Masonry (Ref. 3, 4, 5, & 6)

Weight Class*	Finish**	Weight psf, (kg/m^2)	STC	Reference
L	0	21 (103)	40	Ref. 4
L	0	25 (122)	44	Ref. 4
L	0	36 (176)	45	Ref. 4
L	0	39 (190)	49	Ref. 4
L	0	43 (210)	49	Ref. 4
L	1	22 (107)	43	Ref. 4
L	1	28 (137)	46	Ref. 4
L	1	36 (176)	46	Ref. 4
L	1	32 (156)	43	Ref. 3
L	1	73 (356)	55	Ref. 4
L	2	28 (137)	43	Ref. 3
L	2	30 (146)	48	Ref. 4
L	2	32 (156)	49	Ref. 4
L	2	38 (186)	52	Ref. 4
L	2	42 (205)	50	Ref. 5
L	2	49 (239)	55	Ref. 4
L	2	54 (264)	52	Ref. 4
L	2	67 (327)	56	Ref. 4
L	2	79 (386)	56	Ref. 4
L	2	81 (395)	58	Ref. 4
N	1	29 (142)	44	Ref. 4
N	1	33.5 (164)	48	Ref. 4
N	1	39 (190)	48	Ref. 4
N	2	27 (132)	45	Ref. 5
N	2	42 (205)	50	Ref. 4
N	2	92 (449)	56	Ref. 4
N	2	54 (264)	52	Ref. 5
N	0	26.5 (129)	41	Ref. 4
N	0	48.4 (236)	50	Ref. 6
N	0	53 (259)	52	Ref. 4

*Weight class
 L=Lightweight
 N=Normal weight

**Finish:
 0=bare
 1=paint
 2=plaster

Commentary to TMS Standard TMS 0302-12

Table 7.3-3 — Calculated *STC* Ratings for Clay Masonry Walls[1]

Wall Thickness[2] in. (mm)	Hollow Units Weight psf, (kg/m^2)	STC	Grout Filled Weight psf, (kg/m^2)	STC	Sand Filled Weight psf, (kg/m^2)	STC	Solid Units Weight psf, (kg/m^2)	STC
3 (75)	Not applicable		Not applicable		Not applicable		30 (146)	43
4 (100)	20 (98)	39	38 (186)	45	32 (156)	43	35 (171)	44
6 (150)	32 (156)	43	63 (308)	51	50 (244)	48	55 (269)	49
8 (200)	42 (205)	46	86 (420)	55	68 (332)	52	75 (366)	53
10 (250)	53 (259)	49	109 (532)	58	86 (420)	55	95 (464)	56
12 (300)	62 (303)	51	132 (644)	60	104 (508)	57	115 (561)	58

[1] Based on the smaller specified unit dimension minus the specified tolerance, Clay density of 120 lb/ft^3 (586 kg/m^3); Grout density of 144 lb/ft^3 (703 kg/m^3), Sand density of 100 lb/ft^3 (488 kg/m^3). *STC* values for grout filled and sand filled units assume the materials completely occupy all void areas in and around the units. *STC* values for solid units are based on bed and head joints solidly filled with mortar.

[2] Dimensions in this column reflect equivalent nominal metric unit sizes as opposed to direct SI conversion.

STANDARD METHOD FOR DETERMINING SOUND TRANSMISSION RATINGS FOR MASONRY WALLS (TMS 0302-12)

Commentary to TMS Standard TMS 0302-12

Table 7.3-4 — Calculated *STC* Values for Concrete Masonry Walls[1]

Nominal Unit Size[2] in. (mm)	Density pcf (kg/m³)	STC Hollow Unit	STC Grout Filled	STC Sand Filled	STC Solid Units
4 (100)	80 (1281)	40	45	43	43
6 (150)	80 (1281)	41	51	48	48
8 (200)	80 (1281)	44	54	51	51
10 (250)	80 (1281)	46	58	54	54
12 (300)	80 (1281)	47	60	57	56

Nominal Unit Size[2] in. (mm)	Density pcf (kg/m³)	STC Hollow Unit	STC Grout Filled	STC Sand Filled	STC Solid Units
4 (100)	85 (1362)	40	45	44	44
6 (150)	85 (1362)	42	51	48	48
8 (200)	85 (1362)	44	55	52	52
10 (250)	85 (1362)	46	58	55	55
12 (300)	85 (1362)	48	61	57	57

Nominal Unit Size[2] in. (mm)	Density pcf (kg/m³)	STC Hollow Unit	STC Grout Filled	STC Sand Filled	STC Solid Units
4 (100)	90 (1442)	41	45	44	44
6 (150)	90 (1442)	42	51	49	49
8 (200)	90 (1442)	45	55	52	52
10 (250)	90 (1442)	47	58	55	55
12 (300)	90 (1442)	48	61	57	58

Nominal Unit Size[2] in. (mm)	Density pcf (kg/m³)	STC Hollow Unit	STC Grout Filled	STC Sand Filled	STC Solid Units
4 (100)	95 (1522)	41	46	44	45
6 (150)	95 (1522)	43	51	49	49
8 (200)	95 (1522)	45	55	52	53
10 (250)	95 (1522)	48	58	55	56
12 (300)	95 (1522)	49	61	58	58

Nominal Unit Size[2] in. (mm)	Density pcf (kg/m³)	STC Hollow Unit	STC Grout Filled	STC Sand Filled	STC Solid Units
4 (100)	100 (1602)	42	46	45	45
6 (150)	100 (1602)	43	52	49	50
8 (200)	100 (1602)	46	56	53	54
10 (250)	100 (1602)	48	59	56	56
12 (300)	100 (1602)	49	61	58	59

Nominal Unit Size[2] in. (mm)	Density pcf (kg/m³)	STC Hollow Unit	STC Grout Filled	STC Sand Filled	STC Solid Units
4 (100)	105 (1682)	42	46	45	46
6 (150)	105 (1682)	44	52	50	50
8 (200)	105 (1682)	46	56	53	54
10 (250)	105 (1682)	49	59	56	57
12 (300)	105 (1682)	50	62	58	60

Nominal Unit Size[2] in. (mm)	Density pcf (kg/m³)	STC Hollow Unit	STC Grout Filled	STC Sand Filled	STC Solid Units
4 (100)	110 (1762)	43	47	45	46
6 (150)	110 (1762)	44	52	50	51
8 (200)	110 (1762)	47	56	53	55
10 (250)	110 (1762)	49	59	56	58
12 (300)	110 (1762)	51	62	59	60

Nominal Unit Size[2] in. (mm)	Density pcf (kg/m³)	STC Hollow Unit	STC Grout Filled	STC Sand Filled	STC Solid Units
4 (100)	115 (1842)	43	47	46	46
6 (150)	115 (1842)	45	52	50	51
8 (200)	115 (1842)	47	56	54	55
10 (250)	115 (1842)	50	59	57	58
12 (300)	115 (1842)	51	62	59	61

Nominal Unit Size[2] in. (mm)	Density pcf (kg/m³)	STC Hollow Unit	STC Grout Filled	STC Sand Filled	STC Solid Units
4 (100)	120 (1922)	43	47	46	47
6 (150)	120 (1922)	45	53	50	52
8 (200)	120 (1922)	48	57	54	56
10 (250)	120 (1922)	50	60	57	59
12 (300)	120 (1922)	52	62	59	61

Nominal Unit Size[2] in. (mm)	Density pcf (kg/m³)	STC Hollow Unit	STC Grout Filled	STC Sand Filled	STC Solid Units
4 (100)	125 (2002)	44	48	46	47
6 (150)	125 (2002)	45	53	51	52
8 (200)	125 (2002)	48	57	54	56
10 (250)	125 (2002)	50	60	57	59
12 (300)	125 (2002)	52	63	60	62

Nominal Unit Size[2] in. (mm)	Density pcf (kg/m³)	STC Hollow Unit	STC Grout Filled	STC Sand Filled	STC Solid Units
4 (100)	130 (2082)	44	48	47	48
6 (150)	130 (2082)	46	53	51	53
8 (200)	130 (2082)	49	57	55	57
10 (250)	130 (2082)	51	60	57	60
12 (300)	130 (2082)	52	63	60	62

Nominal Unit Size[2] in. (mm)	Density pcf (kg/m³)	STC Hollow Unit	STC Grout Filled	STC Sand Filled	STC Solid Units
4 (100)	135 (2162)	45	48	47	48
6 (150)	135 (2162)	46	53	51	53
8 (200)	135 (2162)	49	57	55	57
10 (250)	135 (2162)	51	60	58	60
12 (300)	135 (2162)	53	63	60	63

Nominal Unit Size[2] in. (mm)	Density pcf (kg/m³)	STC Hollow Unit	STC Grout Filled	STC Sand Filled	STC Solid Units
4 (100)	140 (2243)	45	48	47	48
6 (150)	140 (2243)	46	54	51	54
8 (200)	140 (2243)	49	58	55	57
10 (250)	140 (2243)	52	61	58	61
12 (300)	140 (2243)	53	63	60	63

Nominal Unit Size[2] in. (mm)	Density pcf (kg/m³)	STC Hollow Unit	STC Grout Filled	STC Sand Filled	STC Solid Units
4 (100)	145 (2323)	45	49	48	49
6 (150)	145 (2323)	47	54	52	54
8 (200)	145 (2323)	50	58	55	58
10 (250)	145 (2323)	52	61	58	61
12 (300)	145 (2323)	54	64	61	64

Nominal Unit Size[2] in. (mm)	Density pcf (kg/m³)	STC Hollow Unit	STC Grout Filled	STC Sand Filled	STC Solid Units
4 (100)	150 (2403)	46	49	48	49
6 (150)	150 (2403)	47	54	52	54
8 (200)	150 (2403)	50	58	56	58
10 (250)	150 (2403)	53	61	59	62
12 (300)	150 (2403)	54	64	61	64

[1] Based on grout density of 140 lb/ft³ (2243 kg/m³), sand density of 90 lb/ft³ (1442 kg/m³), mortar density of 130 lb/ft³ (2082 kg/m³). Percentage solid of units used from mold manufacturers' literature for typical masonry units 4 in. (100 mm) (73.8% solid), 6 in. (150 mm) (55.0% solid); 8 in. (200 mm) (53.0% solid); 10 in. (250 mm) (51.7% solid); 12 in. (300 mm) (48.7% solid). *STC* values for grout filled and sand filled units assume the materials completely occupy all void areas in and around the units. *STC* values for solid units are based on bed and head joints solidly filled with mortar.

[2] Dimensions in this column reflect equivalent metric unit sizes as opposed to direct SI conversions.

STANDARD METHOD FOR DETERMINING SOUND TRANSMISSION RATINGS FOR MASONRY WALLS (TMS 0302-12)

7.3.1 Effect of Drywall on *STC* Ratings of Concrete Masonry Walls

7.3.1.1 The effect of drywall attached directly to the surface of normal weight concrete masonry without a furring space has very little effect on the sound transmission class (*STC*) rating of the wall assembly. Drywall directly attached to lightweight concrete masonry generally improves the *STC* rating by partially sealing of the surface. The more porous the masonry, the better the improvement in *STC* ratings. The amount of improvement is not quantifiable at this time and therefore is not included in the calculated *STC* rating procedure. (Ref. 7 & 8).

7.3.1.2 Significant increases in *STC* ratings in a concrete masonry wall can be achieved by adding gypsum board and sound insulation in the furring space. Three factors govern the amount of improvement in *STC*:
- The method of support. The best method of support for the drywall is the use of independent studs that have no direct connection to the concrete masonry. Resilient metal furring may also be used by itself or in combination with wood furring.
- The depth of the furring space (distance between the drywall and the concrete masonry surfaces).
- The use of sound absorbing material in the furring space. (Ref.7).

Mass-air-mass resonance at low frequencies and narrow furring spaces can cause the STC ratings to drop particularly if that condition exists on both sides of the concrete masonry wall. Under these conditions vibrational energy transfers from the gypsum board through the air space to the wall more effectively than it does through the bare concrete masonry wall. Table 7.3-5 presents the results of Eq. 3-6 for various furring spaces with and without sound-absorbing material in the furring space.

Table 7.3-5—Increase in *STC* Using the Furring Space Depth Indicated and a Single Layer of Drywall

Furring Space Condition	Sides	Furring Space, in. (mm)							
		0.5 (13)	0.8 (19)	1 (25)	1.5 (38)	2 (51)	2.5 (64)	3 (76)	3.5 (89)
No sound-absorbing material in the furring space	one	0.2	0.9	1.6	3.0	4.4	5.8	7.2	8.6
	both	-1.0	-0.1	0.8	2.6	4.4	6.2	8.0	9.8
Furring space filled with sound absorbing material*	one	3.4	4.1	4.9	6.4	7.9	9.4	10.9	12.4
	both	-1.8	1.0	3.8	9.4	15.0	20.6	26.2	31.8

*Fibrous materials, such as cellulose fiber, glass fiber or rock wool insulation, are good materials for absorbing sound; closed-cell materials, such as expanded polystyrene, are not, as they do not significantly absorb sound.

8 — Outdoor-Indoor Transmission Class Ratings

8.1 Laboratory testing
Representative masonry materials need not be from the same manufacturer.

8.2 Field testing
No commentary.

8.3 Calculation
Many ASTM E 90 sound transmission loss tests have been performed on a wide variety of concrete masonry walls. Outdoor-Indoor Transmission Class, *OITC* values for some of these walls have been calculated in accordance with ASTM E 1332 from E 90 test data, and are presented in Table 9.3-2. In general, for masonry walls, heavier walls have higher *OITC* values. Note that the ASTM E 1332 *OITC* calculation requires transmission loss, *STL*, test data from 80 Hz to 4,000 Hz, while ASTM E 90 test reports often do not include *STL* values at 80 Hz. Test reports which do include 80 Hz show that the *STL* value of masonry walls at 80 Hz is typically about the same or higher than that at 100 Hz. For the purposes of this Standard, where *STL* values at 80 Hz were not reported, the 80 Hz *STL* was assumed equal to the 100 Hz *STL*.

OITC data of clay masonry walls (Ref. 2) are plotted against wall weight, *W*, in Figure 8.3-1. The equation for the curve best fitting the data is $OITC = 17.4\, W^{0.224}$ ($OITC = 12.2\, W^{0.224}$) with a correlation coefficient of 0.8453.

OITC data of concrete masonry walls (Ref. 3, 4, 5, and 6) are plotted against wall weight, *W*, in Figure 8.3-2. The equation for the curve best fitting the data is $OITC = 14.7\, W^{0.290}$ ($OITC = 9.28\, W^{0.290}$) with a correlation coefficient of 0.8024.

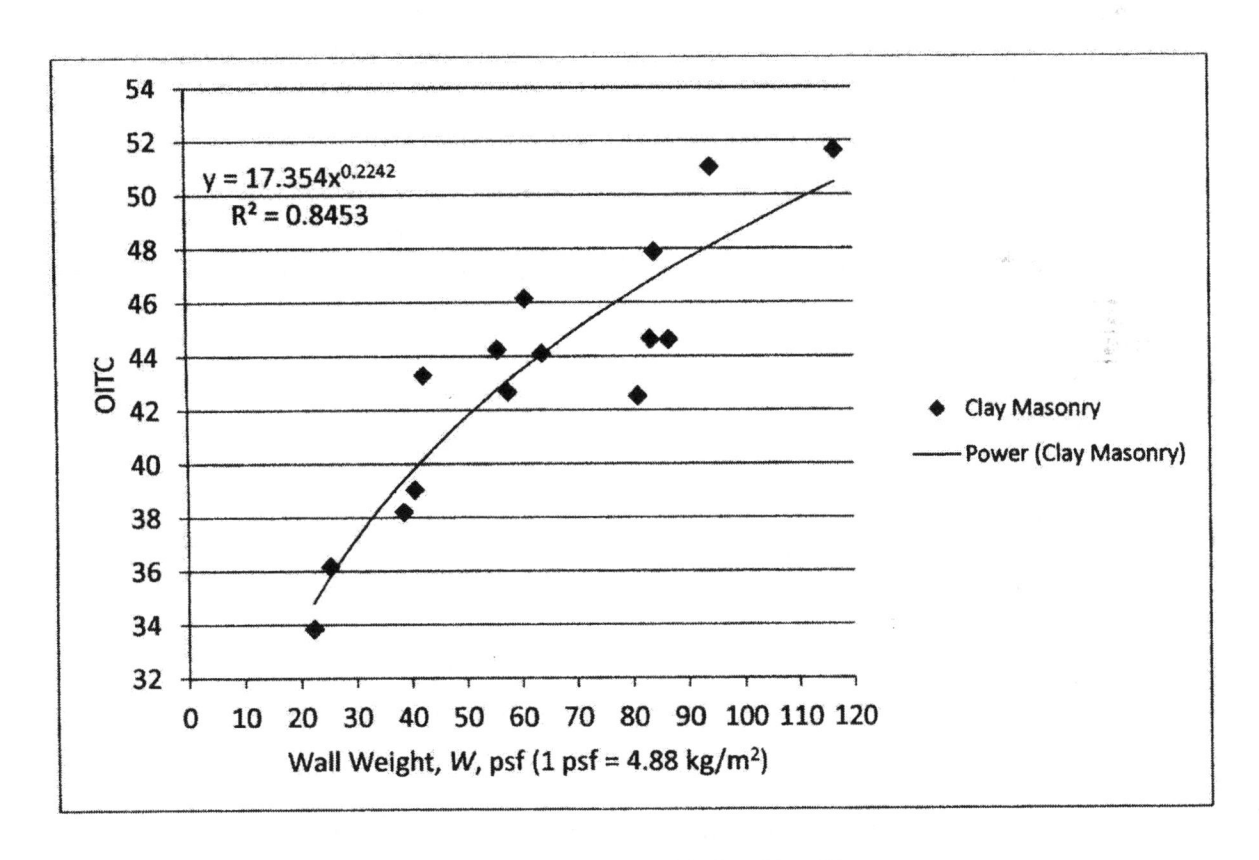

Figure 8.3-1 — *OITC* Curve Fit for Clay Masonry

Table 8.3-1 — *OITC* Data for Clay Masonry (Ref. 2)

Weight psf, (kg/m^2)	OITC
22.3 (109)	34
25.3 (124)	36
38.7 (189)	38
40.6 (198)	39
42.4 (207)	43
55.8 (272)	44
57.7 (282)	43
60.8 (297)	46
63.8 (311)	44
81 (395)	43
83.3 (407)	45
84.1 (411)	48
86.7 (423)	45
94.2 (460)	51
116.7 (570)	52

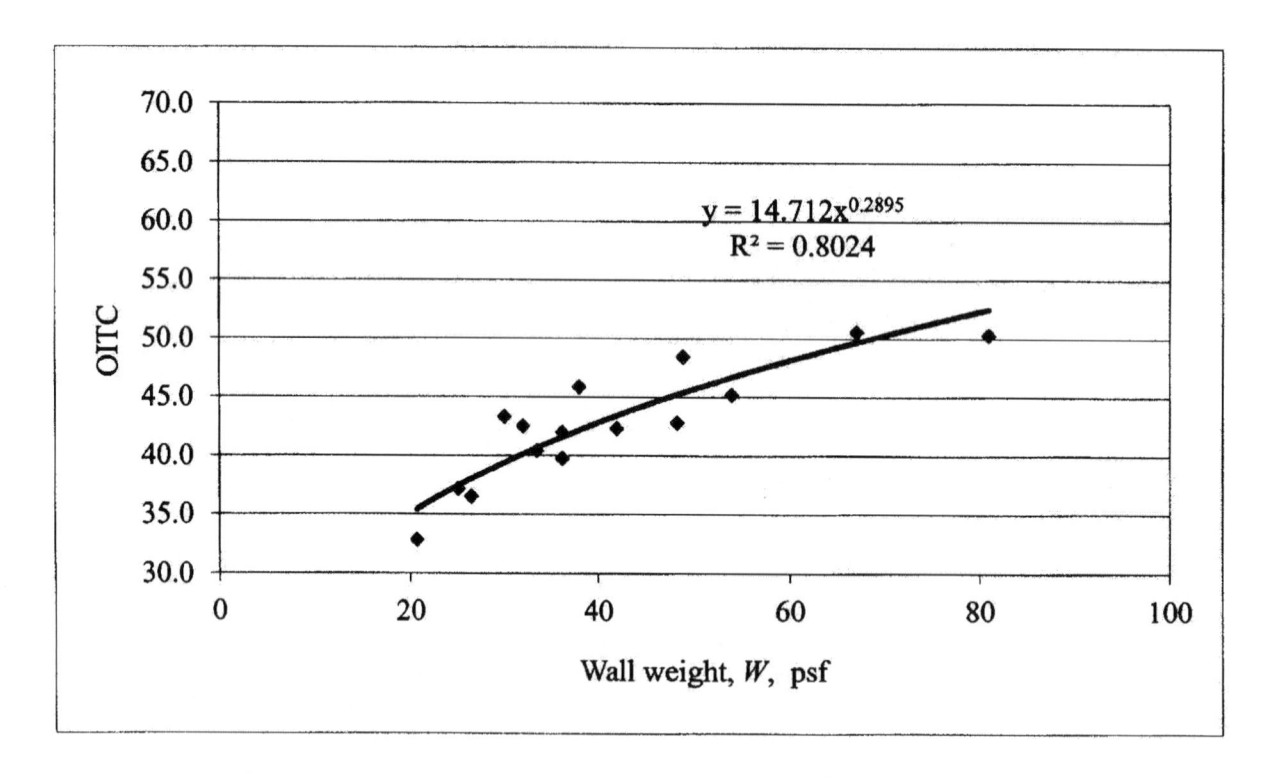

$$y = 14.712 x^{0.2895}$$
$$R^2 = 0.8024$$

Figure 8.3-2 — *OITC* Curve Fit for Concrete Masonry

STANDARD METHOD FOR DETERMINING SOUND TRANSMISSION RATINGS FOR MASONRY WALLS (TMS 0302-12)

Table 8.3-2 — *OITC* Data for Concrete Masonry (Ref. 4 & 6)

Weight Class*	Finish**	Weight psf, (kg/m^2)	OITC	Reference
L	0	20.7 (101)	32	Ref. 4
N	1	26.5 (129)	36	Ref. 4
N	2	32.0 (156)	42	Ref. 4
N	2	42.0 (205)	42	Ref. 4
L	2	36.2 (177)	43	Ref. 4
L	0	25.1 (123)	37	Ref. 4
L	2	54.0 (264)	45	Ref. 4
L	0	36.2 (177)	39	Ref. 4
N	1	33.5 (164)	40	Ref. 4
L	1	36.2 (177)	42	Ref. 4
N	0	48.4 (236)	42	Ref. 6
L	2	38.0 (186)	45	Ref. 4
L	2	67.0 (327)	50	Ref. 4
L	2	49.0 (239)	48	Ref. 4
L	2	81.0 (395)	50	Ref. 4

*Weight class
L=Lightweight
N=Normal weight

**Finish:
0=bare
1=paint
2=plaster

References

1. Berendt, R. D. & Winzer, G. E., "Airborne, Impact and Structural Borne Noise," U. S. Government Printing Office, Washington, D. C., September 1967.

2. "Sound Insulation-Clay Masonry Walls," *Technical Notes on Brick Construction*, No. 5 A, Brick Industry Association (formerly known as the Brick Institute of America), Reston, VA, June 1970.

3. "Sound Transmission Class Ratings for Concrete Masonry Walls," *NCMA TEK*, 13-1, National Concrete Masonry Association, Herndon, VA, 1990.

4. *A Guide to Selecting Concrete Masonry Walls for Noise Reduction*. National Concrete Masonry Association, TR81, Herndon, VA, 1970.

5. *Sound Transmission Loss Through Concrete and Concrete Masonry Walls*. Portland Cement Association, Skokie, IL, 1978.

6. *Sound Transmission Loss Measurements on 190 and 140 mm Single Wythe Concrete Block Walls and on 90 mm Cavity Block Walls*, Report for Ontario Concrete Block Association. National Research Center of Canada Report No. CR-5588.1, 1989.

7. *Controlling Sound Transmission through Concrete Block Walls*, Construction Technology Update No. 13. National Research Council of Canada, 1998.

8. *Sound Transmission Loss Measurements Through 190 mm and 140 mm Blocks with Added Drywall and Through Cavity Block Walls*, Internal Report No. 586. National Research Council of Canada, 1990.

STANDARD METHOD FOR DETERMINING SOUND TRANSMISSION RATINGS FOR MASONRY WALLS (TMS 0302-12)

STANDARD METHOD FOR DETERMINING SOUND TRANSMISSION RATINGS FOR MASONRY WALLS (TMS 0302-12)

TEXT FROM

"*Masonry Inspection Checklist*, copyright © 2004, is reproduced with the permission of the publisher The Masonry Society."

The Masonry Society
105 South Sunset Street, Suite Q
Longmont, CO 80501-6172
www.masonrysociety.org

PREFACE

The Construction Practices Committee of The Masonry Society has prepared this checklist for use during construction. The checklist can be used in whole or in part to suit the needs of a masonry project. The completed document may become part of the job records for that contract. The checklist is meant to serve as a guide for the masonry inspector/observer's work throughout the project.

The Masonry Society (TMS) will appreciate receiving suggestions for improving the checklist. Comments may be submitted by mail to TMS, 105 South Sunset Street, Suite Q, Longmont, CO 80501-6172, by fax at 303-541-9215, or by e-mail at info@masonrysociety.org.

Construction Practices Committee
(Current and past members who assisted on this checklist)

Paul D. Hoggatt[+]
Current Chairman

Turner Smith[*]
Past Chairman

Richard B. Allen[+]	Pam Jergenson[+]	Ed Reed[+]
William Bailey	Lawrence Kahn	Jacob W. Ribar
Robert V. Barnes, Jr.[+]	Robert J. Kudder[+]	Joseph C. Rustic[+]
Christine Beall[+]	John Landry	Daniel Shapiro[+]
Olene L. Bigelow[+]	Richard H. Lauber[+]	Ava Shypula[+]
John M. Bufford[+]	Hugh MacDonald[*]	David M. Sovinski
Jerry Carrier	Theodore W. Marotta	H. Tepper
Robert W. Crooks[+]	Ronald L. Mayes	Al Tomassetti[+]
James T. Darcy[+]	Donald G. McMican[+]	Corey S. Torres[+]
Larry Darling[+]	C. Colin Munro	Gerald Travis[+]
Howard L. Droz[+]	Daniel R. Murray[+]	Donald A. Wakefield
Richard Felice[+]	Charles J. Nacos	Richard A. Weber[+]
Richard Filloramo	William R. Nash[+]	Joseph C. Welte[+]
David Gastgeb	Robert L. Nelson[+]	A. Rhett Whitlock
Edgar F. Glock, Jr.[+]	Jerry M. Painter[+]	William A. Wood[+]
Donald C. Grant	William D. Palmer	Daniel Zechmeister[+]
Clayford T. Grimm[*]	Davis G. Parsons II[+]	Gary L. Zwayer[+]
Mark K. Hottman[+]	Ronald S. Pringle	

[*]deceased
[+]Current Member

Copyright © 2004, The Masonry Society.

All rights reserved including rights of reproduction and use in any form or by any means, including the making of copies by any photo process, or by any electronic or mechanical device, printed, written or oral, or recording for sound or visual reproduction or for any use in any knowledge retrieval system or device, unless permission in writing is obtained from The Masonry Society.

ISBN# 1-929081-21-9
TMS Order No. TMS-5201-04

THE MASONRY SOCIETY
MASONRY INSPECTION CHECKLIST

Masonry Inspection Checklist

INDEX

Introduction

I. Preconstruction
   A. Contract Documents
   B. Mock-up Requirements

II. Preparation for Field Work
   A. Storage of Materials
   B. Substrates
   C. Submittal Compliance
   D. Preconstruction Conference
   E. Sample Panel/Mock-up

III. Quality Assurance
   A. Specification Requirements
   B. Nonconformance Resolution

IV. Construction
   A. Change Orders
   B. Workmanship
   C. Flashing
   D. Weeps
   E. Reinforcement and Connectors
   F. Grout Placement
   G. Insulation
   H. Field Tests
   I. Movement, Control and Expansion Joints
   J. Cleaning

V. Closeout
   A. Operations and Maintenance Manual Requirements
   B. Maintenance Recommendations

Attachments
   Exhibit A Non-Conformance Report
   Exhibit B Non-Conformance Log

THE MASONRY SOCIETY
MASONRY INSPECTION CHECKLIST

INTRODUCTION

This document serves as a tool for masonry construction inspectors, mason contractors, general contractors, architects and engineers to assist them in assuring compliance with the contract documents. Only those items that relate to masonry construction are included. Items that should be checked are listed without explanation. Not all items will be used on every project, and some projects may use items not listed.

The checklist is divided into five parts:

I. Preconstruction
II. Preparation for Field Work
III. Quality Assurance
IV. Construction
V. Closeout

This document is not an inspector's guide to masonry construction, nor is it a training manual. Rather, it can serve as a basis for the inspector/observer's daily log. It is to be used as a tool to aid masonry inspectors/observers in the performance of their duties.

THE MASONRY SOCIETY
MASONRY INSPECTION CHECKLIST

INSPECTION ITEM	REVIEWED		QUALITY ASSURANCE ITEM	COMMENTS
	YES	NO		

I. PRECONSTRUCTION

A. Review Contract Documents

1. **General Conditions and Construction Contract**
   a. Inspector's scope of responsibility ☐ ☐ ☐
   b. Approved alternatives and revisions ☐ ☐ ☐
   c. Addenda ☐ ☐ ☐
   d. Change orders ☐ ☐ ☐

2. **Specifications**
   a. Name and date of applicable building code and its supplement ☐ ☐ ☐
   b. References to standards ☐ ☐ ☐
   c. Specified compressive strength, f'_m ☐ ☐ ☐
   d. Submittals requirements, log ☐ ☐ ☐
   e. Testing requirements ☐ ☐ ☐
   f. Protection of stored materials ☐ ☐ ☐
   g. Special weather conditions ☐ ☐ ☐

Notes: _____

THE MASONRY SOCIETY
MASONRY INSPECTION CHECKLIST

Inspection Checklist

INSPECTION ITEM	REVIEWED YES	REVIEWED NO	QUALITY ASSURANCE ITEM	COMMENTS
h. Mortar type, mix design	☐	☐	☐	
i. Admixtures	☐	☐	☐	
j. Pigments	☐	☐	☐	
k. Grout type, mix design	☐	☐	☐	
l. Metal corrosion protection	☐	☐	☐	
m. Masonry units: size, color, texture, grade, type, strength	☐	☐	☐	
n. Special shapes	☐	☐	☐	
o. Types of accessories	☐	☐	☐	
p. Dampproofing and moisture barriers	☐	☐	☐	
q. Flashing system and weeps	☐	☐	☐	
r. Coatings	☐	☐	☐	
s. Mortar mixing and application procedures	☐	☐	☐	
t. Grout placement procedures	☐	☐	☐	
u. Bond pattern	☐	☐	☐	
v. Mortar joint types, size, alignment of head joints	☐	☐	☐	
w. Filled head and bed joints	☐	☐	☐	
x. Tooling of mortar joints	☐	☐	☐	
y. Wall bracing	☐	☐	☐	
z. Construction tolerances	☐	☐	☐	
aa. Protection of completed work	☐	☐	☐	
bb. Cleaning procedures	☐	☐	☐	

Notes: _____

THE MASONRY SOCIETY
MASONRY INSPECTION CHECKLIST

INSPECTION ITEM	REVIEWED YES	REVIEWED NO	QUALITY ASSURANCE ITEM	COMMENTS
3. Drawings / Design				
a. Name, date, and applicable building code	☐	☐	☐	
b. Compressive strength of masonry, f'_m	☐	☐	☐	
c. Details	☐	☐	☐	
d. Location of expansion and control joints	☐	☐	☐	
e. Compressible fillers, adjustable anchors, and other movement control items	☐	☐	☐	
f. Flashing details and locations	☐	☐	☐	
g. Splices, end dams, and laps	☐	☐	☐	
h. Weepholes and vents	☐	☐	☐	
i. Details and locations of reinforcement	☐	☐	☐	
j. Anchors and wall ties: type, size, and location	☐	☐	☐	
k. Lintels and shelf angles	☐	☐	☐	
l. Insulation size and type	☐	☐	☐	
m. Air barriers	☐	☐	☐	
n. Air space details / drainage accessories	☐	☐	☐	
o. Sealants: location and type	☐	☐	☐	
p. Sealant primers	☐	☐	☐	
q. Foundation and at-grade details	☐	☐	☐	

Notes:

THE MASONRY SOCIETY
MASONRY INSPECTION CHECKLIST

INSPECTION ITEM	REVIEWED		QUALITY ASSURANCE ITEM	COMMENTS
	YES	NO		

4. Submittals
   a. Review approved submittals ☐ ☐ ☐

5. Meetings
   a. Meetings required ☐ ☐ ☐
   b. Attendance ☐ ☐ ☐
   c. Agenda ☐ ☐ ☐

6. Schedules
   a. Construction start date ☐ ☐ ☐
   b. Daily work hours ☐ ☐ ☐
   c. Overtime hours ☐ ☐ ☐
   d. Inspections ☐ ☐ ☐
   e. Testing ☐ ☐ ☐
   f. Weather delay procedures ☐ ☐ ☐
   g. Completion date ☐ ☐ ☐
   h. Punch-lists ☐ ☐ ☐
   i. Substantial completion procedure ☐ ☐ ☐
   j. Warranties ☐ ☐ ☐

Notes:

THE MASONRY SOCIETY
MASONRY INSPECTION CHECKLIST

INSPECTION ITEM	REVIEWED YES	REVIEWED NO	QUALITY ASSURANCE ITEM	COMMENTS
B. Mock-up Requirements				
1. Materials				
a. Specification compliance	☐	☐	☐	
b. Submittals	☐	☐	☐	
2. Construction				
a. Mortar joint size, alignment	☐	☐	☐	
b. Movement joints	☐	☐	☐	
c. Tooling	☐	☐	☐	
d. Penetration details	☐	☐	☐	
e. Flashings, weeps and vents	☐	☐	☐	
f. Cavity details, insulation	☐	☐	☐	
g. Backup details	☐	☐	☐	
h. Anchors, ties and fasteners	☐	☐	☐	
i. Reinforcement	☐	☐	☐	
j. Dampproofing and moisture barriers	☐	☐	☐	
k. Air barriers	☐	☐	☐	
l. Grouting	☐	☐	☐	

Notes: _____

THE MASONRY SOCIETY
MASONRY INSPECTION CHECKLIST

INSPECTION ITEM	OBSERVED/REVIEWED		QUALITY ASSURANCE ITEM	COMMENTS
	YES	NO		

II. PREPARATION FOR FIELD WORK

A. Storage of Materials
1. Covers and protection from weather damage ☐ ☐ ☐
2. Set on pallets or dunnage to prevent contact with the ground or other substrate. ☐ ☐ ☐
3. Isolated from other materials to prevent contamination ☐ ☐ ☐

B. Substrates
1. Clean and within moisture limits ☐ ☐ ☐
2. Structural frames within tolerances ☐ ☐ ☐
3. Footing and ledge dimensions within tolerances ☐ ☐ ☐
4. Grade and wall elevations within tolerances ☐ ☐ ☐
5. Penetrations ☐ ☐ ☐
6. Reinforcement, dowels, anchors ☐ ☐ ☐

C. Submittal Compliance
1. Certificates of Compliance compared to materials ☐ ☐ ☐
2. Notifications of non-conformance ☐ ☐ ☐
3. Removal of rejected materials from site ☐ ☐ ☐

Notes: _____

THE MASONRY SOCIETY
MASONRY INSPECTION CHECKLIST

INSPECTION ITEM	OBSERVED/REVIEWED YES	NO	QUALITY ASSURANCE ITEM	COMMENTS
D. Preconstruction Conference				
1. Conference with representatives of Architect/Engineer, Masonry Contractor, Masonry Inspector, Masonry Suppliers, Testing Agency as needed.	☐	☐	☐	_____
2. Review specifications, drawings and inspection program; resolve ambiguities; confirm scheduling.	☐	☐	☐	_____
3. Review mock-up and/or sample panel.	☐	☐	☐	_____
4. Review submittals, discrepancies, substitutions	☐	☐	☐	_____
5. Review testing requirements, laboratory/field	☐	☐	☐	_____
E. Sample Panel / Mock-up				
1. Displays typical workmanship and materials including masonry units and mortar joints	☐	☐	☐	_____
2. Meeting				
a. Compliance/non-compliance	☐	☐	☐	_____
b. Testing results	☐	☐	☐	_____
c. Quality assurance procedures	☐	☐	☐	_____
d. Observations	☐	☐	☐	_____
e. Evaluations	☐	☐	☐	_____
f. Recommendations	☐	☐	☐	_____
g. Approval	☐	☐	☐	_____
3. Water leakage testing	☐	☐	☐	_____
4. Determine brick IRA	☐	☐	☐	_____

Notes: _____

THE MASONRY SOCIETY
MASONRY INSPECTION CHECKLIST

INSPECTION ITEM	OBSERVED/ REVIEWED		QUALITY ASSURANCE ITEM	COMMENTS
	YES	NO		

III. QUALITY ASSURANCE
   A. Specification Requirements May Include
      1. Certificates for delivered materials indicating compliance with specification ☐ ☐
      2. Verify materials and source match certificate ☐ ☐
      3. Verify f_m prior to construction ☐ ☐
      4. Verify mortar proportions ☐ ☐
      5. Observe brick placement and mortar joint construction ☐ ☐
      6. Verify location of reinforcement, connectors ☐ ☐
      7. In prestress work
         a. Verify tendon size, location, anchorage ☐ ☐
         b. Verify grout proportions for anchorages and bonded tendons ☐ ☐
      8. Verify grout space, reinforcement, mix proportions ☐ ☐
      9. Observe prism, mortar, and grout test specimens ☐ ☐
     10. Verify f_m during construction ☐ ☐

   B. Nonconformance Resolution
      1. Review specified reporting procedures ☐ ☐
      2. Document noncompliance areas ☐ ☐
      3. Review rework, repair and documentation ☐ ☐

Notes:

THE MASONRY SOCIETY
MASONRY INSPECTION CHECKLIST

INSPECTION ITEM	OBSERVED/ REVIEWED		QUALITY ASSURANCE ITEM	COMMENTS
	YES	NO		

IV. CONSTRUCTION

A. Change Orders ☐ ☐ ☐ _____

B. Workmanship
1. Specified masonry materials
   a. Units ☐ ☐ ☐ _____
   b. Mortar ☐ ☐ ☐ _____
   c. Grout ☐ ☐ ☐ _____
2. Preparation
   a. Mortar batching ☐ ☐ ☐ _____
   b. Preparation of units
      1) Wetting ☐ ☐ ☐ _____
      2) Cutting ☐ ☐ ☐ _____
3. Placement of units
   a. Mortar ☐ ☐ ☐ _____
   b. Masonry unit ☐ ☐ ☐ _____
   c. Head joint ☐ ☐ ☐ _____
   d. Tooling ☐ ☐ ☐ _____
4. Tolerances ☐ ☐ ☐ _____
5. Built-in Items ☐ ☐ ☐ _____
6. Hot and Cold weather provisions ☐ ☐ ☐ _____

Notes: _____

THE MASONRY SOCIETY
MASONRY INSPECTION CHECKLIST

Inspection Checklist

INSPECTION ITEM	OBSERVED/REVIEWED YES / NO	QUALITY ASSURANCE ITEM	COMMENTS
7. Wall protection			
a. Stability/bracing	☐ ☐	☐	
b. Covering of walls	☐ ☐	☐	
c. Protection of completed work	☐ ☐	☐	
C. Flashing			
1. Material	☐ ☐	☐	
2. Placement	☐ ☐	☐	
D. Weeps			
1. Material	☐ ☐	☐	
2. Placement	☐ ☐	☐	
E. Reinforcement and Connectors			
1. Reinforcing steel			
a. Size, grade	☐ ☐	☐	
b. Bends	☐ ☐	☐	
c. Laps/joints	☐ ☐	☐	
d. Placement	☐ ☐	☐	

Notes:

THE MASONRY SOCIETY
MASONRY INSPECTION CHECKLIST

INSPECTION ITEM	OBSERVED/ REVIEWED		QUALITY ASSURANCE ITEM	COMMENTS
	YES	NO		

2. Joint reinforcement
   a. Size
   b. Type
   b. Placement
   c. Splices

3. Anchors
   a. Size
   b. Type
   b. Placement

4. Ties
   a. Size
   b. Type
   c. Placement

5. Fasteners
   a. Size
   b. Type
   b. Placement

F. Grout Placement
   1. Verify minimum grout dimensions
   2. Check slump
   3. Verify consolidation
   4. Verify reconsolidation

Notes:

**THE MASONRY SOCIETY
MASONRY INSPECTION CHECKLIST**

INSPECTION ITEM	OBSERVED/REVIEWED		QUALITY ASSURANCE ITEM	COMMENTS
	YES	NO		

G. Insulation
   1. Cavity Insulation
      a. Type ☐ ☐ ☐
      b. Dimension ☐ ☐ ☐
      c. Orientation ☐ ☐ ☐
      d. Fastening/attachment/spacing ☐ ☐ ☐
      e. Joint treatment ☐ ☐ ☐
   2. Unit Cell Insulation
      a. Verify placement ☐ ☐ ☐
      b. Cut masonry unit installation ☐ ☐ ☐
   3. Expanding Foam
      a. Observation holes ☐ ☐ ☐

H. Field Tests
   1. Mortar/Grout
      a. Frequency requirements met ☐ ☐ ☐
      b. Observe specimen preparation, sampling, storage ☐ ☐ ☐
   2. Composite prisms
      a. Frequency requirements met ☐ ☐ ☐
      b. Observe specimen preparation, sampling, storage ☐ ☐ ☐

Notes: _____

THE MASONRY SOCIETY
MASONRY INSPECTION CHECKLIST

INSPECTION ITEM	OBSERVED/ REVIEWED YES / NO	QUALITY ASSURANCE ITEM	COMMENTS
I. Movement Joints – Control & Expansion Joints			
1. Material			
a. Preformed size and type	☐ ☐	☐	
b. Backer rod	☐ ☐	☐	
c. Sealant	☐ ☐	☐	
d. Bond breaker	☐ ☐	☐	
2. Location and Placement			
a. Verify locations horizontal and vertical	☐ ☐	☐	
b. Verify clear, open joint	☐ ☐	☐	
c. Sealant preparation and installation	☐ ☐	☐	

Notes: _____

THE MASONRY SOCIETY
MASONRY INSPECTION CHECKLIST

Inspection Checklist

INSPECTION ITEM	OBSERVED/REVIEWED YES / NO	QUALITY ASSURANCE ITEM	COMMENTS
J. Cleaning			
1. Dry cleaning mortar residue and droppings	☐ ☐	☐	
2. Wet cleaning after minimum curing time	☐ ☐	☐	
3. Compatibility of specified cleaning materials with			
a. Units and mortar	☐ ☐	☐	
b. Sealants	☐ ☐	☐	
c. Windows, doors, louvers	☐ ☐	☐	
d. Flashings and roofing	☐ ☐	☐	
4. Procedure			
a. Test area / mock-up of cleaning method for each material; units, stone, precast and adjoining construction. Suggest sample panel/mock-up	☐ ☐	☐	
b. Protective measures			
1) Personal Protective Equipment	☐ ☐	☐	
2) Protection of adjoining materials	☐ ☐	☐	
c. Manufacturer's recommendations for product application	☐ ☐	☐	
d. Temperature range	☐ ☐	☐	

Notes: _____

THE MASONRY SOCIETY
MASONRY INSPECTION CHECKLIST

INSPECTION ITEM	OBSERVED/REVIEWED YES	OBSERVED/REVIEWED NO	QUALITY ASSURANCE ITEM	COMMENTS
V. CLOSEOUT				
A. Operations and Maintenance Manual Requirements				
1. Provide manuals to the General Contractor and the Owner	☐	☐	☐	
2. List materials; model number, suppliers, name, address				
a. Units	☐	☐	☐	
b. Mortar – mix	☐	☐	☐	
c. Reinforcement	☐	☐	☐	
d. Ties, anchors, and fasteners	☐	☐	☐	
e. Grout – mix	☐	☐	☐	
f. Sealants and backer rod	☐	☐	☐	
g. Water repellent	☐	☐	☐	
h. Vapor barrier	☐	☐	☐	
i. Weeps	☐	☐	☐	
j. Flashings	☐	☐	☐	
k. Lintels, sills, and shelf angles	☐	☐	☐	

Notes: _____

THE MASONRY SOCIETY
MASONRY INSPECTION CHECKLIST

Inspection Checklist

INSPECTION ITEM	OBSERVED/ REVIEWED YES / NO	QUALITY ASSURANCE ITEM	COMMENTS
3. As built drawings	☐ ☐	☐	_____
4. Warranties	☐ ☐	☐	_____
5. Testing data			_____
a. Laboratory - units, mortar, and grout	☐ ☐	☐	_____
b. Field testing	☐ ☐	☐	_____
B. Maintenance recommendations	☐ ☐	☐	_____

Notes: _____

THE MASONRY SOCIETY
MASONRY INSPECTION CHECKLIST

Exhibit A
NON-CONFORMANCE REPORT

Copy Safety _____Yes _____No   Page_____of_____

| Subcontractor/Vendor | Subcontractor Purchase Order No. | Date: | Non-conformance No. |

| Non-conforming Item No: | | Location: | Specification/Drawing |

Description of Non-Conformance _____
_____
_____

PROPOSED DISPOSITION

Proposed Disposition:

_____Rework   _____Repair   _____Scrap   _____Use as is   _____Return to Supplier

_____
_____

| Report Prepared by | Date | Owner | Date |
| _____ | _____ | _____ | _____ |

| QC Approval | Date | Owner | Date |
| _____ | _____ | _____ | _____ |

COMPLETION OF APPROVED DISPOSITION

Re-inspection / Verification Results:

_____Accept   Inspected Contractor QC Manager: _____

_____Reject   Inspected (Owner QC): _____

Comments: _____

FINAL ACCEPTANCE

Project Manager:_____   Date:_____

Resident Engineer:_____   Date:_____

THE MASONRY SOCIETY
MASONRY INSPECTION CHECKLIST

Exhibit B
Non-Conformance Log[1]

Non-Conformance Log No.	Date	Related Non-Conformance Report	Date of Report	Spec. Section Drawing No.	Description of Non-Conformance	Status/Date (see notes below)

Notes: 1 = Item does not conform or is deficient.
2 = Correction required/in process.
3 = Non-conformance corrected/resolved

THE MASONRY SOCIETY
MASONRY INSPECTION CHECKLIST

BIA TECHNICAL NOTES on Brick Construction

SUBJECT INDEX (April 2014)

Copies of BIA Technical Notes can be purchased directly from Brick Industry Association, 1850 Centennial Park Drive, Reston, Virginia 20191-1525, Tel: (703) 620-0010, Fax: (703) 620-3928 or individual copies are available to view and download free of charge from BIA's website at http://www.gobrick.com.

SUBJECT	NUMBER
A	
ACI 530/ASCE 5/TMS 402	
Building Code	3
ADA	14E
Adhered Veneer	28C
Admixtures in Mortar	1, 8
Anchor Bolt	44
Arches	31-31C
Construction	31
Flashing	31
Semi-circular	31C
Structural Design	31A
ASTM International Standards	
Anchors and Ties	7A, 28B, 44B
Brick	9A
Mortar	8
Pavers	9A, 14
Testing	39 Series
B	
Barrier Walls	7
Beams	17B
Bearing Walls	
(See Engineered Brick Masonry)	
Bonds and Patterns in Brickwork	30
Paving Patterns	14, 29
Bond Breaks	18A, 21B
Bond - Mortar	8 Series
Reinforced Brick Masonry	17, 17A
Brick Sizes	9B, 10
C	
Calculated Fire Resistance	16
Caps	36A
Cavity Walls	21-21C
Construction	21C
Detailing	21B
Glazed Brick	13
Materials	21A
Passive Solar Heating	43
Properties	21

SUBJECT	NUMBER
Chimneys	19-19C
Classification of Brick	9A
Cleaning	20
Efflorescence	20
Coatings for Brick	6, 6A
Cold Weather Construction	1
Guide Specifications	11A
Color - Brick	9
Mortar	8
Columns and Pilasters	3B
Compressive Strength	
Brick Masonry	3A, 39A, 42
Brick Units	9A, 39
Mortar	8 Series
Pavers	9A, 14
Walls	39A, 42
Condensation	28B, 47
Control Joints (See Expansion Joints)	
Copings	36
Corbels and Racking	36A
Corrosion	
Metal Ties	7A, 44B
Shelf Angles	7A
Steel Lintels	7A, 31B
Coursing Tables for Brick	10
Cracking	18
Curtain and Panel Walls	17L
D	
Dampproofing	46
Differential Movement	18, 18A
Bond Breaks	18A
Expansion Joints in Paving	14 Series
Expansion Joints in Walls	18A
Flexible Anchorage	18A
Material Properties	18
Structures without Shelf Angles	18A
Volume Changes and Effects	18
Dimensioning	10
Direct Gain, Passive Solar Heating	43
Drainage Walls	7

BIA TEK NOTES

SUBJECT	NUMBER
E	
Efflorescence	
Identification and Prevention	23
Causes and Prevention	23A
Removal	20
Empirical Design of Brick Masonry	42
Energy	
Codes	4B
Embodied	48
Heat Transmission Coefficients	4
Engineered Brick Masonry	
Allowable Design Stresses	3A, 39A
Bearing Wall	24
Building Code Requirements	3, 16
Construction	24F
Detailing	24G
Guide Specifications	11 Series
Material Properties	3A
Quality Control	39B
Section Properties	3B
Shear Wall Design	24C
Testing	39 Series
Wall Types and Properties	3B
Equivalent Thickness	16
Estimating Material Quantities	10
Expansion Joints	18A
Paving	14 Series
F	
Fasteners for Brick Masonry	44A
Fences	29A
Field Panels	9B
Fireplaces, Residential	19-19E
Contemporary Projected Corner, Rumford, Multi-faced	19C
Details and Construction	19B
Finnish Style Masonry Heater	19E
Russian Style Masonry Heater	19D
Fire Resistance	16
Flashing, Types and Selection	7A
Arches	31
Details	7
Replacement	46
Flexible Paving Systems	14
Floor-Wall Connections, Bearing Wall	26
Freeze Thaw Durability	7A, 9A, 9B
Freezing, Protection from	1

SUBJECT	NUMBER
G	
Garden Walls	29A
Girders, Reinforced Brick Masonry	17M
Glazed Brick	
Specifications	9A
Walls	13
Glossary	2
Green Building	48
Grout	17, 17A
Properties	3A
Testing	39
Guide Specifications	11 Series
H	
Heat Transmission Coefficients	4
High-Lift Grouting	17A
Hollow Brick Masonry	41
Reinforced	17 Series
Hot Weather Construction	1
I	
Inspection	46
Reinforced Brick Masonry	17A
Initial Rate of Absorption	7A, 8B, 9A, 9B, 39
L	
Landscape Architecture	29-29B
Accessibility	14E
Garden Walls	29A
Miscellaneous Applications	29B
Paving	14 Series
Pedestrian Applications	29
Lateral Forces, Shear Wall Design	24C
LEED	14D, 48
Lintels	
Reinforced Brick	17B
Structural Steel	31B
Loadbearing Brick Homes	26
M	
Maintenance	46
Cleaning	20
Manufacturing of Brick	9
Material Properties	3A
Masonry Heaters	19D, 19E
Modular Brick Masonry	10
Moisture Control	
Barrier Walls	7

BIA TEK NOTES

SUBJECT	NUMBER
Caps and Copings	7, 36A
Condensation	47
Corrosion	7A, 31B, 44B
Drainage Walls	7
Flashing	7A
Glazed Brick Walls	13
Maintenance	46
Mortar	8
Rain Screen Wall	27
Repointing	46
Water Repellent Coatings	6A
Weeps	7
Moisture Expansion	18
Mortars for Brickwork	8 Series
Cold Weather Construction	1
Efflorescence	8, 23 Series
Estimating Quantities	10
Guide Specifications	11E
Joints	7B, 21C
Materials	8
Mixing	8B
Paving Systems	14, 14C
Quality Assurance	8B
Reinforced Masonry	17A
Repointing	46
Selection	8B
Movement (see Differential Movement)	

N

SUBJECT	NUMBER
Noise Barrier Walls	45
Structural Design	45A

P

SUBJECT	NUMBER
Painting Brick Masonry	6
Parapets	7, 18 Series, 36A
Passive Solar	
Cooling	43C
Details	43G
Heating	43
Materials	43D
Patterns	30
Paving Systems	14 Series
Accessibility	14, 14E
Adhesives	14B
Bases	14
Bituminous Setting Bed	14B
Clay Pavers	9A, 14
Cleaning	20
Coatings	6A
Details	14 Series
Drainage	14 Series
Edge Restraint	14 Series
Expansion Joints	14 Series
Ice and Snow Removal	14
Infiltration Rate	14D
Installation	14 Series
Interlock	14A
Joint Sand	14A, 14B
Maintenance	14 Series
Mortar Setting Bed	14C
Patterns	14, 29
Permeable Pavements	14, 14D
Root Control	14E
Runoff Coefficient	14D
Sand Setting Bed	14A
Tack Coat	14B
Traffic	14
Piers and Pilasters	3B
Portland Cement/Lime Mortar	8 Series
Prefabricated Brick Masonry	
Introduction	40
Thin Brick	28C
Pressure-Equalized Rain Screen Wall	27

R

SUBJECT	NUMBER
Rain Penetration (see Moisture Control)	
Rain Screen wall	27
Recycled Content	48
Reinforced Brick Masonry	
Beams	17B
Curtain and Panel Walls	17L
Flexural Design	17B
Girders	17M
High-Lift Grouted	17A
History	17
Hollow Brick Masonry	26, 41
Inspection	17A
Lintel Design	17B
Materials	17A
Mortar and Grout	17A
Specifications	11 Series
Workmanship	17A
Repointing	46
Retrofit	28A
Rigid Paving Systems	14, 14C
Rumford Fireplaces	19C
R-Values	4, 4B

BIA TEK NOTES

SUBJECT	NUMBER

S

Salvaged Brick 15
Sealers (See Water Repellents)
Sealants...18A, 28
Section Properties. 3B
Selection of Brick 9B
Serpentine Walls..29A
Shelf Angles
    Typical Details...7, 28B
    Corrosion Resistance..7A
Single-Wythe Bearing Walls 26
Sills 36
Sizes of Brick 9B, 10
Slip/Skid Resistance 14, 14E
Soffits 36
Solar Energy (see Passive Solar Systems)
Sound Barriers (see Noise Barrier Walls)
Sound Insulation...5A
Spalling 46
Specifications, General ....11 Series
    ACI 530.1/ASCE 6/TMS 602 3
    Brick 9A
    Cold and Hot Weather Construction 1
    Mortars 8, 11E
    Pavers 9A, 14
Stains
    Identification and Prevention..................................23
    Removal 20
Steel Studs 28B
Steps and Ramps 29
Sustainability 48
Sustainable Development 29

T

Terminology 2
Terraces 29
Testing of Brick and Mortar ....39 Series
    Allowable Design Stresses....................................3A, 39A
    Quality Control 39B
Thermal Expansion of Walls ....18 Series
Thermal Storage Walls, Passive Solar Heating........43
Thermal Transmission Coefficients 4
Thin Brick 28C
    Modular Panel 28C
    Precast Concrete 28C
    Prefabricated Panel 28C
    Thick Set 28C
    Thin Set 28C
    Tilt-up Concrete 28C
Ties and Reinforcement
    Adjustable 44B

SUBJECT	NUMBER

    Corrosion Resistance 7A, 44B
    Joint Reinforcement 44B
    Specifications 11A
Tolerances 9A, 11C
Tooling 7B
Tuckpointing (see Repointing)

U

Used Brick...15
U-Values 4, 4B

V

Veneer Construction...28 Series
    Existing Construction 28A
    Hollow Brick 41
    Steel Studs 28B
    Thin Brick Veneer 28C
    Wood Studs 28

W

Wall Ties..44B
Water Penetration (see Moisture Control)
Water Repellent Coatings..6A
Weeps 7, 46
Winter Construction
    (see Cold Weather Construction)
Wood Studs 28
Workmanship...7B, 21C
    Reinforced Brick Masonry 17A
    Specifications ....11 Series

BIA TEK NOTES

TEK MANUAL
for Concrete Masonry Design and Construction

Copies of NCMA Technical Notes can be purchased directly from National Concrete Masonry Association, 13750 Sunrise Valley Drive, Herndon, Virginia 20171-4662, Tel: (703) 713-1900, Fax: (703) 713-1910 or through http://www.ncma.org.

Table of Contents

1 – Building Codes & Specifications

- **1-01F** ASTM Specifications for Concrete Masonry Units (2012)
- **1-02C** Specification for Masonry Structures (2010)
- **1-03D** Building Code Requirements for Concrete Masonry (2011)
- **1-04** Glossary of Concrete Masonry Terms (2004)

2 – C/M Unit Properties, Shapes, & Sizes

- **2-01A** Typical Sizes and Shapes of Concrete Masonry Units (2002)
- **2-02B** Considerations for Using Specialty Concrete Masonry Units (2010)
- **2-03A** Architectural Concrete Masonry Units (2001)
- **2-04B** Segmental Retaining Wall Units (2008)
- **2-05B** New Concrete Masonry Unit Configurations Under ASTM C90 (2012)
- **2-06** Density-Related Properties of Concrete Masonry Assemblies (2008)

3 – Construction

- **3-01C** All-Weather Concrete Masonry Construction (2002)
- **3-02A** Grouting Concrete Masonry Walls (2005)
- **3-03B** Hybrid Concrete Masonry Construction Details (2009)
- **3-04C** Bracing Concrete Masonry Walls Under Construction (2014)
- **3-05A** Surface Bonded Concrete Masonry Construction (1998)
- **3-06C** Concrete Masonry Veneers (2012)
- **3-07A** Concrete Masonry Fireplaces (2003)
- **3-08A** Concrete Masonry Construction (2001)
- **3-09A** Strategies for Termite Resistance (2000)
- **3-10A** Metric Concrete Masonry Construction (2008)
- **3-11** Concrete Masonry Basement Wall Construction (2001)
- **3-12** Construction of High-Rise Concrete Masonry Buildings (1998)
- **3-13** Construction of Low-Rise Concrete Masonry Buildings (2005)
- **3-14** Post-Tensioned Concrete Masonry Wall Construction (2002)

4 – Costs, Estimating

- **4-01A** Productivity and Modular Coordination in Concrete Masonry Construction (2002)
- **4-02A** Estimating Concrete Masonry Materials (2004)

5 – Details

- **5-01B** Concrete Masonry Veneer Details (2003)
- **5-02A** Clay and Concrete Masonry Banding Details (2002)
- **5-02CA** CAN-TEK Clay and Concrete Masonry Banding Details (2014)
- **5-03A** Concrete Masonry Foundation Wall Details (2003)
- **5-04B** Concrete Masonry Residential Details (2002)
- **5-05B** Integrating Concrete Masonry Walls with Metal Building Systems (2011)
- **5-06A** Curtain and Panel Walls of Concrete Masonry (2001)
- **5-07A** Floor and Roof Connections to Concrete Masonry Walls (2001)
- **5-08B** Detailing Concrete Masonry Fire Walls (2005)
- **5-09A** Concrete Masonry Corner Details (2004)
- **5-10A** Concrete Masonry Radial Wall Details (2006)
- **5-11** Residential Details for High Wind Areas (2003)
- **5-12** Modular Layout of Concrete Masonry (2008)
- **5-13** Rolling Door Details for Concrete Masonry Construction (2007)
- **5-14** Concrete Masonry Hurricane and Tornado Shelters (2008)
- **5-15** Details for Half-High Concrete Masonry Units (2010)
- **5-16** Aesthetic Design with Concrete Masonry (2011)

NCMA TEK NOTES

6 – Energy

6-01C R-Values of Multi-Wythe Concrete Masonry Walls (2013)
6-02C R-Values and U-Factors for Single Wythe Concrete Masonry Walls (2013)
6-03 Shifting Peak Energy Loads with Concrete Masonry Construction (1991)
6-04B Energy Code Compliance Using COMcheck (2012)
6-05A Passive Solar Design Strategies (2006)
6-06B Determining the Recycled Content of Concrete Masonry Products (2009)
6-07A Earth Sheltered Buildings (2006)
6-09C Concrete Masonry & Hardscape Products in LEED 2009 (2009)
6-10A Concrete Masonry Radiant Heating/Cooling Systems (2006)
6-11A Insulating Concrete Masonry Walls (2010)
6-12C International Energy Conservation Code (2006 Ed.) and Concrete Masonry (2007)
6-12D Concrete Masonry in the 2009 Edition of the IECC (2012)
6-12E Concrete Masonry in the 2012 Edition of the IECC (2012)
6-13B Thermal Bridges in Wall Construction (2010)
6-14A Control of Air Leakage in Concrete Masonry Walls (2011)
6-15A Radon-Resistant Concrete Masonry Foundation Walls (2006)
6-16A Heat Capacity (HC) Values for Concrete Masonry Walls (2008)
6-17B Condensation Control in Concrete Masonry Walls (2011)

7 – Fire Resistance

7-01C Fire Resistance Rating of Concrete Masonry Assemblies (2009)
7-02 Balanced Design Fire Protection for Multifamily Housing (2008)
7-03A Firestopping for Concrete Masonry Walls (2010)
7-04A Foam Plastic Insulation in Concrete Masonry Walls (2013)
7-05A Evaluating Fire-Exposed Concrete Masonry Walls (2006)
7-06A Steel Column Fire Protection (2009)

8 – Maintenance & Cleaning

8-01A Maintenance of Concrete Masonry Walls (2004)
8-02A Removal of Stains from Concrete Masonry (2005)
8-03A Control and Removal of Efflorescence (2003)
8-04A Cleaning Concrete Masonry (2005)

9 – Mortar, Grout, & Stucco

9-01A Mortars for Concrete Masonry (2004)
9-02B Self Consolidating Grout for Concrete Masonry (2007)
9-03A Plaster and Stucco for Concrete Masonry (2002)
9-04A Grout for Concrete Masonry (2005)

10 – Movement Control

10-01A Crack Control in Concrete Masonry Walls (2005)
10-02C Control Joints for Concrete Masonry Walls - Empirical Method (2010)
10-03 Control Joints for Concrete Masonry Walls - Alternative Engineered Method (2003)
10-04 Crack Control for Concrete Brick and Other Concrete Masonry Veneers (2001)

11 – Articulated Concrete Block (ACBs)

(Note: Paver TEK Discountinued)
11-09B Articulated Concrete Block for Erosion Control (2014)
11-12A ACB Revetment Design - Factor of Safety Method (2011)
11-13 Articulating Concrete Block (ACB) Installation (2006)

12 – Reinforcement & Connectors

12-01B Anchors and Ties for Masonry (2011)
12-02B Joint Reinforcement for Concrete Masonry (2005)
12-03C Design of Anchor Bolts Embedded in Concrete Masonry (2013)
12-04D Steel Reinforcement for Concrete Masonry (2006)
12-05 Fasteners for Concrete Masonry (2005)
12-06 Splices, Development and Standard Hooks for Concrete Masonry, 2003 & 2006 IBC (2007)
12-06A Splices, Development and Standard Hooks for Concrete Masonry Based on the 2009 & 2012 IBC (2013)

13 – Sound

13-01C Sound Transmission Class Ratings for Concrete Masonry Walls (2012)
13-02A Noise Control with Concrete Masonry (2007)

NCMA TEK NOTES

13-03A Concrete Masonry Highway Sound Barriers (1999)
13-04A Outdoor-Indoor Transmission Class of Concrete Masonry Walls (2012)

14 – Structural - General

14-01B Section Properties of Concrete Masonry Walls (2007)
14-02 This TEK has been replaced by TEK 14-4A and TEK 14-7A as appropriate
14-03A Designing Concrete Masonry Walls for Wind Loads (1995)
14-04B Strength Design Provisions for Concrete Masonry (2008)
14-05A Loadbearing Concrete Masonry Wall Design (2006)
14-06 Concrete Masonry Bond Patterns (2004)
14-07B Allowable Stress Design of Concrete Masonry (2006 & 2009 IBC) (2009)
14-07C Allowable Stress Design of Concrete Masonry Based on the 2012 IBC & 2011 MSJC (2013)
14-08B Empirical Design of Concrete Masonry Walls (2008)
14-09A Hybrid Concrete Masonry Design (2009)
14-10B Impact Resistance of Concrete Masonry for Correctional Facilities (2003)
14-11B Strength Design of Concrete Masonry Walls for Axial Load & Flexure (2003)
14-12B Seismic Design Forces on Concrete Masonry Buildings (2005)
14-13B Concrete Masonry Wall Weights (2008)
14-14 Concrete Masonry Arches (1994)
14-15B Allowable Stress Design of Pier and Panel Highway Sound Barrier Walls (2004)
14-16B Concrete Masonry Fence Design (2007)
14-17A Software for the Structural Design of Concrete Masonry (2010)
14-18B Seismic Design and Detailing Requirements for Masonry Structures (2009)
14-19A Allowable Stress Design Tables for Reinforced Concrete Masonry Walls (2005)
14-19B ASD Tables for Reinforced Concrete Masonry Walls Based on the 2012 IBC & 2011 MSJC (2011)
14-20A Post-Tensioned Concrete Masonry Wall Design (2002)
14-21A Design of Concrete Masonry Walls for Blast Loading (2014)
14-22 Design and Construction of Dry-Stack Masonry Walls (2003)
14-23 Design of Concrete Masonry Infill (2012)

15 – Structural - Foundation & Retaining Walls

15-01B Allowable Stress Design of Concrete Masonry Foundation Walls (2001)
15-02B Strength Design of Reinforced Concrete Masonry Foundation Walls (2004)
15-03A Roles and Responsibilities on Segmental Retaining Wall Projects (2010)
15-04B Segmental Retaining Wall Global Stability (2010)
15-05B Segmental Retaining Wall Design (2010)
15-06 Concrete Masonry Gravity Retaining Walls (1995)
15-07B Concrete Masonry Cantilever Retaining Walls (2005)
15-08A Guide to Segmental Retaining Walls (2009)
15-09A Seismic Design of Segmental Retaining Walls (2010)

16 – Structural - Multi-Wythe Walls

16-01A Multiwythe Concrete Masonry Walls (2005)
16-02B Structural Design of Unreinforced Composite Masonry (2001)
16-03B Reinforced Concrete Masonry Composite Walls (2006)
16-04A Design of Concrete Masonry Noncomposite (Cavity) Walls (2004)

17 – Structural - Beams, Columns & Lintels

17-01D ASD of Concrete Masonry Lintels Based on the 2012 IBC/2011 MSJC (2011)
17-02A Precast Lintels for Concrete Masonry Construction (2000)
17-03A Allowable Stress Design of Concrete Masonry Columns (2001)
17-04B Allowable Stress Design of Concrete Masonry Pilasters (2000)

18 – Quality Assurance, Inspection, & Testing

18-01B Evaluating the Compressive Strength of Concrete Masonry based on 2012 IBC/2011 MSJC (2011)
18-01C Evaluating the Compressive Strength of Concrete Masonry based on 2015 IBC/2013 MSJC (2014)
18-02C Sampling and Testing Concrete Masonry Units (2014)
18-03B Concrete Masonry Inspection (2006)

18-04A Creep Properties of Post-Tensioned and High-Rise Concrete Masonry (2000)
18-05B Masonry Mortar Testing (2014)
18-06 Structural Testing of Concrete Masonry Assemblages (1997)
18-07 Compressive Strength Testing Variables for Concrete Masonry Units (2004)
18-08B Grout Quality Assurance (2005)
18-09A Evaluating Existing Concrete Masonry Construction (2003)
18-10 Sampling and Testing Segmental Retaining Wall Units (2005)
18-11B Inspection Guide for Segmental Retaining Walls (2012)

19 – Water Penetration Resistance

19-01 Water Repellents for Concrete Masonry Walls (2006)
19-02B Design for Dry Single-Wythe Concrete Masonry Walls (2012)
19-03B Preventing Water Penetration in Below-Grade Concrete Masonry Walls (2012)
19-04A Flashing Strategies for Concrete Masonry Walls (2008)
19-05A Flashing Details for Concrete Masonry Walls (2008)
19-06 Joint Sealants for Concrete Masonry Walls (2014)
19-07 Characteristics of Concrete Masonry Units with Integral Water Repellent (2008)

NCMA TEK NOTES

INDEX

A

AAC Masonry	47, 369
Acceptable, Accepted	337
Acceptance Relative to Strength Requirements	111
Access for Special Inspection	34
Accessories	382
Asphalt Emulsion	382
Joint Fillers	382
Masonry Cleaner	382
Acoustical Leaks	569
Adhered Masonry Veneer	23, 26
Exterior Adhered Masonry Veneer	26
Exterior Adhered Masonry Veneer—Porcelain Tile	27
Interior Adhered Masonry Veneers	27
Adhered Veneer	274
Adhered Veneer Requirements	355
Administration and Enforcement	5
Adobe Construction	47, 52
Air-Borne Sound	21
Masonry	21
Allowable Compressive Stresses	295
Allowable Compressive Stresses for Empirical Design of Masonry	296
Allowable Flexural Tensile Stresses for Clay and Concrete Masonry	192
Allowable Shear on Bolts in Adobe Masonry	53
Allowable Story Height	64
Allowable Stress Design	50
Allowable Stress Design of Masonry	175
Reinforced Masonry	196
Unreinforced Masonry	188
Allowable Stress Gross Cross-Sectional Area for Dry-Stacked, Surface-Bonded Concrete Masonry Walls	52
Allowable Stresses	52
Alternative Materials, Design and Methods of Construction and Equipment	5
Research Reports	5
Tests	5
Anchor	94
Anchor Bolts	151, 245, 405
Anchor Bolts Embedded in Grout	176, 205
Anchor Pullout	94
Anchorage	29, 302
Structural Walls	29
Anchorage Design	160
Anchored Masonry Veneer	23, 25
Seismic Requirements	25
Tolerances	25
Anchored Veneer	268
Anchors, Ties, and Accessories	375
Application and Measurement of Prestressing Force	411
Approval of Special Systems of Design or Construction	84

Architect/Engineer	337
Architectural Conditions	431, 500
Area, Gross Cross-Sectional	47, 94, 337
Area, Net Cross-Sectional	47, 94, 337
Area, Net Shear	94
Ash Dump Cleanout	54
Asphalt Emulsion	382
Autoclaved Aerated Concrete (AAC)	47, 94, 337
Autoclaved Aerated Concrete (AAC) Masonry	94, 338
Autoclaved Aerated Concrete Masonry Veneer	267
Axial Compression and Flexure	188, 196, 236
Axial Tension	194, 212, 239, 249

B

B Bar Reinforcement	496
Backing	23, 94
Basic Load Combinations	29
Base Surface Treatment	282
Basic Wind Speed Modification for Topographic Wind Speed Up Effect	430
Beams	137
Bearing Area	123
Bearing Stresses	178, 240
Bearing Wall Structures	30
Bed Joint	47, 94
BIA Technical Notes on Brick Construction	603-606
Bond	301
Bond Beam	95, 338
Bond Pattern	391
Bonded Prestressing Tendon	96, 339
Bonding with Masonry Headers	301
Bonding with Wall Ties or Joint Reinforcement	301
Bounding Frame	96
Bracing of Masonry	396
Brick	47
Building Code Requirements and Specification for Masonry Structures	67
Building Official	96
Burning and Welding Operations	412

C

Calculated Fire Resistance	9
Calculated STC Ratings for Clay Masonry Walls	574
Calculated STC Values for Concrete Masonry Walls	575
Calculation	293, 565, 566, 570
Calculation of Fire Resistance	556
Calculation Procedure	556
Calculation for Fire-Exposed Side	556
Calculation for Non-Fire-Exposed Side	556
Fire Exposure	556
Minimum Fire Resistance Provided by Concrete or Masonry	557

INDEX

Cantilever Walls	288, 299
Cast Stone	47
Cavity Wall	96
Cell (Masonry)	47
Ceramic Tile Mortar Compositions	49
Changes in Dimension	58
Chases and Recesses	288, 305
Chimney	47
Factory-Built Chimney	47
Masonry Chimney	47
Metal Chimney	47
Chimney Caps	58
Chimney Clearances	61
Chimney Fireblocking	61
Chimney Types	47
High-Heat Appliance Type	47
Low-Heat Appliance Type	48
Masonry Type	48
Medium-Heat Appliance Type	48
Clay Brick and Tile Masonry	13, 553
Masonry Walls	13
Multiwythe Walls	15
Reinforced Clay Masonry Columns	15
Reinforced Clay Masonry Lintels	15
Clay Brick and Tile Masonry Wall Assemblies	554
Effects of Finish Materials on Fire Resistance	554
Filled and Unfilled Clay Brick and Tile Masonry	554
Multi-Wythe Walls	554
Single-Wythe Walls	554
Clay Flue Lining (Installation)	59
Cleaning	412
Cleanouts	339, 390
Clearance(s)	56
Coatings for Corrosion Protection	376
Code Requirements for Determining Fire Resistance of Concrete and Masonry Construction Assemblies	545
Coefficients for Plaster	15
Coefficients of Creep	120
Coefficients of Moisture Expansion for Clay Masonry	120
Coefficients of Shrinkage	120
Coefficients of Thermal Expansion	120
Cold Weather Construction	363
Collar Joint	48, 96, 339
Column(s)	96, 141
Combustible Framing in Fire Walls	7
Commentary to Direct Design Handbook for Masonry Structures	493
Composite Action	96
Composite Masonry	96
Compressive Strength Determination	348
Compressive Strength of Masonry	96, 339
Compressive Strength of Masonry Based on the Compressive Strength of Clay Masonry Units and Type of Mortar Used in Construction	349
Compressive Strength of Masonry Based on the Compressive Strength of Concrete Masonry Units and Type of Mortar Used in Construction	351
Compressive Strength Requirements	348
Compressive Stress Requirements	295

INDEX

Concealed Spaces----------8
Concentrated Loads----------131
Concrete and Masonry Chimneys for High-Heat Appliances----------58
Concrete and Masonry Chimneys for Medium-Heat Appliances----------59
Concrete and Masonry Foundation Walls----------38
Concrete Masonry----------9, 549
    Concrete Masonry Columns----------13
    Concrete Masonry Lintels----------13
    Concrete Masonry Walls----------9
    Equivalent Thickness----------9
    Multiwythe Masonry Walls----------12
Concrete Masonry Wall Assemblies----------550
    Effects of Finish Materials on Fire Resistance----------551
    Expansion or Contraction Joints----------550
    Multi-Wythe Wall Assemblies----------550
    Single-Wythe Wall Assemblies----------550
Confinement----------407
Connection of Braced Wall Panels----------65
Connection to Structural Frames----------124
Connector----------96
Consolidation----------409
Construction----------50, 564, 568
    Sealing Openings and Joints----------564, 568
    Surface Coatings----------564, 570
Construction Considerations----------111
Construction Loads----------363
Content of Statement of Special Inspections----------34
Continuous, Special Inspection----------33
Contract Documents----------96, 339
Contract Documents and Calculations----------82
Contractor----------339
Contractor Responsibility----------34
Contractor's Services and Duties----------361
Control Joint Detail at C Bar Locations----------491
Conventional Light-Frame Construction----------63
Conversion of Inch-Pound Units to SI Units----------329
Corbel(s)----------96, 143, 247
Corbeling----------58
Corrosion Protection for Tendons----------377
Cover, Grout----------96, 339
Cover, Masonry----------96, 339
Cover, Mortar----------97, 339
C_s Values for Risk Category I and II Structures----------445
C_s Values for Risk Category III Structures----------445
C_s Values for Risk Category IV Structures----------446
C_w Values----------444

D

Damper----------55
Dampproofing----------37
    Walls----------37
Dampproofing and Waterproofing----------37

INDEX

Data for Clay Masonry---571, 578
Data for Concrete Masonry---573, 579
Debris---390
Deep Beam(s)---97, 139
Definition(s)---23, 33, 47, 56, 94, 337, 425, 547, 564
Deflection---240
Deformation Requirements---204, 244
Delivery, Storage, and Handling---362
Depth---97
Design Assumptions---188, 196, 210, 214, 250
Design Criteria---188, 210
Design Lateral Soil Loads---39
Design Methods---47, 234
    Masonry Veneer---47
Design of Adhered Veneer---267
Design of Anchored Veneer---265
Design of Beams, Piers, and Columns---220, 253
Design of Frame Elements with Participating Infills for In-Plane Loads---312
Design of Masonry Infill---307
    Non-Participating Infills---309
    Participating Infills---310
Design of Non-Participating Infills for Out-of-Plane Loads---309
Design of Participating Infills for In-Plane Forces---311
Design of Participating Infills for Out-of-Plane Forces---313
Design of Partition Walls---283
Design Requirements---115
Design Story Drift---97
Design Strength---48, 97, 175, 203, 243, 308
Details---487, 533
Details of Designated and Non-Designated Shear Flanged Wall Intersections---489
Details of Flanged Wall Intersections---534
Details of Reinforcement and Metal Accessories---147
Determination of Minimum Area---60
Development of Bonded Tendons---241
Development of Reinforcement Embedded in Grout---179
Diaphragm---97
Diaphragm Chord Reinforcement (C Bars)---472
Diaphragm Length-to-Width Ratios---293
Dimension Stone---267
Dimensions---43, 48, 97
    Nominal---48, 97, 339
    Specified---48, 97, 339
Direct Design Handbook for Masonry Structures---417
Direct Design Procedure---435, 511
Drift Limits---160
Duties and Powers of Building Official---5

E

Earthquake Loads---30
Effective Compressive Width per Bar---130
Effective Embedment Length for Bent-Bar Anchor Bolts---156
Effective Embedment Length for Headed Anchor Bolts---156

INDEX

Effective Height — 97
Effective Prestress — 97, 235
Effects of Finish Materials on Fire Resistance — 551, 554, 556
Elastic Moduli — 118
Element Interaction — 160
Element Classification — 162
Embedded Conduits, Pipes and Sleeves — 112
Embedded Items and Accessories — 395
Embedded Posts and Poles — 40
    Limitations — 40
Embedment — 147
Empirical Design of Masonry — 51, 289
    Anchorage — 302
    Bond — 301
    Compressive Stress Requirements — 295
    Height — 293
    Lateral Stability — 293
    Lateral Support — 298
    Miscellaneous Requirements — 305
    Thickness of Masonry — 299
Empirically Designed Masonry, Glass Unit Masonry and Masonry Veneer in Risk Category IV — 36
Engineered Design Methods — 175
Equation Conversions — 317
Equivalent Thickness — 9, 550, 553
    Air Spaces and Cells Filled with Loose Fill Material — 550, 554
    Solid Grouted Construction — 550, 554
    Ungrouted or Partially Grouted Construction — 550, 554
Example Building Elevation(s) — 515, 523
Example Roof Plan — 513
Examples of Designations for East-West Wind and Seismic Loading — 513
Examples of Designations for North-South Wind and Seismic Loading — 513
Execution — 389
Expansion Joints — 282
Expansion or Contraction Joints — 550, 554
Exterior Adhered Masonry Veneer — 26
Exterior Adhered Masonry Veneers—Porcelain Tile — 27
Exterior Air — 56
    Clearance — 56
    Exterior Air Intake — 56
    Factory-Built Fireplaces — 56
    Masonry Fireplaces — 56
    Outlet — 56
    Passageway — 56
Exterior Insulation and Finish Systems (EIFS) — 36
    Water-Resistive Barrier Coating — 36
Exterior Wall(s) — 23
Exterior Wall Covering — 23
Exterior Wall Envelope — 23

F

Fabrication — 387
    Prefabricated Masonry — 388

INDEX

Reinforcement	387
Factory-Built Fireplaces	56
Fiberboard	63
Field Quality Control	412
Field Testing	565, 566
Filled and Unfilled Clay Brick and Tile Masonry	554
Fire and Smoke Protection Features	7
Fire Exposure	556
Fire Resistance	24
Fire Resistance of Clay-Masonry-Protected Steel Columns	19, 555
Fire Resistance of Concrete-Masonry-Protected Steel Columns	17, 552
Fire Resistance Periods of Clay Masonry Walls	13
Fire-Resistance-Rated Walls	7
Through Penetrations	8
Fire Resistance Rating for Bearing Steel Frame Brick Veneer Walls or Partitions	14
Fire-Resistance Rating of Concrete Masonry Assemblies	550
Fire-Resistance Rating of Structural Members	7
Fire Resistance Requirements for Plaster	9
Fire-Resistant Penetrations and Joints	36
Fire-Resistant Joint Systems	36
Penetration Firestops	36
Fire Walls	7
Firebox Dimensions	55
Firebox Walls	54
Steel Fireplace Units	54
Fireplace	48
Fireplace Clearance	55
Fireplace Drawings	54
Fireplace Fireblocking	56
Fireplace Throat	48
Flashing	25
Masonry	25
Flexural Cracking	249
Flexural Strength of Unreinforced (Plain) AAC Masonry Members	249
Flood Hazard Areas	5
Flood Resistance	24
Flood Resistance for Coastal High-Hazard Areas and Coastal A Zones	24
Floor and Roof Anchorage	304
Floor Decks	63
Floor Framing	63, 64
Floor Joist	64
Girders	64
Flue Area	60
Determination of Minimum Area	60
Minimum Area	60
Flue Lining (Material)	58
Concrete and Masonry Chimneys for High-Heat Appliances	59
Concrete and Masonry Chimneys for Medium-Heat Appliances	59
Residential-Type Appliances (General)	58
Flue Sizes for Masonry Chimneys	60
Footings and Foundations	54, 57
Ash Dump Cleanout	54
Foundation Pier(s)	48, 97, 300
Foundation Wall Construction	301
Foundation Walls	37, 299

INDEX

Concrete and Masonry Foundation Walls---38
    Prescriptive Design of Concrete and Masonry Foundation Walls---38
    Rubble Stone Foundation Walls---38
Foundation Walls, Retaining Walls and Embedded Posts and Poles---37
Foundations---42
Foundations and Footings---64
  Foundation Plates and Sills---64

G

General Analysis and Design Considerations---115
  Connection to Structural Frames---124
  Loading---115
  Masonry Not Laid in Running Bond---125
  Material Properties---117
  Section Properties---121
General Beam Design---137
General Column Design---141
General Construction Requirements---63
General Design Requirements---29, 267
General Requirements---77
  Approval of special systems of design or construction---84
  Standards Cited in this Code---84
Glass Unit Masonry---24, 48, 53, 97, 277, 339, 369
  Base Surface Treatment---282
  Expansion Joints---282
  Mortar---282
  Panel Size---277
  Reinforcement---282
  Support---280
Glass Unit Masonry Panel Anchors---407
Grout---50, 97, 339, 386, 564
Grout Demonstration Panel---362
Grout Lift---340
Grout Materials---370
Grout Placement---407
  Alternate Grout Placement---410
  Confinement---407
  Consolidation---409
  Grout for AAC Masonry---410
  Grout Key---410
  Grout Lift Height---409
  Grout Pour Height---408
  Placing Time---407
Grout Pour---340
Grout, Self-Consolidating---98, 339
Grout Space Requirements---113, 408
Grouting Bonded Tendons---412
Grouting, Minimum Spaces---111
Gypsum Wallboard---557

INDEX

H

Head Joint — 48, 98
Header (Bonder) — 98
Hearth and Hearth Extension — 55
    Hearth Extension Thickness — 55
    Hearth Thickness — 55
Hearth Extension Dimensions — 55
Height — 293
Horizontal Assemblies — 8
    Membrane Penetrations — 8
    Through Penetrations — 8
Horizontal Reinforcing — 54, 57
Hot Weather Construction — 366

I

Illustration of Exception Three Chimney Clearance Provision — 61
Illustration of Exception to Fireplace Clearance Provision — 56
Increase in STC Using the Furring Space Depth Indicated and a Single Layer of Drywall — 576
Infill — 98
infill, Net Thickness — 98
Infill, Non-Participating — 98
Infill, Participating — 98
In-Plane Connection Requirements for Participating Infills — 310
In-Plane Isolation Joints for Non-Participating Infills — 309
Inspection(s) — 389
Inspection Agency's Services and Duties — 361
Inspection, Continuous — 98, 340
Inspection, Periodic — 98, 340
Installation — 57
Installation of Finishes — 557
    Gypsum Wallboard — 557
    Plaster and Stucco — 557
Installation of Wall Coverings — 24
Interior Adhered Masonry Veneers — 27
Interior Environment — 21
Intersecting Walls — 127, 288, 302

J

Jacking Force — 234
Joint Fillers — 382
Joint Reinforcement — 375
Joint Sealants — 564

INDEX

619

L

Laboratory Testing	565, 566
Lap Splices	51, 288
Lap Splices of Reinforcement	489, 534
Lateral Force Coefficients for Ordinary Plain (Unreinforced) Masonry Shear Walls	464
Lateral Force Coefficients for Ordinary Reinforced Masonry Shear Walls	464-468
Lateral Force Coefficients for Special Reinforced Masonry Shear Walls	469-472
Lateral Load Distribution	116
Lateral Load Resistance	115
Lateral Stability	293
Lateral Support	298
Laterally Restrained Prestressing Tendon	98
Laterally Unrestrained Prestressing Tendon	98
Level A Quality Assurance	105, 107, 357
Level B Quality Assurance	106, 108, 358
Level C Quality Assurance	106, 110, 360
LFRS Options	443
Licensed Design Professional	99
Lightly Loaded Columns	142
Limit Design Method	315
Mechanism Deformation	316
Mechanism Strength	316
Yield Mechanism	315
Limitations Based on Building Height and Basic Wind Speed	291
Limitations of Concrete or Masonry Veneer	65
Lintel and Throat	55
Damper	55
Lintel Design Tables	473-480
Lintels	288, 305
List of Lateral Force Coefficients Tables	463
Load-bearing Corbels	143
Load Combinations	29
Load Combinations using Allowable Stress Design	29
Basic Load Combinations	29
Load, Dead	99
Load, Live	99
Load Path	160
Load, Service	99
Loading	115
Loading Conditions	432, 505
Longitudinal Reinforcement	99
Long-Term Loading	63

M

Mandatory Requirements Checklist	414
Masonry	24, 47, 48
Glass Unit Masonry	48
Plain Masonry	48
Reinforced Masonry	48
Solid Masonry	48

INDEX

Unreinforced (Plain) Masonry	48
Masonry Assemblies	127
Masonry Breakout	99
Masonry Chimney	7, 57
Masonry Chimney Cleanout Openings	61
Masonry Cleaner	382
Masonry Construction	36, 50
Empirically Designed Masonry, Glass Unit Masonry and Masonry Veneer in Risk Category IV	36
Molded Cornices	50
Support on Wood	50
Vertical Masonry Foundation Elements	36
Masonry Construction Materials	49
Masonry Erection	391
Bond Pattern	391
Bracing of Masonry	396
Embedded Items and Accessories	395
Placing Adhered Veneer	395
Placing Mortar and Units	391
Site Tolerances	396
Masonry Fireplaces	54, 56
Masonry Heater Clearance	57
Masonry Heaters	56
Masonry Inspection Checklist	581
Masonry Lintels	487, 533
Masonry not Laid in Running Bond	125
Masonry, Partially Grouted	99, 340
Masonry Partition Walls	283
Anchorage	288
Lateral Support	286
Miscellaneous Requirements	288
Prescriptive Design of Partition Walls	283
Masonry Protection	363
Masonry Unit(s)	48, 49, 564
Second-Hand Units	49
Hollow	48, 99, 340
Solid	48, 99, 340
Masonry-Unit Footings	43
Dimensions	43
Offsets	43
Masonry Unit Materials	371
Masonry Veneer	47
Masonry Veneer Chimneys	58
Masonry Walls	13
Material and Construction Requirements	433, 506
Material Properties	117, 207, 245
Materials	24, 564
Grout	564
Joint Sealants	564
Masonry Units	564
Mortar	564
Maximum Bar Size	51
Maximum Horizontal Spans for Walls Above and Below Openings for Seismic Conditions	481
Maximum Horizontal Spans for Walls Above and Below Openings for Wind Conditions	481
Maximum l/t and h/t	286, 298
Maximum l/t or h/t for 5 psf Lateral Load	287

INDEX

Maximum l/t or h/t for 10 psf Lateral Load	287
Maximum Vertical Spans for Wall Segments	447-454
Maximum Vertical Spans for Walls without Openings Constructed using Lightweight Concrete Masonry Units for Seismic Conditions and for Ground Snow Loads	454-456
Maximum Vertical Spans for Walls without Openings Constructed using Medium and Normal Weight Concrete Masonry Units for Seismic Conditions and for Ground Snow Loads	457-458
Maximum Vertical Spans without Openings for Walls Constructed of Lightweight Concrete Masonry Units with Veneer Cladding for Seismic Conditions for Ground Snow Loads	459-460
Maximum Vertical Spans without Openings for Walls Constructed using Medium and Normal Weight Concrete Masonry Units with Veneer Cladding for Seismic Conditions for Ground Snow Loads	461-462
Mean Daily Temperature	340
Mechanism Deformation	316
Mechanism Strength	316
Metal Reinforcement and Accessories	50
Minimum Bend Diameter for Reinforcing Bars	149
Minimum Cover of Longitudinal Reinforcement in Fire-Resistance-Rated Reinforced Concrete Masonry Lintels	13
Minimum Daily Temperature	340
Minimum Diameters of Bend	149
Minimum Dimension of Concrete Masonry Columns	13
Minimum Equivalent Thickness of Bearing or Nonbearing Concrete Masonry Walls	12
Minimum Fire Resistance Provided by Concrete or Masonry	557
Minimum/Maximum (not Less Than....not More Than)	340
Minimum Permissible Effective Embedment Length	157
Minimum Prescriptive Seismic Reinforcement for Ordinary Reinforced Masonry Shear Walls as Required by the MSJC Code	450, 534
Minimum Prescriptive Seismic Reinforcement for Special Reinforced Masonry Shear Walls as Required by the MSJC Code	490, 535
Minimum Requirements	81
Minimum Standards and Quality	63
Minimum Thickness	299
Minimum Thickness of Weather Coverings	25
Mixing	385
Grout	386
Mortar	385
Thin-Bed Mortar for AAC	387
Modulus of Elasticity	99
Modulus of Rigidity	99
Modulus of Rupture	208
Molded Cornices	50
Mortar	48, 49, 282, 385, 564
Masonry Mortar	49
Mortar for Adhered Masonry Veneer	50
Mortars for Ceramic Wall and Floor Tile	49
Surface-bonding Mortar	49
Mortar Materials	367
AAC Masonry	369
Glass Unit Masonry	369
Mortar, Surface-Bonding	48
Multiple Flues	59
Multiplying Factor for Finishes on Non-Fire-Exposed Side of Concrete Slabs and Concrete and Masonry Walls	556
Multiplying Factor for Finishes on Nonfire-Exposed Side of Wall	11
Multiwythe Masonry Elements	133
Multiwythe Masonry Walls	12
Multi-Wythe Wall Assemblies	550
Multiwythe Walls	15, 554

INDEX

N

Natural or Cast Stone — 302
Net Cross-Sectional Area of Round Flue Sizes — 61
Net Cross-Sectional Area of Square and Rectangular Flue Sizes — 61
New Materials — 33
Nominal Axial Strength of Unreinforced (Plain) AAC Masonry Members — 249
Nominal Bearing Strength — 207, 246
Nominal Dimension — 48
Nominal Flexural and Axial Strength — 210
Nominal Shear Strength — 213
Nominal Shear Strength of Unreinforced (Plain) AAC Masonry Members — 249
Nominal Strength — 48, 99
Noncombustible Furring — 9
Non-Load-Bearing Corbels — 143
Nonparticipating Elements — 162
Non-Participating Infills — 309
Non-Participating Wall Detailing Option — 524
Notations — 49, 87, 426 497, 546, 563

O

Offsets — 43, 58
Outlet — 56
Outdoor-Indoor Transmission Class Ratings — 566, 577
    Calculation — 566, 577
    Field Testing — 566
    Laboratory Testing — 566, 577
Out-of-Plane Connection Requirements for Participating Infills — 310

P

Panel Size — 277
Participating Elements — 162
Participating Infills — 310
Partition Wall — 99, 340
Penetration Firestops — 36
Penetrations — 7
Performance Requirements — 23
Performance Specification for Corrosion-Inhibiting Coating — 381
Periodic, Special Inspection — 33
Permissible Stresses in Prestressing Tendons — 234
Permitted Mortar for Seismic Design Category — 433
Pier and Curtain Wall Foundations — 43
Piers — 99
Pilasters — 143
Placement — 151
Placement of Reinforcement — 147
Placement Tolerances for Reinforcement — 488, 533
Placing Adhered Veneer — 395

INDEX

Placing Mortar and Units ---381
Placing Time ---407
Plain Masonry ---48
Plain Masonry Foundation Walls ---39
Plaster and Stucco ---557
Post-Tensioned Masonry Members ---234
Post-Tensioning ---100, 341
Prefabricated Masonry ---388
Preparation ---390
    Cleanouts ---390
    Debris ---390
    Reinforcement ---390
    Wetting Masonry Units ---390
Prescriptive Design Methods ---263
Prescriptive Design of Concrete and Masonry Foundation Walls ---38
Prescriptive Design of Partition Walls ---283
Prescriptive Footings for Light-Frame Construction ---43
Prescriptive Footings Supporting Walls of Light-Frame Construction ---43
Prescriptive Fire Resistance ---9
Prescriptive Requirements for Adhered Masonry Veneer ---274
Prescriptive Requirements for Anchored Masonry Veneer ---268
Prestressed Masonry ---48, 100, 233, 341
    Axial Compression and Flexure ---236
    Axial Tension ---239
    Deflection ---240
    Design Methods ---234
    Development of Bonded Tendons ---241
    Permissible Stresses in Prestressing ---234
    Prestressing Tendon Anchorages, Couplers, and End Blocks ---240
    Protection of Prestressing Tendons and Accessories ---241
    Shear ---239
Prestressing Anchorages, Couplers, and End Blocks ---380
Prestressing Grout ---100, 341
Prestressing Steel ---121
Prestressing Tendon(s) ---100, 341, 375
Prestressing Tendon Anchorages, Couplers, and End Blocks ---240
Prestressing Tendon Installation and Stressing Procedure ---411
    Application and Measurement of Prestressing Force ---411
    Burning and Welding Operations ---412
    Grouting Bonded Tendons ---412
    Site Tolerances ---411
Pretensioning ---100, 341
Prism ---100, 341
Prism Test Method ---354
Project Conditions ---363
    Cold Weather Construction ---363
    Construction Loads ---363
    Hot Weather Construction ---366
    Masonry Protection ---363
Project Drawings ---100, 341
Project Specifications ---100, 341
Projected Area for Axial Tension ---153
Projected Area for Shear ---154
Protection of Prestressing Tendons and Accessories ---241
Protection of Reinforcement and Metal Accessories ---148

INDEX

Q

Quality and Construction---105
    Construction Considerations---111
    Quality Assurance Program---105
Quality Assurance---50, 100, 341, 357
    Contractor's Services and Duties---361
    Grout Demonstration Panel---362
    Inspection Agency's Services and Duties---361
    Sample Panels---362
    Testing Agency's Services and Duties---357
Quality Assurance Program---105

R

Radius of Gyration---123
Rain Caps---58
Rated Fire-Resistance Periods for Various Walls and Partitions---10
Reference Standards---343, 549, 563
Reinforced AAC Masonry---250
Reinforced Clay Masonry Columns---15, 553
Reinforced Clay Masonry Lintels---15, 553
Reinforced Concrete Masonry Columns---551
Reinforced Concrete Masonry Lintels---551
Reinforced Masonry---48, 196, 214
Reinforced Masonry Columns---550
Reinforced Masonry Lintels---15, 550
Reinforcement---100, 282, 341, 387, 398
Reinforcement, Metal Accessories, and Anchor Bolts---147
    Anchor Bolts---151
    Details of Reinforcement and Metal Accessories---147
Reinforcement, Prestressing Tendons, and Metal Accessories---374
    Anchors, Ties, and Accessories---375
    Coatings for Corrosion Protection---376
    Corrosion Protection for Tendons---377
    Joint Reinforcement---375
    Prestressing Anchorages, Couplers, and End Blocks---380
    Prestressing Tendons---375
    Reinforcing bars---374
    Stainless Steel---376
Reinforcement Requirements and Details---215, 251
Reinforcement, Tie, and Anchor Installation---398
    Anchor Bolts---405
    Basic Requirements---398
    Glass Unit Masonry Panel Anchors---407
    Reinforcement---398
    Veneer Anchors---406
    Wall Ties---404
Reinforcing Bars---374
Report Requirement---34
Required Special Inspections and Tests---35
Required Strength---49, 100, 203, 243, 308

INDEX

Research Reports ---5
Residential-Type Appliances (General) ---58
Retaining Walls ---39
    Design Lateral Soil Loads ---39
    Safety Factor ---39
Roofs ---293
Rubble Stone Foundation Walls ---38
Running Bond ---48, 100, 341

S

Sample Panels ---362
Scope and Administration ---5
Sealing Around Penetrations and Fixtures ---569
Sealing Openings and Joints ---564, 568
Second-Hand Units ---49
Section Properties ---121
Seismic Anchorage ---54, 57
Seismic Design ---50
Seismic Design Category A Requirements ---170
Seismic Design Category B Requirements ---170
Seismic Design Category C Requirements ---171
Seismic Design Category D Requirements ---173
Seismic Design Category E and F Requirements ---174
Seismic Design Category (SDC) for Site Class A ---441
Seismic Design Category (SDC) for Site Class B ---442
Seismic Design Category (SDC) for Site Class C ---442
Seismic Design Category (SDC) for Site Class D ---443
Seismic Design Requirements ---159
    Element Classification ---162
    General Analysis ---160
    Seismic Design Category Requirements ---170
Seismic Design Requirements for Masonry ---50
Seismic Reinforcement ---54, 57
    Horizontal Reinforcing ---54, 57
    Vertical Reinforcing ---54, 57
Seismic Requirements in the Statement of Special Inspections ---34
Service Load Requirements ---237
Shaft Enclosures ---7
Shallow Foundations ---43
Shear ---194, 199, 239
Shear Strength in Multiwythe Masonry Elements ---207
Shear Stress in Multiwythe Masonry Elements ---178
Shear Wall(s) ---101, 293
Shear Wall, Detailed Plain (Unreinforced) AAC Masonry ---101
Shear Wall, Detailed Plain (Unreinforced) Masonry ---101
Shear Wall, Intermediate Reinforced Masonry ---101
Shear Wall, Intermediate Reinforced Prestressed Masonry ---101
Shear Wall Layout at Corners ---503
Shear Wall, Ordinary Plain (Unreinforced) AAC Masonry ---101
Shear Wall, Ordinary Plain (Unreinforced) Masonry ---101
Shear Wall, Ordinary Plain (Unreinforced) Prestressed Masonry ---101
Shear Wall, Ordinary Reinforced AAC Masonry ---101

INDEX

Shear Wall, Ordinary Reinforced Masonry	101
Shear Wall, Special Reinforced Masonry	101
Shear Wall, Special Reinforced Prestressed Masonry	101
Single-Wythe Wall Assemblies	550
Single-Wythe Walls	554
Site Conditions	429, 499
Site Tolerances	396, 411
Size of Reinforcement	147
Slab-Type Veneer	26
Slump Flow	102, 341
Smoke Chamber Walls	55
Smoke Chamber Dimensions	55
Soils and Foundations	37
Solid Grouted Construction	550
Solid Masonry	48
Sound Transmission	21
Sound Transmission Class Ratings	565, 570
Calculation	565, 570
Field Testing	565
Laboratory Testing	565, 570
Space Around Lining	59
Spark Arrestors	58
Special Boundary Elements	102
Special Inspection	33, 47
Continuous, Special Inspection	33
Periodic, Special Inspection	33
Special Inspections and Tests	33
Access for Special Inspection	34
Report Requirement	34
Special Inspector Qualifications	33
Statement of Special Inspections	34
Special Inspections and Tests, Contractor Responsibility and Structural Observations	33
Special Inspections for Seismic Resistance	36
Special Inspector	33
Specification for Masonry Structures	331
Specified Dimension	48
Specified Compressive Strength of AAC Masonry, f'_{AAC}	102
Specified Compressive Strength of Masonry, f'_m	48, 102, 341
Splices	51
Splices of Reinforcement	51
Stainless Steel	376
Standard Hooks	149
Standard Method for Determining Sound Transmission Ratings for Masonry Walls	559
Standards Cited in this Code	84, 425
Statement of Special Inspections	34
Content of Statement of Special Inspections	34
Seismic Requirements in the Statement of Special Inspections	34
Wind Requirements in the Statement of Special Inspections	34
Steel Fireplace Units	54
Steel Reinforcement—Allowable Stresses	196
Stiffness	122
Stirrup	102
Stone Masonry	48, 102, 342
Ashlar Stone Masonry	102, 342
Rubble Stone Masonry	102, 342

INDEX

Stone Veneer — 25
Strength — 29, 48
    Design Strength — 48
    Nominal Strength — 48
    Required Strength — 49
    Strength Design — 29
Strength Contribution from Reinforcement — 248
Strength Design of Autoclaved Aerated Concrete (AAC) Masonry — 243
    Reinforced AAC Masonry — 250
    Unreinforced (Plain) AAC Masonry — 248
Strength Design of Masonry — 51, 203
    Reinforced Masonry — 214
    Unreinforced (Plain) Masonry — 210
Strength for Resisting Loads — 248
Strength of Joints — 243
Strength-Reduction Factors — 102, 203, 244, 308
Strength Requirements — 240
Stress Calculations — 121
Structural Design — 29
Structural Elements — 127
    Beams — 137
    Columns — 141
    Corbels — 143
    Masonry Assemblies — 127
    Pilasters — 143
Structural Integrity — 30
Structural Observation(s) — 33, 35
    Structural Observations for Seismic Resistance — 35
    Structural Observations for Wind Requirements — 35
Structural Steel Columns Protected by Clay Masonry — 556
    Calculation of Fire Resistance — 556
Structural Steel Columns Protected by Concrete Masonry — 551
Structural Walls — 29
Submittals — 356
Submittals to the Building Official — 35
Support Elements — 288, 299
Support on Wood — 50
Surface-Bonded Walls — 52
    Construction — 52
    Strength — 52
Surface-Bonding Mortar — 49
Surface Coatings — 564, 570
System Description — 348
    Adhered Veneer Requirements — 355
    Compressive Strength Determination — 348
    Compressive Strength Requirements — 348

T

TEK Manual for Concrete Masonry Design and Construction — 607-610
Tendon Anchorage — 102, 342
Tendon Coupler — 102, 342
Tendon Jacking Force — 102, 342

INDEX

Termination---58
    Chimney Caps---58
    Rain Caps---58
    Spark Arrestors---58
Terra Cotta---26
Testing Agency's Services and Duties---357
Testing Prism from Constructed Masonry---355
Tests---5
Thermal Conductivity of Concrete or Clay Masonry Units---16
Thickness of Masonry---299
Thin-Bed Mortar---103
Thin-Bed Mortar for AAC---387
Tie, Lateral---103
Tie, Wall---49, 103
Tile, Structural Clay---49
Time Assigned to Finish Materials on Fire-Exposed Side of Concrete and Masonry Walls---557
Time Assigned to Finish Materials on Fire-Exposed Side of Wall---11
Transverse Reinforcement---103

U

Unbonded Prestressing Tendon---103, 342
Ungrouted or Partially Grouted Construction---550
Unit Masonry Protection---9
Units---277
Unreinforced Masonry---188
Unreinforced (Plain) AAC Masonry---248
Unreinforced (Plain) Masonry---48, 103, 210

V

Veneer---23, 263
    Adhered Veneer---274
    Anchored Veneer---268
Veneer, Adhered---103, 342
Veneer, Anchored---103
Veneer Anchors---406
Veneer, Masonry---103
Vertical and Lateral Flame Propagation---24
Vertical Masonry Foundation Elements---36, 42
Vertical Openings---7
Vertical Reinforcing---54, 57
Visual Stability Index (VSI)---103, 342

INDEX

629

W

- Wall(s) ---- 49, 103, 342
  - Cavity Wall ---- 49
  - Dry-Stacked, Surface-Bonded Walls ---- 49
  - Parapet Wall ---- 49
- Wall Design for In-Plane Loads ---- 227, 260
- Wall Design for Out-of-Plane Loads ---- 224, 257
- Wall Lateral Support Requirements ---- 298
- Wall, Load-Bearing ---- 103, 342
- Wall, Masonry Bonded Hollow ---- 103, 342
- Wall Thickness ---- 58
  - Masonry Veneer Chimneys ---- 58
- Wall Ties ---- 404
- Walls Adjoining Structural Framing ---- 304
- Water-Resistive Barrier ---- 23
- Water-Resistive Barrier Coating ---- 36
- Waterproofing ---- 37
  - Walls ---- 37
- Weather Protection ---- 23, 24
- Wetting Masonry Units ---- 390
- Width ---- 103
- Wind-Borne Debris Protection Fastening Schedule for Wood Structural Panels ---- 30
- Wind Loads ---- 30
- Wind Requirements in the Statement of Special Inspections ---- 34
- Wood ---- 63
- Wythe ---- 49, 103, 342

INDEX

ICC INTERNATIONAL CODE COUNCIL

People Helping People Build a Safer World®

NEW!
INSPECTOR SKILLS
The first book to address the "other" skills necessary for the people side of inspection

When coupled with a solid technical knowledge of codes and construction practices, effective "soft skills" elicit cooperation, generate respect and credibility, and improve the image of inspectors and code safety departments. Soft skills are non-technical traits and behaviors that enhance an inspector's ability to interact with others and to successfully carry out job duties.

Topics include:
- Approaches to inspection
- Getting along
- Customer service
- Ethics
- Effective communication
- And much more

ICC's new **Inspector Skills** raises awareness of the importance of soft skills and provides guidance for improving those skills. The book is ideal for inspectors in all disciplines and can also benefit students, permit technicians, plan reviewers and building officials.

SOFT COVER #7104S

NEW!
STRENGTHEN STRUCTURES THROUGH SPECIAL INSPECTION
Special Inspection Manual, 2012 Edition

ICC's newest title details how specialized oversight and increased inspection during construction can help to prevent significant structural damage.

Features:
- Contains detailed information about the IBC requirements for special inspection
- Supports preparation for special inspection certification exams
- Aids building departments in their creation of procedures
- Provides forms to run a special inspection overview program
- Defines responsibilities of the special inspector, building official, project owner, engineer or architect of record and contractor

SOFT COVER #4019S12

ORDER YOURS TODAY! 1-800-786-4452 | www.iccsafe.org/books

TMS Seminar Offerings

The Masonry Society (TMS) has several excellent seminar series that have been highly rated by attendees and that are reasonably priced. Listed below are some of the more popular seminar presentations TMS offers. TMS's seminars are professionally presented and can be structured to provide continuing education credits. In addition to the seminars below, TMS can also provide shorter presentations on topics as diverse as code updates, moisture penetration resistance, fire resistance, sound control, and sustainable design.

University Professors Masonry Workshop (UPMW)

The UPMW is a forum for faculty who are teaching masonry, or will be teaching masonry, to learn about the design, specification, construction, and evaluation of masonry. Effective and innovative ways to teach this information to their students are discussed. The goal of the UPMW is simple: to assist professors teaching masonry so that students, who will be future designers and construction managers, are comfortable using masonry on their projects. For mor information visit www.masonrysociety.org/html/education/workshops/UPMW

Designing Masonry to the MSJC and IBC

This intense seminar is aimed at practicing designers reviewing the masonry design and specification provisions in the *Building Code Requirements and Specification for Masonry Structures* (TMS 402/ACI 530/ASCE 5) which serves as the primary reference for the International Building Code (IBC) for the design and construction of masonry. Instructors provide background on major Quality Assurance issues and also highlight the masonry requirements contained in the IBC. Several practical design examples of typical masonry elements are reviewed using the TMS 402 and revisions to the design examples based on modifications by the IBC are also discussed.

Special Inspection of Masonry Construction

This comprehensive seminar is aimed at helping inspectors, contractors, engineers, and building officials identify and check key quality assurance items on the jobsite. The seminar uses the IBC and the TMS Masonry Standard as primary references which is the basis for the current certification examination for Structural Masonry Inspectors. The seminar also overviews related ASTM standards and inspection guides. Attendees learn code inspection requirements, material requirements for clay brick, concrete block, mortar, grout, and reinforcement, proper placement of materials, how to determine the compressive strength of masonry, and appropriate severe weather procedures.

The Role of Masonry in LEED™ and Sustainable Design

Increasingly, environmental considerations are a part of the design process for buildings, and it is therefore vital that designers, contractors, and manufacturers are aware of the attributes of material and construction methods. TMS masonry experts introduce attendees to the U.S. Green Building Council's LEED™ Rating System and the associated certification process. Seminar leaders overview sustainable design and the LEED™ Rating System, address LEED™ credit categories, and identify how masonry products can contribute to LEED™ credits. This presentation gives insight on the environmental building marketplace, as well as covering criteria and considerations used in selecting masonry products for LEED™ projects.

Direct Design of Masonry Structures – A Simplified Approach for Single Story CMU Buildings

In 2010, The Masonry Society (TMS) published a new design standard (TMS 403 – *Direct Design Handbook for Masonry Structures*) for the design of single-story, loadbearing, concrete masonry structures. This updated handbook, based on ASCE 7 load requirements, and the strength design procedures of TMS 402/ACI 530/ASCE 5, provides a simple and rapid procedure for the design of single-story concrete masonry structures. This seminar reviews this new handbook, its basis, the direct design procedure, and the limitations on its use. Instructors discuss how the procedure can be used to produce efficient, safe designs, more rapidly than by using conventional methods.

For additional information on these or other TMS Seminar Offerings, contact TMS at 303-939-9700 or info@masonrysociety.org

THE MASONRY SOCIETY
Advancing the knowledge of masonry

105 South Sunset Street, Suite Q
Longmont, CO 80501
www.masonrysociety.org

MASONRY INSTITUTE OF AMERICA

VISIT OUR WEBSITE

www.masonryinstitute.org

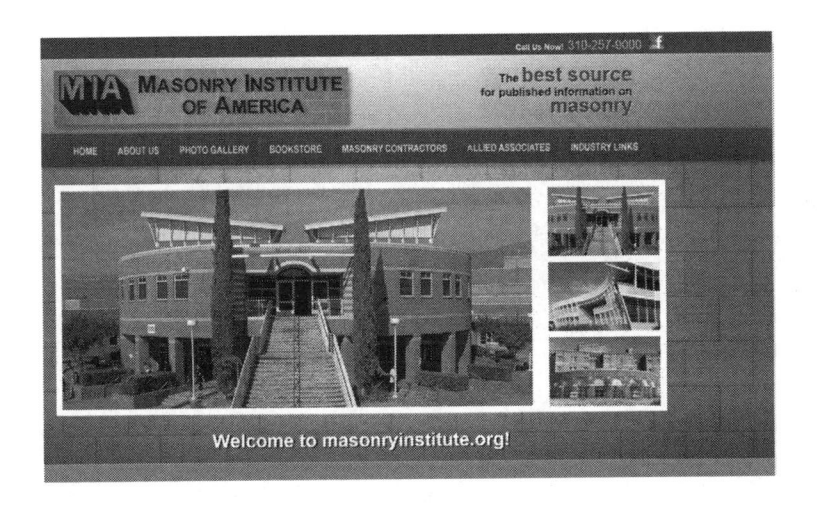

- Industry News
- Photo Gallery
- Bookstore
- How to Become a Contractor
- How to Become an Inspector
- Masonry Contractors
- Allied Associates
- Technical Questions
- Industry Links
- Request Information

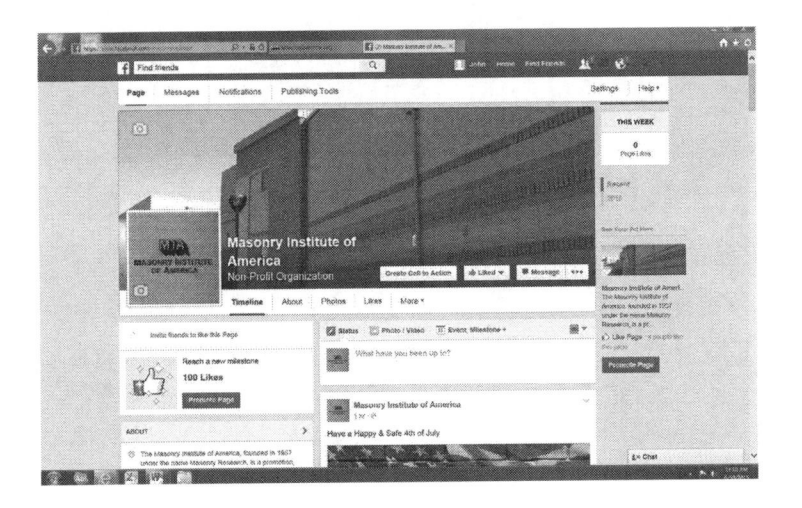

Be Social and Learn Masonry from the
Best Source
for Published Information on
Masonry

Like us on facebook